Graduate Texts in Mathematics

(continued after index)

Peter Abramenko
Kenneth S. Brown

Buildings

Theory and Applications

 Springer

Peter Abramenko
Department of Mathematics
University of Virginia
Charlottesville, VA 22904
USA
pa8e@virginia.edu

Kenneth S. Brown
Department of Mathematics
Cornell University
Ithaca, NY 14853
USA
kbrown@cornell.edu

ISBN: 978-1-4419-2701-9   e-ISBN: 978-0-387-78835-7

Mathematics Subject Classification (2000): (Primary) 51E24, 20E42, 20F55, 51F15, (Secondary) 14L35, 20F05, 20F65, 20G15, 20J05, 22E40, 22E65

© 2010 Springer Science+Business Media, LLC

Based in part on *Buildings*, Kenneth S. Brown, 1989, 1998

Printed on acid-free paper.

9 8 7 6 5 4 3 2 1

springer.com

To our wives, Monika and Susan

To our wives, Monika and Susan

Preface

This text started out as a revised version of *Buildings* by the second-named author [53], but it has grown into a much more voluminous book. The earlier book was intended to give a short, friendly, elementary introduction to the theory, accessible to readers with a minimal background. Moreover, it approached buildings from only one point of view, sometimes called the "old-fashioned" approach: A building is a simplicial complex with certain properties.

The current book includes all the material of the earlier one, but we have added a lot. In particular, we have included the "modern" (or "W-metric") approach to buildings, which looks quite different from the old-fashioned approach but is equivalent to it. This has become increasingly important in the theory and applications of buildings. We have also added a thorough treatment of the Moufang property, which occupies two chapters. And we have added many new exercises and illustrations. Some of the exercises have hints or solutions in the back of the book. A more extensive set of solutions is available in a separate solutions manual, which may be obtained from Springer's Mathematics Editorial Department.

We have tried to add the new material in such a way that readers who are content with the old-fashioned approach can still get an elementary treatment of it by reading selected chapters or sections. In particular, many readers will want to omit the optional sections (marked with a star). The introduction below provides more detailed guidance to the reader.

In spite of the fact that the book has almost quadrupled in size, we were still not able to cover all important aspects of the theory of buildings. For example, we give very little detail concerning the connections with incidence geometry. And we do not prove Tits's fundamental classification theorems for spherical and Euclidean buildings. Fortunately, the recent books of Weiss [281, 283] treat these classification theorems thoroughly.

Applications of buildings to various aspects of group theory occur in several chapters of the book, starting in Chapter 6. In addition, Chapter 13 is devoted to applications to the cohomology theory of groups, while Chapter 14 sketches a variety of other applications.

Most of the material in this book is due to Jacques Tits, who originated the theory of buildings. It has been a pleasure studying Tits's work. We were especially pleased to learn, while this book was in the final stages of production, that Tits was named as a corecipient of the 2008 Abel prize. The citation states:

> Tits created a new and highly influential vision of groups as geometric objects. He introduced what is now known as a Tits building, which encodes in geometric terms the algebraic structure of linear groups. The theory of buildings is a central unifying principle with an amazing range of applications. . . .

We hope that our exposition helps make Tits's beautiful ideas accessible to a broad mathematical audience.

We are very grateful to Pierre-Emmanuel Caprace, Ralf Gramlich, Bill Kantor, Bernhard Mühlherr, Johannes Rauh, Hendrik Van Maldeghem, and Richard Weiss for many helpful comments on a preliminary draft of this book. We would also like to thank all the people who helped us with the applications of buildings that we discuss in Chapter 14; their names are mentioned in the introduction to that chapter.

Charlottesville, VA, and Ithaca, NY   *Peter Abramenko*
June 2008   *Kenneth S. Brown*

Contents

List of Figures

Introduction

Buildings were introduced by Jacques Tits in order to provide a unified geometric framework for understanding semisimple complex Lie groups and, later, semisimple algebraic groups over an arbitrary field. The definition evolved gradually during the 1950s and 1960s and reached a mature form in about 1965. Tits outlined the theory in a 1965 Bourbaki Seminar exposé [243] and gave a full account in [247]. At that time, Tits thought of a building as a simplicial complex with a family of subcomplexes called *apartments*, subject to a few axioms that will be stated in Chapter 4. Each apartment is made up of *chambers*, which are the top-dimensional simplices. This viewpoint is sometimes called the "old-fashioned approach" to buildings, but we will use the more neutral phrase *simplicial approach*.

In the more "modern" approach, introduced by Tits in a 1981 paper [255], one forgets about all simplices except the chambers, and one forgets about apartments. The definition is recast entirely in terms of objects called *chamber systems*. For lack of a better term, we will refer to this as the *combinatorial approach* to buildings. The definition from this point of view also evolved over a period of years, and it reached a mature form in the late 1980s. An important catalyst was the theory of twin buildings, which was being developed by Ronan and Tits. The final version of the combinatorial definition can be found in [261], where a building is viewed as a set C (the *chambers*), together with a *Weyl-group-valued distance function* subject to a few axioms.

A third way of thinking about buildings, which we call the *metric approach*, is gotten by taking geometric realizations of the structures described in the previous two paragraphs. It turns out that this can always be done so as to obtain a metric space with nice geometric properties. The possibility of doing this has been known for a long time in special cases where the apartments are spheres or Euclidean spaces. But M. Davis [88] discovered much more recently that it can be done in general. In this approach to buildings, apartments again play a prominent role but are viewed as metric spaces rather than simplicial complexes.

The three approaches to buildings are distinguished by how one thinks of a chamber. In the simplicial approach, chambers are maximal simplices. In the combinatorial approach, chambers are just elements of an abstract set, or vertices of a graph. And in the metric approach, chambers are metric spaces.

Our goal in this book is to treat buildings from all three of these points of view. The various approaches complement one another and are all useful. On the other hand, we recognize that some readers may prefer one particular viewpoint. We have therefore tried to create more than one path through the book so that, for example, the reader interested only in the combinatorial approach can learn the basics without having to spend too much time studying buildings as simplicial complexes. More detailed guidance is given in Section 0.9.

The remainder of this introduction is intended to provide an overview of the various ways of thinking about buildings, as well as a guide to the rest of the book. The reader need not be concerned about unexplained terminology or notation; we will start from scratch in Chapter 1.

All three approaches to buildings start with Coxeter groups.

0.1 Coxeter Groups and Coxeter Complexes

A *Coxeter group* of rank n is a group generated by n elements of order 2, subject to relations that give the orders of the pairwise products of the generators. Thus the group of order 2 is a Coxeter group of rank 1, and the dihedral group D_{2m} of order $2m$ $(m \geq 2)$ is a Coxeter group of rank 2, with presentation

$$D_{2m} = \left\langle \, s, t \; ; \; s^2 = t^2 = (st)^m = 1 \, \right\rangle .$$

[*Warning*: Some mathematicians, following the standard notation of crystallography, write D_m instead of D_{2m}.] The infinite dihedral group

$$D_\infty = \left\langle \, s, t \; ; \; s^2 = t^2 = 1 \, \right\rangle$$

is also a rank-2 Coxeter group; there is no relation for the product st, because it has infinite order. Readers who have studied Lie theory have seen *Weyl groups*, which are the classical examples of (finite) Coxeter groups. For example, the symmetric group S_3 on 3 letters, which is the same as the dihedral group of order 6, is the Weyl group of type A_2. And the symmetric group S_4 on 4 letters is the Weyl group of type A_3, with presentation

$$S_4 = \left\langle \, s, t, u \; ; \; s^2 = t^2 = u^2 = (st)^3 = (tu)^3 = (su)^2 = 1 \, \right\rangle .$$

Certain infinite Coxeter groups also arise in Lie theory, as *affine Weyl groups*. For example, D_∞ is the affine Weyl group of type $\tilde{A}_1$, and the Coxeter group W with presentation

$$W = \left\langle \, s, t, u \; ; \; s^2 = t^2 = u^2 = (st)^3 = (tu)^3 = (su)^3 = 1 \, \right\rangle \qquad (0.1)$$

is the affine Weyl group of type $\tilde{A}_2$.

The given set S of generators of order 2 should be viewed as part of the structure, but one often suppresses it for simplicity. When we need to be precise, we will talk about the *Coxeter system* (W, S) rather than the Coxeter group W. The system (W, S) is said to be *reducible* if S admits a partition $S = S' \amalg S''$ such that all elements of S' commute with all elements of S''. In this case W splits as a direct product $W' \times W''$ of two Coxeter groups of lower rank.

Every finite Coxeter group can be realized in a canonical way as a group of orthogonal transformations of Euclidean space, with the generators of order 2 acting as reflections with respect to hyperplanes. Thus D_{2m} acts on the plane, with s and t acting as reflections through lines that meet at an angle of π/m. And S_4 admits a reflection representation on 3-dimensional space. To construct this representation, let S_4 act on $\mathbb{R}^4$ by permuting the coordinates, and then restrict to the 3-dimensional subspace $x_1 + x_2 + x_3 + x_4 = 0$. More geometrically, we get this action by viewing S_4 as the group of symmetries of a regular tetrahedron.

Given a finite Coxeter group W and its reflection representation on Euclidean space, consider the set of hyperplanes whose reflections belong to W. If we cut the unit sphere by these hyperplanes, we get a cell decomposition of the sphere. The cells turn out to be spherical simplices, and we obtain a simplicial complex $\Sigma = \Sigma(W)$ (or $\Sigma(W, S)$) triangulating the sphere. This is called the *Coxeter complex* associated with W.

For D_{2m} acting on the plane, Σ is a circle decomposed into $2m$ arcs. For the action of S_4 on $\mathbb{R}^3$ mentioned above, Σ is the triangulated 2-sphere shown in Figure 0.1.* There are 6 reflecting hyperplanes, which cut the sphere into 24 triangular regions. Combinatorially, Σ is the barycentric subdivision of the boundary of a tetrahedron, as indicated in the picture. (One face of an inscribed tetrahedron is visible.) The vertex labels in the picture will be explained in the next section.

A similar but more complicated construction yields a Coxeter complex associated with an arbitrary Coxeter group W. For example, $\Sigma(D_\infty)$ is a triangulated line, with the generators s and t acting as affine reflections with respect to the endpoints of an edge. And the Coxeter complex for the group W defined in equation (0.1) is the Euclidean plane, tiled by equilateral triangles. [The generators s, t, u act as reflections with respect to the sides of one such triangle.]

We will give a detailed treatment of Coxeter groups in Chapters 1 and 2. The Coxeter complex associated with a finite Coxeter group will arise naturally from our discussion, and this will motivate the general theory of Coxeter complexes to be given in Chapter 3.

---

* Figure 0.1 was drawn by Bill Casselman for the article [54]. We are grateful to him for permission to reproduce it here.

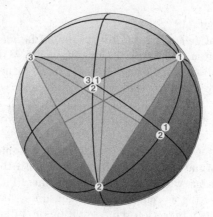

Fig. 0.1. The Coxeter complex of type A_3 (drawn by Bill Casselman).

0.2 Buildings as Simplicial Complexes

We begin with the canonical example of a building: Let k be a field and let $\Delta = \Delta(k^n)$ be the abstract simplicial complex whose vertices are the nonzero proper subspaces of the vector space k^n and whose simplices are the chains

$$V_1 < V_2 < \cdots < V_r$$

of such subspaces. Every simplex σ is contained in a subcomplex, called an apartment, which is isomorphic to the Coxeter complex associated with the symmetric group on n letters. To find such an apartment, choose a basis $e_1, e_2, \ldots, e_n$ of k^n such that every subspace V_i that occurs in σ is spanned by some subset of the basis vectors. We then get an apartment containing σ by taking *all* simplices whose vertices are spanned by subsets of the basis vectors.

Figure 0.1 shows an apartment for the case $n = 4$. The labels on the vertices indicate which basis vectors span the corresponding subspace. Thus the vertex labeled 2 is the line spanned by e_2, the vertex labeled by both 1 and 2 is the plane spanned by e_1 and e_2, and the vertex labeled by 1, 2, and 3 is the 3-dimensional space spanned by e_1, e_2, e_3. These three subspaces form a chain, so they span a 2-simplex in Δ.

For a second example of a building, take any simplicial tree with no endpoints (i.e., every vertex is incident to at least two edges). Any copy of the real line in the tree is an apartment, isomorphic to the Coxeter complex associated with the infinite dihedral group.

As these examples suggest, a building is a simplicial complex that is the union of "apartments," each of which is a Coxeter complex. There are axioms that specify how the apartments are glued together. They are easy to write down; the reader anxious to see them can look ahead to Section 4.1. But they are not easy to grasp intuitively until one works with them for a while.

The simplices of top dimension are called *chambers*, while those of codimension 1 are called *panels*. Note that a Coxeter complex is itself a building, with a single apartment. In fact, Coxeter complexes are precisely the *thin* buildings. (This means that every panel is a face of exactly two chambers.) The more interesting buildings are the *thick* buildings, i.e., those in which every panel is a face of at least three chambers. The building $\Delta(k^n)$ described above is thick, and a tree is a thick building if and only if every vertex is incident to at least three edges.

There is a well-defined Coxeter system (W, S) associated with a building Δ, such that the apartments are all isomorphic to $\Sigma(W, S)$. One says that Δ is a building of type (W, S), and one calls W the *Weyl group* of Δ. Much of the terminology (rank, reducibility, ...) from the theory of Coxeter groups is carried over to buildings. In addition, one says that a building is *spherical* if W is finite (in which case the apartments are triangulated spheres). Thus $\Delta(k^n)$ is an irreducible spherical building of rank $n - 1$, while a tree with no endpoints is an irreducible building of rank 2 that is not spherical. (As we will see later, it is an example of a *Euclidean building*.)

0.3 Buildings as W-Metric Spaces

Let Δ be a building of type (W, S) as above, and let $\mathcal{C} = \mathcal{C}(\Delta)$ be its set of chambers. It turns out that there is a natural way to define a W-valued *distance function*

$$\delta \colon \mathcal{C} \times \mathcal{C} \to W$$

that describes the relative position of any two chambers. Intuitively, $\delta(C, D)$ for $C, D \in \mathcal{C}$ is something like a vector pointing from C to D. The definition of δ will be given in Section 4.8; all we will say at the moment is that $\delta(C, D)$ contains information about the totality of minimal galleries from C to D. Here a *gallery* is a finite sequence of chambers such that any two consecutive ones have a common panel, and it is *minimal* if there is no shorter gallery with the same first and last chambers. Minimal galleries are combinatorial analogues of geodesics.

It turns out that one can completely reconstruct the building Δ from the data consisting of the Coxeter system (W, S), the set $\mathcal{C}$ of chambers, and the function δ. Moreover, one can write down simple axioms that these data must satisfy in order that they come from a building. The axioms can be found in Section 5.1.1. Some of them resemble the axioms for an ordinary metric space, so we will sometimes refer to $(\mathcal{C}, \delta)$ as a *W-metric space*.

From the combinatorial point of view, then, a building of type (W, S) is simply a W-metric space. The most striking thing about this approach is that the axioms contain nothing resembling the existence of apartments. Indeed, in proving that the combinatorial definition is equivalent to the simplicial one, the key step is to use the axioms to prove the existence of apartments. All of this will be carried out in Chapter 5.

The combinatorial approach is both more abstract and more elementary than the simplicial approach. It is more abstract because the geometric intuition is gone. Thus one no longer visualizes chambers as regions cut out by hyperplanes, and one no longer visualizes apartments as simplicial complexes associated with reflection groups. But it is more elementary because the underlying mathematical object can be boiled down to nothing more than a graph with colored edges. [The vertices of the graph are the elements of $\mathcal{C}$, and two such vertices C, D are connected by an edge with "color" $s \in S$ if and only if $\delta(C, D) = s$.]

0.4 Buildings and Groups

We mentioned above that Tits introduced buildings because of his interest in Lie groups and algebraic groups. The connection between buildings and groups is provided by Tits's theory of *BN-pairs*, which we treat in Chapter 6. Given a group G with a pair of subgroups B, N satisfying certain axioms, one constructs a building on which G operates as a group of automorphisms. Conversely, a sufficiently nice action of a group on a building yields a BN-pair.

This theory provides a very good illustration of the usefulness of thinking of buildings as W-metric spaces, so we have chosen to take this point of view in Chapter 6. An alternative treatment of BN-pairs based on the simplicial approach can be found in the earlier book by the second author [53], and an outline is given in Exercise 6.54 in the present book.

0.5 The Moufang Property and the Classification Theorem

One of Tits's greatest achievements is the classification of thick, irreducible, spherical buildings of rank at least 3, proved in [247]. Roughly speaking, the result is that such buildings correspond to classical groups and simple algebraic groups (of relative rank at least 3) defined over an arbitrary field. The rank restriction cannot be avoided. The buildings of rank 2, for example, include those of type A_2, which are essentially the same as projective planes. And there is no hope of classifying projective planes, even the finite ones.

If, however, one imposes a certain symmetry condition on the buildings (the so-called *Moufang property*), then the classification extends to rank 2. This result is due to Tits and Weiss [262]. In rank ≥ 3 the Moufang property does not need to be added as a hypothesis because Tits proved that all thick, irreducible, spherical buildings of rank ≥ 3 have the Moufang property. We

will study the Moufang property in Chapter 7 for spherical buildings and in
Chapter 8 for more general buildings.

The proof of the classification theorem is long and involved. In this book
(Chapter 9) we only give a rough statement of the theorem, with some pointers
to the literature for readers who want more details.

0.6 Euclidean Buildings

We have seen that spherical buildings arise in connection with algebraic groups
over an arbitrary field. If the field comes equipped with a discrete valuation,
then (under suitable hypotheses on the group) there is a second building, in
which the apartments are Euclidean spaces instead of spheres. Such buildings
are called *Euclidean* buildings, or buildings of *affine type*. We study them
in Chapter 11 after laying the foundations by treating Euclidean reflection
groups and Euclidean Coxeter complexes in Chapter 10. Along the way we
develop the theory of affine Weyl groups alluded to in Section 0.1 above.
Chapter 10 also includes a brief outline of the theory of hyperbolic reflection
groups.

Euclidean buildings, as we will see, admit a canonical metric, which restricts to a Euclidean metric on each apartment. This was introduced and
exploited by Bruhat and Tits [59], who showed further that every Euclidean
building is a CAT(0) space (although they did not use that terminology).
This means that it has metric properties analogous to those of complete simply connected Riemannian manifolds of nonpositive curvature.

0.7 Buildings as Metric Spaces

M. Davis [88] made the surprising discovery that for *every* building there
is a geometric realization that admits a CAT(0) metric. This is primarily
of interest in the nonspherical case, so assume that Δ is a building of type
(W, S) with W infinite. We then have a Coxeter complex $\Sigma = \Sigma(W, S)$, as we
mentioned above. In the Euclidean case, Σ triangulates a Euclidean space, as
in Section 0.6. In general, however, it has no natural geometric structure.

But Davis found a way to truncate Σ so as to obtain a different "geometric realization" of (W, S), which we denote by Σ_d, and which has a natural
CAT(0) metric. It is a subspace of the geometric realization of the simplicial
complex Σ. Returning to our building Δ, we then get a CAT(0) geometric
realization by replacing each apartment by a copy of Σ_d. This is the metric-
space approach to buildings mentioned at the beginning of this introduction.
Davis's theory will be treated in Chapter 12, where we also give a general
procedure for constructing metric realizations of buildings. For example, if
the Weyl group is a hyperbolic reflection group, then there is a realization in
which every apartment is a hyperbolic space.

0.8 Applications of Buildings

Buildings have many uses beyond those originally envisaged by Tits. For example, the authors of this book first got interested in buildings because of applications to the cohomology theory of groups. We survey the applications to group cohomology in Chapter 13, and we mention a variety of other applications in Chapter 14.

0.9 A Guide for the Reader

Chapters 1–4, 6, 10, 11, and 13 constitute a revised and enlarged version of the original book [53]. Readers who want an elementary introduction in the spirit of that book, with buildings always viewed as simplicial complexes, can concentrate on those chapters (with perhaps an occasional glance at Chapter 5 for terminology). Such readers may also want to omit Sections 2.5 and 3.6 on first reading, returning to them later as necessary. In addition, there are several optional sections (marked with a star) that may be omitted.

At the other extreme, readers who are primarily interested in the combinatorial approach to buildings can read Chapters 1, 2, and 5, with an occasional glance at Chapters 3 and 4 for motivation or terminology.

Chapters 7 and 8 treat the Moufang property. Chapter 7 covers the spherical case and requires only the simplicial approach. Chapter 8, on the other hand, is more advanced. It requries both the simplicial and the combinatorial approaches, and it relies on some of the earlier starred sections that are not needed elsewhere in the book. That whole chapter can be viewed as optional and may safely be omitted. We have tried, nevertheless, to write that chapter at the same level as the rest of the book in order to make this material accessible. Chapter 9 (on the classification theorem) should make sense to readers who know either the simplicial approach or the combinatorial approach to buildings. Chapter 11 (Euclidean buildings), on the other hand, requires familiarity with the simplicial approach.

Chapter 12, on metric realizations of buildings, can technically be approached either from the simplicial or the combinatorial viewpoint. As a practical matter, however, this chapter makes use of Euclidean buildings for motivation, so it may be difficult reading for someone who knows only the combinatorial approach. Finally, Chapters 13 and 14 (on applications of buildings) definitely require the simplicial approach.

1

Finite Reflection Groups

This book is about connections between groups and geometry. We begin by considering groups of isometries of Euclidean space generated by hyperplane reflections. In order to avoid technicalities in this introductory chapter, we confine our attention to *finite* groups and we require our reflections to be with respect to *linear* hyperplanes (i.e., hyperplanes passing through the origin). We will generalize this in Chapter 10, replacing "finite" by "discrete" and "linear" by "affine."

1.1 Definitions

Let V be a Euclidean vector space, i.e., a finite-dimensional real vector space with an inner product.

Definition 1.1. A *hyperplane* in V is a subspace H of codimension 1. The *reflection* with respect to H is the linear transformation $s_H \colon V \to V$ that is the identity on H and is multiplication by -1 on the (1-dimensional) orthogonal complement $H^\perp$ of H. If α is a nonzero vector in $H^\perp$, so that $H = \alpha^\perp$, we will sometimes write s_α instead of s_H.

Example 1.2. Let $s \colon \mathbb{R}^n \to \mathbb{R}^n$ interchange the first two coordinates, i.e.,

$$s(x_1, x_2, x_3, \dots, x_n) = (x_2, x_1, x_3, \dots, x_n).$$

Equivalently, s transposes the first two standard basis vectors e_1, e_2 and fixes the others. Then s is the identity on the hyperplane $x_1 - x_2 = 0$, which is the orthogonal complement of $\alpha := e_1 - e_2$, and $s(\alpha) = -\alpha$. So s is the reflection s_α.

For future reference, we derive a formula for s_α. Given $x \in V$, write $x = h + \lambda\alpha$ with $h \in H$ and $\lambda \in \mathbb{R}$. Taking the inner product of both sides with α, we obtain $\lambda = \langle \alpha, x \rangle / \langle \alpha, \alpha \rangle$, where the angle brackets denote the inner product in V. Then $s_\alpha(x) = h - \lambda\alpha = x - 2\lambda\alpha$, and hence

$$s_\alpha(x) = x - 2\frac{\langle \alpha, x \rangle}{\langle \alpha, \alpha \rangle}\alpha \, . \qquad (1.1)$$

In words, this says that the mirror image of x with respect to $\alpha^\perp$ is obtained by subtracting twice the component of x in the direction of α, thereby changing the sign of that component. Suppose, for instance, that $\alpha = e_1 - e_2$ as in Example 1.2; then $\langle \alpha, \alpha \rangle = 2$, so equation (1.1) becomes

$$s_\alpha(x) = x - \langle \alpha, x \rangle \alpha \, .$$

Definition 1.3. A *finite reflection group* is a finite group W of invertible linear transformations of V generated by reflections s_H, where H ranges over a set of hyperplanes.

The group law is of course composition. We will sometimes refer to the *pair* (W, V) as a finite reflection group when it is necessary to emphasize the vector space V on which W acts.

The requirement that W be finite is a very strong one. Suppose, for instance, that $\dim V = 2$ and that W is generated by two reflections $s := s_H$ and $s' := s_{H'}$. Then the rotation $ss' \in W$ has infinite order (and hence W is infinite) unless the angle between the lines H and H' is a rational multiple of π. The following criterion is often used to verify that a given group generated by reflections is finite:

Lemma 1.4. *Let Φ be a finite set of nonzero vectors in V, and let W be the group generated by the reflections s_α ($\alpha \in \Phi$). If Φ is invariant under the action of W, then W is finite.*

Proof. We will show that W is isomorphic to a group of permutations of the finite set Φ. Let V_1 be the subspace of V spanned by Φ, and let V_0 be its orthogonal complement. Then $V_0 = \bigcap_{\alpha \in \Phi} \alpha^\perp$, which is the fixed-point set $V^W := \{v \in V \mid wv = v \text{ for all } w \in W\}$. In view of the orthogonal decomposition $V = V_0 \oplus V_1$, it follows that an element of W is completely determined by its action on V_1 and hence by its action on Φ. $\qquad \square$

The group W defined in the lemma will be denoted by W_Φ. Such groups arise classically in the theory of Lie algebras, where Φ is the *root system* associated with a complex semisimple Lie algebra and W_Φ is the corresponding *Weyl group*. (This explains the use of the letter W for a finite reflection group.)

We will not need the precise definition of "root system," but the interested reader can find it in Appendix B. For now, we need to know only that a root system satisfies the hypotheses of Lemma 1.4 as well as an integrality condition that forces W_Φ to leave a lattice invariant. It will be convenient to have a name for sets Φ as in the lemma that are not necessarily root systems in the classical sense.

Definition 1.5. A set Φ satisfying the hypotheses of Lemma 1.4 will be called a *generalized root system*. The elements of Φ will be called *roots*. We will always assume (without loss of generality) that our generalized root systems are *reduced*, in the sense that $\pm\alpha$ (for $\alpha \in \Phi$) are the only scalar multiples of α that are again roots. Thus there is exactly one pair $\pm\alpha$ for each generating reflection in the statement of Lemma 1.4.

To emphasize the distinction between generalized root systems and the classical ones that leave a lattice invariant, we will sometimes refer to the classical ones as *crystallographic* root systems.

It is also convenient to have some terminology for the sort of decomposition of V that arose in the proof of Lemma 1.4. Let W be a group generated by reflections s_H ($H \in \mathcal{H}$), where $\mathcal{H}$ is a set of hyperplanes. Let V_0 be the fixed-point set

$$V^W = \bigcap_{H \in \mathcal{H}} H .$$

Definition 1.6. We call V_0 the *inessential* part of V, and we call its orthogonal complement V_1 the *essential* part of V. The pair (W, V) is called *essential* if $V_1 = V$, or, equivalently, if $V_0 = 0$. The dimension of V_1 is called the *rank* of the finite reflection group W.

The study of a general (W, V) is easily reduced to the essential case. Indeed, V_1 is W-invariant since V_0 is, and clearly $(V_1)^W = 0$; so we have an orthogonal decomposition $V = V_0 \oplus V_1$, where the action of W is trivial on the first summand and essential on the second. We may therefore identify W with a group acting on V_1, and as such, W is essential (and still generated by reflections). If W is the group W_Φ associated with a generalized root system, then W is essential if and only if Φ spans V.

Exercise 1.7. Show that every finite reflection group W has the form W_Φ for some generalized root system Φ.

1.2 Examples

There are two classical families of examples of finite reflection groups. The first, as we have already indicated, consists of Weyl groups of (crystallographic) root systems. The second consists of symmetry groups of regular solids. We will not assume that the reader knows anything about either of these two subjects. But it will be convenient to use the language of root systems or regular solids informally as we discuss examples. It is a fact that *all* finite reflection groups can be explained in terms of one or both of these theories; we will return to this in the next section.

Example 1.8. The group W of order 2 generated by a single reflection s_α is a finite reflection group of rank 1. After passing to the essential part of V, we may identify W with the group $\{\pm 1\}$ acting on $\mathbb{R}$ by multiplication. It is the group of symmetries of the regular solid $[-1, 1]$ in $\mathbb{R}$. It is also the Weyl group of the root system $\Phi := \{\pm\alpha\}$, which is called the root system of type A_1.

Example 1.9. Let V be 2-dimensional, and choose two hyperplanes (lines) that intersect at an angle of π/m for some integer $m \geq 2$. Let s and t be the corresponding reflections and let W be the group $\langle s, t \rangle$ they generate. [Here and throughout this book we use angle brackets to denote the group generated by a given set.] Then the product $\rho := st$ is a rotation through an angle of $2\pi/m$ and hence is of order m. Moreover, s conjugates ρ to $s(st)s = ts = \rho^{-1}$ and similarly for t, so the cyclic subgroup $C := \langle \rho \rangle$ of order m is normal in W. Finally, the quotient W/C is easily seen to be of order 2; hence W is indeed a finite reflection group, of order $2m$.

This group W is called the *dihedral group* of order $2m$, and we will denote it by D_{2m}. If $m \geq 3$, W is the group of symmetries of a regular m-gon. If $m = 3$, 4, or 6, then W can also be described as the Weyl group of a root system Φ, said to be of type A_2, B_2, or G_2, respectively. The root system of type A_2 ($m = 3$) consists of 6 equally spaced vectors of the same length, as shown in Figure 1.1, which also shows the three reflecting hyperplanes (lines). There are two oppositely oriented root vectors for each hyperplane. To get

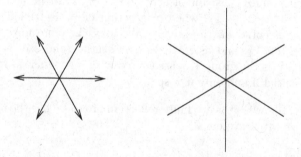

Fig. 1.1. The root system of type A_2 and the reflecting hyperplanes.

B_2 and G_2 ($m = 4$ and $m = 6$), we can take m equally spaced unit vectors together with the sum of any two cyclically consecutive ones, as shown in Figure 1.2

Of course, we can always get D_{2m} from the generalized root system consisting of $2m$ equally spaced unit vectors; but this is not crystallographic for $m > 3$.

Example 1.10. Let W be the group of linear transformations of $\mathbb{R}^n$ ($n \geq 2$) that permute the standard basis vectors $e_1, e_2, \ldots, e_n$. Thus W is isomorphic to the symmetric group S_n on n letters and can be identified with the group

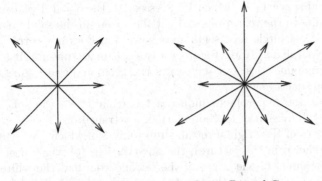

Fig. 1.2. The root systems of type B_2 and G_2.

of $n \times n$ permutation matrices. It is generated by the $\binom{n}{2}$ transpositions s_{ij} $(i < j)$, where s_{ij} interchanges the ith and jth coordinates, so it is a finite reflection group (see Example 1.2). Note that $(W, \mathbb{R}^n)$ is not essential. In fact V^W is the line $x_1 = x_2 = \cdots = x_n$ spanned by the vector $e := (1, 1, \ldots, 1)$. So the subspace V_1 of $\mathbb{R}^n$ on which W is essential is the $(n-1)$-dimensional subspace $e^\perp$ defined by $\sum_{i=1}^n x_i = 0$, whence W has rank $n - 1$.

The interested reader can verify that W is the group of symmetries of a regular $(n-1)$-simplex in V_1. [Hint: The convex hull σ of $e_1, \ldots, e_n$ is a regular $(n-1)$-simplex in the affine hyperplane $\sum x_i = 1$, which is parallel to V_1. The desired regular simplex in V_1 is now obtained from σ via the translation $x \mapsto x - b$, where b is the barycenter of σ.] W is also the Weyl group of a root system in V_1, called the root system of type A_{n-1}. It consists of the $n(n-1)$ vectors $e_i - e_j$ $(i \neq j)$.

When $n = 2$, this example reduces to Example 1.8; when $n = 3$, it reduces to Example 1.9 with $m = 3$ (after we pass to the essential part), i.e., W is dihedral of order 6. For $n = 4$, Figure 1.3 shows the unit sphere in the

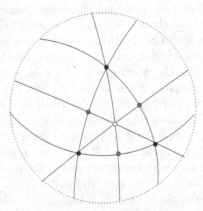

Fig. 1.3. The hyperplanes for the reflection group of type A_3.

3-dimensional space V_1 on which W is essential; the $6 = \binom{4}{2}$ planes $x_i = x_j$ (corresponding to the reflections s_{ij}) cut the sphere in the solid great circles. The dotted great circle represents an equator and does not correspond to a reflecting hyperplane. [Note: Figure 1.3 is a schematic picture; it accurately shows the combinatorics, but it distorts the geometry. See Figure 0.1 for a more accurate picture.]

Note that the hyperplanes induce a triangulation of the sphere as the barycentric subdivision of the boundary of a tetrahedron. The black vertices are the vertices of the original tetrahedron (only 3 of which are visible in the hemisphere shown in the picture); the gray vertices (of which 3 are visible) are the barycenters of the edges of the tetrahedron; and the white vertices (of which one is visible) are the barycenters of the 2-dimensional faces of the tetrahedron. We will see later that this generalizes to arbitrary n: the reflecting hyperplanes triangulate the sphere in V_1 as the barycentric subdivision of the boundary of an $(n-1)$-simplex. See Exercise 1.112.

Example 1.11. Let W be the group of linear transformations of $\mathbb{R}^n$ $(n \geq 1)$ leaving invariant the set $\{\pm e_i\}$ of standard basis vectors and their negatives. In terms of matrices, W can be viewed as the group of $n \times n$ monomial matrices whose nonzero entries are ± 1. [Recall that a *monomial matrix* is one with exactly one nonzero element in every row and every column.] Elements of W are sometimes called "signed permutations." The group W is generated by transpositions s_{ij} as above, together with reflections $t_1, \ldots, t_n$, where t_i changes the sign of the ith coordinate (i.e., t_i is the reflection in the hyperplane $x_i = 0$). Hence W is a finite reflection group of order $2^n n!$, and this time it is essential.

Once again, the interested reader is invited to verify that W is the group of symmetries of a regular solid in $\mathbb{R}^n$, which one can take to be the n-cube $[-1, 1]^n$. Alternatively, take the solid to be the convex hull of the $2n$ vectors $\{\pm e_i\}$; this is a "hyperoctahedron." [The hyperoctahedron is the *dual* of the cube, which means that it is the convex hull of the barycenters of the faces of the cube. Since a solid and its dual have the same symmetry group, it makes no difference which one we choose. We had no reason to mention this in our previous examples because the dual of a regular m-gon is again a regular m-gon, and the dual of a regular simplex is again a regular simplex.]

And once again, W is the Weyl group of a root system, called the root system of type B_n, consisting of the vectors $\pm e_i \pm e_j$ $(i \neq j)$ together with the vectors $\pm e_i$. Alternatively, W can be described as the Weyl group of the root system of type C_n, consisting of the vectors $\pm e_i \pm e_j$ $(i \neq j)$ together with the vectors $\pm 2e_i$; this is dual to type B_n. [Every root system Φ has a "dual," as we explain in Appendix B. A root system and its dual have the same Weyl group. The root systems mentioned in Examples 1.8–1.10, like the regular solids, are self-dual, so this issue did not arise.]

When $n = 1$ this example reduces to Example 1.8; when $n = 2$ it reduces to Example 1.9 with $m = 4$, i.e., W is dihedral of order 8.

Exercise 1.12. Implicit in the last example is the fact that W contains the reflection s_α, where $\alpha := e_i + e_j$ $(i \neq j)$. Verify this by giving an explicit description of s_α in terms of coordinates and/or in terms of its effect on the standard basis vectors.

Example 1.13. Let Φ be the set of vectors $\pm e_i \pm e_j$ $(i \neq j)$ in $\mathbb{R}^n$ $(n \geq 2)$. This is the root system of type D_n. The corresponding Weyl group W_Φ is a subgroup of index 2 in the group W of Example 1.11. If we think of the elements of W as monomial matrices whose nonzero elements are ± 1, then W_Φ consists of those elements with an even number of minus signs.

When $n = 2$, this example reduces to Example 1.9 with $m = 2$, i.e., W_Φ is dihedral of order 4. When $n = 3$, W_Φ is isomorphic to the Weyl group of type A_3; see Exercise 1.99.

We close this section by mentioning an uninteresting way of constructing new examples of finite reflection groups from given ones:

Exercise 1.14. Given finite reflection groups (W', V') and (W'', V''), show that the direct product $W := W' \times W''$ can be realized as a finite reflection group acting on the orthogonal direct sum $V := V' \oplus V''$.

Definition 1.15. A finite reflection group (W, V) is called *reducible* if it decomposes as in the exercise, with V' and V'' nontrivial, and it is called *irreducible* otherwise.

We will see later that an essential finite reflection group always admits a canonical decomposition into "irreducible components" (Exercise 1.100). For example, the Weyl group of type D_2 decomposes as a product of two copies of the Weyl group of type A_1.

1.3 Classification

Finite reflection groups (W, V) have been completely classified up to isomorphism. In this section we list them briefly; see Bourbaki [44], Grove–Benson [124], or Humphreys [133] for more details. We will confine ourselves to the reflection groups that are *essential*, *irreducible*, and *nontrivial*; all others are obtained from these by taking direct sums and, possibly, adding an extra summand on which the group acts trivially.

First, we list three infinite families of reflection groups:

- Type A_n $(n \geq 1)$: Here W is the symmetric group on $n + 1$ letters, acting as in Example 1.10 on a certain n-dimensional subspace of $\mathbb{R}^{n+1}$. This group is the group of symmetries of a regular n-simplex, and it can also be described as the Weyl group of the root system of type A_n.

- Type C_n, also called type B_n ($n \geq 2$): This is the group W of signed permutations acting on $\mathbb{R}^n$ as in Example 1.11. (We require $n \geq 2$ because Example 1.11 with $n = 1$ gives the group of type A_1 again.) The group W is the group of symmetries of the n-cube (or n-dimensional hyperoctahedron); it is also the Weyl group of the root system of type B_n and the root system of type C_n. Following a common convention in the theory of buildings, we will usually call W the reflection group of type C_n, but we may occasionally call it the reflection group of type B_n.

- Type D_n ($n \geq 4$): This is the Weyl group of the root system of type D_n that we saw in Example 1.13. It does not correspond to any regular solid.

Next, there are seven exceptional groups:

- Type E_n ($n = 6, 7, 8$): This is the Weyl group of a root system of the same name. It does not correspond to any regular solid.

- Type F_4: This is the Weyl group of a root system of the same name; it is also the group of symmetries of a certain self-dual 24-sided regular solid in $\mathbb{R}^4$ whose (3-dimensional) faces are solid octahedra.

- Type G_2: This is the Weyl group of the root system of the same name that we saw in Example 1.9. It is dihedral of order 12, so we can also describe it as the group of symmetries of a hexagon.

- Type H_n ($n = 3, 4$): This does not correspond to any root system, but it is the symmetry group of a regular solid X. When $n = 3$, X is the dodecahedron (which has 12 pentagonal faces) or, dually, the icosahedron (which has 20 triangular faces). When $n = 4$, X is a 120-sided solid in $\mathbb{R}^4$ (with dodecahedral faces) or, dually, a 600-sided solid (with tetrahedral faces).

Finally, we have the dihedral groups D_{2m} (not to be confused with the groups of type D_n listed above). If $m = 2$, the group is reducible (it is $\{\pm 1\} \times \{\pm 1\}$ acting on $\mathbb{R} \oplus \mathbb{R}$). The cases $m = 3$ and 4 correspond, respectively, to the groups of type A_2 and C_2. And the case $m = 6$ corresponds to the group of type G_2. This leaves:

- Type $I_2(m)$ ($m = 5$ or $m \geq 7$): The group W is the dihedral group of order $2m$. It is the symmetry group of a regular m-gon, but it does not correspond to any root system.

Remarks 1.16. (a) The subscript in the notation for each type is the rank of the reflection group.

(b) There is also a classification of the irreducible root systems. They are the root systems mentioned in the discussion above, except that C_n should be included only for $n \geq 3$ in order to avoid repetition; see Exercise 1.17.

(c) To learn more about the regular solids mentioned above, see Coxeter [86] or Lyndon [156] or further references cited therein.

Exercise 1.17. Show that the root system of type C_2 is the same as that of type B_2, up to a rotation and a rescaling of the metric.

1.4 Cell Decomposition

Let (W, V) be an essential finite reflection group. The hyperplanes H with $s_H \in W$ cut V into polyhedral pieces, which turn out to be cones over simplices. Intersecting these cones with the unit sphere, one obtains a simplicial decomposition of the sphere. These assertions will be proved in Section 1.5. (We have already seen an example of it in Figure 1.3.) The purpose of the present section is to lay the groundwork by studying the polyhedral decomposition of Euclidean space induced by an arbitrary finite set $\mathcal{H}$ of hyperplanes. Following standard terminology, we will also say that $\mathcal{H}$ is a *hyperplane arrangement*. In this chapter hyperplane arrangements will always be assumed to be *finite* and to consist of *linear* hyperplanes. We will have occasion to consider infinite arrangements of affine hyperplanes in Chapter 10.

This section is long because it develops from scratch some basic facts about polyhedral geometry. Readers who are already familiar with these facts or are willing to accept them as "intuitively obvious" can read the first few subsections quickly for notation and terminology. Section 1.4.6, however, is likely to be new for many readers.

Throughout this section V will denote a finite-dimensional real vector space, and $\mathcal{H} = \{H_i\}_{i \in I}$ will denote a hyperplane arrangement in V indexed by a finite set I. We assume the hyperplanes are listed without repetition, i.e., $H_i \neq H_j$ for $i \neq j$.

1.4.1 Cells

For each $i \in I$, let $f_i \colon V \to \mathbb{R}$ be a nonzero linear function such that H_i is defined by $f_i = 0$. The function f_i is uniquely determined by H_i, up to multiplication by a nonzero scalar.

Definition 1.18. A *cell* in V with respect to $\mathcal{H}$ is a nonempty set A obtained by choosing for each $i \in I$ a sign $\sigma_i \in \{+, -, 0\}$ and specifying $f_i = \sigma_i$. [Here "$f_i = +$" means $f_i > 0$, and similarly for "$f_i = -$."] Thus A is defined by homogeneous linear equalities or strict inequalities, one for each hyperplane. In more geometric language, we have

$$A = \bigcap_{i \in I} U_i, \tag{1.2}$$

where U_i is either H_i or one of the open half-spaces of V determined by H_i. The sequence $\sigma := (\sigma_i)_{i \in I}$ that encodes the definition of A is called the *sign sequence* of A and is denoted by $\sigma(A)$. The cells such that $\sigma_i \neq 0$ for all i are called *chambers*.

Note that the chambers are nonempty convex open sets that partition the complement $V \smallsetminus \bigcup_{i \in I} H_i$, so they are the connected components of the complement. In general, a cell A is open relative to its *support*, which is defined to be the subspace

$$\operatorname{supp} A := \bigcap_{\sigma_i(A)=0} H_i$$

of V. Equivalently, $\operatorname{supp} A$ is the subspace defined by the equalities $f_i = 0$ that occur in the description of A. Since A is open in $\operatorname{supp} A$, we can also describe $\operatorname{supp} A$ as the linear span of A. The *dimension* of A is, by definition, the dimension of its support.

Definition 1.19. We denote by $\Sigma(\mathcal{H})$ the set of all cells and by $\mathcal{C}(\mathcal{H})$ the subset consisting of all chambers.

The cells A form a partition of V into disjoint convex cones, where a *cone* is a subset closed under multiplication by positive scalars. Figure 1.4 shows a simple example, where $\mathcal{H}$ consists of three lines in the plane, numbered 1, 2, and 3. There are 13 cells: 6 chambers (open sectors), 6 open rays, and the cell consisting of the origin. Sign sequences for the chambers are indicated, based on the assumption that the f_i are chosen to be positive on the chamber labeled $+++$. The reader is advised to fill in the sign sequences for the lower-dimensional cells; for example, the ray separating the chambers $+++$ and $+-+$ has sign sequence $+0+$.

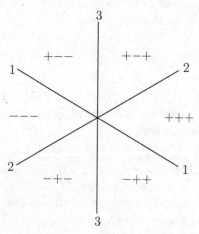

Fig. 1.4. Three lines in the plane.

For another example, let $\mathcal{H}$ consist of the 3 coordinate planes in $\mathbb{R}^3$. There are 27 cells, one for each possible sign sequence: 8 open orthants, which are the chambers, 12 open sectors (4 orthants in each coordinate plane), 6 open rays (two in each coordinate line), and the origin. A second rank-3 example was drawn in Figure 1.3; there are 6 planes, and the picture shows the intersections of the cells with the unit sphere. There are 24 chambers, corresponding to triangular regions on the sphere; 36 two-dimensional faces, corresponding to edges on the sphere; 14 rays, corresponding to the vertices on the sphere; and the origin, corresponding to the empty subset of the sphere.

1.4.2 Closed Cells and the Face Relation

We begin by defining a partial order of the set $\Sigma := \Sigma(\mathcal{H})$ of cells, so that Σ becomes a *poset* (partially ordered set).

Definition 1.20. Given cells $A, B \in \Sigma$, we say that B is a *face* of A, and we write $B \le A$, if for each $i \in I$ either $\sigma_i(B) = 0$ or $\sigma_i(B) = \sigma_i(A)$. More concisely, the ordering on cells is given by the coordinatewise ordering on sign sequences, where we make the convention that $+$ and $-$ are bigger than 0. In terms of linear equalities and inequalities, $B \le A$ if and only if the description of B is obtained from that of A by changing zero or more inequalities to equalities.

For example, the chamber $+++$ in Figure 1.4 has four faces: the chamber itself, the rays $+0+$ and $0++$, and the origin (with sign sequence 000).

Definition 1.21. Given a cell A, let $\overline{A}$ be the set obtained by replacing the open half-spaces that occur in equation (1.2) by the corresponding closed half-spaces. Equivalently, replace the strict inequalities $f_i > 0$ or $f_i < 0$ in the description of A by the corresponding weak inequalities $f_i \ge 0$ or $f_i \le 0$. We call $\overline{A}$ the *closed cell* associated to A. The cell A itself, by contrast, will often be called an *open cell*, even though it is not in general an open subset of V. [The more common term is *relatively open* cell, since, as we have already noted, A is open in its support and hence in $\overline{A}$.]

For example, the closed cell corresponding to each of the chambers in Figure 1.4 above is a closed sector.

Remark 1.22. Readers familiar with cell complexes may find the term "cell" confusing, since our closed cells are not topological balls. Whenever confusion might arise, we will call the cells defined here *conical cells*. If we assume that $\mathcal{H}$ is *essential*, by which we mean that $\bigcap_{H \in \mathcal{H}} = \{0\}$, then every closed conical cell is in fact the cone over a topological ball, gotten by intersecting the cell with a sphere. The proof is left to the interested reader. (See also Section A.2.3 below, where a more precise result will be proved.)

It is immediate from the definitions that

$$\overline{A} = \bigcup_{B \le A} B . \tag{1.3}$$

Since the open cells are disjoint, it follows that the face relation can be characterized in terms of the closed cells:

$$B \leq A \iff \overline{B} \subseteq \overline{A}.$$

This shows, in particular, that $B = A$ if and only if $\overline{B} = \overline{A}$. Hence:

Proposition 1.23. *The function $A \mapsto \overline{A}$ is a bijection from the open cells to the closed cells.* □

We will find it helpful to have a geometric description of the correspondence between open cells and closed cells that does not refer to $\mathcal{H}$:

Proposition 1.24. *Let A be an open cell.*

(1) *$\overline{A}$ is the closure of A in V (in the sense of point-set topology).*
(2) *Let L be the linear span of $\overline{A}$. Then A is the interior of $\overline{A}$ in L, i.e., the largest open subset of L contained in $\overline{A}$.*

Proof. (1) Clearly $\overline{A}$ is closed in V, so it contains the closure of A. Conversely, given $y \in \overline{A}$, choose $x \in A$ and consider the closed line segment from x to y, denoted by $[x, y]$. Each equality in the description of A holds on the whole line segment; and each strict inequality holds on the half-open segment $[x, y)$. So $[x, y) \subseteq A$ and hence y is in the closure of A.

(2) Note first that $L = \operatorname{supp} A$; for $\operatorname{supp} A$ contains $\overline{A}$ and is spanned by A, so it is also spanned by $\overline{A}$. We therefore have $A \subseteq \operatorname{int}_L(\overline{A})$ (the latter being the interior of $\overline{A}$ in L) since A is open in its support. Conversely, suppose $y \in \overline{A} \smallsetminus A$ and consider the segment $[x, y]$ again. Since $y \notin A$, there must be an inequality in the description of A, say $f_i > 0$, such that $f_i(y) = 0$. So if the line segment is continued past y, we immediately have $f_i < 0$, which means we have left $\overline{A}$ (but stayed in L). Hence $y \notin \operatorname{int}_L(\overline{A})$. □

Our next observation is that we can give a direct definition of what it means to be a closed cell, independent of the notion of *open cell*. Recall that a closed cell is defined by equalities or weak inequalities, one for each $i \in I$. Conversely, suppose X is an arbitrary set defined by specifying for each i the equality $f_i = 0$ or one of the weak inequalities $f_i \geq 0$ or $f_i \leq 0$; we will show that X is a closed cell:

Proposition 1.25. *Let X be a set defined by equalities or weak inequalities as above. Then X is a closed cell with respect to $\mathcal{H}$.*

Proof. Let σ_i be 0 if $f_i = 0$ on X. Otherwise, either $f_i \geq 0$ on X or $f_i \leq 0$ on X, and we take σ_i to be $+$ or $-$, accordingly. [*Caution:* It is possible that our original description of X involved an inequality, say $f_i \geq 0$, but that nevertheless $f_i = 0$ on X; so σ_i is 0 in this case.] Let A be the set defined by the signs σ_i. If A is nonempty, then it is a cell and $X = \overline{A}$. To prove $A \neq \emptyset$, choose for each i with $\sigma_i \in \{+, -\}$ a vector $x_i \in X$ with $f_i(x_i) \neq 0$. Let x be the sum of these vectors (or 0 if there are none). Then $x \in A$. □

Corollary 1.26. *An intersection of closed cells is a closed cell.*   □

We turn, finally, to the geometric meaning of the face relation. If one visualizes a cell A in dimension 2 or 3, one sees easily what its faces are, without knowing the particular system of equalities and inequalities by which A was defined. Roughly speaking, the faces are the flat pieces into which the boundary of A decomposes. The following proposition states this precisely:

Proposition 1.27. *Let A be a cell. Then two distinct points $y, z \in \overline{A}$ lie in the same face of A if and only if there is an open line segment containing both y and z and lying entirely in $\overline{A}$. Consequently, the partition of $\overline{A}$ into faces depends only on A as a subset of V, and not on the arrangement $\mathcal{H}$.*

Proof. Suppose y and z are in the same face $B \leq A$. For each condition $f_i = \sigma_i$ in the description of B, we can extend the segment $[y, z]$ slightly in both directions without violating the condition. Since there are only finitely many such conditions, it follows that B contains an open segment containing y and z; hence so does $\overline{A}$.

Suppose now that y and z are in different faces of A. Then there is some i such that y and z behave differently with respect to f_i, say $f_i(y) > 0$ and $f_i(z) = 0$. If we now continue the segment $[y, z]$ past z, we immediately have $f_i < 0$, so we leave $\overline{A}$; hence there is no open segment in $\overline{A}$ containing both y and z.   □

The significance of this for us is that if we want to understand the polyhedral structure of a particular cell A, then we can replace $\mathcal{H}$ by any other hyperplane arrangement for which A is still a cell. We record this for future reference:

Corollary 1.28. *Let A be a cell with respect to $\mathcal{H}$. If A is also a cell with respect to an arrangement $\mathcal{H}'$, then the faces of A defined using $\mathcal{H}'$ are the same as those defined using $\mathcal{H}$.*   □

In practice, we will want to take a minimal set of hyperplanes for a given A. In the next subsection we spell out exactly how to do this in case A is a chamber.

Exercise 1.29. Given $A \in \Sigma$, show that $\bigcup_{B \geq A} B$ is a convex open subset of V. [Suggestion: First draw a picture to see why this is plausible.]

1.4.3 Panels and Walls

Definition 1.30. A cell A with exactly one 0 in its sign sequence is called a *panel*. This is equivalent to saying that supp A is a hyperplane, which is then necessarily in $\mathcal{H}$. If the panel A is a face of a chamber C, then we will also say that A is a *panel of C* and that its support hyperplane H is a *wall* of C.

In low-dimensional examples like the one in Figure 1.4, one sees easily that every chamber is defined by the inequalities corresponding to its walls; the other inequalities are redundant. We will show that this is always the case. Fix a chamber C. We say that C is *defined* by a subset $\mathcal{H}' \subseteq \mathcal{H}$ if C is defined by the conditions $f_i = \sigma_i$, where i ranges over the indices such that $H_i \in \mathcal{H}'$.

Lemma 1.31. *If $H \in \mathcal{H}$ is not a wall of C, then C is defined by $\mathcal{H}' := \mathcal{H} \smallsetminus \{H\}$.*

Proof. Assume, to simplify the notation, that C is defined by the inequalities $f_i > 0$ for all i, and let j be the index such that $H = H_j$. Suppose C is not defined by $\mathcal{H}'$. Then removing the inequality $f_j > 0$ results in a set C' strictly bigger than C. Choose $y \in C' \smallsetminus C$ and $x \in C$. Since $f_j(x) > 0$ and $f_j(y) \leq 0$, there is a point $z \in (x, y]$ such that $f_j(z) = 0$. This point z is then in a panel A of C supported by H, so H is a wall of C. $\square$

Proposition 1.32. *Let C be a chamber and let $\mathcal{H}_C$ be its set of walls. Then C is defined by $\mathcal{H}_C$, and $\mathcal{H}_C$ is the smallest subset of $\mathcal{H}$ with this property.*

Proof. If C is defined by $\mathcal{H}' \subseteq \mathcal{H}$, then we can use $\mathcal{H}'$ to determine the walls of C by Corollary 1.28; hence $\mathcal{H}' \supseteq \mathcal{H}_C$. It remains to show that C is defined by $\mathcal{H}_C$. If $\mathcal{H}$ contains any H that is not a wall of C, then we can remove it by Lemma 1.31 to get a smaller defining set $\mathcal{H}'$. Now C is still a chamber with respect to $\mathcal{H}'$, and replacing $\mathcal{H}$ by $\mathcal{H}'$ does not change the walls. So we may repeat the process to remove another nonwall, and so on. Since $\mathcal{H}$ is finite, we arrive at $\mathcal{H}_C$ after finitely many steps. $\square$

The proof we just gave made crucial use of the fact that the notion of "wall" does not depend on the particular defining set of hyperplanes. Here is a simple intrinsic characterization of the walls:

Proposition 1.33. *Let C be a chamber and let H be a linear hyperplane in V. Then H is a wall of C if and only if C lies on one side of H and $\overline{C} \cap H$ has nonempty interior in H.*

Proof. If H is the support of a panel A of C, then certainly C lies on one side of H and $\overline{C} \cap H$ contains A, which is a nonempty open subset of H. Conversely, suppose H is a hyperplane such that C lies on one side of H and $\overline{C} \cap H$ has nonempty interior in H. Then C is still a chamber with respect to $\mathcal{H}^+ := \mathcal{H} \cup \{H\}$, so we can use $\mathcal{H}^+$ to determine the faces of C. By Proposition 1.25, $\overline{C} \cap H$ is a closed cell $\overline{A}$ with respect to $\mathcal{H}^+$, and the corresponding open cell A is a face of C because $\overline{A} \subseteq \overline{C}$. Since $\overline{A}$ is contained in H and has nonempty interior in H, the support of A must be H. Thus A is a panel of C and its support H is therefore a wall of C. $\square$

Exercises

1.34. This exercise outlines a more direct proof that any chamber is defined by its walls; fill in the missing details.

Let C and $\mathcal{H}_C$ be as in Proposition 1.32, and let C' be the $\mathcal{H}_C$-chamber containing C. Suppose $C' \neq C$. Choose $y \in C' \smallsetminus C$ and $x \in C$, and consider the line segment $[x, y]$. By moving y slightly if necessary, we may assume $y \notin \overline{C}$, so that the segment $[x, y]$ crosses at least one $H \in \mathcal{H}$. And by moving x slightly if necessary, we may assume that the segment never crosses more than one H at a time. The first H that is crossed as we traverse the segment starting at x is then a wall of C, contradicting the definition of C.

1.35. Assume that $\mathcal{H}$ is essential, as defined in Remark 1.22. Show that every closed cell $\overline{A}$ is the closed convex cone generated by the 1-dimensional faces of A, i.e., every $x \in \overline{A}$ can be expressed as $x = \sum_{k=1}^{m} y_j$, where each y_k is in a 1-dimensional face of A. [Note: These 1-dimensional faces are rays. They therefore correspond to vertices if we think of cells in terms of their intersections with a sphere as in Remark 1.22.]

1.4.4 Simplicial Cones

Let C be a fixed but arbitrary chamber and let $\mathcal{H}'$ be its set of walls. It will be convenient to take the index set I for $\mathcal{H}$ to be $\{1, 2, \ldots, m\}$ for some m. For simplicity of notation we will assume that the elements of $\mathcal{H}'$ are the hyperplanes $f_i = 0$ for $1 \leq i \leq r$ and that $f_i > 0$ on C for $1 \leq i \leq m$.

Let $V_0 := \bigcap_{i=1}^{m} H_i$. We call $\mathcal{H}$ *essential* if $V_0 = 0$. There is no loss of generality in restricting attention to the essential case. For if we set $V_1 := V/V_0$, then the linear functions f_i pass to the quotient V_1 and define an essential set of hyperplanes there. And the cells determined by these hyperplanes in V_1 are in 1–1 correspondence with the cells in V. More precisely, the cells in V are the inverse images in V of the cells in V_1. [Geometrically, then, the cells in V are simply the cells in V_1 "fattened up" by a factor $\mathbb{R}^d$, where $d := \dim V_0$.]

Note that V_0 is itself a cell, with sign sequence $(0, 0, \ldots, 0)$. It is the smallest cell, in the sense that it is a face of every cell, so $\mathcal{H}$ is essential if and only if the smallest cell is a point. Note that V_0 is also the smallest face of C. Since the faces of C can be determined by using $\mathcal{H}'$ instead of $\mathcal{H}$ (Section 1.4.2), it follows that $V_0 = \bigcap_{i=1}^{r} H_i$.

Assume now that $\mathcal{H}$ is essential. Then our last observation says that $\bigcap_{i=1}^{r} H_i = 0$. It follows that $r \geq n := \dim V$. It is easy to visualize examples in which inequality holds (e.g., C could be the cone over an open square, in which case $r = 4 > 3 = \dim V$). We will now prove that equality holds if and only if the cone C is *simplicial*, by which we mean that for some basis $e_1, \ldots, e_n$ of V, C consists of the linear combinations $\sum_{i=1}^{n} \lambda_i e_i$ with all $\lambda_i > 0$. [In other words, C is the interior of the cone over the simplex with vertices $e_1, \ldots, e_n$.]

Proposition 1.36. *Assume that* $\mathcal{H}$ *is essential. Then the following conditions on the chamber C are equivalent:*

(i) C *is a simplicial cone.*
(ii) C *has exactly n panels, i.e., $r = n$.*
(iii) $f_1, \ldots, f_r$ *are linearly independent.*
(iv) $f_1, \ldots, f_r$ *form a basis for the dual space V^* of V.*

Proof. As we noted above, the assumption that $\mathcal{H}$ is essential implies that $\bigcap_{i=1}^{r} H_i = 0$, i.e., that the equations $f_1 = 0, \ldots, f_r = 0$ have only the trivial solution. The equivalence of (ii), (iii), and (iv) follows easily from this by elementary linear algebra.

Suppose now that (ii)–(iv) hold, and let $(e_i)_{1 \le i \le n}$ be the basis of V dual to (f_i). Then the description "$f_i > 0$ for $1 \le i \le n$" of C implies that C consists of the positive linear combinations of the e_i, which proves (i).

Conversely, (i) implies that C is defined by $x_i > 0$ for $1 \le i \le n$, where x_i is the ith coordinate function with respect to some basis for V. We can use this description of C to determine its walls, which are easily seen to be the coordinate hyperplanes $x_i = 0$; this proves (ii)–(iv).   □

1.4.5 A Condition for a Chamber to Be Simplicial

The result of this subsection will be used later to show that the chambers associated to an essential finite reflection group are always simplicial cones.

We continue with the notation of the previous subsection. Assume further that V has an inner product $\langle -, - \rangle$. Then the linear function f_i is given by $\langle e_i, - \rangle$ for a unique vector $e_i \in V$. Replacing f_i by a scalar multiple, we may assume $\|e_i\| = 1$; thus e_i is one of the two unit vectors perpendicular to H_i. Whenever there is a fixed chamber C under discussion, as there is at the moment, then we can remove this ambiguity by requiring that e_i point toward the side of H_i containing C. This is equivalent to requiring, as above, that $f_i > 0$ on C.

In summary, then, we are now assuming that the chamber C is defined by $\langle e_i, - \rangle > 0$ for $1 \le i \le m$, where the e_i are unit vectors, and that the first r of these inequalities in fact suffice to define C. Moreover, no smaller set of linear inequalities defines C. We repeat, for emphasis, that the collection of vectors $(e_i)_{1 \le i \le r}$ is completely determined by C, up to reindexing. The following proposition gives a sufficient condition for C to be simplicial in terms of the matrix of inner products $\langle e_i, e_j \rangle$ $(1 \le i, j \le r)$, often called the *Gram matrix* of C.

Proposition 1.37. *Assume that* $\mathcal{H}$ *is essential. If $\langle e_i, e_j \rangle \le 0$ for each $i \ne j$ $(i, j \le r)$, i.e., if the angle between e_i and e_j is not acute, then C is a simplicial cone.*

Proof. According to Proposition 1.36, we must show that $e_1, \ldots, e_r$ are linearly independent. If not, we claim that there is a nontrivial linear relation among them with nonnegative coefficients. For let $\sum_{i=1}^{r} \lambda_i e_i = 0$ be an arbitrary nontrivial linear relation. If the nonzero λ_i all have the same sign, the claim follows at once. Otherwise, we can rewrite the relation in the form

$$\sum_{j \in J} \mu_j e_j = \sum_{k \in K} \mu_k e_k \,,$$

with J and K disjoint nonempty subsets of $\{1, \ldots, r\}$ and all coefficients positive. Then the inner product of the left-hand side of this equation with the right-hand side is ≤ 0. But this is the inner product of a vector with itself, so that vector must be 0. Thus both sides of the equation are 0, and the claim is proved.

Note that what we have done so far applies to *any* set of vectors with pairwise nonpositive inner products. But now let's add the additional information that the inequalities $\langle e_i, - \rangle > 0$ $(i \leq r)$ define the (nonempty) chamber C. This is clearly inconsistent with the existence of a nontrivial nonnegative linear relation among the e_i, so we have reached a contradiction. Thus $e_1, \ldots, e_r$ are indeed linearly independent. $\square$

1.4.6 Semigroup Structure

We return now to the general setup. Thus $\mathcal{H} = \{H_i\}_{i \in I}$ is not necessarily essential, and V is not assumed to be equipped with an inner product. We saw in Section 1.4.2 that the set $\Sigma := \Sigma(\mathcal{H})$ of cells is a poset under the face relation. What is less obvious, and perhaps surprising, is that there is a natural way to multiply cells, so that Σ becomes a *semigroup*. This product was introduced by Bland in the early 1970s in connection with linear programming, and it eventually led to one approach to the theory of oriented matroids; see [37]. Tits [247] discovered the product independently (in the setting of Coxeter complexes and buildings), although he phrased his version of the theory in terms of "projection operators" rather than products.

We proceed now to the definition of the product. Given two cells $A, B \in \Sigma$, choose $x \in A$ and $y \in B$, and consider a typical point $p_t := (1 - t)x + ty$ on the line segment $[x, y]$ $(0 \leq t \leq 1)$. For each $i \in I$ and all sufficiently small $t > 0$, the sign of $f_i(p_t)$ is the same as the sign of $f_i(x)$ unless $f_i(x) = 0$, in which case the sign of $f_i(p_t)$ is the same as the sign of $f_i(y)$. Hence there is a cell C that contains p_t for all sufficiently small $t > 0$, and its sign sequence is given by $\sigma_i(C) = \sigma_i(A)$ unless $\sigma_i(A) = 0$, in which case $\sigma_i(C) = \sigma_i(B)$. Note that the sign sequence of C, and hence C itself, depends only on A and B, not on the choice of $x \in A$ and $y \in B$. We will call C the *product* of A and B:

Definition 1.38. Given two cells $A, B \in \Sigma$, their *product* is the cell AB with sign sequence

$$\sigma_i(AB) = \begin{cases} \sigma_i(A) & \text{if } \sigma_i(A) \neq 0, \\ \sigma_i(B) & \text{if } \sigma_i(A) = 0. \end{cases} \qquad (1.4)$$

The cell AB is characterized by the property that if we choose $x \in A$ and $y \in B$, then $(1-t)x + ty$ is in AB for all sufficiently small $t > 0$.

See Figure 1.5 for a simple example, where A and B are half-lines and AB turns out to be a chamber. For a second example, let A' be the half-line opposite A in the same figure; then $AA' = A$. One can easily check from (1.4)

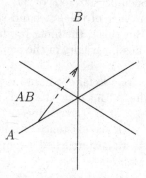

Fig. 1.5. The product of two half-lines.

that the associative law holds:

$$A(BC) = (AB)C \qquad (1.5)$$

for all $A, B, C \in \Sigma$. In fact, the triple product, with either way of associating, can be characterized by the property that $\sigma_i(ABC)$ is $\sigma_i(A)$ unless $\sigma_i(A) = 0$, in which case it is $\sigma_i(B)$ unless $\sigma_i(B) = 0$, in which case it is $\sigma_i(C)$. So Σ is indeed a semigroup. It has an identity, consisting of the cell $\bigcap_{i \in I} H_i$ with sign sequence $(0, 0, \ldots, 0)$.

Following Tits [247], we will often call AB the *projection* of B on A and write

$$AB = \operatorname{proj}_A B \,.$$

This may serve as a reminder of the geometric meaning of the product. We will see, however, that the product notation is quite useful, especially to facilitate application of the associative law. Note that the associative law, in the language of projections, takes the complicated form

$$\operatorname{proj}_A(\operatorname{proj}_B C) = \operatorname{proj}_{\operatorname{proj}_A B} C \,. \qquad (1.6)$$

Equation (1.6) appears (in a slightly different context from ours) in Tits's appendix [249] to Solomon's paper [221] on the descent algebra, and the observation that (1.6) is actually an associative law can be used to give a much simpler treatment; see [55].

The geometry of projections is especially clear when the second factor is a chamber. In order to state the result, we introduce a metric on the set $\mathcal{C} := \mathcal{C}(\mathcal{H})$ of chambers. We will temporarily denote this metric by $d_{\mathcal{H}}(-, -)$; later, after showing that $d_{\mathcal{H}}$ coincides with another naturally defined metric, we will drop the subscript $\mathcal{H}$.

Definition 1.39. The *distance* $d_{\mathcal{H}}(C, D)$ between two chambers C, D is the number of hyperplanes in $\mathcal{H}$ separating C and D. Equivalently, $d_{\mathcal{H}}(C, D)$ is the number of positions at which the sign sequences of C and D differ.

The following result justifies the term "projection."

Proposition 1.40. *Given a cell A and a chamber C, the product AC (or the projection of C on A) is a chamber having A as a face; among the chambers having A as a face, it is the unique one at minimal distance from C.*

Proof. To minimize the distance to C of a chamber $D \geq A$, we must maximize the number of indices i such that $\sigma_i(D) = \sigma_i(C)$. We have no choice about $\sigma_i(D)$ whenever $\sigma_i(A) \neq 0$, so the best we can do is make $\sigma_i(D) = \sigma_i(C)$ whenever $\sigma_i(A) = 0$. This is precisely what the definition of AC in (1.4) achieves. $\quad\square$

Finally, since Σ is now both a poset and a semigroup, it is natural to ask how these structures interact. We record a few simple results in the following proposition, whose proof is routine and is left to the reader.

Proposition 1.41. *Let A and B be arbitrary cells.*

(1) $A \leq AB$, with equality if and only if $\operatorname{supp} B \leq \operatorname{supp} A$.
(2) $A \leq B$ if and only if $AB = B$.
(3) $\operatorname{supp} A = \operatorname{supp} B$ if and only if $AB = A$ and $BA = B$.
(4) AB and BA have the same support, which is the intersection of the hyperplanes in $\mathcal{H}$ containing both A and B. $\quad\square$

Exercises

1.42. Prove the following more precise version of Proposition 1.40: For any chamber $D \geq A$,

$$d_{\mathcal{H}}(C, D) = d_{\mathcal{H}}(C, AC) + d_{\mathcal{H}}(AC, D) . \tag{1.7}$$

In the language of Dress–Scharlau [97], this says that the set $\mathcal{C}_{\geq A}$ of chambers $D \geq A$ is a *gated* subset of the metric space of chambers. Here AC is the "gate" through which one enters $\mathcal{C}_{\geq A}$ to get from C to an arbitrary chamber $D \geq A$. See Figure 1.6 for a schematic illustration.

1.43. We say that cells $A, B, \ldots$ are *joinable* if they have an upper bound in the poset Σ. Show that this holds if and only if they commute with one another in the semigroup Σ, in which case their product is their least upper bound.

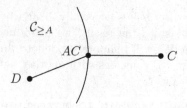

Fig. 1.6. The gate property.

1.44. If A and B have the same support, show that left multiplication by A gives a bijection $\Sigma_{\geq B} \to \Sigma_{\geq A}$, with inverse given by multiplication by B. This holds, for example, if A and B are *opposite*, i.e., $A = -B$.

1.45. Given $A \in \Sigma$, show that the poset $\Sigma_{\geq A}$ is isomorphic to the set of cells of a hyperplane arrangement.

1.4.7 Example: The Braid Arrangement

Let $\mathcal{H}$ be the arrangement in $\mathbb{R}^n$ consisting of the $\binom{n}{2}$ hyperplanes $x_i = x_j$ ($i \neq j$). This has already occurred implicitly in the discussion of Example 1.10. This arrangement, or its essential version in the $(n-1)$-dimensional subspace $x_1 + \cdots + x_n = 0$, is called the *braid arrangement* for reasons that are explained in [182]. It is also called, for more transparent reasons, the *reflection arrangement of type* A_{n-1}. A chamber with respect to $\mathcal{H}$ is a nonempty set defined by inequalities $x_i - x_j > 0$ or $x_i - x_j < 0$ ($i < j$), i.e., it is a set defined by specifying an ordering of the coordinates. Thus there are $n!$ chambers, one for each possible ordering. A typical chamber is given by

$$x_{\pi(1)} > x_{\pi(2)} > \cdots > x_{\pi(n)},$$

where π is a permutation of $\{1, 2, \ldots, n\}$. Figure 1.7 shows the correspondence between chambers and permutations when $n = 4$. Here a permutation π is represented by its list of values $\pi(1)\pi(2) \cdots \pi(n)$. (Recall that this is a rank-3 example, and that cells can be represented by their intersections with the unit sphere; see Figure 1.3.)

Faces are gotten by changing zero or more inequalities to equalities. They correspond to *compositions* $B = (B_1, B_2, \ldots, B_k)$ of the set $\{1, 2, \ldots, n\}$, also called *ordered partitions*. Here the blocks B_i form a set partition in the usual sense, and their order matters. The set composition B encodes the ordering of the coordinates and which coordinates are equal to one another. For example, the common face between the chambers 1234 and 1324 in Figure 1.7 is given by

$$x_1 > x_2 = x_3 > x_4,$$

and it corresponds to the set composition $(\{1\}, \{2, 3\}, \{4\})$. Notice that the chambers can be identified with the set compositions in which all blocks are singletons.

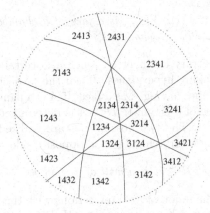

Fig. 1.7. Chambers correspond to permutations.

One can verify from equation (1.4) the following interpretation of the product in terms of set compositions: Take (nonempty) intersections of the blocks in lexicographic order; more precisely, if $B = (B_1, \ldots, B_l)$ and $C = (C_1, \ldots, C_m)$, then

$$BC = (B_1 \cap C_1, \ldots, B_1 \cap C_m, \ldots, B_l \cap C_1, \ldots, B_l \cap C_m)\hat{}\,,$$

where the hat means "delete empty intersections." More briefly, BC is obtained by using C to refine B.

This product, at least when the second factor is a chamber, has an interesting interpretation in terms of card shuffling. See [55] and further references cited there.

Remark 1.46. Although chambers correspond to permutations, their product in the semigroup Σ of cells has nothing to do with the usual product of permutations. In fact, the product of two chambers is always equal to the first factor.

1.4.8 Formal Properties of the Poset of Cells

Recall from Section 1.4.2 that the poset $\Sigma := \Sigma(\mathcal{H})$ of (open) cells is isomorphic to the poset of *closed* cells, where the latter is ordered by inclusion. (See Proposition 1.23 and the paragraph preceding it.) Recall, also, that any intersection of closed cells is a closed cell (Corollary 1.26); consequently:

Proposition 1.47. *Any two elements of Σ have a greatest lower bound.*   □

We will denote by $A \cap B$ the greatest lower bound of two open cells A and B. It is, of course, *not* the set-theoretic intersection of A and B, this intersection being empty unless $A = B$; it is, rather, the open cell whose closure is the intersection of $\overline{A}$ and $\overline{B}$.

Proposition 1.48. *Any cell $A \in \Sigma$ is a face of a chamber. If A is a panel, then it is a face of exactly two chambers.*

Proof. Choose an arbitrary chamber C. [Such a C certainly exists: V is not the union of finitely many hyperplanes.] Then the projection AC defined in Section 1.4.6 is a chamber having A as a face. If A is a panel with support H_i, then the sign sequence of a chamber $D \geq A$ is determined except for $\sigma_i(D)$, which is either $+$ or $-$. The two possibilities are realized by AC and AC', where C and C' are arbitrary chambers on opposite sides of H_i. $\square$

Corollary 1.49. *Every $H \in \mathcal{H}$ is a wall of a chamber.*

Proof. H cannot be the union of its intersections with the other hyperplanes, so there is at least one panel A with support H. Hence H is a wall of each of the chambers $C > A$. $\square$

1.4.9 The Chamber Graph

Definition 1.50. Two chambers C and C' are *adjacent* if they are distinct and have a common panel A.

Note that C and C' are then the two chambers having A as a face, and their sign sequences differ in exactly one position. Thus the hyperplane $H := \operatorname{supp} A$ is the unique element of $\mathcal{H}$ separating C from C'; in particular, $d_{\mathcal{H}}(C, C') = 1$, where $d_{\mathcal{H}}$ is our metric on the set $\mathcal{C} := \mathcal{C}(\mathcal{H})$ of chambers (Definition 1.39). Moreover, $A = C \cap C'$. [One can prove this last assertion by a dimension argument, or by looking at sign sequences, or simply by checking the definition of $C \cap C'$ above.] We will often say, in this situation, that "C and C' are adjacent along the wall H."

Example 1.51. If $\mathcal{H}$ is the braid arrangement (Section 1.4.7), then chambers are labeled by permutations, viewed as lists of numbers. Two chambers are adjacent if and only if the lists differ by the interchange of two consecutive elements. See Figure 1.7.

Definition 1.52. The *chamber graph* associated with $\mathcal{H}$ is the graph whose vertex set is the set $\mathcal{C}$ of chambers, with an edge joining two chambers C, C' if and only if they are adjacent.

We can visualize the chamber graph by putting a dot in each chamber and an edge cutting across each panel, as in Figure 1.8. We will sometimes draw the schematic diagram

$$C \underset{H}{\rule{0pt}{1.5em}\!\!-\!\!\!-\!\!\!-\!\!} C' \tag{1.8}$$

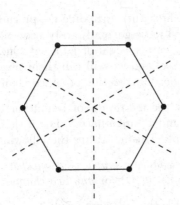

Fig. 1.8. The chamber graph is a hexagon.

to indicate that C and C' are adjacent along H. The horizontal line is intended to suggest an edge in the chamber graph, and the dashed vertical line represents the wall that is crossed in going from C to C'.

There is a canonical metric on the vertices of any graph, where the distance between two vertices is the minimal length of a path joining them. The usual convention is that the distance is ∞ if the two vertices are in different connected components. But we will see below that the chamber graph is in fact connected and that moreover, the graph metric on the set of chambers coincides with the metric $d_{\mathcal{H}}$ of Definition 1.39. Before proceeding to this, we introduce some terminology that we will be using throughout the book.

Definition 1.53. A path in the chamber graph is called a *gallery*. Thus a gallery is a sequence of chambers $\Gamma = (C_0, C_1, \dots, C_l)$ such that consecutive chambers C_{i-1} and C_i $(i = 1, \dots, l)$ are adjacent. The integer l is called the *length* of Γ. We will write

$$\Gamma: C_0, \dots, C_l$$

and say that Γ is a gallery from C_0 to C_l, or that Γ *connects* C_0 and C_l. The minimal length l of a gallery connecting two chambers C, D is called the *gallery distance* between C and D and is denoted $d(C, D)$. Finally, a gallery $C = C_0, \dots, C_l = D$ of minimal length $l = d(C, D)$ is called a *minimal gallery* from C to D. This is the same as what is commonly called a *geodesic* in the chamber graph.

Once we have proven that $d = d_{\mathcal{H}}$, we will no longer need the notation $d_{\mathcal{H}}$, nor will we need to refer to the distance as "gallery distance," though we may still do so occasionally for emphasis.

We sometimes represent a gallery schematically by means of a diagram

$$\Gamma: C_0 \text{———} C_1 \text{———} C_2 \text{———} \cdots \text{———} C_l \ ,$$

which may be further decorated with hyperplanes as in the diagram (1.8).

Warning. In some of the literature, including the precursor [53] of the present book, galleries are defined more generally to be sequences as above in which consecutive chambers are either equal or adjacent. Such sequences do come up naturally, as we will see, and we will call them *pregalleries*. A pregallery can be converted to a gallery by deleting repeated chambers.

We noted above that the metric $d_\mathcal{H}$ of Definition 1.39 has the property that $d_\mathcal{H}(C, C') = 1$ if C and C' are adjacent, i.e., if they are connected by an edge in the chamber graph. This motivates the following:

Proposition 1.54. *The chamber graph is connected, and the gallery distance $d(C, D)$ is equal to $d_\mathcal{H}(C, D)$ for any two chambers C, D.*

The crux of the proof is the following result:

Lemma 1.55. *For any two chambers $C \neq D$, there is a chamber C' adjacent to C such that $d_\mathcal{H}(C', D) = d_\mathcal{H}(C, D) - 1$.*

Proof. Since C is defined by its set of walls (Proposition 1.32), there must be a wall of C that separates C from D. [Otherwise, we would have $D \subseteq C$, contradicting the fact that distinct chambers are disjoint.] Let A be the corresponding panel of C, and let C' be the projection AD (Section 1.4.6). Then C' is adjacent to C, and $d_\mathcal{H}(C', D) = d_\mathcal{H}(C, D) - 1$.   □

Proof of the proposition. Given two chambers C, D, we may apply the lemma finitely many times to obtain a gallery of length $d_\mathcal{H}(C, D)$ from C to D. In particular, the chamber graph is connected and $d \leq d_\mathcal{H}$. To prove the opposite inequality, consider a gallery

$$C = C_0, C_1, \ldots, C_l = D$$

of minimal length $l = d(C, D)$. Then $d_\mathcal{H}(C_{i-1}, C_i) = 1$ for $i = 1, \ldots, l$, whence $d_\mathcal{H}(C, D) \leq l$.   □

Given a minimal gallery $C = C_0, \ldots, C_l = D$, let $H_1, \ldots, H_l \in \mathcal{H}$ be the hyperplanes such that C_{i-1} and C_i are adjacent along H_i. [*Warning:* This notation has nothing to do with our original indexing of the elements of $\mathcal{H}$ as $\{H_i\}_{i \in I}$; we will have no further need for that indexing.] We will refer to the H_i as the "walls crossed" by the gallery. Since exactly one component of the sign sequence changes as we move from one chamber to the next, and since exactly $l = d(C, D)$ signs must change altogether, it is clear that $H_1, \ldots, H_l$ are distinct and are precisely the elements of $\mathcal{H}$ that separate C from D. Conversely, suppose we have a gallery from C to D that does not cross any wall more than once. If k is the length of the gallery, then exactly k signs change, so $k = l$ and the gallery is minimal. This proves the following:

Proposition 1.56. *A gallery from C to D is minimal if and only if it does not cross any wall more than once. In this case the walls that it crosses are precisely those that separate C from D.*   □

Since the set $\mathcal{C} = \mathcal{C}(\mathcal{H})$ of chambers is a metric space, it has a well-defined *diameter*, which we will also refer to as the diameter of Σ; by definition, it is the maximum distance $d(C, D)$ between two chambers C, D. The following result is immediate from the interpretation of the metric on $\mathcal{C}$ as $d_{\mathcal{H}}$:

Proposition 1.57. *The diameter of $\mathcal{C}$ is $m := |\mathcal{H}|$. For any chamber C, there is a unique chamber D with $d(C, D) = m$, namely, the opposite chamber $D = -C$.* □

Observe that for any chambers C and D,

$$d(C, D) + d(D, -C) = m. \tag{1.9}$$

Indeed, every hyperplane in $\mathcal{H}$ separates D from either C or $-C$, but not both. Thus if we concatenate a minimal gallery from C to D with a minimal gallery from D to $-C$, we get a minimal gallery from C to $-C$. Consequently:

Corollary 1.58. *For any chambers C, D, there is a minimal gallery from C to $-C$ passing through D.* □

We have confined ourselves so far to distances and galleries between chambers. But it is also possible to consider distances and galleries involving cells other than chambers. The basic facts about these are easily deduced from the chamber case via the theory of projections (Section 1.4.6); see Exercises 1.61 and 1.62 below.

Exercises

1.59. Let C be a chamber.

(a) If A is a cell that is not a chamber, show that AC is not opposite to C.
(b) Conversely, if D is any chamber not opposite C, then $D = AC$ for some panel A of D.

1.60. Arguing as in the proof of Proposition 1.54, prove the following criterion for recognizing the distance function on a graph. Let G be a graph with vertex set $\mathcal{V}$, and let $\delta \colon \mathcal{V} \times \mathcal{V} \to \mathbb{Z}_+$ be a function, where $\mathbb{Z}_+$ is the set of nonnegative integers. Call two vertices *incident* if they are connected by an edge. Assume:

(1) $\delta(v, v) = 0$ for all vertices v.
(2) If v and v' are incident, then $|\delta(v, w) - \delta(v', w)| \leq 1$ for all vertices w.
(3) Given vertices $v \neq w$, there is a vertex v' incident to v such that $\delta(v', w) < \delta(v, w)$.

Then G is connected, and δ is the graph metric.

1.61. Given $A, C \in \Sigma$ with C a chamber, consider galleries

$$C_0, \dots, C_l = C$$

with $A \leq C_0$. Such a gallery will be said to *connect* A to C. Show that a gallery from A to C of minimal length must start with $C_0 = AC$. Deduce that the minimal length $d(A, C)$ of such a gallery is $|\mathcal{S}(A, C)|$, where $\mathcal{S}(A, C)$ is the set of hyperplanes in $\mathcal{H}$ that strictly separate A from C. [A hyperplane is said to *strictly separate* two subsets if they are contained in opposite open half-spaces.]

1.62. More generally, given any two cells $A, B \in \Sigma$, consider galleries Γ of the form

$$C_0, \dots, C_l$$

with $A \leq C_0$ and $B \leq C_l$. In other words, Γ is a path in the chamber graph starting in $\mathcal{C}_{\geq A}$ and ending in $\mathcal{C}_{\geq B}$. Show that the minimal length $d(A, B)$ of such a gallery is $|\mathcal{S}(A, B)|$, where $\mathcal{S}(A, B)$ has the same meaning as in the previous exercise. More concisely,

$$d(\mathcal{C}_{\geq A}, \mathcal{C}_{\geq B}) = |\mathcal{S}(A, B)|,$$

where the left side denotes the usual distance between subsets of a metric space. Moreover, the chambers C_0 that can start a minimal gallery are precisely those having AB as a face.

A glance at Dress–Scharlau [97] is illuminating in connection with the previous exercise.

1.63. Generalize Corollary 1.58 as follows: For any cell A and chamber D, there is a minimal gallery from A to $-A$ passing through D.

1.64. Proposition 1.57 can be viewed as giving a characterization of the chamber $-C$ opposite a given chamber C in terms of the metric on $\mathcal{C}$. In this exercise we extend that characterization to arbitrary cells.

(a) Fix a cell $A \in \Sigma$, and consider the maximum value of $d(A, B)$, as B varies over all cells. Show that $d(A, B)$ achieves this maximum value if and only if $B \geq -A$.

(b) Deduce that for any $B \in \Sigma$, we have $B = -A$ if and only if $\dim B \leq \dim A$ and $d(A, B) = \max \{d(A, B') \mid B' \in \Sigma\}$.

1.65. Let $\mathcal{D}$ be a nonempty set of chambers. Show that the following conditions are equivalent:

(i) For any $D, D' \in \mathcal{D}$, every minimal gallery from D to D' is contained in $\mathcal{D}$.

(ii) $\mathcal{D}$ is the set of chambers in an intersection of half-spaces bounded by hyperplanes in $\mathcal{H}$.

We say that $\mathcal{D}$ is *convex* if the equivalent conditions (i) and (ii) are satisfied.

1.66. Given two chambers C, D, show that their convex hull (the smallest convex set of chambers containing both of them) consists of the chambers E such that $d(C, D) = d(C, E) + d(E, D)$. In other words, it consists of the chambers that can occur in a minimal gallery from C to D. In particular, the convex hull of any two opposite chambers $C, -C$ is the entire set $\mathcal{C}$ of chambers.

1.67. For any $A \in \Sigma$, show that the set $\mathcal{C}_{\geq A}$ of chambers having A as a face is convex.

1.68. Let $\Sigma' \subseteq \Sigma$ be a nonempty set of cells closed under passage to faces. Let $|\Sigma'|$ be the corresponding subset of V, i.e., $|\Sigma'| := \bigcup_{A \in \Sigma'} A$. Prove that the following three conditions are equivalent:

(i) Σ' is a subsemigroup of Σ.
(ii) Σ' is the set of cells in an intersection of closed half-spaces bounded by hyperplanes in $\mathcal{H}$.
(iii) $|\Sigma'|$ is a convex subset of V.

If Σ' contains at least one chamber, show that (i)–(iii) are equivalent to:

(iv) The maximal elements of Σ' are chambers, and the set of chambers in Σ' is convex.
(v) Given $A, C \in \Sigma'$ with C a chamber, Σ' contains every minimal gallery from A to C.

1.5 The Simplicial Complex of a Reflection Group

We return, finally, to the setup at the beginning of the chapter, where V is assumed to have an inner product, W is a finite reflection group acting on V, and $\mathcal{H}$ is a set of hyperplanes such that the reflections s_H $(H \in \mathcal{H})$ generate W. We assume further that $\mathcal{H}$ is W-invariant. Such an $\mathcal{H}$ certainly exists. For example, we can take $\mathcal{H}$ to consist of all hyperplanes H with $s_H \in W$; the W-invariance of this set follows from the easily verified identity $s_{wH} = w s_H w^{-1}$. Or if W is defined via a generalized root system Φ, then we can take $\mathcal{H} = \{\alpha^{\perp}\}_{\alpha \in \Phi}$. We will see (Corollary 1.72 below) that there is a unique $\mathcal{H}$, so the two choices just described actually coincide; but this is not obvious a priori.

Throughout this section we denote by Σ, or $\Sigma(W, V)$, the set $\Sigma(\mathcal{H})$ of cells in V with respect to $\mathcal{H}$ that we studied in Section 1.4.

1.5.1 The Action of W on $\Sigma(W, V)$

Since Σ is defined in terms of $\mathcal{H}$ and the linear structure on V, it is clear that W permutes the cells and preserves all of the structure that we introduced in Section 1.4: the face relation, products, adjacency,.... In particular, W acts on Σ as a group of poset automorphisms and a group of semigroup automorphisms, and W acts on the chamber graph as a group of graph automorphisms.

Let C be a fixed but arbitrary chamber, called the *fundamental chamber*, and let S be the set of reflections with respect to the walls of C. The elements of S are called the *fundamental reflections* or *simple reflections*. For $s \in S$ let H_s be the hyperplane fixed by s, and let A_s be the panel of C with support H_s. Then we have $A_s = sA_s < sC$, so C and sC are the two chambers having A_s as a face (Proposition 1.48). Thus the chambers adjacent to C are the chambers sC for $s \in S$. Using the W-action, we deduce that for any $w \in W$, the chambers adjacent to wC are the chambers wsC for $s \in S$. Note that wC and wsC are adjacent along the wall wH_s, which is the fixed hyperplane of the reflection wsw^{-1}. Schematically:

$$
C \underset{H_s}{\rule{0pt}{1em}\!-\!-\!} sC \quad \xrightarrow{\ w\ } \quad wC \underset{wH_s}{\rule{0pt}{1em}\!-\!-\!} wsC \tag{1.10}
$$

It follows that galleries starting at C are in 1–1 correspondence with *words* in the alphabet S; the gallery

$$
C \rule[0.3em]{2em}{0.4pt} s_1 C \rule[0.3em]{2em}{0.4pt} s_1 s_2 C \rule[0.3em]{1em}{0.4pt} \cdots \rule[0.3em]{1em}{0.4pt} s_1 s_2 \cdots s_l C \tag{1.11}
$$

corresponds to the word $s_1 s_2 \cdots s_l$.

The reader may find it helpful to trace out some galleries for the reflection group of type A_2 pictured in Figure 1.9. Here W is the dihedral group of order 6, generated by two reflections s, t whose product is a rotation of order 3. In the figure, w_0 is the element $sts = tst$ of W.

We can now prove the main result of this section, after which we will say more about galleries.

Theorem 1.69.

(1) *The set S of fundamental reflections generates W.*

(2) *The action of W is simply transitive on the set $\mathcal{C}$ of chambers. Thus there is a 1–1 correspondence between W and $\mathcal{C}$ given by $w \leftrightarrow wC$, where C is the fundamental chamber. In particular, the number of chambers is $|W| :=$ the order of W.*

Proof. The proof will proceed in several steps.

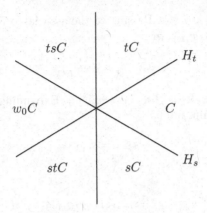

Fig. 1.9. The chambers for $W = \langle\, s, t \,;\, s^2 = t^2 = (st)^3 = 1 \,\rangle$.

(a) We first show that the subgroup $W' := \langle S \rangle$ generated by S acts transitively on $\mathcal{C}$. Given $D \in \mathcal{C}$ we can choose a gallery from C to D since the chamber graph is connected (Proposition 1.54). This gallery has the form (1.11), so $D = wC$ with $w := s_1 s_2 \cdots s_l \in W'$.

(b) Next we prove assertion (1) of the theorem, which says that $W = W'$. It suffices to show that W' contains the generators s_H of W ($H \in \mathcal{H}$). Given $H \in \mathcal{H}$, take a chamber D having H as a wall (Corollary 1.49). Then $D = wC$ for some $w \in W'$ by (a), so $H = wH_s$ for some wall H_s of C, and $s_H = wsw^{-1} \in W'$.

(c) To complete the proof, we must show that $wC \neq C$ for $w \neq 1$ in W. We will prove that in fact,

$$d(C, wC) = l(w) , \qquad (1.12)$$

where the right-hand side is the minimal length of an S-word representing w, i.e., the smallest $l \geq 0$ such that

$$w = s_1 s_2 \cdots s_l \qquad (1.13)$$

with $s_i \in S$. Thus we must show that the gallery from C to wC corresponding to (1.13) as in (1.11) is minimal if $l = l(w)$. Write the gallery as

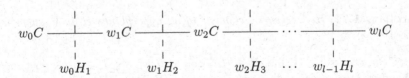

where $w_i := s_1 \cdots s_i$ and $H_i := H_{s_i}$ (see (1.10), and note that $w_i = w_{i-1}s_i$ for $i \geq 1$). Here $w_0 := 1$. If the gallery is not minimal, then two of the walls indicated above must coincide (Proposition 1.56). Thus $w_{i-1}H_i = w_{j-1}H_j$ for some i, j with $1 \le i < j \le l$. Passing to the associated reflections, we obtain $w_{i-1} s_i w_{i-1}^{-1} = w_{j-1} s_j w_{j-1}^{-1}$, or

$$w_i w_{i-1}^{-1} = w_j w_{j-1}^{-1}.$$

This can be rewritten as $w_{i-1}^{-1} w_{j-1} = w_i^{-1} w_j$. Expanding both sides in terms of $s_1, \ldots, s_l$, we obtain, finally,

$$s_i \cdots s_{j-1} = s_{i+1} \cdots s_j.$$

Hence

$$
\begin{aligned}
w &= s_1 \cdots s_{i-1}(s_i \cdots s_{j-1}) s_j \cdots s_l \\
&= s_1 \cdots s_{i-1}(s_{i+1} \cdots s_j) s_j \cdots s_l \\
&= s_1 \cdots \hat{s}_i \cdots \hat{s}_j \cdots s_l,
\end{aligned}
$$

where the hats indicate deleted letters. This contradicts the minimality of l and completes the proof. □

The miracle that occurred at the end of the proof leads to the following surprising result:

Corollary 1.70. *Let $w = s_1 s_2 \cdots s_m$ with $s_i \in S$. If there exists a shorter expression for w as a product of elements of S, then there are indices $i < j$ such that*

$$w = s_1 \cdots \hat{s}_i \cdots \hat{s}_j \cdots s_m.$$

Proof. The hypothesis implies that $d(C, wC) < m$, so the gallery corresponding to the given expression for w is not minimal. The conclusion now follows from the proof of (c) above. □

Remark 1.71. We will explore the algebraic consequences of this remarkable property of (W, S), which we call the *deletion condition*, in a more general setting in Chapter 2. And in Chapter 3, again in a more general setting, we will give an alternative proof of it that seems less magical. See Lemma 3.70 and the paragraph following the proof of Lemma 3.71.

Next, here is the promised uniqueness of $\mathcal{H}$.

Corollary 1.72. *$\mathcal{H}$ necessarily consists of all hyperplanes H in V such that $s_H \in W$.*

Proof. Suppose $s_H \in W$ but $H \notin \mathcal{H}$. Then $H \not\subseteq \bigcup_{H' \in \mathcal{H}} H'$, so H must meet a chamber D. Since the element $w := s_H$ of W fixes H, it follows that wD meets D and hence that $wD = D$, contradicting the theorem. □

We close this subsection by summarizing what we now know about the connection between S-words and galleries. This is most easily stated in the language of Cayley graphs. We recall the definition of the latter in the form that is most convenient for our purposes; there are slight variants of the definition in the literature.

Definition 1.73. Let G be a group and let S be a symmetric set of generators of G that does not contain the identity. Here "symmetric" means that $S = S^{-1}$. Then the *Cayley graph* of (G, S) is the (undirected) graph whose vertex set is G and whose edges are the unordered pairs $\{g, h\}$ such that $h = gs$ for some $s \in S$.

Note that the left-translation action of G on itself induces a left action of G on the Cayley graph, since the edges are defined using right translation. Note further that paths from 1 to g in the Cayley graph correspond to decompositions of g as a word in the elements of S. In particular, the distance from 1 to g is the minimal length l of an expression

$$g = s_1 s_2 \cdots s_l \tag{1.14}$$

of g as a product of generators.

Definition 1.74. We call the minimal length l of a decomposition as in (1.14) the *length of g with respect to S*, and we write

$$l = l_S(g) .$$

We omit the subscript S if it is clear from the context. A minimal-length decomposition (1.14) is called a *reduced decomposition* of g.

The following result is little more than a restatement of our earlier analysis of galleries, combined with assertion (2) of Theorem 1.69.

Corollary 1.75. *The chamber graph of $\Sigma(W, V)$ is isomorphic, as a graph with W-action, to the Cayley graph of (W, S). For any $w \in W$, there is a 1–1 correspondence between galleries from C to wC and decompositions of w as an S-word. It associates to the decomposition $w = s_1 s_2 \cdots s_l$ the gallery pictured in (1.11). Consequently, minimal galleries from C to wC correspond to reduced decompositions of w, and*

$$d(C, wC) = l(w) . \tag{1.15}$$

Proof. The bijection $wC \leftrightarrow w$ sets up the isomorphism on the level of vertices. The remaining details should be clear at this point and are left to the reader. $\square$

In working with Cayley graphs, one often labels the edge from g to gs by the generator s. (Cayley [78] thought of the label as representing a color, and he called the graph a "colourgroup.") Following this convention, we will often write

$$wC \xrightarrow{\quad s \quad} wsC \tag{1.16}$$

and say that wC is *s-adjacent* to wsC.

Warning. If C_1 and C_2 are adjacent chambers, then the generator s that labels the edge in the chamber graph joining them is *not* in general the reflection that takes C_1 to C_2. Indeed, the reflection taking wC to wsC is wsw^{-1}, which is generally different from s.

Exercise 1.76. Deduce from equation (1.15) that $l(ws) = l(w) \pm 1$ for all $w \in W$ and $s \in S$. Deduce further that $l(sw) = l(w) \pm 1$ for all w, s.

Note. The essential content of this is that one cannot have $l(ws) = l(w)$. One can prove this purely algebraically by a determinant argument. It is *not*, however, a general property of length functions on groups. Consider, for example, the direct product of two groups of order 2, with the three nontrivial elements as generators.

1.5.2 The Longest Element of W

Recall from the general theory of hyperplane arrangements that $-C$ is the unique chamber at maximal distance from the fundamental chamber C (Proposition 1.57). This leads to the following results about W and its generating set S.

Proposition 1.77. *Let (W, V) be a finite reflection group.*

(1) *W has a unique element w_0 of maximal length. Its length is given by $l(w_0) = |\mathcal{H}|$.*
(2) *The element w_0 is characterized by the property that $w_0 C = -C$, where C is the fundamental chamber.*
(3) *$l(w w_0) = l(w_0) - l(w)$ for all $w \in W$.*
(4) *$w_0^2 = 1$, and w_0 normalizes the set S of fundamental reflections.*

Proof. By Theorem 1.69, there is a unique $w_0 \in W$ such that $w_0 C = -C$. Parts (1) and (2) now follow at once from Proposition 1.57 and equation (1.15).
    (3) We have

$$\begin{aligned}
l(w w_0) &= d(C, w w_0 C) \\
&= d(C, w(-C)) \\
&= d(-C, wC) \\
&= |\mathcal{H}| - d(C, wC) \\
&= l(w_0) - l(w),
\end{aligned}$$

where the second-to-last equality follows from equation (1.9).

(4) $w_0^2 C = w_0(-C) = -w_0(C) = C$, so Theorem 1.69 implies that $w_0^2 = 1$. Note next that $-C$ has the same walls as C, so S is the set of reflections with respect to the walls of $-C$. On the other hand, $w_0 S w_0^{-1}$ [$= w_0 S w_0$] is the set of reflections with respect to the walls of $-C = w_0 C$. So $w_0 S w_0^{-1} = S$.   □

It follows from (4) that conjugation by w_0 induces an involution of S (possibly trivial), which we denote by σ_0. We will give a geometric interpretation of this involution in Proposition 1.130.

Exercise 1.78. Give an algebraic proof that w_0 normalizes S by using (3) to calculate $l(w_0 s w_0)$ for $s \in S$.

1.5.3 Examples

Example 1.79. Suppose that (W, V) is essential and of rank 2. One could simply give a direct analysis of this situation, but it will be instructive to see what Theorem 1.69 says about it. Let $m := |\mathcal{H}|$. Then $m \geq 2$, and the m lines in $\mathcal{H}$ divide the plane V into $2m$ chambers, each of which is a sector determined by two rays. The transitivity of W on the set of sectors implies that they are all congruent, so each sector must have angle $2\pi/2m = \pi/m$. In view of assertion (1) of Theorem 1.69, W is generated by two reflections in lines L_1 and L_2 that intersect at an angle of π/m. In other words, W is *dihedral of order $2m$ and (W, V) looks exactly like Example 1.9.*

Let us also record, for future reference, the following fact about this example: Let L_1 and L_2 be the walls of one of the chambers C, and let e_i ($i = 1, 2$) be the unit normal to L_i pointing to the side of L_i containing C; see Figure 1.10. Then the inner product of e_1 and e_2 is given by

$$\langle e_1, e_2 \rangle = -\cos \frac{\pi}{m}.$$

[To understand the sign, note that the angle between e_1 and $-e_2$ is π/m.]

Example 1.80. This is a trivial generalization of the previous example, but it will be useful to have it on record. Assume that (W, V) has rank 2 but is not necessarily essential. In other words, if we write $V = V_0 \oplus V_1$ as in Section 1.1, then $\dim V_1 = 2$. By the previous example applied to (W, V_1), we have $W \cong D_{2m}$ for some $m \geq 2$. Moreover, if C_1 is a chamber in V_1 with walls L_i and normals e_i as above, then $V_0 \times C_1$ is a chamber in V with walls $V_0 \oplus L_i$ and the same normals e_i. In particular, it is still true that a chamber C has two walls and that the corresponding unit normals (pointing toward the side containing C) satisfy

$$\langle e_1, e_2 \rangle = -\cos \frac{\pi}{m}. \tag{1.17}$$

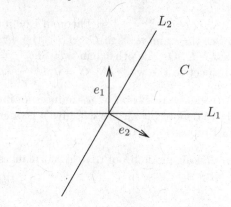

Fig. 1.10. The canonical unit normals associated with a chamber.

Example 1.81. (Type A_{n-1}) Let W be the symmetric group on n letters acting on $\mathbb{R}^n$ as in Example 1.10. Thus a permutation π acts by $\pi(v_i) = v_{\pi(i)}$ for $1 \leq i \leq n$, where $v_1, \ldots, v_n$ is the standard basis for $\mathbb{R}^n$. [These basis vectors were called e_i in Example 1.10, but we now call them v_i in order to avoid confusion with the canonical unit vectors associated to a chamber.] In terms of coordinates, the action is given by $\pi(x_1, \ldots, x_n) = (y_1, \ldots, y_n)$ with

$$x_i = y_{\pi(i)} \quad \text{for } 1 \leq i \leq n . \tag{1.18}$$

Indeed, we have $\pi\left(\sum_{i=1}^n x_i v_i\right) = \sum_{i=1}^n x_i \pi(v_i) = \sum_{i=1}^n x_i v_{\pi(i)}$, whence (1.18).

To analyze this example we can take $\mathcal{H}$ to be the braid arrangement discussed in Section 1.4.7; this set is clearly W-invariant, and the corresponding reflections generate W. We already saw in Section 1.4.7 that the chambers are in 1–1 correspondence with elements of W. The correspondence given there is identical with the one predicted by Theorem 1.69, if we take the fundamental chamber C to be given by

$$x_1 > x_2 > \cdots > x_n . \tag{1.19}$$

To see this, just observe that by (1.18), πC is defined by

$$x_{\pi(1)} > x_{\pi(2)} > \cdots > x_{\pi(n)} . \tag{1.20}$$

The set of inequalities (1.19) is a minimal set of defining inequalities for C, so the latter has $n-1$ walls, the ith of which is the hyperplane $H_{i,i+1}$ given by $x_i = x_{i+1}$ ($i = 1, \ldots, n-1$). [*Note*: $n-1$ is the "right" number of walls, since this example has rank $n - 1$.] The reflection with respect to the ith wall is the transposition $s_i := s_{i,i+1}$ that interchanges i and $i+1$, so assertion (1) of Theorem 1.69 reduces to the well-known fact that the $n-1$ pairwise adjacent transpositions generate the symmetric group.

We remark in passing that equation (1.15) also reduces to a well-known fact about the symmetric group. Recall first that an *inversion* of a permutation

$\pi \in W$ is a pair (i, j) with $1 \leq i < j \leq n$ and $\pi(i) > \pi(j)$. Then the well-known fact is that the length $l_S(\pi)$ is equal to the number of inversions of π. To derive this from (1.15), observe that by (1.20), a wall $x_i = x_j$ with $i < j$ separates C from πC if and only if i occurs later than j in the list of numbers $\pi(1), \ldots, \pi(n)$, i.e., if and only if $\pi^{-1}(i) > \pi^{-1}(j)$. Thus (1.15) says that $l(\pi)$ is equal to the number of inversions of π^{-1}; since $l(\pi) = l(\pi^{-1})$, this proves our assertion.

Let's compute, now, the canonical unit vectors $e_1, \ldots, e_{n-1}$ associated to C. If we let $v_1, \ldots, v_n$ be the standard basis vectors for $V := \mathbb{R}^n$ as above, then the ith inequality defining C can be written $\langle v_i - v_{i+1}, x \rangle > 0$, so the unit vector e_i perpendicular to the ith wall and pointing toward the side containing C is given by

$$e_i = \frac{v_i - v_{i+1}}{\sqrt{2}} .$$

In particular, we can calculate the inner product

$$\langle e_i, e_j \rangle = \begin{cases} 1 & \text{for } j = i, \\ -1/2 & \text{for } j = i+1, \\ 0 & \text{for } j > i+1. \end{cases}$$

Note that $1 = -\cos(\pi/1)$, $-1/2 = -\cos(\pi/3)$, and $0 = -\cos(\pi/2)$. Hence the inner product calculation can be written in the more concise form

$$\langle e_i, e_j \rangle = -\cos \frac{\pi}{m_{ij}} ,$$

where m_{ij} is the order of $s_i s_j$ (or, equivalently, $2m_{ij}$ is the order of the dihedral subgroup generated by s_i and s_j). This formula should not be surprising, in view of (1.17).

Example 1.82. (Type C_n) Let W be the signed permutation group acting on $\mathbb{R}^n$ as in Example 1.11. Then $\mathcal{H}$ consists of the hyperplanes $x_i - x_j = 0$ $(i \neq j)$, $x_i + x_j = 0$ $(i \neq j)$, and $x_i = 0$. To describe a chamber, one has to say which coordinates are positive and which are negative, and one has to specify an ordering of the absolute values of the coordinates. It follows that there are $2^n n!$ chambers, each defined by n inequalities of the form

$$\epsilon_1 x_{\pi(1)} > \epsilon_2 x_{\pi(2)} > \cdots > \epsilon_n x_{\pi(n)} > 0$$

with $\epsilon_i \in \{\pm 1\}$ and $\pi \in S_n$. As fundamental chamber we can take

$$x_1 > x_2 > \cdots > x_n > 0 .$$

The interested reader can work out the fundamental reflections, the canonical unit vectors, and so on, as in Example 1.81. The reader might further want to work out the poset/semigroup of cells, as we did for type A_{n-1} in Section 1.4.7.

Example 1.83. (Type D_n) Let W be the subgroup of the signed permutation group consisting of elements that change an even number of signs (Example 1.13). Then $\mathcal{H}$ consists of the hyperplanes $x_i - x_j = 0$ and $x_i + x_j = 0$ $(i \neq j)$. To figure out what the chambers look like, consider two coordinates, say x_1 and x_2. From the fact that x_1 is comparable to both x_2 and $-x_2$ on any given chamber C, one can deduce that one of the coordinates is bigger than the other in absolute value and that this coordinate has a constant sign. In other words, we have an inequality of the form $\epsilon x_1 > |x_2|$ or $\epsilon x_2 > |x_1|$ on C, where $\epsilon = \pm 1$. It follows that there are $2^{n-1}n!$ chambers, each defined by inequalities of the form

$$\epsilon_1 x_{\pi(1)} > \epsilon_2 x_{\pi(2)} > \cdots > \epsilon_{n-1} x_{\pi(n-1)} > |x_{\pi(n)}| \qquad (1.21)$$

with $\epsilon_i \in \{\pm 1\}$ and $\pi \in S_n$. Note that the last inequality is equivalent to two linear inequalities, $\epsilon_{n-1} x_{\pi(n-1)} > x_{\pi(n)}$ and $\epsilon_{n-1} x_{\pi(n-1)} > -x_{\pi(n)}$, so we have n linear inequalities in all.

As fundamental chamber we take

$$x_1 > x_2 > \cdots > x_{n-1} > |x_n|,$$

with walls $x_1 = x_2$, $x_2 = x_3$, ..., $x_{n-1} = x_n$, and $x_{n-1} = -x_n$. Further analysis is left to the interested reader.

Example 1.84. This final example is intended to provide some geometric intuition. Several statements will be made without proof, and the reader is advised not to worry too much about this.

Let W be the reflection group of type H_3, i.e., the group of symmetries of a regular dodecahedron in $V := \mathbb{R}^3$. It is convenient to restrict the action of W to the unit sphere S^2 and to think of W as a group of isometries of this sphere. As such, it is the group of symmetries of the regular tessellation of the sphere obtained by radially projecting the faces of the dodecahedron onto the sphere. Let P be one of the 12 spherical pentagons that occur in this tessellation. It has interior angles $2\pi/3$, since there are 3 pentagons at each vertex.

The circles of symmetry of this tessellation (corresponding to the planes of symmetry of the dodecahedron) barycentrically subdivide P, thereby cutting it into 10 spherical triangles. A typical such triangle T has angles $\pi/2$, $\pi/3$, and $\pi/5$. The angle $\pi/5 = 2\pi/10$ occurs at the center of P; the angle $\pi/3$, which is half of the interior angle $2\pi/3$ of P, occurs at a vertex of P; and the angle $\pi/2$ occurs at the midpoint of an edge of P, where the line from the center of P perpendicularly bisects that edge. See Figure 1.11.* Finally, a typical chamber C in V is simply the cone over such a triangle T. There are $12 \cdot 10 = 120$ such chambers, so $|W| = 120$. Thus the dodecahedral group W is

---

* Figure 1.11 first appeared in Klein–Fricke [145, p. 106] and is reprinted from a digital image provided by the Cornell University Library's Historic Monograph Collection.

Fig. 1.11. The dodecahedral tessellation, barycentrically subdivided.

a group of order 120 generated by 3 reflections. The calculation of the angles of T above makes it easy to compute the orders of the pairwise products of the generating reflections. One has, for a suitable numbering s_1, s_2, s_3 of these reflections,

$$(s_1 s_2)^3 = (s_2 s_3)^5 = (s_1 s_3)^2 = 1 \, .$$

Exercises

1.85. Recall from the discussion near the end of Section 1.5.1 that we can distinguish various types of adjacency. Spell out what that means in Examples 1.81, 1.82, and 1.83. (For the A_{n-1} case, see Example 1.51.)

1.86. Find w_0 and the induced involution of S in Examples 1.81, 1.82, and 1.83, where w_0 is the element of maximal length (Section 1.5.2).

1.87. In Example 1.84, W is a familiar group of order 120. Which one is it?

1.5.4 The Chambers Are Simplicial

Let (W, V) be a finite reflection group, and let the notation be as in Section 1.5.1. Thus we have a fundamental chamber C with walls H_s ($s \in S$). Let e_s be the unit normal to H_s pointing to the side of H_s containing C. The *Gram matrix* of C is the matrix of inner products $\langle e_s, e_t \rangle$, whose rows and columns are indexed by S (see Section 1.4.5).

The reader who has worked through the examples in Section 1.5.3 will not be surprised by the following computation of the Gram matrix:

Theorem 1.88. *With the notation above, we have*

$$\langle e_s, e_t \rangle = -\cos \frac{\pi}{m(s,t)} \tag{1.22}$$

for $s, t \in S$, where $m(s, t)$ is the order of st. In particular, $\langle e_s, e_t \rangle \le 0$ for $s \neq t$. Consequently, C is a simplicial cone if (W, V) is essential.

The proof will make use of the following lemma:

Lemma 1.89. *Given $s \neq t$ in S, let W' be the group generated by s and t. Then W' is a rank-2 reflection group, and C is contained in a W'-chamber C' having H_s and H_t as its walls.*

Proof. We have $V^{W'} = H_s \cap H_t = (\mathbb{R}e_s \oplus \mathbb{R}e_t)^\perp$, so (W', V) has rank 2. Let $\mathcal{H}' \subseteq \mathcal{H}$ be the set of hyperplanes of the form $w'H_s$ or $w'H_t$ with $w' \in W'$. Then $\mathcal{H}'$ is W'-invariant, and the reflections with respect to the elements of $\mathcal{H}'$ are in W' and generate it. Hence $\mathcal{H}'$ is the set of W'-walls, i.e., the set of hyperplanes that define the W'-cells. Since C is convex and is disjoint from all the elements of $\mathcal{H}'$, it is contained in a W'-chamber C'. The rank calculation shows that C' has two walls. To see that these are H_s and H_t, note that $\overline{C'} \cap H_s \supseteq \overline{C} \cap H_s$, which has nonempty interior in H_s, and similarly for H_t. So H_s and H_t are walls of C' by Proposition 1.33. $\square$

Proof of the theorem. The last assertion follows from Proposition 1.37, so we need only prove the first assertion. We may assume $s \neq t$. Let W' and C' be as in the lemma. Then $\langle e_s, - \rangle$ and $\langle e_t, - \rangle$ are positive on $C \subseteq C'$, so e_s and e_t are the canonical unit normals to the walls of C'. The inner product formula (1.22) now follows from Example 1.80. $\square$

One immediate consequence of the theorem is the following criterion for reducibility:

Corollary 1.90. *Assume that (W, V) is essential. Then (W, V) is reducible if and only if there is a partition of S into (nonempty) subsets S', S'' such that $m(s, t) = 2$ for all $s \in S'$ and $t \in S''$.*

Proof. Suppose there is such a partition. Let W' and W'' be the subgroups $\langle S' \rangle$ and $\langle S'' \rangle$, and let V' (respectively V'') be the subspace of V spanned by the e_s with $s \in S'$ (respectively $s \in S''$). Then we have an orthogonal decomposition $V = V' \oplus V''$, and W can be identified with $W' \times W''$ acting on this direct sum. Thus (W, V) is reducible. The converse is equally easy and is left to the reader. $\square$

The next corollary is more interesting. Recall from Section 1.5.2 that W has a unique longest element w_0 and that w_0 normalizes S.

Corollary 1.91. *Assume that (W, V) is essential and irreducible.*

(1) *w_0 is the only nontrivial element of W that normalizes S.*
(2) *The center of W is trivial unless W contains -1, where -1 denotes $-\mathrm{id}_V$. In this case $w_0 = -1$ and the center is $\{\pm 1\} = \{1, w_0\}$.*

Proof. (1) We will give two proofs, one algebraic and one geometric. Both are instructive.

Algebraic proof. Suppose $w \in W$ normalizes S, and set $s' := wsw^{-1}$ for any $s \in S$. Since we_s is a unit vector orthgonal to $wH_s = H_{s'}$, we have $we_s = \epsilon_s e_{s'}$ with $\epsilon_s = \pm 1$. Let $S_+ := \{s \in S \mid \epsilon_s = 1\}$ and $S_- := \{s \in S \mid \epsilon_s = -1\}$. We claim that $m(s,t) = 2$ for $s \in S_+$ and $t \in S_-$. By irreducibility (and the previous corollary), this will imply that either S_+ or S_- is all of S, i.e., that ϵ_s is independent of s.

To prove the claim, note that

$$\begin{aligned}
\langle e_{s'}, e_{t'} \rangle &= -\langle we_s, we_t \rangle && \text{because } s \in S_+ \text{ and } t \in S_- \\
&= -\langle e_s, e_t \rangle && \text{because } w \text{ is orthogonal} \\
&= \cos(\pi/m) && \text{by (1.22)},
\end{aligned}$$

where $m = m(s,t)$. On the other hand, (1.22) also implies

$$\langle e_{s'}, e_{t'} \rangle = -\cos(\pi/m') \, ,$$

where $m' = m(s', t')$. But $m = m'$ because the function $s \mapsto s'$ is the restriction of a group automorphism, so we must have $\cos(\pi/m) = 0$, i.e., $m(s,t) = 2$. This proves the claim.

We now know that ϵ_s is independent of s. So either $wC = C$ or $wC = -C = w_0 C$. Since W is simply transitive on the chambers, we conclude that $w = 1$ or $w = w_0$.

Geometric proof. Assume that $w \neq 1$ and that $wSw^{-1} = S$, and consider the chamber $D := wC$. The reflections with respect to its walls are the reflections in $wSw^{-1} = S$, so D has the same walls H_s as C. The crux of the proof is now the following claim: Given $s, t \in S$ with $m(s,t) > 2$, either H_s and H_t both separate D from C or else neither of them separates D from C. The claim implies, by irreducibility and the fact that $D \neq C$, that every H_s separates D from C. Hence D is defined by $\langle e_s, - \rangle < 0$ for all $s \in S$, i.e., $D = -C = w_0 C$ and $w = w_0$.

To prove the claim, we set $W' := \langle s, t \rangle$ and apply Lemma 1.89 to get a W'-chamber $C' \supseteq C$ such that H_s and H_t are the walls of C'. We can also apply the lemma with C replaced by D to get a W'-chamber $D' \supseteq D$ such that H_s and H_t are the walls of D'. Now we know exactly what rank-2 reflection groups look like, and since $m(s,t) > 2$, there are only two W'-chambers with H_s and H_t as walls, namely, C' and $-C'$. Hence $D' = \pm C'$, and the claim follows at once.

(2) The center of W normalizes S, so if there is a nontrivial center then w_0 is central and is the unique nontrivial element of the center. But if w_0 is central, then $w_0 e_s = -e_s$ for all s; hence $w_0 = -1$. $\square$

Exercises

1.92. Calculate the integers $m(s,t)$ for the reflection groups of type C_n and D_n (Examples 1.82 and 1.83) and verify (1.22) by direct calculation.

1.93. Suppose (W,V) is essential but reducible. Show that the normalizer of S is bigger than $\{1, w_0\}$.

1.5.5 The Coxeter Matrix

We continue to assume that (W,V) is a finite reflection group with a fundamental chamber C and corresponding set S of fundamental reflections. We assume further that (W,V) is essential. Then $|S| = n := \dim V$ by Theorem 1.88 [since the simplicial cone C has exactly n walls], and the vectors e_s ($s \in S$) form a basis for V. This fact, when combined with the calculation (1.22) of the Gram matrix, has the following important consequence:

Corollary 1.94. *Assume that (W,V) is essential. Then (W,V) is completely determined, up to isomorphism, by the matrix $M := \big(m(s,t)\big)_{s,t \in S}$.*

Proof. Given M, we can recover (W,V) as follows: V can be identified with $\mathbb{R}^S$, the vector space of "S-tuples" $(x_s)_{s \in S}$, with standard basis $(e_s)_{s \in S}$. We give $\mathbb{R}^S$ the inner product defined by (1.22), and we can then identify W with the group of linear automorphisms of $\mathbb{R}^S$ generated by the orthogonal reflections with respect to the hyperplanes $e_s{}^\perp$. □

Definition 1.95. The matrix $M = (m(s,t))_{s,t \in S}$ is called the *Coxeter matrix* associated to W. More precisely, M is associated to W together with a choice of fundamental chamber. It is an $n \times n$ matrix whose rows and columns are indexed by the set S of fundamental reflections.

The short explanation of Corollary 1.94 is that the Coxeter matrix determines the Gram matrix and the Gram matrix determines (W,V). Note that we have the following explicit formula for s in terms of the inner product, and hence in terms of the Coxeter matrix:

$$s(x) = x - 2\langle e_s, x \rangle e_s .\tag{1.23}$$

This is simply formula (1.1) specialized to the case that α is a unit vector.

Remark 1.96. The Coxeter matrix has the following formal properties: It is a symmetric matrix of integers $m(s,t)$, with $m(s,s) = 1$ and $m(s,t) \geq 2$ for $s \neq t$. But not every such matrix can be the Coxeter matrix of a finite reflection group. A further necessary (and, as we will see in Section 2.5.4, sufficient) condition is that the matrix $A := \big(-\cos(\pi/m(s,t))\big)_{s,t \in S}$ must be positive definite. This fact, together with Corollary 1.94, is the basis for the classification result stated in Section 1.3. Indeed, the proof of that result in Bourbaki [44], Grove–Benson [124], and Humphreys [133] consists in analyzing the possibilities for M, given that A is positive definite.

Exercise 1.97. What happens to M if we change the choice of C?

1.5.6 The Coxeter Diagram

Instead of working directly with the Coxeter matrix M, one usually works with a diagram called the *Coxeter diagram*, which encodes all the information in M. The diagram has n vertices, one for each $s \in S$, and the vertices corresponding to distinct elements s, t are connected by an edge if and only if $m(s,t) \geq 3$. If $m(s,t) \geq 4$, then there is more than one convention in the literature as to how to indicate this in the diagram; the one we will follow is simply to label the edge with the number $m(s,t)$. In summary, a labeled edge (with label necessarily at least 4) indicates the value of the corresponding $m(s,t)$; an unlabeled edge indicates that $m(s,t) = 3$; and the lack of an edge joining s and t indicates that $m(s,t) = 2$.

The Coxeter diagrams for all of the irreducible finite reflection groups are shown in Table 1.1. Based on the examples we have given (and Exercise 1.92), the reader should be able to check that the diagrams are correct for the cases A_n, C_n, D_n, G_2, H_3, and $I_2(m)$.

Remarks 1.98. (a) Note that the diagrams that occur in this table are very special. For example, the graphs are all trees; there is very little branching in these trees; and the edge labels are rarely necessary (i.e., the numbers $m(s,t)$ are rarely bigger than 3). One does not need the full force of the classification theorem in order to know these properties; in fact, these properties are among the first few observations that occur in the *proof* of the classification theorem given in the cited references.

(b) Readers who have studied Lie theory will be familiar with the *Dynkin diagram* of a root system. The Dynkin diagram is similar to the Coxeter diagram, but it contains slightly more information; in particular, it contains enough information to distinguish the root system of type B_n from that of type C_n for $n \geq 3$, even though these root systems have the same Weyl group.

(c) In the diagrams corresponding to root systems (all but the last three diagrams in Table 1.1), the only edge labels that occur are 4 and 6. According to a common convention different from the one we have adopted, one omits these labels and instead draws a double bond (two parallel edges) when $m(s,t) = 4$ and a triple bond (three parallel edges) when $m(s,t) = 6$.

Exercises

1.99. Compute the Coxeter diagrams for the reflection groups of type D_2 and D_3. Why aren't these listed in the table?

1.100. Show that an essential finite reflection group (W, V) is irreducible if and only if the graph underlying its Coxeter diagram is connected. Deduce, in the reducible case, a canonical decomposition

$$(W, V) \cong (W_1 \times \cdots \times W_k, V_1 \oplus \cdots \oplus V_k)$$

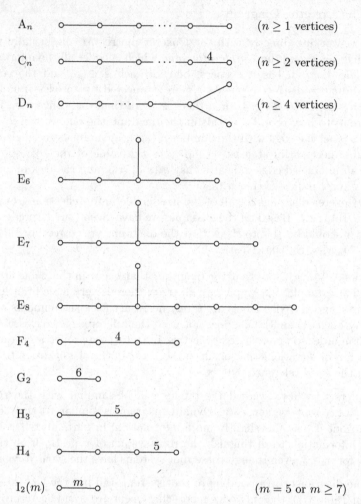

Table 1.1. Coxeter diagrams of the irreducible finite reflection groups.

into "irreducible components," one for each connected component of the Coxeter diagram.

1.101. Let (W, V) be an essential irreducible finite reflection group. The purpose of this exercise is to show that (W, V) is also irreducible in the sense of representation theory, i.e., the only W-invariant subspaces of V are $\{0\}$ and V. Let V' be a W-invariant subspace.

(a) For each $s \in S$, show that either V' contains e_s or V' is contained in the hyperplane $H_s := e_s^\perp$.

(b) If V' contains e_s for some $s \in S$, show that V' contains e_s for all $s \in S$.

(c) Deduce from (a) and (b) that $V' = \{0\}$ or V. Thus the action of W on V is irreducible in the sense of representation theory.

(d) Show that the only linear endomorphisms of V that commute with all elements of W are the scalar-multiplication operators. [This implies that the action of W on V is *absolutely irreducible*.]

1.102. Use Exercise 1.101 to give a new proof of Corollary 1.91(2): The center of an essential, irreducible finite reflection group W is trivial unless W contains -1, in which case the center is $\{\pm 1\}$.

1.5.7 Fundamental Domain and Stabilizers

When studying the action of a group on a set, one wants to know how many orbits there are and what the stabilizers are at typical points of these orbits. Both of these questions have extremely simple answers in the case of W acting on V. We need one bit of terminology.

Definition 1.103. If a group G acts on a space X, then we call a subset $Y \subseteq X$ a *strict fundamental domain* if Y is closed and is a set of representatives for the G-orbits in X.

Theorem 1.104. *Let (W, V) be a finite reflection group, C a chamber, and S the set of reflections with respect to the walls of C. Then $\overline{C}$ is a strict fundamental domain for the action of W on V. Moreover, the stabilizer W_x of a point $x \in \overline{C}$ is the subgroup $\langle S_x \rangle$ generated by $S_x := \{s \in S \mid sx = x\}$. In particular, W_x fixes every point of $\overline{A}$, where A is the cell containing x.*

Proof. Since W is transitive on the chambers, it is clear that every point of V is W-equivalent to a point of $\overline{C}$. Everything else in the theorem will follow if we prove the following claim: For $x, y \in \overline{C}$ and $w \in W$, if $wx = y$ then $x = y$ and $w \in \langle S_x \rangle$. We argue by induction on the length $l := l(w)$ of w with respect to S.

If $l = 0$ there is nothing to prove, so assume $l > 0$ and choose a reduced decomposition $w = s_1 \cdots s_l$. Since the corresponding gallery from C to wC is minimal (Corollary 1.75), we know that C and wC are separated by the wall H_1 fixed by s_1. We therefore have

$$wx = y \in \overline{C} \cap w\overline{C} \subseteq H_1 \,.$$

So if we apply s_1 to both sides of the equation $wx = y$, we obtain

$$w'x = s_1 y = y \,,$$

where $w' := s_1 w = s_2 \cdots s_l$. By the induction hypothesis, it follows that $x = y$ [whence $s_1 \in S_x$] and that $w' \in \langle S_x \rangle$. So $w = s_1 w'$ is also in $\langle S_x \rangle$, and the proof is complete. $\qquad\square$

Corollary 1.105. *For any cell A, the stabilizer W_A of A (as a set) fixes A pointwise.*

Proof. We may assume that A is a face of the fundamental chamber and hence that $A \subseteq \overline{C}$. Then no two distinct points of A are W-equivalent, and the result follows at once.   □

Exercise 1.106. Let A and B be cells, and let AB be their product (Section 1.4.6). Show that $W_{AB} = W_A \cap W_B$.

1.5.8 The Poset Σ as a Simplicial Complex

The fact that every chamber is a simplicial cone in the essential case suggests that the hyperplanes in $\mathcal{H}$ cut the unit sphere in V into (spherical) simplices. Thus it seems intuitively clear that the poset $\Sigma := \Sigma(W, V)$ of cells can be identified with the poset of simplices of a simplicial complex that triangulates a sphere of dimension $\mathrm{rank}(W, V) - 1$. In this subsection we prove this statement rigorously. Before proceeding, the reader might find it helpful to look at the first few paragraphs of Appendix A, where we explain our conventions regarding simplicial complexes. In particular, the statement of the following proposition has to be understood in terms of Definition A.1.

Proposition 1.107. *The poset Σ is a simplicial complex.*

Proof. We may assume that (W, V) is essential, since Σ remains unchanged, up to canonical isomorphism, if we pass to the essential part. According to Definition A.1, we must check two conditions. Condition (a) is that any two elements of Σ have a greatest lower bound; this has already been proved in Proposition 1.47. As to Condition (b), concerning the poset $\Sigma_{\leq A}$ of faces of a cell $A \in \Sigma$, we know that A is a face of a chamber, so it suffices to consider the case that A is a chamber. But it is a trivial matter to compute the poset of faces of a simplicial cone, and this poset is indeed isomorphic to the set of subsets of $\{1, \ldots, n\}$.   □

We gave this somewhat abstract proof of the proposition in order to introduce the unorthodox terminology that we use regarding simplicial complexes; this will be useful later. But it is easy to chase through the discussion in Section A.1.1 in order to describe in more conventional terms how, in the essential case, Σ can be identified with an abstract simplicial complex (in which the simplices are certain finite subsets of a set of "vertices"):

Every 1-dimensional cell $A \in \Sigma$ is a ray $\mathbb{R}_+^* v$, where $\mathbb{R}_+^*$ is the set of positive reals and v is a unit vector; the unit vectors v that arise in this way are the *vertices* of our simplicial complex. For each $(q + 1)$-dimensional cell $A \in \Sigma$ ($q \geq -1$), there is a *q-simplex* $\{v_0, \ldots, v_q\}$ in our complex, where the v_i are the unit vectors in the 1-dimensional faces of A. It should be clear that we do indeed obtain a simplicial complex in this way and that Σ can be identified with the poset of simplices of this complex. Notice that we have allowed $q = -1$ above. The cell A is $\{0\}$ in this case, and it corresponds to the empty set of vertices. [Our convention, as explained in Section A.1.1, is that the empty set is always included as a simplex of an abstract simplicial complex.]

Proposition 1.108. *The geometric realization $|\Sigma|$ is canonically homeomorphic to a sphere of dimension* $\mathrm{rank}(W, V) - 1$.

Proof. Again we may assume that (W, V) is essential, in which case we will exhibit a homeomorphism from $|\Sigma|$ to the unit sphere in V. Recall from Section A.1.1 that $|\Sigma|$ consists of certain convex combinations $\sum_v \lambda_v v$, where v ranges over the vertices of Σ, viewed as basis vectors of an abstract vector space. Now the vertices $v_0, \ldots, v_q$ of any $A \in \Sigma$ can also be viewed as unit vectors in V, and as such, they are linearly independent. Hence we have a map $|\Sigma| \to V \smallsetminus \{0\}$, given by $\sum \lambda_v v \mapsto \sum \lambda_v v$. Composing this with radial projection, we obtain a continuous map $\phi \colon |\Sigma| \to S^{n-1}$. Since ϕ takes $|A| \subset |\Sigma|$ bijectively to $A \cap S^{n-1} \subset V$, it is bijective and therefore a homeomorphism (by compactness of $|\Sigma|$). $\qquad\square$

In view of the results of this section, an essential finite reflection group of rank n is also called a *spherical reflection group* of *dimension* $n - 1$.

Exercise 1.109. Suppose W is the group of symmetries of a regular solid X. Make an intelligent guess as to how to describe Σ directly in terms of X.

1.5.9 A Group-Theoretic Description of Σ

We started the chapter with a "concrete" group W, given to us as a group of linear transformations (or, in more geometric language, as a group of isometries of Euclidean space, or, even better, as a group of isometries of a sphere). The geometry gave us, after we chose a fundamental chamber C, a set S of generators of W. The geometry also gave us a simplicial complex $\Sigma := \Sigma(W, V)$, constructed by means of hyperplanes and half-spaces. We will prove below, however, that if we forget the geometry and just view W as an abstract group with a given set S of generators, then we can reconstruct Σ by pure group theory. This observation will have far-reaching consequences. For simplicity, we assume in this subsection that (W, V) is *essential*.

Consider first the subcomplex $\Sigma_{\leq C}$ consisting of the faces of the fundamental chamber C. To every face $A \leq C$, we associate its stabilizer W_A. In view of Theorem 1.104 and Corollary 1.105, W_A is also the stabilizer of any point $x \in A$, and it fixes $\overline{A}$ pointwise. The theorem also says that W_A is generated by a subset of our given generating set S. Subgroups of this form have a name:

Definition 1.110. A subgroup of W is called a *standard parabolic subgroup*, or simply a *standard subgroup*, if it is generated by a subset of S. Any conjugate of such a subgroup will be called *parabolic*, without the adjective "standard."

Thus we have a function ϕ from $\Sigma_{\leq C}$ to the set of standard subgroups of W, and we will show that ϕ is a bijection. In fact, we can construct the inverse ψ of ϕ by taking fixed-point sets: Let W' be a standard subgroup of W, generated by a set $S' \subseteq S$; then the fixed-point set of W' in $\overline{C}$ is obtained by intersecting $\overline{C}$ with the walls of C corresponding to the reflections in S'. So this fixed-point set is equal to $\overline{A}$ for some $A \leq C$, and we can set $\psi(W') := A$. Using the stabilizer calculation in Section 1.5.7, one can easily check that ψ is inverse to ϕ.

Note next that ϕ and its inverse ψ are order-reversing. For ψ, this is immediate from the definition. In the case of ϕ, the assertion follows from the fact that W_A fixes $\overline{A}$ pointwise and hence stabilizes every face of A. We therefore have a poset isomorphism

$$\Sigma_{\leq C} \cong (\text{standard subgroups})^{\mathrm{op}}, \qquad (1.24)$$

where "op" indicates that we are using the opposite of the usual inclusion order. We will also describe the poset on the right in (1.24) as the poset of standard subgroups, ordered by reverse inclusion. Figure 1.12 illustrates this isomorphism when $n = 3$ and $S = \{s, t, u\}$. Here C is the cone over a triangle, and we have drawn a slice T of C (or, equivalently, the intersection of C with the unit sphere). The figure shows the stabilizer of almost every face of T, the one exception being the empty face, which is not visible in the picture. The empty face corresponds to the cell $A := \{0\}$, which would appear in the picture if we drew the whole chamber C instead of just T. It is the smallest face of T, and its stabilizer is the largest standard subgroup of W, namely, W itself. Similarly, the largest face is T itself, whose stabilizer is the smallest standard subgroup $\{1\}$ (generated by $\emptyset \subset S$).

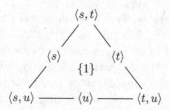

Fig. 1.12. The standard parabolic subgroups as stabilizers.

Returning now to the general case, we can use the W-action to extend our isomorphism to one from the whole poset Σ to the set of *standard cosets* in W, i.e., the cosets wW' of standard subgroups. Indeed, we can send a typical element $wA \in \Sigma$ ($w \in W$, $A \leq C$) to the coset wW_A. It is a routine matter to deduce the following result from what we did above for $\Sigma_{\leq C}$:

Theorem 1.111. *There is a poset isomorphism*

$$\Sigma \cong (\text{standard cosets})^{\text{op}}$$

that is compatible with the W-action, where W acts on the cosets by left translation.   □

We can express the theorem more briefly by saying that Σ is W-isomorphic to the poset of standard cosets in W, ordered by reverse inclusion.

Exercise 1.112. Let W be the reflection group of type A_{n-1} (symmetric group on n letters), with the standard choice of fundamental chamber. Thus S is the set $\{s_1, \ldots, s_{n-1}\}$ of basic transpositions, where s_i interchanges i and $i + 1$. We stated without proof in the discussion of Example 1.10 that the complex Σ associated to W is the barycentric subdivision of the boundary of an $(n - 1)$-simplex. (See also Exercise 1.109.) Prove this rigorously.

1.5.10 Roots and Half-Spaces

Finally, having reconstructed the complex $\Sigma := \Sigma(W, V)$ from the algebraic data (W, S), we wish to do the same for some other geometric concepts: walls, reflections, roots, and half-spaces. The set $\mathcal{H}$ of walls and the corresponding set T of reflections are easy to describe algebraically:

Proposition 1.113. *Let T be the set of reflections in W. Then T is the set of conjugates of elements of S, and there is a bijection $\mathcal{H} \to T$ given by $H \mapsto s_H$. This bijection is W-equivariant, where W acts on T by conjugation.*

Proof. The fact that T is the set of conjugates of elements of S was proved in step (b) of the proof of Theorem 1.69, and the bijection with $\mathcal{H}$ follows from Corollary 1.72. Finally, W-equivariance is simply the familiar formula $s_{wH} = w s_H w^{-1}$, which we have already used several times.   □

Turning next to roots, suppose that W is the reflection group W_Φ associated with a generalized root system Φ. Note, then, that the open half-spaces determined by $\mathcal{H}$ are in 1–1 correspondence with roots: To a half-space U bounded by H we associate the root α that is orthogonal to H and points toward U. [Recall from Section 1.1 that our root systems are assumed to be reduced, so there is only one such α.] Thus U is defined by $\langle \alpha, - \rangle > 0$.

Definition 1.114. We call the open half-space U and the corresponding root α *positive* if U contains the fundamental chamber C, and we call them *negative* otherwise. A positive root α is called *simple* if $\alpha^\perp$ is a wall of C.

(Thus the simple roots give, after normalization, the unit vectors e_s that we discussed earlier.)

We wish to recover Φ, as a set with W-action, from (W, S). Now we have already described $\mathcal{H}$ algebraically, and there is an obvious bijection $\Phi \leftrightarrow \mathcal{H} \times \{\pm 1\}$ where $\mathcal{H} \times \{+1\}$ corresponds to the positive roots and $\mathcal{H} \times \{-1\}$ corresponds to the negative roots. To work out the W-action, we need to know when an element $w \in W$ transforms a positive root to a negative root.

Lemma 1.115. *Let α be a positive root with corresponding hyperplane $H :=$ $\alpha^\perp$. For any $w \in W$, $w\alpha$ is a negative root if and only if H separates C from $w^{-1}C$.*

Proof. Let U be the half-space corresponding to α. Then

$$w\alpha \text{ is negative} \iff wU \text{ is negative}$$
$$\iff wU \text{ does not contain } C$$
$$\iff U \text{ does not contain } w^{-1}C$$
$$\iff H \text{ separates } C \text{ from } w^{-1}C. \qquad \square$$

Let's specialize to the case that w is a fundamental reflection $s \in S$. Then the wall H_s fixed by s is the only wall that separates C from sC. So the lemma in this case says that $s\alpha$ is negative if and only if $\alpha^\perp = H_s$. This gives the following interpretation of Φ as a set with W-action:

Proposition 1.116. *There is a W-equivariant bijection $\Phi \leftrightarrow \mathcal{H} \times \{\pm 1\}$, where the action of a generator s on $\mathcal{H} \times \{\pm 1\}$ is given by*

$$(H, \epsilon) \mapsto \begin{cases} (sH, \epsilon) & \text{if } H \neq H_s, \\ (H, -\epsilon) & \text{if } H = H_s. \end{cases}$$

$\square$

Propositions 1.113 and 1.116 have a purely group-theoretic consequence, whose significance will become clear in the next chapter:

Corollary 1.117. *Let T be the set of conjugates of elements of S. Then there is an action of W on $T \times \{\pm 1\}$ such that a generator $s \in S$ acts by*

$$(t, \epsilon) \mapsto \begin{cases} (sts, \epsilon) & \text{if } t \neq s, \\ (s, -\epsilon) & \text{if } t = s. \end{cases}$$

$\square$

Finally, we will describe algebraically the half-spaces corresponding to the fundamental reflections. Here we identify a half-space with the set of chambers it contains, and we use the bijection $wC \leftrightarrow w$ of Theorem 1.69 to relate this to a subset of W. Our task, then, is to describe the set of $w \in W$ such that $wC \subseteq U_+(s)$, where $U_+(s)$ is the positive half-space bounded by H_s.

Proposition 1.118. *For all $s \in S$ and $w \in W$, $wC \subseteq U_+(s)$ if and only if $l(sw) > l(w)$.*

Proof. We have

$$wC \subseteq U_+(s) \iff H_s \text{ does not separate } C \text{ from } wC$$
$$\iff d(sC, wC) > d(C, wC)$$
$$\iff d(C, swC) > d(C, wC)$$
$$\iff l(sw) > l(w),$$

where we have used the W-invariance of $d(-, -)$ to write $d(sC, wC) = d(C, swC)$. □

Example 1.119. We illustrate the concepts in this section by applying them to the case that W is the symmetric group on n letters acting on $\mathbb{R}^n$ as in Examples 1.10 and 1.81. Recall that a permutation π acts on $\mathbb{R}^n$ by $\pi e_i = e_{\pi(i)}$, where $e_1, \ldots, e_n$ is the standard basis for $\mathbb{R}^n$.

(a) *Walls and reflections.* There is one wall H_{ij} for each *unordered* pair i, j of integers with $1 \le i, j \le n$ and $i \ne j$; it is the hyperplane given by $x_i = x_j$. The corresponding reflection is the transposition $s_{ij} \in W$ that interchanges i and j.

(b) *Roots and half-spaces.* We can take our root system Φ to be the set of vectors $\alpha_{ij} := e_i - e_j$. Thus there is one root for each *ordered* pair of integers i, j with $1 \le i, j \le n$ and $i \ne j$; the corresponding half-space is given by $x_i > x_j$. Recalling that the fundamental chamber C is defined by $x_1 > \cdots > x_n$, we see that α_{ij} is a positive root if and only if $i < j$. So the bijection $\Phi \leftrightarrow \mathcal{H} \times \{\pm 1\}$ is given by $\alpha_{ij} \leftrightarrow (H_{ij}, \epsilon)$, where $\epsilon = +1 \iff i < j$.

Let's take the analysis one step further and identify the roots with subsets of W. Here if α is a root and U is the associated half-space $\langle \alpha, - \rangle > 0$, the corresponding subset of W is $\{w \in W \mid wC \subseteq U\}$. Thinking of elements of W as permutations π, we claim that

$$\alpha_{ij} \leftrightarrow \{\pi \in W \mid \pi^{-1}(i) < \pi^{-1}(j)\} . \tag{1.25}$$

The condition $\pi^{-1}(i) < \pi^{-1}(j)$ has a concrete interpretation if we represent a permutation π by its list of values $\pi(1)\pi(2) \cdots \pi(n)$ as in Section 1.4.7. Namely, it says that i occurs before j in the list. To prove the claim, we need only recall that πC is the chamber given by $x_{\pi(1)} > \cdots > x_{\pi(n)}$. Clearly this chamber is contained in the half-space $x_i > x_j$ if and only if i precedes j in the list representing π.

The reader might find it instructive to verify that the sets of permutations corresponding to α_{14} and α_{41} in Figure 1.7 do indeed form a pair of opposite hemispheres.

(c) *The action of W on roots.* It is immediate from the definitions that $w\alpha_{ij} = \alpha_{w(i)w(j)}$ for any i, j and any permutation $w \in W$. One can easily verify by direct calculation that this is consistent with the correspondence in (1.25), i.e., that left multiplication by w maps the set on the right side of (1.25) to $\{\pi \in W \mid \pi^{-1}(w(i)) < \pi^{-1}(w(j))\}$.

(d) *The simple roots.* Finally, we illustrate Proposition 1.118 in this example. The fundamental reflections are the transpositions $s_i := s_{i,i+1}$, where $1 \leq i \leq n-1$. The corresponding root is $e_i - e_{i+1}$, and the corresponding subset of W, according to (b), is the set of permutations π such that $\pi^{-1}(i) < \pi^{-1}(i+1)$. Recall now that the length of an element $w \in W$ is the number of inversions (see Example 1.81). The interpretation of Proposition 1.118, then, is the following assertion, which one can easily check directly: i precedes $i+1$ in the list $\pi(1) \cdots \pi(n)$ if and only if interchanging i and $i+1$ increases the number of inversions.

Exercises

1.120. Show that every positive root is a nonnegative linear combination of simple roots.

1.121. Show that every root is W-equivalent to a simple root.

1.122. Write down the simple roots for the root systems of type A_n, B_n, C_n, and D_n, based on the fundamental chambers given in Section 1.5.3.

1.123. Given $s \in S$, let α_s be the corresponding simple root. For any $w \in W$, show that $w\alpha_s$ is a positive root if and only if $l(ws) > l(w)$.

1.124. A restatement of Proposition 1.118, in view of Exercise 1.76, is that $wC \not\subseteq U_+(s)$ if and only if $l(sw) < l(w)$, i.e., if and only w admits a reduced decomposition starting with s. Interpret this geometrically in terms of galleries.

1.125. For any $w \in W$, show that $l(w)$ is the number of positive roots α such that $w\alpha$ is negative. [Note that this proves, again, that the length of a permutation is the number of inversions.] Deduce that the longest element $w_0 \in W$ (Section 1.5.2) is characterized by the property that it takes every positive root to a negative root.

1.126. With w_0 as in the previous exercise, show that the action of w_0 on the simple roots α_s ($s \in S$) is given by

$$w_0 \alpha_s = -\alpha_{\sigma_0(s)} \, ,$$

where σ_0 is the involution of S introduced at the end of Section 1.5.2.

1.6 Special Properties of Σ

We close this chapter by mentioning three properties of the simplicial complex $\Sigma := \Sigma(W, V)$ associated to a finite reflection group:

- Σ is a flag complex (Section 1.6.1).
- Σ is a colorable chamber complex (Section 1.6.2).
- Σ is determined by its associated chamber system (Section 1.6.3).

The terminology is explained in Appendix A; see Sections A.1.2, A.1.3, and A.1.4.

The importance of these properties will not become clear until later in the book. The reader may wish to skip ahead to Chapter 2 and return to the present section as needed.

1.6.1 Σ Is a Flag Complex

In many of our examples of finite reflection groups, we have remarked that Σ is a barycentric subdivision. And we note in Section A.1.2 that barycentric subdivisions are always flag complexes. The following result is therefore not surprising:

Proposition 1.127. *The simplicial complex Σ associated to a finite reflection group is a flag complex.*

Proof. In view of the characterization of flag complexes given in Proposition A.7, it suffices to note that every set of pairwise joinable simplices is joinable. This is in fact true in greater generality. Indeed, if Σ is the poset of cells associated to an arbitrary hyperplane arrangement, then every set of pairwise joinable elements of Σ is joinable. This follows from the criterion for joinability given in Exercise 1.43. □

1.6.2 Σ Is a Colorable Chamber Complex

It is immediate that the simplicial complex Σ associated to a finite reflection group is a chamber complex as defined in Section A.1.3. And, using barycentric subdivisions as motivation again, we already know that Σ is often colorable. In fact, it is always colorable.

Proposition 1.128. *The chamber complex Σ associated with a finite reflection group W is colorable.*

Proof. We will use the criterion in terms of retractions stated at the end of Section A.1.3. Choose a chamber C. Then we can define $\phi \colon \Sigma \to \Sigma_{\leq C}$ by letting $\phi(A)$ be the unique face of C that is W-equivalent to A (see Theorem 1.104). It is easy to check that ϕ is a well-defined chamber map and a retraction. □

It is clear from the proof that two simplices have the same type (or color) if and only if they are in the same W-orbit. In particular:

Corollary 1.129. *The action of W on Σ is type-preserving.* □

Recall from Section A.1.3 that the (essentially unique) type function on Σ is completely determined once one assigns types to the vertices of a "fundamental" chamber C. There is a canonical choice in which the set of colors is the set S of fundamental reflections. Namely, for each vertex v of C, the panel of C not containing v is fixed by a unique reflection s, and we declare v to have type s. More succinctly, the panel of C fixed by s has cotype s, i.e., it is an s-panel. See Figure 1.13.

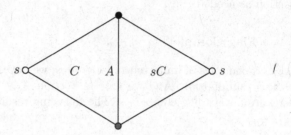

Fig. 1.13. A is the s-panel of C.

We can also describe the type function by means of the correspondence between simplices and standard cosets (Theorem 1.111): The simplex corresponding to a coset $w\langle S'\rangle$ has cotype S'.

Figure 1.14 shows the canonical type function when W is the reflection group of type A_2 (see also Figure 1.9). Here Σ is combinatorially a hexagon. Our definition implies that the white vertex of the fundamental chamber has type s; hence all of the white vertices have type s. Similarly, all the black vertices have type t.

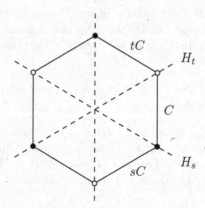

Fig. 1.14. The canonical type function; $\circ = s$, $\bullet = t$.

The reader might find it instructive to work out the types of the three vertices of the fundamental chamber in Figure 1.12. For example, the vertex with stabilizer $\langle s, t \rangle$ has type u.

The fact that the set S plays a dual role, being both a subset of W and a set of types of vertices of Σ, is potentially confusing. In practice, however, it turns out to be quite useful. As an illustration of the dual role, consider the *opposition involution* of $\Sigma = \Sigma(W, V)$, which is by definition the simplicial automorphism $A \mapsto -A$ for $A \in \Sigma$. We denote it by op_Σ. Thinking of S as the set of types of vertices, we get an induced *type-change involution* $(\mathrm{op}_\Sigma)_*$ (possibly trivial) by Proposition A.14. On the other hand, thinking of S as a subset of W, we have an involution σ_0 of S, given by conjugation by the longest element w_0 (Section 1.5.2). These two involutions turn out to coincide:

Proposition 1.130. *The type-change map* $(\mathrm{op}_\Sigma)_*$ *is* σ_0.

Proof. Fix $s \in S$ and let A be the panel of C of cotype s, i.e., the panel of C fixed by s. We have to show that the cotype of the panel $-A$ of $-C$ is $\sigma_0(s)$. In view of Corollary 1.129, the cotype of $-A$ is the same as that of $w_0(-A) = -w_0(A)$. The latter is the panel of C fixed by $w_0 s w_0 = \sigma_0(s)$, so it does indeed have cotype $\sigma_0(s)$. $\square$

Corollary 1.131. *If* (W, V) *is essential, then the following conditions are equivalent:*

(i) *The involution* σ_0 *is trivial.*
(ii) w_0 *is central in* W.
(iii) *The opposition involution of* $\Sigma = \Sigma(W, V)$ *is given by the action of* w_0.
(iv) $w_0 = -1$.
(v) W *contains* -1.
(vi) $w_0 D = -D$ *for every chamber* D.

If (W, S) *is irreducible, these conditions are also equivalent to:*

(vi) W *has a nontrivial center.*

Proof. From the original definition of σ_0, we see that it is trivial if and only if w_0 commutes with each $s \in S$. Hence (i) and (ii) are equivalent. On the other hand, Proposition 1.130 shows that (i) holds if and only if op_Σ is type-preserving. But this holds if and only if w_0 acts on Σ as op_Σ. [If -1 is type-preserving, then it agrees with w_0 on the vertices of the fundamental chamber; moving out along galleries, one deduces that it agrees with w_0 on the entire chamber complex Σ.] Thus (i) and (iii) are equivalent.

Next, the fact that (W, V) is essential implies that if w_0 and op_Σ agree as simplicial maps, then $w_0 = -1$ as a linear map on V. This follows, for example, from the fact that the vertices of Σ can be identified with a set of unit vectors that span V (see Section 1.5.8). Hence (iii) and (iv) are equivalent. And (iv) is equivalent to (v) because w_0 is the unique element of W that takes C to $-C$.

Turning now to (vi), we can argue that it is equivalent to (iii) because two simplicial automorphisms that agree on all chambers must agree on all panels and hence on all vertices. Alternatively, (vi) is equivalent to (ii) because (vi) says that w_0 is independent of the choice of fundamental chamber and hence is invariant under conjugation.

Finally, if (W, S) is irreducible, then we have already shown in Corollary 1.91 (see also Exercise 1.102) that the center of W is nontrivial if and only if (v) holds. $\square$

Exercises

1.132. Give an example to show that we cannot drop the assumption that (W, V) is essential in the corollary.

1.133. Using the classification of finite reflection groups (Section 1.3), find the involution σ_0 for as many of them as you can. [See Section 5.7 for the complete list.]

1.6.3 Σ Is Determined by Its Chamber System

We continue to let Σ be the chamber complex $\Sigma(W, V)$ associated with a finite reflection group. Choose a fundamental chamber C and let S be the set of fundamental reflections. Then, as we just saw in the previous subsection, we have a canonical type function on Σ with values in S. According to Section A.1.4, this yields a "chamber system," consisting of the set of chambers together with "s-adjacency" relations, one for each $s \in S$. For example, sC and C are s-adjacent for any $s \in S$, since they have the same s-panel by definition of the canonical type function. The action of W on Σ being type-preserving (Corollary 1.129), it follows that wsC is s-adjacent to wC for all $w \in W$ and $s \in S$. Thus the present s-adjacency relations are the same as those defined at the end of Section 1.5.1.

Proposition 1.134. *Σ satisfies the hypotheses of Proposition A.20. In particular, Σ is completely determined by its chamber system.*

This is a special case of the following:

Proposition 1.135. *Let Σ be the poset of cells associated to a hyperplane arrangement. For any $A \in \Sigma$ and any chambers $C, D \in \Sigma_{\geq A}$, every minimal gallery joining C to D lies entirely in $\Sigma_{\geq A}$. In particular, if Σ is simplicial (and hence a chamber complex), then $\mathrm{lk}_\Sigma A$ is a chamber complex.*

(Here $\mathrm{lk}_\Sigma A$ is the *link* of A in Σ; see Definition A.19.)

Proof. This has already been proved in Exercise 1.67, but here is an independent proof. Given chambers C, D, let $\Gamma\colon C = C_0, \ldots, C_l = D$ be a minimal gallery. Then the walls $H_1, \ldots, H_l$ crossed by Γ separate C from D (Proposition 1.56). For each $i = 1, \ldots, l$, it follows that A is contained in both closed half-spaces bounded by H_i; hence $A \subseteq H_i$. Assuming inductively that $A \leq C_{i-1}$, we conclude that $A \subseteq \overline{C}_{i-1} \cap H_i = \overline{C}_{i-1} \cap \overline{C}_i$; hence $A \leq C_i$. $\square$

We close with a trivial exercise, designed to force the reader to read the definition of "residue" in Section A.1.4.

Exercise 1.136. Let Σ again be the simplicial complex associated with a finite reflection group. Choose a fundamental chamber so that the set of chambers can be identified with W. Show that the residues in the chamber system of Σ are precisely the standard cosets in W. Thus we recover from Proposition A.20(4) our order-reversing bijection between Σ and the set of standard cosets.

2

Coxeter Groups

Let W be a group generated by a set S of elements of order 2. In case W is a finite reflection group and S is the set of reflections with respect to the walls of some fixed chambers, there is a rich geometric theory that can be constructed from (W, S) by pure group theory. For example, the standard cosets, ordered by reverse inclusion, form a simplicial complex that triangulates a sphere (see Section 1.5.9). In this chapter and the next, we try to develop a similar geometric theory for more general pairs (W, S).

Two things will result from this study. First, we will discover some interesting facts about the combinatorial group theory of finite reflection groups. Second, we will discover a much larger class of groups W that deserve to be called "reflection groups" (or, more precisely, *discrete* reflection groups). The study of these groups W and their associated simplicial complexes was initiated by Tits [240], who called the groups *Coxeter groups* and the complexes *Coxeter complexes*.

2.1 The Action on Roots

We continue to denote by W be an arbitrary group generated by a set S of elements of order 2. If W is to be a "reflection group," then certain elements have to be singled out as "reflections," and they should be in 1–1 correspondence with "walls." Each wall should determine two "half-spaces" (or "roots"), and there should be an action of W on these. We discussed this action in detail for finite reflection groups in Section 1.5.10; see, in particular, Corollary 1.117. This leads to the following condition that (W, S) ought to satisfy if it is to behave like a reflection group. We will call it condition (A) for "action":

(A) *Let T be the set of conjugates of elements of S. There is an action of W on $T \times \{\pm 1\}$ such that a generator $s \in S$ acts as the involution ρ_s given by*

$$\rho_s(t, \epsilon) = \begin{cases} (sts, \epsilon) & \text{if } t \neq s, \\ (s, -\epsilon) & \text{if } t = s. \end{cases}$$

Definition 2.1. We will call the elements of T *reflections.*

In the next chapter we will see that condition (A) is sufficient to let us construct a simplicial complex $\Sigma = \Sigma(W, S)$ on which W acts, with the elements of T being reflections in a sense that will be made precise. For now, however, we will simply use the word "reflection" as an aid to the intuition, and we will explore the algebraic consequences of (A) using Chapter 1 as a guide.

We begin by studying decompositions of elements of W as words in the generating set S. Given $w \in W$ and a decomposition $w = s_1 \cdots s_l$ with $s_i \in S$, consider the sequence of reflections given by

$$t_i := w_{i-1} s_i w_{i-1}^{-1} \tag{2.1}$$

for $i = 1, \ldots, l$, where $w_i := s_1 \cdots s_i$. We saw this sequence in Section 1.5.1; in that setting, the decomposition of w corresponds to a gallery Γ from C to wC (where C is the fundamental chamber) and the t_i are the reflections with respect to the walls crossed by Γ. This motivates the following result:

Lemma 2.2. *Suppose (W, S) satisfies* (A). *Then one can associate to each $w \in W$ a finite subset $T(w) \subseteq T$ with the following properties:*

(1) *$|T(w)| = l(w)$, where $l := l_S$ is the length function on W with respect to S.*

(2) *For any reduced decomposition $w = s_1 \cdots s_l$, the reflections t_i defined in (2.1) are distinct and are precisely the elements of $T(w)$.*

(3) *Consider an arbitrary decomposition $w = s_1 \cdots s_l$. For any $t \in T$, one has $t \in T(w)$ if and only if t occurs an odd number of times in the sequence t_i defined in (2.1).*

Proof. It should be clear that heuristically, $T(w)$ is supposed to be the set of reflections with respect to the "walls that separate C from wC." For finite reflection groups, Lemma 1.115 says that a wall H separates C from wC if and only if the action of w^{-1} on the roots takes the positive root determined by H to a negative root. With this as motivation, we define $T(w)$ by

$$T(w) := \left\{ t \in T \mid w^{-1} \cdot (t, 1) = (w^{-1}tw, -1) \right\} .$$

Consider now an arbitrary decomposition $w = s_1 \cdots s_l$ with $s_i \in S$. Since $w^{-1} = s_l \cdots s_1$, we can compute $w^{-1} \cdot (t, 1)$ (for $t \in T$) by first applying s_1, then applying s_2, and so on. After applying $s_1, \ldots, s_{i-1}$, we will have an element of the form $(w_{i-1}^{-1} t w_{i-1}, \epsilon)$, where $w_{i-1} = s_1 \cdots s_{i-1}$ as above, and we must apply s_i to this element.

In view of the definition of ρ_{s_i}, the application of s_i will change ϵ to $-\epsilon$ if and only if $w_{i-1}^{-1} t w_{i-1} = s_i$, i.e., if and only if $t = t_i$. Hence $w^{-1} \cdot (t, 1) = (w^{-1}tw, (-1)^p)$, where p is the number of i such that $t = t_i$. By the definition of $T(w)$, we therefore have $t \in T(w)$ if and only if p is odd. This proves (3).

Parts (1) and (2) follow from (3) together with the following claim: Given a reduced decomposition $w = s_1 \cdots s_l$, the associated reflections t_i are distinct. To prove the claim, suppose $t_i = t_j$ with $1 \leq i < j \leq l$. Then a simple computation gives, exactly as in the proof of Theorem 1.69,

$$w = s_1 \cdots \hat{s}_i \cdots \hat{s}_j \cdots s_l$$

for some indices $i < j$. This contradicts the assumption that we started with a reduced decomposition, so the lemma is proved.   □

This proof has the same amazing consequence that we observed for finite reflection groups in Corollary 1.70. In order to state it, we introduce a second condition, which we call the *deletion condition*, that a general pair (W, S) may or may not satisfy:

(D) *If* $w = s_1 \cdots s_m$ *with* $m > l(w)$, *then there are indices* $i < j$ *such that* $w = s_1 \cdots \hat{s}_i \cdots \hat{s}_j \cdots s_m$.

Corollary 2.3. *If* (W, S) *satisfies* (A), *then it satisfies* (D).

Proof. Suppose $w = s_1 \cdots s_m$ with $m > l(w)$, and consider the corresponding sequence of reflections t_i $(i = 1, \ldots, m)$. These cannot all be distinct, since this would imply, by assertion (3) of Lemma 2.2, $|T(w)| = m$, contradicting (1). Hence $t_i = t_j$ for some $i < j$, and we can delete two letters as above.   □

2.2 Examples

2.2.1 Finite Reflection Groups

It is obvious that a finite reflection group as in Chapter 1 satisfies (A) (where S is the set of reflections with respect to the walls of a fixed fundamental chamber). Indeed, we arrived at the formulation of (A) by writing down a condition that was already known for finite reflection groups (Corollary 1.117).

Conversely, it is true (but *not* obvious) that every finite group satisfying (A) is a finite reflection group. We will prove this in Section 2.5.4 below. In view of this fact, our remaining examples will necessarily be infinite groups.

2.2.2 The Infinite Dihedral Group

Let W be the *infinite dihedral group* D_∞. By definition, this is the group defined by the presentation

$$W := \left\langle s, t \; ; \; s^2 = t^2 = 1 \right\rangle.$$

For readers not familiar with this notation for group presentations, it simply means that we start with the free group $F := F(s, t)$ on two generators s, t and then divide out by the smallest normal subgroup containing s^2 and t^2. Note that the finite dihedral groups D_{2m} are quotients of W. It follows that the generators s, t of F map to distinct nontrivial elements of W, so no confusion will result if we use the same letters s, t to denote those elements of W. It also follows that st has infinite order in W and hence that W is infinite.

Let $S := \{s, t\} \subseteq W$. We will explain from three different points of view why (W, S) satisfies (A).

(i) Combinatorial group theory

The definition of W via the presentation above makes it easy to define homomorphisms from W to another group. One need only specify two elements of the target group whose squares are trivial, and there is then a homomorphism taking s and t to these elements. In particular, if we want W to act on some set, it suffices to specify involutions ρ_s and ρ_t of that set, and then we can make s and t act as ρ_s and ρ_t, respectively. Condition (A) is now evident.

(ii) Euclidean geometry

We make W act as a group of isometries of the real line L by letting s act as the reflection about 0 ($x \mapsto -x$) and t act as the reflection about 1 ($x \mapsto 2-x$). Note, then, that W acts as a group of *affine* transformations $x \mapsto ax + b$. This action has an associated "chamber geometry," entirely analogous to what we saw in Chapter 1 for finite (linear) reflection groups. It is illustrated in Figure 2.1, where C denotes the open unit interval. The vertices in the picture are the integers. The two colors, black and white, indicate the two W-orbits of vertices.

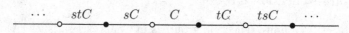

$\cdots \quad stC \quad\quad sC \quad\quad C \quad\quad tC \quad\quad tsC \quad \cdots$

Fig. 2.1. The chambers for D_∞; affine version.

One can now check that the set T in the statement of (A) is the set of elements of W that act as reflections about integers, and one can identify $T \times \{\pm 1\}$ with the set of half-lines whose endpoint is an integer. The action of W on L induces an action of W on this set of half-lines, and condition (A) follows.

(iii) Linear algebra

There is a standard method for "linearizing" affine objects by embedding the affine space in question as an affine hyperplane (i.e., a translate of a linear hyperplane) in a vector space of one higher dimension. In the present case, we do this by identifying the line L above with the affine line $y = 1$ in the plane $V = \mathbb{R}^2$. The affine action of W on L extends to a linear action of W on V.

Explicitly, since we want $s(x,1) = (-x,1)$, we can set $s(x,y) = (-x,y)$; in other words, we can make s act via the matrix

$$\begin{pmatrix} -1 & 0 \\ 0 & 1 \end{pmatrix}.$$

Similarly, to make $t(x,1) = (2 - x,1)$, we can set $t(x,y) := (2y - x, y)$; thus t acts via the matrix

$$\begin{pmatrix} -1 & 2 \\ 0 & 1 \end{pmatrix}.$$

The picture of W acting on V is shown in Figure 2.2. It is simply the cone over the picture of W acting on L. (C now denotes the *cone* over the unit interval in the line $y = 1$.) The set T is now the set of reflections with respect to the walls of the chambers shown in the picture, and we may identify $T \times \{\pm 1\}$ with the set of half-planes determined by these walls. Condition (A) now follows easily from the action of W on these half-planes.

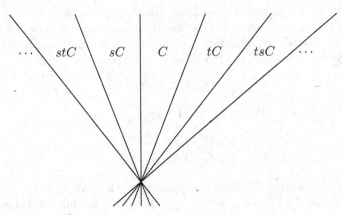

$$\cdots \quad stC \quad sC \quad C \quad tC \quad tsC \quad \cdots$$

Fig. 2.2. The chambers for D_∞; linear version.

Let's compare this situation with that of Chapter 1. As in that context, s and t act as linear reflections on V, provided we interpret this term suitably:

Definition 2.4. If V is a real vector space, not necessarily endowed with an inner product, then a *linear reflection* on V is a linear map that is the identity on a (linear) hyperplane H and is multiplication by -1 on some complement of H, i.e., a 1-dimensional subspace H' such that $V = H \oplus H'$. The reflections considered in Chapter 1, where V has an inner product and $H' = H^\perp$, will be called *orthogonal* reflections from now on to distinguish them from the more general linear reflections that we have just defined. Note that a linear reflection is *not* uniquely determined by its hyperplane H of fixed points.

In the present example it is still true, as in Chapter 1, that W is generated by linear reflections whose associated hyperplanes are the two walls of a "fundamental chamber" C. And it is still true that $\overline{C}$ is a strict fundamental domain for the action of W on $\bigcup_{w \in W} w\overline{C}$. But this union is not the whole vector space V. It is, rather, the convex cone consisting of the open upper half-plane together with the origin. This is a very general phenomenon, as we will see in Section 2.6.

Note that the chamber geometry for W acting on V is very similar to the chamber geometry for finite reflection groups. For example, the chamber graph can be identified with the Cayley graph of (W, S), and the analogue of Proposition 1.118 remains valid (with the same proof). We record this explicitly for future reference. Let H_s and H_t be the fixed hyperplanes of s and t, respectively, and let $U_\pm(s)$ and $U_\pm(t)$ be the corresponding open half-spaces, where the positive half-space is the one containing C.

Lemma 2.5. *For all $w \in W$, $wC \subseteq U_+(s)$ if and only if $l(sw) > l(w)$, and $wC \subseteq U_+(t)$ if and only if $l(tw) > l(w)$.* □

Finally, we will show that our representation of W as a "linear reflection group" admits a description resembling the description of a finite reflection group in terms of its Coxeter matrix (Section 1.5.5). Since we have no natural inner product on V, we introduce the dual space V^* and use inner-product notation for the canonical pairing $V^* \times V \to \mathbb{R}$, i.e.,

$$\langle f, v \rangle := f(v)$$

for $f \in V^*$ and $v \in V$. Define $e_s, e_t \in V^*$ by

$$\langle e_s, (x, y) \rangle := x \,,$$
$$\langle e_t, (x, y) \rangle := y - x \,.$$

With these definitions, the fixed hyperplane for s is given by $\langle e_s, - \rangle = 0$, the fixed hyperplane for t is given by $\langle e_t, - \rangle = 0$, and the fundamental chamber C is given by $\langle e_s, - \rangle > 0$ and $\langle e_t, - \rangle > 0$.

We still have a Coxeter matrix M specifying the orders of the pairwise products of the generators. It is given by

$$M = \begin{pmatrix} 1 & \infty \\ \infty & 1 \end{pmatrix} .$$

The corresponding Coxeter diagram is $\circ \overset{\infty}{\rule{1.2cm}{0.4pt}} \circ$. Imitating equation (1.22), we now put a symmetric bilinear form on V^* by setting

$$B(e_s, e_s) = B(e_t, e_t) = -\cos\frac{\pi}{1} = 1$$

and

$$B(e_s, e_t) = B(e_t, e_s) = -\cos\frac{\pi}{\infty} = -1 \,.$$

Next, define linear reflections s', t' on V^* by using this bilinear form as in equation (1.23):

$$s'(f) = f - 2B(e_s, f)e_s ,$$
$$t'(f) = f - 2B(e_t, f)e_t .$$

Note that $s'(e_s) = -e_s$ and s' fixes the hyperplane $B(e_s, -) = 0$, which is spanned by $e_s + e_t$; so s' is indeed a reflection. Similarly, t' is a reflection, with the same fixed hyperplane.

It turns out that s' and t' are the reflections s^* and t^* on V^* induced by s and t. To check this, one can simply compute s^* and t^* on e_s and e_t. For example,

$$\langle s^*(e_s), (x,y) \rangle = \langle e_s, s(x,y) \rangle = \langle e_s, (-x,y) \rangle = -x = \langle -e_s, (x,y) \rangle ,$$

so $s^*(e_s) = -e_s = s'(e_s)$. The remaining computations are equally easy and are left to the reader.

In summary, our reflection representation of D_∞ on V could have been obtained as follows: Start with an abstract vector space $\mathbb{R}e_s \oplus \mathbb{R}e_t$ [which is our V^*] and define a linear action of D_∞ on it by copying the formulas from the finite case, using the Coxeter matrix. Now pass to the dual space $(\mathbb{R}e_s \oplus \mathbb{R}e_t)^*$ [which is our V] to obtain an action in which we have the familiar sort of chamber geometry.

Remark 2.6. It is natural to ask whether we had to pass to the dual space in order to obtain the chamber geometry. The answer is yes—our two fundamental reflections acting on $\mathbb{R}e_s \oplus \mathbb{R}e_t$ have the same fixed hyperplane, so they do not determine a chamber in that space. See Exercise 2.7 below for further insight into the difference between the D_∞-action on V and its action on V^*. In the finite case, on the other hand, the duality was hidden because, in the presence of a W-invariant inner product, there is a canonical identification of V with its dual. We will return to this circle of ideas in Section 2.5 below.

Exercise 2.7.

(a) If s is a linear reflection on a 2-dimensional vector space V, show that the only s-invariant affine lines not passing through the origin are those parallel to the (-1)-eigenspace.

(b) Deduce that two linear reflections s, t of V have a common invariant line not passing through the origin if and only if they have the same (-1)-eigenspace.

(c) Suppose s and t have the same $(+1)$-eigenspace. Show that the induced reflections s^* and t^* of V^* have the same (-1)-eigenspace.

2.2.3 The Group $\mathrm{PGL}_2(\mathbb{Z})$

Let $\mathrm{GL}_2(\mathbb{Z})$ be the group of 2×2 invertible matrices over the ring $\mathbb{Z}$ of integers. Let $\mathrm{PGL}_2(\mathbb{Z})$ be the quotient of $\mathrm{GL}_2(\mathbb{Z})$ by the central subgroup of order two generated by -1 (= the negative of the identity matrix). Thus $\mathrm{PGL}_2(\mathbb{Z})$ is obtained from $GL_2(\mathbb{Z})$ by identifying a matrix with its negative. We denote a typical element of $GL_2(\mathbb{Z})$ by

$$\begin{pmatrix} a & b \\ c & d \end{pmatrix}$$

and its image in $PGL_2(\mathbb{Z})$ by

$$\begin{bmatrix} a & b \\ c & d \end{bmatrix}.$$

It is easy to check that $W := PGL_2(\mathbb{Z})$ is generated by the set $S = \{s_1, s_2, s_3\}$ of elements of order 2 defined by

$$s_1 = \begin{bmatrix} 0 & 1 \\ 1 & 0 \end{bmatrix}, \qquad s_2 = \begin{bmatrix} -1 & 1 \\ 0 & 1 \end{bmatrix}, \qquad s_3 = \begin{bmatrix} -1 & 0 \\ 0 & 1 \end{bmatrix}.$$

(One can see this by thinking about elementary row operations.) We now show that condition (A) is satisfied. We will use three different methods, analogous to those used for D_∞. In each of the three cases, however, we will have to use one or more nontrivial facts that will be stated without proof. Readers who are not familiar with these facts are advised to just read the discussion casually for the main ideas.

(i) Combinatorial group theory

A simple computation shows that the products s_1s_2, s_1s_3, and s_2s_3 have orders 3, 2, and ∞, respectively. It is also true (but not obvious) that W admits a presentation in which the defining relations simply specify the orders of these pairwise products:

$$W = \left\langle s_1, s_2, s_3 \; ; \; s_1^2 = s_2^2 = s_3^2 = (s_1s_2)^3 = (s_1s_3)^2 = 1 \right\rangle.$$

[Some readers will be familiar with the fact that W has a subgroup $PSL_2(\mathbb{Z})$ of index 2 that admits a presentation $\langle u, v \; ; \; u^3 = v^2 = 1 \rangle$; see Serre [217, Section I.4.2] or Lehner [152, Sections IV.5H and VII.2F]. It is not too hard to deduce the presentation for W stated above from this presentation for $PSL_2(\mathbb{Z})$.]

To verify (A), now, we need only check that the involutions $\rho_i := \rho_{s_i}$ that occur in the statement of (A) satisfy the defining relations for W. Consider, for instance, the relation $(\rho_1\rho_2)^3 = 1$. Let $S' := \{s_1, s_2\}$ and let W' be the dihedral group of order 6 generated by S'. The set T' of reflections in W' (i.e., the W'-conjugates of s_1 and s_2), form a subset of the set T of reflections in W (the W-conjugates of the elements of S).

Suppose, now, that we apply $(\rho_1\rho_2)^3$ to $(t, \epsilon) \in T \times \{\pm 1\}$. Since $(s_1s_2)^3 = 1$ in W, we will get $(t, \pm\epsilon)$. Clearly the only thing we have to worry about is the possibility of sign changes in the second factor as we successively apply the ρ_i. But no sign changes will ever occur unless there is an element $w' \in W'$ with $w'tw'^{-1} \in S'$, in which case we have $t \in T'$. Thus we are reduced to showing that $(\rho_1\rho_2)^3$ is the identity on $T' \times \{\pm 1\}$, which follows from the fact that W' is a finite reflection group and hence is already known to satisfy (A). [Alternatively, we could complete the proof by doing an easy computation in the dihedral group D_6.]

Remark 2.8. Note, for future reference, that this proof works whenever W admits a presentation of the form

$$W = \left\langle\, S\,;\, (st)^{m(s,t)} = 1 \,\right\rangle,$$

where $m(s,t)$ is the order of st and there is one relation for each pair s, t such that $m(s,t) < \infty$. We will return to this in Section 2.4.

(ii) Hyperbolic geometry

There is a famous tessellation of the hyperbolic plane by ideal hyperbolic triangles (i.e., triangles having their vertices on the circle at infinity). Figure 2.3 shows this tessellation in the unit disk model of the hyperbolic plane.[*]

Fig. 2.3. A tessellation of the hyperbolic plane.

The (hyperbolic) lines of symmetry of this tessellation barycentrically subdivide it; see Figure 2.4. The group of symmetries of the original tessellation is the group of hyperbolic isometries generated by the reflections with respect

---

[*] Figures 2.3 and 2.4 first appeared in Klein–Fricke [145, pp. 111 and 112]. Figure 2.5 is based on a picture in [145, p. 106]. We are grateful to Cornell University Library's Historic Monograph Collection for providing digital images of these pictures.

Fig. 2.4. The same tessellation, subdivided by the lines of symmetry.

to the lines of symmetry, and it is, in fact, precisely the group W. In order to explain this in slightly more detail, we switch to the upper-half-plane model of the hyperbolic plane. Figure 2.5 shows the barycentric subdivision in this model. To relate the two models of the hyperbolic plane, one should think of the vertices of the big triangle in Figure 2.3 as corresponding to the points 0, 1, and ∞ in Figure 2.5. The barycenter of this big triangle is shown as a heavy dot in Figure 2.5. The action of W on the upper half-plane is given by

$$
\begin{bmatrix} a & b \\ c & d \end{bmatrix} \cdot z = \begin{cases} \dfrac{az+b}{cz+d} & \text{if } ad - bc = 1, \\[2ex] \dfrac{a\bar{z}+b}{c\bar{z}+d} & \text{if } ad - bc = -1, \end{cases}
$$

where $\bar{z}$ is the complex conjugate of z. Many readers will be familiar with this action restricted to $\mathrm{PSL}_2(\mathbb{Z})$, where, of course, complex conjugation does not arise. Complex conjugation is necessary for the full group W, however, because elements of negative determinant acting by linear fractional transformations interchange the upper and lower half-planes.

Now under this action, the generating set S of W is the set of reflections in the three sides of one of the "chambers" C, as indicated in Figure 2.5. Moreover, it is known that $\overline{C}$ is a strict fundamental domain for the action of W.

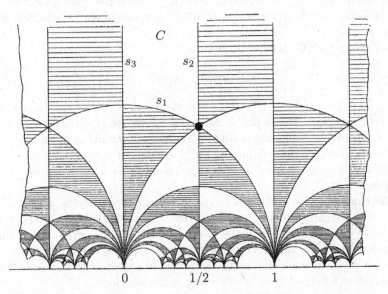

Fig. 2.5. The barycentric subdivision in the upper-half-plane model.

A proof of this can be found in almost any book that discusses modular forms. [Actually it is more likely that the analogous fact about $\mathrm{PSL}_2(\mathbb{Z})$ is proved: $\overline{C} \cup s_3\overline{C}$ is a fundamental domain (but not a strict fundamental domain) for this group. See, for instance, Serre [215, Section VII.1.2] or Lehner [152, Section IV.5H].]

Readers who have followed all of this can probably complete the geometric proof that (W, S) satisfies condition (A). Just identify $T \times \{\pm 1\}$ with the set of hyperbolic half-planes determined by the hyperbolic lines in Figure 2.5, and use the action of W on these half-planes.

(iii) Linear algebra

As was the case with the group D_∞, the linear algebra approach will take the longest to explain. But it is quite instructive and worth at least reading through, without necessarily checking all the details. It is based on a 3-dimensional linear representation of W that has been studied extensively, starting with Gauss.

The vector space V on which W acts is the space of real quadratic forms q in two variables, i.e., the space of functions $q \colon \mathbb{R}^2 \to \mathbb{R}$ given by $q(x) = ax_1^2 + 2bx_1x_2 + cx_2^2$, where $x = (x_1, x_2)$. Note that we can also write $q(x) = \beta(x, x)$, where β is the bilinear form on $\mathbb{R}^2$ with matrix

$$A := \begin{pmatrix} a & b \\ b & c \end{pmatrix} .$$

Thus we can, when it is convenient, identify V with the space of symmetric bilinear forms on $\mathbb{R}^2$, or, equivalently, with the space of real, symmetric 2×2 matrices.

The group $G = \mathrm{GL}_2(\mathbb{R})$ acts on V by

$$(g \cdot q)(x) := q(xg)$$

for $g \in G$, $q \in V$, and $x \in \mathbb{R}^2$, where x is viewed as a row vector on the right side of the equation. This action is said to be by *change of variable*, since $g \cdot q$ is obtained from q by replacing x_1 and x_2 by linear functions of x_1 and x_2 (with coefficients given by the columns of g). In terms of the symmetric matrix A corresponding to q, the action of g is given by $A \mapsto gAg^t$, where g^t is the transpose of g.

The elements $q \in V$ fall into exactly six orbits under the action of G. First, there are three types of nondegenerate forms: positive definite (G-equivalent to $x_1^2 + x_2^2$); negative definite (G-equivalent to $-x_1^2 - x_2^2$); and indefinite (G-equivalent to $x_1^2 - x_2^2$). Next, there are the nonzero degenerate forms, which are either positive semidefinite (G-equivalent to x_1^2) or negative semidefinite (G-equivalent to $-x_1^2$). And finally, there is the zero form.

It is easy to visualize this partition of V into G-orbits. Let $Q \colon V \to \mathbb{R}$ be given by

$$Q(q) := -\det A = b^2 - ac \,,$$

where A is the matrix corresponding to q as above. (Thus Q is a quadratic form on the 3-dimensional space V of quadratic forms.) Then the degenerate forms q are the points of the quadric surface $Q = 0$ in V. If we introduce new coordinates x, y, z in V by setting

$$b = x \,,$$
$$a = z + y \,,$$
$$c = z - y \,,$$

then Q becomes $x^2 + y^2 - z^2$, so the quadric surface of degenerate forms is the double cone $z^2 = x^2 + y^2$. [Draw a picture!] The exterior of the cone is given by $Q > 0$ and consists of the indefinite forms. And the interior $Q < 0$ has two components, the upper half ($z > 0$), consisting of the positive definite forms, and the lower half, consisting of the negative definite forms.

The action of $G = \mathrm{GL}_2(\mathbb{R})$ on V is really an action of the quotient $G/\{\pm 1\} = \mathrm{PGL}_2(\mathbb{R})$, so we may restrict the action to $W = \mathrm{PGL}_2(\mathbb{Z}) \leq \mathrm{PGL}_2(\mathbb{R})$. This is the desired 3-dimensional representation of W. Here are the basic facts about this representation:

First, the W-action leaves the form Q invariant, i.e., $Q(wq) = Q(q)$ for $w \in W$ and $q \in V$. This follows from the fact that every $g \in \mathrm{GL}_2(\mathbb{Z})$ has $\det g = \pm 1$, so that

$$\det gAg^t = (\det g)^2 \det A = \det A$$

for any symmetric 2×2 matrix A. So W also leaves invariant the symmetric bilinear form B on V such that $Q(q) = B(q, q)$. One can easily compute B explicitly; in terms of symmetric matrices, we have

$$B(A, A') = bb' - \frac{1}{2}(ac' + a'c),$$

where

$$A = \begin{pmatrix} a & b \\ b & c \end{pmatrix} \quad \text{and} \quad A' = \begin{pmatrix} a' & b' \\ b' & c' \end{pmatrix}.$$

The next observation is that the generators s_i of W act on V as linear reflections. In fact, computing the (± 1)-eigenspaces of s_i, one finds that s_i has a 1-dimensional (-1)-eigenspace $\mathbb{R}e_i$ and that s_i fixes the hyperplane $H_i := e_i^{\perp}$, where $e_i^{\perp}$ is defined with respect to our bilinear form $B(-, -)$. One can take the e_i, which are determined up to scalar multiplication, to be the following symmetric matrices:

$$e_1 = \begin{pmatrix} 1 & 0 \\ 0 & -1 \end{pmatrix}, \qquad e_2 = \begin{pmatrix} -1 & -1 \\ -1 & 0 \end{pmatrix}, \qquad e_3 = \begin{pmatrix} 0 & 1 \\ 1 & 0 \end{pmatrix}.$$

And the fixed hyperplanes H_i are given, respectively, by $a = c$, $c = 2b$, and $b = 0$.

We chose the eigenvectors e_i above so that they would satisfy $Q(e_i) = 1$; this determines them up to sign. It then follows as a formal consequence that the reflections s_i are given by the usual formula:

$$s_i q = q - 2B(e_i, q)e_i;$$

for the map defined by this formula is the identity on $e_i^{\perp}$ and sends e_i to $-e_i$.

We now focus on the action of W on the cone P of positive definite forms, and we look for a fundamental domain for this action. Concretely, this means that we are looking for canonical representatives for the positive definite forms q under integral change of variable. Gauss found the following fundamental domain. Let C be the simplicial cone in V defined by the inequalities $a > c > 2b > 0$. Then $C \subseteq P$, and $\overline{C}$ is (more or less) a fundamental domain for the action of W on P.

The qualifier "more or less" here refers to the fact that $\overline{C}$ touches the boundary of P. For if one computes the vertices of C (i.e., the rays that are 1-dimensional faces of C), one finds that they are represented by the forms x_1^2, $x_1^2 + x_2^2$, and $x_1^2 + x_1 x_2 + x_2^2$, the first of which is degenerate. So the correct statement is the following: Let X be the convex cone in V consisting of the positive definite forms together with the forms $\lambda(ax_1 + bx_2)^2$ with $\lambda \geq 0$ and $a, b \in \mathbb{Z}$. Then $X = \bigcup_{w \in W} w\overline{C}$, and $\overline{C}$ is a strict fundamental domain for the action of W on X; moreover, the open simplicial cones wC are disjoint from one another.

Note that the walls of C are precisely the fixed hyperplanes H_i of the reflections s_i. So we have, once again, the usual sort of chamber geometry, and it is possible to verify condition (A) by identifying $T \times \{\pm 1\}$ with the half-spaces in V determined by the walls of the chambers wC. Details are omitted.

One final comment: We normalized the e_i above so that we would have $B(e_i, -) > 0$ on C. In view of Chapter 1 and the infinite dihedral group example, it is therefore to be expected that

$$B(e_i, e_j) = -\cos \frac{\pi}{m_{ij}},$$

where m_{ij} is the order of $s_i s_j$. This is indeed the case, as direct computation shows. Thus our representation of W acting on V is what we should now be ready to call the "canonical linear representation" of W. Note also, for future reference, that the bilinear form B in this example is nondegenerate, although not positive definite. Indeed, we showed above that Q could be written as $x^2 + y^2 - z^2$ after a change of coordinates in V, so B has signature $(2, 1)$ [there are 2 plus signs and 1 minus sign].

Exercise 2.9. What is the connection between the points of view in (ii) and (iii)?

2.3 Consequences of the Deletion Condition

We return to the general theory, which is much easier than the examples. Thus W is an arbitrary group with a set S of generators of order 2. We saw at the end of Section 2.1 that if the action condition (A) holds then so does the deletion condition (D). We now explore some consequences of (D). We begin by giving two reformulations of it.

2.3.1 Equivalent Forms of (D)

We will need to formalize the concept of "word," which we have already used informally. By a *word* in the generating set S we mean a sequence $\mathbf{s} = (s_1, \ldots, s_d)$ of elements of S. We will often be less formal and simply say that the "expression" $s_1 \cdots s_d$ is a word. But we must be careful to distinguish a word $\mathbf{s}$ from the *element* $w = s_1 \cdots s_d \in W$ that it represents. Whenever there is danger of confusion, we will be more precise and revert to the sequence notation $(s_1, \ldots, s_d)$.

Definition 2.10. A word $\mathbf{s} = (s_1, \ldots, s_d)$ is called *reduced* if the corresponding element $w := s_1 \cdots s_d$ has length $l(w) = d$, i.e., if it cannot be represented by a shorter word. We will also say, in this situation, that $\mathbf{s}$ is a *reduced decomposition* of w, or, less formally, that the equation $w = s_1 \cdots s_d$ is a reduced decomposition of w.

We can now state the first consequence of condition (D). It is called the *exchange condition*:

(E) *Given $w \in W$, $s \in S$, and any reduced decomposition $w = s_1 \cdots s_d$ of w, either $l(sw) = d + 1$ or else there is an index i such that*

$$w = ss_1 \cdots \hat{s}_i \cdots s_d \,.$$

The proof that (D) implies (E) is immediate. For if $l(sw) < d+1$, then (D) implies that sw is equal to $ss_1 \cdots s_d$ with two letters deleted. Since $l(w) = d$, one of the deleted letters must be the initial s; multiplying by s, we obtain the conclusion of (E).

In order to put (E) into perspective, note that for general (W, S), we have the following three possibilities for $l(sw)$: (a) $l(sw) = l(w) + 1$; this happens if and only if we get a reduced decomposition of sw by putting s in front of a reduced decomposition of w. (b) $l(sw) = l(w) - 1$; this happens if and only if w admits a reduced decomposition starting with s. (c) $l(sw) = l(w)$.

[Possibility (c) might seem counterintuitive at first, but easy examples show that it can happen. See the note following Exercise 1.76.]

The content of (E), then, is the following: First, possibility (c) is prohibited. Second, if (b) holds then we can always find a reduced decomposition of w starting with s by taking an arbitrary reduced decomposition $w = s_1 \cdots s_d$ and then "exchanging" a suitable letter s_i for an s in front.

Note that (E) seems to be asymmetric, in that it involves only *left* multiplication by elements of S; but if (E) holds, then we can apply it to w^{-1} to deduce the analogous fact about right multiplication. We will use this observation without comment whenever it is convenient.

Next we record a consequence of (E). It will be called the *folding condition* for reasons that will become clear in Section 3.3.3 (see Remark 3.40).

(F) *Given $w \in W$ and $s, t \in S$ such that $l(sw) = l(w)+1$ and $l(wt) = l(w)+1$, either $l(swt) = l(w) + 2$ or else $swt = w$.*

To see that (E) implies (F), take a reduced decomposition $w = s_1 \cdots s_d$. Then the word $s_1 \cdots s_d t$ is a reduced decomposition of wt. Applying (E) to s and wt, we conclude that either $l(swt) = d+2$ or else we can exchange one of the letters in $s_1 \cdots s_d t$ for an s in front. Now the letter exchanged for s cannot be an s_i, since that would contradict the assumption that $l(sw) = d + 1$; so the letter must be the final t. Thus $wt = sw$; hence $swt = w$.

Finally, we show that (F) implies (D): Suppose $w = s_1 \cdots s_d$ with $d > l(w)$. Assuming (F), we will show by induction on d that we can delete two letters. We may assume that the words $s_1 \cdots s_{d-1}$ and $s_2 \cdots s_d$ are both reduced; otherwise, we are done by the induction hypothesis. Set $w' = s_2 \cdots s_{d-1}$. (This makes sense, because we necessarily have $d \geq 2$.) Then $l(s_1 w') = l(w') + 1 = l(w' s_d)$, and $l(s_1 w' s_d) < l(w') + 2$; so (F) implies that $s_1 w' s_d = w'$, i.e., that $w = \hat{s}_1 s_2 \cdots s_{d-1} \hat{s}_d$.

In summary, we now know the following relations among the four conditions that we have introduced:

Proposition 2.11. *Given a pair (W, S) consisting of a group W and a set S of generators of order 2, we have*

$$(A) \implies (D) \iff (E) \iff (F) .$$ □

2.3.2 Parabolic Subgroups and Cosets

Assume throughout this subsection that (W, S) satisfies the equivalent conditions (D), (E), and (F). We will derive some important consequences, mostly involving standard parabolic subgroups and cosets. These are defined exactly as in the case of finite reflection groups:

Definition 2.12. For any subset $J \subseteq S$, we denote by W_J the subgroup $\langle J \rangle$ generated by J. We call W_J a *standard parabolic subgroup*, or simply a *standard subgroup*. Any coset wW_J will be called a *standard coset*.

Proposition 2.13. *The function $J \mapsto W_J$ is a poset isomorphism from the set of subsets of S to the set of standard subgroups of W, where both sets are ordered by inclusion. The inverse is given by $W' \mapsto W' \cap S$.*

Proof. Consider the map from standard subgroups to subsets of S given by $W' \mapsto W' \cap S$ for any standard subgroup W'. It is clear that $W' = \langle W' \cap S \rangle$ if W' is a standard subgroup. It is also clear that $J \subseteq W_J \cap S$ for any $J \subseteq S$. To prove the opposite inclusion, suppose $s \in W_J \cap S$. Then we can express s as a J-word and repeatedly apply the deletion condition until the word's length has been reduced to 1; thus $s \in J$. Hence $J = W_J \cap S$, and our two maps are inverses of one another. Finally, both maps clearly preserve inclusions, so they are poset isomorphisms.   □

Our next observation is that when dealing with elements w of standard subgroups, we can write $l(w)$ without ambiguity.

Proposition 2.14. *Let W_J be a standard subgroup, where $J \subseteq S$. For any $w \in W_J$,*

$$l_J(w) = l_S(w) .$$

Proof. Suppose we have a J-reduced decomposition

$$w = s_1 \cdots s_d. \tag{2.2}$$

Thus $s_i \in J$ for each i and there is no shorter J-word representing w. We must show that there is no shorter S-word representing w. If there were a shorter S-word representing w, then we could get one by deleting two letters in (2.2). But this would contradict the assumption that the decomposition in (2.2) is J-reduced.   □

Here is another easy, but very useful, consequence of the deletion condition:

Lemma 2.15. *Given $J \subseteq S$, $w \in W_J$, and $s \in S \smallsetminus J$, we have $l(sw) = l(w) + 1$.*

Proof. Choose a reduced decomposition $w = s_1 \cdots s_l$ with $s_i \in J$ for all i. Suppose $l(sw) < l(w)$. Then $w = ss_1 \cdots \hat{s}_i \cdots s_l$ for some i by the exchange condition. This implies that $s \in W_J \cap S = J$, where the equality follows from Proposition 2.13. But this contradicts our hypothesis that $s \notin J$, so we must have $l(sw) \geq l(w)$ and hence $l(sw) = l(w) + 1$. $\qquad\square$

This leads to the following useful result:

Proposition 2.16. *For any $w \in W$ there is a subset $S(w) \subseteq S$ such that all reduced decompositions of w involve precisely the letters in $S(w)$. Moreover, $S(w)$ is the smallest subset $J \subseteq S$ with $w \in W_J$.*

Proof. Both assertions will follow if we prove the following: Given two decompositions $w = s_1 \cdots s_l = t_1 \cdots t_r$ with the one on the left reduced, each s_i is equal to some t_j. We argue by induction on $l = l(w)$, which may be assumed > 0. Let $J = \{t_1, \ldots, t_r\}$. Since $w \in W_J$ and $l(s_1 w) < l(w)$, the lemma implies that $s_1 \in J$. Now $s_2 \cdots s_l = s_1 w \in W_J$, so we also have $s_2, \ldots, s_l \in J$ by the induction hypothesis. $\qquad\square$

Proposition 2.17. *Fix $w \in W$ and let $J := \{s \in S \mid l(sw) < l(w)\}$. Then every reduced J-word can occur as an initial subword of a reduced decomposition of w. Hence*

$$l(w'w) = l(w) - l(w') \qquad (2.3)$$

for every $w' \in W_J$. In particular, the length function is bounded on W_J.

Proof. Let $t_1 \cdots t_l$ be a reduced J-word. Arguing by induction on l, we may assume that we have a reduced decomposition

$$w = t_2 \cdots t_l s_1 \cdots s_r.$$

Since $l(t_1 w) < l(w)$, we can exchange one of the letters in this decomposition for a t_1 in front. The exchanged letter cannot be a t_i, since that would contradict the assumption that the word $t_1 \cdots t_l$ is reduced. So it must be an s_i; hence

$$w = t_1 \cdots t_l s_1 \cdots \hat{s}_i \cdots s_r.$$

This proves the first assertion of the proposition. Applying it to a J-reduced decomposition of w'^{-1}, we obtain (2.3). Finally, (2.3) shows that the length function on W_J is bounded by $l(w)$. $\qquad\square$

Note that the last assertion of the proposition implies that W_J is finite if J is finite. Of course J is automatically finite if S is finite, which it is in most applications of the theory. But Proposition 2.16 implies that J is finite even if S is infinite, since $J \subseteq S(w)$, and the latter is obviously finite. Proposition 2.17 therefore has the following consequence:

Corollary 2.18. *With J as in Proposition 2.17, the group W_J is finite.* $\square$

Here is an important special case:

Corollary 2.19. *W is finite if and only if it has an element w_0 such that $l(sw_0) \leq l(w_0)$ for all $s \in S$. In this case w_0 has maximal length and is the unique element of maximal length, and it has order 2. Moreover,*

$$l(ww_0) = l(w_0) - l(w) \tag{2.4}$$

for all $w \in W$.

Proof. If W is finite, it obviously has an element w_0 of maximal length, and then necessarily $l(sw_0) \leq l(w_0)$ for all $s \in S$. Conversely, if w_0 is an element such that $l(sw_0) \leq l(w_0)$ for all s, then in fact $l(sw_0) < l(w_0)$ for all s (see Section 2.3.1), so W is finite by Corollary 2.18, and equation (2.4) follows from (2.3). Taking $w = w_0$ in (2.4), we see that $w_0^2 = 1$. And taking $w \neq w_0$ in (2.4), we see that $l(w) < l(w_0)$, so w_0 has maximal length and is the unique element of maximal length. $\square$

We have already encountered w_0 several times in Chapter 1, starting in Section 1.5.2. See Exercise 1.59 for a geometric explanation of the fact that w_0 is characterized by the inequality $l(sw_0) < l(w_0)$ for all s. [Note that this can also be written as $l(w_0 s) < l(w_0)$ for all s.]

Next, we show how the deletion condition leads to an interesting result about standard cosets.

Proposition 2.20. *Let W_J be a standard subgroup $(J \subseteq S)$. Then every left coset wW_J has a unique representative w_1 of minimal length. It is characterized by the property*

$$l(w_1 s) = l(w_1) + 1 \tag{2.5}$$

for all $s \in J$. Moreover,

$$l(w_1 w_J) = l(w_1) + l(w_J) \tag{2.6}$$

for all $w_J \in W_J$.

Proof. Choose w_1 of minimal length in the coset. Then $l(w_1 s) \geq l(w_1)$ for all $s \in J$; hence (2.5) holds. To prove (2.6) (which implies the uniqueness of w_1), choose a reduced decomposition $w_1 = s_1 \cdots s_l$, and consider an arbitrary element $w_J \in W_J$ and an arbitrary reduced J-decomposition $w_J = t_1 \cdots t_r$ $(t_i \in J)$. We must show that the word $s_1 \cdots s_l t_1 \cdots t_r$ is reduced.

If $s_1 \cdots s_l t_1 \cdots t_r$ is not reduced, then we can delete two letters. Neither deleted letter can be an s_i, since that would yield an element of $w_1 W_J$ shorter than w_1. On the other hand, the deleted letters cannot both be t_j's, since that would yield a shorter decomposition of w_J. So we have a contradiction, and the decomposition $w_1 w_J = s_1 \cdots s_l t_1 \cdots t_r$ is indeed reduced.

Finally, if $w_J \neq 1$, then $r > 0$ and $l(w_1 w_J t_r) < l(w_1 w_J)$. Thus w_1 is the unique element of the coset satisfying (2.5) for all $s \in J$. $\square$

Definition 2.21. Given $s \in S$, we say that an element $w \in W$ is (right) *s-reduced* if $l(ws) = l(w)+1$. Given $J \subseteq S$, we say that w is (right) *J-reduced* if it is s-reduced for all $s \in J$. One defines "left s-reduced" and "left J-reduced" elements similarly. Finally, given two subsets $J, K \subseteq S$, we say that an element $w \in W$ is (J, K)-*reduced* if it is left J-reduced and right K-reduced.

Thus the proposition says that the minimal-length representative of a left W_J-coset is the unique right J-reduced element of that coset. Left-reduced elements are similarly related to right cosets $W_J w$, and, as we will see, (J, K)-reduced elements are related to double cosets $W_J w W_K$.

Remark 2.22. To get some geometric intuition for Proposition 2.20, suppose W is a finite reflection group. Then standard cosets wW_J correspond to simplices A, and coset representatives correspond to chambers $D \geq A$. The representative of minimal length corresponds to the chamber $D_1 \geq A$ that is closest to the fundamental chamber C, i.e., $D_1 = AC$ (see Section 1.4.6). Moreover, equation (2.6) is a restatement of the gate property of Exercise 1.42; see Exercise 2.27 below.

It turns out that general products AB (with B not necessarily a chamber) are related to double cosets. We will explain this in the next chapter in a more general setting where the groups are not necessarily finite. Our treatment will make use of the following generalization of Proposition 2.20 to double cosets:

Proposition 2.23. *Let W_J and W_K be standard subgroups $(J, K \subseteq S)$. Then every (W_J, W_K)-double coset $W_J w W_K$ has a unique representative w_1 of minimal length. It is (J, K)-reduced and is the unique (J, K)-reduced element of the double coset. Moreover, every element w of the double coset $W_J w_1 W_K$ can be written as*

$$w = w_J w_1 w_K \tag{2.7}$$

with $w_J \in W_J$, $w_K \in W_K$, and

$$l(w) = l(w_J) + l(w_1) + l(w_K) . \tag{2.8}$$

The proof will use the following consequence of (D):

Lemma 2.24. *Let J and K be subsets of S, and let w_1 be an element of minimal length in its double coset $W_J w_1 W_K$. Suppose $u \in W_J$ and $v \in W_K$ are elements such that $l(uw_1v) < l(u) + l(w_1) + l(v)$. Given reduced decompositions $u = s_1 \cdots s_l$ and $v = t_1 \cdots t_r$, we have $uw_1v = u'w_1v'$, where $u' = s_1 \cdots \hat{s}_i \cdots s_l$ and $v' = t_1 \cdots \hat{t}_j \cdots t_r$ for some indices $1 \leq i \leq l$ and $1 \leq j \leq r$.*

Proof. Consider the decomposition of uw_1v obtained by combining the given decompositions of u and v with a reduced decomposition of w_1. By hypothesis this is not reduced, so we can delete two letters. The assumption on w_1 implies that neither of the deleted letters can involve the w_1-part of the word. Since we used reduced decompositions of u and v, the only possibility is that one deleted letter is an s_i and the other is a t_j. □

Proof of the proposition. Choose w_1 of minimal length in the double coset. Then w_1 is trivially (J, K)-reduced. Next, repeated applications of Lemma 2.24 show that any element $w \in W_J w_1 W_K$ can be written as in (2.7) in such a way that (2.8) holds; this implies that w_1 is the unique element of minimal length. Finally, if w_J in (2.7) is nontrivial then w is not left J-reduced, and if w_K is nontrivial then w is not right K-reduced. Hence w_1 is the only (J, K)-reduced element of the double coset.   □

We close with a technical result that will be needed in Chapter 5. It will actually fall out of our treatment of products in Section 3.6.4 (see Exercise 3.114), but we present here a purely group-theoretic proof.

Lemma 2.25. *With the notation of Proposition 2.23,*

$$W_J \cap w_1 W_K w_1^{-1} = W_{J_1} \, ,$$

where $J_1 := J \cap w_1 K w_1^{-1}$.

Proof. Given $u \in W_J \cap w_1 W_K w_1^{-1}$, we must show that $u \in W_{J_1}$. Equivalently, given $u \in W_J$ and $v \in W_K$ such that $u w_1 v = w_1$, we must show that $u \in W_{J_1}$. Note first that by repeated applications of Lemma 2.24, we have $l(u) = l(v)$. Write $u = s_1 \cdots s_n$ with $s_1, \ldots, s_n \in J$ and $n = l(u) = l(v)$. We show by induction on n that $s_i \in w_1 W_K w_1^{-1}$ for all i.

We have $l(s_n w_1 v) < l(s_n) + l(w_1) + l(v) = 1 + l(w_1) + n$, since otherwise $u w_1 v = (s_1 \cdots s_{n-1})(s_n u v)$ would have length $\geq l(s_n w_1 v) - (n-1) = l(w_1) + 2$. We can now apply Lemma 2.24 again to get $s_n w_1 v = w_1 v'$ for some $v' \in W_K$ and hence $s_n \in w_1 W_K w_1^{-1}$. But we also have $w_1 = u w_1 v = (s_1 \cdots s_{n-1}) s_n w_1 v = (s_1 \cdots s_{n-1}) w_1 v'$. So we can apply the induction hypothesis to deduce that also $s_1, \ldots, s_{n-1} \in w_1 W_K w_1^{-1}$. Finally, we observe that $s = w_1 x w_1^{-1}$ with $s \in J$ and $x \in W_K$ implies $l(x) = 1$ by the argument at the beginning of the proof ($s w_1 x^{-1} = w_1$). So indeed we have $s_i \in J_1$ for all i; hence $u \in W_{J_1}$.   □

Exercises

Assume throughout these exercises that (W, S) satisfies the deletion condition.

2.26. Given two standard subgroups W_J, W_K $(J, K \subseteq S)$, show that their intersection $W_J \cap W_K$ is the standard subgroup $W_{J \cap K}$. Generalize to an arbitrary family of standard subgroups.

2.27. If W is a finite reflection group, rewrite equation (2.6) to get the gate property of Exercise 1.42.

2.28. Let $w = s_1 \cdots s_n$, where the s_i are distinct elements of S. Show that $l(w) = n$.

2.29. Assume that the Coxeter diagram of (W, S) has no isolated nodes. Show that every proper standard subgroup of W has index at least 3.

2.30. Suppose that the elements of S can be enumerated as $s_1, \ldots, s_n$ so that $m(s_i, s_{i+1}) > 2$ for $i = 1, \ldots, n-1$. Thus the Coxeter diagram contains a path of length $n - 1$. Set $w_i := s_1 s_2 \cdots s_i$ for $i = 0, \ldots, n$.

(a) Show that the word defining w_i is reduced, i.e., $l(w_i) = i$.
(b) If $j \neq i$, show that $l(w_i s_j) = i + 1$, i.e., w_i is right s_j-reduced in the sense of Definition 2.21.

2.31. Let W be the finite reflection group of type A_{n-1} (symmetric group on n letters) with its standard generators $s_1, \ldots, s_{n-1}$, and set $J := S \smallsetminus \{s_1\}$.

(a) Show that the right J-reduced elements are the n elements $w_i :=$ $s_i \cdots s_2 s_1$ $(i = 0, \ldots, n-1)$.
(b) List the left J-reduced elements of W and the (J, J)-reduced elements.
(c) Generalize to the case $n = \infty$.

2.3.3 The Word Problem

We continue to assume that (W, S) satisfies the equivalent conditions (D), (E), and (F). The *word problem* seeks an algorithm that does the following: Given two S-words $\mathbf{s} = (s_1, \ldots, s_d)$ and $\mathbf{t} = (t_1, \ldots, t_e)$, decide whether they represent the same element of W. Let's begin with the case of a dihedral group D_{2m} generated by two elements s, t such that st has order m $(2 \leq m \leq \infty)$.

It is obvious, first of all, that we may confine our attention to *alternating* words $\mathbf{s}$ and $\mathbf{t}$, i.e., words with no consecutive s's or t's. Secondly, we may assume that both words have length at most m. For the relation $(st)^m = 1$ (if m is finite) can be rewritten as

$$stst \cdots = tsts \cdots \,, \tag{2.9}$$

where both sides have length m. So in any alternating word of length $> m$, we can take a subword $(s, t, \ldots)$ of length m and replace it by the word $(t, s, \ldots)$ of length m, thereby creating an (s, s) or (t, t) that can be deleted. Finally, the word problem for alternating words of length $\leq m$ has the following simple solution: The only two distinct alternating words of length $\leq m$ that represent the same element of D_{2m} are the two of length m (when $m < \infty$), as in (2.9). The proof is an easy computation, which is left to the reader. [Alternatively, think about what galleries look like when the plane is divided into $2m$ chambers by m lines through the origin (if m is finite) or when the line is divided into infinitely many intervals (if m is infinite).]

Returning now to a general (W, S), not necessarily dihedral, consider the *Coxeter matrix*

$$M := \big(m(s, t)\big)_{s, t \in S} \,,$$

where $m(s, t)$ is the order of st.

Definition 2.32. By an *elementary M-operation* on a word we mean an operation of one of the following two types:

(I) Delete a subword of the form (s, s).

(II) Given $s, t \in S$ with $s \neq t$ and $m(s, t) < \infty$, replace an alternating subword $(s, t, \dots)$ of length $m = m(s, t)$ by the alternating word $(t, s, \dots)$ of length m.

Call a word *M-reduced* if it cannot be shortened by any finite sequence of elementary M-operations.

It is not hard to see that one can algorithmically enumerate all possible words obtainable from a given one by elementary M-operations. [For examples, see Exercises 2.38 and 2.39 below.] In particular, one can decide whether a word is M-reduced. Similarly, one can decide whether a word **s** can be converted to a given word **t** by means of elementary M-operations. Consequently, the following theorem of Tits [246] solves the word problem.

Theorem 2.33.

(1) *A word is reduced if and only if it is M-reduced.*

(2) *If* **s** *and* **t** *are reduced, then they represent the same element of W if and only if* **s** *can be transformed to* **t** *by elementary M-operations of type (II).*

Proof. We begin with (2), for which it suffices to prove the "only if" part. Suppose $\mathbf{s} = (s_1, \dots, s_d)$ and $\mathbf{t} = (t_1, \dots, t_d)$ are reduced words representing the same element $w \in W$. We will show by induction on $d = l(w)$ that **s** can be changed to **t** by operations of type (II). Let $s = s_1$ and $t = t_1$. There are two cases.

(a) $s = t$. Then we can cancel the first letter from each side of the equation

$$s_1 \cdots s_d = t_1 \cdots t_d,$$

and we are done by the induction hypothesis.

(b) $s \neq t$. Since $l(sw) < l(w)$ and $l(tw) < l(w)$, Proposition 2.17 implies that $m := m(s, t)$ is finite and that there is a reduced decomposition **u** of w starting with the alternating word $(s, t, s, t, \dots)$ of length m. Let **u'** be the word obtained from **u** by replacing this initial segment of length m by the word $(t, s, t, s, \dots)$ of length m. We can then get from **s** to **t** by

$$\mathbf{s} \to \mathbf{u} \to \mathbf{u'} \to \mathbf{t},$$

where the first and third arrows are given by case (a) and the second is an operation of type (II). This completes the proof of (2).

Turning now to (1), we must show that if $\mathbf{s} = (s_1, \dots, s_d)$ is not reduced, then it can be shortened by M-operations. We argue by induction on d. If the subword $\mathbf{s'} := (s_2, \dots, s_d)$ is not reduced, then we are done by the induction hypothesis. So assume that **s'** is reduced and let $w' = s_2 \cdots s_d$.

Since $l(s_1w') \neq l(w') + 1$, there is a reduced decomposition of w' starting with s_1, say $\mathbf{t}' = (s_1, t_1, \ldots, t_{d-2})$. By statement (2), which we have already proved, we can transform $\mathbf{s}'$ to $\mathbf{t}'$ by M-operations; hence we can transform $\mathbf{s}$ to $(s_1, s_1, t_1, \ldots, t_{d-2})$ by M-operations. But this can then be shortened to $\mathbf{t} := (t_1, \ldots, t_{d-2})$ by an operation of type (I).   □

Remark 2.34. It is immediate from the theorem that all reduced decompositions of a given element $w \in W$ involve the same generators. This was already proved in Proposition 2.16, but the theorem probably gives a conceptually clearer explanation of why it is true.

Note that the solution of the word problem in Theorem 2.33 gives a complete description of the elements of W in terms of the Coxeter matrix. Consequently, we have the following generalization of a result that we already knew for finite reflection groups:

Corollary 2.35. *W is determined up to isomorphism by its Coxeter matrix.*
    □

We can make this more precise, in a way that gives us new information even for finite reflection groups:

Corollary 2.36. *W admits the presentation*

$$W = \left\langle \, S \, ; \, (st)^{m(s,t)} = 1 \, \right\rangle ,$$

where there is one relation for each pair s, t with $m(s,t) < \infty$.

Proof. Let $\tilde{W}$ be the abstract group defined by this presentation, and consider the canonical surjection $\tilde{W} \twoheadrightarrow W$. By Theorem 2.33, an element $\tilde{w}$ in the kernel can be represented by a word $\mathbf{s}$ that is reducible to the empty word by M-operations. But M-operations do not change the element of $\tilde{W}$ represented by a word, so $\tilde{w} = 1$.   □

We close by giving, as a sample application of Theorem 2.33, a lemma that will be useful in the next subsection. Recall from Proposition 2.16 that $S(w)$ for $w \in W$ denotes the set of generators $s \in S$ that occur in some (every) reduced decomposition of w.

Lemma 2.37. *If $w \in W$ and $s \in S \setminus S(w)$ satisfy $l(sws) < l(w) + 2$, then s commutes with all elements of $S(w)$.*

Proof. We have $l(sw) = l(w) + 1 = l(ws)$ by Lemma 2.15. Therefore, in view of the folding condition (Section 2.3.1), the hypothesis $l(sws) < l(w) + 2$ is equivalent to the equation $sw = ws$. We now show by induction on $l := l(w)$ that s commutes with all elements of $S(w)$. We may assume $l \geq 1$. Choose a reduced decomposition $w = s_1 \cdots s_l$, and consider the equation

$$ss_1 \cdots s_l = s_1 \cdots s_l s \ .$$

By Theorem 2.33, we can perform M-operations of type (II) to the word on the left in order to convert it to the word on the right. One of these operations must involve the initial s. Prior to applying this operation, we have a word of the form $st_1 \cdots t_l$ with $w = t_1 \cdots t_l$ and all $t_i \in S \smallsetminus \{s\}$; so the operation is possible only if $m(s, t_1) = 2$. Thus s commutes with t_1, hence also with $t_1 w = t_2 \cdots t_l$, and an application of the induction hypothesis completes the proof. □

Exercises

2.38. Suppose that $|S| = 3$ and that the generators s, t, u satisfy $m(s, t) > 2$ and $m(t, u) > 2$. Show that the word $utstu$ is reduced.

2.39. Let W be the reflection group of type A_3 [symmetric group on four letters]. Find all reduced decompositions of the longest element w_0, which is $s_1 s_3 s_2 s_1 s_3 s_2$.

2.40. Give an example to show that the hypothesis $s \notin S(w)$ in Lemma 2.37 cannot be replaced by the weaker hypothesis $l(sw) = l(w) + 1 = l(ws)$.

2.41. Given $w \in W$ and $s \in S \smallsetminus S(w)$ with $w^{-1} sw \in S$, show that s commutes with all elements of $S(w)$.

2.42. Recall from the classification of irreducible finite reflection groups that their Coxeter diagrams have a number of special properties, including the following: The edge labels are never greater than 5 if there are least 3 vertices; the graph is a tree; it branches at at most one vertex, which is then necessarily of degree 3; if it branches, there are no labeled edges (i.e., there are no $m(s, t) > 3$); if it does not branch, there is at most one labeled edge. As we mentioned in Remark 1.98, these facts are proved in the course of proving the classification theorem. Show that they all follow from Tits's solution of the word problem.

*2.3.4 Counting Cosets

We continue to assume that (W, S) satisfies the deletion condition. We begin with the following result of Deodhar [96, Proposition 4.2], who says that the result was also known to Howlett.

Proposition 2.43. *Assume that (W, S) is irreducible (i.e., its Coxeter diagram is connected) and that W is infinite. If J is a proper subset of S, then W_J has infinite index in W.* □

The proof will use the following lemma:

Lemma 2.44. *Given $s, t \in S$ with $m(s, t) > 2$, let $J := S \smallsetminus \{s\}$ and $K :=$ $J \smallsetminus \{t\} = S \smallsetminus \{s, t\}$. Suppose w is a right K-reduced element of W_J. Then ws is (right) J-reduced.*

Proof. Note first that $l(ws) = l(w) + 1$ by Lemma 2.15. We must show that $l(wsr) = l(w) + 2$ for $r \in S \smallsetminus \{s\}$. If $r \neq t$, then $r \in K$ and w is r-reduced. Since w is also s-reduced, we have $l(wsr) = l(w) + l(sr) = l(w) + 2$ by Proposition 2.20. Suppose now that $r = t$. Assuming, as we may, that $l(w) > 0$, we can write $w = vt$ where v is t-reduced. [Recall that w is a K-reduced element of W_J, so every reduced decomposition of it ends in t.] Since v is also s-reduced, Proposition 2.20 yields

$$l(wst) = l(vtst) = l(v) + 3 = l(w) + 2,$$

as required. □

Proof of Proposition 2.43. We will give the proof under the assumption that S is finite. See Exercise 2.47 below for the case of infinite S. We may assume $J = S \smallsetminus \{s\}$ for some $s \in S$, and we argue by induction on $|S|$. The result is trivial if W_J is finite, so assume W_J is infinite. Our task is to produce infinitely many (right) J-reduced elements of J. Now (W_J, J) might be reducible, so we cannot directly apply the induction hypothesis. But we can decompose the Coxeter graph of (W_J, J) into connected components with vertex sets $J_1, J_2, \ldots$, and at least one of these (say J_1) must correspond to an infinite group W_{J_1}. So we can apply the induction hypothesis to the latter.

Choose $t \in J_1$ with $m(s, t) > 2$; such a t exists by irreducibility of (W, S). By induction, W_{J_1} contains infinitely many $(J_1 \smallsetminus \{t\})$-reduced elements w. To complete the proof, observe that for each such w, the element ws is (right) J-reduced: It is J_1-reduced by Lemma 2.44, and it is $(J \smallsetminus J_1)$-reduced by Lemma 2.15. □

We now generalize the proposition to double cosets. To the best of our knowledge, this generalization has not previously appeared in the literature.

Proposition 2.45. *Assume that (W, S) is irreducible, S is finite, and W is infinite. If I and J are proper subsets of S, then $W_I \backslash W / W_J$ is infinite.*

Remark 2.46. In contrast to the situation for ordinary cosets, one cannot avoid the assumption that S is finite. A counterexample was given in Exercise 2.31(c), and we briefly recall it here: The group W has infinitely many generators $s_1, s_2, \ldots$, where $m(s_i, s_{i+1}) = 3$ and $m(s_i, s_j) = 2$ if $j > i+1$; thus the Coxeter diagram is an infinite path with unlabeled edges. Set $J := S \smallsetminus \{s_1\}$. Then straightforward computations show that there are only two (J, J)-reduced elements, 1 and s_1, so $|W_J \backslash W / W_J| = 2$. The conceptual explanation is that W is a doubly transitive permutation group on the set $\mathbb{N}$ of natural numbers, and W_J is the stabilizer of $1 \in \mathbb{N}$. Then W / W_J is infinite because $\mathbb{N}$ is infinite, but $|W_J \backslash W / W_J| = 2$ because W_J acts transitively on $\mathbb{N} \smallsetminus \{1\}$. See Exercise 2.31 and its solution for more details.

Proof of Proposition 2.45. We argue by induction on $|S|$. We may assume that $I = S \setminus \{s'\}$ and $J = S \setminus \{s\}$ for some $s', s \in S$. We may also assume, in view of Proposition 2.43, that W_I and W_J are both infinite. Our task is to show that W contains infinitely many (I, J)-reduced elements.

Case I, $s = s'$. This is very similar to the proof of Proposition 2.43. As in that proof, we can find $J_1 \subseteq J$ and $t \in J_1$ such that $W_1 := W_{J_1}$ is infinite and irreducible and $m(s, t) > 2$. By the induction hypothesis, W_1 contains infinitely many (K, K)-reduced elements w, where $K := J_1 \setminus \{t\}$. We will show that sws is (J, J)-reduced for each such w with $l(w) > 1$. By symmetry it suffices to show that sws is right J-reduced, i.e., that $l(swsr) = l(sws) + 1$ for all $r \in J$. Note first that $t \in S(w)$, so

$$l(sws) = l(w) + 2 \tag{2.10}$$

by Lemma 2.37. [Recall that $m(s, t) > 2$, i.e., s does not commute with t.] Consider now the following three possibilities for r.

(a) $r \in J \setminus J_1$. Then $l(swsr) = l(sws) + 1$ by Lemma 2.15.

(b) $r \in K$. Then sw is right r-reduced since $l(wr) = l(w) + 1$ and $s \notin J_1$, and it is right s-reduced by (2.10). We therefore have $l(swsr) = l(sw) + l(sr)$ by Proposition 2.20; hence $l(swsr) = l(sw) + 2 = l(sws) + 1$.

(c) $r = t$. Note that we necessarily have $w = tut$ with $u \in W_1$ and $l(w) = l(u) + 2$. (Recall that w is (K, K)-reduced and that $l(w) > 1$.) We must show that $l(stutst) = l(stuts) + 1$. Observe that stu is right s-reduced by Lemma 2.37 and is right t-reduced since $(stu)t = sw$ and $s \notin J$. Hence Proposition 2.20 yields $l(stutst) = l(stu) + l(tst) = l(stu) + 3 = l(stuts) + 1$.

Case II, $s \neq s'$. Denote by J_1 the connected component of $J = S \setminus \{s\}$ that contains s'. Suppose W_{J_1} is finite. Then there must exist another connected component J_2 of J such that W_{J_2} is infinite. (Recall from the beginning of the proof that W_I and W_J are infinite.) It follows that $I' := J_2 \cup \{s\}$ is connected, $W_{I'}$ is infinite, and I' is contained in $I = S \setminus \{s'\}$. So if we denote by I_1 the connected component of I that contains s, then W_{I_1} is infinite. We may therefore interchange the roles of I and J if necessary and assume from the beginning that $W_1 := W_{J_1}$ is infinite.

As in Case I, choose $t \in J_1$ with $m(s, t) > 2$. Set $K := J_1 \setminus \{t\}$ and $H := J_1 \setminus \{s'\}$. (Possibly $t = s'$ and $K = H$.) The induction hypothesis gives us infinitely many (H, K)-reduced elements $w \in W_1$, and we again consider only those w with $l(w) > 1$. Any such w must (being (H, K)-reduced) have a "reduced decomposition" $w = s'ut$ with $u \in W_1$, where "reduced" means that $l(w) = l(u) + 2$.

We claim that for each such w, the element ws is (I, J)-reduced. We show first that $l(rws) > l(ws)$ for all $r \in I = S \setminus \{s'\}$. This is clear for $r \in J \setminus J_1$. It is also clear for $r = s$ in view of Lemma 2.37, since $t \in S(w)$ and $m(s, t) > 2$. Finally, if $r \in H = J_1 \setminus \{s'\}$, then $l(rw) > w$ by the assumption on w, and $l(rws) > l(rw)$ since $s \notin J_1$, so $l(rws) = l(w) + 2 > l(ws)$.

Next we show that $l(wsr) > l(ws)$ for all $r \in J = S \smallsetminus \{s\}$. This is again trivial if r is in $J \smallsetminus J_1$. For $r \in K = J_1 \smallsetminus \{t\}$ we know that w is right r-reduced as well as right s-reduced; hence $l(wsr) = l(w) + 2 > l(ws)$. It remains to consider the case $r = t$. In this case $s'u$ is right t-reduced (since $s'ut$ is a reduced decomposition of w) and right s-reduced (since $s'u \in W_1$). Recalling again that $m(s,t) > 2$, we conclude that $l(wst) = l(s'utst) = l(s'u) + 3 = l(w) + 2 > l(ws)$. $\square$

Exercises

2.47. Prove Proposition 2.43 if S is infinite.

2.48. What goes wrong if we try to prove Proposition 2.45 for infinite S by imitating the solution to Exercise 2.47?

2.4 Coxeter Groups

We return now to an arbitrary (W, S), where W is a group and S is a set of generators of W of order 2. We have seen (Corollary 2.36) that if (W, S) satisfies (D) then it admits a presentation in which the relations simply specify the orders of the pairwise products of the generators. Tits [240] initiated the systematic study of groups with such a presentation. He called them *Coxeter groups*, since Coxeter [85] had earlier studied finite groups with a presentation of this type. We will therefore call the following condition on (W, S) the *Coxeter condition*:

(C) *W admits the presentation*

$$\left\langle\, S\,;\, (st)^{m(s,t)} = 1 \,\right\rangle,$$

where $m(s,t)$ is the order of st and there is one relation for each pair s, t with $m(s,t) < \infty$.

We have now introduced five conditions on (W, S), and we have shown that they are related as follows:

$$(\text{A}) \implies (\text{D}) \iff (\text{E}) \iff (\text{F}) \implies (\text{C}).$$

On the other hand, a Coxeter presentation as in (C) is precisely what we used in Section 2.2.3 to give a proof by combinatorial group theory that $\mathrm{PGL}_2(\mathbb{Z})$ satisfies (A). As we noted in Remark 2.8, this proof goes through with no change to show, in general, that (C) implies (A). Thus we have come full circle:

Theorem 2.49. *The conditions* (A), (C), (D), (E), *and* (F) *are equivalent.*
                    $\square$

It now seems safe to conclude that we have found the right class of groups that deserve to be called "reflection groups." We follow Tits's terminology, however, and call them Coxeter groups:

Definition 2.50. We say that W is a *Coxeter group* (or, more precisely, that the pair (W, S) is a *Coxeter system*) if the equivalent conditions of Theorem 2.49 are satisfied. The matrix $M = (m(s,t))$ will be called the *Coxeter matrix* of (W, S), and the cardinality $|S|$ will be called the *rank* of (W, S).

Exercise 2.51. Let (W, S) be a Coxeter system, and let W_J be a standard subgroup ($J \subseteq S$). Show that (W_J, J) is a Coxeter system.

2.5 The Canonical Linear Representation

Starting in this section, and throughout the rest of the book, **we assume that the generating set S is finite** unless we explicitly state that S might be infinite. The finite case is the most important one for the theory of buildings, and many of the arguments are simpler in that case. Some of what we do, however, would in fact generalize to the case of infinite S.

We emphasized in our discussion of examples in Section 2.2 that the Coxeter groups in those examples all admit a canonical linear representation that can be described in terms of the Coxeter matrix. And of course, this linear representation was there from the start in the case of finite reflection groups (Chapter 1). We show now that such a representation exists for *every* Coxeter group. As a byproduct of the discussion we will obtain an answer to the following natural question, which may have already occurred to the reader when we formulated the Coxeter condition (C): Which matrices M can occur as the Coxeter matrix of a Coxeter group?

Let $M = (m(s,t))_{s,t \in S}$ be a matrix with $m(s,t) \in \mathbb{Z} \cup \{\infty\}$. For the moment, S is just an index set, i.e., there is no group W yet.

Definition 2.52. We call M a *Coxeter matrix* if

$$m(s,s) = 1 \quad \text{and} \quad 2 \le m(s,t) = m(t,s) \le \infty \text{ for } s \ne t .$$

We denote by W_M the group defined by the presentation

$$W_M := \left\langle S ; (st)^{m(s,t)} = 1 \right\rangle ,$$

where, as usual, the relation occurs only if $m(s,t) < \infty$.

Note that the image of S in $W := W_M$ consists of elements of order 2 (i.e., $s \ne 1$ in W for each $s \in S$). This follows from the fact that there is a homomorphism $W \to \{\pm 1\}$ with $s \mapsto -1$ for each $s \in S$. But it is not obvious that S injects into W or that st has order precisely $m(s,t)$. [It is conceivable, a priori, that the order of st in W is a proper divisor of $m(s,t)$.] So we cannot immediately assert that W is a Coxeter group with Coxeter matrix M. But we will prove this to be the case using the *canonical linear representation* that we are about to construct.

2.5.1 Construction of the Representation

As in Section 1.5.5, we introduce the vector space $V := \mathbb{R}^S$ of S-tuples, with its standard basis $(e_s)_{s \in S}$. Let B be the symmetric bilinear form on V such that

$$B(e_s, e_t) = -\cos \frac{\pi}{m(s,t)} \, .$$

We wish to make W act on V as a "reflection group," where the bilinear form B plays the role of the inner product that was available in Chapter 1. Note first that if $\alpha \in V$ is any vector such that $B(\alpha, \alpha) \neq 0$, then we have a decomposition $V = \mathbb{R}\alpha \oplus \alpha^\perp$, where $\alpha^\perp := \{x \in V \mid B(\alpha, x) = 0\}$. There is therefore a linear reflection σ on V that sends α to $-\alpha$ and is the identity on $\alpha^\perp$. It is clear from this description that σ is *orthogonal* with respect to B, i.e., $B(\sigma(x), \sigma(y)) = B(x, y)$ for all $x, y \in V$. Moreover, σ is given by the familiar formula from Chapter 1: Assuming, for simplicity, that $B(\alpha, \alpha) = 1$, we have

$$\sigma(x) = x - 2B(\alpha, x)\alpha \, . \tag{2.11}$$

We now try to make W act on V so that a generator $s \in S$ acts as the reflection σ_s with respect to e_s. In other words, we want

$$s(x) = \sigma_s(x) = x - 2B(e_s, x)e_s \tag{2.12}$$

for $s \in S$ and $x \in V$. To get a well-defined action, we must show that

$$(\sigma_s \sigma_t)^m = \mathrm{id}_V$$

if $m := m(s, t) < \infty$. This is clear if $s = t$, so assume $s \neq t$.

Note that the bilinear form B is positive definite on the subspace $V_1 := \mathbb{R}e_s \oplus \mathbb{R}e_t$. So V_1 can be viewed as a Euclidean plane, and σ_s and σ_t induce orthogonal reflections on that plane. Since the angle between e_s and e_t is $\pi - \pi/m$, the angle between the fixed lines of σ_s and σ_t is π/m. Using standard Euclidean geometry, we conclude that the product $\sigma_s \sigma_t$ acts on V_1 as a rotation of order m. Moreover, σ_s and σ_t act trivially on the orthogonal complement $V_0 := V_1^\perp$ with respect to B, and we have a decomposition $V = V_1 \oplus V_0$ because B is nondegenerate on V_1. Hence $\sigma_s \sigma_t$ has order m as an automorphism of V. This proves that we do in fact get the desired action of W on V and that st has order $m(s, t)$ whenever the latter is finite.

Given s, t with $m(s, t) = \infty$, it is still true that σ_s and σ_t restricted to $V_1 := \mathbb{R}e_s \oplus \mathbb{R}e_t$ generate a faithful linear representation of D_∞, which is the same as the one we discussed in Section 2.2.2 (where the vector space with basis e_s, e_t was called V^*), so $\sigma_s \sigma_t$ has infinite order; hence st has infinite order. [Alternatively, one can simply write out the matrices of σ_s and σ_t on V_1 and check that the product has 1 as an eigenvalue of multiplicity two; since the product is nontrivial, this implies that it has infinite order.]

The discussion in the previous two paragraphs shows that the various σ_s are distinct from one another, so S injects into W. We may therefore view S as a subset of W, and we have proven the following result:

Theorem 2.53. *The elements of S are distinct and of order 2 in W_M, and st has order precisely $m(s,t)$ in W_M. Hence (W_M, S) is a Coxeter system with Coxeter matrix M.* □

It is now clear that a Coxeter matrix in the sense of Definition 2.52 is the same as what we called a Coxeter matrix in Section 2.4. Moreover, the action of $W := W_M$ on V provides a *canonical linear representation* for any Coxeter group. It was first introduced by Tits [240]. Our next goal is to show that it is faithful, i.e., that W injects into the group of linear automorphisms of V; thus every Coxeter group can be viewed as a linear reflection group.

Remark 2.54. It is sometimes convenient to write the formula (2.12) for the action of W on V in a way that resembles a familiar formula from the theory of root systems (see equation (B.3) in Appendix B). For $s \in S$, let $e_s^\vee \in V^*$ be given by

$$\langle e_s^\vee, x \rangle := 2B(e_s, x)$$

for $x \in V$, where, as in Section 2.2.2, we use angle brackets for the natural pairing between V and V^*. Then (2.12) becomes

$$s(x) = x - \langle e_s^\vee, x \rangle e_s . \tag{2.13}$$

Exercises

2.55. Let M be a matrix as at the beginning of this section, but suppose we drop the symmetry assumption $m(s,t) = m(t,s)$.

(a) If M is not symmetric, show that there are elements $s, t \in S$ such that the order of the image of st is not $m(s,t)$. Where does the proof of Theorem 2.53 break down?
(b) Let S' be the image of S in $W = W_M$. Show that (W, S') is still a Coxeter system.
(c) Give a procedure for determining the set S' and the Coxeter matrix M' of (W, S').

2.56. This is a generalization of Exercise 1.101. Let (W, S) be an irreducible Coxeter system.

(a) Show that every proper W-invariant subspace of V is contained in the radical of B, the latter being $\{x \in V \mid B(x,y) = 0 \text{ for all } y \in V\}$. In particular, the canonical linear representation is irreducible if B is nondegenerate.
(b) Show that the only linear endomorphisms of V that commute with all elements of W are the scalar-multiplication operators.

2.5.2 The Dual Representation

The proof that the canonical linear representation is faithful will be based on chamber geometry. The reader who has worked through the discussion of D_∞ in Section 2.2.2 will not be surprised that the natural place to look for chambers is in the dual space V^* of V. Note first that the action of W on V induces an action of W on V^*. It is defined by

$$\langle w\xi, x \rangle = \langle \xi, w^{-1}x \rangle$$

or, equivalently,

$$\langle w\xi, wx \rangle = \langle \xi, x \rangle$$

for $w \in W$, $\xi \in V^*$, and $x \in V$. Using this definition and equation (2.13), one can check that the action of a generator $s \in S$ is given by

$$s(\xi) = \xi - \langle \xi, e_s \rangle e_s^\vee \tag{2.14}$$

for $\xi \in V^*$. In particular, s acts on V^* as a linear reflection whose fixed hyperplane H_s is given by $\langle -, e_s \rangle = 0$. Let C be the simplicial cone in V^* defined by the inequalities $\langle -, e_s \rangle > 0$ for $s \in S$. We call C the *fundamental chamber*. Its walls are the hyperplanes H_s.

Remark 2.57. The fact that the vectors e_s form a basis for V, whereas the $e_s^\vee$ do not in general form a basis for V^*, explains why V^* is the natural place to look for chambers rather than V.

We claim now that

$$wC \cap C = \emptyset \text{ for } 1 \neq w \in W. \tag{2.15}$$

The idea behind the proof of (2.15) is that if we choose $s \in S$ such that $l(sw) < l(w)$, then H_s ought to separate C from wC (see Section 3.3.3). This motivates the following lemma, which implies (2.15):

Lemma 2.58. *Fix $s \in S$ and let $U_+(s)$ and $U_-(s)$ be the open half-spaces in V^* defined, respectively, by $\langle -, e_s \rangle > 0$ and $\langle -, e_s \rangle < 0$. Then for any $w \in W$ we have*

$$wC \subseteq \begin{cases} U_+(s) & \text{if } l(sw) = l(w) + 1, \\ U_-(s) & \text{if } l(sw) = l(w) - 1. \end{cases}$$

Proof. We argue by induction on $l(w)$. If $l(sw) < l(w)$, then we may apply the induction hypothesis to the element sw to get $swC \subseteq U_+(s)$; multiplying by s, we obtain $wC \subseteq sU_+(s) = U_-(s)$, as required. Suppose now that $l(sw) > l(w)$. We may assume $w \neq 1$, so there is a $t \in S$ (necessarily different from s) such that $l(tw) < l(w)$. Choose a reduced decomposition of w starting with as long a subword as possible involving just s and t. This yields a factorization $w = w'w''$ with $w' \in W' := \langle s, t \rangle$, $l(w) = l(w') + l(w'')$, $l(sw'') > l(w'')$, and $l(tw'') > l(w'')$. By the induction hypothesis, we have $w''C \subseteq C' :=$ $U_+(s) \cap U_+(t)$. Now C' is essentially the fundamental chamber for the dual of the canonical representation of W'. More precisely, the dual of the canonical representation of W' is a 2-dimensional quotient of V^*, and C' is the inverse image of the fundamental chamber. (See Exercise 2.60 below.) But we have studied the canonical representation of W' and its dual in detail, whether W' is finite or infinite, and we know that its chamber geometry behaves in the expected way. [See Proposition 1.118 for the finite case and Lemma 2.5 for the infinite case.] In particular, since $l(sw') > l(w')$, it follows that $w'C' \subseteq U_+(s)$; hence $wC = w'w''C \subseteq w'C' \subseteq U_+(s)$. $\qquad\qquad\square$

The following result is an immediate consequence of (2.15):

Theorem 2.59. *The action of W on V is faithful. Moreover, W is a discrete subgroup of the topological group $\mathrm{GL}(V)$ of linear automorphisms of V.* $\quad\square$

Exercise 2.60. Let W_J be a standard subgroup ($J \subseteq S$). Then we have a canonical linear representation of W_J on a vector space V_J with basis $(e_s)_{s \in J}$. Show that there is a W_J-equivariant surjection $V^* \twoheadrightarrow V_J{}^*$ with kernel $\bigcap_{s \in J} H_s$, which is the fixed-point set of W_J in V^*. Informally, then, we can say that the action of W_J on V^* is essentially the dual of the canonical linear representation of W_J.

2.5.3 Roots, Walls, and Chambers

Definition 2.61. The vectors we_s ($w \in W$, $s \in S$) in V will be called *roots*, the hyperplanes wH_s ($w \in W$, $s \in S$) in V^* *walls*, and the simplicial cones wC ($w \in W$) in V^* *chambers*. We denote by Φ the set of all roots, by $\mathcal{H}$ the set of all walls, and by $\mathcal{C}$ the set of all chambers.

We emphasize once again that the roots are in V, while the chambers and walls are in V^*. Note that in view of (2.15), W acts simply transitively on $\mathcal{C}$.

Let T be the set of reflections in W as defined in Section 2.1. Then the elements of T act on V as orthogonal reflections with respect to the bilinear form B. More precisely, for any $t \in T$ there are two unit vectors $\pm\alpha$ (with respect to B) in the (-1)-eigenspace of t acting on V, and the action of t can then be written as in equation (2.11). There are also analogues of equations (2.13) and (2.14). We will write $t = s_\alpha$ when t is associated to $\pm\alpha$ in this way. Note that we have $\pm\alpha \in \Phi$. Moreover, one can easily check the equation

$$ws_\alpha w^{-1} = s_{w\alpha}$$

for $w \in W$ and $\alpha \in \Phi$.

It is clear from this discussion that there is a bijection between T and the set of pairs $\pm\alpha$ of opposite roots, under which $\pm\alpha$ corresponds to the reflection $s_\alpha \in T$. We therefore obtain, by duality, a bijection between T and $\mathcal{H}$, which associates to $t \in T$ the fixed hyperplane of t acting on V^*. To verify this, observe that if $t = s_\alpha$, the fixed hyperplane of t in V^* is given by $\langle -, \alpha \rangle = 0$. [This follows, for instance, from the analogue of (2.14) with e_s replaced by α.]

Turning now to chambers and walls, Lemma 2.58 implies that every chamber $D = wC$ lies on one side of each wall; hence D has a well-defined *sign sequence*

$$\sigma(D) = \left(\sigma_H(D) \right)_{H \in \mathcal{H}}$$

with $\sigma_H(D) \in \{+, -\}$; here we could arbitrarily choose the positive and negative sides of H, but we follow the usual convention that the positive side is the one containing C. A chamber D is clearly determined by its sign sequence, since D is a simplicial cone whose walls form a subset of $\mathcal{H}$. Even though our sign sequences are infinite in general, we claim that the sign sequences of any two chambers differ in only finitely many places. The following lemma is the key step in the proof of this:

Lemma 2.62. *For any $s \in S$, H_s is the only wall separating C from sC.*

Proof. We will use the theory of (finite) hyperplane arrangements developed in Section 1.4. Let $\mathcal{H}_0 \subseteq \mathcal{H}$ be any finite subset containing the walls of C and sC. Then C and sC are $\mathcal{H}_0$-chambers with a common wall $H_s = sH_s$. Moreover, if A_s is the panel of C with support H_s, then A_s is fixed by s and hence is also the panel of sC with support H_s. Thus C and sC are adjacent in the sense of Section 1.4.9 (applied to $\mathcal{H}_0$), and H_s is the unique element of $\mathcal{H}_0$ that separates them. Since $\mathcal{H}_0$ could have been chosen to contain any given wall, the lemma follows. $\square$

Consider now an arbitrary chamber wC ($w \in W$), and choose a decomposition $w = s_1 \cdots s_n$ ($s_i \in S$). Since $s_1 \cdots s_{i-1}C$ and $s_1 \cdots s_iC$ are separated by only one wall, the sign sequences of C and $wC = s_1 \cdots s_nC$ differ in at most n positions. Using the W-action, we obtain the result claimed above, which we record for future reference:

Lemma 2.63. *Any two chambers are separated by only finitely many walls.*
    $\square$

Remark 2.64. We began our development of the theory of Coxeter groups in Section 2.1 by introducing the set $T \times \{\pm 1\}$ and suggesting that its elements should be thought of intuitively as roots. It is now clear that this abstract set can be identified with the set of roots Φ introduced above. One can in fact develop the entire theory of Coxeter groups by starting with the canonical linear representation and using its action on Φ systematically. This is the approach taken in the book by Humphreys [133], based on ideas of Deodhar [96].

2.5.4 Finite Coxeter Groups

We show here that if the Coxeter group W is finite, then it is a finite reflection group in the sense of Chapter 1. Roughly speaking, then, finite Coxeter groups are exactly the same thing as finite reflection groups.

Theorem 2.65. *If W is finite, then the bilinear form B is positive definite and W is a finite reflection group acting on V, with S as the set of reflections with respect to the chamber defined by $B(e_s, -) > 0$ for all $s \in S$.*

Proof. Let $V_1 = V^*$, and let's temporarily forget that V_1 is the dual of V. By Theorem 2.59, W can be identified with a finite group of linear transformations of V_1 generated by linear reflections. We can make them orthogonal reflections by putting a W-invariant inner product on V_1. To this end, start with an arbitrary inner product $(-, -)$ on V_1, and construct a W-invariant inner product $\langle -, - \rangle$ by "averaging":

$$\langle x, y \rangle := \sum_{w \in W} (wx, wy) \,.$$

For each $s \in S$, the (± 1)-eigenspaces of s are orthogonal to one another with respect to this W-invariant inner product, so s indeed acts on V_1 as an orthogonal reflection. Thus (W, V_1) is a finite reflection group in the sense of Chapter 1. The set $\mathcal{H}$ defined in Definition 2.61 satisfies the conditions of Section 1.5, so it is the set of walls of (W, V_1). Our fundamental chamber C is therefore a chamber for (W, V_1), and S is a set of fundamental reflections.

Let f_s be the unit vector in $H_s^\perp$ pointing to the side of H_s containing C. Then we know from Section 1.5.5 that the usual formulas hold: $s(x) = x - 2\langle f_s, x \rangle f_s$ for $x \in V_1$, and $\langle f_s, f_t \rangle = -\cos(\pi/m(s,t))$. Thus V_1, with its W-action and inner product, is isomorphic to V, with its W-action and bilinear form B. Everything we know about (W, V_1) can now be transported to (W, V). $\square$

Remarks 2.66. (a) Combining Theorem 2.65 with the classification of finite reflection groups (Section 1.3), we recover Coxeter's list [85] of the finite Coxeter groups.

(b) We can now apply to arbitrary finite Coxeter groups all of the results of Chapter 1. For example, the result about the normalizer of S (Corollary 1.91) is valid for every finite Coxeter group.

Next, we wish to show that the converse of the first assertion of Theorem 2.65 is also true.

Proposition 2.67. *If B is positive definite, then W is finite.*

Proof. W is a subgroup of the orthogonal group consisting of all linear transformations of V that leave B invariant. Since B is positive definite, this orthogonal group is compact. In view of Theorem 2.59, W is a discrete subgroup of a compact group; hence it is finite. $\square$

Combining Theorem 2.65, Proposition 2.67, and the results of Chapter 1, we obtain the following:

Corollary 2.68. *The following conditions on a Coxeter system* (W, S) *are equivalent:*

(i) W *is finite.*
(ii) W *can be realized as a finite reflection group, with* S *as the set of reflections with respect to the walls of a fundamental chamber.*
(iii) *The canonical bilinear form* B *on* $V := \mathbb{R}^S$ *is positive definite.* $\square$

Finally, we add several more useful criteria for a Coxeter group to be finite. Recall that C is the fundamental chamber, $\mathcal{C}$ is the set of all chambers, $\mathcal{H}$ is the set of walls, T is the set of reflections, and Φ is the set of roots.

Proposition 2.69. *The following conditions on a Coxeter system* (W, S) *are equivalent:*

(i) W *is finite.*
(ii) $-C$ *is a chamber.*
(iii) $\mathcal{H}$ *is finite.*
(iv) T *is finite.*
(v) Φ *is finite.*
(vi) $\mathcal{C}$ *is finite.*

Proof. It is clear from the discussion in Section 2.5.3 that conditions (iii), (iv), and (v) are all equivalent. So it suffices to show

$$(\mathrm{i}) \implies (\mathrm{ii}) \implies (\mathrm{iii}) \implies (\mathrm{vi}) \implies (\mathrm{i}) \,.$$

If W is finite, then it is a finite reflection group, so $-C$ is a chamber. Hence (i) $\implies$ (ii). Next, (ii) $\implies$ (iii) by Lemma 2.63 since every wall separates $-C$ from C. Finally, (iii) $\implies$ (vi) trivially, and (vi) $\implies$ (i) because W acts simply transitively on $\mathcal{C}$. $\square$

2.5.5 Coxeter Groups and Geometry

In this short subsection we make some remarks about three classes of Coxeter groups that are related to classical geometry.

Definition 2.70. We will say that a Coxeter system (W, S) of rank n is *spherical*, of dimension $n-1$, if it satisfies the equivalent conditions of Corollary 2.68.

The motivation for this definition should be clear from Chapter 1. The examples in Section 2.2 were designed to suggest that there are also reasonable notions of *Euclidean* and *hyperbolic* Coxeter system. We will study these in Chapter 10, but by way of preview, we state some facts that have already been illustrated in the examples. First, we introduce some standard terminology.

Definition 2.71. The bilinear form B and the Coxeter system (W, S) are said to be of *positive type* if B is positive semidefinite, i.e., if $B(x, x) \geq 0$ for all $x \in V$.

Note that we may have $B(x, x) = 0$ for some $x \neq 0$; this holds if and only if B is degenerate (Exercise 2.72).

Now suppose that (W, S) is an *irreducible* Coxeter system of rank $n := |S|$. Then:

(1) (W, S) is spherical (of dimension $n-1$) if and only if B is positive definite.
(2) (W, S) is Euclidean (of dimension $n - 1$) if and only if B is of positive type but degenerate.
(3) (W, S) is hyperbolic (of dimension $n - 1$) if B is nondegenerate of signature $(n - 1, 1)$ and $B(x, x) < 0$ for all x in the fundamental chamber C.

We have already proven (1), and we will return to (2) and (3) in Chapter 10, where in particular, we will define the terms "Euclidean Coxeter system" and "hyperbolic Coxeter system." But statement (3) needs some explanation before we move on.

Recall first that the *signature* of a nondegenerate bilinear form on a finite-dimensional real vector is the pair of integers (p, q) such that when the form is diagonalized it has p positive entries and q negative entries on the diagonal. Secondly, since B is nondegenerate in (3), we can use it to identify V with its dual. So the fundamental chamber C that we defined in V^* can be identified with the chamber (still called C) in V defined by $B(e_s, -) > 0$ for all $s \in S$. Thus it makes sense to talk about the value of $B(x, x)$ for $x \in C$. Thirdly, note that there is an "if" but not an "only if" in (3). The reason is that the condition in (3) characterizes the special class of hyperbolic Coxeter groups that act on hyperbolic space with a simplex as fundamental domain. But it turns out that the fundamental domain for a hyperbolic reflection group can be a more complicated polyhedron. We will explain this further in Section 10.3.

Exercise 2.72. Let B be a positive semidefinite symmetric bilinear form on a real vector space. If $B(x, x) = 0$ for some x, show that x is in the radical of B, i.e., $B(x, y) = 0$ for all y.

2.5.6 Applications of the Canonical Linear Representation

Returning to the general case, we first use the canonical linear representation to calculate the normalizer of the generating set S. We already know what this is for a finite Coxeter group (Remark 2.66(b)), so we will know it for every Coxeter group if we treat the infinite irreducible case. In the finite case we gave two proofs, one algebraic and one geometric. We are now in a position to give an analogue of the algebraic proof; in the next chapter we will see that the geometric proof also generalizes (see Exercise 3.122).

Proposition 2.73. *If (W, S) is an irreducible Coxeter system with W infinite, then the normalizer of S is trivial. In particular, the center of W is trivial.*

(This result first appeared in print in [44, Section V.4, Exercise 3]; see also [96, Proposition 4.1].)

Proof. The proof is almost the same as the algebraic proof of Corollary 1.91, except that the bilinear form $B(-,-)$ replaces the inner product $\langle -,- \rangle$, and one has to work in both V, which contains the set Φ of roots, and V^*, which contains the chambers. Here are the details. Suppose $w \in W$ normalizes S. For each $s \in S$, $s' := wsw^{-1} \in T$ is the reflection corresponding to the roots $\pm we_s \in \Phi$. Hence $we_s = \pm e_{s'}$. Irreducibility now implies, exactly as in the proof of Corollary 1.91, that the ambiguous sign is independent of s. So either $we_s = e_{s'}$ for all $s \in S$ or $we_s = -e_{s'}$ for all $s \in S$. Considering now the action of w on V^*, we conclude that either $wC = C$ or $wC = -C$. The second case is impossible by Proposition 2.69, so $wC = C$ and hence $w = 1$ by simple transitivity. □

As a second application, we prove the following result of Niblo and Reeves (cf. [180, Lemma 3]):

Proposition 2.74. *If W is infinite, then W contains two reflections t, t' whose product tt' has infinite order.*

Remark 2.75. Our standing assumption that S is finite is crucial here. The proposition is obviously false, for example, if S is infinite but W_J is finite for every finite $J \subseteq S$.

The proof will make use of two lemmas. The first will be proved later, but we state it here for ease of reference.

Lemma 2.76. *Let $W' \le W$ be a subgroup generated by two reflections. If W' is finite, then W' is contained in a finite parabolic subgroup.*

This is a special case of Proposition 2.87, which we will prove in the next section using the Tits cone. Alternatively, there is a direct combinatorial proof of this special case that will arise naturally in our study of Coxeter complexes in the next chapter; see Corollary 3.167.

The next lemma is the crucial one. The "if" part of the first assertion can be found in [49, Proposition 1.4], where it is attributed to Dyer [99].

Lemma 2.77. *Let α and β be roots with $\alpha \neq \pm\beta$, and let s_α and s_β be the corresponding reflections. Then $s_\alpha s_\beta$ has finite order if and only if $|B(\alpha, \beta)| < 1$. Moreover, the set of real numbers*

$$E := \{B(\alpha, \beta) \mid \alpha, \beta \in \Phi, |B(\alpha, \beta)| < 1\}$$

is finite.

Proof. Let W' be the dihedral group generated by s_α and s_β; it is finite if and only if $s_\alpha s_\beta$ has finite order. Suppose first that $|B(\alpha, \beta)| < 1$, and let V' be the 2-dimensional subspace of V generated by α and β. The hypothesis implies that the form B is positive definite on V'. In particular, it is nondegenerate, so we have an orthogonal decomposition $V = V' \oplus V''$, where V'' is the orthogonal complement of V' with respect to B. The dihedral group W' acts trivially on V'', so we may view it as a subgroup of the orthogonal group acting on V'. Since W' is discrete and the orthogonal group is compact, it follows that W' is finite.

Conversely, suppose W' is finite. By Lemma 2.76, we then have $wW'w^{-1} \leq W_J$ for some $w \in W$ and some $J \subseteq S$ such that W_J is finite. Replacing α and β by $w\alpha$ and $w\beta$, we may assume that $W' \leq W_J$. By Corollary 2.68, B is positive definite on the subspace V_J spanned by the e_s with $s \in J$, so the inequality $|B(\alpha, \beta)| < 1$ will follow if we can show that α and β are in V_J. To this end, note that, as above, we have $V = V_J \oplus (V_J)^\perp$, and W_J acts trivially on $(V_J)^\perp$. So V_J must contain the (-1)-eigenspaces of s_α and s_β; hence $\alpha, \beta \in V_J$.

Finally, we prove the finiteness of E. By the arguments above, E is the set of numbers $B(\alpha, \beta)$ (with $\alpha, \beta \in \Phi$, $\alpha \neq \pm\beta$) such that s_α and s_β are contained in a finite standard subgroup W_J. Now there are only finitely many possibilities for J, and for each J, there are only finitely many possibilities for the pair $s_\alpha, s_\beta \in W_J$. Since a root α is determined up to sign by the reflection s_α, the finiteness of E follows. $\square$

Proof of Proposition 2.74. With E as in the lemma, set $N := (|E|+1)^{|S|}$. We claim that any subset of T with more than N elements must contain two reflections whose product has infinite order. Since T is infinite by Proposition 2.69, this implies the proposition.

In view of Lemma 2.77, an equivalent formulation of the claim is that any set $\Psi \subseteq \Phi$ with more than N elements and with $\Psi \cap -\Psi = \emptyset$ must contain two distinct roots α, β such that $|B(\alpha, \beta)| \geq 1$. Suppose, to the contrary, that $|B(\alpha, \beta)| < 1$ for all $\alpha, \beta \in \Psi$ with $\alpha \neq \beta$. Let $\Psi' \subseteq \Psi$ be a basis for the subspace V' of V spanned by Ψ (so Ψ' has at most $|S|$ elements). For any $\alpha \in \Psi$ there are at most N possibilities for the sequence $(B(\alpha, \gamma))_{\gamma \in \Psi'}$, since each component is either 1 or is in E. Since $|\Psi| > N$, there must exist two distinct elements $\alpha, \beta \in \Psi$ such that $B(\alpha, -) = B(\beta, -)$ as linear functions on V'. But this yields the contradiction

$$1 = B(\alpha, \alpha) = B(\beta, \alpha) < 1 .   \square$$

Exercise 2.78. Give an alternative proof that the center of an infinite irreducible Coxeter group is trivial using the method of Exercise 1.102.

*2.6 The Tits Cone

We continue to assume that (W, S) is an arbitrary Coxeter system with S finite. In Section 2.5 we constructed a representation of W on the vector space $V = \mathbb{R}^S$. We introduced a set $\mathcal{C}$ of *chambers* in V^*, determined by a set $\mathcal{H}$ of *walls*. Here we carry the chamber geometry further and show that a great deal of what we did in Chapter 1 for finite reflection groups extends to general Coxeter groups. The main results first appeared in an unpublished paper of Tits [240]. Published accounts later appeared in Bourbaki [44], Vinberg [270], and Humphreys [133]. We have marked this subsection as optional because, while it contains an extremely useful tool for the study of Coxeter groups, it is not really needed in the rest of this book.

2.6.1 Cell Decomposition

We assume here that the reader is familiar with the elementary geometry of polyhedral sets defined by finitely many linear equalities and inequalities, as developed in Section 1.4. In particular, we will make use of the fact that such a set has well-defined *faces*. These can be determined using a collection of defining equalities and inequalities as in Definition 1.20, and they can also be characterized intrinsically (Proposition 1.27).

We apply this first to the fundamental chamber C, which is a simplicial cone. It has one face A for each subset $J \subseteq S$, defined by $\langle -, e_s \rangle = 0$ for $s \in J$ and $\langle -, e_s \rangle > 0$ for $s \in S \setminus J$. We use these faces and the W-action to define the cells that will be of interest to us.

Definition 2.79. The transforms wA ($w \in W$, $A \leq C$) will be called *cells*. The *Tits cone* X is defined to be the union of all the cells. Equivalently,

$$X = \bigcup_{w \in W} w\overline{C}.$$

Note that every cell is a polyhedral set of the sort discussed above, defined by finitely many linear equalities and inequalities.

Theorem 2.80. *The cone X is convex. For any $x, y \in X$, the line segment $[x, y]$ crosses only finitely many walls and is contained in a finite union of cells. Moreover:*

(1) *$\overline{C}$ is a strict fundamental domain for the action of W on X.*
(2) *The stabilizer of any $x \in \overline{C}$ is the standard subgroup of W generated by $S_x := \{ s \in S \mid sx = x \}$.*
(3) *For each cell A and wall H, A is contained either in H or in one of the open half-spaces determined by H.*

Proof. To prove the first part of the theorem, we may assume $x \in \overline{C}$ and $y \in w\overline{C}$ for some $w \in W$. Then $[x, y]$ crosses only finitely many walls by Lemma 2.63, since any wall that it crosses separates C from wC. We will prove by induction on $l(w)$ that $[x, y]$ is contained in a finite union of cells (and hence, in particular, it is contained in X). Let z be the point such that $[x, z] = \overline{C} \cap [x, y]$; see Figure 2.6. Then $[x, z]$ is contained in the union of the faces of C, so it is enough to show that $[z, y]$ is contained in a finite union of cells. We may assume $y \neq z$. For each $s \in S$, we have $z \in U_+(s)$ or $z \in H_s$, and

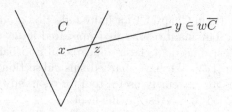

Fig. 2.6. Proof of convexity.

there must be at least one s with $z \in H_s$ and $y \in U_-(s)$; otherwise, we could move a positive distance from z toward y without leaving $\overline{C}$. Lemma 2.58 now implies that $w = sw'$ with $l(w') = l(w) - 1$, and then $[z, y] = s[z, sy]$ with $sy \in w'\overline{C}$, so we are done by the induction hypothesis.

We now proceed with (1)–(3). For (1) and (2) we must show that if $wx = y$ with $x, y \in \overline{C}$, then $x = y$ and $w \in \langle S_x \rangle$. We argue by induction on $l(w)$, which may be assumed > 0. Write $w = sw'$ with $l(w) = l(w') + 1$. Then $w'x = sy$. The left side is in $\overline{U_+(s)}$ by Lemma 2.58, and the right side is in $\overline{U_-(s)}$. So $w'x = sy \in H_s$, and hence $w'x = y \in H_s$. We now have $x = y$ and $w' \in \langle S_x \rangle$ by the induction hypothesis, and finally $w = sw' \in \langle S_x \rangle$ because $s \in S_y = S_x$.

(3) We may assume $A \leq C$ and $H = wH_s$. We know that A is contained in at least one of the two closed half-spaces bounded by H, since C is in one of the open half-spaces. So we must show that if A meets H then A is contained in H. In other words, if t is the reflection wsw^{-1} whose fixed-point set is H, we must show that if $tx = x$ for some $x \in A$ then $tx = x$ for all $x \in A$. This follows immediately from (2), since S_x corresponds to the walls of C containing x, which are the same as the walls of C containing A. $\square$

It follows from (3) that every cell A has a well-defined sign sequence $\sigma(A)$, generalizing the sign sequences for chambers, where now the possibility $\sigma_H(A) = 0$ is allowed. Moreover, A is defined by the equalities and inequalities corresponding to the signs. [It suffices to check this when A is a face of the fundamental chamber, in which case the result is trivial.] In particular, distinct cells are disjoint. It follows easily that the face relation, which makes sense a priori because each cell is defined by finitely many linear equalities and inequalities, has the usual interpretation:

$$B \leq A \iff \overline{B} \subseteq \overline{A} \iff \sigma(B) \leq \sigma(A) ,$$

where the ordering on sign sequences is the same as in Definition 1.20.

Remarks 2.81. (a) Since the cells are disjoint, the statement that every closed line segment in X is contained in only finitely many cells can be strengthened: Every closed line segment meets only finitely many cells.

(b) Even though $\mathcal{H}$ is infinite, it still determines a partition of V^* into sets determined by sign sequences, exactly as in Section 1.4. For lack of a better name, we call these sets $\mathcal{H}$-*cells*. What we have just shown, then, is that the cells of the Tits cone are in fact $\mathcal{H}$-cells. One must be careful in what follows to distinguish *cells* as in Definition 2.79 from the more general $\mathcal{H}$-*cells*.

(c) Although each cell of the Tits cone is determined by finitely many of the hyperplanes in $\mathcal{H}$, this is not necessarily true of arbitrary $\mathcal{H}$-cells. The reader might find it instructive to find all the $\mathcal{H}$-cells if W is the infinite dihedral group (Figure 2.2).

We can now carry the theory further.

Proposition 2.82.

(1) *Given cells A and B, there is a (unique) cell AB such that*

$$\sigma_H(AB) = \begin{cases} \sigma_H(A) & \text{if } \sigma_H(A) \neq 0, \\ \sigma_H(B) & \text{if } \sigma_H(A) = 0. \end{cases} \tag{2.16}$$

For any $x \in A$ and $y \in B$, we have $(1 - \epsilon)x + \epsilon y \in AB$ for all sufficiently small $\epsilon > 0$. The product $(A, B) \mapsto AB$ makes the set of cells a semigroup.
(2) *X is the entire space V^* if and only if W is finite.*

Proof. (1) This is proved as in Section 1.4.6, the essential point being that a line segment in X crosses only finitely many walls.

(2) If W is finite then we saw in the previous section that W can be identified with a finite reflection group acting on V^*, so $X = \bigcup w\overline{C} = V^*$ by Chapter 1. Conversely, suppose $X = V^*$. Then $-C$ is contained in X. Since $-C$ is obviously an $\mathcal{H}$-cell, it follows that $-C$ is a cell of X, hence a chamber of X, so W is finite by Proposition 2.69.　　　　　　　　　　　　　　□

Remark 2.83. Using the fact that the chambers are simplicial cones, one can show as in Section 1.5.8 that the poset Σ of cells is a simplicial complex, whose vertices are the cells that are rays. Moreover, one can easily check, as in Section 1.5.9, that this simplicial complex is isomorphic to the poset $\Sigma(W, S)$ of standard cosets, ordered by reverse inclusion. This simplicial complex, called the *Coxeter complex* associated to (W, S), will be the main object of study in the next chapter.

2.6.2 The Finite Subgroups of W

As an application of the Tits cone, we will use it to analyze the finite subgroups of W. As above, let Σ denote the set of cells in X. Fix $A \in \Sigma$, and let $\mathcal{H}_A$ be the set of walls containing A. Then $\mathcal{H}_A$ determines a partition of V^* into $\mathcal{H}_A$-cells. This is coarser than the partition into $\mathcal{H}$-cells, i.e., every $\mathcal{H}_A$-cell is a union of $\mathcal{H}$-cells. As in Exercise 1.45, we can prove the following:

Lemma 2.84. *Let $\Sigma_{\geq A}$ be the set of cells of X having A as a face, and let Σ_A be the set of $\mathcal{H}_A$-cells that meet X. For any cell $B \in \Sigma_{\geq A}$, let $f(B)$ be the $\mathcal{H}_A$-cell containing B. Then $f: \Sigma_{\geq A} \to \Sigma_A$ is a bijection.*

Proof. On the level of sign sequences, f just picks out the components of $\sigma(B)$ corresponding to the hyperplanes in $\mathcal{H}_A$. It is 1–1 because the remaining components of $\sigma(B)$ are the same as those of $\sigma(A)$. To prove that f is surjective, start with an $\mathcal{H}_A$-cell B' that meets X, choose a cell B of X contained in B', and form the product AB. Then a consideration of sign sequences shows that $f(AB) = B'$. $\square$

Lemma 2.85. *Let W_A be the stabilizer of A. Then W_A is finite if and only if $\mathcal{H}_A$ is finite.*

Proof. If W_A is finite, then it contains only finitely many reflections, so $\mathcal{H}_A$ is finite. Conversely, if $\mathcal{H}_A$ is finite, then there are only finitely many $\mathcal{H}_A$-cells, and hence Σ_A is finite. Lemma 2.84 now implies that $\Sigma_{\geq A}$ contains only finitely many chambers. So W_A is finite, since it acts simply transitively on those chambers. [Given chambers $D, E \geq A$, we know that there is a unique $w \in W$ such that $wD = E$. Then wA and A are W-equivalent faces of E; hence $wA = A$ by Theorem 2.80.] $\square$

Lemma 2.86. *Let X_f be the set of points $x \in X$ whose stabilizer W_x is finite. Given $x \in X_f$ and $y \in X$ with $x \neq y$, the half-open line segment $[x, y)$ is contained in X_f. In particular, X_f is convex.*

Proof. In view of Lemma 2.85, X_f consists of the points $x \in X$ such that x is contained in only finitely many walls. The result now follows from the fact that $[x, y]$ crosses only finitely many walls. $\square$

We can now prove the main result of this subsection.

Proposition 2.87. *Every finite subgroup of W is contained in a finite parabolic subgroup.*

Proof. Note that X_f is W-invariant and that by Theorem 2.80, the stabilizers W_x for $x \in X_f$ are precisely the finite parabolic subgroups of W. So our task is to show that every finite subgroup W' of W fixes a point of X_f. The latter being convex by Lemma 2.86, we can prove this by averaging: Start with an arbitrary $x \in X_f$, and then $\sum_{w \in W'} wx$ is a point of X_f fixed by W'. $\square$

Remark 2.88. See Bourbaki [44, Section V.4, Exercise 2(d)] or Brink and Howlett [49, Proposition 1.3] for other proofs of the proposition.

Exercises

2.89. Let A be a cell and W_A its stabilizer. If W_A is finite, show that every $\mathcal{H}_A$-cell meets X, so the map f of Lemma 2.84 is a bijection from $\Sigma_{\geq A}$ to the set of all $\mathcal{H}_A$-cells.

2.90. (a) Show that X_f is open in V^* and hence is the interior of X in V^*.
(b) Show that the action of W on X_f is proper. Since the stabilizers W_x for $x \in X_f$ are finite by definition, the content of this is that every $x \in X_f$ has a W_x-invariant neighborhood U such that $wU \cap U = \emptyset$ if $w \notin W_x$.

2.6.3 The Shape of X

Suppose (W, S) is irreducible. If W is neither spherical nor Euclidean, then a result of Vinberg [270, p. 1112, Lemma 15] says that the Tits cone X is strictly convex, i.e., its closure does not contain any lines through the origin. (This is obviously false in the spherical case. It is also false in the Euclidean case, where the closure of the Tits cone is a closed half-space. We have seen this in the case of the infinite dihedral group in Section 2.2.2, and the assertion in general follows from some results that we will prove in Section 10.2.2.) Our goal in this subsection is to prove the following weak form of Vinberg's result, which is valid in the Euclidean case also; see Krammer [149, Theorem 2.1.6] for a different proof.

Proposition 2.91. *If W is infinite and irreducible, then the Tits cone X does not contain any lines through the origin.*

Our proof will be based on the following lemma:

Lemma 2.92. *Suppose W is infinite and irreducible. For any $x \neq 0$ in X, there are infinitely many walls not containing x.*

Proof. We may assume that the cell A containing x is a face of the fundamental chamber C and hence that its stabilizer is W_J for some $J \subsetneq S$. Suppose x (and hence A) is contained in all but finitely many walls. Then there is an upper bound on the gallery distance $d(A, D)$, where D ranges over the chambers wC ($w \in W$). [Here $d(A, D)$ is defined as in Exercise 1.61, and, as in that exercise, it is equal to the number of walls that strictly separate A from D.] Equivalently, there is an upper bound on $d(wA, C)$ for $w \in W$. This implies that the W-orbit of A is finite and hence that W_J has finite index in W, contradicting Proposition 2.43.   □

(See also Exercise 3.83(b), where the same result is stated and proved from a combinatorial point of view.)

Proof of Proposition 2.91. Suppose X contains a pair of opposite points $\pm x$ with $x \neq 0$. Since x and $-x$ are strictly separated by all walls that do not contain them, it follows from Theorem 2.80 that x is contained in all but finitely many walls. This contradicts Lemma 2.92.   □

*2.7 Infinite Hyperplane Arrangements

In the previous section we encountered a possibly infinite hyperplane arrangement for which we were nevertheless able to carry out much of the theory of Section 1.4. The purpose of the present (optional) section is to axiomatize this situation and carry the theory further.

Throughout this section we denote by $\mathcal{H}$ an arbitrary collection of linear hyperplanes in a finite-dimensional real vector space V. For convenience, we assume that we have chosen for each $H \in \mathcal{H}$ a linear function $f_H \colon V \to \mathbb{R}$ such that H is defined by the equation $f_H = 0$. Exactly as in Definition 1.18, we then obtain a partition of V into "cells" A, each of which is defined by equalities or strict inequalities, one for each $H \in \mathcal{H}$. We again encode the definition of a cell A by its *sign sequence* $\sigma(A) = (\sigma_H(A))_{H \in \mathcal{H}}$. Explicitly, we define $\sigma(A)$ to be $\sigma(x)$ for any $x \in A$, where $\sigma_H(x)$ is the sign of $f_H(x)$. The set of cells is a poset under the *face relation* defined in terms of sign sequences as in Definition 1.20.

Continuing as in Section 1.4.2, we can also work with the *closed cells* $\overline{A}$. Here $\overline{A}$ is the set obtained by replacing the strict inequalities in the definition of A by weak inequalities; it is also the topological closure of $\overline{A}$ in V. We have

$$\overline{A} = \bigcup_{B \le A} B \, .$$

This implies the following characterization of the partial order on cells, which does not explicitly refer to $\mathcal{H}$:

$$B \le A \iff \overline{B} \subseteq \overline{A} \, . \tag{2.17}$$

We define the *support* of a cell A, denoted by $\operatorname{supp} A$, to be its linear span, and we define the *dimension* of A by $\dim A := \dim(\operatorname{supp} A)$. It is immediate from the definitions that if $B < A$ then $\dim A < \dim B$, since A is contained in at least one hyperplane that does not contain B. In particular, every nonempty collection of cells has a maximal element.

Assume throughout the rest of the section that we are given a nonempty set Σ of cells, and set

$$X := \bigcup_{A \in \Sigma} A \, .$$

We can now state our axioms.

(H0) X *linearly spans* V.

This axiom is harmless; if it failed, we could simply replace V by the span V' of X, and we could replace $\mathcal{H}$ by $\mathcal{H}' := \{H \cap V' \mid H \in \mathcal{H}, H \not\supseteq X\}$.

(H1) Σ *is closed under passage to faces; equivalently, X is a union of closed cells.*

The next two axioms are more serious and can be viewed as finiteness properties. For any $A, B \in \Sigma$, let $\mathcal{S}(A, B)$ be the set of hyperplanes $H \in \mathcal{H}$ that strictly separate A and B, i.e., that satisfy $\sigma_H(A) = -\sigma_H(B) \ne 0$.

(H2) *For any $A, B \in \Sigma$, the set $\mathcal{S}(A, B)$ is finite.*

(H3) *For any $A \in \Sigma$ there is a finite subset $\mathcal{H}_A \subseteq \mathcal{H}$ that defines A.*

Here, as in Section 1.4.3, the statement means that A is defined by the conditions $f_H = \sigma_H(A)$ for $H \in \mathcal{H}_A$. It follows that each cell $A \in \Sigma$ is a polyhedral cone of the sort studied in Section 1.4.

One consequence of (H3) is that A is open in its support. More precisely, we have

$$\operatorname{supp} A \ \subseteq \bigcap_{\substack{H \in \mathcal{H} \\ H \supseteq A}} H \ \subseteq \bigcap_{\substack{H \in \mathcal{H}_A \\ H \supseteq A}} H \,, \tag{2.18}$$

and A is open in the last of these spaces. Hence the latter is spanned by A, and the three spaces are equal.

Before proceeding further, we need to resolve a potential ambiguity. Given $A \in \Sigma$, we can talk about the faces of A as defined at the beginning of this section; let's call these the $\mathcal{H}$-*faces* of A. But if (H3) holds, then it would seem more natural to consider the $\mathcal{H}_A$-faces of A, which can in fact be intrinsically defined, without reference to the set $\mathcal{H}_A$ (see Proposition 1.27). It is this second notion of "face" that we used in the case of the Tits cone. Fortunately, there is no conflict:

Lemma 2.93. *Suppose that axiom* (H3) *holds. Given $A \in \Sigma$ and $\mathcal{H}_A$ as in* (H3), *the $\mathcal{H}_A$-faces of A are the same as the $\mathcal{H}$-faces of A.*

Proof. Since $\mathcal{H}_A \subseteq \mathcal{H}$, the partition of $\overline{A}$ into $\mathcal{H}$-cells refines the partition into $\mathcal{H}_A$-cells. It therefore suffices to show that every $\mathcal{H}_A$-face of A is contained in an $\mathcal{H}$-cell. Let B be an $\mathcal{H}_A$-face of A, and consider any $H \in \mathcal{H}$. Suppose, for instance, that $f_H \geq 0$ on $\overline{A}$. Then $f_H \geq 0$ on B; hence, since B is open in its support, either $f_H > 0$ on B or $f_H \equiv 0$ on B. Thus every f_H has a constant sign on B, so B is contained in an $\mathcal{H}$-cell. $\square$

We turn now to the most interesting axioms, involving products and convexity. Given two sign sequences $\sigma = (\sigma_H)_{H \in \mathcal{H}}$ and $\tau = (\tau_H)_{H \in \mathcal{H}}$, we define their *product* $\sigma\tau$ to be the sign sequence given by

$$(\sigma\tau)_H = \begin{cases} \sigma_H & \text{if } \sigma_H \neq 0, \\ \tau_H & \text{if } \sigma_H = 0, \end{cases}$$

for $H \in \mathcal{H}$. Consider now the following two conditions:

(H4) *For any two cells $A, B \in \Sigma$, there is a cell $AB \in \Sigma$ such that $\sigma(AB) = \sigma(A)\sigma(B)$.*

(H5) X *is a convex subset of V.*

These two conditions are in fact equivalent:

Proposition 2.94. *In the presence of* (H1) *and* (H2), *axioms* (H4) *and* (H5) *are equivalent to one another. When these axioms are satisfied, the product of cells can be characterized as follows: Given $x \in A$ and $y \in B$, the cell AB contains $(1 - t)x + ty \in AB$ for all sufficiently small $t > 0$.*

Proof. Suppose (H4) holds. Given $x, y \in X$, let A (resp. B) be the cell containing x (resp. y). We show by induction on $|\mathcal{S}(A, B)|$ that the open segment (x, y) is contained in X. If $\mathcal{S}(A, B) \neq \emptyset$, let z be the first point (i.e., the point closest to x) where (x, y) crosses a hyperplane in $\mathcal{H}$. Otherwise, set $z = y$. Let F be the cell containing z. Then every point in (x, z) has sign sequence equal to $\sigma(A)\sigma(B)$, so $(x, z) \subseteq AB \subseteq X$. We also have $F \leq AB$, so F is in Σ and z is in X. If $z = y$, we are done. Otherwise, $\mathcal{S}(F, B) \subsetneqq \mathcal{S}(A, B)$; hence $(x, y) = (x, z) \cup \{z\} \cup (z, y) \subseteq X$ by the induction hypothesis. This proves (H5).

Conversely, suppose that (H5) holds. Given $A, B \in \Sigma$, choose $x \in A$ and $y \in B$, and consider $z_t := (1 - t)x + ty$ for $0 < t \leq 1$. By hypothesis, $z_t \in X$ for all t. If t is small enough, $\sigma_H(z_t) = \sigma_H(A) \neq 0$ for all $H \in \mathcal{S}(A, B)$; hence $\sigma_H(z_t) = \bigl(\sigma(A)\sigma(B)\bigr)_H$ for such H. And if $H \in \mathcal{H} \smallsetminus \mathcal{S}(A, B)$, then $\sigma_H(z_t) = \bigl(\sigma(A)\sigma(B)\bigr)_H$ for *all* $t \in (0, 1)$. There is therefore a cell in Σ with sign sequence $\sigma(A)\sigma(B)$, and it contains z_t for sufficiently small $t > 0$. Thus (H4) holds, as does the last assertion of the proposition. $\qquad\square$

Assume from now on that X and Σ satisfy axioms (H0)–(H5). (In particular, X could be the Tits cone associated to a Coxeter group.) It is then easy to extend to Σ most of the concepts and results of Section 1.4. We will briefly run through some of these.

(1) Any two cells $A, B \in \Sigma$ have a greatest lower bound $A \cap B$, whose corresponding closed cell is the intersection $\overline{A} \cap \overline{B}$.

Use a finite subset of $\mathcal{H}$ that defines both A and B, and then appeal to the corresponding fact about finite hyperplane arrangements.

(2) A *chamber* of Σ is a cell $C \in \Sigma$ such that $\sigma_H(C) \neq 0$ for all $H \in \mathcal{H}$. These are precisely the maximal elements of Σ.

Indeed, a chamber is trivially maximal. Conversely, suppose $C \in \Sigma$ is maximal and consider any $H \in \mathcal{H}$. In view of (H0), there is a cell $A \in \Sigma$ with $\sigma_H(A) \neq 0$. But then $C = CA$ by maximality; hence $\sigma_H(C) = \sigma_H(CA) \neq 0$, and C is a chamber.

(3) A *panel* is a cell $P \in \Sigma$ with exactly one 0 in its sign sequence. Equivalently, it is a cell in Σ of dimension equal to $\dim V - 1$. Every panel is a face of at least one chamber and at most two.

(4) Two distinct chambers $C, D \in \Sigma$ are *adjacent* if they have a common panel. One can now define galleries in the obvious way and prove that any two chambers C, D can be connected by a gallery; moreover, the minimal length of such a gallery is $|\mathcal{S}(C, D)|$.

(5) More generally, we can consider galleries connecting two arbitrary cells $A, B \in \Sigma$ as in Exercise 1.62. The solution to that exercise goes through

without change to show that the minimal length $d(A, B)$ of such a gallery is $|\mathcal{S}(A, B)|$. Moreover, the chambers that can start a minimal gallery from A to B are precisely those having AB as a face. In particular, every minimal gallery from a cell A to a chamber C starts with AC.

We turn now to subcomplexes, which we did *not* have occasion to consider in the setting of finite hyperplane arrangements. By a *subcomplex* of Σ we mean a nonempty subset Σ' that is closed under passage to faces. Let Σ' be a subcomplex and let $X' := \bigcup_{A \in \Sigma'} A$. Then we can study Σ' by viewing it as a set of $\mathcal{H}'$-cells in the linear span V' of X', where

$$\mathcal{H}' := \{H \cap V' \mid H \in \mathcal{H}, \, H \not\supseteq X'\} \ .$$

Viewed in this way, Σ' satisfies all of the axioms of this section except possibly the (equivalent) axioms (H4) and (H5).

Definition 2.95. We say that a subcomplex Σ' of Σ is *convex* if Σ' satisfies (H4) and (H5), i.e., if $X' := \bigcup_{A \in \Sigma'} A$ is a convex subset of V or, equivalently, if Σ' is a subsemigroup of Σ.

Thus we can apply to convex subcomplexes all of the results that we have proven about Σ. To state one explicitly, assume for simplicity that the chambers of Σ are simplicial cones (as in the setting of Section 2.6). We make this assumption only so that we can apply the language of Section A.1.3. Then the results of this section show that Σ is a chamber complex in which any panel is a face of at most two chambers. Consequently:

Proposition 2.96. *Suppose that the chambers of Σ are simplicial cones. Then every convex subcomplex Σ' of Σ is a chamber complex in which every panel is a face of at most two chambers.* □

(Of course we might have $\dim \Sigma' < \dim \Sigma$, so Σ' is not in general a *chamber subcomplex* of Σ.)

We close this section by giving some useful characterizations of convex subcomplexes. Note first that there is an obvious way of constructing convex subcomplexes of Σ using half-spaces. Namely, for any $H \in \mathcal{H}$ we have a convex subcomplex $\Sigma_+(H)$ (resp. $\Sigma_-(H)$) consisting of the cells in Σ on which $f_H \geq 0$ (resp. $f_H \leq 0$). Further examples can be obtained from these by taking intersections. For example, the convex subcomplex $\Sigma_0(H) := \{A \in \Sigma \mid \sigma_H(A) = 0\}$ is the intersection $\Sigma_+(H) \cap \Sigma_-(H)$. The following proposition, which should be compared with Exercise 1.68, implies that *every* convex subcomplex can be obtained as an intersection of such "halves" of Σ.

Let $\mathcal{D}$ be a nonempty set of chambers in Σ. We say that $\mathcal{D}$ is *convex* if for all $C, D \in \mathcal{D}$, every minimal gallery in Σ from C to D is contained in $\mathcal{D}$.

Proposition 2.97. *Let Σ' be a subcomplex of Σ. Then the following two conditions are equivalent:*

(i) Σ' *is a convex subcomplex of Σ.*

(ii) Σ' *is an intersection of subcomplexes of the form $\Sigma_\pm(H)$ $(H \in \mathcal{H})$.*

If Σ' contains at least one chamber, then (i) *and* (ii) *are equivalent to each of the following conditions:*

(iii) *The maximal elements of Σ' are chambers of Σ, and the set of chambers in Σ' is convex.*

(iv) *Given $A, C \in \Sigma'$ with C a chamber, Σ' contains every minimal gallery in Σ from A to C.*

Proof. The implication (ii) $\implies$ (i) is trivial, since an intersection of convex subcomplexes is a convex subcomplex. [Note that it is automatically nonempty because it contains the smallest cell, which is $\bigcap_{H \in \mathcal{H}} H$.] To prove the converse, assume first that Σ' contains a chamber, in which case we will prove (i) $\implies$ (iv) $\implies$ (iii) $\implies$ (ii).

(i) $\implies$ (iv): Consider a minimal gallery $A \le C_0, C_1, \ldots, C_l = C$ in Σ. Set $A_0 := A$ and $A_i := C_{i-1} \cap C_i$ for $i = 1, \ldots, l$. Then $C_i, C_{i+1}, \ldots, C_l$ is a minimal gallery from A_i to C for $0 \le i \le l$; hence $C_i = A_i C$. So if (i) holds and $A, C \in \Sigma'$, it follows inductively that $C_i \in \Sigma'$ for all i.

(iv) $\implies$ (iii): This is trivial.

(iii) $\implies$ (ii): Let Σ'' be the intersection of the subcomplexes $\Sigma_\pm(H)$ that contain Σ'. If (iii) holds, we claim that Σ' and Σ'' have the same chambers. This implies that they are equal, since Σ'' is a subsemigroup of Σ containing a chamber, and hence every maximal cell of Σ'' is a chamber. To prove the claim, let $\mathcal{C}$ be the set $\mathcal{C}(\Sigma)$ of chambers of Σ, and set $\mathcal{D} := \mathcal{C}(\Sigma')$. We must show that for any $C \in \mathcal{C} \smallsetminus \mathcal{D}$ there is a hyperplane $H \in \mathcal{H}$ that separates C from $\mathcal{D}$.

Choose $D \in \mathcal{D}$ at minimal distance from C, and let $D, D', \ldots, C$ be a minimal gallery from D to C. Then $D' \notin \mathcal{D}$, and the hyperplane H separating D from D' also separates D from C. We will show that all chambers in $\mathcal{D}$ are on the D-side of H. Given $E \in \mathcal{D}$, we have $d(E, D) = d(E, D') \pm 1$. The sign cannot be $+$, because then there would be a minimal gallery from E to D passing through D', contradicting the convexity of $\mathcal{D}$. So the sign is $-$, which means that E and D are on the same side of H. This completes the proof that (iii) $\implies$ (ii) and hence that all four conditions are equivalent when Σ' contains a chamber.

Suppose now that Σ' does not necessarily contain a chamber. To prove (i) $\implies$ (ii), note that (i) implies that all maximal cells of Σ' have the same support U. [If A and B are maximal, then $A = AB$ and $B = BA$; now use the fact that AB and BA have the same zeros in their sign sequences.] Let $\mathcal{H}_U := \{H \cap U \mid H \in \mathcal{H}, H \not\supseteq U\}$, and let Σ_U be the set of elements of Σ contained in U. Then Σ_U is a set of $\mathcal{H}_U$-cells in U satisfying all of our axioms, and Σ' is a convex subcomplex of Σ_U whose maximal simplices are chambers of Σ_U. Moreover, the "halves" of Σ_U are the subcomplexes $\Sigma_U \cap \Sigma_\pm(H)$ for $H \in \mathcal{H}$, $H \not\supseteq U$. By the case already treated, it follows that Σ' is an intersection of such subcomplexes. This implies (ii), since

$$\Sigma_U = \bigcap_{H \supseteq U} \Sigma_0(H) = \bigcap_{H \supseteq U} \Sigma_+(H) \cap \Sigma_-(H) \,. \qquad\qquad \square$$

Exercise 2.98. Show that in the presence of the other axioms, (H3) can be replaced by the following apparently weaker condition:

(H3′) *Any chamber $C \in \Sigma$ can be defined by finitely many linear inequalities of the form $f > 0$.*

3

Coxeter Complexes

Let (W, S) be a Coxeter system (Definition 2.50). Recall from Section 2.6 that S *is assumed to be finite*. In Chapter 2 we proved some algebraic results about (W, S), guided by our geometric intuition from Chapter 1. We now develop the corresponding geometric theory. Our point of view in this chapter is purely combinatorial, though we often indicate alternative proofs that make use of the (optional) Sections 2.6 and 2.7. We will also refer to those sections in some exercises, which the reader may omit.

3.1 The Coxeter Complex

Recall from Definition 2.12 that a standard coset in W is a coset of the form wW_J with $w \in W$ and $W_J := \langle J \rangle$ for some subset $J \subseteq S$. The results of Section 1.5.9 [and Section 2.6] motivate the following definition.

Definition 3.1. Let $\Sigma(W, S)$ be the poset of standard cosets in W, ordered by reverse inclusion. Thus $B \le A$ in Σ if and only if $B \supseteq A$ as subsets of W, in which case we say that B is a *face* of A. We call $\Sigma(W, S)$ the *Coxeter complex* associated to (W, S).

It is worth reemphasizing the motivating example:

Definition 3.2. The Coxeter complex $\Sigma(W, S)$ is called *spherical* if it is finite or, equivalently, if W is finite.

The terminology comes from the fact that by Theorem 2.65, a spherical Coxeter complex is isomorphic to the complex $\Sigma(W, V)$ associated to a finite reflection group (Section 1.5); hence it is a simplicial complex triangulating a sphere by Proposition 1.108. In the more general setting of Definition 3.1, the word "complex" will be justified in Theorem 3.5 below, where we will prove that $\Sigma := \Sigma(W, S)$ is indeed a simplicial complex. [See also Remark 2.83.] Anticipating this result, we proceed with some further terminology.

Definition 3.3. The elements of Σ are called *simplices*. The maximal simplices, which are the singletons $\{w\}$, are called *chambers* and are identified with the elements of W. The simplices of the form $w\langle s\rangle = \{w, ws\}$ (with $w \in W$ and $s \in S$) are called *panels*. We set $C := 1$ and call it the *fundamental chamber*. Each panel $w\langle s\rangle$ is a face of exactly two chambers, w and ws, which are said to be *s-adjacent*.

Note that there is an action of W on Σ by left translation, and the action on the chambers is simply transitive. Note further that we can construct "galleries" in Σ in the way that is familiar from Chapter 1: Given $w \in W$ and a decomposition $w = s_1 \cdots s_l$ (with $s_i \in S$), we have a sequence of chambers

$$\Gamma: C = C_0, C_1, \ldots, C_l = wC, \tag{3.1}$$

where $C_i = s_1 \cdots s_i C$, and C_{i-1} is s_i-adjacent to C_i for $i = 1, \ldots, l$.

At this point the reader may need to refer to Section A.1.3 in Appendix A for the terminology regarding chamber complexes and type functions. Let's add one more bit of terminology:

Definition 3.4. A chamber complex is called *thin* if every panel is a face of exactly two chambers.

Theorem 3.5. *The poset* $\Sigma := \Sigma(W, S)$ *is a simplicial complex. Moreover, it is a thin chamber complex of rank equal to* $|S|$*, it is colorable, and the action of* W *on* Σ *is type-preserving.*

Proof. To show that Σ is simplicial, there are two things we must verify (see Definition A.1):

(a) Any two elements $A, B \in \Sigma$ have a greatest lower bound.
(b) For any $A \in \Sigma$, the poset $\Sigma_{\leq A}$ is a Boolean lattice.

For (a) we can use the W-action on Σ to reduce to the case that one of the two elements is a face of the fundamental chamber C, i.e., is a standard subgroup. What we must prove, then, is that a standard subgroup W_J and a standard coset wW_K (where $J, K \subseteq S$) have a least upper bound in the set of standard cosets, with respect to the ordering by inclusion. Now any standard coset containing the two given ones contains the identity and hence is a standard subgroup. Moreover, it contains w and hence also $W_K = w^{-1}(wW_K)$. So the upper bounds of our two standard cosets are the standard subgroups containing J, K, and w. In view of Proposition 2.16, there is indeed a smallest upper bound, namely, the standard subgroup W_L, where $L := J \cup K \cup S(w)$.

To prove (b), we may assume that A is the fundamental chamber C. In this case, $\Sigma_{\leq C}$ is the set of standard subgroups of W (ordered by reverse inclusion). By Proposition 2.13 we have

$$\Sigma_{\leq C} \cong (\text{subsets of } S)^{\text{op}} \cong (\text{subsets of } S),$$

where the second isomorphism is given by $J \mapsto S \setminus J$ for $J \subseteq S$. This proves (b) and completes the proof that Σ is simplicial. The proof also shows that all maximal simplices have the same rank, equal to $|S|$. And the discussion surrounding (3.1) above implies that any two of them can be connected by a gallery and that any panel is a face of exactly two chambers. So Σ is a thin chamber complex.

Finally, we can define a W-invariant type function τ on Σ, with values in S, by setting $\tau(wW_J) := S \setminus J$.   □

We will continue to denote by τ the type function constructed in the proof. For emphasis, we repeat the definition:

Definition 3.6. $\Sigma(W, S)$ has a *canonical type function* with values in S, defined by
$$\tau(wW_J) = S \setminus J$$
for $w \in W$ and $J \subseteq S$. Equivalently, the simplex wW_J has *cotype J*.

We have already seen the canonical type function in Section 1.6.2 in the context of finite reflection groups, where we also saw examples illustrating it. Here is one more:

Example 3.7. Let W be the group of isometries of the plane generated by the (affine) reflections with respect to the sides of an equilateral triangle. This is an example of a Euclidean reflection group. Although we will not treat the theory of such groups systematically until Chapter 10, the reader should find it plausible that W is the Coxeter group
$$\left\langle\, s, t, u \; ; \; s^2 = t^2 = u^2 = (st)^3 = (tu)^3 = (su)^3 = 1 \,\right\rangle$$
and that the Coxeter complex $\Sigma(W, \{s, t, u\})$ is the plane tiled by equilateral triangles. We will give an ad hoc proof of this in Section 3.4.2 below (Example 3.76); in the meantime, the reader is advised to take the assertion on faith. Figure 3.1 shows the panels of the fundamental chamber C labeled by the reflections that fix them, or, equivalently, by their cotypes. The black vertex of C is not in the panel fixed by s, so it is of type s and hence all black vertices are of type s. Similar remarks apply to the other two types.

Exercises

Throughout these exercises (W, S) is a Coxeter system and $\Sigma := \Sigma(W, S)$ is the associated Coxeter complex.

3.8. Give an alternative proof that Σ is simplicial based on Exercise A.3.

3.9. The canonical type function yields a notion of *s-adjacency* for any $s \in S$ (Section A.1.4). Show that this is consistent with Definition 3.3.

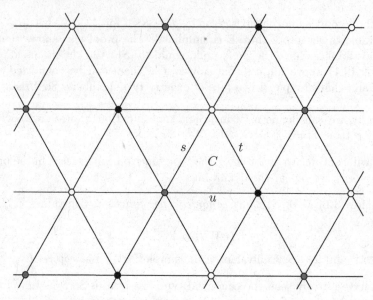

Fig. 3.1. The canonical type function; black $= s$, gray $= t$, white $= u$.

3.10. For every simplex $A \in \Sigma$, show that

$$A = \bigcap_{C \geq A} C \,,$$

where C ranges over the chambers $\geq A$.

3.11. (a) Let C and D be chambers of Σ such that $d(C, D') \leq d(C, D)$ for every chamber D' adjacent to D. Show that Σ is spherical and that C and D are opposite. [Recall that if Σ is spherical, then Σ can be identified with the complex associated to a finite reflection group, so "opposite" makes sense.]

  (b) Deduce (or show directly) that Σ is spherical if and only if it has finite diameter, where the *diameter* of a chamber complex is the supremum of the gallery distances between its chambers.

3.12. Assume that (W, S) is irreducible and W is infinite. The content of Proposition 2.43, then, is that Σ has infinitely many vertices of each type s. [Take $J = S \smallsetminus \{s\}$ in the proposition.]

  (a) Deduce that for each chamber C and each $s \in S$, the distance $d(C, y)$ is unbounded as y ranges over the vertices of type s. Equivalently, $d(C, y)$ is unbounded for a fixed vertex y of type s as C ranges over all chambers. Here $d(-, -)$ denotes the gallery distance defined in Section A.1.3.

  (b) Use Proposition 2.45 to prove the following stronger result: For each vertex $x \in \Sigma$ and each $s \in S$, the distance $d(x, y)$ is unbounded as y ranges over the vertices of type s.

3.13. For any simplex $A \in \Sigma$, show that the stabilizer W_A of A in W acts transitively on the set $\mathcal{C}(\Sigma)_{\geq A}$ of chambers having A as a face.

Warning. There is potential confusion between the notation W_A for the stabilizer of a simplex and W_J for a standard subgroup. But it should always be clear from the context which one is intended.

3.14. We have followed the conventions of Section A.1.1 in describing Σ as a poset. Convert this to a more conventional description as follows: There is one vertex for each maximal (proper) standard coset, and a finite collection of such cosets forms a simplex if and only if their intersection is nonempty. More succinctly, one expresses this by saying that Σ is the *nerve* of the covering of W by its maximal standard cosets.

3.15. Give a bijection between the geometric realization $|\Sigma|$ and equivalence classes of points of the Tits cone under multiplication by positive scalars.

3.2 Local Properties of Coxeter Complexes

We continue to denote by Σ the Coxeter complex $\Sigma(W, S)$ associated to a fixed Coxeter system (W, S). By "local properties" of Σ we mean properties of the *links* of simplices (Definition A.19). For example, it is of interest to know whether these links are chamber complexes.

Proposition 3.16. *Given $A \in \Sigma$, let $J := S \smallsetminus \tau(A)$ be its cotype. Then $\operatorname{lk}_\Sigma A$ is isomorphic to the Coxeter complex $\Sigma(W_J, J)$ associated to the Coxeter system (W_J, J). In particular, this link is a chamber complex.*

(Note that the statement makes sense because (W_J, J) is indeed a Coxeter system by Exercise 2.51.)

Proof. We may assume that A is a face of the fundamental chamber. Then A is the standard subgroup W_J. Recall now that there is a poset isomorphism $\operatorname{lk}_\Sigma A \cong \Sigma_{\geq A}$; hence the link of A is isomorphic to the set of standard cosets in W that are contained in W_J, ordered by reverse inclusion. But the standard cosets that are contained in W_J are precisely the same as the standard cosets associated to the Coxeter system (W_J, J). Thus $\Sigma_{\geq A} = \Sigma(W_J, J)$. $\square$

It follows that Σ satisfies the hypotheses of Proposition A.20. Consequently:

Corollary 3.17. *Σ is completely determined by its underlying chamber system. More precisely, the simplices of Σ are in 1–1 correspondence with the residues in $\mathcal{C}(\Sigma)$, ordered by reverse inclusion. Here a simplex A corresponds to the residue $\mathcal{C}(\Sigma)_{\geq A}$, consisting of the chambers having A as a face.* $\square$

Remark 3.18. We have included this corollary only to force the reader to learn the terminology associated with chamber systems, especially the concept of *residue*. But the statement of the corollary is in fact a complete tautology in view of the definition of Σ in terms of standard cosets. Indeed, if one identifies chambers with elements of W, then it is immediate from the definitions that the residues *are* the standard cosets.

Proposition 3.16 has a simple interpretation in terms of Coxeter matrices. Recall, first, that the Coxeter system (W, S) is determined by its Coxeter matrix $M = \big(m(s,t)\big)_{s,t \in S}$. So we may think of Σ as a simplicial complex associated to M. Next, note that the rows and columns of M are indexed by S, which is also the set of types of the vertices of Σ. What the proposition says, then, is that lk A is the Coxeter complex associated to the matrix M_J obtained from M by selecting the rows and columns belonging to the cotype J of A.

This becomes even easier to use if we translate it into the language of *Coxeter diagrams*. Recall that the diagram of (W, S) is a graph D with labels on some edges. There is one vertex for each $s \in S$, with s joined to t if $m(s,t) \geq 3$, and with a label over that edge if $m(s,t) \geq 4$. The passage from M to M_J above, and hence the passage from Σ to lk A, corresponds to passing to the *induced subdiagram* D_J with vertex set J equal to the cotype of A. In other words, we retain the vertices in the cotype (and all edges between them). Equivalently, we delete all vertices in $\tau(A)$ (and all edges touching them).

Consider, for example, the group $W = \mathrm{PGL}_2(\mathbb{Z})$ studied in Section 2.2.3. Its diagram is

$$\circ \!\!-\!\!-\!\!-\!\!- \circ \overset{\infty}{-\!\!-\!\!-} \circ \, .$$

The Coxeter complex Σ has rank 3 (dimension 2), so there are three types of vertices. Let's compute the link of each type of vertex.

According to the recipe above, we must delete one vertex at a time from the Coxeter diagram of W. This yields the Coxeter diagrams of the dihedral groups D_{2m}, where $m = \infty$, 2, and 3, respectively. Now it is easy to figure out what the Coxeter complex associated to D_{2m} looks like, and in fact, we have already seen it in Chapters 1 and 2. Namely, it is a $2m$-gon; in other words, it is a triangulated circle with $2m$ edges if $m < \infty$, and it is a triangulated line if $m = \infty$. So our three links in this example are a line, a quadrilateral, and a hexagon.

Exercise 3.19. Look at Figure 2.5. Can you find the three types of links in the picture?

This example illustrates a general principle, valid for all Coxeter complexes: The link of a codimension-2 simplex of cotype $\{s, t\}$ (with $s \neq t$) is a $2m$-gon, where $m = m(s,t)$. This fact yields a geometric interpretation of the Coxeter matrix M:

Corollary 3.20. *The Coxeter matrix M of (W, S) can be recovered from Σ as follows: For any $s, t \in S$ with $s \neq t$, $m(s, t)$ is the unique number m $(2 \leq m \leq \infty)$ such that the link of every simplex of cotype $\{s, t\}$ is a $2m$-gon.*
$\square$

This shows, in particular, that the Coxeter group W is determined up to isomorphism by Σ. We will see this again in the next section, from a different point of view.

Remark 3.21. Note that a $2m$-gon has diameter m, where the *diameter* of a chamber complex is the supremum of the gallery distances between its chambers. So we can also write the geometric interpretation of M as

$$m(s, t) = \operatorname{diam}(\operatorname{lk} A) \, ,$$

where A has cotype $\{s, t\}$ as above. The result in this form is valid even when $s = t$. [In this case the link has exactly two chambers, which are adjacent, so the diameter is indeed $1 = m(s, s)$.]

We can use this corollary, together with Tits's solution to the word problem for Coxeter groups, to give a simple answer to a question that might seem, a priori, to be very difficult: How can one describe the totality of minimal galleries connecting two given chambers? This is easy in the 1-dimensional case, where Σ is a $2m$-gon: Minimal galleries are unique unless $m < \infty$ and the two given chambers C_1 and C_2 are at maximum distance m from each other, i.e., they are *opposite*. In this case there are exactly two minimal galleries connecting C_1 to C_2.

Translating this result to the link of a simplex A of codimension 2 in an arbitrary Coxeter complex, we obtain a similar description of the minimal galleries in the subposet $\Sigma_{\geq A}$. Visualize, for example, the case that Σ is 2-dimensional and A is a vertex v whose link is finite. Then for some $m < \infty$, $\Sigma_{\geq A}$ contains $2m$ chambers that form a solid $2m$-gon centered at v. The only nonuniqueness of minimal galleries in this subposet arises from the fact that there are two ways of going around the $2m$-gon to get from a given chamber to the opposite chamber.

Since galleries correspond to words, we can use the solution to the word problem (Section 2.3.3) to analyze the general case. The answer, roughly, is that the nonuniqueness of minimal galleries in a Coxeter complex can be explained entirely in terms of the obvious nonuniqueness that occurs in links of codimension-2 simplices. To state this precisely, we need some terminology.

Definition 3.22. If $\Gamma \colon C_0, \dots, C_d$ is a gallery, then the *type* of Γ is the sequence $\mathbf{s} := (s_1, \dots, s_d)$ such that C_{i-1} is s_i-adjacent to C_i for $i = 1, \dots, d$.

(This notion of "type of a gallery" makes sense in any colorable chamber complex; we will use it again later.)

Suppose Γ has a subgallery of type $(s, t, s, t, \dots)$ and of length $m = m(s, t) < \infty$, where $s \neq t$. Then this subgallery lies in $\Sigma_{\geq A}$ for some codimension-2 simplex A of cotype $\{s, t\}$, and we may replace it by the other minimal gallery in $\Sigma_{\geq A}$ with the same extremities. This produces a new gallery Γ' from C_0 to C_d.

Definition 3.23. The gallery Γ' is said to be obtained from Γ by an *elementary homotopy*. Two galleries are said to be *homotopic* if there is a finite sequence of elementary homotopies transforming one to the other.

Figure 3.2 shows an elementary homotopy from a gallery of type (u, t, s, t, u) to one of type (u, s, t, s, u).

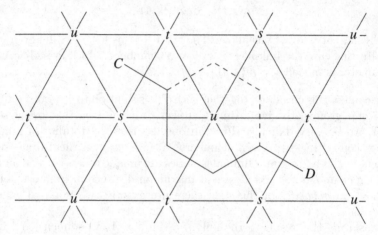

Fig. 3.2. An elementary homotopy.

The following result is now immediate from our earlier observations:

Proposition 3.24. *Any two minimal galleries with the same extremities are homotopic.* □

Remark 3.25. We have framed our discussion in terms of links. We could equally well have used the language of residues. Indeed, if A is a codimension-2 simplex of cotype $\{s, t\}$ as above, then the set of chambers in $\Sigma_{\geq A}$ is a residue of type $\{s, t\}$. So our elementary homotopies all take place in rank-2 residues. Here the *rank* of a residue is the cardinality of its type, which is the same as the codimension of the corresponding simplex A.

Exercise 3.26. Give a method for using homotopies to decide whether a given gallery is minimal and if not, to obtain a minimal gallery from it.

Finally, we can use our calculation of links to answer another natural question, at least to readers who are familiar with combinatorial topology:

When is Σ a manifold? This question arises naturally because triangulated manifolds (without boundary) are the canonical examples of thin chamber complexes (Example A.9). We already know the answer if W is finite: In this case Σ is a sphere (see the remarks following Definition 3.2); in particular, Σ is a manifold.

What happens if W and Σ are infinite? There is an obvious necessary condition. Namely, manifolds are locally compact, hence *locally finite*, i.e., every nonempty simplex A is a face of only finitely many chambers. In other words, the link of A must be finite. Conversely, if the link of every nonempty simplex is finite, then it is in fact a sphere (since it is a finite Coxeter complex). We leave it as an exercise for the interested reader to deduce that Σ is then a manifold. This proves the following:

Corollary 3.27. *The following conditions are equivalent:*

(i) *Σ is a manifold.*
(ii) *Σ is locally finite.*
(iii) *Every proper standard subgroup of W is finite.* $\square$

For example, the Coxeter complex associated to $\mathrm{PGL}_2(\mathbb{Z})$ is not a manifold. One can see the nonmanifold points in Figures 2.3 and 2.5: They are the cusps.

Exercise 3.28. If (iii) holds and W is infinite, show that (W, S) is irreducible.

Remark 3.29. Condition (iii) is quite restrictive. One can show that it holds only in the following three cases: (a) W is finite; (b) W is an irreducible Euclidean reflection group; (c) W is a hyperbolic reflection group whose fundamental domain is a closed simplex contained entirely in the interior of the hyperbolic space. See Chapter 10 for definitions of the terms used in (b) and (c) and for more information.

Even though we have not yet officially discussed Euclidean reflection groups, most readers probably have some intuition about them. For example, D_∞ is a Euclidean reflection group acting on the line, and the group of Example 3.7 is a Euclidean reflection group acting on the plane. It is natural to wonder why *reducible* Euclidean reflection groups were excluded in Remark 3.29 (and in Exercise 3.28): Given Euclidean reflection groups W_1 and W_2 acting on Euclidean spaces E_1 and E_2, isn't their product W a Euclidean reflection group acting on $E = E_1 \times E_2$, which is a Euclidean space and hence a manifold? And doesn't Σ triangulate this manifold? The answer is "yes" to the first question, but "no" to the second. The following exercise explains what happens.

Exercise 3.30. Let (W', S') and (W'', S'') be Coxeter systems, and let (W, S) be their product (with $W := W' \times W''$ and $S := S' \cup S''$). Show that

$$\Sigma(W, S) \cong \Sigma(W', S') * \Sigma(W'', S'') \,,$$

where the asterisk denotes the join operation. [Recall that the *join* Δ of two simplicial complexes Δ' and Δ'' with vertex sets $\mathcal{V}'$ and $\mathcal{V}''$ has vertex set equal to the disjoint union $\mathcal{V}' \amalg \mathcal{V}''$ and has one simplex $A' \cup A''$ for every $A' \in \Delta'$ and $A'' \in \Delta''$. From the poset point of view, then, Δ is simply the Cartesian product of Δ' and Δ''. But its geometric realization $|\Delta|$ is *not* the Cartesian product $|\Delta'| \times |\Delta''|$; in fact, Δ does not even have the right dimension for this to be true.]

Returning to the question whether a reducible Coxeter group can yield a Coxeter complex that is a manifold, the essential point is that the join of two manifolds that are not spheres is generally not a manifold. [But the join of two spheres is a sphere.]

This discussion suggests that the Coxeter complex $\Sigma(W, S)$ is not always the "best" geometric model for a Coxeter group W. For example, it would seem more reasonable to use a product of Euclidean spaces rather than a join of Euclidean spaces in the case of a reducible Euclidean reflection group. The result is a cell complex whose cells are products of simplices rather than simplices. We will return to this circle of ideas in Chapter 12.

3.3 Construction of Chamber Maps

We continue to assume that (W, S) is a Coxeter system and that $\Sigma = \Sigma(W, S)$ is the associated Coxeter complex.

3.3.1 Generalities

In studying Σ, it is quite easy to work with the chambers and the adjacency relations. It is awkward, on the other hand, to work with the vertices, which correspond to the maximal (proper) standard cosets wW_J with $J = S \smallsetminus \{s\}$ for some $s \in S$ (see Exercise 3.14). It is therefore useful that we never have to think about the vertices, i.e., that Σ is determined by its associated *chamber system*, consisting of the set of chambers [which correspond to the elements of W] together with the adjacency relations [given by right multiplication by elements of S]. See Corollary 3.17 and Remark 3.18. The specific consequence of this that we will need is that if we want to construct an endomorphism of Σ (i.e., a chamber map $\phi\colon \Sigma \to \Sigma$), then we need only give a function ϕ' on the chambers that is compatible with the adjacency relations.

We could deduce this from general considerations involving chamber systems, but we prefer to give a direct proof. In order to motivate the precise statement, let's think about what "compatible" should mean in the rough statement given above. If we take this to mean "preserving s-equivalence for all s," then we are dealing only with type-preserving endomorphisms of Σ. To handle the general case we must specify, in addition to ϕ', a permutation ϕ'' of S that describes how ϕ mixes up the vertex types (see Proposition A.14).

The compatibility condition, then, is that ϕ' takes s-equivalent chambers to $\phi''(s)$-equivalent chambers.

Here, now, is the precise result:

Lemma 3.31. *Endomorphisms ϕ of Σ are in 1–1 correspondence with pairs (ϕ', ϕ''), where ϕ' is a function $W \to W$, ϕ'' is a permutation of S, and $\phi'(ws) = \phi'(w)$ or $\phi'(w)\phi''(s)$ for all $w \in W$ and $s \in S$.*

Proof. Let ϕ be an endomorphism of Σ. Then the restriction of ϕ to the chambers yields a function $\phi' \colon W \to W$. (Recall that the chambers are the singleton standard cosets and are identified with the elements of W.) We also have a type-change map ϕ_* (Proposition A.14), which is a bijection $\phi'' := \phi_* \colon S \to S$. Then ϕ takes s-adjacent chambers to $\phi''(s)$-equivalent chambers, i.e., $\phi'(ws) = \phi'(w)$ or $\phi'(w)\phi''(s)$.

Note that ϕ is completely determined by the pair (ϕ', ϕ''). For if $A := wW_J$ is an arbitrary simplex of Σ, then A is the face of cotype J of the chamber w; so $\phi(A)$ must be the face of $\phi'(w)$ of cotype $\phi''(J)$; in other words, $\phi(wW_J) = \phi'(w)W_{\phi''(J)}$.

Finally, we must show that every pair (ϕ', ϕ'') as in the statement of the lemma arises from an endomorphism ϕ. To this end we simply define ϕ, as we must, by $\phi(wW_J) = \phi'(w)W_{\phi''(J)}$. It is easy to check that ϕ is a well-defined chamber map that induces ϕ' on the chambers and ϕ'' on the types.   □

The next two subsections illustrate the lemma.

3.3.2 Automorphisms

Recall that the W-action on Σ is simply transitive on the chambers; in particular, this action is faithful, in the sense that the corresponding homomorphism $W \to \operatorname{Aut} \Sigma$ is injective. Here $\operatorname{Aut} \Sigma$ denotes the group of simplicial automorphisms of Σ.

Proposition 3.32. *The image of $W \hookrightarrow \operatorname{Aut} \Sigma$ is the normal subgroup $\operatorname{Aut}_0 \Sigma$ consisting of the type-preserving automorphisms of Σ.*

(This shows, for the second time, that W is determined up to isomorphism by its Coxeter complex Σ.)

Proof. We already know that W acts as a group of type-preserving automorphisms of Σ. Conversely, suppose ϕ is an arbitrary type-preserving automorphism, and let ϕ' and ϕ'' be its "components" as in Lemma 3.31. Then ϕ'' is the identity, so $\phi'(ws) = \phi'(w)s$ for all w and s. [The possibility $\phi'(ws) = \phi'(w)$ is excluded because ϕ is an automorphism.] It follows easily that $\phi'(w) = \phi'(1)w$ for all w, so ϕ' is left multiplication by $w_1 := \phi'(1)$, and hence ϕ is given by the action of w_1. This proves everything except the normality of $\operatorname{Aut}_0 \Sigma$, which is left as an exercise.   □

There is a second obvious source of automorphisms of Σ. Namely, there is a homomorphism $\mathrm{Aut}(W, S) \to \mathrm{Aut}\,\Sigma$, where $\mathrm{Aut}(W, S)$ is the group of automorphisms of W stabilizing S; for such an automorphism takes standard cosets to standard cosets and hence induces an automorphism of Σ.

Proposition 3.33. *The homomorphism* $\mathrm{Aut}(W, S) \to \mathrm{Aut}\,\Sigma$ *just defined is injective, and its image is the group* $\mathrm{Aut}(\Sigma, C)$ *consisting of the automorphisms of* Σ *that stabilize the fundamental chamber* $C = 1$.

Proof. Given $f \in \mathrm{Aut}(W, S)$, its image $\phi \in \mathrm{Aut}\,\Sigma$ has components $\phi' = f$ and $\phi'' := f|_S$. This shows that the homomorphism is injective. And ϕ stabilizes C because $f(1) = 1$. Conversely, suppose we are given $\phi \in \mathrm{Aut}(\Sigma, C)$, and let ϕ', ϕ'' be its components. Then ϕ' is a bijection satisfying $\phi'(1) = 1$ and $\phi'(ws) = \phi'(w)\phi''(s)$. It follows that $\phi'(s_1 \cdots s_d) = \phi''(s_1) \cdots \phi''(s_d)$ for all $s_1, \ldots, s_d \in S$. This implies that ϕ' is a homomorphism, hence an automorphism, and that $\phi'(s) = \phi''(s)$ for all $s \in S$. Thus ϕ' is in $\mathrm{Aut}(W, S)$ and ϕ is its image in $\mathrm{Aut}(\Sigma, C)$. $\qquad\qquad\square$

Remark 3.34. The group $\mathrm{Aut}(W, S)$ is quite easy to understand, in view of the Coxeter presentation of W: An element of this group is determined by giving a permutation π of S that is compatible with the Coxeter matrix, in the sense that $m(\pi(s), \pi(t)) = m(s, t)$ for all $s, t \in S$. More concisely, $\mathrm{Aut}(W, S)$ is simply the group of automorphisms of the Coxeter diagram of (W, S).

Exercises

3.35. Show that the full automorphism group of Σ is the semidirect product $\mathrm{Aut}_0\,\Sigma \rtimes \mathrm{Aut}(\Sigma, C)$. Hence $\mathrm{Aut}\,\Sigma \cong W \rtimes \mathrm{Aut}(W, S)$.

3.36. Suppose W is an irreducible finite reflection group. By looking at the list in Section 1.5.6 of possible Coxeter diagrams, show that with one exception, $\mathrm{Aut}(W, S)$ is either trivial or of order 2. [The exception is the group of type D_4.] So, with one exception, W is either the full automorphism group of Σ or a subgroup of index 2.

3.37. Specialize now to the case that W is the group of symmetries of a regular solid X, and note (again by looking at the list) that $\mathrm{Aut}(W, S)$ is of order 2 if and only if X is self-dual. Explain this geometrically. More precisely, explain why an isomorphism from X to its dual induces a "type-reversing" automorphism of Σ.

3.3.3 Construction of Foldings

As a final illustration of Lemma 3.31, we will construct maps that, intuitively, "fold Σ onto a half-space along a wall." The significance of this will become clear in the next section.

Proposition 3.38. *Let C_1 and C_2 be adjacent chambers of $\Sigma = \Sigma(W, S)$. Then there is an endomorphism ϕ of Σ with the following properties:*

(1) *ϕ is a retraction onto its image α.*
(2) *Every chamber in α is the image of exactly one chamber not in α.*
(3) *$\phi(C_2) = C_1$.*

To construct ϕ, we may assume that C_1 is the fundamental chamber C, in which case C_2 is necessarily sC for some $s \in S$. Before beginning the proof based on Lemma 3.31, we remark that there is a very short proof that uses the Tits cone instead of the proposition. Namely, identify Σ with the set of cells in the latter, and let α_+ (resp. α_-) be the set of cells in the closed half-space $\overline{U_+(s)}$ (resp. $\overline{U_-(s)}$), where the notation is that of Sections 2.5 and 2.6. Then we can take ϕ to be the map given by the reflection s on α_- and by the identity on α_+. It is well defined because s is the identity on $\alpha_+ \cap \alpha_-$, which consists of the cells in H_s.

But we will give a purely combinatorial proof using Lemma 3.31. The crux of the proof is the next lemma, which constructs the ϕ' component of the desired ϕ (still assuming that $C_1 = C$ and $C_2 = sC$). Recall, for motivation, that there are two possibilities for an element $w \in W$: either $l(sw) = l(w) - 1$ or $l(sw) = l(w) + 1$. In the first case, w admits a reduced decomposition starting with s, so there is a minimal gallery of the form $C, sC, \ldots, wC$. We therefore expect that there is a "wall" that separates C from sC, and this wall should also separate C from wC. Thus we should have $wC \notin \alpha$ in this case. In the second case, there is a minimal gallery of the form $C, sC, \ldots, swC$. So we expect that swC is not in α but that its "mirror image" wC is in α. These considerations motivate the following lemma and its proof:

Lemma 3.39. *Fix $s \in S$. Then there is a function $\phi_s : W \to W$ with the following properties:*

(1) *ϕ is a retraction onto its image α_s, which consists of the elements $w \in W$ such that $l(sw) = l(w) + 1$.*
(2) *Each element of α_s is the image under ϕ_s of exactly one element of the complement α'_s.*
(3) *The left-translation action of s on W interchanges the sets α_s and α'_s.*
(4) *For each $t \in S$, ϕ_s takes t-adjacent elements of W to elements that are either equal or t-adjacent.*

Proof. It is clear how we should define ϕ_s:

$$\phi_s(w) := \begin{cases} w & \text{if } l(sw) = l(w) + 1, \\ sw & \text{if } l(sw) = l(w) - 1. \end{cases}$$

And it is immediate from this definition that (1)–(3) hold. It remains to verify (4). We will prove a more precise result, which should be plausible in view of the "folding" interpretation: Consider two t-adjacent elements w and wt for some $w \in W$. Then we claim:

(a) If w and wt are both in α_s or both in α'_s, then $\phi_s(wt) = \phi_s(w)t$.

(b) If w is in α_s and wt is in α'_s, then $\phi_s(w) = w = \phi_s(wt)$.

Assertion (a) is immediate from the definition of ϕ_s. To prove (b), note that the assumptions imply that $l(sw) = l(w) + 1$ and $l(swt) = l(wt) - 1$. This implies, first, that $l(wt) = l(w) + 1$. For we have

$$l(wt) = l(swt) + 1 \geq l(sw) = l(w) + 1 .$$

We can now apply the folding condition (F) of Section 2.3.1 to conclude that $swt = w$; hence $\phi_s(wt) = swt = w$, as claimed.   □

Remark 3.40. The proof explains why we called condition (F) the *folding condition*.

Proof of Proposition 3.38. Assuming still that $C_1 = C$ and $C_2 = sC$, we can set $\phi' = \phi_s$ and $\phi'' = \mathrm{id}_S$. Everything should be clear now, except perhaps for (1), which can be expressed by saying that ϕ is *idempotent*, i.e., that $\phi^2 = \phi$. But ϕ^2 and ϕ are type-preserving chamber maps that agree on chambers; hence they agree on all simplices.   □

3.4 Roots

We are ready, finally, to complete the circle of ideas begun in Chapter 2. Recall that our treatment of Coxeter groups, starting in Section 2.1, was based on the intuition that W should be a "reflection group" and that there should be a pair of "opposite roots" for each reflection. We justified this intuition by means of the canonical linear repesentation in Section 2.5.3. We now describe an alternative approach to roots from a purely combinatorial point of view.

   We will begin by developing, following Tits [247], a theory of roots and reflections in an arbitrary thin chamber complex; this theory is based on the notion of "folding" that we have already introduced informally. Once the basic properties of foldings have been laid out, it will be evident that a Coxeter complex $\Sigma(W, S)$ does indeed possess a rich supply of roots and that W is generated by reflections of Σ. Finally, we will prove a theorem of Tits that characterizes the Coxeter complexes as the thin chamber complexes with a "rich supply" of roots.

   A note on terminology: We will not have root *vectors* in this context, but we will have the analogues of half-spaces, and, following Tits, these will be called "roots." In the setting of finite reflection groups, they are of course in canonical 1–1 correspondence with root vectors (see Section 1.5.10). Roots are sometimes called *half-apartments* because Coxeter complexes are called *apartments* in the theory of buildings, as we will see in Chapter 4. As a reminder of the heuristic connection with the root vectors of Chapter 1, the roots in the present chapter will be denoted by lowercase Greek letters $\alpha, \beta, \ldots$.

3.4.1 Foldings

Let Σ be an arbitrary thin chamber complex. Recall that an endomorphism ϕ of Σ is called *idempotent* if $\phi^2 = \phi$ or, equivalently, if ϕ is a retraction onto its image.

Definition 3.41. A *folding* of Σ is an idempotent endomorphism ϕ such that for every chamber $C \in \phi(\Sigma)$ there is exactly one chamber $C' \in \Sigma \smallsetminus \phi(\Sigma)$ with $\phi(C') = C$.

Proposition 3.38 gives many examples of foldings. More concretely, the reader can easily visualize foldings in the finite Coxeter complexes of Chapter 1 (where one folds along a hyperplane) or in the plane tiled by equilateral triangles (Example 3.7 above).

Let ϕ be a folding and let α be its image $\phi(\Sigma)$. It is easy to see that α is a chamber complex in its own right, since ϕ takes galleries to pregalleries. Let α' be the subcomplex of Σ generated by the chambers not in α; thus α' consists of all such chambers and their faces. By the definition of "folding," then, ϕ induces a bijection

$$\mathcal{C}(\alpha') \overset{\sim}{\longrightarrow} \mathcal{C}(\alpha) \,,$$

where $\mathcal{C}(\alpha)$ (resp. $\mathcal{C}(\alpha')$) denotes the set of chambers in α (resp. α'). For brevity, we will temporarily refer to α and α' as the *halves* of Σ determined by ϕ.

We now define a function ϕ' on $\mathcal{C}(\Sigma)$ by taking $\phi'|_{\mathcal{C}(\alpha')}$ to be the identity and $\phi'|_{\mathcal{C}(\alpha)}$ to be the inverse of the bijection above. Intuitively, ϕ' is the "folding opposite to ϕ"; but ϕ' is not really a folding, since it is defined only on chambers. We do not know whether, in the present generality, ϕ' can be extended to an endomorphism of Σ. Nevertheless, the following is true.

Lemma 3.42. ϕ' *takes adjacent chambers to chambers that are equal or adjacent.*

Proof. Let C and D be adjacent chambers. If they are both in α', there is nothing to prove. So assume that at least one of them, say C, is in α. Then $\phi'(C)$ is the unique chamber $C' \in \alpha'$ such that $\phi(C') = C$. Let $A := C \cap D$ be the common panel of C and D, and let A' be the panel of C' such that $\phi(A') = A$. Finally, let D' be the chamber adjacent to C' along A'. See Figure 3.3. The figure shows the picture we expect if C and D are both in α; the dashed vertical line in the middle is intended to suggest the "fold line," i.e., the "wall separating α from α'."

Since $\phi(D')$ is a chamber having A as a face, we must have either $\phi(D') = C$ or $\phi(D') = D$. Suppose first that $D' \in \alpha'$, as suggested by the picture. Then we cannot have $\phi(D') = C$, since then C' and D' would be distinct chambers in α' mapping to C. So we must have $\phi(D') = D$, which implies that $D \in \alpha$ and that $\phi'(D) = D'$. Thus $\phi'(D)$ is adjacent to $\phi'(C)$ in this case.

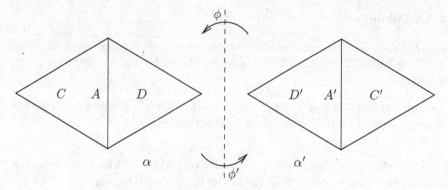

Fig. 3.3. Proof of Lemma 3.42, first case.

The other possibility is that $D' \in \alpha$. In this case the correct picture is presumably as in Figure 3.4, but we must prove this rigorously. Since D' is in α, so is its face A'. Hence $A = \phi(A') = A'$. Thus all four of our chambers have the common face A. The thinness of Σ now implies that $\{C, D\} = \{C', D'\}$.

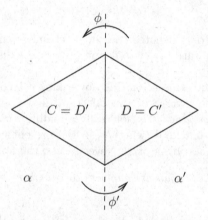

Fig. 3.4. Proof of Lemma 3.42, second case.

Since $C \neq C'$ [because one is in α and the other is in α'], the only possibility is that $C = D' \in \alpha$ and $D = C' \in \alpha'$. Thus $\phi'(D) = D = C' = \phi'(C)$. □

Note that as a consequence of this lemma, ϕ' takes galleries to pregalleries. In particular, it follows that α' is a chamber complex. We now proceed to develop the basic properties of our folding ϕ and the associated function ϕ' and subcomplexes α and α'.

Lemma 3.43. *There exists a pair C, C' of adjacent chambers with $C \in \alpha$ and $C' \in \alpha'$. For any such pair, we have $\phi(C') = C$ and $\phi'(C) = C'$.*

Proof. Since $\mathcal{C}(\alpha)$ and $\mathcal{C}(\alpha')$ are both nonempty, there is a gallery Γ that starts in α and ends in α'. Then Γ must cross from α to α' at some point, whence the first assertion. Suppose, now, that C and C' are as in the statement of the lemma, and let $A := C \cap C'$. Then $A < C \in \alpha$, so A is fixed by ϕ and hence $\phi(C')$ has A as a face. By thinness, we must have $\phi(C') = C$ or $\phi(C') = C'$. But the second possibility would imply $C' \in \alpha$, so $\phi(C') = C$. It now follows from the definition of ϕ' that $\phi'(C) = C'$. $\qquad\square$

Lemma 3.44. α and α' are convex subcomplexes of Σ, in the sense that if Γ is a minimal gallery in Σ with both extremities in α (resp. α'), then Γ lies entirely in α (resp. α').

Proof. Suppose Γ is a minimal gallery with both extremities in α. If Γ is not contained in α, then it must cross from α to α' at some point. Thus there is a pair of consecutive chambers in Γ to which we can apply Lemma 3.43. But then the pregallery $\phi(\Gamma)$ has a repetition. We can therefore get a shorter gallery with the same extremities as Γ, contradicting the minimality. A similar argument (using ϕ') works for α'. $\qquad\square$

Lemma 3.45. *Let C and C' be as in Lemma 3.43. Then*

$$\mathcal{C}(\alpha) = \{D \in \mathcal{C}(\Sigma) \mid d(D, C) < d(D, C')\}$$

and

$$\mathcal{C}(\alpha') = \{D \in \mathcal{C}(\Sigma) \mid d(D, C) > d(D, C')\}.$$

In particular, no chamber of Σ is equidistant from C and C'.

(Note that the last assertion is not vacuous, i.e., there are thin chamber complexes in which a chamber D is equidistant from two adjacent chambers C, C'. The intuitive reason for the impossibility of this in the present context is that the "wall" separating C from C' would have to cut through D, contradicting the fact that our two halves α and α' are subcomplexes.)

Proof. Note that the right-hand sides of the two equalities to be proved are disjoint sets of chambers. Consequently, since α and α' partition the chambers of Σ, it suffices to prove that the left-hand sides are contained in the right-hand sides. Suppose, then, that we are given a chamber $D \in \alpha$, and let Γ be a minimal gallery from D to C'. Then, as before, Γ must cross from α to α' at some point, so we may fold it (i.e., apply ϕ to it) to obtain a pregallery from D to $\phi(C') = C$ that has a repetition. Hence $d(D, C) < d(D, C')$, as required. A similar argument using ϕ' proves the second inclusion. $\qquad\square$

Lemma 3.46. *Suppose C and C' are adjacent chambers such that $\phi(C') = C$. Then ϕ is the unique folding taking C' to C.*

Proof. Note first that we have $C \in \phi(\Sigma) = \alpha$ and $C' \in \alpha'$ [because $\phi(C') \neq C'$]. So Lemma 3.45 is applicable and yields a description of the two halves α and α' of Σ determined by ϕ. If ψ is a second folding with $\psi(C') = C$, then we can similarly apply Lemma 3.45 to obtain the same description of the halves of Σ determined by ψ. In particular, it follows that ψ, like ϕ, is the identity on α and maps $\mathcal{C}(\alpha')$ bijectively to $\mathcal{C}(\alpha)$. We must show that ψ agrees with ϕ on all vertices of α'.

To begin with, we know that the two foldings both take C' to C and fix all vertices of the panel $C \cap C'$ of C'; hence they agree pointwise on C' (i.e., they agree on all vertices of C'). We will complete the proof by showing that ϕ and ψ continue to agree pointwise as we move away from C' along a gallery Γ in α'. It suffices to show that if ϕ and ψ agree pointwise on a chamber $D \in \alpha'$ then they agree pointwise on any chamber $E \in \alpha'$ that is adjacent to D.

Let A be the common panel $D \cap E$. Let $D_1 := \phi(D) = \psi(D)$, let $A_1 := \phi(A) = \psi(A)$, and let E_1 be the unique chamber distinct from D_1 and having A_1 as a face; see Figure 3.5. Then necessarily $\phi(E) = E_1 = \psi(E)$; for the only

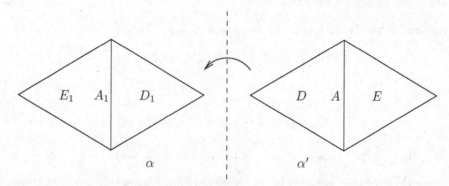

Fig. 3.5. Uniqueness of foldings.

other possibility is that ϕ or ψ maps E to D_1, contradicting the injectivity of ϕ and ψ on $\mathcal{C}(\alpha')$. And ϕ and ψ must agree pointwise on E, since they are already known to agree on all but one vertex of E. $\square$

Remark 3.47. The argument used in the previous two paragraphs will be called the *standard uniqueness argument*. It will be used repeatedly as we proceed. For pedagogical reasons, we prefer not to formalize the argument, since we think readers will benefit from thinking it through several more times. The basic idea to remember is the following: If a chamber map is known on all the vertices of one chamber, then one can often figure out what it has to do as one moves away from that chamber along a gallery. We have already used this idea twice before: once in the proof of uniqueness of type functions (Section A.1.3) and once in the proof of Corollary 1.131.

Definition 3.48. We will say that the folding ϕ is *reversible* if the function ϕ' defined above on chambers extends to a folding. Note that if C and C' are as in Lemma 3.46, then we have $\phi'(C) = C'$; so if ϕ is reversible, then the extension of ϕ' to a folding is unique: It is the unique folding of Σ taking C to C'. We will use the same symbol ϕ' for this extension, and we will call it the folding *opposite* to ϕ.

Lemma 3.49. *Let C and C' be adjacent chambers with $\phi(C') = C$. Then ϕ is reversible if and only if there exists a folding taking C to C'. In this case there is an automorphism s of Σ such that $s|_\alpha = \phi'$ and $s|_{\alpha'} = \phi$. This automorphism is of order 2, and it can be characterized as the unique nontrivial automorphism of Σ that fixes $C \cap C'$ pointwise. Finally, the set of simplices of Σ fixed by s is the subcomplex $\alpha \cap \alpha'$ of Σ.*

Proof. We have already seen that if ϕ is reversible then the opposite folding ϕ' takes C to C'. Conversely, suppose there is a folding ϕ_1 such that $\phi_1(C) = C'$. Then we can apply Lemma 3.45 to ϕ_1 to deduce that ϕ_1 determines the same halves α and α' as ϕ (but with their roles reversed, i.e., α' is the image of ϕ_1). In particular, ϕ and ϕ_1 are both the identity on $H := \alpha \cap \alpha'$, so there is a well-defined endomorphism s of Σ with $s|_\alpha = \phi_1$ and $s|_{\alpha'} = \phi$. Note that H is the full fixed-point set of s; for if $A \notin H$, say $A \notin \alpha$, then $s(A) = \phi(A) \neq A$.

It is clear that s maps $\mathcal{C}(\alpha)$ bijectively to $\mathcal{C}(\alpha')$, and vice versa, so s is bijective on $\mathcal{C}(\Sigma)$. Hence s^2 is bijective on $\mathcal{C}(\Sigma)$. Since s^2 fixes C pointwise, the standard uniqueness argument is applicable and shows that s^2 is the identity. In particular, s is an automorphism.

We now prove that $\phi_1|_{\mathcal{C}(\Sigma)} = \phi'$, and hence that ϕ is reversible. Since ϕ' and ϕ_1 are both the identity on $\mathcal{C}(\alpha')$, it suffices to consider chambers $D \in \alpha$. For any such D we have $D = s^2(D) = \phi(\phi_1(D))$, so $\phi_1(D)$ is the (unique) chamber in α' that is mapped by ϕ to D. Hence $\phi_1(D) = \phi'(D)$ by the definition of the latter.

Finally, to prove the characterization of s stated in the lemma, suppose that t is another nontrivial automorphism fixing $C \cap C'$ pointwise. Then t must interchange C and C'; for otherwise t would have to fix them pointwise, and the standard uniqueness argument would show that t is trivial. Thus t agrees with s (pointwise) on C, and both are bijective on $\mathcal{C}(\Sigma)$. We can therefore apply the standard uniqueness argument yet again to deduce that $s = t$. $\square$

We now introduce geometric language and summarize some of the results above in this language.

Definition 3.50. A *root* of Σ is a subcomplex α that is the image of a reversible folding ϕ. In view of Lemmas 3.43 and 3.46, the folding ϕ is uniquely determined by α. The subcomplex α' generated by the chambers not in α is again a root, being the image of the opposite folding ϕ'; it is called the root *opposite* to α. We will often write $\alpha' = -\alpha$. The intersection $\partial\alpha := \alpha \cap -\alpha$ of two opposite roots will be called the *wall* bounding $\pm\alpha$.

Note that we can recover the pair of roots $\pm\alpha$ from the wall $H = \partial\alpha$ and, in fact, from any panel A of Σ that is contained in H. To see this, it suffices to describe the foldings $\{\phi, \phi'\}$ in terms of A: Let C_1 and C_2 be the chambers having A as a face. Note that they are necessarily in opposite roots, because α and α' each contain a chamber $\geq A$. There is a unique folding ϕ_1 (resp. ϕ_2) such that $\phi_1(C_2) = C_1$ (resp. $\phi_2(C_1) = C_2$), and $\{\phi, \phi'\} = \{\phi_1, \phi_2\}$.

A wall H determines an automorphism $s := s_H$ by Lemma 3.49. It fixes H pointwise and interchanges the two roots determined by H. For any panel $A \in H$ as in the previous paragraph, we can characterize s as the unique nontrivial automorphism of Σ that fixes A pointwise; in particular, s is the unique nontrivial automorphism fixing every simplex of H.

Definition 3.51. Given a wall H, we call s_H the *reflection* of Σ with respect to H.

Finally, two chambers $C, C' \in \Sigma$ will be said to be *separated* by the wall H if one is in α and the other is in $-\alpha$. If the two chambers are adjacent, Lemmas 3.43 and 3.46 imply that H is then the *unique* wall separating them.

In case Σ is a Coxeter complex $\Sigma(W, S)$, Proposition 3.38 shows that every pair C_1, C_2 of adjacent chambers is separated by a wall. For we have a folding taking C_2 to C_1 and also one taking C_1 to C_2; these foldings are therefore opposite to one another by Lemma 3.49 and determine a wall separating C_1 from C_2. If C_1 and C_2 are C and sC for some $s \in S$, where C is the fundamental chamber, it is easy to see that the reflection associated to this wall is given by the action of s. It follows that the reflections of Σ determined by all possible walls are precisely the elements of W that we called *reflections* in Chapter 2.

For future reference, let's explicitly spell out what the roots look like in this example, starting with the "simple roots." As in Definition 3.3, we identify the chambers of $\Sigma = \Sigma(W, S)$ with the elements of W. For each $s \in S$ there is then a unique root α_s with $1 \in \alpha_s$ and $s \notin \alpha_s$. In view of Lemma 3.45, its set of chambers is given by

$$\mathcal{C}(\alpha_s) = \{w \in W \mid l(sw) > l(w)\} \ . \tag{3.2}$$

See also Lemma 3.39, where the folding onto α_s was explicitly constructed. The general root, then, is gotten from roots of the form α_s by using the W-action. Explicitly, if we are given a pair of adjacent chambers w, ws with $w \in W$ and $s \in S$, then $w\alpha_s$ is the root containing w but not ws.

The reader who prefers to think of all this in terms of the Tits cone can easily reformulate the definitions from that point of view. For example, the roots correspond in the obvious way to the closed half-spaces associated with the hyperplanes that were called walls in Sections 2.5 and 2.6; see the paragraph following the statement of Proposition 3.38.

Example 3.52. Let W be the symmetric group on n letters with its standard generating set $S = \{s_1, \ldots, s_{n-1}\}$, where s_i is the transposition that interchanges i and $i + 1$. In Example 1.119 we studied roots in $\Sigma(W, S)$ from the point of view of finite reflection groups. We show now how the present combinatorial approach leads to the same conclusions. We need some preliminary observations.

For any permutation $\pi \in W$, let $\iota(\pi)$ be the number of *inversions* of π, i.e., the number of ordered pairs (i, j) with $1 \leq i < j \leq n$ and $\pi(i) > \pi(j)$. For any $1 \leq i \leq n - 1$ one easily checks that

$$\pi(i) < \pi(i+1) \implies \iota(\pi s_i) = \iota(\pi) + 1,$$
$$\pi(i) > \pi(i+1) \implies \iota(\pi s_i) = \iota(\pi) - 1.$$

Indeed, if one identifies a permutation π with the list $\pi(1), \ldots, \pi(n)$, then the effect of right-multiplying by s_i is to interchange the elements in positions i and $i + 1$. Our assertions follow at once. One can very quickly deduce that $\iota(\pi) = l(\pi)$, a fact that we have already proven twice by other methods (see Example 1.81 and Exercise 1.125). Consequently,

$$l(\pi s_i) > l(\pi) \iff \pi(i) < \pi(i+1) . \tag{3.3}$$

Returning now to roots, we can use (3.3) and (3.2) to get the following concrete description of the simple root $\alpha_i := \alpha_{s_i}$ for $1 \leq i \leq n - 1$:

$$\mathcal{C}(\alpha_i) = \left\{ \pi \in W \mid \pi^{-1}(i) < \pi^{-1}(i+1) \right\} . \tag{3.4}$$

To see this, note that

$$
\begin{aligned}
\pi \in \alpha_i &\iff l(s_i \pi) > l(\pi) && \text{by (3.2)} \\
&\iff l(\pi^{-1} s_i) > l(\pi^{-1}) && \text{since } l(w) = l(w^{-1}) \\
&\iff \pi^{-1}(i) < \pi^{-1}(i+1) && \text{by (3.3) .}
\end{aligned}
$$

An arbitrary root α has the form $w\alpha_i$ for some $w \in W$ and $1 \leq i \leq i + 1$, and one checks that its chamber set is $\{\pi \in W \mid \pi^{-1}(w(i)) < \pi^{-1}(w(i+1))\}$. Now $w(i)$ and $w(i+1)$ can be any pair of integers $i \neq j$ with $1 \leq i, j \leq n$. So we have arrived at the same conclusion as in Example 1.119: Σ has one root α_{ij} for each ordered pair of integers i, j with $1 \leq i, j \leq n$ and $i \neq j$. Its chamber set is given by

$$\mathcal{C}(\alpha_{ij}) = \left\{ \pi \in W \mid \pi^{-1}(i) < \pi^{-1}(j) \right\} .$$

We close this subsection by recording two simple but useful facts about roots and walls. The first is that one cannot have nested roots in a spherical Coxeter complex. (In fact, we will see in Section 3.6.8 that this property *characterizes* spherical Coxeter complexes.)

Lemma 3.53. *If α and β are distinct roots of a spherical Coxeter complex Σ, then $\alpha \not\subseteq \beta$.*

Proof. This is obvious from the point of view of Chapter 1, where roots correspond to half-spaces whose bounding hyperplanes pass through the origin. Alternatively, the lemma follows from the fact that α and β have the same finite number of chambers, equal to half the number of chambers in Σ. See Exercise 3.56 for a third proof.   □

For the final observation, recall that a subcomplex Δ' of a simplicial complex Δ is said to be *full* if it contains every simplex of Δ whose vertices are all in Δ'.

Lemma 3.54. *Roots and walls in a thin chamber complex Σ are full subcomplexes.*

Proof. For roots, this follows from Lemma A.15. For walls the result follows from the fact that a wall is an intersection of two roots. Alternatively, one can use the fact that a wall is the fixed-point set of a reflection.   □

Exercises

3.55. Let $\pm\alpha$ be a pair of opposite roots with bounding wall $H := \alpha \cap -\alpha$, and let C, C' be chambers with $C \in \alpha$ and $C' \in -\alpha$. Recall that one can speak of the gallery distance $d(-,-)$ between arbitrary simplices (Section A.1.3). Prove the following generalization of Lemma 3.45:

$$\alpha \smallsetminus H = \{A \in \Sigma \mid d(A, C) < d(A, C')\} \ ,$$
$$-\alpha \smallsetminus H = \{A \in \Sigma \mid d(A, C) > d(A, C')\} \ ,$$
$$H = \{A \in \Sigma \mid d(A, C) = d(A, C')\} \ .$$

3.56. Give a proof of Lemma 3.53 based on the opposition involution (Section 1.6.2).

3.57. Let α be a root and s the associated reflection. If C and C' are chambers in α, show that $d(C, sC') > d(C, C')$.

3.58. This is a continuation of Example 3.52. Thus W is the symmetric group on n letters with its standard generating set S. According to Exercise 1.112 and its solution, $\Sigma := \Sigma(W, S)$ is isomorphic to the flag complex Σ' of proper, nonempty subsets of $\{1, 2, \ldots, n\}$. We wish to describe the roots α_{ij} from this point of view. Given indices $1 \le i, j \le n$ with $i \ne j$, let α'_{ij} be the root of Σ' corresponding to α_{ij} under the canonical isomorphism between Σ and Σ'.

(a) Show that the vertices of α'_{ij} are the proper nonempty subsets $X \subset \{1, \ldots, n\}$ such that $j \in X \implies i \in X$.
(b) Show that the vertices of $\partial\alpha'_{ij}$ are the proper nonempty subsets $X \subset \{1, \ldots, n\}$ such that $i \in X \iff j \in X$.
(c) Show that the interior vertices of α'_{ij} (i.e., the vertices in $\alpha'_{ij} \smallsetminus \partial\alpha'_{ij}$) are the proper nonempty subsets $X \subset \{1, \ldots, n\}$ such that $i \in X$ and $j \notin X$.

3.59. Let (W, S) be a Coxeter system.

(a) Prove the following *strong form of the exchange condition*: Given $w \in W$, suppose t is a reflection such that $l(tw) < l(w)$. Then for any decomposition $w = s_1 \cdots s_d$, there is an index i such that $tw = s_1 \cdots \hat{s}_i \cdots s_d$. (Note that if $t \in S$, this essentially reduces to the exchange condition stated in Section 2.3.1, except that there we also required the decomposition of w to be reduced.)

(b) Given $w, w' \in W$, show that the following conditions are equivalent:
   (i) For every decomposition $w = s_1 \cdots s_d$, there is an index i such that $w' = s_1 \cdots \hat{s}_i \cdots s_d$.
   (ii) For some reduced decomposition $w = s_1 \cdots s_d$, there is an index i such that $w' = s_1 \cdots \hat{s}_i \cdots s_d$.
   (iii) $l(w') < l(w)$, and there is a reflection t such that $w' = tw$.

(c) The *Bruhat graph* of (W, S) is the directed graph with vertex set W and with a directed edge $w' \to w$ whenever w' and w satisfy the equivalent conditions in (b). Show that this graph is acyclic (i.e., there are no directed cycles). Consequently, there is a partial order on W, called the *Bruhat order*, with $w' \le w$ if and only if there is a directed path

$$w' = w_0 \to w_1 \to \cdots \to w_k = w$$

from w' to w $(k \ge 0)$.

(d) Given $w \in W$ and $s \in S$, note that sw and w are always comparable in the Bruhat order: We have $w < sw$ if $l(sw) > l(w)$, and $sw < w$ otherwise. So the expression "$\max\{sw, w\}$" is meaningful. Prove now that if $w' < w$ then $sw' \le \max\{sw, w\}$ for any $s \in S$. [Note that we might have $sw' = w = \max\{sw, w\}$, so equality can definitely hold. If $l(sw) > l(w)$, however, then length considerations show that we must have $sw' < \max\{sw, w\} = sw$.]

(e) Given $w, w' \in W$, show that the following conditions are equivalent:
   (i) $w' < w$ in the Bruhat order.
   (ii) For every decomposition of w as a product of elements of S, there is a decomposition of w' obtained by deleting one or more letters.
   (iii) For some reduced decomposition of w, there is a decomposition of w' obtained by deleting one or more letters.

Remark 3.60. The Bruhat order was introduced by Chevalley in a widely circulated unpublished manuscript in the late 1950s. It arose in connection with his study of inclusion relations among "Schubert varieties." A slightly edited version of Chevalley's paper finally appeared in print in 1994; see [82]. In a foreword to that paper, Borel pointed out that the name "Chevalley order" would be more appropriate than "Bruhat order," and he proposed "Bruhat–Chevalley order" as a compromise. This suggestion does not seem to have gained wide acceptance, probably because there is too much existing literature referring to the Bruhat order. See Humphreys [133, Sections 5.9–5.11] for more information about the Bruhat order and for further references.

3.61. Let $\pm\alpha$ be a pair of opposite roots, let C, C' be adjacent chambers with $C \in \alpha$ and $C' \in -\alpha$, and let v be the vertex of C not in the common panel $C \cap C'$. Show that $v \notin -\alpha$.

3.62. Let Σ be the Coxeter complex $\Sigma(W, S)$. For any simplex $A \in \Sigma$, show that the stabilizer W_A of A in W is generated by the reflections s_H, where H ranges over the walls containing A.

3.63. You have now seen the standard uniqueness argument applied several times. Try to write down a lemma that includes all of these applications. [*Warning*: Unless you have incredible foresight, you can expect to have to modify your lemma one or more times as you see further applications of the argument. In fact, this might even happen in the next few pages.]

3.4.2 Characterization of Coxeter Complexes

It should now be clear that Coxeter complexes possess a good theory of roots. Our next goal is to show that this property characterizes Coxeter complexes among the thin chamber complexes. Since we will be considering chamber complexes that are not necessarily given to us as $\Sigma(W, S)$, it is convenient to slightly expand our previous terminology.

Definition 3.64. A simplicial complex Σ is called a *Coxeter complex* if it is isomorphic to $\Sigma(W, S)$ for some Coxeter system (W, S). It is called a *spherical* Coxeter complex if it is finite.

This differs from our previous use of the term "Coxeter complex" in that we do not assume that $\Sigma = \Sigma(W, S)$. In fact, we do not assume that we are given a specific isomorphism $\Sigma \xrightarrow{\sim} \Sigma(W, S)$ as part of the structure of Σ. In particular, no chamber of Σ has been singled out as "fundamental."

The following theorem of Tits says, roughly speaking, that Coxeter complexes can be characterized as the thin chamber complexes with "enough" roots.

Theorem 3.65. *A thin chamber complex Σ is a Coxeter complex if and only if every pair of adjacent chambers is separated by a wall.*

(We can restate the condition of the theorem as follows: For every ordered pair C, C' of adjacent chambers, there is a folding ϕ of Σ with $\phi(C') = C$. We do not need to specify here that ϕ is reversible; for this follows, as we saw above in the case of $\Sigma(W, S)$, from the existence of a folding taking C' to C.)

Proof of Theorem 3.65 (start). We have already proven the "only if" part. For the converse, assume that every pair of adjacent chambers is separated by a wall. Choose an arbitrary chamber C, called the *fundamental chamber*, and let S be the set of reflections determined by the panels of C. Let $W \le \operatorname{Aut} \Sigma$ be the subgroup generated by S. We will prove that (W, S) is a Coxeter system and that $\Sigma \cong \Sigma(W, S)$.

We could simply repeat, essentially verbatim, the arguments that led to the analogous results for finite reflection groups in Chapter 1. For the sake of variety, however, we will use a different method. This is actually a little longer, but it adds some geometric insight that we would not get by repeating the previous arguments. In particular, it gives a simple geometric explanation of the deletion condition.

We now proceed with a sequence of lemmas, after which we can complete the proof.

Lemma 3.66. *W acts transitively on the chambers of Σ.*

Proof. This is identical to the proof given in Chapter 1 for finite reflection groups (Theorem 1.69). $\qquad\square$

Lemma 3.67. *Σ is colorable.*

Proof. Let $\overline{C}$ be the subcomplex $\Sigma_{\le C}$. It suffices to show that $\overline{C}$ is a retract of Σ. The idea for showing this is to construct a retraction ρ by folding and folding and folding..., until the whole complex Σ has been folded up onto $\overline{C}$.

To make this precise, let $C_1, \dots, C_n$ be the chambers adjacent to C, and let $\phi_1, \dots, \phi_n$ be the foldings such that $\phi_i(C_i) = C$. Let ψ be the composite $\phi_n \circ \cdots \circ \phi_1$. We claim that $d(C, \psi(D)) < d(C, D)$ for any chamber $D \ne C$. To prove this, let $\Gamma \colon C, C', \dots, D$ be a minimal gallery from C to D; we will show that $\psi(\Gamma)$ has a repetition. If $\phi_1(\Gamma)$ has a repetition, we are done. Otherwise, the standard uniqueness argument shows that ϕ_1 fixes all the chambers of Γ pointwise. In this case, repeat the argument with ϕ_2, and so on. Eventually we will be ready to apply the folding ϕ_i that takes C' to C. If the previous foldings did not already produce a repetition in Γ, then they have fixed Γ pointwise, and the application of ϕ_i yields a pregallery with a repetition. This proves the claim.

It follows that for any chamber D, $\psi^k(D) = C$ for k sufficiently large. Since ψ fixes C pointwise, this implies that the "infinite iterate" $\rho := \lim_{k \to \infty} \psi^k$ is a well-defined chamber map that retracts Σ onto $\overline{C}$. $\qquad\square$

It will be convenient to choose a fixed type function τ with S as the set of types, analogous to the canonical type function that we used earlier in the chapter. To this end we assign types to the vertices of the fundamental chamber C by declaring that the panel fixed by the reflection $s \in S$ is an s-panel. We then extend this to all of Σ by means of a retraction ρ of Σ onto $\overline{C}$. Note that this type function τ has a property that by now should be very familiar: For any $s \in S$, the chambers C and sC are s-adjacent.

Lemma 3.68. *Foldings and reflections are type-preserving; hence all elements of W are type-preserving. Consequently, wC and wsC are s-adjacent for any $w \in W$ and $s \in S$.*

Proof. A folding ϕ fixes at least one chamber pointwise, so the type-change map ϕ_* is the identity (see Proposition A.14). This proves that foldings are type-preserving, and everything else follows from this. $\qquad\square$

If $\Gamma\colon C_0, \ldots, C_d$ is a gallery and H_i is the wall separating C_{i-1} from C_i, then, as usual, we will say that $H_1, \ldots, H_d$ are the *walls crossed by* Γ.

Lemma 3.69. *If $\Gamma\colon C_0, \ldots, C_d$ is a minimal gallery, then the walls crossed by Γ are distinct and are precisely the walls separating C_0 from C_d. Hence the distance between two chambers is equal to the number of walls separating them.*

Proof. Suppose H is a wall separating C_0 from C_d. Let $\pm\alpha$ be the corresponding roots, say with $C_0 \in \alpha$ and $C_d \in -\alpha$. Then there must be some i with $1 \le i \le d$ such that $C_{i-1} \in \alpha$ and $C_i \in -\alpha$. Since α and $-\alpha$ are convex (Lemma 3.44), it follows that we have $C_0, \ldots, C_{i-1} \in \alpha$ and $C_i, \ldots, C_d \in -\alpha$. In other words, Γ crosses H exactly once. Now suppose H is a wall that does not separate C_0 from C_d. Then C_0 and C_d are both in the same root α, so the convexity of α implies that Γ does not cross H. $\qquad\square$

The crux of the proof of Lemma 3.69, obviously, is the convexity of roots, which in turn was based on the idea of using foldings to shorten galleries. We can now use this same idea to prove a geometric analogue of the deletion condition. The statement uses the notion of *type* of a gallery (Definition 3.22).

Lemma 3.70. *Let Γ be a gallery of type $\mathbf{s} = (s_1, \ldots, s_d)$. If Γ is not minimal, then there is a gallery Γ' with the same extremities as Γ such that Γ' has type $\mathbf{s}' = (s_1, \ldots, \hat{s}_i, \ldots, \hat{s}_j, \ldots, s_d)$ for some $i < j$.*

Proof. Since Γ is not minimal, Lemma 3.69 implies that the number of walls separating C_0 from C_d is less than d. Hence the walls crossed by Γ cannot all be distinct; for if a wall is crossed exactly once by Γ, then it certainly separates C_0 from C_d. We can therefore find a root α and indices i, j, with $1 \le i < j \le d$, such that C_{i-1} and C_j are in α but $C_k \in -\alpha$ for $i \le k < j$; see Figure 3.6. Let ϕ be the folding with image α. If we modify Γ by applying ϕ to the portion $C_i, \ldots, C_{j-1}$, we obtain a pregallery with the same extremities that has exactly two repetitions:

$$C_0, \ldots, C_{i-1}, \phi(C_i), \ldots, \phi(C_{j-1}), C_j, \ldots, C_d.$$

So we can delete C_{i-1} and C_j to obtain a gallery Γ' of length $d - 2$. The type $\mathbf{s}'$ of Γ' is $(s_1, \ldots, \hat{s}_i, \ldots, \hat{s}_j, \ldots, s_d)$ because ϕ is type-preserving. $\qquad\square$

Lemma 3.71. *The action of W is simply transitive on the chambers of Σ.*

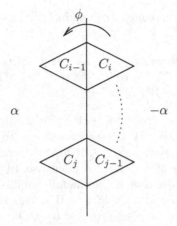

Fig. 3.6. A geometric proof of the deletion condition.

Proof. We have already noted that the action is transitive. To prove that the stabilizer of C is trivial, note that if $wC = C$ then w fixes C pointwise, since w is type-preserving. But then $w = 1$ by the standard uniqueness argument. ☐

It follows from Lemma 3.71 that we have a bijection $W \to \mathcal{C}(\Sigma)$ given by $w \mapsto wC$. This yields the familiar 1–1 correspondence between galleries starting at C and words $\mathbf{s} = (s_1, \ldots, s_d)$, where the gallery (C_i) corresponding to $\mathbf{s}$ is given by $C_i := s_1 \cdots s_i C$ for $i = 0, \ldots, d$. In view of Lemma 3.68, the type of this gallery is the word $\mathbf{s}$ that we started with. So a direct translation of Lemma 3.70 into the language of group theory yields the deletion condition for (W, S). Consequently:

Lemma 3.72. (W, S) *is a Coxeter system.* ☐

Remark 3.73. Another way to prove that (W, S) is a Coxeter system is to verify condition (A) of Chapter 2 by using the action of W on the set of roots of Σ. Indeed, Lemma 3.66 implies that every panel of Σ is W-equivalent to a face of C. Hence every reflection of Σ is W-conjugate to an element of S. This shows that the "reflections" in W, in the sense of Definition 2.1, are precisely the reflections of Σ obtained from the theory of foldings. We can therefore identify the set T used in Chapter 2 with the set of reflections of Σ, and we can identify $T \times \{\pm 1\}$ with the set of roots of Σ. The action of W on the roots therefore yields an action of W on $T \times \{\pm 1\}$ with the properties required for condition (A). Details are left to the interested reader.

For the next lemma, we need a simplicial analogue of the concept of "strict fundamental domain" (Definition 1.103).

Definition 3.74. If a group G acts on a simplicial complex Δ, then we call a set of simplices $\Delta' \subseteq \Delta$ a *simplicial fundamental domain* if Δ' is a subcomplex of Δ and is a set of representatives for the G-orbits of simplices.

(This yields a strict fundamental domain $|\Delta'|$ for the action of G on the geometric realization $|\Delta|$.)

Lemma 3.75. *The subcomplex $\overline{C} := \Sigma_{\leq C}$ is a simplicial fundamental domain for the action of W on Σ. Moreover, the stabilizer of the face of C of cotype J is the standard subgroup W_J of W.*

Proof. The first assertion follows from the transitivity of W on the chambers, together with the fact that W is type-preserving. To prove the second, let A be a face of C and let $\tau(A) = S \smallsetminus J$. It follows from the definition of τ that J is the set of elements of S that fix A pointwise. In particular, the subgroup W_J stabilizes A. To prove that W_J is the full stabilizer, suppose $wA = A$. We will show by induction on $l(w)$ that $w \in W_J$. We may assume $w \neq 1$, so we can write $w = sw'$ with $s \in S$ and $l(w') < l(w)$. Our correspondence between words and galleries now implies that there is a minimal gallery of the form $C, sC, \ldots, wC$. By Lemma 3.69, then, the wall H corresponding to s separates C from wC.

Let α be the root bounded by H that contains C. Then $wC \in -\alpha = s\alpha$, so we have $w'C \in \alpha$. The equation $wA = A$ now yields

$$w'A = sA \in \alpha \cap s\alpha = H \ ,$$

hence $A \in H$ and $w'A = A$. We therefore have $s \in J$ [because s fixes A pointwise] and $w' \in W_J$ by induction; thus $w = sw' \in W_J$. $\square$

We have now done all the work required to complete the proof of the theorem.

Proof of Theorem 3.65 (end). Recall that we have assumed that every pair of adjacent chambers in Σ is separated by a wall, and we are trying to prove that Σ is a Coxeter complex. By Lemma 3.72, we have a Coxeter system (W, S), and Lemma 3.75 easily yields an isomorphism $\Sigma \cong \Sigma(W, S)$. Thus Σ is a Coxeter complex. $\square$

Example 3.76. Let Σ be the plane tiled by equilateral triangles. It is geometrically evident that we can construct, for any adjacent chambers C, C', a folding taking C' to C. So Σ is indeed a Coxeter complex, as claimed in Example 3.7. To see that the Coxeter group W is the one given in that example, one can compute the orders of pairwise products of fundamental reflections, or one can observe that the link of every vertex is a hexagon.

The last assertion of Lemma 3.69 is the analogue of a fact that we used many times in Chapter 1, giving two different ways of computing the distance between two chambers. The final result of this section generalizes this to arbitrary simplices. Recall that one can talk about the gallery distance $d(A, B)$ between arbitrary simplices (Section A.1.3).

Definition 3.77. We say that a wall H *strictly separates* two simplices if they are in opposite roots determined by H and neither is in H. We denote by $\mathcal{S}(A, B)$ the set of walls that strictly separate two simplices A and B.

Proposition 3.78. *For any two simplices A, B in a Coxeter complex Σ, we have*
$$d(A, B) = |\mathcal{S}(A, B)| \,,$$
i.e., $d(A, B)$ is equal to the number of walls H that strictly separate A from B. More precisely, the walls crossed by any minimal gallery from A to B are distinct and are precisely the walls in $\mathcal{S}(A, B)$.

Proof. A proof from the point of view of the Tits cone was sketched in Section 2.7. Here is a combinatorial proof: Let $\Gamma \colon C_0, \ldots, C_d$ be a minimal gallery from A to B. Then it is also a minimal gallery from C_0 to C_d, so it crosses d distinct walls, and these are the walls separating C_0 from C_d. It is immediate that $\mathcal{S}(A, B) \subseteq \mathcal{S}(C_0, C_d)$, so Γ crosses all the walls in $\mathcal{S}(A, B)$. We must show, conversely, that every wall H crossed by Γ is in $\mathcal{S}(A, B)$. Suppose not. Then there is a root α bounded by H that contains both A and B. But then we can get a shorter gallery from A to B by applying the folding of Σ onto α. This contradicts the minimality of Γ. $\qquad\square$

We close this section by making some remarks that will be useful later, concerning links. Given a simplex A in a Coxeter complex Σ, recall that its link $\Sigma' := \mathrm{lk}_{\Sigma} A$ is again a Coxeter complex (Proposition 3.16). We wish to explicitly describe its walls and roots. Suppose H is a wall of Σ containing A, and let $\pm\alpha$ be the corresponding roots. Then one checks immediately from the definitions that $H' := H \cap \Sigma'$ is a wall of Σ', with associated roots $\pm\alpha' := \pm\alpha \cap \Sigma'$.

Proposition 3.79. *The function $H \mapsto H' := H \cap \Sigma'$ is a bijection from the set of walls of Σ containing A to the set of walls of Σ'. Similarly, the function $\alpha \mapsto \alpha' := \alpha \cap \Sigma'$ is a bijection from the set of roots of Σ whose boundary contains A to the set of roots of Σ'.*

Proof. It suffices to prove the first assertion. Since a wall of Σ' is completely determined by any panel that it contains, we can reformulate the assertion as follows: For any panel P' of Σ', there is a unique wall H of Σ with $A \in H$ and $P' \in H \cap \Sigma'$. Equivalently, there is a unique wall of Σ containing the simplex $P := P' \cup A$. [The equivalence follows from the fact that walls are full subcomplexes by Lemma 3.54.] Since P is a panel of Σ, the proposition is now immediate. $\qquad\square$

Remark 3.80. Recall that we may identify Σ' with $\Sigma_{\geq A}$ via $B' \mapsto B' \cup A$ for $B' \in \Sigma'$, and $B \mapsto B \smallsetminus A$ for $B \in \Sigma_{\geq A}$. If we make this identification, then the bijections in the proposition are still given by intersection. In other words, if H is a wall of Σ containing A and $H' := H \cap \Sigma'$, then

$$\{B \in \Sigma_{\geq A} \mid (B \smallsetminus A) \in H'\} = H \cap \Sigma_{\geq A}\,,$$

and similarly for roots.

Exercises

Assume throughout these exercises that Σ is a Coxeter complex.

3.81. Let H be a wall with associated roots $\pm\alpha$, and let A be an arbitrary simplex. Show that $A \in H$ if and only if there are chambers $C, C' \geq A$ with $C \in \alpha$ and $C' \in -\alpha$.

3.82. With the notation of the previous exercise, if $A \in H$ show that the chambers C, C' can be taken to be adjacent. Thus there is a panel P in Σ such that $A \leq P \in H$.

3.83. Assume that Σ is infinite.

(a) Show that Σ has infinitely many walls.
(b) Assume that Σ is *irreducible* (i.e., its Coxeter diagram is connected). For every vertex x of Σ, show that there are infinitely many walls not containing x. [See Lemma 2.92 for the same result expressed in terms of the Tits cone.]

3.5 The Weyl Distance Function

In this section we introduce an important tool, whose usefulness will become more and more apparent as we develop the theory of buildings. Let Σ be a Coxeter complex. Choose a type function on Σ with values in a set S, which is not necessarily given to us as the set of generators of a Coxeter group.

Definition 3.84. The *Coxeter matrix* of Σ is the matrix $M = \big(m(s,t)\big)_{s,t\in S}$ defined by

$$m(s,t) := \operatorname{diam}(\operatorname{lk} A)\,,$$

where A is any simplex of cotype $\{s,t\}$ (see Remark 3.21). Note that if $\Sigma = \Sigma(W, S)$ for some Coxeter system (W, S) and we use the canonical type function, then M is the Coxeter matrix of (W, S). It follows that M is well defined in general. The *Weyl group* of Σ is defined to be the Coxeter group W_M defined by M. It has generating set S and defining relations $(st)^{m(s,t)} = 1$.

Note that if Σ is given to us as $\Sigma(W, S)$ (with its canonical type function), then W_M is the group W that we started with. The following result is therefore not surprising:

Proposition 3.85. *There is a type-preserving isomorphism $\Sigma \cong \Sigma(W_M, S)$, where $\Sigma(W_M, S)$ is given its canonical type function with values in S.*

Proof. By definition, there is a simplicial isomorphism $\phi\colon \Sigma \overset{\sim}{\longrightarrow} \Sigma(W', S')$ for some Coxeter system (W', S'). Let $\phi_*\colon S \to S'$ be the induced type-change bijection (Proposition A.14), where $\Sigma(W', S')$ is given its canonical type function. For any $s, t \in S$ and any simplex A of cotype $\{s, t\}$, the image $A' := \phi(A)$ has cotype $\{s', t'\}$, where $s' := \phi_*(s)$ and $t' := \phi_*(t)$. Since ϕ induces an isomorphism $\mathrm{lk}_\Sigma A \overset{\sim}{\longrightarrow} \mathrm{lk}_{\Sigma'} A'$, it follows that $m(s, t) = m'(s', t')$, where $m'(-, -)$ denotes the Coxeter matrix of (W', S'). Hence ϕ_* extends to an isomorphism $(W_M, S) \overset{\sim}{\longrightarrow} (W', S')$ of Coxeter systems, which in turn induces an isomorphism $\psi\colon \Sigma(W_M, S) \overset{\sim}{\longrightarrow} \Sigma(W', S')$. Note that the induced type-change bijection $\psi_*\colon S \to S'$ is equal to ϕ_*. It follows that the composite isomorphism $\psi^{-1} \circ \phi\colon \Sigma \overset{\sim}{\longrightarrow} \Sigma(W_M, S)$ is type-preserving. $\square$

This motivates the following terminology.

Definition 3.86. Let (W, S) be a Coxeter system with Coxeter matrix M. We say that a Coxeter complex Σ is of *type* (W, S) (or of *type* M) if Σ comes equipped with a type function having values in S such that the Coxeter matrix of Σ is M or, equivalently, such that there is a type-preserving isomorphism $\Sigma \overset{\sim}{\longrightarrow} \Sigma(W, S)$. We can then identify W with the *Weyl group* W_M of Σ.

We now wish to define a function $\delta\colon \mathcal{C}(\Sigma) \times \mathcal{C}(\Sigma) \to W_M$, called the *Weyl distance function*, such that

$$d(C_1, C_2) = l\big(\delta(C_1, C_2)\big) \tag{3.5}$$

for any two chambers C_1, C_2. Intuitively, $\delta(C_1, C_2)$ is something like a vector pointing from C_1 to C_2; it tells us the distance from C_1 to C_2 as well as what "direction" to go in to get from C_1 to C_2.

To define $\delta(C_1, C_2)$, choose an arbitrary gallery from C_1 to C_2, let $(s_1, s_2, \dots, s_d)$ be its type, and set

$$\delta(C_1, C_2) := s_1 s_2 \cdots s_d \in W_M . \tag{3.6}$$

To see that the right-hand side is independent of the choice of gallery, we may assume that $\Sigma = \Sigma(W, S)$ with its canonical type function. Then we can identify $\mathcal{C}(\Sigma)$ with W, and a gallery of type $(s_1, \dots, s_d)$ from a chamber w_1 to a chamber w_2 has the form $w_1, w_1 s_1, \dots, w_1 s_1 \cdots s_d = w_2$. Hence the right-hand side of (3.6) is equal to $w_1^{-1} w_2$, which is indeed independent of the choice of gallery. See Figure 3.2 for an example, where $\delta(C, D) = utstu = ustsu$.

This discussion gives us a concrete interpretation of δ as a "difference" map $W \times W \to W$, sending (w_1, w_2) to $w_1^{-1} w_2$, when $\Sigma = \Sigma(W, S)$. Equivalently,

$$\delta(w_1 C, w_2 C) = w_1^{-1} w_2 , \tag{3.7}$$

where C is the fundamental chamber. In particular,

$$\delta(C, wC) = w . \tag{3.8}$$

One can also deduce from the discussion that for arbitrary Σ, galleries from C_1 to C_2 are in 1–1 correspondence with decompositions of $\delta(C_1, C_2)$, and minimal galleries correspond to reduced decompositions.

Finally, returning to an arbitrary Coxeter complex of type (W, S) we show that δ extends in a natural way to a function on arbitrary pairs of simplices. Let A be a simplex of cotype J and let B be a simplex of cotype K ($J, K \subset S$). Consider the set of elements $\delta(C, D)$, where C and D are chambers with $C \geq A$ and $D \geq B$. We claim that this set is a double coset $W_J w W_K$. To see this, we may assume $\Sigma = \Sigma(W, S)$ with its canonical type function. Thus A is a coset $w_1 W_J$, B is a coset $w_2 W_K$, C corresponds to an arbitrary element of A, and D corresponds to an arbitrary element of B. The set of elements $\delta(C, D)$ is then the set of differences $A^{-1}B := \{a^{-1}b \mid a \in A, \, b \in B\}$, which is the double coset $(w_1 W_J)^{-1}(w_2 W_K) = W_J w_1^{-1} w_2 W_K$, whence the claim.

We can now define $\delta(A, B)$ to be the element of minimal length in the double coset (see Proposition 2.23). Note that pairs C, D with $C \geq A$, $D \geq B$, and $\delta(C, D) = \delta(A, B)$ are precisely those pairs such that there is a minimal gallery from A to B of the form

$$C = C_0, \ldots, C_l = D \, .$$

We have proved the following:

Proposition 3.87. *Let Σ be a Coxeter complex of type (W, S), and let A and B be arbitrary simplices. Then there is an element $\delta(A, B) \in W$ such that*

$$\delta(A, B) = \delta(C_0, C_l)$$

for any minimal gallery $C_0, \ldots, C_l$ from A to B. In particular,

$$d(A, B) = l\big(\delta(A, B)\big) \, . \qquad \square$$

Note that reduced decompositions of $\delta(A, B)$ are *not* necessarily in 1–1 correspondence with minimal galleries from A to B, since in general there is more than one possible C_0 that can start such a gallery. The reader can easily find examples of this in Figure 3.2. We will get a clearer understanding of this phenomenon in the next section.

Exercise 3.88. Prove the following strong version of the triangle inequality: Given three chambers C_1, C_2, C_3, we have $\delta(C_1, C_3) = \delta(C_1, C_2)\delta(C_2, C_3)$.

3.6 Products and Convexity

This section gives analogues for Coxeter complexes of some of the results of Chapter 1 on hyperplane arrangements. The results should all be believable because of this analogy, but there are technicalities. The reader anxious to get

to buildings may want to skip ahead to Chapter 4 and return to the present
section as needed.

Throughout this section, Σ denotes an arbitrary Coxeter complex in the
sense of Definition 3.64, and $\mathcal{H}$ denotes its set of walls (Definition 3.50). For
each pair $\pm\alpha$ of opposite roots, we arbitrarily declare one of them to be
positive and the other negative. The most common convention is to choose a
"fundamental chamber" C and declare a root to be positive if it contains C.

3.6.1 Sign Sequences

Let A be a simplex of Σ and let H be a wall with its associated pair of
roots $\pm\alpha$, where α is the positive one. We have three possibilities: A is in α
but not $-\alpha$, A is in $-\alpha$ but not α, or A is in H. We set $\sigma_H(A) = +, -,$ or 0,
accordingly. The resulting family

$$\sigma(A) := \big(\sigma_H(A)\big)_{H\in\mathcal{H}}$$

is the *sign sequence* of A.

Remark 3.89. If $\Sigma = \Sigma(W, S)$, the sign sequence just defined can be identified with the sign sequence introduced in our study of the Tits cone (Section 2.6).

The first observation is that the sign sequence determines the face relation
in the expected way. As in Definition 1.20, we order sign sequences coordinatewise, with the convention that $0 < +$ and $0 < -$.

Proposition 3.90. *Given simplices $A, B \in \Sigma$, we have $B \leq A$ if and only if
$\sigma(B) \leq \sigma(A)$. In particular, $A = B$ if and only if $\sigma(A) = \sigma(B)$, i.e., a simplex
is uniquely determined by its sign sequence.*

Proof. If one wants to use the Tits cone, the result is already contained in
Section 2.6 (see the paragraph following Definition 2.79). But here is a purely
combinatorial proof.

Suppose $B \leq A$. Then $\sigma(B) \leq \sigma(A)$, since every root containing A must
contain B (roots are subcomplexes). Conversely, suppose $\sigma(B) \leq \sigma(A)$. Then
Proposition 3.78 implies that $d(A, B) = 0$. In other words, there is a chamber C having both A and B as faces. We now show that every vertex v of B
is also a vertex of A. Let P be the panel of C not containing v, and let H be
the wall containing P. Then $v \notin H$ (Exercise 3.61), so $\sigma_H(B) \neq 0$ and hence
$\sigma_H(A) \neq 0$. Thus A is not a face of the panel P, which means that v is a
vertex of A. $\square$

Exercise 3.91. Show that a simplex A is a chamber if and only if $\sigma_H(A) \neq 0$
for all $H \in \mathcal{H}$.

We will make extensive use of sign sequences, primarily in connection
with products (Section 3.6.4 below). But first we pause to give two easier
applications, the first involving convexity and the second involving supports.

3.6.2 Convex Sets of Chambers

Definition 3.92. Let Δ be a chamber complex, and let $\mathcal{C} := \mathcal{C}(\Delta)$ be its set of chambers. A subset $\mathcal{D} \subseteq \mathcal{C}$ is called *convex* if it is nonempty and for all $D, D' \in \mathcal{D}$, every minimal gallery in $\mathcal{C}$ from D to D' is contained in $\mathcal{D}$.

For example, $\mathcal{C}(\alpha)$ is a convex subset of $\mathcal{C}(\Sigma)$ for any root α of our Coxeter complex Σ (Lemma 3.44). We can get further examples from this one, since an intersection of convex sets is convex (if it is nonempty). For example:

Proposition 3.93. *For any simplex $A \in \Sigma$, the set $\mathcal{C}(\Sigma)_{\geq A}$ of chambers having A as a face is convex. More concisely, residues are convex.*

Proof. Proposition 3.90 implies that $\mathcal{C}(\Sigma)_{\geq A}$ is an intersection of convex sets of the form $\mathcal{C}(\alpha)$ for various roots α, one for each wall H such that $\sigma_H(A) \neq 0$. This proves the first assertion, and the second assertion is simply a restatement of the first (see Corollary 3.17). $\square$

We will study convexity systematically in Section 3.6.6, but we give here one further result, since we have just been talking about intersections of roots.

Proposition 3.94. *Let $\mathcal{D}$ be a nonempty set of chambers in Σ. Then $\mathcal{D}$ is a convex subset of $\mathcal{C}(\Sigma)$ if and only if $\mathcal{D}$ is an intersection of sets $\mathcal{C}(\alpha)$ for some family of roots α.*

Proof. It suffices to prove the "only if" part. The proof is essentially the same as the solution to Exercise 1.65 [see also Proposition 2.97], but for variety, we will say it in a slightly different way that will yield extra information (see Exercise 3.97).

By a *boundary panel* of $\mathcal{D}$ we mean a panel A such that one of the two chambers having A as a face is in $\mathcal{D}$ and the other is not. Let these two chambers be denoted by D, D', with $D \in \mathcal{D}$, and let $\alpha = \alpha_A$ be the root containing D but not D'. Assume now that $\mathcal{D}$ is convex. We claim that $\mathcal{D} \subseteq \alpha$. For suppose $E \in \mathcal{D}$ but $E \notin \alpha$. Then $d(E, D) = d(E, D') + 1$ by Lemma 3.45, and hence there is a minimal gallery from E to D passing through D'. By convexity this implies $D' \in \mathcal{D}$, contradicting our assumptions.

To complete the proof we will show that

$$\mathcal{D} = \bigcap_A \mathcal{C}(\alpha_A),$$

where A ranges over the boundary panels of $\mathcal{D}$. If C is a chamber of Σ not in $\mathcal{D}$, we must find a boundary panel A with $C \notin \alpha_A$. Choose $D \in \mathcal{D}$ at minimal distance from C, and let $D, D', \ldots, C$ be a minimal gallery from D to C. Then $D' \notin \mathcal{D}$, so $A := D \cap D'$ is a boundary panel. Since $d(C, D') < d(C, D)$, we have $D \notin \alpha_A$ by Lemma 3.45, as required. $\square$

Exercises

3.95. Give an alternative proof of Proposition 3.93 using properties of reduced words in Coxeter groups.

3.96. Let α be a root, and let C be a chamber not in α but adjacent to a chamber in α. Show that $\mathcal{C}(\alpha)$ is maximal among the convex sets of chambers not containing C.

3.97. Let $\mathcal{D}$ be a nonempty convex set of chambers in Σ, and consider the expression $\mathcal{D} = \bigcap_A \mathcal{C}(\alpha_A)$ that occurred in the proof of Proposition 3.94. There may be redundancy in this expression because a root α could be α_A for more than one boundary panel A. To remedy this, we index α_A by the wall $H := \alpha_A \cap -\alpha_A$, said to be a *wall of* $\mathcal{D}$. We then set $\alpha_H := \alpha_A$; it is independent of the choice of A because, of the two roots bounded by H, it is the one containing $\mathcal{D}$. In summary, we have

$$\mathcal{D} = \bigcap_H \mathcal{C}(\alpha_H) \, ,$$

where H ranges over the walls of $\mathcal{D}$.

(a) Show that this expression is irredundant, in the sense that if any $\mathcal{C}(\alpha_H)$ is deleted, then the resulting intersection is strictly bigger than $\mathcal{D}$.

(b) Let H_1 and H_2 be distinct walls of $\mathcal{D}$, let A_i be a boundary panel of $\mathcal{D}$ contained in H_i for $i = 1, 2$, let α_i be the corresponding root $\alpha_{H_i} = \alpha_{A_i}$, and let D_i, D_i' be the chambers $> A_i$, with $D_i \in \mathcal{D}$. Show that

$$d(D_1', D_2') = d(D_1, D_2) + 2 \, ;$$

in other words, there is a minimal gallery of the form $D_1', D_1, \ldots, D_2, D_2'$, where the inner part is a minimal gallery from D_1 to D_2 in $\mathcal{D}$.

3.6.3 Supports

Definition 3.98. The *support* of a simplex $A \in \Sigma$, denoted by $\operatorname{supp} A$, is the intersection of the walls containing A.

We record two results about supports that will be needed later. They both have easy proofs using the Tits cone (which we leave to the interested reader), and they also have easy combinatorial proofs that we will give.

Proposition 3.99. *A is a maximal simplex of its support.*

Proof. If $A < B$ in Σ, then Proposition 3.90 implies that there is a wall H with $\sigma_H(A) = 0$ and $\sigma_H(B) \neq 0$; hence $B \notin \operatorname{supp} A$. $\square$

Proposition 3.100. *If $\Sigma = \Sigma(W, S)$ with its natural W-action, then the stabilizer W_A of any simplex A fixes* supp A *pointwise, i.e., it fixes every simplex of* supp A. *Moreover,* supp A *is the full fixed-point set of W_A.*

Proof. This is an immediate consequence of the fact that W_A is generated by the reflections that fix A, i.e., the reflections s_H with $A \in H$; see Exercise 3.62.
□

There is a lot more to say about supports, but it fits most naturally into the setting of convex subcomplexes, which we cannot treat properly until we have introduced products.. We will therefore leave the subject now and return to it in Section 3.6.6.

Exercises

3.101. If $\Sigma = \Sigma(W, S)$ and we identify Σ with the set of cells in the Tits cone, how is the present definition of support related to the support as defined in Section 2.7?

3.102. What is the support of a vertex if Σ is 1-dimensional?

3.6.4 Semigroup Structure

Given $A, B \in \Sigma$, we wish to define their *product AB* as in Section 1.4.6. Thus AB should be the simplex with sign sequence given by

$$\sigma_H(AB) = \begin{cases} \sigma_H(A) & \text{if } \sigma_H(A) \neq 0, \\ \sigma_H(B) & \text{if } \sigma_H(A) = 0, \end{cases} \tag{3.9}$$

for any wall $H \in \mathcal{H}$. Of course, one has to prove the existence of such a simplex.

The existence proof was quite easy in the setting of Chapter 1, and we have already given the analogous easy proof in general using the Tits cone (Proposition 2.82, part (1)). But we will give an independent proof here, which is purely combinatorial. It is longer, but it is instructive, and it will generalize to buildings in the next chapter. In the course of the proof we will see that the chambers having AB as a face are precisely those that can start a minimal gallery from A to B, as we would expect from Exercise 1.62 [and Section 2.7]. This completely characterizes the desired AB, since a simplex is determined by the set of chambers having it as a face; see, for instance, Corollary 3.17 or Exercise 3.10.

The existence of the desired AB is quite easy to prove if B is a chamber (in which case (3.9) forces AB to be a chamber), so we begin with that case. The result should be compared with Proposition 1.40.

Proposition 3.103. *Given a simplex A and a chamber C, there is a (unique) chamber AC such that for any $H \in \mathcal{H}$,*

$$\sigma_H(AC) = \begin{cases} \sigma_H(A) & \text{if } \sigma_H(A) \neq 0, \\ \sigma_H(C) & \text{if } \sigma_H(A) = 0. \end{cases} \tag{3.10}$$

It has A as a face, and among the chambers having A as a face, it is the unique one at minimal distance from C. Every minimal gallery from A to C starts with AC.

Proof. Choose a minimal gallery $\Gamma \colon C_0, \dots, C_d = C$ from A to C. Given $H \in \mathcal{H}$, if $\sigma_H(A) \neq 0$, then $\sigma_H(C_0) = \sigma_H(A)$, since $C_0 \geq A$. If $\sigma_H(A) = 0$, then $H \notin \mathcal{S}(A, C)$ [see Definition 3.77], so Proposition 3.78 implies that Γ does not cross H; hence $\sigma_H(C_0) = \sigma_H(C)$. Thus C_0 is the desired chamber AC. The last two assertions of the proposition follow from the existence proof, since we started with an arbitrary minimal gallery from A to C. Alternatively, the second-to-last assertion follows from (3.10) as in the proof of Proposition 1.40, and the last assertion is simply a restatement of it. □

Definition 3.104. Given simplices $A, C \in \Sigma$ with C a chamber, their *product* is the chamber AC described in Proposition 3.103. The product is also denoted by $\mathrm{proj}_A C$ and called the *projection* of C onto A.

Equation (3.10) leads to the following important property of AC, which we call the *gate property*.

Proposition 3.105. *For any simplex A and any chambers C, D with $D \geq A$,*

$$d(C, D) = d(C, AC) + d(AC, D) . \tag{3.11}$$

Proof. Partition the walls H separating C from D into two subsets according to whether or not $\sigma_H(A) = 0$. Those with $\sigma_H(A) = 0$ are precisely the walls separating AC from D, while those with $\sigma_H(A) \neq 0$ are the walls separating AC from C. Equation (3.11) now follows from the fact that we can compute distances by counting separating walls (Lemma 3.69). □

We already saw the gate property in the context of hyperplane arrangements in Chapter 1; see Exercise 1.42 and Figure 1.6. As we noted there, it says that $\mathcal{C}_{\geq A}$ is a "gated subset" of the metric space $\mathcal{C} = \mathcal{C}(\Sigma)$ in the sense of Dress–Scharlau [97].

We now proceed to the existence of AB for arbitrary B. We will use ideas borrowed from the Dress–Scharlau theory, with some simplifications achieved in the present context by the use of sign sequences. Our first goal will be to find a simplex AB such that the chambers $\geq AB$ are precisely those that can start a minimal gallery from A to B. We will then be able to check that (3.9) holds.

Let $\mathcal{C} = \mathcal{C}(\Sigma)$, and for any simplex A, let $\mathcal{C}_A \subseteq \mathcal{C}$ be the set $\mathcal{C}_{\geq A}$ of chambers $\geq A$. For any two simplices A, B, let $\mathcal{C}_{A,B} \subseteq \mathcal{C}_A$ be the set of chambers that can start a minimal gallery from A to B.

Proposition 3.106.

(1) $\mathcal{C}_{A,B}$ is the image of the projection map $\mathcal{C}_B \to \mathcal{C}_A$ given by $D \mapsto AD$.
(2) The projection maps $\mathcal{C}_{A,B} \rightleftarrows \mathcal{C}_{B,A}$, given by $C \mapsto BC$ and $D \mapsto AD$ $(C \in \mathcal{C}_{A,B},\ D \in \mathcal{C}_{B,A})$, define mutually inverse bijections.
(3) Given any minimal gallery $C_0, \dots, C_l$ from A to B, we have $C_0 = AC_l$ and $C_l = BC_0$. In other words, C_0 and C_l correspond to one another under the bijections in (2).

Proof. A minimal gallery $C_0, \dots, C_l$ from A to B is also minimal from A to C_l, so $C_0 = AC_l$ by Proposition 3.103. Similarly, $C_l = BC_0$. This proves (3) and shows that $\mathcal{C}_{A,B}$ is contained in the image of the projection map $\mathcal{C}_B \to \mathcal{C}_A$, which is part of (1). To prove the opposite inclusion, consider any $D \in \mathcal{C}_B$. Then one sees by checking sign sequences that $\mathcal{S}(AD, B(AD)) = \mathcal{S}(A, B)$; hence $d(AD, B(AD)) = d(A, B)$ by Proposition 3.78. Thus there is a minimal gallery from A to B starting with AD. This proves (1). It follows from (1) that the projection maps do define maps $\mathcal{C}_{A,B} \rightleftarrows \mathcal{C}_{B,A}$, and it is easy to check (using sign sequences) that these maps are inverse to one another, whence (2). $\square$

We need one more simple observation before we can complete the analysis of $\mathcal{C}_{A,B}$.

Lemma 3.107. The projection $\mathcal{C} \to \mathcal{C}_A$ takes adjacent chambers to chambers that are equal or adjacent. Consequently, $\mathcal{C}_{A,B}$ is a connected subset of the chamber graph of Σ, i.e., any two elements of $\mathcal{C}_{A,B}$ can be connected by a gallery in $\mathcal{C}_{A,B}$.

Proof. The first assertion is immediate if one calculates projections in terms of sign sequences (equation (3.10)). The second assertion now follows from part (1) of Proposition 3.106 because $\mathcal{C}_B$ is connected (by Proposition 3.16 or Proposition 3.93). $\square$

We can now prove the main result of this subsection. Choose a type function on Σ with values in a set S, so that we can define the Weyl group $W = W_M$ and the Weyl distance function δ as in Section 3.5. Recall that the abstract set S then becomes a set of generators of W.

It will be convenient to use residue terminology in what follows. Recall that by Corollary 3.17, the residues are the sets of the form $\mathcal{C}_A$, one for each simplex A. The main point in what follows is to show that $\mathcal{C}_{A,B}$ is again a residue, so that we can define AB to be the corresponding simplex.

Theorem 3.108. Let A be a simplex of cotype J, let B be a simplex of cotype K, and let $w = \delta(A, B)$. Then $\mathcal{C}_{A,B}$ is a residue of type $J_1 := J \cap wKw^{-1}$. In other words, there is a simplex AB of cotype J_1 such that $\mathcal{C}_{A,B} = \mathcal{C}_{AB}$, i.e., the chambers that can start a minimal gallery from A to B are precisely those having AB as a face. Moreover, the sign sequence of AB is given by equation (3.9).

Proof. To show that $\mathcal{C}_{A,B}$ is contained in a residue of type J_1 it suffices, by the lemma, to prove that any two adjacent chambers in $\mathcal{C}_{A,B}$ are J_1-equivalent. Let $C_1, C_2 \in \mathcal{C}_{A,B}$ be s-adjacent, and let $D_1 = BC_1$ and $D_2 = BC_2$. Then D_1 and D_2 are distinct by part (2) of Proposition 3.106 and are adjacent by Lemma 3.107. Let $t \in S$ be the element such that D_1 and D_2 are t-adjacent. For $i = 1, 2$ there is a minimal gallery from A to B starting with C_i and, necessarily, ending with D_i (see part (3) of Proposition 3.106). So $\delta(C_1, D_1) = w = \delta(C_2, D_2)$. Computing $\delta(C_1, D_2)$ in two different ways, we conclude that $sw = wt$. See Figure 3.7, which should be viewed as a schematic picture of the chamber graph. We have $s \in J$ because $C_1, C_2 \geq A$, and similarly $t \in K$.

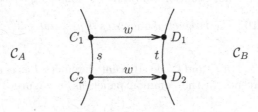

Fig. 3.7. $sw = wt$.

So $s = wtw^{-1} \in J \cap wKw^{-1}$, whence C_1 and C_2 are J_1-equivalent.

To show that $\mathcal{C}_{A,B}$ is an entire residue of type J_1, it suffices to prove that if C_1 is in $\mathcal{C}_{A,B}$ and C_2 is s-adjacent to C_1 for some $s \in J_1$, then C_2 is in $\mathcal{C}_{A,B}$. Write $sw = wt$ with $t \in J$, and let $D_1 = BC_1$. Then $\delta(C_2, D_1) = sw = wt$; see Figure 3.8. Since decompositions of $\delta(C_2, D_1)$ correspond to galleries from

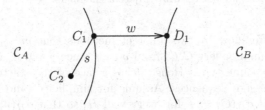

Fig. 3.8. $\delta(C_2, D_1) = sw$.

C_2 to D_1, it follows that there is a chamber D_2 such that $\delta(C_2, D_2) = w$ and $\delta(D_2, D_1) = t$. In other words, we have achieved the situation in Figure 3.7, where $D_2 \in \mathcal{C}_B$ because $t \in K$. Since $d(C_2, D_2) = l(w) = d(A, B)$, it follows that $C_2 \in \mathcal{C}_{A,B}$, as required.

Now let AB be the simplex such that $\mathcal{C}_{A,B} = \mathcal{C}_{AB}$. We must calculate the sign sequence of AB. We have $AB \geq A$ since $\mathcal{C}_{AB} \subseteq \mathcal{C}_A$; so $\sigma_H(AB) = \sigma_H(A)$ if $\sigma_H(A) \neq 0$. Suppose $\sigma_H(A) = 0$. Then $H \notin \mathcal{S}(A, B)$, so the minimal galleries $\Gamma: C_0, \ldots, C_l$ from A to B do not cross H, i.e., C_0 and C_l are on the same side of H. If $\sigma_H(B) \neq 0$, it follows that $\sigma_H(C_0) = \sigma_H(B)$. In other words, every chamber $C \geq AB$ satisfies $\sigma_H(C) = \sigma_H(B)$; hence $\sigma_H(AB) = \sigma_H(B)$ by Exercise 3.81. If $\sigma_H(B) = 0$, on the other hand, then Γ is in one root α associated to H, and we can fold onto $-\alpha$ to get another minimal gallery from A to B. Thus there are elements of $\mathcal{C}_{A,B} = \mathcal{C}_{AB}$ on both sides of H, and $\sigma_H(AB) = 0$. This proves (3.9). $\square$

Definition 3.109. Given simplices $A, B \in \Sigma$, their *product* is the chamber AB described in Theorem 3.108 and characterized by equation (3.9). The product is also denoted by $\mathrm{proj}_A B$ and called the *projection* of B onto A.

As in Section 1.4.6, equation (3.9) has the following consequence:

Corollary 3.110. *The product of simplices is associative. Hence Σ is a semigroup.* $\square$

Example 3.111. Let C and C' be adjacent chambers. Let v, v' be the vertices of C, C' that are not in the common panel, as in Figure 3.9. Consider the

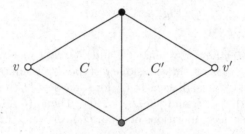

Fig. 3.9. Adjacent chambers.

product vv'. We show by two different methods that $vv' \leq C$. Method 1: There is a minimal gallery C, C' from v to v' since v and v' are not joinable. [They have the same type.] Hence vv' is a face of the starting chamber C. Method 2: Use sign sequences. Assume for simplicity (and without loss of generality) that $\sigma_H(C) = +$ for every wall H, so that $\sigma_H(v) \geq 0$ for all H. We then have $\sigma_H(C') = +$ for all H except the one containing $P := C \cap C'$; hence $\sigma_H(v') \geq 0$ for all H except the one containing P. Since $\sigma_H(v) = +$ for that exceptional wall, it follows that $\sigma_H(vv') \geq 0$ for all H and hence that $vv' \leq C$.

We close this subsection by recording some connections between the poset and semigroup structures on Σ as in Proposition 1.41 and Exercise 1.44. The proofs are easy via sign sequences and are left to the reader

Proposition 3.112. *Let A and B be arbitrary simplices in Σ.*

(1) $A \leq AB$, with equality if and only if $\mathrm{supp}\, B \leq \mathrm{supp}\, A$.

(2) $A \leq B$ if and only if $AB = B$.

(3) supp $A =$ supp B if and only if $AB = A$ and $BA = B$.

(4) If supp $A =$ supp B, then left multiplication by B and A defines mutually inverse bijections $\Sigma_{\geq A} \rightleftarrows \Sigma_{\geq B}$. In particular, $\dim A = \dim B$.

(5) AB and BA have the same support, which is the intersection of the walls containing both A and B. $\square$

Corollary 3.113. *For any simplices $A, B \in \Sigma$, $\dim AB = \dim BA$. Consequently, $\dim AB \geq \max \{\dim A, \dim B\}$.*

Proof. The first assertion follows immediately from parts (5) and (4) of the proposition. For the second, we have $\dim AB \geq \dim A$ trivially because $A \leq AB$, and similarly $\dim BA \geq \dim B$; now use the fact that $\dim BA = \dim AB$.
            $\square$

Exercises

3.114. Use Theorem 3.108 to give a new proof of Lemma 2.25.

3.115. Show that every root is a subsemigroup, and hence every intersection of roots is a subsemigroup. In particular, this applies to the *support* of any simplex (Definition 3.98).

3.116. Show that (finitely many) simplices $A, B, \ldots, C$ are joinable if and only if they commute with one another in the semigroup Σ, in which case their product is their least upper bound (see Exercise 1.43). Deduce, as in the proof of Proposition 1.127, that Σ is a flag complex.

3.117. Given simplices $A_1, A_2, B \in \Sigma$ with $A_1 \leq A_2$, show that $d(A_1, B) \leq d(A_2, B)$, with equality if $A_2 \geq A_1 B$. In particular, $d(A, B) = d(AB, B)$ for any two simplices A, B.

3.118. Figure 3.9 suggests that $vv' = C$. Give examples to show that this is not necessarily the case. For instance, vv' could be a vertex or an edge.

3.119. Recall that the link $L_A := \mathrm{lk}_\Sigma A$ of any simplex A is again a Coxeter complex; hence it has a semigroup structure. Is it a subsemigroup of Σ? If not, how is the product on L_A related to the product in Σ?

The remaining exercises are intended to show how the use of products can sometimes replace arguments based on the Tits cone. The intent of the exercises, then, is that they should be solved combinatorially, without the Tits cone. Given a chamber C and a panel P of C, the wall containing P will be called a *wall of C*. Thus every chamber has exactly $n + 1$ walls if $\dim \Sigma = n$.

3.120. Fix a chamber C and let $\mathcal{H}_C$ be its set of walls.

(a) Show that C is defined by $\mathcal{H}_C$; in other words, if D is a chamber such that $\sigma_H(D) = \sigma_H(C)$ for all $H \in \mathcal{H}_C$, then $D = C$.

(b) Suppose A is a simplex such that $\sigma_H(A) \le \sigma_H(C)$ for all $H \in \mathcal{H}_C$. Show that $A \le C$.

(c) If A and B are faces of C, show that $A \le B$ if and only if $\sigma_H(A) \le \sigma_H(B)$ for all $H \in \mathcal{H}_C$.

(d) If $A \le C$, show that A is defined by $\mathcal{H}_C$; in other words, if B is a simplex such that $\sigma_H(B) = \sigma_H(A)$ for all $H \in \mathcal{H}_C$, then $B = A$.

3.121. (a) Let C be a chamber, and let s and t be reflections with respect to two distinct walls of C, denoted by H_s and H_t. Let m be the order of st, and assume $m \ge 3$. If D is another chamber that also has H_s and H_t as two of its walls, show that either H_s and H_t both separate C from D or else neither of them separates C from D.

(b) Give an example to show that we cannot drop the assumption that $m \ge 3$ in (a).

(c) Generalize (a) as follows. Let $H_1, \ldots, H_k$ be walls of a chamber C such that the corresponding reflections s_i generate an irreducible Coxeter group. If D is another chamber having $H_1, \ldots, H_k$ as walls, show that either every H_i separates C from D or else no H_i separates C from D.

3.122. Use the previous exercise to give a combinatorial proof of the following fact, which we have proven earlier by different methods: If (W, S) is irreducible and $wSw^{-1} = S$ for some $w \ne 1$ in W, then W is finite and w is the longest element. [We gave two proofs of this for finite reflection groups, one algebraic and one geometric; see the proof of Corollary 1.91. And we generalized the algebraic proof to the infinite case in the proof of Proposition 2.73. The point of the exercise is that we now have the tools to generalize the geometric proof.]

3.6.5 Applications of Products

In this brief subsection we use products to prove two results that will be needed later. Both proofs make use of the following lemma.

Lemma 3.123. *Let $\Sigma = \Sigma(W, S)$, and let W_A for $A \in \Sigma$ be the stabilizer of A in W. Then for any two simplices $A, B \in \Sigma$ we have $W_{AB} = W_A \cap W_B$.*

Proof. It is clear that $W_A \cap W_B \le W_{AB}$ and that $W_{AB} \le W_A$. [For the latter, note that $A \le AB$ and W is type-preserving.] So all that remains to show is that W_{AB} fixes B. This follows from Proposition 3.100 because $B \in \operatorname{supp} BA = \operatorname{supp} AB$. $\square$

We can now prove a finiteness result that, a priori, is far from obvious:

Proposition 3.124. *Let A and B be arbitrary simplices of $\Sigma(W, S)$. Choose a chamber $C \ge AB$. Then every minimal gallery from A to B is equivalent under $W_A \cap W_B$ to one that starts with C. In particular, there are only finitely many $(W_A \cap W_B)$-orbits of minimal galleries from A to B.*

Proof. By the lemma and Exercise 3.13, $W_A \cap W_B$ is transitive on $\mathcal{C}_{AB}$. This implies the first assertion. For the second assertion, we need only recall that a minimal gallery from A to B starting with C must end with BC, so there are only finitely many of these, one for each reduced decomposition of $\delta(C, BC)$ $[= \delta(A, B)]$. □

Our next result is taken from Tits [247, Lemma 12.12].

Proposition 3.125. *Let Σ be an irreducible Coxeter complex, and let H be a wall of Σ. Then Σ contains a chamber C that is disjoint from H, in the sense that none of the vertices of C are in H. More generally, every simplex A disjoint from H is a face of a chamber disjoint from H.*

Note that we cannot drop the irreducibility assumption. For example, suppose Σ is of type $A_1 \times A_1$, i.e., Σ is the poset of cells associated with the reflection group $\{\pm 1\} \times \{\pm 1\}$ acting on $\mathbb{R} \times \mathbb{R}$. Then there are two walls and four chambers, and each chamber has a nontrivial face in each wall.

Proof of the proposition. Let A be disjoint from H, and choose a maximal simplex $B \geq A$ disjoint from H. Clearly B is not the empty simplex, since H cannot contain every vertex of Σ. We will show that B is a chamber. Let τ be a type function on Σ with values in a set S. By irreducibility of Σ, it suffices to show that $m(s, t) = 2$ for all $s \in \tau(B)$ and $t \in S \smallsetminus \tau(B)$. [This will imply that $\tau(B) = S$, so that B is a chamber.]

Choose a panel $P \geq B$ of cotype t, and let C and C' be the chambers having P as a face. Let v (resp. v') be the vertex of C (resp. C') not in P. Thus $\tau(v) = \tau(v') = t$. Let F (resp. F') be the panel of C (resp. C') of cotype s. Thus the vertex of C not in F, which is the same as the vertex of C' not in F', has type s and hence is a vertex of B. Note further that v and v' are both joinable with B, so we must have $v, v' \in H$ by the maximality of B. Figure 3.10 summarizes some of the notation. We have drawn the picture so

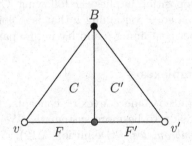

Fig. 3.10. $m(s, t) = 2$; $\bullet = s$, $\circ = t$.

as to suggest that supp $F = $ supp F', since the proof will show that this is in fact the case.

We now assume, without loss of generality, that $\Sigma = \Sigma(W, S)$ with its canonical type function and that C is the fundamental chamber. Then t is the fundamental reflection that fixes $P = C \cap C'$, and s is the fundamental reflection that fixes F. Consider the product vv'. By Example 3.111 we have $vv' \leq C$. We also have $vv' \in H$ because v and v' are both in H. Since the vertex of C not in F is in B and hence not in H, it follows that $vv' \leq F$.

Thus vv' is fixed by the reflection s. In view of Lemma 3.123, it follows that s fixes v'. But s also fixes $F \cap F'$, which contains every vertex of F' except v'. So s fixes F'. On the other hand, we have $F' = tF$, so the reflection fixing F' is the conjugate tst of s. Thus $tst = s$ and $m(s, t) = 2$. □

Note that a great deal of the proof remains valid if we drop the irreducibility assumption, but the conclusion becomes more complicated. Namely, we can no longer prove that B is a chamber, but we can say that $\tau(B)$ is a union of one or more connected components of the Coxeter diagram. This yields the following more precise result:

Proposition 3.126. *Let Σ be a Coxeter complex with a type function τ. If H is a wall of Σ and A is a simplex disjoint from H, then there is a simplex $B \geq A$ disjoint from H such that $\tau(B)$ is a union of connected components of the Coxeter diagram of Σ.* □

This has the following immediate consequence, which we will have occasion to use in Chapter 7:

Corollary 3.127. *Let Σ be a Coxeter complex, and let r be the minimal cardinality of a connected component of its Coxeter diagram. If H is a wall of Σ and A is a simplex disjoint from H, then there is a simplex $B \geq A$ disjoint from H such that $\operatorname{rank} B \geq r$. Equivalently, there is a chamber $C \geq A$ such that the maximal face of C in H has codimension $\geq r$.* □

Example 3.128. Suppose the Coxeter diagram of Σ has no isolated nodes. Then $r \geq 2$, and the proposition implies the following: Given a wall H and a vertex $v \notin H$, there is an edge containing v that is disjoint from H. Equivalently, there is a chamber containing v and having no panel in H.

3.6.6 Convex Subcomplexes

We have already introduced some convexity concepts, but with the aid of products, we are now ready to be more systematic. We begin by discussing convexity for *chamber subcomplexes* (Definition A.12).

Definition 3.129. Let Δ be a chamber complex and let Δ' be a chamber subcomplex. We say that Δ' is a *convex subcomplex* of Δ if its set of chambers $\mathcal{C}(\Delta')$ is a convex subset of $\mathcal{C}(\Delta)$, i.e., if every minimal gallery in Δ joining two chambers of Δ' is contained in Δ'.

In this subsection we will be concerned exclusively with the case that Δ is a Coxeter complex Σ. In this case, as we will see, convexity is closely related to the product in Σ. The following lemma gives the first hint of this.

Lemma 3.130. *Let Σ be a Coxeter complex.*

(1) *Every convex chamber subcomplex of Σ is a subsemigroup of Σ.*
(2) *Let Σ' be a subcomplex of Σ that is a subsemigroup. If $\Sigma' \cap \mathcal{C}(\Sigma)$ is a convex subset of $\mathcal{C}(\Sigma)$, then Σ' is a convex chamber subcomplex of Σ.*
(3) *Let Σ' be an intersection of convex chamber subcomplexes of Σ. If Σ' contains at least one chamber, then Σ' is a convex chamber subcomplex of Σ.*

Proof. (1) Let Σ' be a convex chamber subcomplex of Σ. Given $A, B \in \Sigma'$, we must show that $AB \in \Sigma'$. Choose a chamber $C \in \Sigma'$ with $C \geq B$. By checking sign sequences, one sees that $AB \leq AC$; so it suffices to prove $AC \in \Sigma'$. To this end choose a chamber $D \geq A$ and note that by the gate property (3.11), there is a minimal gallery from D to C passing through AC. This gallery is contained in Σ' by convexity; hence $AC \in \Sigma'$.

(2) The only thing that needs to be proved is that Σ' is a chamber subcomplex of Σ. The convexity assumption implies that $\Sigma' \cap \mathcal{C}(\Sigma)$ is nonempty and that any two of its elements can be joined by a gallery in Σ'. So we need only show that every maximal simplex A of Σ' is a chamber of Σ. Choose an arbitrary chamber $C \in \Sigma' \cap \mathcal{C}(\Sigma)$. Then we have $A \leq AC \in \Sigma'$, so $A = AC$ and A is indeed a chamber of Σ.

(3) It is obvious that $\mathcal{C}(\Sigma')$ is a convex subset of $\mathcal{C}(\Sigma)$. Moreover, (1) implies that Σ' is an intersection of subsemigroups and hence is itself a subsemigroup. The result therefore follows from (2). $\square$

We can now give several characterizations of convex chamber subcomplexes.

Theorem 3.131. *Let Σ be a Coxeter complex and Σ' a subcomplex containing at least one chamber. Then the following conditions are equivalent:*

(i) *Σ' is a convex chamber subcomplex of Σ.*
(ii) *Σ' is an intersection of roots.*
(iii) *Σ' is a subsemigroup of Σ.*
(iv) *Given $A, C \in \Sigma'$ with C a chamber of Σ, every minimal gallery from A to C in Σ is contained in Σ'.*

Proof. The equivalence of (i) and (ii) is almost immediate from Proposition 3.94, but one must be a little careful: Suppose (i) holds, and let Σ'' be the intersection of all roots containing Σ'. Then Σ'' is a chamber subcomplex of Σ by part (3) of the lemma, and $\mathcal{C}(\Sigma'') = \mathcal{C}(\Sigma')$ by Proposition 3.94. So $\Sigma'' = \Sigma'$ and (ii) holds. Conversely, (ii) $\implies$ (i) by part (3) of the lemma again, since roots are convex chamber subcomplexes by Lemma 3.44.

We already know from part (1) of Lemma 3.130 that (i) $\implies$ (iii), and it is easy to check that (iv) $\implies$ (i). So it remains to prove (iii) $\implies$ (iv). Suppose (iii) holds, and let A, C be as in (iv). Consider a minimal gallery $A \leq C_0, C_1, \ldots, C_l = C$ from A to C. Recall from Proposition 3.103 that $C_0 = AC$. Moreover, if A_i is the common panel between C_{i-1} and C_i for $i = 1, \ldots, n$, then we have $C_1 = A_1 C$, $C_2 = A_2 C$, and so on, as one sees by checking sign sequences or by using the fact that C is closer to C_i than to C_{i-1}. Since Σ' is a subsemigroup and a subcomplex, it follows inductively that all $C_i \in \Sigma'$. □

Remark 3.132. The equivalence between (i) and (iii) has an intuitive interpretation. In the setting of hyperplane arrangements, for example, knowing the product AB of two cells A, B is equivalent to knowing the beginning of a line segment joining a point of A to a point of B. An analogue of Theorem 3.131 in that context was given in Exercise 1.68 [see also Proposition 2.97]. In our present combinatorial setting, we can therefore think of products as a substitute for line segments (or geodesics). Of course minimal galleries also provide a substitute for geodesics, and it is reassuring that they yield the same notion of convexity for chamber subcomplexes.

Examples 3.133. (a) Given two chambers C, D, we define their *convex hull* $\Gamma(C, D)$ to be the smallest convex chamber subcomplex containing C and D or, equivalently, the intersection of all roots containing C and D. One easily checks by counting separating walls (cf. Exercise 1.66) that the chambers $E \in \Gamma(C, D)$ are precisely those such that

$$d(C, D) = d(C, E) + d(E, D) ,$$

i.e., they are the chambers that can occur in a minimal gallery from C to D. See Figure 3.11 for an example. Note that in Figure 3.11, one could also speak of the convex hull of C and D in the usual sense of Euclidean geometry, and this is strictly smaller than the geometric realization of $\Gamma(C, D)$. Whenever this could lead to confusion, we will call $\Gamma(C, D)$ the *combinatorial convex hull* of C and D.

(b) For any simplex $A \in \Sigma$, the *star* of A, denoted by st A or $\mathrm{st}_\Sigma A$, is the set of simplices joinable to A. It consists of all the chambers $\geq A$ and their faces, so it is a convex chamber subcomplex of Σ by Proposition 3.93.

(c) The concept of "convex hull" that we introduced in (a) generalizes in an obvious way. Given an arbitrary collection of simplices containing at least one chamber, we define their *convex hull* to be the smallest convex chamber subcomplex containing them or, equivalently, the intersection of all roots containing them. For example, one can speak of the convex hull $\Gamma(A, C)$ of a simplex A and a chamber C. It is an intersection of roots, where there is one root for each wall that does not strictly separate A from C.

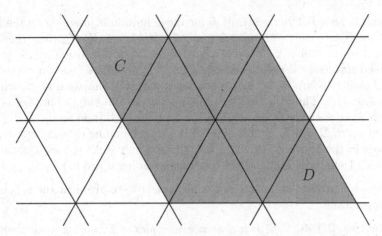

Fig. 3.11. The (combinatorial) convex hull of C and D.

(d) Here is a specific instance of (c) that will be useful in the theory of buildings. Suppose Σ is spherical. Let P and P' be opposite panels, and let C' be a chamber having P' as a face. Then the wall of Σ containing P and P' is the only wall that does not strictly separate P from C', so the convex hull $\Gamma(P, C')$ is the (unique) root containing P and C'. See Figure 3.12.

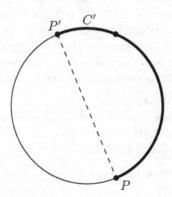

Fig. 3.12. The convex hull of P and C' is a root.

We turn now to convexity for subcomplexes that are not necessarily chamber subcomplexes. It is not clear that there is a sensible way to define this concept in the generality of arbitrary chamber complexes. For Coxeter complexes, however, Theorem 3.131 and Remark 3.132 motivate the following generalization of convexity to arbitrary subcomplexes:

Definition 3.134. A subcomplex Σ' of a Coxeter complex Σ is called a *convex subcomplex* if it is a subsemigroup of Σ.

Remark 3.135. Further motivation for the definition is provided by the theory of the Tits cone (Sections 2.6 and 2.7). If $\Sigma = \Sigma(W, S)$, then the Tits cone is a convex cone X in a real vector space V of dimension $|S|$. It is decomposed into conical cells by a collection of hyperplanes, one for each wall of Σ. These cells are in 1–1 correspondence with the simplices of Σ, with a dimension shift. Thus the cell $\{0\}$ corresponds to the empty simplex of Σ, the cells that are rays correspond to the vertices of Σ, and so on. If Σ' is a subcomplex of Σ and X' is the union of the corresponding cells in X, then Σ' is convex in the sense of Definition 3.134 if and only if X' is a convex subset of V. (See Definition 2.95 and the discussion leading up to it.)

The main results about convex subcomplexes are given in the following two propositions:

Proposition 3.136. *If Σ' is a convex subcomplex of Σ, then Σ' is a chamber complex in which every panel is a face of at most two chambers.*

Proposition 3.137. *Let Σ' be a subcomplex of a Coxeter complex Σ. Then the following conditions are equivalent:*

  (i) *Σ' is convex.*
  (ii) *Σ' is an intersection of roots.*
  (iii) *Σ' is an intersection of convex chamber subcomplexes.*

It is trivial that an intersection of convex subcomplexes is convex. So the essential content of Proposition 3.137 is the implication (i) $\implies$ (ii). This has a concrete interpretation in terms of sign sequences. It says that every convex subcomplex is defined by conditions of the form $\sigma_H(A) \geq 0$, $\sigma_H(A) \leq 0$, or $\sigma_H(A) = 0$, where there is at most one such condition for each $H \in \mathcal{H}$. We will discuss the proofs of the propositions after a remark and a few examples.

Remark 3.138. A useful consequence of Proposition 3.137 and Lemma 3.54 is that convex subcomplexes of a Coxeter complex are always full subcomplexes. One can also derive this directly from the definition of "convex subcomplex," together with Exercise 3.116.

Examples 3.139. (a) Every wall is an intersection of two roots and hence is convex.

(b) The support of any simplex is an intersection of walls and hence is convex.

(c) The *convex hull* of a collection of simplices is the smallest convex subcomplex containing them or, equivalently, the intersection of all roots containing them. For example, we can speak of the convex hull $\Gamma(A, B)$ of two simplices A, B. It is defined by one condition for each wall $H \notin \mathcal{S}(A, B)$: If $\sigma_H(A) = \sigma_H(B) = 0$, the condition is $\sigma_H = 0$; otherwise, $\sigma_H(A)$ and $\sigma_H(B)$ are either both ≥ 0 or both ≤ 0, and the condition is $\sigma_H \geq 0$ or $\sigma_H \leq 0$ accordingly. The reader is encouraged to draw some examples of $\Gamma(A, B)$ in Figure 3.11. See Exercises 3.148 and 3.149 below for more information about $\Gamma(A, B)$.

Propositions 3.136 and 3.137 both have fairly short proofs based on the Tits cone, which we have already given in the optional Section 2.7 (see Propositions 2.96 and 2.97). We wish to give combinatorial proofs also. The treatment that follows is based largely on [8, Section 1]. We begin with some easy consequences of the definition.

Lemma 3.140. *Let Σ' be a convex subcomplex of Σ.*

(1) *All maximal simplices of Σ' have the same dimension.*
(2) *If A is a maximal simplex of Σ', then $\Sigma' \subseteq \operatorname{supp} A$.*

Proof. (1) If A and B are maximal simplices of Σ', then $AB = A$ and $BA = B$, so A and B have the same dimension by Corollary 3.113.

(2) For any simplex $B \in \Sigma'$, we have $AB = A$ by maximality of A; hence $\operatorname{supp} A = \operatorname{supp} AB = \operatorname{supp} BA$ (see Proposition 3.112(5)). Since $B \leq BA$, it follows that $B \in \operatorname{supp} A$. $\square$

This has some useful consequences for supports:

Corollary 3.141.

(1) *For any simplex $A \in \Sigma$, we have $\dim A = \dim(\operatorname{supp} A)$.*
(2) *Let A and B be simplices in Σ with $\dim A = \dim B$. If $B \in \operatorname{supp} A$, then $\operatorname{supp} A = \operatorname{supp} B$.*

Proof. (1) The subcomplex $\operatorname{supp} A$ is convex, so all of its maximal simplices have the same dimension by the lemma; now apply Proposition 3.99.

(2) B is a maximal simplex of $\operatorname{supp} A$, so $\operatorname{supp} A \subseteq \operatorname{supp} B$ by the lemma. The opposite inclusion is immediate from the definition of "support." $\square$

Lemma 3.142. *If Σ' is a convex subcomplex of Σ, then any two maximal simplices $A, B \in \Sigma'$ can be connected by a Σ'-gallery.*

Proof. We argue by induction on $d(A, B)$, where the latter denotes the gallery distance between A and B *in* Σ (see Section A.1.3). If $d(A, B) = 0$, then A and B are joinable in Σ, and one sees immediately by using sign sequences that $AB = BA$ (see Exercise 3.116); hence $A = B$ by maximality, and there is nothing to prove.

Assume now that $d(A, B) > 0$, and choose a minimal Σ-gallery $C_0, C_1, \ldots$ from A to B. Then $A \nleq C_1$, so $A' := A \cap C_1$ has codimension 1 in A, and $d(A', B) < d(A, B)$. Now observe that $A'B$ is again a maximal simplex of Σ' since $\dim A'B = \dim BA' = \dim B$, where the first equality follows from Corollary 3.113. Moreover, $d(A'B, B) = d(A', B)$ by Exercise 3.117, so $d(A'B, B) < d(A, B)$. By the induction hypothesis, there is a Σ'-gallery from $A'B$ to B; since A and $A'B$ have the common codimension-1 face A', it follows that there is a Σ'-gallery from A to B. $\square$

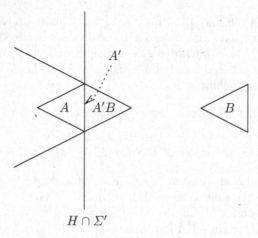

$$H \cap \Sigma'$$

Fig. 3.13. A wall separating A from B.

One can understand the proof intuitively by considering the wall H of Σ separating C_0 from C_1. Then $A' \in H$, so we can think of $H \cap \Sigma'$ as a "wall of A in Σ'" separating A from B; see Figure 3.13. Readers of Section 2.7 will note that this intuition can be made precise via the theory of hyperplane arrangements.

Lemma 3.143. *Let x be a vertex of Σ. Then* $\operatorname{supp} x$ *is 0-dimensional and either has x as its only vertex or else has exactly two vertices x, y.*

The combinatorial proof of this is somewhat tricky. To avoid disrupting the flow of ideas, we postpone the proof to the next subsection.

Remarks 3.144. (a) Lemma 3.143 is obvious from the point of view of the Tits cone. [The vertex x corresponds to a cell that is a ray, and $\operatorname{supp} x$ corresponds to the cells in the line spanned by that ray.] Moreover, one obtains from this point of view a sharper result: If $\operatorname{supp} x$ contains a vertex $y \neq x$, then the sign sequence of y is opposite to that of x (i.e., $\sigma_H(y) = -\sigma_H(x)$ for every wall H). In more geometric language, this says that every wall of Σ either contains both x and y or else strictly separates x from y. We will say that x and y are *opposite vertices* in this situation.

If Σ is spherical, then every vertex x has an opposite vertex $-x$. It turns out that this is essentially the only situation in which opposite vertices exist. See Proposition 2.91 and Exercise 3.156.

(b) It may seem counterintuitive at first that a 2-vertex 0-dimensional simplicial complex, which is not even connected, can be viewed as convex. But the interpretation in terms of the Tits cone explains this, since, as we just saw in (a), that 0-dimensional complex corresponds to a line in a vector space. And a line is indeed convex in the usual sense.

We can now prove our two propositions:

Proof of Proposition 3.136. Lemmas 3.140 and 3.142 show that Σ' is a chamber complex, so it remains to show that every Σ'-panel P is a face of at most two Σ'-chambers. Let A be a Σ'-chamber having P as a face. Since $\Sigma' \subseteq \operatorname{supp} A$ and the two have a common maximal simplex, they have the same dimension. We may therefore assume that $\Sigma' = \operatorname{supp} A$. Moreover, we may replace Σ by $\operatorname{lk}_\Sigma P$ (see Proposition 3.79) and thereby reduce to the case that P is the empty simplex and A is a vertex. The result now follows from Lemma 3.143. □

Proof of Proposition 3.137. It is immediate that (ii) $\implies$ (iii) $\implies$ (i), so it suffices to show that (i) $\implies$ (ii). Assume that Σ' is convex. Given $B \in \Sigma$ such that $B \notin \Sigma'$, we must find a root α of Σ that contains Σ' but not B. Choose a Σ'-chamber A that minimizes $d(A, B)$, the latter being the gallery distance in Σ as in the proof of Lemma 3.142. Then $\Sigma' \subseteq \operatorname{supp} A$ by Lemma 3.140, and the two have the same dimension. If $B \notin \operatorname{supp} A$, then we are done because $\operatorname{supp} A$ is an intersection of roots. So assume $B \in \operatorname{supp} A$.

Next, note that $d(A, B) > 0$, since otherwise we would have $AB = BA$; this would imply $B \le AB = A$, contradicting the assumption that $B \notin \Sigma'$. Now apply the proof of Lemma 3.142 to the convex subcomplex $\operatorname{supp} A$. We obtain a codimension-1 face A' of A such that $A'B$ is a $(\operatorname{supp} A)$-chamber with $d(A'B, B) < d(A, B)$. Moreover, we get a wall H of Σ such that $A' \in H$ and $H \in \mathcal{S}(A, B)$ (see Proposition 3.78). This is illustrated in Figure 3.13, which should now be viewed as a picture of $\operatorname{supp} A$. Since $\sigma_H(A'B) = \sigma_H(B)$, we conclude that $H \in \mathcal{S}(A'B, B)$. To complete the proof, we show that Σ' is contained in the root α bounded by H that contains A.

It suffices to show that α contains every Σ'-chamber D. Consider the product $A'D \in \Sigma'$. Then $A'D$ is a chamber of Σ' (and $\operatorname{supp} A$), since $\dim A'D \ge \dim D = \dim \Sigma' = \dim \operatorname{supp} A$. Now A and $A'B$ are the only two $(\operatorname{supp} A)$-chambers having A' as a face by Proposition 3.136, and $A'B \notin \Sigma'$ by the choice of A and the fact that $d(A'B, B) < d(A, B)$; so we must have $A'D = A$. Thus $\sigma_H(D) = \sigma_H(A)$, and hence $D \in \alpha$. □

Exercises

We continue to assume that Σ is a Coxeter complex.

3.145. Find the convex hull of C and D in Figure 3.2.

3.146. This is a variant of Example 3.133(d). Suppose that Σ is spherical and set $d := \operatorname{diam} \Sigma$. If C and D are chambers such that $d(C, D) = d - 1$, show that $\Gamma(C, D)$ is a root.

3.147. If Σ is spherical, show that every root is a maximal (proper) convex subcomplex.

3.148. Show that $\Gamma(A, B)$ is the intersection of the subcomplexes $\Gamma(C, D)$ with $C \in \mathcal{C}_{A,B}$ and $D = BC$.

3.149. Show that AB is a maximal simplex in $\Gamma(A, B)$. Show further that it is the unique maximal simplex of $\Gamma(A, B)$ having A as a face. So we have yet another characterization of AB.

3.150. Show that the convex hull of a finite set of simplices is finite.

3.151. What is the convex hull of a single simplex? Show directly, without using Proposition 3.137, that it is an intersection of roots.

3.152. Given chambers $C, D \in \Sigma$, let $I(C, D)$ be the intersection of all roots α with $C \in \alpha$ and $D \notin \alpha$. Prove that $I(C, D)$ is a convex chamber subcomplex of Σ whose chambers are those $E \in \mathcal{C}(\Sigma)$ such that

$$d(D, E) = d(D, C) + d(C, E) \, .$$

Intuitively, these are the chambers that one can reach by moving on a geodesic from D to C and continuing past C.

3.153. Let Σ' be a convex subcomplex of Σ and H a wall not containing Σ'.

(a) Show that $\dim(H \cap \Sigma') < \dim \Sigma'$.
(b) Now specialize to the case that $\Sigma' = \operatorname{supp} A$ for some simplex A. If H contains a codimension-1 face P of A, show that $H \cap \operatorname{supp} A = \operatorname{supp} P$.

3.154. Let C be a chamber and A a face of C. Show that $\operatorname{supp} A$ is the intersection of the walls of C containing A.

3.155. Let (W, S) be a Coxeter system and let $\Sigma = \Sigma(W, S)$ with its natural W-action. Given $A \in \Sigma$, let W_A be the stabilizer of A and let N_A be the normalizer of W_A in W. Show that N_A/W_A acts simply transitively on the set of simplices in $\operatorname{supp} A$ having the same type as A.

3.156. (a) Suppose Σ is infinite and irreducible (i.e., its Coxeter diagram is connected). If x is a vertex of Σ, show that x does not have an opposite vertex as defined in Remark 3.144(a). Thus the support of x consists only of x and the empty simplex. [Proposition 2.91 proves this result from the point of view of the Tits cone. The present exercise is asking for a combinatorial proof.]

(b) In the general case, decompose Σ as a join $\Sigma_1 * \cdots * \Sigma_n$ according to the connected components of the Coxeter diagram (Exercise 3.30). Show that a vertex x has an opposite vertex y if and only if the factor Σ_i containing x is spherical.

3.157. Let (W, S) be a Coxeter system, and let $J \subseteq S$ satisfy $|J| = |S| - 1$. Let N_J be the normalizer of W_J in W.

(a) Show that W_J has index 1 or 2 in N_J.

(b) Show that $W_J = N_J$ if (W, S) is irreducible and W is infinite.

3.158. Assume that $\Sigma = \Sigma(W, S)$, and identify Σ with the set of cells in the Tits cone. Let Σ_f be the set of simplices in Σ with finite stabilizer. (Equivalently, Σ_f is the poset of finite standard cosets ordered by reverse inclusion, or the set of simplices with finite link, or the set of simplices whose link is a sphere.) It follows trivially from Lemma 3.123 that Σ_f is a subsemigroup of Σ. Does this yield an alternative proof that X_f is convex in Lemma 2.86?

3.6.7 The Support of a Vertex

In this subsection we give a combinatorial proof of Lemma 3.143, which was needed in the combinatorial proofs of Propositions 3.136 and 3.137. This turns out to be less straightforward than one would expect, given how obvious the result is from the point of view of the Tits cone.

Let Σ be a Coxeter complex. We start by considering 1-dimensional convex subcomplexes of Σ and using products to construct "geodesic paths." In the following lemma, $d(-, -)$ denotes gallery distance in Σ.

Lemma 3.159. *Let Σ' be a 1-dimensional convex subcomplex of Σ, and let x and y be two distinct vertices of Σ'. For any edge F of Σ' such that $yx \leq F$, there is a sequence $x = x_0, \ldots, x_n = y$ of vertices of Σ' with the following properties:*

(1) *$E_i := \{x_{i-1}, x_i\}$ is an edge of Σ' for all $1 \leq i \leq n$, and $E_n = F$.*

(2) *$yx_i \leq F$ for all i.*

(3) *$E_i = x_{i-1}F$ and $yE_i = F$ for all $1 \leq i \leq n$.*

(4) *$d(x_i, y) = d(x_i, F) = d(E_{i+1}, F) = d(E_{i+1}, y)$ for all $0 \leq i < n$.*

(5) *$d(x_i, y) < d(x_{i-1}, y)$ for all $0 < i < n$.*

Proof. Note that the construction of the x_i is forced on us by properties (1)–(5): We must have $E_1 = xF$ by (3), and then x_1 must be the vertex of E_1 different from $x = x_0$. If $x_1 = y$, we are done; otherwise, we must have $E_2 = x_1 F$, and so on. The essential content of the lemma, then, is that this process terminates and yields a sequence of vertices in Σ' satisfying all the stated properties. We now give the formal proof, arguing by induction on $d(x, y)$.

If $d(x, y) = 0$, then x and y are joinable in Σ. The edge joining them is the product $xy = yx$ (Exercise 3.116), and this is in Σ' by convexity. So $F = \{x, y\}$ and we can take $n = 1$, $x_0 = x$, and $x_1 = y$. Properties (1)–(5) hold trivially.

Suppose $d(x, y) > 0$. Set $E_1 := xF$ and note that this is an edge of Σ'. Denote by x_1 the vertex of E_1 different from $x = x_0$. We have $yE_1 = yxF = F$, since $yx \leq F$, and then $yx_1 \leq yE_1 = F$. Thus (2) and (3) hold for $i = 1$. We can now prove (4) for $i = 0$ by three applications of Exercise 3.117. Indeed, the three equations

$$d(x, y) = d(x, F) = d(E_1, F) = d(E_1, y)$$

follow, respectively, from the relations $yx \leq F$, $xF = E_1$, and $yE_1 = F$. So we will be set up to apply the induction hypothesis as soon as we verify that $d(x_1, y) < d(x, y)$. (Note that $x_1 \neq y$, since $d(x, y) > 0$.)

Let $C_0, C_1, \ldots, C_m$ be a minimal gallery from x to y in Σ. We can take $C_0 \geq E_1$, since $E_1 = xF \geq xy$. Then minimality of the gallery implies that x cannot be a vertex of C_1, so x_1 must be a vertex of C_1. Thus $d(x_1, y) < m = d(x, y)$, and we can apply the induction hypothesis to x_1 and y to complete the proof.   □

We can now prove Lemma 3.143. Recall the statement of the lemma: *If x is a vertex of the Coxeter complex Σ, then* $\operatorname{supp} x$ *is 0-dimensional and either has x as its only vertex or else has exactly two vertices.*

Proof. We already know the first assertion (Corollary 3.141), so we need only show that $\operatorname{supp} x$ cannot have three vertices. We argue by induction on $\dim \Sigma$, which may be assumed > 0. Suppose y is a vertex different from x in $\operatorname{supp} x$, so that $xy = x$ and $yx = y$. Let Σ' be the support of an edge containing x. Then Σ' is a 1-dimensional convex subcomplex of Σ containing x and y. Applying the induction hypothesis to links as in the proof of Proposition 3.136, we conclude that Σ' is a 1-dimensional chamber complex in which every panel is a face of at most two chambers. In other words, Σ' is either a circle or a (possibly infinite) line segment.

Suppose now that there is a third vertex $z \in \operatorname{supp} x$. If Σ' is a line segment, we may assume that z lies between x and y. Since there is no backtracking in the path $x = x_0, x_1, \ldots, x_n = y$ constructed in Lemma 3.159, this path must be the unique geodesic between x and y in Σ'. We therefore have $z = x_i$ for some $0 < i < n$. Now $d(E_{i+1}, y) = d(x_i, y) < d(x_{i-1}, y) = d(E_i, y)$. But E_{i+1} and E_i are edges containing $x_i = z = zy$, where the second equation comes from the assumption that $z \in \operatorname{supp} x = \operatorname{supp} y$. So $d(E_{i+1}, y) = d(z, y) = d(E_i, y)$. This contradiction shows that z cannot exist if Σ' is a line segment. (Side remark: It also follows from this argument that x and y are endpoints of Σ'.)

Now suppose that Σ' is a circle. Then y is contained in precisely two edges F and F' of Σ'. By Lemma 3.159, these two edges (which contain $yx = y$) give rise to two paths $P: E_0, \ldots, E_{n-1} = F$ and $P': E_0', \ldots, E_{n'-1}' = F'$, both starting at x and ending at y. Since there is no backtracking in these paths, their union must be the full circle Σ'. By the same argument as in the previous paragraph, no interior vertex of P or P' can be in $\operatorname{supp} y$, again contradicting the existence of z.   □

Remarks 3.160. (a) This combinatorial proof of Lemma 3.143 is an instructive illustration of the use of products, but it is considerably trickier than the (almost trivial) proof using the Tits cone. This illustrates the power and usefulness of the Tits cone.

(b) Although we have succeeded, with some effort, in giving a combinatorial proof of Lemma 3.143, we do not know how to prove combinatorially the stronger result stated in Remark 3.144(a).

Exercise 3.161. Let x, y, and Σ' be as in the proof. Thus either Σ' is a circle or else Σ' is a line segment with x and y as endpoints. The first case occurs whenever x and y are opposite vertices of a spherical Coxeter complex. Give an example to show that the second case can occur also.

3.6.8 Links Revisited; Nested Roots

We continue to denote by Σ an arbitrary Coxeter complex. In this final subsection we use the theory of products and convexity as an aid in proving that, as we stated before Lemma 3.53, Σ contains nested roots if it is infinite. The proof will use some facts about links that will also be needed in Chapter 7, so we begin with those.

Fix a simplex A of Σ, and set $\Sigma' := \mathrm{lk}_\Sigma A$. Recall from Proposition 3.79 that we have a bijection $H \mapsto H' := H \cap \Sigma'$ from the set of walls of Σ containing A to the set of walls of Σ'. Similarly, there is a bijection $\alpha \mapsto \alpha' := \alpha \cap \Sigma'$ from the set of roots α of Σ with $A \in \partial\alpha$ to the set of roots of Σ'. Using products, one can describe the inverses of these bijections. The full result is stated in Exercise 3.168 below. We will need only part of this result, which we give in the following lemma:

Lemma 3.162. Let A and Σ' be as above, let α be a root of Σ with $A \in \partial\alpha$, and let $\alpha' := \alpha \cap \Sigma'$. Then

$$\mathcal{C}(\alpha) = \{C \in \mathcal{C}(\Sigma) \mid (AC \smallsetminus A) \in \alpha'\} \ .$$

Proof. Let C be a chamber of Σ. Using the convexity of roots and Remark 3.80, we have

$$C \in \alpha \implies AC \in \alpha \cap \Sigma_{\geq A} \implies (AC \smallsetminus A) \in \alpha' \ .$$

Similarly,

$$C \in -\alpha \implies AC \in -\alpha \cap \Sigma_{\geq A} \implies (AC \smallsetminus A) \in (-\alpha)' = -\alpha' \ .$$

The lemma follows at once.   □

This has the following immediate consequence. Note that the "if" part would not be obvious without the lemma.

Corollary 3.163. Let α and β be roots of Σ with $A \in \partial\alpha \cap \partial\beta$, and let $\alpha' := \alpha \cap \Sigma'$ and $\beta' := \beta \cap \Sigma'$. Then $\alpha \subseteq \beta$ if and only if $\alpha' \subseteq \beta'$.   □

We turn now to nested roots. We say that a pair of distinct roots $\{\alpha, \beta\}$ is *nested* if $\alpha \subseteq \beta$ or $\beta \subseteq \alpha$.

Lemma 3.164. *Let α and β be roots of Σ with $\alpha \neq \pm\beta$. If the pairs $\{\alpha, \beta\}$ and $\{-\alpha, \beta\}$ are both non-nested, then $\partial\alpha \cap \partial\beta$ is a chamber complex of codimension 2 in Σ, and $\mathrm{lk}_\Sigma A$ is spherical for every maximal simplex $A \in \partial\alpha \cap \partial\beta$.*

Proof. The hypothesis says that the intersection $(\pm\alpha) \cap (\pm\beta)$ contains a chamber for each of the four possible choices of sign. Considering a minimal gallery from a chamber in $\alpha \cap \beta$ to a chamber in $\alpha \cap -\beta$, we obtain a panel $P \in \alpha \cap \partial\beta$. Similarly, there is a panel $Q \in (-\alpha) \cap \partial\beta$. Recall now that $\partial\beta$ is a chamber complex (see Proposition 3.136) and that P and Q are maximal simplices in it. We can therefore choose P and Q to be $\partial\beta$-adjacent [consider a $\partial\beta$-gallery joining them]. Then $P \cap Q$ is in $\partial\alpha \cap \partial\beta$ and has codimension 2 in Σ, so $\mathrm{codim}(\partial\alpha \cap \partial\beta) \leq 2$. Equality must hold, since $\partial\alpha$ and $\partial\beta$ cannot have a panel of Σ in common. And $\partial\alpha \cap \partial\beta$ is a chamber complex by Proposition 3.136 again. Finally, let A be a maximal simplex of $\partial\alpha \cap \partial\beta$. Then $\Sigma' := \mathrm{lk}_\Sigma A$ is a rank-2 Coxeter complex, so either it is spherical or else it is a triangulated line. Let $\alpha' := \alpha \cap \Sigma'$ and $\beta' := \beta \cap \Sigma'$. Then the pairs $\{\alpha', \beta'\}$ and $\{-\alpha', \beta'\}$ are both non-nested by Corollary 3.163, so Σ' cannot be a line and is therefore spherical.   □

For any root α, we denote by s_α the reflection with respect to $\partial\alpha$.

Proposition 3.165. *Let α and β be roots of Σ with $\alpha \neq \pm\beta$. Then the following conditions are equivalent:*

(i) *Either $\{\alpha, \beta\}$ is nested or $\{-\alpha, \beta\}$ is nested.*
(ii) *The product $s_\alpha s_\beta$ has infinite order.*
(iii) *For every simplex $A \in \partial\alpha \cap \partial\beta$, the link $\mathrm{lk}_\Sigma A$ is not spherical.*

Proof. (i) $\implies$ (ii): Suppose one of the pairs is nested, say $\alpha \subsetneqq \beta$. Setting $w := s_\alpha s_\beta$, we then have

$$w\alpha \subsetneqq w\beta = s_\alpha(-\beta) \subsetneqq s_\alpha(-\alpha) = \alpha \,.$$

There is therefore an infinite descending chain

$$\alpha \supsetneqq w\alpha \supsetneqq w^2\alpha \supsetneqq \cdots ,$$

implying (ii).

(ii) $\implies$ (iii): We may assume that $\Sigma = \Sigma(W, S)$. The element $s_\alpha s_\beta \in W$ fixes every simplex $A \in \partial\alpha \cap \partial\beta$, so (ii) implies that the stabilizer W_A of A in Σ is infinite. Since $\mathrm{lk}_\Sigma A$ is the Coxeter complex associated to the Coxeter group W_A, it follows that the link is not spherical.

(iii) $\implies$ (i): This follows from Lemma 3.164.   □

Corollary 3.166. *Σ has nested roots if and only if it is not spherical.*

Proof. We already know that spherical Coxeter complexes do not have nested roots (Lemma 3.53). If Σ is not spherical, on the other hand, then it has a pair of reflections whose product has infinite order by Proposition 2.74, so it has a pair of nested roots by Proposition 3.165.   □

As a second corollary, we can give a new proof of a special case of Proposition 2.87, and we can even give a sharper result in this case:

Corollary 3.167. *If (W, S) is a Coxeter system and $W' \leq W$ is a finite subgroup generated by two reflections, then W' is contained in a finite parabolic subgroup of W of rank 2.*

Proof. Let $\Sigma := \Sigma(W, S)$. Combining the proposition with Lemma 3.164, we see that W' stabilizes a codimension-2 simplex with spherical link. The stabilizer of this link is a finite parabolic subgroup of W of rank 2. $\qquad\square$

Exercise 3.168. Let A and Σ' be as in the beginning of this subsection.

(a) Let H be a wall of Σ containing A, and let $H' := H \cap \Sigma'$. Then

$$H = \{B \in \Sigma \mid (AB \smallsetminus A) \in H'\} . \tag{3.12}$$

(b) Let α be a root of Σ with $A \in \partial\alpha$, and let $\alpha' := \alpha \cap \Sigma'$. Then

$$\alpha = \{B \in \Sigma \mid (AB \smallsetminus A) \in \alpha'\} . \tag{3.13}$$

Buildings as Chamber Complexes

As we stated in the introduction, there is more than one approach to buildings. The point of view in this chapter is that buildings are simplicial complexes satisfying certain axioms. These are quite easy to state, but not so easy to motivate. We will not attempt to explain how Tits came up with them now, but we will make some historical remarks in Chapter 6 that should make the definition seem less mysterious.

The terminology used in this subject is attributed by Tits to Bourbaki (see [248]). In order to understand where it comes from, one should interpret the word "chamber" that we have been using as meaning "room." Thus Coxeter complexes are divided up into rooms by walls, and they are therefore called "apartments." Buildings, then, are complexes that are built by putting apartments together. We now state the axioms, which specify the rules for putting the apartments together.

4.1 Definition and First Properties

Definition 4.1. A *building* is a simplicial complex Δ that can be expressed as the union of subcomplexes Σ (called *apartments*) satisfying the following axioms:

(B0) *Each apartment Σ is a Coxeter complex.*

(B1) *For any two simplices $A, B \in \Delta$, there is an apartment Σ containing both of them.*

(B2) *If Σ and Σ' are two apartments containing A and B, then there is an isomorphism $\Sigma \to \Sigma'$ fixing A and B pointwise.*

(Recall that a map fixes a simplex A *pointwise* if it fixes every vertex of A.)

Note that we can take both A and B to be the empty simplex in (B2); hence any two apartments are isomorphic. This implies, in particular, that

Δ is finite-dimensional, its dimension being the common dimension of its apartments. Note also that Δ is a chamber complex. For if C and C' are maximal simplices, then they are also maximal simplices of some apartment Σ by (B1), so they have the same dimension and are connected by a gallery.

Any collection $\mathcal{A}$ of subcomplexes Σ satisfying the axioms will be called a *system of apartments* for Δ. Thus a building is a simplicial complex that admits a system of apartments. Note that we do *not* require that a building be equipped, as part of its structure, with a specific system of apartments. The reason for this is that it turns out that a building always admits a canonical system of apartments. And in the important special case that the apartments are *finite* Coxeter complexes, it is even true that there is a *unique* system of apartments. We will prove both of these assertions later in the chapter (Sections 4.5 and 4.7, respectively).

Remark 4.2. The complexes we have called buildings are sometimes called *weak buildings* in the literature, the term "building" being reserved for the case in which Δ is *thick*. This means, by definition, that every panel is a face of at least three chambers. With our definition, by contrast, a building can even be thin. Indeed, a Coxeter complex is a thin building with a single apartment. If we confine ourselves to the thick case, then axiom (B0) can be considerably weakened. Namely, we need only assume that the apartments Σ are thin chamber complexes, and it then follows from (B1) and (B2) that they are in fact Coxeter complexes. The proof of this will be given in Section 4.13.

Remark 4.3. Axiom (B2) can be replaced by the following weaker axiom, which is simply the special case of (B2) in which one of the two simplices is a chamber. Some care is needed in the precise formulation, since in the absence of (B2), we do not yet know that all apartments have the same dimension; thus we need to avoid ambiguity in our use of the word "chamber."

(B2') *Let Σ and Σ' be apartments containing simplices A, C, where C is a chamber of Σ. Then there is an isomorphism $\Sigma \xrightarrow{\sim} \Sigma'$ fixing A and C pointwise.*

To see that this implies (B2) (in the presence of (B0) and (B1)), consider an arbitrary pair of simplices A, B contained in two apartments Σ and Σ'. Choose a chamber $C \geq A$ in Σ and a chamber $D \geq B$ in Σ', and choose an apartment Σ'' containing C and D. Assuming (B2'), we have isomorphisms

$$\Sigma \xrightarrow{\sim} \Sigma'' \xrightarrow{\sim} \Sigma',$$

where the first isomorphism fixes C and B pointwise and the second isomorphism fixes A and D pointwise. The composite is then an isomorphism $\Sigma \xrightarrow{\sim} \Sigma'$ fixing A and B pointwise, so (B2) holds.

Remark 4.4. Axiom (B2'), in turn, is equivalent to the following axiom, which appears at first glance to be stronger:

(B2″) *Let Σ and Σ' be two apartments containing a simplex C that is a chamber of Σ. Then there is an isomorphism $\Sigma \xrightarrow{\sim} \Sigma'$ fixing every simplex in $\Sigma \cap \Sigma'$.*

For suppose that (B2′) holds, and let Σ, Σ', and C be as in (B2″). Then we have, for each $A \in \Sigma \cap \Sigma'$, an isomorphism $\phi_A \colon \Sigma \xrightarrow{\sim} \Sigma'$ fixing A and C pointwise. But our standard uniqueness argument (Section 3.4.1) shows that there is at most one isomorphism from Σ to Σ' fixing C pointwise. So all the ϕ_A are equal to a single isomorphism ϕ, which therefore fixes the entire intersection $\Sigma \cap \Sigma'$.

Remark 4.5. One can strengthen (B2″) still further, by dropping the assumption that the two apartments have a common chamber. In other words, the isomorphisms in (B2) can always be taken to fix every simplex in the intersection. We will be able to prove this later in the chapter; see Proposition 4.101 and Exercise 4.108.

Assume, for the remainder of this section, that Δ is a building and that $\mathcal{A}$ is a fixed system of apartments.

Proposition 4.6. *Δ is colorable. Moreover, the isomorphisms $\Sigma \xrightarrow{\sim} \Sigma'$ in axiom (B2) can be taken to be type-preserving.*

Proof. Fix an arbitrary chamber C, and assign types to its vertices arbitrarily. If Σ is any apartment containing C, then [since Coxeter complexes are colorable] the assignment of types on C extends uniquely to a type function τ_Σ of Σ. For any two such apartments Σ, Σ', the type functions τ_Σ and $\tau_{\Sigma'}$ agree on $\Sigma \cap \Sigma'$; this follows from the fact that $\tau_{\Sigma'}$ can be constructed as $\tau_\Sigma \circ \phi$, where $\phi \colon \Sigma' \xrightarrow{\sim} \Sigma$ is the isomorphism fixing $\Sigma \cap \Sigma'$ as in (B2″). The various τ_Σ therefore fit together to give a type function τ defined on the union of the apartments containing C. But this union is all of Δ by (B1), so the first assertion of the proposition is proved.

To prove the second assertion, it suffices to consider the isomorphisms that occur in axiom (B2′). But such an isomorphism is automatically type-preserving, since it fixes a chamber pointwise. □

Choose a fixed type function τ on Δ with values in a set S. In view of the essential uniqueness of type functions, nothing we do will depend in any serious way on this choice. For any apartment Σ, the function τ yields a *Coxeter matrix* $M := \big(m(s,t)\big)_{s,t \in S}$, defined by

$$m(s,t) := \mathrm{diam}(\mathrm{lk}_\Sigma A) ,$$

where A is any simplex in Σ of cotype $\{s,t\}$ (see Section 3.2). Since any two apartments are isomorphic in a type-preserving way, M does not depend on Σ:

Proposition 4.7. *All apartments have the same Coxeter matrix M.* □

We will therefore call M the *Coxeter matrix* of Δ. Similarly, we can speak of the *Coxeter diagram* of Δ; it is a graph with one vertex for each $s \in S$. Strictly speaking, we should be talking about the Coxeter matrix and diagram of the pair $(\Delta, \mathcal{A})$; but we will show in Section 4.4 that the matrix and diagram are really intrinsically associated to Δ and do not depend on the system of apartments $\mathcal{A}$.

The importance of the Coxeter matrix, of course, is that it completely determines the isomorphism type of the apartments. Let's spell this out in detail: Let W_M be the Coxeter group associated to M, with generating set S and relations $(st)^{m(s,t)} = 1$. In the language of Section 3.5, W_M is the Weyl group of every apartment. Let Σ_M be the Coxeter complex $\Sigma(W_M, S)$. It has a canonical type function with values in S. We can now state the following consequence of Propositions 4.7 and 3.85:

Corollary 4.8. *For any apartment Σ, there is a type-preserving isomorphism $\Sigma \cong \Sigma_M$. Thus Σ, endowed with the type function $\tau|_\Sigma$, is a Coxeter complex of type M, or of type (W_M, S), in the sense of Definition 3.86.* □

Finally, we record one more simple consequence of the axioms. Recall that the study of local properties of Coxeter complexes consisted of a single result, which said that the link of a simplex in a Coxeter complex is again a Coxeter complex (Proposition 3.16). The situation for buildings is similar:

Proposition 4.9. *If Δ is a building, then so is* lk A *for any $A \in \Delta$. In particular, the link is a chamber complex.*

Proof. Choose a fixed system of apartments $\mathcal{A}$ for Δ. Given $A \in \Delta$, let $\mathcal{A}'$ be the family of subcomplexes of $\mathrm{lk}_\Delta\, A$ of the form $\mathrm{lk}_\Sigma\, A$, where Σ is an element of $\mathcal{A}$ containing A. Any such subcomplex is a Coxeter complex by the result cited above. So it remains to verify (B1) and (B2). Given $B, B' \in \mathrm{lk}_\Delta\, A$, we can join them with A to obtain simplices $A \cup B$ and $A \cup B'$ in Δ. Since Δ satisfies (B1), there is an apartment Σ containing both of these simplices. Hence $\mathrm{lk}_\Sigma\, A$ is an element of $\mathcal{A}'$ containing B and B'. This proves that $\mathcal{A}'$ satisfies (B1), and the proof of (B2) is similar. □

Remark 4.10. The proof, together with the discussion in Section 3.2, tells us how to get the Coxeter diagram of lk A from that of Δ: If A has cotype $J \subseteq S$ and D is the Coxeter diagram of Δ, then the Coxeter diagram of lk A is the induced diagram D_J with vertex set J.

As in the case of Coxeter complexes, we can immediately apply the results of Section A.1.4 involving residues:

Corollary 4.11. *Δ is completely determined by its underlying chamber system. More precisely, the simplices of Δ are in 1–1 correspondence with the residues in $\mathcal{C} := \mathcal{C}(\Delta)$, ordered by reverse inclusion. Here a simplex A corresponds to the residue $\mathcal{C}_{\geq A}$, consisting of the chambers having A as a face.* □

Exercises

4.12. Show that every thin building is a Coxeter complex.

4.13. Let Δ be a building. For any simplex $A \in \Delta$, show that the residue $\mathcal{C}(\Delta)_{\geq A}$ is a convex subset of $\mathcal{C}(\Delta)$ in the sense of Definition 3.92.

4.14. (a) Given a simplex A in a building Δ, one could try to define the *support* of A by choosing an apartment Σ containing A and declaring $\operatorname{supp} A$ to be $\operatorname{supp}_{\Sigma} A$, where the latter is the support of A in Σ (Definition 3.98). Show that this does not work. In other words, if Σ and Σ' are two apartments containing A, $\operatorname{supp}_{\Sigma} A$ need not equal $\operatorname{supp}_{\Sigma'} A$.

(b) Show, on the other hand, that the relation "$\operatorname{supp} A = \operatorname{supp} B$" is a well-defined relation on the simplices of Δ, i.e., if A and B have the same support in one apartment containing them, then they have the same support in every apartment containing them.

4.2 Examples

Almost all of the examples in this section will be defined as flag complexes, so readers may need to consult Section A.1.2 for the terminology before proceeding.

Let P be a set with an "incidence" relation as in the section just cited. Assume, in addition, that P is partitioned into nonempty subsets $P_0, P_1, \ldots, P_{n-1}$. Elements of P_i are said to have *type i*. Or, to use more intuitive language, elements of $P_0, P_1, P_2 \ldots$ might be called points, lines, planes, etc. If the incidence relation has the property that two elements of the same type are never incident unless they are equal, then we will call P (together with the partition and incidence relation) an *incidence geometry of rank n*. (In some of the literature one requires, in addition, that every maximal flag include an element from each of the sets $P_0, P_1, \ldots, P_{n-1}$.)

If $n = 1$, then we just have a set of points, with no further structure; one can think of it as a "line." If $n = 2$, then P is a "plane" consisting of points and lines, with some points declared to be incident to some lines. If $n = 3$, there are points, lines, and planes. And so on.

In practice, of course, one is interested in incidence geometries that are subject to certain axioms, such as the axioms for projective geometry or some other kind of geometry. We will see below that different types of geometry correspond to different types of buildings (where the *type* of a building is determined by its Coxeter matrix).

We proceed now to the examples, starting with a case that is trivial but nonetheless instructive.

Example 4.15. Suppose Δ is a building of rank 1 (dimension 0). Then every apartment must be a 0-sphere S^0, since this is the only Coxeter complex of rank 1. In particular, Δ must have at least two vertices. Conversely, a rank-1 simplicial complex with at least two vertices is a building (with every 2-vertex subcomplex as an apartment). Thus the rank-1 buildings are precisely the flag complexes of the rank-1 incidence geometries with at least 2 points. [It is, of course, reasonable to demand that a 1-dimensional space, or "line," have at least 2 points. In fact, one often even demands that there be at least 3 points, which is equivalent to requiring the flag complex Δ to be thick.]

Example 4.16. Suppose Δ is a building of rank 2 (dimension 1). Then an apartment Σ must be a $2m$-gon for some m ($2 \leq m \leq \infty$). We will draw the Coxeter diagram as

$$\circ \underset{}{\overset{m}{\text{———}}} \circ$$

which should be interpreted as

$$\circ \qquad \circ$$

if $m = 2$ and as

$$\circ\text{———}\circ$$

if $m = 3$.

Let's begin with the case $m = 2$. Then every apartment is a quadrilateral:

(As usual, the two colors, black and white, represent the two types of vertices.) It follows easily from the building axioms that every vertex of type $\bullet$ is connected by an edge to every vertex of type $\circ$, i.e., Δ is a *complete bipartite graph*. In the language of incidence geometry, Δ is the flag complex of a "plane" in which every point is incident to every line. Conversely, the flag complex of such a plane is always a rank-2 building (with $m = 2$), provided that there are at least two points and at least two lines.

Note that we can also describe Δ as the join of two rank-1 buildings. This suggests a general fact:

Exercise 4.17. If Δ is a building whose Coxeter diagram is disconnected, show that Δ is canonically the join of lower-dimensional buildings, one for each connected component of the diagram.

Returning now to Example 4.16, suppose next that $m = 3$. Then every apartment is a hexagon, which we may draw as the barycentric subdivision of a triangle:

This picture suggests a configuration of three lines in a plane (one line for each ○), whose pairwise intersections yield three points of the plane (one for each ●). The existence of many such apartments, as guaranteed by (B1), makes it plausible that Δ is the flag complex of a projective plane. Recall the definition of the latter:

Definition 4.18. A *projective plane* is a rank-2 incidence geometry satisfying the following three axioms:

(1) Any two points are incident to a unique line.
(2) Any two lines are incident to a unique point.
(3) There exist three noncollinear points.

With this definition, it is indeed the case that our building Δ is the flag complex of a projective plane. You may find it instructive to try to prove this as an exercise. [The exercise is not entirely routine; if you get stuck, you will see it again in Exercise 4.46.] Conversely, the flag complex of a projective plane is a building, with one apartment for every triangle in the projective plane. This converse *is* a routine exercise.

The most familiar example of a projective plane is the projective plane over a field k. By definition, the set P_0 of "points" is the set of 1-dimensional subspaces of the 3-dimensional vector space k^3; the set P_1 of "lines" is the set of 2-dimensional subspaces of k^3; and "incidence" is given by inclusion, i.e., a point $x \in P_0$ is incident to a line $L \in P_1$ if $x < L$ as subspaces of k^3.

It is now easy to construct concrete examples of buildings. Let P be the projective plane over $\mathbb{F}_2$, for instance, where $\mathbb{F}_2$ is the field with two elements. This plane is also called the *Fano* projective plane. It has 7 points (each on exactly 3 lines) and 7 lines (each containing exactly 3 points). The resulting flag complex Δ is a thick building with 14 vertices and 21 edges. We will see in Exercise 4.23 below that the points of P can be put in 1–1 correspondence with the 7th roots of unity ζ^j ($\zeta = e^{2\pi i/7}$, $j = 0, \ldots, 6$) in such a way that the lines of P are the triples $\{\zeta^j, \zeta^{j+1}, \zeta^{j+3}\}$, $j = 0, \ldots, 6$. This leads to the picture of Δ shown in Figure 4.1. The interested reader can locate some of the apartments (there are 28 of them) and verify some cases of the building axioms.

Remark 4.19. This picture is misleading in one respect; namely, it fails to reveal how much symmetry Δ has. One can see from the picture that Δ admits an action of the dihedral group D_{14}, but in fact $\operatorname{Aut} \Delta$ is of order 336. The subgroup $\operatorname{Aut}_0 \Delta$ of type-preserving automorphisms is $\operatorname{GL}_3(\mathbb{F}_2)$, which is the simple group of order 168.

Continuing with Example 4.16, one could analyze in a similar way the buildings corresponding to $m = 4, 5, 6, \ldots$. Each value of m corresponds to a particular type of plane geometry.

Definition 4.20. An incidence plane P is called a *generalized m-gon* if its flag complex is a building of type $I_2(m)$.

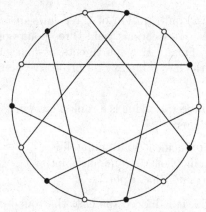

Fig. 4.1. The incidence graph of the Fano plane.

The terminology comes from the fact that P has formal properties analogous to those of the geometry consisting of the vertices and edges of an m-gon. See Van Maldeghem [264] or Tits–Weiss [262] for more information. See also Proposition 4.44 below, where we spell out precisely what it means for the flag complex to be a building of type $I_2(m)$.

Generalized 3-gons are also called *generalized triangles*. Thus a generalized triangle is the same thing as a projective plane. Generalized 4-gons are also called *generalized quadrangles* or *polar planes*. A generalized quadrangle has no triangles, but there do exist lots of quadrilaterals. Every quadrilateral yields an apartment in the flag complex, this apartment being an octagon (or barycentrically subdivided quadrilateral). See Exercise 4.24 below for a concrete example of a generalized quadrangle.

Finally, in case $m = \infty$, buildings of type $I_2(\infty)$ are simply trees with no endpoints (where an endpoint of a tree is a vertex that is on only one edge). To see that such a tree is a building, simply take the apartments to be all possible subcomplexes that are lines (i.e., ∞-gons); the verification of the building axioms is a routine matter. The converse, that every building of this type is a tree, is more challenging. We will treat it in Proposition 4.44 below, along with our characterization of generalized m-gons for $m < \infty$.

The two remaining examples are intended to provide a brief glimpse of some higher-dimensional buildings from the point of view of incidence geometry. Details (including definitions of some of the terms), will be omitted; these can be found in Tits [247]. (See also Scharlau [207].) We will, however, give many details for the most important case in the next section. And we will show in Chapter 6 how to construct further examples of buildings via group theory rather than incidence geometry.

Examples 4.21. (a) If P is an n-dimensional projective space, then its flag complex is a rank-n building of type A_n, i.e., having Coxeter diagram

$$\circ\!\!-\!\!-\!\!\circ\!\!-\!\!-\!\!\circ\!\!-\cdots-\!\!\circ\!\!-\!\!-\!\!\circ \quad (n \text{ vertices}).$$

Every apartment is isomorphic to the barycentric subdivision of the boundary of an n-simplex, and there is one such apartment for every frame in the projective space (where a frame is a set of $n + 1$ points in general position). Conversely, every building of type A_n is the flag complex of a projective space.

When $n = 2$, this example reduces to the case $m = 3$ of Example 4.16.

(b) If P is an n-dimensional polar space, then its flag complex is a rank-n building of type C_n, i.e., having Coxeter diagram

$$\circ\!\!-\!\!-\!\!\circ\!\!-\!\!-\!\!\circ\!\!-\cdots-\!\!\circ\!\!\overset{4}{-\!\!-\!\!}\circ \quad (n \text{ vertices}).$$

Every apartment is isomorphic to the barycentric subdivision of the boundary of an n-cube (or n-dimensional hyperoctahedron), and there is one such apartment for every "polar frame" in the given polar space. Conversely, every building of type C_n is the flag complex of a polar space.

When $n = 2$, this example reduces to the case $m = 4$ of Example 4.16.

Remark 4.22. It is no accident that all of the examples in this section (except trees) have been defined as flag complexes. Indeed, we will see in Exercise 4.50 below that *every* building is a flag complex.

Exercises

4.23. (a) Let V be a 3-dimensional vector space over $\mathbb{F}_2$, and let $P = P(V)$ be the projective plane in which the points are the nonzero vectors in V and the lines are the triples $\{u, v, w\}$ with $u + v + w = 0$. Show that P is isomorphic to the projective plane over $\mathbb{F}_2$.

(b) Let $\Delta(V)$ be the flag complex of $P(V)$, with its canonical type function. If V^* is the dual of V, show that the correspondence between subspaces of V and subspaces of V^* induces a type-reversing isomorphism $\Delta(V) \cong \Delta(V^*)$. Consequently, any isomorphism $V \xrightarrow{\sim} V^*$ induces a type-reversing automorphism of $\Delta(V)$. If the isomorphism $V \xrightarrow{\sim} V^*$ comes from a nondegenerate symmetric bilinear form on V, show that the resulting automorphism of $\Delta(V)$ is an involution.

(c) Let V be the field $\mathbb{F}_8$, viewed as a vector space over $\mathbb{F}_2$. Show that there is a 7th root of unity $\zeta \in \mathbb{F}_8$ such that the lines in $P = P(V)$ are the triples $L_i = \{\zeta^i, \zeta^{i+1}, \zeta^{i+3}\}$, $i \in \mathbb{Z}/7\mathbb{Z}$.

(d) With $V = \mathbb{F}_8$ as in (c), recall that there is a nondegenerate symmetric bilinear form on V given by $\langle x, y \rangle = \operatorname{tr}(xy)$, where $\operatorname{tr} \colon \mathbb{F}_8 \to \mathbb{F}_2$ is the trace. This induces a type-reversing involution σ of $\Delta = \Delta(V)$ by (b). Show that σ is given on vertices by $\zeta^i \leftrightarrow L_{6-i}$. Describe σ in terms of the picture of Δ in Figure 4.1.

4.24. In this exercise we will use some standard algebraic terminology concerning bilinear forms. Readers not familiar with this terminology can look ahead at Section 6.6, where all the terms are defined. Let V be a 4-dimensional vector space over the field $\mathbb{F}_2$, with basis e_1, e_2, f_1, f_2. There is a nondegenerate alternating bilinear form $\langle -, - \rangle$ on V such that

$$\langle e_i, f_i \rangle = \langle f_i, e_i \rangle = 1$$

for $i = 1, 2$ and all other "inner products" $\langle u, v \rangle$ of basis vectors are 0. The purpose of this exercise is to construct a generalized quadrangle Q from V and $\langle -, - \rangle$. The *points* of Q are defined to be the nonzero vectors in V, and the *lines* of Q are the 2-dimensional totally isotropic subspaces of V. A point p is *incident* to a line L if $p \in L$.

(a) Show that there are 15 points, each incident to 3 lines, and 15 lines, each incident to 3 points. Thus the flag complex Δ of Q has 30 vertices and 45 edges.

(b) Show how to use the four given basis vectors to construct a quadrilateral in Q and hence an octagon in Δ. This octagon will be called the *standard apartment*. More generally, any four vectors in V with inner products like those of the basis vectors give rise to an octagon in Δ called an *apartment*. [Four vectors of this form are said to form a *symplectic basis* of V.]

(c) Show that Δ is a building of type $I_2(4)$.

(d) Call two vertices of Δ *opposite* if there is an apartment Σ containing them and they are opposite in Σ in the obvious sense (recall that Σ is an octagon). Show that two vertices are opposite if and only if (i) they are noncollinear points of Q or (ii) they are nonintersecting lines of Q.

(e) Show that Δ contains 5 vertices (but not 6) that are pairwise opposite.

4.3 The Building Associated to a Vector Space

Let V be a vector space of finite dimension $n \geq 2$ over an arbitrary field k.

Definition 4.25. The *projective space* associated to V consists of the nonzero proper subspaces of V, two such being called *incident* if one is contained in the other. (This is an example of a projective space of rank $(n-1)$.) Let $\Delta = \Delta(V)$ be the flag complex of this projective space; thus the simplices are chains

$$V_1 < V_2 < \cdots < V_k$$

of nonzero proper subspaces of V.

The maximal simplices of Δ are the chains

$$V_1 < V_2 < \cdots < V_{n-1} \tag{4.1}$$

with $\dim V_i = i$. In what follows we will find it notationally convenient to set $V_0 = 0$ and $V_n = V$. We will call the maximal simplices *chambers*, even though we do not yet know that Δ is a chamber complex.

The purpose of this section is to outline a proof that Δ is a building, as claimed in Example 4.21(a) above. By a *frame* in V we will mean a set $\mathcal{F} = \{L_1, \ldots, L_n\}$ of 1-dimensional subspaces of V such that $V = L_1 \oplus \cdots \oplus L_n$. Given such a frame, consider the set of subspaces V' of V such that V' is spanned by a nonempty proper subset of $\mathcal{F}$. Let $\Sigma(\mathcal{F})$ be the subcomplex of Δ consisting of flags of such subspaces. Call a subcomplex of this form an *apartment*. When $n = 3$, for example, every apartment is a hexagon, as shown in Figure 4.2. We now proceed with an outline of a proof that Δ is a building.

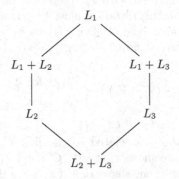

Fig. 4.2. An apartment when $n = 3$.

Let Σ_0 be the flag complex of the poset of nonempty proper subsets of $\{1, 2, \ldots, n\}$. This is a Coxeter complex of type A_{n-1}, as we explained in the solution to Exercise 1.112. To verify axiom (B0), one checks that every apartment $\Sigma(\mathcal{F})$ is isomorphic to Σ_0. This is trivial to verify, but it will be useful for us to make the isomorphism explicit.

Choose a chamber C in $\Sigma(\mathcal{F})$ with vertices V_i as in (4.1). Then the elements of $\mathcal{F}$ can be numbered $L_1, L_2, \ldots, L_n$ in such a way that $V_i = L_1 + \cdots + L_i$ for all i (including $i = 0$ and $i = n$). We then have an isomorphism $\phi = \phi_{\mathcal{F},C} \colon \Sigma_0 \xrightarrow{\sim} \Sigma(\mathcal{F})$ given on vertices by $I \mapsto \bigoplus_{i \in I} L_i$ for any nonempty proper subset $I \subset \{1, \ldots, n\}$. The inverse of ϕ is the map $\psi = \psi_{\mathcal{F},C} \colon \Sigma(\mathcal{F}) \xrightarrow{\sim} \Sigma_0$ defined as follows on vertices: Given a nonzero proper subspace U of V that is a vertex of $\Sigma(\mathcal{F})$, consider the filtration $(U \cap V_i)_{0 \le i \le n}$ of U induced by C. Then ψ records the indices at which this filtration jumps; more precisely, ψ is given by

$$\psi(U) := \{i \ge 1 \mid U \cap V_i > U \cap V_{i-1}\} \ .$$

Note that the definition of ψ makes no reference to $\mathcal{F}$, only to C. In particular, if $\mathcal{F}'$ is a second frame whose apartment $\Sigma(\mathcal{F}')$ contains C, then $\psi_{\mathcal{F},C}$ and $\psi_{\mathcal{F}',C}$ agree on $\Sigma(\mathcal{F}) \cap \Sigma(\mathcal{F}')$. Axiom (B2'') is now immediate: $\phi_{\mathcal{F}',C} \circ \psi_{\mathcal{F},C}$ is an isomorphism $\Sigma(\mathcal{F}) \xrightarrow{\sim} \Sigma(\mathcal{F}')$ fixing the intersection pointwise.

Turning now to (B1), consider two chambers C, C' of Δ, with vertex sets

$$V_1 < \cdots < V_{n-1} \quad \text{and} \quad V_1' < \cdots < V_{n-1}',$$

respectively. View $\{V_i\}_{0 \le i \le n}$ and $\{V_i'\}_{0 \le i \le n}$ as composition series for V. According to the Jordan–Hölder theorem, there is a permutation π of $\{1, \ldots, n\}$ such that $V_i'/V_{i-1}' \cong V_j/V_{j-1}$ if $j = \pi(i)$. This is of course a triviality in the present context of vector spaces; but we need to review how the proof of the Jordan–Hölder theorem yields a canonical π and canonical isomorphisms $V_i'/V_{i-1}' \cong V_j/V_{j-1}$.

For each $i \in \{1, \ldots, n\}$, the composition series $\{V_j\}_{0 \le j \le n}$ induces a filtration of V_i'/V_{i-1}'. [First intersect with V_i'; then take images mod V_{i-1}'.] Since V_i'/V_{i-1}' is 1-dimensional, this filtration must be trivial, i.e., only one of the successive quotients is nontrivial. Define $\pi(i)$ to be the index j such that the jth quotient is nontrivial. Equivalently, $j = \pi(i)$ is characterized by the property that

$$V_{i-1}' + (V_i' \cap V_k) = \begin{cases} V_{i-1}' & \text{for } k < j, \\ V_i' & \text{for } k \ge j. \end{cases}$$

The resulting function $\pi = \pi(C, C') \colon \{1, \ldots, n\} \to \{1, \ldots, n\}$ is called the *Jordan–Hölder permutation* associated to the pair C, C'. To see that π is indeed a permutation, one shows that $\pi(C, C')$ and $\pi(C', C)$ are inverses of one another. More precisely, one shows that if $\pi(i) = j$ there are isomorphisms

$$\frac{V_i'}{V_{i-1}'} \xleftarrow{\sim} \frac{V_i' \cap V_j}{(V_{i-1}' \cap V_j) + (V_i' \cap V_{j-1})} \xrightarrow{\sim} \frac{V_j}{V_{j-1}}$$

induced by inclusions.

All of this can be extracted from the proof of the Jordan–Hölder theorem as given in many standard texts, such as Jacobson [139, Section 3.3]. But one can also just prove it directly. To show, for instance, that $V_{i-1}' \cap V_j \le V_{j-1}$, suppose this is false; then $V_j = V_{j-1} + (V_{i-1}' \cap V_j)$. Intersect both sides with V_i' to obtain $V_i' \cap V_j = (V_i' \cap V_{j-1}) + (V_{i-1}' \cap V_j) \le V_{i-1}'$, contradicting the definition of j.

It is now easy to verify (B1): Given maximal simplices C, C', one can find a frame $\mathcal{F}$ such that $\Sigma(\mathcal{F})$ contains C and C' by choosing, for each i, j as above, a 1-dimensional subspace $L_j \le V_i' \cap V_j$ whose image in both V_i'/V_{i-1}' and V_j/V_{j-1} is nontrivial.

This completes our sketch of the proof that $\Delta(V)$ is a building. As a byproduct of the proof, we have obtained a function $\mathcal{C}(\Delta) \times \mathcal{C}(\Delta) \to W$, where $W = S_n$, i.e., W is the Coxeter group of type A_{n-1}. The reader who has read Section 3.5 may be able to guess the geometric meaning of this function. Exercise 4.28 below provides a further clue. We will return to this in Exercise 4.93.

Remark 4.26. Since a vector space is determined up to isomorphism by its dimension, this section essentially contains one example for each $n \ge 2$ and each field k. But, as we have already seen in Exercise 4.23 and will see again in Exercises 4.31 and 4.32, it can be conceptually helpful to consider arbitrary vector spaces that are not assumed to be given as k^n.

Exercises

4.27. Show that $\Delta(V)$ is thick.

4.28. What is $\pi(C, C')$ if C and C' are i-adjacent, i.e., if $V_j = V'_j$ for $j \neq i$ and $V_i \neq V'_i$?

4.29. Let C and C' be chambers with vertices V_i and V'_i as above. Call C and C' *opposite* if there is a frame $\mathcal{F} = \{L_1, \ldots, L_n\}$ such that $V_i = L_1 + L_2 + \cdots + L_i$ and $V'_i = L_n + L_{n-1} + \cdots + L_{n-i+1}$ for all i. Show that $\mathcal{F}$ is then unique. In words, two opposite chambers are contained in a unique apartment. See Theorem 4.70 below for a generalization.

4.30. As we noted above, the map $\psi_{\mathcal{F},C}$ depends only on C, and not on $\mathcal{F}$. Show that we obtain in this way a well-defined chamber map $\psi_C \colon \Delta \to \Sigma_0$. Deduce that every apartment $\Sigma(\mathcal{F})$ is a retract of Δ.

4.31. Observe that the building $\Delta(V)$ comes equipped with a canonical coloring, having values in $\{1, \ldots, n-1\}$, where the "color" of a vertex is its dimension as a subspace of V. Show that there is a canonical type-reversing isomorphism $\phi \colon \Delta(V) \xrightarrow{\sim} \Delta(V^*)$, where V^* is the vector space dual to V. Here "type-reversing" means that a vertex of type i is sent to a vertex of type $n - i$. In other words, ϕ induces the unique nontrivial automorphism of the Coxeter diagram of type A_{n-1} [assuming $n \geq 3$]. Deduce, as in Exercise 4.23(b), that the buildings constructed in this section always admit type-reversing automorphisms.

4.32. In this section we have assumed that V is a vector space over an arbitrary field k. Fields, by convention, are usually assumed to be commutative. Suppose, however, that we allow k to be a *division ring* (also called a *skew field*), i.e., a possibly noncommutative ring with $1 \neq 0$ in which every nonzero element is invertible. Does anything change?

4.4 Retractions

Retractions, as we saw in Chapter 3, can be quite useful technical tools. In this section we establish the existence and formal properties of retractions of a building onto its apartments. Exercise 4.30 has already given us an example of such a retraction.

Assume throughout this section that Δ is a building and that $\mathcal{A}$ is an arbitrary system of apartments.

Proposition 4.33. *Every apartment is a retract of Δ.*

Proof. This is very similar to the construction of a type function. Fix a chamber C of the given apartment Σ, and consider all the apartments Σ' that contain C. For any such Σ' there is a unique isomorphism $\phi_{\Sigma'} : \Sigma' \xrightarrow{\sim} \Sigma$ that fixes C pointwise, where the existence is given by axiom (B2) and the uniqueness is proved by the standard argument. For any two such apartments Σ', Σ'', the isomorphisms $\phi_{\Sigma'}$ and $\phi_{\Sigma''}$ agree on $\Sigma' \cap \Sigma''$. This follows from the fact that we can construct $\phi_{\Sigma''}$ by composing $\phi_{\Sigma'}$ with the isomorphism $\Sigma'' \xrightarrow{\sim} \Sigma'$ that fixes $\Sigma' \cap \Sigma''$ pointwise [see (B2″)]. The various isomorphisms $\phi_{\Sigma'}$ therefore fit together to give a chamber map $\rho : \Delta \to \Sigma$, and ρ is a retraction since ϕ_Σ is the identity.   □

One useful consequence of this is that combinatorial distances between chambers of Δ can be computed in terms of the distance functions on apartments, which we understand reasonably well in terms of separating walls or reduced words:

Corollary 4.34. *Let C and D be chambers of Δ, and let Σ be any apartment containing C and D. Then $d_\Delta(C, D) = d_\Sigma(C, D)$. Consequently, the diameter of Δ is equal to the diameter of any apartment.*

Proof. Suppose Γ is a minimal gallery in Σ from C to D. Then Γ is also minimal in Δ; for if there were a shorter gallery in Δ, then we could get a shorter one in Σ by applying a retraction. This proves the first assertion. As an immediate consequence, we have diam $\Sigma \leq$ diam Δ. To prove the opposite inequality, let C' and D' be arbitrary chambers of Δ and let Σ' be an apartment containing them. Then we have

$$d_\Delta(C', D') = d_{\Sigma'}(C', D') \leq \text{diam } \Sigma'.$$

But $\Sigma \cong \Sigma'$, so diam $\Sigma' =$ diam Σ and hence diam $\Delta \leq$ diam Σ.   □

Remark 4.35. The statement and proof of the corollary remain valid if C and D are replaced by arbitrary simplices. (Recall from Section A.1.3 that $d(A, B)$ makes sense for any simplices A, B.)

Corollary 4.34 can be used to prove the important fact that the Coxeter matrix M of Δ, as defined in Section 4.1, really is an invariant of Δ:

Corollary 4.36. *The Coxeter matrix M depends only on Δ, not on the system of apartments. It is given by*

$$m(s, t) = \text{diam}(\text{lk}_\Delta A) ,$$

where A is any simplex of cotype $\{s, t\}$.

Proof. Suppose A has cotype $\{s, t\}$, and let Σ be any apartment containing A. Then we have $m(s, t) = \text{diam}(\text{lk}_\Sigma A)$ by definition. But $\text{lk}_\Sigma A$ is an apartment in the building $\text{lk}_\Delta A$, so it has the same diameter as the latter by Corollary 4.34. This proves the second assertion, and the first assertion follows at once.   □

As an example, suppose that Δ is 1-dimensional and that the apartments are $2m$-gons as in Example 4.16. Then *every* system of apartments in Δ must consist of $2m$-gons (for the same $m = \operatorname{diam}\Delta$).

In view of Corollary 4.36, the building Δ has a well-defined *type* and a well-defined *Weyl group* $W = W_M$. We have already used this language informally in connection with some of the examples in Section 4.2. Let's make this more precise, as in Definition 3.86.

Definition 4.37. Let (W, S) be a Coxeter system with Coxeter matrix M. We say that a building Δ is of *type* (W, S) (or of *type* M) if Δ comes equipped with a type function having values in S such that the Coxeter matrix of Δ is M. We then say that W is the *Weyl group* of Δ.

Returning now to the general study of retractions, note that the proof of Proposition 4.33 actually yields, for any apartment Σ and any chamber $C \in \Sigma$, a canonical retraction of Δ onto Σ.

Definition 4.38. Given an apartment Σ and a chamber $C \in \Sigma$, there is a *canonical retraction* $\rho = \rho_{\Sigma,C} \colon \Delta \to \Sigma$. It is called the retraction onto Σ *centered* at C. It can be characterized as the unique chamber map $\Delta \to \Sigma$ that fixes C pointwise and maps every apartment containing C isomorphically onto Σ.

This characterization makes it appear that ρ depends on the apartment system $\mathcal{A}$. But part (3) of the following proposition gives a different characterization, which shows that ρ depends only on Σ and C, not on $\mathcal{A}$.

Proposition 4.39. *The retraction* $\rho = \rho_{\Sigma,C}$ *has the following properties.*

(1) *For any face* $A \leq C$, $\rho^{-1}(A) = \{A\}$.
(2) ρ *preserves distances from* C, *i.e.,* $d(C, \rho(D)) = d(C, D)$ *for any chamber* $D \in \Delta$. *More generally,* $d(A, \rho(B)) = d(A, B)$ *for any* $A \leq C$ *and any* $B \in \Delta$.
(3) ρ *is the unique chamber map* $\Delta \to \Sigma$ *that fixes* C *pointwise and preserves distances from* C.

Proof. (1) Suppose $B \in \Delta$ is a simplex such that $\rho(B) = A \leq C$. Choose an apartment Σ' containing both B and C. Then $\rho|_{\Sigma'}$ is an isomorphism, and it maps both A and B to A. Hence $B = A$.

(2) Given $A \leq C$ and any simplex B, choose an apartment Σ' containing C and B. Since $\rho|_{\Sigma'}$ is an isomorphism, we have $d_\Sigma(A, \rho(B)) = d_{\Sigma'}(A, B)$. In view of Remark 4.35 above, we can delete the subscripts to obtain $d(A, \rho(B)) = d(A, B)$.

(3) Suppose $\phi \colon \Delta \to \Sigma$ is another chamber map that fixes C pointwise and preserves distances from C. Then ϕ and ρ must each take any minimal gallery in Δ starting at C to a pregallery of the same length in Σ. The pregallery is therefore a gallery, and this is all that is needed to make the standard uniqueness argument go through; hence $\phi = \rho$. $\qquad\square$

Property (2) above enables one, in practice, to figure out what ρ looks like. Suppose, for example, that Δ is a tree as in Example 4.16. Then Σ is a triangulated line and C is an edge of Σ. Let v and w be the vertices of C. Then it follows easily from (2) that ρ simply "flattens Δ out" onto Σ, with the part of Δ closer to v than to w going to the corresponding part of Σ, and similarly for the part closer to w. See Figure 4.3.

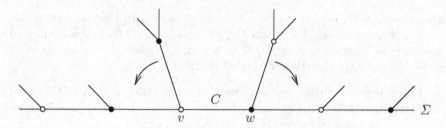

Fig. 4.3. The retraction $\rho_{\Sigma,C}$.

As an application of retractions, we show that every apartment Σ is a convex chamber subcomplex of Δ in the sense of Definition 3.129.

Proposition 4.40. *Every apartment Σ is a convex chamber subcomplex of Δ. More precisely, given $C, A \in \Sigma$ with C a chamber, every minimal gallery in Δ between C and A is contained in Σ.*

Proof. Let $\Gamma: C_0, \ldots, C_d$ be a minimal gallery from C to A. If Γ is not contained in Σ, then there is an index $i \geq 1$ with $C_{i-1} \in \Sigma$ and $C_i \notin \Sigma$. Let D be the chamber of Σ distinct from C_{i-1} and having $C_{i-1} \cap C_i$ as a face, and let ρ be the retraction $\rho_{\Sigma,D}$. Then $\rho(C_i) = C_{i-1}$ (see Figure 4.4), so the pregallery $\rho(\Gamma)$ from C to A has a repetition, contradicting the minimality of Γ. □

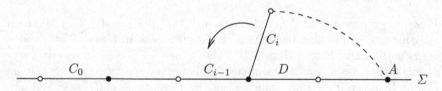

Fig. 4.4. ρ shortens Γ.

As a second application of retractions, we give a criterion for recognizing minimal galleries in a building. Recall that we have a type function on Δ with values in a set S and that S can also be viewed as the set of generators of the Weyl group $W = W_M$ of Δ.

Proposition 4.41. *Let $\Gamma\colon C_0,\ldots,C_d$ be a gallery of type $\mathbf{s} = (s_1,\ldots,s_d)$. Then Γ is minimal if and only if $\mathbf{s}$ is a reduced word.*

Proof. If Γ is minimal, then it is contained in an apartment by Proposition 4.40; $\mathbf{s}$ is then reduced by the connection between words in a Coxeter group and galleries in the associated Coxeter complex. Conversely, suppose $\mathbf{s}$ is reduced. We may assume by induction that the subgallery $C_1,\ldots,C_d$ is minimal and hence is contained in an apartment Σ. Then $\rho_{\Sigma,C_1}(\Gamma)$ is a gallery in Σ with the same type $\mathbf{s}$ as Γ, so it is minimal. It follows that Γ is minimal, since its image under a chamber map is minimal.   □

Corollary 4.42. *Let C and D be chambers, and let $\mathbf{s}$ be a reduced word. If there is a gallery from C to D of type $\mathbf{s}$, then there is a unique such gallery.*

Proof. Choose an apartment Σ containing C and D. Any gallery from C to D of type $\mathbf{s}$ is minimal and hence is contained in Σ. The result now follows from the thinness of Σ.   □

We close this section by using retractions to complete the discussion of rank-2 buildings that we began in Example 4.16. We need some terminology.

Definition 4.43. Let Δ be a graph, i.e., a 1-dimensional simplicial complex. The *girth* of Δ is the smallest integer $k \geq 3$ such that Δ contains a k-gon, provided such an integer k exists. Otherwise, Δ is a tree, and we define its girth to be ∞.

Proposition 4.44. *Let Δ be a connected bipartite graph in which every vertex is a face of at least two edges. Then Δ is a building if and only if Δ has diameter m and girth $2m$ for some m with $2 \leq m \leq \infty$. In this case Δ has type $I_2(m)$.*

("Diameter" here means the supremum of the distances $d(u,v)$ between vertices u,v. Here $d(-,-)$ denotes the usual distance between two vertices, not the gallery distance defined in Section A.1.3. Note that when $m = \infty$, the content of the proposition is that a building of type o—∞—o is the same thing as a tree without endpoints, as claimed in Example 4.16.)

Proof. Suppose that Δ has diameter m and girth $2m$. Then one easily checks the following properties for any two vertices u,v:

- If $d(u,v) < m$, then there is a unique path from u to v of length $\leq m$ with no backtracking. ["No backtracking" means that any three consecutive vertices are distinct.]
- If $0 < d(u,v) < m$, then there is a unique vertex v' adjacent to v such that $d(u,v') = d(u,v) - 1$. For all other vertices v' adjacent to v, $d(u,v') = d(u,v) + 1$.
- If $d(u,v) = m$ (and hence $m < \infty$), then $d(u,v') = m - 1$ for all vertices v' adjacent to v.

It is now a routine exercise to show that Δ is a building of type $I_2(m)$, with the apartments being all the $2m$-gons in Δ.

Conversely, suppose that Δ is a building of type $I_2(m)$. Since each apartment is a $2m$-gon and any two vertices are contained in an apartment, it suffices to show that Δ does not contain a k-gon for $k < 2m$. Let Z be a k-gon in Δ with $3 \le k < \infty$. Let C be a chamber in Z, with vertices v and w, let Σ be any apartment containing C, and let ρ be the retraction $\rho_{\Sigma,C}$. As we traverse Z starting at C (thought of as oriented from v to w, say), the image under ρ is a closed curve in Σ passing through the vertices $v, w, \dots$ and never traversing C again before returning to v [see part (1) of Proposition 4.39]. Since Σ is a $2m$-gon, this is possible only if $k \ge 2m$. $\square$

Remark 4.45. In view of Definition 4.20, the proposition can be viewed as a characterization of generalized m-gons; this characterization is often taken as the definition of a generalized m-gon.

Exercises

4.46. Deduce from Proposition 4.44 with $m = 3$ that every building of type A_2 is the flag complex of a projective plane, as claimed in Example 4.16.

4.47. Give an alternative proof that a building of type $I_2(m)$ has girth $2m$ by considering types of galleries.

4.48. Let Δ be the incidence graph of a generalized m-gon (i.e., a building of type $\circ\!\!\underline{m}\!\!\circ$). Given two vertices u, v of Δ, a path of minimal length from u to v is called a *geodesic*. As we noted in the proof of Proposition 4.44, there is a unique such geodesic if $d(u, v) < m$. If $d(u, v) = m < \infty$, however, there are at least two geodesics from u to v.

(a) Let Δ' be a convex chamber subcomplex of Δ. If u and v are vertices of Δ' such that $d(u, v) < m$ in Δ, show that the (unique) geodesic joining them is contained in Δ'.

(b) If $d(u, v) = m < \infty$, show that there is a convex chamber subcomplex Δ' containing u and v but only one of the geodesics joining them.

4.49. (a) Where in the proof of Proposition 4.40 did we use the assumption that C is a chamber?

(b) Give an example to show that this assumption cannot be dropped.

4.50. Show that every building is a flag complex.

4.51. Given an apartment Σ, a chamber $C \in \Sigma$, and a chamber $D \in \Delta$, show that there is a unique type-preserving chamber map $\rho\colon \Delta \to \Sigma$ such that $\rho(D) = C$ and ρ maps every apartment containing D isomorphically onto Σ. For lack of a better name, we will call ρ the *canonical map* $\Delta \to \Sigma$ such that $\rho(D) = C$.

4.52. (a) Let Δ be a thick building, let C be a chamber, and let $\mathcal{A}'$ be the set of apartments containing C. Show that C is the only chamber in $\bigcap_{\Sigma \in \mathcal{A}'} \Sigma$. (In more intuitive language, every closed chamber is an intersection of apartments. The interested reader can phrase this precisely in terms of the geometric realization.)

(b) Give an example to show that the thickness assumption cannot be dropped.

4.53. (a) Recall that the notions of *elementary homotopy* and *homotopy* were defined for galleries in Coxeter complexes in Section 3.2. Generalize to buildings.

(b) Given a nonminimal gallery, how can one modify it to obtain a minimal gallery with the same extremities?

4.5 The Complete System of Apartments

We have seen several cases in which something that seemed a priori to depend on a choice of apartment system $\mathcal{A}$ turned out to be independent of $\mathcal{A}$. The next theorem is the ultimate result of this type; it can be viewed as saying that all systems of apartments in a given building are compatible with one another. Although the statement of the result does not refer to colorings, we will continue to assume that Δ comes equipped with a fixed type function having values in a set S. In particular, we have a Coxeter matrix M, which will be used in the proof of the theorem.

Theorem 4.54. *If Δ is a building, then the union of any family of apartment systems is again an apartment system. Consequently, Δ admits a largest system of apartments.*

Proof. It is obvious that (B0) and (B1) hold for the union, so the only problem is to prove (B2). We will work with the variant (B2″). Suppose, then, that Σ and Σ' are apartments in different apartment systems and that $\Sigma \cap \Sigma'$ contains at least one chamber. We must find an isomorphism $\Sigma' \xrightarrow{\sim} \Sigma$ that fixes $\Sigma \cap \Sigma'$ pointwise.

Choose an arbitrary chamber $C \in \Sigma \cap \Sigma'$. There are then two obvious candidates for the desired isomorphism $\Sigma' \xrightarrow{\sim} \Sigma$. On the one hand, we know by Corollary 4.36 that Σ and Σ' have the same Coxeter matrix M, so we can find a type-preserving isomorphism $\phi \colon \Sigma' \xrightarrow{\sim} \Sigma$ by Corollary 4.8. And we can certainly choose ϕ such that $\phi(C) = C$, since the group of type-preserving automorphisms of Σ is transitive on the chambers. It then follows that ϕ fixes C pointwise. Unfortunately, it is not obvious that ϕ fixes $\Sigma \cap \Sigma'$ pointwise.

The other candidate is provided by the theory of retractions. Namely, let ρ be the retraction $\rho_{\Sigma,C}$ and let $\psi \colon \Sigma' \to \Sigma$ be the restriction of ρ to Σ'. Then ψ obviously fixes $\Sigma \cap \Sigma'$ pointwise, simply because ρ is a retraction onto Σ.

But it is not obvious that ψ is an isomorphism. [Readers who are tempted to say that ψ is an isomorphism by the construction of ρ should recall that we do not know that Σ' is part of an apartment system containing Σ; indeed, that is what we are trying to prove!]

To complete the proof, we will show by the standard uniqueness argument that ϕ and ψ are in fact the same map, which therefore has all the required properties. Since ϕ and ψ both fix C pointwise, the standard argument will go through if we can show that if Γ is a minimal gallery in Σ' starting at C, then the pregalleries $\phi(\Gamma)$ and $\psi(\Gamma)$ are galleries. This is clear for $\phi(\Gamma)$ since ϕ is an isomorphism. And it is true for $\psi(\Gamma)$ because of two facts proved in the previous section: (a) Γ is still minimal when viewed as a gallery in Δ; and (b) ρ preserves distances from C.   □

Definition 4.55. The maximal apartment system will be called the *complete* system of apartments. It consists, then, of all subcomplexes $\Sigma \subseteq \Delta$ such that Σ is in some apartment system $\mathcal{A}$.

Remark 4.56. This description of the complete apartment system is not very informative. Here are two characterizations that are more useful. Let Σ be a chamber subcomplex of Δ (Definition A.12), and let $\mathcal{A}$ be the complete system of apartments. Then:

(1) Σ is in $\mathcal{A}$ if and only if Σ is isomorphic to Σ_M.
(2) Σ is in $\mathcal{A}$ if and only if Σ is a thin, convex chamber subcomplex of Δ.

We have already proved the necessity of each of the conditions. The sufficiency of (2) will be proved in Section 4.8 using the Weyl distance function (see Theorem 4.86). The sufficiency of (1) is easier and will be proved in Proposition 4.59 below. The proof will use the following two lemmas. The first is a simple combinatorial fact.

Lemma 4.57. *Let* $\mathbf{m} = (m_i)_{i \in I}$ *and* $\mathbf{m}' = (m'_i)_{i \in I}$, *where* I *is a finite index set and the elements* m_i, m'_i *are in some totally ordered set. Suppose that* $\mathbf{m}'$ *is a permutation of* $\mathbf{m}$ *in the sense that there is a bijection* $f: I \to I$ *such that* $m'_i = m_{f(i)}$ *for all* $i \in I$. *If* $m_i \leq m'_i$ *for all* i, *then* $m_i = m'_i$ *for all* i.

(In our application of this, $\mathbf{m}$ and $\mathbf{m}'$ will be Coxeter matrices, and the totally ordered set will be $\mathbb{N} \cup \{\infty\}$, where $\mathbb{N}$ is the set of natural numbers.)

Sketch of proof. Let m be the smallest element that occurs in $\mathbf{m}$, and let $J := \{i \in I \mid m_i = m\}$. Then we have $m'_i \geq m$ for $i \in J$ and $m'_i > m$ for $i \notin J$. But m must occur in $\mathbf{m}'$ as many times as it occurs in $\mathbf{m}$, so $m'_i = m = m_i$ for all $i \in J$. Now consider the second smallest element that occurs in $\mathbf{m}$, and so on.   □

We return now to our building Δ. Recall that we have a type function τ on Δ with values in a set S, which gives us a Coxeter matrix M and a Weyl group $W := W_M$. Suppose Σ is a chamber subcomplex of Δ and is known to be a Coxeter complex. Then τ restricts to a type function on Σ, and it then makes sense to talk about the Coxeter matrix $M' = \big(m'(s,t)\big)_{s,t \in S}$ of Σ, with $m'(s,t) = \operatorname{diam} \operatorname{lk}_\Sigma A$ for any simplex $A \in \Sigma$ of cotype $\{s,t\}$. Easy examples show that Σ need not be isomorphic to the apartments of Δ. In particular, the matrix M' is not necessarily equal to the Coxeter matrix M of Δ. We will prove in the next lemma, however, that $M' = M$ if Σ is isomorphic to Σ_M. Note that this is not at all obvious, since we do not know, a priori, that there is a *type-preserving* isomorphism between Σ and Σ_M (with respect to the type function $\tau|_\Sigma$ on Σ and the canonical type function τ_M on Σ_M).

Lemma 4.58. *Let Σ be a subcomplex of Δ that is isomorphic to Σ_M, and let M' be its Coxeter matrix as above. Then $M' = M$. Consequently, there is a type-preserving isomorphism between Σ and Σ_M.*

Proof. We know from Proposition 3.85 that there is a type-preserving isomorphism between Σ and $\Sigma_{M'}$. So the second assertion follows from the first. To prove the first assertion, let $\phi \colon \Sigma \overset{\sim}{\longrightarrow} \Sigma_M$ be an isomorphism, and let $f \colon S \to S$ be the type-change map ϕ_* (Proposition A.14). Thus $\tau_M(\phi(A)) = f(\tau(A))$ for every simplex $A \in \Sigma$. In view of the interpretation of Coxeter matrices in terms of links (Corollary 3.20), it follows that $m(f(s), f(t)) = m'(s,t)$ for all $s,t \in S$. On the other hand, we claim that $m(s,t) \le m'(s,t)$ for all $s,t \in S$. To see this, consider $s \ne t$ in S and let A be a simplex in Σ of cotype $\{s,t\}$. Then $\operatorname{lk}_\Sigma A$ is a $2m'(s,t)$-gon and is a chamber subcomplex of $\operatorname{lk}_\Delta A$. The latter is a graph of girth $2m(s,t)$ by Proposition 4.44, so $m(s,t) \le m'(s,t)$, as claimed. Lemma 4.57 now implies that $M = M'$. $\qquad\square$

Proposition 4.59. *If Σ is a subcomplex of Δ that is isomorphic to Σ_M, then Σ is an apartment in the complete system of apartments.*

Proof. It suffices to show that if Σ is adjoined to an apartment system $\mathcal{A}$, then axiom (B2″) still holds. The proof is essentially the same as the proof of Theorem 4.54. The given complex Σ plays the role of the complex Σ' that occurred in that proof, and the only extra ingredient required is that one needs to appeal to Proposition 4.41 and Lemma 4.58 to show that every minimal gallery in Σ is still minimal in Δ. $\qquad\square$

Remark 4.60. It follows from Proposition 4.59 that the *intrinsic* property of a chamber subcomplex of a building that it is isomorphic to Σ_M implies the strong *relative* property of being convex in Δ. This is a remarkable fact about buildings that is not at all obvious a priori.

Exercise 4.61. A subcomplex α of a building Δ is called a *root* if there is an apartment Σ (in the complete system of apartments) such that $\alpha \subseteq \Sigma$ and α is a root in Σ.

(a) If α is a root in Δ, show that it is a root in every apartment containing it.

(b) If α is a root in Δ, show that it has a well-defined boundary $\partial \alpha$, equal to its boundary in every apartment containing it.

4.6 Subbuildings

Definition 4.62. Let Δ be a building and Δ' a chamber subcomplex of Δ. We say that Δ' is a *subbuilding* of Δ if Δ' is a building in its own right and every apartment of Δ' is an apartment of Δ. Here "apartment" refers to the complete systems of apartments in Δ' and Δ.

The condition on apartments in Definition 4.62 can be restated as a condition on Coxeter matrices if we assume that Δ comes equipped with a type function. So suppose Δ is a building of type M in the sense of Definition 4.37. The given type function on Δ restricts to a type function on any chamber subcomplex Δ' of Δ. So Δ', if it is a building, has a Coxeter matrix M' indexed by the same set $S \times S$ that indexes M. The following statement therefore makes sense:

Proposition 4.63. *Let Δ be a building of type M. A chamber subcomplex Δ' of Δ is a subbuilding if and only if Δ' is a building in its own right and its Coxeter matrix is M.*

Proof. The "only if" part follows immediately from the fact that the Coxeter matrix of a building is the same as the Coxeter matrix of any of its apartments. Conversely, suppose Δ' is a chamber subcomplex of Δ and is a building with Coxeter matrix M. Then every apartment of Δ' is isomorphic to Σ_M, so it is an apartment of Δ by Proposition 4.59. Thus Δ' is a subbuilding of Δ. $\square$

Example 4.64. Let $\Delta = \Delta(k^n)$ be the building associated to an n-dimensional vector space over a field k (Section 4.3). If k' is a subfield of k, then there is an obvious embedding of $\Delta' := \Delta((k')^n)$ as a subcomplex of Δ, and this subcomplex is a subbuilding. One can see this directly from Definition 4.62 and the construction of apartments in Section 4.3. Alternatively, it follows at once from Proposition 4.63.

There is another characterization of subbuildings, for which we need some terminology.

Definition 4.65. A chamber complex is said to be *weak* if every panel is a face of at least two chambers.

This somewhat strange terminology is motivated by the concept of "weak building" mentioned in Remark 4.2.

Theorem 4.66. *Let Δ be a building and Δ' a chamber subcomplex of Δ. Then Δ' is a subbuilding if and only if it is weak and is convex in Δ.*

In view of Proposition 4.63, we obtain the following nonobvious consequence: Suppose a building Δ' of the same type as Δ is embedded in Δ as a simplicial subcomplex; then Δ' is automatically *convex* in Δ (cf. Remark 4.60).

The "if" part of the theorem requires tools that will be introduced in the next chapter, so we defer its proof; see Proposition 5.94. (The hard part is to use the hypothesis that Δ' is weak and convex in order to produce apartments.) The "only if" part is easier, however, and we can prove it now.

Proof of the "only if" part. Suppose Δ' is a subbuilding of Δ. Then Δ' is trivially weak, since every panel is contained in an apartment. To prove that Δ' is convex in Δ, let C and D be two chambers of Δ', and let Σ be an apartment of Δ' containing them. Then Σ is an apartment of Δ and hence is convex in Δ. Consequently, every minimal gallery in Δ from C to D is contained in Σ and therefore in Δ'. $\square$

4.7 The Spherical Case

Our treatment of spherical buildings in this section will use the theory of finite reflection groups whenever it is convenient to do so. The reader who prefers a purely combinatorial approach will find one in the next chapter (Section 5.7). Recall from Definitions 3.2 and 3.64 that a Coxeter complex Σ is called *spherical* if it is finite or, equivalently, if it is isomorphic to the complex $\Sigma(W, V)$ associated to a finite reflection group (Section 1.5). Another equivalent condition is that Σ have finite diameter (see Exercise 3.11(b)).

Definition 4.67. A building is called *spherical* if its apartments are spherical Coxeter complexes.

Since the diameter of a building is equal to the diameter of any apartment (Corollary 4.34), a building is spherical if and only if it has finite diameter. The next definition is motivated by Proposition 1.57.

Definition 4.68. We say that two chambers C, C' in a spherical building Δ are *opposite*, and we write C op C', if $d(C, C') = \operatorname{diam} \Delta$.

We record, for ease of reference, the following result (Corollary 1.58):

Lemma 4.69. *Let C and C' be opposite chambers in a spherical building, and let Σ be any apartment containing C and C'. Then every chamber of Σ occurs in some minimal gallery from C to C'.* $\square$

[This is the combinatorial analogue of the following geometric fact: Given two opposite points x, x' of a sphere, the geodesics (great semicircles) from x to x' cover the entire sphere.]

Since we know that any apartment Σ is convex, it follows from the lemma that Σ is the convex hull of $\{C, C'\}$ for any pair of opposite chambers

$C, C' \in \Sigma$, i.e., Σ is the smallest convex chamber subcomplex of Δ containing C and C'. This simple observation leads to the following theorem, which shows that the nature of apartment systems in a spherical building is considerably simpler than in the general case:

Theorem 4.70. *A spherical building Δ admits a unique system of apartments. The apartments are precisely the convex hulls of pairs C, C' of opposite chambers.*

Proof. Let $\mathcal{A}$ be an arbitrary system of apartments, and let $\mathcal{A}'$ be the set of convex hulls of pairs of opposite chambers. Every apartment $\Sigma \in \mathcal{A}$ contains a pair of opposite chambers and is their convex hull, as we observed above; so $\mathcal{A} \subseteq \mathcal{A}'$. For the opposite inclusion, consider a pair C, C' of opposite chambers in Δ. There is an apartment $\Sigma \in \mathcal{A}$ containing them both, and Σ is equal to their convex hull; hence this convex hull is indeed in $\mathcal{A}$. $\square$

Remarks 4.71. (a) The theorem shows that the opposition relation contains a lot of information about the building Δ. In the thick case, one can in fact reconstruct Δ as a simplicial complex from the set of chambers together with the opposition relation on it. This is a result of Abramenko and Van Maldeghem [15]. The key step is to characterize adjacency in terms of opposition.

(b) In nonspherical buildings there can definitely exist apartment systems other than the complete one. We will see in Chapter 6 that such apartment systems arise naturally from group theory. But there is an easy example available now, namely, the case that Δ is a tree. The complete apartment system $\mathcal{A}$ in this case consists of all lines in Δ, but it easy to see that one usually does not need to take all of the lines as apartments in order to satisfy the building axioms. Readers familiar with ends of trees (see Serre [217, Sections I.2.2 and II.1.1]) can understand the situation as follows: The set $\mathcal{A}$ of lines is in 1–1 correspondence with the set of unordered pairs of distinct ends of Δ; in particular, $\mathcal{A}$ has a natural topology. To get a system of apartments, one need only take a dense subset of $\mathcal{A}$.

For future reference, we remark that the concept of opposite chamber extends to arbitrary simplices in a spherical building. Recall first that by definition, any apartment Σ is isomorphic to the complex $\Sigma(W, V)$ associated to a finite reflection group. It is not hard to check that this yields a well-defined simplicial automorphism op_Σ of Σ, called the *opposition involution*, corresponding to the automorphism $A \mapsto -A$ of $\Sigma(W, V)$. (See Section 1.6.2.) We can characterize it without reference to reflection groups by the property that A and $\mathrm{op}_\Sigma(A)$ have opposite sign sequences, i.e., they are contained in the same walls of Σ and are strictly separated by every wall of Σ that does not contain them.

Definition 4.72. We say that $\mathrm{op}_\Sigma(A)$ is the simplex *opposite* to A in Σ. We call two simplices $A, B \in \Delta$ *opposite*, and we write A op B, if they are opposite in some (or every) apartment containing them.

See Exercises 4.79 and 4.80 below for other characterizations of the opposition relation.

The notion of "opposite chamber" has many uses. It arises, for instance, if one attempts to analyze the homotopy type of a spherical building Δ (i.e., the homotopy type of the geometric realization $|\Delta|$). Here is a sketch of how that can be done, following [220].

Fix a chamber C of Δ and let Δ' be the subcomplex obtained by deleting all chambers opposite to C. We claim that Δ' is contractible. Now a contractible subcomplex can be collapsed to a point without affecting the homotopy type [129, p. 11; 224, 7.1.5 and 7.6.2]. So the claim yields the following theorem of Solomon and Tits [220]:

Theorem 4.73. *If Δ is a spherical building of rank n, then $|\Delta|$ has the homotopy type of a bouquet of $(n-1)$-spheres, where there is one sphere for every apartment containing a fixed chamber C.* $\square$

It remains to say something about the claim. Note first that every apartment Σ containing C admits a canonical type-preserving isomorphism to Σ_M, with C going to the fundamental chamber of Σ_M. Now $|\Sigma_M|$ can be identified with the unit sphere in the vector space V on which the reflection group W_M acts. Hence Σ admits a "spherical geometry"; in particular, the punctured sphere $|\Sigma \cap \Delta'|$ admits a canonical contracting homotopy that contracts it to the barycenter of $|C|$ along geodesics, i.e., arcs of great circles. One can show that the various homotopies, one for each apartment containing C, are compatible with one another; hence they fit together to give a well-defined contracting homotopy on Δ'.

Remarks 4.74. (a) The idea of introducing geodesics (and other geometric notions) into the study of buildings is extremely useful. We will return to it in Chapters 11 and 12. In particular, the spherical case will be treated in Section 12.2, Example 12.39.

(b) In Section 4.12 we give a purely combinatorial proof, also based on [220], of the contractibility of Δ'. That proof, while not as close to the geometric intuition as the proof sketched above, has the advantage that the method extends to nonspherical buildings and enables one to analyze their homotopy type as well. The result is that every nonspherical building is contractible.

(c) Readers familiar with Cohen–Macaulay complexes [36, Section 11] can easily deduce from our study of the homotopy type of a building that every building is a homotopy-Cohen–Macaulay complex. [The point here is that links in buildings are again buildings, so we also understand the homotopy type of any link.]

Exercises

4.75. Show that a spherical building is finite if and only if every panel is a face of only finitely many chambers.

4.76. Let Δ be a spherical building, and let α be a root of Δ as defined in Exercise 4.61. If Σ and Σ' are two distinct apartments containing α, show that $\Sigma \cap \Sigma' = \alpha$.

4.77. Let Δ be an arbitrary building, and let C and D be chambers. Show that $d(C, A) < d(C, D)$ for every panel A of D if and only if Δ is spherical and D is opposite C.

4.78. Let Δ be the complex $\Delta(V)$ associated to an n-dimensional vector space V, as in Section 4.3. Let C and C' be chambers with vertex sets $V_1 < \cdots < V_{n-1}$ and $V_1' < \cdots < V_{n-1}'$, respectively. Show that C and C' are opposite if and only if $V = V_i \oplus V_{n-i}'$ for all i.

4.79. Prove the following characterization of opposite simplices, which does not refer to apartments. Two simplices A, B in a spherical building Δ are opposite if and only they satisfy the following condition:

(Op) *For every chamber $C \geq A$ in Δ there is a chamber $D \geq B$ that is opposite C, and for every chamber $D \geq B$ there is a chamber $C \geq A$ that is opposite C.*

4.80. Here is another characterization of the opposition relation that does not refer to apartments. Given a simplex A in a spherical building Δ, show that for any $B \in \Delta$, A op B if and only if $\dim B \leq \dim A$ and $d(A, B) = \max \{d(A, B') \mid B' \in \Delta\}$.

4.8 The Weyl Distance Function

We return to an arbitrary building Δ. As usual, we assume that Δ has been provided with a type function having values in a set S. Recall that Δ then has a Coxeter matrix M and a Weyl group $W = W_M$, generated by S. In the terminology of Definition 4.37, Δ is a *building of type* (W, S). By Section 3.5, each apartment Σ has a well-defined Weyl distance function $\delta_\Sigma \colon \mathcal{C}(\Sigma) \times \mathcal{C}(\Sigma) \to W$, which contains information about types of galleries in Σ. Using the convexity of apartments, we will extend these functions to a function $\delta = \delta_\Delta$ defined on $\mathcal{C}(\Delta) \times \mathcal{C}(\Delta)$.

Proposition 4.81. *There is a function $\delta \colon \mathcal{C}(\Delta) \times \mathcal{C}(\Delta) \to W$ with the following properties:*

(1) *Given a minimal gallery $\Gamma \colon C_0, \ldots, C_d$ of type $\mathbf{s}(\Gamma) = (s_1, \ldots, s_d)$, $\delta(C_0, C_d)$ is the element $w = s_1 \cdots s_d$ represented by $\mathbf{s}(\Gamma)$.*

(2) *Let C and D be chambers, and let $w = \delta(C, D)$. The function $\Gamma \mapsto \mathbf{s}(\Gamma)$ gives a 1–1 correspondence between minimal galleries from C to D and reduced decompositions of w.*

Proof. Given chambers C, D, choose an apartment Σ containing them and set $\delta(C, D) = \delta_\Sigma(C, D)$. This is independent of the choice of Σ by Proposition 4.6. Assertions (1) and (2) now follow from the convexity of apartments and the corresponding facts about δ_Σ (Section 3.5). $\square$

Definition 4.82. Let Δ be a building of type (W, S), and let $\mathcal{C} = \mathcal{C}(\Delta)$. The function $\delta \colon \mathcal{C} \times \mathcal{C} \to W$ of Proposition 4.81 is called the *Weyl distance function* associated to Δ.

Recall from Proposition 4.41 that a gallery is minimal if and only if its type is reduced. So we can characterize δ as follows:

Proposition 4.83. *Let C and D be chambers, and suppose there is a gallery from C to D whose type $(s_1, \ldots, s_d)$ is a reduced word; then $\delta(C, D) = s_1 \cdots s_d$.* $\square$

Next we record some formal properties of δ.

Proposition 4.84. *The Weyl distance function $\delta \colon \mathcal{C}(\Delta) \times \mathcal{C}(\Delta) \to W$ has the following properties:*

(1) $\delta(C, D) = 1$ *if and only if* $C = D$.
(2) $\delta(D, C) = \delta(C, D)^{-1}$.
(3) *If $\delta(C', C) = s \in S$ and $\delta(C, D) = w$, then $\delta(C', D) = sw$ or w. If, in addition, $l(sw) = l(w) + 1$, then $\delta(C', D) = sw$.*
(4) *If $\delta(C, D) = w$, then for any $s \in S$ there is a chamber C' such that $\delta(C', C) = s$ and $\delta(C', D) = sw$. If $l(sw) = l(w) - 1$, then there is a unique such C'.*

Proof. (1) and (2) are immediate. To prove (3) and (4), let $\delta(C, D) = w$, let $\mathbf{s} = (s_1, \ldots, s_d)$ be a reduced decomposition of w, and choose a gallery $C = C_0, \ldots, C_d = D$ of type $\mathbf{s}$. If $l(sw) = l(w) + 1$, then for any C' with $\delta(C', C) = s$, the gallery $C', C_0, \ldots, C_d$ has reduced type $(s, s_1, \ldots, s_d)$, so $\delta(C', D) = sw$ by Proposition 4.83. Both (3) and (4) follow in this case. Now suppose $l(sw) = s(w) - 1$. Then we can choose $\mathbf{s}$ such that $s_1 = s$. We then have $\delta(C, C_1) = s$ and $\delta(C_1, D) = sw$. If C' is any chamber distinct from C_1 with $\delta(C', C) = s$, then the three chambers C', C, C_1 are all s-adjacent to one another, and we have a gallery $C', C_1, \ldots, C_d$ of reduced type $\mathbf{s}$. Hence $\delta(C', D) = w$ and the proof of (3) and (4) is complete. $\square$

Properties (1) and (2) are analogues of properties of the distance function d on a metric space $[d(x, y) = 0$ if and only if $x = y$, and $d(y, x) = d(x, y)]$. Similarly, the first sentence of property (3) is similar to the triangle inequality [if $d(x', x) = 1$ and $d(x, y) = l$, then $l - 1 \leq d(x', y) \leq l + 1$]. We will often draw a triangle such as

as a schematic illustration of this "triangle inequality." Property (4) and the second sentence of (3) are more special, but one can get some intuition about them by thinking about trees. Indeed, they are reminiscent of the following property of a tree: If x and y are distinct vertices of a tree, then there is a unique vertex x' adjacent to x such that $d(x', y) = d(x, y) - 1$; for all other vertices x' adjacent to x, $d(x', y) = d(x, y) + 1$. [Think of y as the root of the tree; there are many ways to move away from the root, but only one way to move closer.]

Remarks 4.85. (a) We could equally well have formulated and proved analogues of (3) and (4) with the s-adjacency on the right instead of the left, i.e., for the situation

$$C \xrightarrow{\ w\ } D$$
$$\{ws, w\} \searrow \quad \downarrow s$$
$$D'$$

but there is no need to do so. Indeed, the two versions are formally equivalent in the presence of (2).

(b) Recall that one can compute δ in an apartment using arbitrary galleries, not just minimal ones. As a result, one has a very strong version of the triangle inequality for three chambers in an apartment (see Exercise 3.88): $\delta(C_1, C_3) = \delta(C_1, C_2)\delta(C_2, C_3)$. For three chambers in a building, however, (4) shows that this need not hold. In fact, we will see in Chapter 5 that this equality holds for three chambers in Δ if and only if they are contained in an apartment of the complete apartment system. (See Exercise 5.77.)

We can now complete the circle of ideas begun in Remark 4.56:

Theorem 4.86. *Let Σ be a thin, convex chamber subcomplex of Δ. Then Σ is an apartment in the complete apartment system.*

Proof. By Proposition 4.59, it suffices to show that $\Sigma \cong \Sigma(W, S)$. Note that both Σ and $\Sigma(W, S)$ are determined by their underlying chamber system, as in Section A.1.4. For $\Sigma(W, S)$, we already observed this in Section 3.2. For Σ, the assertion follows from convexity. Indeed, for any simplex $A \in \Sigma$, $\mathcal{C}(\Sigma)_{\geq A} = \mathcal{C}(\Sigma) \cap \mathcal{C}(\Delta)_{\geq A}$, which is convex by Exercise 4.13 and hence is gallery connected. It therefore suffices to show that Σ and $\Sigma(W, S)$ have isomorphic chamber systems. This is where the Weyl distance function is useful.

Fix an arbitrary "fundamental chamber" $C_0 \in \mathcal{C}(\Sigma)$, and define $\phi \colon \mathcal{C}(\Sigma) \to W$ by $\phi(C) := \delta(C_0, C)$ for $C \in \mathcal{C}(\Sigma)$. We show first that ϕ is surjective: Given $w \in W$, choose a reduced decomposition $\mathbf{s}$ of w and consider the gallery of type $\mathbf{s}$ in Σ starting at C_0. It ends at a chamber C such that $\delta(C_0, C) = w$, so $\phi(C) = w$.

Next, we claim that ϕ preserves s-adjacency for each $s \in S$; in other words, if C and C' are s-adjacent chambers of Σ, then $\phi(C') = \phi(C)s$. Let $w = \phi(C) = \delta(C_0, C)$. If $l(ws) = l(w) + 1$, our claim follows from property (3) of δ (and Remark 4.85(a)). If $l(ws) = l(w) - 1$, then there is a minimal gallery from C_0 to C in Δ ending with an s-adjacency. The second-to-last chamber of this gallery is in Σ by convexity; hence it is equal to C' by thinness. Thus $\delta(C_0, C') = ws$, and the claim is proved.

By repeated applications of the claim, we obtain

$$\phi(C') = \phi(C)\delta(C, C') \tag{4.2}$$

for any $C, C' \in \mathcal{C}(\Sigma)$. In particular, ϕ is injective and hence bijective. Moreover, it follows from (4.2) that if $\phi(C)$ is s-adjacent to $\phi(C')$, then C is s-adjacent to C'. Thus ϕ is an isomorphism of chamber systems, and the proof is complete. $\square$

Remark 4.87. Equation (4.2) can also be written as

$$\phi(C)^{-1}\phi(C') = \delta(C, C')$$

or, equivalently,

$$\delta(\phi(C), \phi(C')) = \delta(C, C') , \tag{4.3}$$

where the δ on the left denotes Weyl distance in $\Sigma(W, S)$. Thus ϕ is an "isometry" with respect to Weyl distance. Such isometries play a central role in the approach to buildings that we will take in Chapter 5.

Finally, we return to the general theory and observe that, as in Section 3.5, we can extend the Weyl distance function to arbitrary pairs of simplices.

Proposition 4.88. *Given two simplices* $A, B \in \Delta$, *there is an element* $\delta(A, B) \in W$ *such that*

$$\delta(A, B) = \delta(C_0, C_d) \tag{4.4}$$

for any minimal gallery $C_0, \ldots, C_d$ *from* A *to* B.

Proof. Choose an apartment Σ containing A and B, and set $\delta(A, B) = \delta_\Sigma(A, B)$. This is independent of the choice of Σ by Proposition 4.6. Equation (4.4) follows from Proposition 3.87, since the given gallery is a minimal gallery from C_0 to C_d and hence is contained in an apartment. $\square$

Exercises

4.89. Let Δ be a building of type (W, S) and let Δ' be a subbuilding (Definition 4.62). Show that the Weyl distance function on Δ' is the restriction to $\mathcal{C}(\Delta') \times \mathcal{C}(\Delta')$ of the Weyl distance function on Δ.

4.90. Let Δ be a spherical building. Given a chamber C and an apartment Σ of Δ, show that Σ contains a chamber D that is opposite to C.

4.91. For any apartment Σ and chamber $C \in \Sigma$, show that the retraction $\rho = \rho_{\Sigma,C}$ preserves Weyl distances from C, i.e., $\delta(C, \rho(D)) = \delta(C, D)$ for any chamber $D \in \Delta$.

4.92. Let W_J be a standard subgroup of W, where $J \subseteq S$. Show that two chambers C, D are in the same J-residue if and only if $\delta(C, D) \in W_J$.

4.93. Suppose $\Delta = \Delta(V)$ as in Section 4.3. Show that W can be identified with the symmetric group on n letters and that δ associates to any pair C, C' of chambers the Jordan–Hölder permutation $\pi(C, C')$.

4.94. Given a chamber $C \in \mathcal{C} := \mathcal{C}(\Delta)$, let $\phi = \phi_C : \mathcal{C} \to W$ be the map $\delta(C, -)$.

  (a) Show that ϕ preserves s-equivalence for all $s \in S$ and hence extends to a type-preserving chamber map $\phi : \Delta \to \Sigma(W, S)$.

  (b) Show that the chamber map ϕ maps every apartment Σ containing C isomorphically onto $\Sigma(W, S)$.

  (c) Given an apartment Σ containing C, how is ϕ related to the retraction $\rho_{\Sigma,C}$?

  (d) How can we reconstruct the canonical maps ρ of Exercise 4.51 from maps of the form ϕ_C?

4.9 Projections (Products)

We continue to assume that Δ is a building of type (W, S). Our goal here is to define, for any simplices $A, B \in \Delta$, a *product* $AB \in \Delta$, also called the *projection* of B onto A and denoted by $\mathrm{proj}_A B$. This will extend the product that is already known to exist in any apartment (Section 3.6.4). The reader is warned that the product is generally not associative, although its restriction to any apartment is associative. For this reason, the term "projection" is much more common than "product" in the literature. On the other hand, the product notation has the advantage of leading to simpler-looking formulas. We begin with the important case in which the second factor B is a chamber.

Proposition 4.95. *Given a simplex A and a chamber C, let AC be the product of A and C in any apartment containing A and C. Then AC is independent of the choice of apartment, and every minimal gallery from A to C starts with AC. Moreover, the gate property*

$$d(C, D) = d(C, AC) + d(AC, D) \tag{4.5}$$

holds for any chamber $D \geq A$.

Proof. Choose a minimal gallery $\Gamma \colon C_0, \ldots, C_d$ from A to C. If Σ is any apartment containing A and C, then Γ is contained in Σ by Proposition 4.40; hence C_0 is equal to the product AC defined in Σ (Section 3.6.4). This proves the first assertion of the proposition. To prove the gate property (4.5), choose Σ to contain D and C (hence also A), and use the fact that the gate property is known to hold in Σ (Section 3.6.4, equation (3.11)). $\qquad\square$

We turn now to the general case, in which B is arbitrary, beginning with the generalization to buildings of Proposition 3.106. Let $\mathcal{C} = \mathcal{C}(\Delta)$. As in Section 3.6.4, we set $\mathcal{C}_A := \mathcal{C}_{\geq A}$, and we denote by $\mathcal{C}_{A,B}$ the set of chambers in $\mathcal{C}_A$ that can start a minimal gallery in Δ from A to B. And, as in Section 3.6.4, we know that the sets of the form $\mathcal{C}_A$ are precisely the residues (Corollary 4.11).

Proposition 4.96.

(1) *$\mathcal{C}_{A,B}$ is the image of the projection map $\mathcal{C}_B \to \mathcal{C}_A$ given by $D \mapsto AD$.*
(2) *The projection maps $\mathcal{C}_{A,B} \rightleftarrows \mathcal{C}_{B,A}$, given by $C \mapsto BC$ and $D \mapsto AD$ ($C \in \mathcal{C}_{A,B}$, $D \in \mathcal{C}_{B,A}$), define mutually inverse bijections.*
(3) *Given any minimal gallery $C_0, \ldots, C_d$ from A to B, we have $C_0 = AC_d$ and $C_d = BC_0$. In other words, C_0 and C_d correspond to one another under the bijections in (2).*

Proof. A minimal gallery $C_0, \ldots, C_d$ from A to B is also minimal from A to C_d, so $C_0 = AC_d$ by Proposition 4.95. Similarly, $C_d = BC_0$. This proves (3) and shows that $\mathcal{C}_{A,B}$ is contained in the image of the projection map $\mathcal{C}_B \to \mathcal{C}_A$, which is part of (1). To prove the opposite inclusion, consider any $D \in \mathcal{C}_B$, and choose an apartment Σ containing A and D. By Proposition 3.106, there is a minimal gallery from A to B in Σ starting with AD, and this is still minimal in Δ by Remark 4.35. This proves (1). It follows from (1) that the projection maps do define maps $\mathcal{C}_{A,B} \rightleftarrows \mathcal{C}_{B,A}$, and it is easy to check (using the corresponding fact about Coxeter complexes) that these maps are inverse to one another, whence (2). $\qquad\square$

We can now prove the main result of this section.

Theorem 4.97. *Let A be a simplex of cotype J, let B be a simplex of cotype K, and let w be the Weyl distance $\delta(A, B)$. Then $\mathcal{C}_{A,B}$ is a residue of type $J_1 := J \cap wKw^{-1}$. In other words, there is a simplex AB of cotype J_1 such that $\mathcal{C}_{A,B} = \mathcal{C}_{AB}$, i.e., the chambers that can start a minimal gallery from*

A to B are precisely those having AB as a face. Moreover, the product AB so defined agrees with the product computed in any apartment containing A and B.

Proof. Given $C_1, C_2 \in \mathcal{C}_{A,B}$, let $D_i = BC_i$ for $i = 1, 2$. Then $C_1 = AD_1$ by Proposition 4.96, so any apartment Σ containing C_2 and D_1 also contains C_1 [because it contains A]. Similarly, Σ contains B and hence it contains D_2. Recalling that $d(C_1, D_1) = d(A, B) = d(C_2, D_2)$, we see that C_1 and C_2 are chambers in Σ that start minimal galleries from A to B in Σ. We can therefore apply Theorem 3.108 to Σ to conclude that they are J_1-equivalent. Thus $\mathcal{C}_{A,B}$ is contained in a residue of type J_1, so there is a unique simplex AB of cotype J_1 such that $\mathcal{C}_{A,B} \subseteq \mathcal{C}_{AB}$.

It is obvious that the product AB so defined can be computed in any apartment containing A and B. It remains only to show that every chamber $C \geq AB$ can start a minimal gallery from A to B. But this follows again from Theorem 3.108, applied to any apartment containing C and B.   □

Exercises

4.98. Give an example to show that the product of simplices is not associative in general.

4.99. In Exercise 4.14 we saw that the relation "$\operatorname{supp} A = \operatorname{supp} B$" makes sense in a building even though the support of a simplex is not well defined. Reformulate that relation using projections.

4.100. If $\operatorname{supp} A = \operatorname{supp} B$, show that left multiplication by B and A define mutually inverse bijections $\Delta_{\geq A} \rightleftarrows \Delta_{\geq B}$. In particular, this holds if Δ is spherical and A and B are opposite.

4.10 Applications of Projections

It will become more and more clear as we proceed that projections are a fundamental tool in the theory of buildings. Here we just give a few sample applications, the first of which is a proof of the result stated in Remark 4.5:

Proposition 4.101. *For any two apartments Σ, Σ' there is a type-preserving isomorphism $\phi\colon \Sigma \xrightarrow{\sim} \Sigma'$ fixing $\Sigma \cap \Sigma'$ pointwise.*

Proof. The proof is similar in spirit to the proof of (B2″) in Section 4.1. Let M be a maximal simplex in $\Sigma \cap \Sigma'$. Choose chambers $C \in \Sigma$ and $C' \in \Sigma'$ having M as a face. For each simplex $A \in \Sigma \cap \Sigma'$, we will construct a type-preserving isomorphism $\phi_A\colon \Sigma \xrightarrow{\sim} \Sigma'$ such that ϕ_A fixes M and A and $\phi_A(C) = C'$. By the standard uniqueness argument, all of these ϕ_A are equal and provide the desired ϕ. [Note that the standard uniqueness argument works because all ϕ_A, being type-preserving, agree pointwise on C.]

To construct ϕ_A, start with any type-preserving isomorphism $\psi\colon \Sigma \xrightarrow{\sim} \Sigma'$ fixing M and A (Proposition 4.6). Now let ϕ_A be the composite $w \circ \psi$, where w is the unique type-preserving automorphism of Σ' taking $\psi(C)$ to C'. Note that w, hence also ϕ_A, necessarily fixes M, since M is a face of both $\psi(C)$ and C'. What is less obvious is that w (hence also ϕ_A) fixes A. This is where projections are useful.

Since Σ and Σ' are both closed under products in Δ, the same is true of $K := \Sigma \cap \Sigma'$. In particular, K is a convex subcomplex of Σ' in the sense of Section 3.6.6. Hence $K \subseteq \operatorname{supp}_{\Sigma'} M$, the latter being the support of M in Σ' (Lemma 3.140). Proposition 3.100 now implies that the stabilizer of M in the group of type-preserving automorphisms of Σ' fixes K pointwise, so $wA = A$. $\qquad\square$

Remark 4.102. See Exercise 4.108 below for a variant of this proof, which is conceptually clearer but requires slightly more from the theory of convex subcomplexes of Coxeter complexes.

Here is a second application of projections, which requires only the "easy" case given in Proposition 4.95. Recall from Section 4.7 that two simplices in a spherical building are said to be *opposite* if they are opposite in some (and hence every) apartment containing them.

Proposition 4.103. *Let C be a chamber of a spherical building Δ and let P be a panel in Δ that is opposite some panel of C. Among the chambers $D > P$, there is a unique one that is not opposite C, namely, the chamber $D = PC$.*

Proof. Note first that PC is not opposite C, since in any apartment containing C and P, the wall $\operatorname{supp} P$ containing P does not separate C from PC. Now consider any chamber $D > P$ other than PC. Any apartment Σ containing C and D also contains PC, and the sign sequences of C and PC in Σ are opposite with respect to all walls except $\operatorname{supp} P$. Since D is adjacent to PC along that wall, it follows that the sign sequence of D is opposite to that of C with respect to all walls of Σ. $\qquad\square$

Finally, we use projections to prove one more result along the same lines.

Proposition 4.104. *Let Δ be a thick spherical building. For any two chambers of Δ, there is a chamber that is opposite both of them.*

Remark 4.105. This is a simple application of Proposition 4.103 if the two given chambers C, D are adjacent. To see this, let P be the common panel $C \cap D$, and choose, by thickness, a chamber $E > P$ different from C, D. Choose an apartment Σ containing C and E, and let C' be the chamber of Σ opposite C; see Figure 4.5. Then E is not opposite C', so, by Proposition 4.103, E is the unique chamber $> P$ that is not opposite C'. In particular, D is opposite C', and hence C' is the desired chamber opposite both C and D.

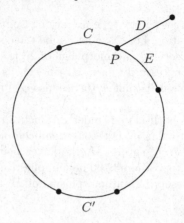

Fig. 4.5. A chamber opposite both C and D.

In the general case we will give a different argument, which is based on the following two lemmas.

Lemma 4.106. *Let C and D be chambers in a spherical building. If C and D are not opposite, then there is a panel $P < C$ such that $C = PD$.*

Proof. This is actually a special case of Exercise 4.77 (see also part (b) of Exercise 1.59), but we will give an independent proof. Let D' be the chamber opposite D in an apartment containing C and D, and choose a minimal gallery from D' to D passing through C (Lemma 4.69). Since $C \neq D'$, there is a chamber E preceding C in this gallery, and $d(E, D) = d(C, D) + 1$. If P is the panel $E \cap C$, it follows easily from the gate property (see Proposition 4.95) that $PD = C$. Indeed, all chambers $> P$ other than PD are at the same distance $d(PD, D) + 1$ from D; so if we encounter two chambers $> P$ at different distances from D, the closer one must be PD.   □

Lemma 4.107. *Let C, D, and E be chambers in a thick spherical building. If E is not opposite D, then there is a chamber E' adjacent to E such that $d(E', D) > d(E, D)$ and $d(E', C) \geq d(E, C)$.*

Proof. By Lemma 4.106, there is a panel $P < E$ such that $E = PD$. By thickness, there is chamber $E' > P$ different from both PC and $PD = E$. Using the gate property, we find that

$$d(E', D) = d(PD, D) + 1 = d(E, D) + 1$$

and

$$d(E', C) = d(PC, C) + 1 \geq d(E, C).   □$$

Proof of Proposition 4.104. Let C and D be the two given chambers, and choose, among the chambers E opposite C, one that is at maximal distance from D. It follows from Lemma 4.107 that E is opposite D, since otherwise we could find a chamber E' that is still opposite C but further away from D.   □

Exercises

4.108. This exercise gives an alternative proof of Proposition 4.101. Let Σ and Σ' be apartments in a building, and let $K := \Sigma \cap \Sigma'$. It is a convex subcomplex of both Σ and Σ'; in particular, it is a chamber complex in which every panel is a face of at most two chambers (Proposition 3.136). Let M be a K-chamber, and let $\phi \colon \Sigma \to \Sigma'$ be an isomorphism fixing M pointwise. The following steps outline a proof that ϕ fixes K pointwise. Fill in the details.

(a) Set $\Lambda := \mathrm{supp}_{\Sigma} M$ and $\Lambda' := \mathrm{supp}_{\Sigma'} M$. Then Λ and Λ' are chamber complexes, of the same dimension as K, in which every panel is a face of at most two chambers. Moreover, $K = \Lambda \cap \Lambda'$. See Figure 4.6.

Fig. 4.6. The intersection of two apartments.

(b) ϕ maps Λ isomorphically onto Λ'.
(c) Since ϕ fixes a K-chamber pointwise, ϕ is the identity on K by the standard uniqueness argument.

4.109. Let α and α' be roots in a building Δ (see Exercise 4.61). If $\alpha \cap \alpha' = \partial\alpha = \partial\alpha'$, show that $\alpha \cup \alpha'$ is an apartment in the complete apartment system of Δ.

4.110. Show that the conclusion of Proposition 4.104 is false if Δ is not thick.

4.111. What happens in Lemma 4.107 if the building is not spherical?

4.112. Let Δ be a thick spherical building. For any panel A, show that the cardinality of $\mathcal{C}_A$ depends only on the type of A. Give an example to show that one cannot drop the sphericity assumption.

4.113. Let Δ be a spherical building. Suppose Δ has one chamber C such that every panel of C is a face of at least 3 chambers. Show that Δ is thick. Give an example to show that one cannot drop the sphericity assumption.

4.114. Let Δ be the flag complex of a generalized quadrangle Q (see Definition 4.20 and the discussion following it). Suppose Q has a thick line (i.e., a line with at least 3 points) and a thick point (i.e., a point that is on at least 3 lines). Show that Δ is thick.

4.11 Convex Subcomplexes

Throughout this section, Δ denotes an arbitrary building.

4.11.1 Chamber Subcomplexes

The theory of convex chamber subcomplexes of Δ (see Definition 3.129) is similar to the corresponding theory for Coxeter complexes, except that there is no longer a characterization of convex subcomplexes as intersections of roots:

Proposition 4.115. *If Δ' is a subcomplex of Δ that contains at least one chamber, then the following conditions are equivalent:*

(i) Δ' *is a convex chamber subcomplex of Δ.*

(ii) Δ' *is closed under products, i.e., for any $A, B \in \Delta'$, the projection* $\mathrm{proj}_A B = AB$ *is in Δ'.*

(iii) *Given $A, C \in \Delta'$ with C a chamber of Δ, every minimal gallery from A to C in Δ is contained in Δ'.*

Moreover, an intersection of convex chamber subcomplexes of Δ is again a convex chamber subcomplex if it contains at least one chamber.

Proof. The arguments given in Section 3.6.6 for Coxeter complexes (see Lemma 3.130 and Theorem 3.131) extend to buildings, with no essential change.   □

In view of the last assertion of the proposition, we can construct convex hulls in the usual way:

Definition 4.116. Given an arbitrary collection of simplices containing at least one chamber, their *convex hull* is the intersection of all convex chamber subcomplexes containing them.

The convex hull is itself a convex chamber subcomplex containing the given simplices, so it is the smallest one. We will also call the convex hull the *combinatorial convex hull* in any context where the term "convex hull" might be ambiguous.

An important example is the convex hull $\Gamma(A, C)$ of a simplex A and a chamber C. Since apartments are convex, $\Gamma(A, C)$ is contained in any apartment Σ containing A and C; hence $\Gamma(A, C)$ coincides with the convex hull of A and C in Σ. In particular, the convex hull $\Gamma(C, D)$ of two chambers C, D is a convex chamber subcomplex whose chambers are precisely the chambers that can occur in a minimal gallery between C and D (see Example 3.133(a)). A special case of this played an important role in Section 4.7.

The following example gives another interesting special case, involving roots. Here a *root* of Δ is a subcomplex that is contained in some apartment Σ and is a root in Σ. It then follows easily from the building axioms that α is a root in every apartment containing it and that α has a well-defined boundary $\partial\alpha$, equal to its boundary in any apartment containing it; see Exercise 4.61.

Example 4.117. Suppose α is a root of a spherical building Δ. Let P be a panel in $\partial\alpha$, let P' be the panel in $\partial\alpha$ opposite P, and let C' be the chamber of α having P' as a face. Then $\Gamma(P, C') = \alpha$. To see this, choose an apartment Σ containing α. As we noted above, $\Gamma(P, C')$ coincides with the convex hull of P and C' in Σ. Our assertion now follows from Example 3.133(d).

As another illustration of convex hulls, we give a characterization of the set of apartments containing a given root of a spherical building. This result will be used in Chapter 7. We need some notation. Given a root α of Δ and a panel $P \in \partial\alpha$, set

$$\mathcal{C}(P, \alpha) := \mathcal{C}_P \smallsetminus \{C\} \ ,$$

where $\mathcal{C}_P = \mathcal{C}(\Delta)_{\geq P}$ and C is the unique chamber in α having P as a face. One can visualize $\mathcal{C}(P, \alpha)$ as the set of chambers of Δ that are "attached to α along P." We denote by $\mathcal{A}(\alpha)$ the set of apartments of Δ containing α.

Lemma 4.118. *Let α be a root in a spherical building Δ, and let P be a panel in $\partial\alpha$. Then there is a canonical bijection from $\mathcal{C}(P, \alpha)$ to $\mathcal{A}(\alpha)$. It associates to any chamber $D \in \mathcal{C}(P, \alpha)$ the convex hull of D and α.*

Proof. Let C be the chamber in α having P as a face, let P' be the panel opposite P in α, and let C' be the chamber in α having P' as a face. By Proposition 4.103, C is the unique chamber in $\mathcal{C}_P$ that is *not* opposite C'. In particular, every chamber $D \in \mathcal{C}(P, \alpha)$ is opposite C'; hence the convex hull $\Gamma(D, C')$ is an apartment Σ'. Now $\alpha = \Gamma(P, C')$ by Example 4.117, so Σ' contains α and hence is the convex hull of D and α. We therefore have a map $\mathcal{C}(P, \alpha) \to \mathcal{A}(\alpha)$ that sends a chamber $D \in \mathcal{C}(P, \alpha)$ to the convex hull Σ' of D and α. It is easily seen to be a bijection; the inverse associates to an apartment $\Sigma' \in \mathcal{A}(\alpha)$ the unique chamber $D \in \Sigma'$ opposite C'. $\square$

Exercise 4.119. Let Δ be a spherical building of diameter d. Show that the roots of Δ are precisely the convex hulls $\Gamma(C, D)$, where C and D are chambers such that $d(C, D) = d - 1$.

*4.11.2 General Subcomplexes

The results of this optional subsection will not be needed later, so we will be brief. As in Section 3.6.6, we can extend the notion of convexity to arbitrary subcomplexes:

Definition 4.120. Let Δ' be a subcomplex of Δ. We say that Δ' is a *convex subcomplex* if it is closed under products.

See Remark 3.132 for the intuition behind this definition. Exercise 4.124 below provides further motivation.

There is a smallest convex subcomplex containing any given collection of simplices, called their *convex hull*. As above, the convex hull $\Gamma(A, B)$ of *two* simplices A, B is the same as their convex hull in any apartment containing them. We record the one useful fact about general convex subcomplexes:

Proposition 4.121. *Every convex subcomplex Δ' of Δ is a chamber complex.*

Proof. Let A and B be maximal simplices of Δ', choose an apartment Σ containing them, and let $\Sigma' := \Sigma \cap \Delta'$. Then Σ' is a convex subcomplex of Σ; hence it is a chamber complex by Proposition 3.136. Since A and B are maximal in Σ', they have the same dimension and are connected by a gallery in Σ'; this is also a gallery in Δ'. $\qquad\square$

Remark 4.122. Our definition of convexity seems to us to be the most useful and intuitive one, for reasons mentioned above. But the reader should be aware that there is a different definition in the literature, due to Tits [247, 1.5], according to which a subcomplex is called convex if it is an intersection of convex chamber subcomplexes. In view of Proposition 4.115, convexity in the sense of Tits (or "T-convexity" for short) implies convexity in our sense. And Proposition 3.137 says that T-convexity is equivalent to convexity if Δ consists of a single apartment. But convexity in our sense does *not* imply T-convexity in general. [The point is that we have no tools in general for constructing "enough" convex chamber subcomplexes. In the case of a single apartment, however, we can use roots.] If fact, H. Van Maldeghem has pointed out to us that there are infinitely many counterexamples. The remainder of this section is devoted to a detailed description of the smallest of these.

In Exercise 4.24 we constructed a generalized quadrangle Q whose associated building Δ of type C_2 has the following properties:

(a) Every vertex is a face of exactly 3 edges.
(b) Δ contains 5 pairwise opposite vertices.

Clearly any set X of pairwise opposite vertices forms a 0-dimensional convex subcomplex of Δ, since $uv = u$ if u and v are opposite (see part (3) of Proposition 3.112). But the following proposition shows that X is not T-convex if $|X|$ is at least 4.

Proposition 4.123. *With Δ as above, let X be a set of 4 pairwise opposite vertices. Then the only convex chamber subcomplex of Δ containing X is Δ itself.*

Proof. Let Δ' be a convex chamber subcomplex containing X. We will show that Δ' is a thick subbuilding of Δ, from which it follows (by (a) above) that $\Delta' = \Delta$. The first step is to show that Δ' contains an apartment of Δ, so that Δ' is a subbuilding by Exercise 4.125 below. The reader is advised to draw a picture as we proceed; the final result is in Figure 4.7.

Assume for definiteness that the elements of X are points of the quadrangle Q, and let x, x' be two of these points. It follows easily from the convexity assumption that Δ' contains a geodesic from x to x', i.e., a path of length 4 (see Exercise 4.48). Denote the vertices along this path by (x, L, p, M, x'). The vertex p is at distance 2 from at most 3 elements of X, since each of the 3 lines of Q containing p can be incident with at most one element of X. So, since $|X| = 4$, there is a $y \in X$ that is opposite p.

Using $d(-,-)$ to denote the graph distance between two vertices, we have $d(y, M) = 3$ and hence, by the exercise just cited, Δ' contains the geodesic (y, L', q, M) from y to M. Note that q is different from p and x' because p and x' are both opposite y; in the case of x', we are using here the assumption that any two elements of X are opposite. For the same reason, $L' \neq M$. It follows that the path (L', q, M, p, L) of length 4 is a geodesic, so that L' and L are opposite [use the fact that Δ has girth 8 by Proposition 4.44.] Hence $d(x, L') = 3$, and Δ' contains the geodesic (x, M', p', L') from x to L'. Combining this geodesic with the path (x, L, p, M, q, L') of length 5, and using again the fact that Δ has girth 8, we obtain an octagon in Δ' as shown in Figure 4.7. This octagon is an apartment of Δ by Proposition 4.59, and we

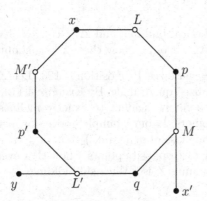

Fig. 4.7. An apartment in Δ'.

are done with step 1 of the proof.

The second and final step is to show that Δ' is thick. By Exercise 4.114 it suffices to show that Δ' contains a thick point and a thick line. Now we already saw, as a byproduct of step 1, that Δ' has thick lines (M and L' in the notation above). So it suffices to find a thick point. Choose $x \in X$ and a line $L \in \Delta'$ incident to x. Denote the points of $X \setminus \{x\}$ by y_1, y_2, y_3, and note that since x is opposite y_i, we have $d(L, y_i) = 3$ for $i = 1, 2, 3$. Hence Δ' contains the geodesic (L, p_i, M_i, y_i) from L to y_i for each i. Now the line L is incident to exactly 3 points, one of which is x (which is different from all the p_i), so at least two of the p_i have to coincide, say $p_1 = p_2 =: p$. We have $M_1 \neq M_2$ since y_1 and y_2 are opposite, so p is the desired thick point, as illustrated in Figure 4.8. $\qquad\square$

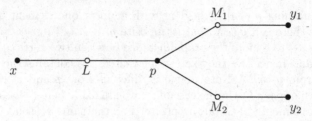

Fig. 4.8. A thick point in Δ'.

Exercises

4.124. Let Δ be an arbitrary building and let Δ' be a subcomplex. Show that Δ' is convex if and only if $\Delta' \cap \Sigma$ is a convex subcomplex of Σ for every apartment Σ.

4.125. Let Δ be a spherical building and Δ' a convex chamber subcomplex. If Δ' contains an apartment of Δ, show that Δ' is a subbuilding of Δ.

4.126. This exercise generalizes Proposition 4.123. Let Δ be the incidence graph of a thick generalized quadrangle. By Exercise 4.112, there are numbers m, n such that every point is incident to exactly m lines and every line is incident to exactly n points. In our example above, for instance, we had $m = n = 3$. Assume (without loss of generality) that $m \geq n$, and let X be a set consisting of $m + 1$ pairwise opposite points. Show that every convex chamber subcomplex of Δ containing X is a thick subbuilding.

*4.12 The Homotopy Type of a Building

The purpose of this section, as stated in Remark 4.74(b), is to outline Solomon's combinatorial method for analyzing the homotopy type of a building. For ease of reference, we restate the result. Fix a chamber $C \in \Delta$.

Theorem 4.127. *If Δ is a spherical building of rank n, then $|\Delta|$ has the homotopy type of a bouquet of $(n-1)$-spheres, where there is one sphere for every apartment containing C. If Δ is a nonspherical building, then $|\Delta|$ is contractible.*

Our outline of the proof will be complete as far as the theory of buildings is concerned, but we will omit some homotopy-theoretic details. The idea is to start with C and then keep track of the homotopy type as one successively adjoins the chambers adjacent to C, then the chambers at distance 2 from C, and so on. The lemma below enables one to figure out what happens each time a new chamber is adjoined (along with its faces). Recall that if A is a simplex in an abstract simplicial complex Δ, then $\overline{A}$ denotes the subcomplex $\Delta_{\leq A}$, whose geometric realization is the closed simplex in $|\Delta|$ corresponding to A.

Lemma 4.128. *Let Δ be an arbitrary building. Fix a chamber C and an integer $d \geq 1$, and let $\mathcal{D}$ be a set of chambers with the following two properties:*

(1) $d(C, D) \leq d$ *for every $D \in \mathcal{D}$.*
(2) $\mathcal{D}$ *contains every chamber $D \in \Delta$ with $d(C, D) < d$.*

Let Δ' be the subcomplex of Δ generated by $\mathcal{D}$, and let D be a chamber of Δ such that $D \notin \mathcal{D}$ and $d(C, D) = d$. Then

$$\overline{D} \cap \Delta' = \bigcup_{A \in \mathcal{P}} \overline{A} \, ,$$

where $\mathcal{P}$ is the set of panels A of D such that $d(C, A) < d$. The set $\mathcal{P}$ contains all the panels of D if and only if Δ is spherical and of diameter d.

Readers familiar with homotopy theory will find it a routine matter to use this lemma to complete the proof of Theorem 4.127. We proceed now to the proof of the lemma, which is a fairly easy consequence of the theory of projections. In fact, we need only the case given by Proposition 4.95, where the second factor is a chamber.

Proof of Lemma 4.128. The first claim is that

$$\overline{D} \cap \Delta' = \{B < D \mid d(C, B) < d\} \, .$$

The right side is trivially contained in the left side by hypothesis (2). To prove the opposite inclusion, suppose $B \in \overline{D} \cap \Delta'$. Then there is a chamber $D' \in \mathcal{D}$ with $B < D'$, and we have

$$d(C, B) \leq d(C, D') \leq d = d(C, D)$$

by hypothesis (1). We cannot have $d(C, B) = d$, since that would imply $d(C, B) = d(C, D') = d(C, D)$ and hence that $D' = BC = D$. So $d(C, B) < d$, as required.

We claim next that if $B < D$ with $d(C, B) < d$, then there is a panel A of D such that $B \leq A$ and $d(C, A) < d$. To see this, recall that

$$d = d(C, D) = d(C, BC) + d(BC, D)$$

by the gate property. In view of the convexity of $\mathcal{C}(\Delta)_{\geq B}$ (Exercise 4.13), it follows that there is a minimal gallery $\Gamma \colon C = C_0, \ldots, C_d = D$ with $B < C_{d-1}$. Setting $A := C_{d-1} \cap D$, we then have $B \leq A$ and $d(C, A) < d$, proving the claim.

Combining the two claims, we have

$$\overline{D} \cap \Delta' = \bigcup_{A \in \mathcal{P}} \overline{A} \, ,$$

where $\mathcal{P}$ is the set of panels A of D such that $d(C, A) < d$. Exercise 4.77 says that $\mathcal{P}$ contains every panel of D if and only if Δ is spherical and $d = \operatorname{diam} \Delta$. $\square$

Remark 4.129. Our approach to the Solomon–Tits theorem is closely related to the theory of shellability. See Björner [35, 36].

Exercise 4.130. Let Δ be a building, and let Δ' be a chamber subcomplex of Δ that is *starlike* from some chamber $C \in \Delta'$, in the sense that every minimal gallery from C to a chamber of Δ' is contained in Δ'. Determine the homotopy type of $|\Delta'|$.

*4.13 The Axioms for a Thick Building

The purpose of this final section of the chapter is to show that axiom (B0) can essentially be eliminated if Δ is thick. We will not actually need this result in what follows, since it will always be clear in our examples that the purported apartments are in fact Coxeter complexes. But the proof is very instructive, being based on a clever use of retractions and the standard uniqueness argument. In addition, it makes (B0) seem much less artificial.

Theorem 4.131. *Let Δ be a thick chamber complex with a family $\mathcal{A}$ of thin chamber subcomplexes Σ satisfying axioms (B1) and (B2). Then every $\Sigma \in \mathcal{A}$ is a Coxeter complex, so Δ is a building and $\mathcal{A}$ is a system of apartments.*

Proof. Note first that much of the theory of retractions developed in Section 4.4 did not use axiom (B0), but only the fact that the apartments Σ are thin chamber complexes. In particular, 4.33, 4.34, and 4.39 remain valid.

We now show that every $\Sigma \in \mathcal{A}$ is a Coxeter complex by constructing foldings (Section 3.4.2). Given adjacent chambers $C, C' \in \Sigma$, we must find a folding $\phi \colon \Sigma \to \Sigma$ such that $\phi(C') = C$. Let A be the common panel $C \cap C'$, let C'' be a third chamber of Δ having A as a face, and let Σ' be an apartment containing C and C''. Let $\phi \colon \Sigma \to \Sigma$ be the restriction to Σ of $\rho_{\Sigma,C'} \circ \rho_{\Sigma',C}$. Then ϕ fixes C pointwise and satisfies $\phi(C') = C$. We will prove that ϕ is a folding. [Draw a picture of the tree case to see why this is plausible.]

In view of Proposition 4.39, ϕ preserves distances from A, i.e., $d(A, \phi(D)) = d(A, D)$ for any chamber $D \in \Sigma$. [Distances here may be computed either in Σ or in Δ, but we will be thinking about Σ-distances when we apply this.] In other words, if Γ is a minimal gallery in Σ from A to D, then $\phi(\Gamma)$ is a minimal gallery from A to $\phi(D)$. In particular, $\phi(\Gamma)$ really is a gallery, not just a pregallery, and this will enable us to apply the standard uniqueness argument.

A first such application shows that if D is a chamber of Σ with $d(A, D) = d(C, D)$ (i.e., if there is a minimal gallery from A to D that starts with C), then ϕ fixes D pointwise. Thus ϕ is the identity on the subcomplex α generated by $\{D \in \mathcal{C}(\Sigma) \mid d(A, D) = d(C, D)\}$. And this subcomplex α is precisely the image of ϕ. For suppose D is any chamber of Σ and Γ is a minimal gallery from A to D; then $\phi(\Gamma)$ is minimal from A to $\phi(D)$ and starts with C, so $\phi(D) \in \alpha$. Thus ϕ is a retraction of Σ onto α.

Everything we have done so far can also be done with the roles of C and C' reversed. Hence there is an endomorphism ϕ' of Σ with $\phi'(C) = C'$ such that ϕ' preserves distances from A and retracts Σ onto the subcomplex α' generated by $\{D \in \mathcal{C}(\Sigma) \mid d(A, D) = d(C', D)\}$.

We show next that α and α' have no chamber in common. Suppose D is a chamber in $\alpha \cap \alpha'$. Then D is fixed pointwise by both ϕ and ϕ'. If Γ is a minimal gallery from D to A, it follows by the standard uniqueness argument that ϕ and ϕ' fix every chamber in Γ pointwise. But this is absurd, since Γ ends with either C or C'.

We now have $\mathcal{C}(\Sigma) = \mathcal{C}(\alpha) \amalg \mathcal{C}(\alpha')$. The proof that ϕ is a folding will be complete if we can show that ϕ maps $\mathcal{C}(\alpha')$ bijectively to $\mathcal{C}(\alpha)$. To this end, consider the composites $\phi\phi'$ and $\phi'\phi$. The first takes C to C and fixes A pointwise, so it fixes C pointwise; it is therefore the identity on α by the standard uniqueness argument. Similarly, the other composite is the identity on α'. Hence ϕ induces an isomorphism $\alpha' \xrightarrow{\sim} \alpha$, with inverse induced by ϕ'.

□

Buildings as W-Metric Spaces

We now present our second approach to buildings, which we called the "combinatorial approach" in the introduction. From a purely logical point of view, the only prerequisite for this chapter is a knowledge of the basic facts about Coxeter groups (Chapter 2). We will, however, make motivational remarks that refer to Chapter 4 (which in turn depends on Chapter 3). And, of course, we will have to use results of Chapter 4 when we prove that the two approaches to buildings are equivalent.

For most of this chapter we do *not* need our standing assumption that the set S of generators of a Coxeter group is finite. We will explicitly state this assumption in the few places where it is needed.

5.1 Buildings of Type (W, S)

In Section 4.8 we introduced a Weyl distance function $\delta \colon \mathcal{C}(\Delta) \times \mathcal{C}(\Delta) \to W$ for a building Δ of type (W, S), and we derived some basic properties of it. In modern building theory, it is important that buildings can be *characterized* by the properties of δ. So we start with a new definition of "building" here as an abstract system $(\mathcal{C}, \delta)$ subject to axioms motivated by Proposition 4.84. We will show in Section 5.6 below that the new definition is equivalent to the one in Chapter 4. The properties of the function δ collected in Proposition 4.84 are redundant, so our definition gives a more economical system of axioms due to Tits [261].

An interesting feature that distinguishes the Weyl distance approach from the simplicial approach is that there is no reference to apartments. These appear only later, as a consequence of a theorem asserting their existence. On the other hand, we need the Weyl group from the beginning, whereas in the simplicial approach this appears only after one chooses a type function. We proceed now to the details.

5.1.1 Definition and Basic Facts

We fix a Coxeter system (W, S) and denote by $l = l_S$ the length function on W with respect to S (see Section 1.5.1).

Definition 5.1. A *building of type* (W, S) is a pair $(\mathcal{C}, \delta)$ consisting of a nonempty set $\mathcal{C}$, whose elements are called *chambers*, together with a map $\delta \colon \mathcal{C} \times \mathcal{C} \to W$, called the *Weyl distance* function, such that for all $C, D \in \mathcal{C}$, the following three conditions hold:

(WD1) $\delta(C, D) = 1$ *if and only if* $C = D$.

(WD2) *If* $\delta(C, D) = w$ *and* $C' \in \mathcal{C}$ *satisfies* $\delta(C', C) = s \in S$, *then* $\delta(C', D) = sw$ *or* w. *If, in addition,* $l(sw) = l(w) + 1$, *then* $\delta(C', D) = sw$.

(WD3) *If* $\delta(C, D) = w$, *then for any* $s \in S$ *there is a chamber* $C' \in \mathcal{C}$ *such that* $\delta(C', C) = s$ *and* $\delta(C', D) = sw$.

As we remarked following Proposition 4.84, the properties of δ that we are now taking as axioms vaguely resemble the axioms for a metric space. We will therefore sometimes call $(\mathcal{C}, \delta)$ a *W-metric space*.

When we use the word "building" from now on, it will usually be clear from the context whether we are talking about a building in the sense of Chapter 4 or a building in the sense of our new definition. But if we need to distinguish between the two concepts, we will use the term *simplicial building* for our original definition and the term *W-metric building* for the new one. We can then express the main result of Section 4.8 by saying that every simplicial building of type (W, S) gives rise to a W-metric building of type (W, S). We will prove that the two concepts are equivalent in Section 5.6.

As in Section 4.8, we will often illustrate (WD2) and (WD3) schematically by drawing triangles, such as

$$
\begin{array}{c}
C' \\
\Big\downarrow s \quad \diagdown^{\{sw, w\}} \\
C \xrightarrow{\ \ w\ \ } D
\end{array}
$$

for (WD2).

We now reintroduce some terminology from previous chapters in the present context.

Definition 5.2. Given $s \in S$, we say that two chambers $C, D \in \mathcal{C}$ are *s-adjacent* if $\delta(C, D) = s$. We say that they are *s-equivalent*, and we write

$$
C \sim_s D ,
$$

if $\delta(C, D) \in \{1, s\}$.

The following lemma shows that, as the terminology suggests, s-adjacency is symmetric and s-equivalence is an equivalence relation.

Lemma 5.3. *Let* $(\mathcal{C}, \delta)$ *be a building of type* (W, S).

(1) *If* $C, D \in \mathcal{C}$ *satisfy* $\delta(C, D) = s \in S$, *then also* $\delta(D, C) = s$.

(2) *If* $C, D, E \in \mathcal{C}$ *satisfy* $\delta(C, D) = \delta(D, E) = s \in S$, *then* $\delta(C, E) \in \{1, s\}$.

Proof. (1) Method 1: Set $w := \delta(D, C)$. By (WD1), $w \neq 1$ and $1 = \delta(C, C)$. By (WD2), we have $1 = \delta(C, C) \in \{sw, w\}$:

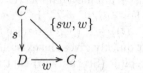

Since $w \neq 1$, we must have $sw = 1$ and hence $\delta(D, C) = w = s$. Method 2: By (WD3), there is a chamber C' such that $\delta(C', C) = s$ and $\delta(C', D) = s^2 = 1$. Then $C' = D$ by (WD1), so $\delta(D, C) = \delta(C', C) = s$.

(2) This is immediate from (WD2), since $\{s^2, s\} = \{1, s\}$:

$\square$

In view of the symmetry of s-adjacency, we will sometimes eliminate the arrowhead between s-adjacent chambers in our diagrams. For example, the triangle illustrating (WD2) can now be drawn as

Definition 5.4. The equivalence classes in $\mathcal{C}$ under s-equivalence are called s-*panels*. A *panel* is by definition an s-panel for some $s \in S$.

The motivation for this terminology comes from Corollary 4.11; we can think intuitively of an s-panel as the set of chambers having a given codimension-1 face of cotype s.

Next, we prove a refinement of (WD3), which is to be expected in view of Proposition 4.84(4).

Lemma 5.5. *The chamber C' in* (WD3) *is uniquely determined if* $l(sw) = l(w) - 1$.

Proof. Let $\mathcal{P}$ be the s-panel containing C, and choose a chamber C' as in (WD3). Then $C' \in \mathcal{P}$, and we are trying to prove that it is the unique chamber in $\mathcal{P}$ such that $\delta(C', D) = w' := sw$. This is immediate from (WD2). For if C'' is a chamber in $\mathcal{P}$ distinct from C', then we have $\delta(C'', C') = s$ and $\delta(C', D) = w'$ with $l(sw') > l(w')$, so $\delta(C'', D) = sw' \neq w'$. $\square$

Remark 5.6. The proof of Lemma 5.5 implies that for any s-panel $\mathcal{P}$ and any chamber D there is a unique chamber $C' \in \mathcal{P}$ that is "closest" to D in the following sense: For any $C \neq C'$ in $\mathcal{P}$, we have $\delta(C, D) = s\delta(C', D)$ and $l\big(\delta(C, D)\big) = l\big(\delta(C', D)\big) + 1$. (See Exercise 5.13.) So the WD axioms are formulated in a way that quickly gives the existence of projections of chambers onto panels. We will see later (Section 5.3) that the axioms are in fact strong enough to yield projections of chambers onto arbitrary "residues."

By (WD3), every panel contains at least two chambers. The building $(\mathcal{C}, \delta)$ is called *thick* (resp. *thin*) if every panel contains at least three (resp. precisely two) chambers. The natural example of a thin building of type (W, S) is the following.

Example 5.7. If we define $\delta_W : W \times W \to W$ by $\delta_W(w_1, w_2) = w_1^{-1} w_2$, then (W, δ_W) is a thin building of type (W, S) (cf. Section 3.5). So here the chambers are elements of W, and each $x \in W$ is s-adjacent to precisely one chamber, namely xs. It is obvious that the axioms (WD1)–(WD3) are satisfied in this example; in addition one *always* has $\delta_W(C', D) = sw$ in (WD2), whether $l(sw) = l(w) + 1$ or not. More generally, for any three chambers $C, D, E \in W$, we have

$$\delta_W(C, E) = \delta_W(C, D)\delta_W(D, E) .$$

We call (W, δ_W) the *standard thin building* of type (W, S). We will see later (Proposition 5.65) that it is the *unique* thin building of type (W, S), up to isomorphism.

Exercises

5.8. Spell out explicitly what a building of type (W, S) is if $|S| = 0$ or 1.

5.9. Show that one gets an equivalent set of axioms for buildings of type (W, S) if one replaces (WD3) by the following two axioms:

(WD3a) *For any $C \in \mathcal{C}$ and any $s \in S$, there exists $C' \in \mathcal{C}$ with $\delta(C', C) = s$.*

(WD3b) *If $\delta(C, D) = w$, then for any $s \in S$ satisfying $l(sw) = l(w) - 1$ there is a chamber $C' \in \mathcal{C}$ such that $\delta(C', C) = s$ and $\delta(C', D) = sw$.*

5.10. Show that one gets an equivalent set of axioms for buildings of type (W, S) if one replaces (WD2) by the following two axioms:

(WD2a) *The relation of s-equivalence (Definition 5.2) is transitive for each* $s \in S$.

(WD2b) *If* $\delta(C', C) = s \in S$, $\delta(C, D) = w$, *and* $l(sw) = l(w) + 1$, *then* $\delta(C', D) = sw$.

5.11. Show that one gets an equivalent set of axioms for buildings of type (W, S) if one deletes the second sentence in (WD2) and adds the following sentence to (WD3):

> *If, in addition,* $l(sw) = l(w) - 1$, *then* C' *is uniquely determined.*

5.12. In this exercise the reader is assumed to be familiar with the Bruhat order (Exercise 3.59). Show that one gets an equivalent set of axioms for buildings of type (W, S) if one replaces (WD2) by the following axiom:

(WD2c) *If* $\delta(C', C) = s \in S$ *and* $\delta(C, D) = w$, *then* $\delta(C', D) \geq sw$ *in the Bruhat order.*

(If the direction of the inequality sign seems surprising in (WD2c), keep in mind that sw might be shorter than w, so it would not be reasonable to expect the opposite inequality to hold. Of course sw might also be longer than w, but in this case neither inequality is surprising.)

5.13. Prove the claim made in Remark 5.6.

5.14. Let T be a tree with no endpoints. Verify directly, without using the fact that T is a simplicial building, that T gives rise to a building of type $(D_\infty, \{s_1, s_2\})$, where $\mathcal{C}$ is the set of edges of T and D_∞ is the infinite dihedral group $\langle s_1, s_2 \; ; \; s_1^2 = s_2^2 = 1 \rangle$.

5.1.2 Galleries and Words

As explained in Section 4.8, each simplicial building Δ gives rise to a W-metric building. There the Weyl distance δ is defined using galleries in Δ. We now reverse the process. We define galleries using the given δ (and the associated adjacency relations), and we prove that δ can be computed in the expected way in terms of galleries. Assume throughout this subsection that $(\mathcal{C}, \delta)$ is a building of type (W, S).

Definition 5.15. We call two chambers $C, D \in \mathcal{C}$ *adjacent* if they are s-adjacent for some $s \in S$. A sequence $\Gamma \colon C_0, \ldots, C_n$ of $n+1$ chambers such that C_{i-1} and C_i are adjacent for all $1 \leq i \leq n$ is called a *gallery* of length n. We say that C_0 and C_n are *connected* by Γ. If there is no gallery of length $< n$ connecting C_0 and C_n, then we say that the *gallery distance* between C_0 and C_n is n and write $d(C_0, C_n) = n$. The gallery Γ is called *minimal* if $d(C_0, C_n) = n$. If $s_i = \delta(C_{i-1}, C_i)$ for $1 \leq i \leq n$, then $\mathbf{s}(\Gamma) := (s_1, \ldots, s_n)$ is called the *type* of the gallery Γ.

We can now derive some basic properties of galleries from the WD axioms.

Lemma 5.16. *Let C and D be chambers and let $w := \delta(C, D)$.*

(1) *If Γ is a gallery of type $\mathbf{s} = (s_1, \ldots, s_n)$ connecting C and D, then there exists a subword $(s_{i_1}, \ldots, s_{i_m})$ of $\mathbf{s}$ such that $w = s_{i_1} \cdots s_{i_m}$, where $0 \leq m \leq n$ and $1 \leq i_1 < \cdots < i_m \leq n$. If, in addition, $\mathbf{s}$ is reduced in the sense of Section 2.3.1, then $w = s_1 \cdots s_n$, and Γ is minimal.*

(2) *If $w = s_1 \cdots s_n$ with $s_1, \ldots, s_n \in S$, then there exists a gallery Γ of type $\mathbf{s} = (s_1, \ldots, s_n)$ connecting C and D. If, in addition, $\mathbf{s}$ is reduced, then this gallery Γ is uniquely determined and minimal.*

Proof. In both parts of the lemma, we proceed by induction on n.

(1) Let $\Gamma \colon C = C_0, C_1, \ldots, C_n = D$ be the given gallery. Let $w' = \delta(C_1, D)$, and apply (WD2) to the triangle

to deduce that $w \in \{s_1 w', w'\}$. By induction, we may assume that w' is the product of the elements occurring in a subword of $(s_2, \ldots, s_n)$. This immediately yields our first claim about w. Note that this part of the proof already implies $d(C, D) \geq l(w)$, since $l(w) \leq m \leq n$, where n is the length of an arbitrary gallery connecting C and D.

Now assume that in addition, $\mathbf{s}$ is reduced. Then also $\mathbf{s}' := (s_2, \ldots, s_n)$ is reduced. So the induction hypothesis yields $w' = \delta(C_1, D) = s_2 \cdots s_n$ in this case. Hence $l(s_1 w') = l(s_1 \cdots s_n) = n = l(w') + 1$. The application of (WD2) to the triangle above therefore yields $w = s_1 w' = s_1 \cdots s_n$. In particular, $n = l(w)$, and by the previous paragraph there is no gallery of length $< l(w)$ connecting C and D. Hence Γ is minimal.

(2) Assume $w = s_1 \cdots s_n$ with $n > 0$. Applying (WD3), we obtain a chamber C_1 that is s_1-adjacent to C with $\delta(C_1, D) = s_1 w = s_2 \cdots s_n$:

By the induction hypothesis there exists a gallery $\Gamma' \colon C_1, \ldots, C_n = D$ of type $\mathbf{s}' := (s_2, \ldots, s_n)$ connecting C_1 and D. Hence $\Gamma \colon C = C_0, C_1, \ldots, C_n = D$ is a gallery of type $\mathbf{s}$ connecting C and D.

Now assume additionally that $\mathbf{s}$ is reduced. Then Γ is minimal by part (1). We want to show that Γ is the only gallery of type $\mathbf{s}$ connecting C and D. Note that the chamber C_1 following $C = C_0$ in Γ has to satisfy $\delta(C, C_1) = s_1$ as well as (by part (1)) $\delta(C_1, D) = s_2 \cdots s_n = s_1 w$. Since $l(s_1 w) = n - 1 = l(w) - 1$, Lemma 5.5 implies that C_1 is uniquely determined. Then by induction the gallery Γ' of reduced type s' connecting C_1 and D is also uniquely determined. Hence Γ is unique. $\square$

Corollary 5.17. *For any two chambers $C, D \in \mathcal{C}$, we have:*

(1) $d(C, D) = l(\delta(C, D))$.
(2) $\delta(D, C) = \delta(C, D)^{-1}$.

Proof. Set $w := \delta(C, D)$ and choose a reduced decomposition $w = s_1 \cdots s_n$ of w. By Lemma 5.16(2), there exists a gallery $\Gamma: C = C_0, \ldots, C_n = D$ of type $s = (s_1, \ldots, s_n)$ connecting C and D. It is minimal since s is reduced, so $d(C, D) = n = l(w)$. This proves (1). Now $D = C_n, \ldots, C_0 = C$ is a gallery of reduced type $(s_n, \ldots, s_1)$ connecting D and C, so Lemma 5.16(1) implies $\delta(D, C) = s_n \cdots s_1 = w^{-1} = \delta(C, D)^{-1}$, proving (2). $\square$

Remark 5.18. Using Corollary 5.17(2), we can deduce the following analogues of (WD2) and (WD3), thereby removing the asymmetry in those axioms:

(WD2′) *If $D' \in \mathcal{C}$ satisfies $\delta(D, D') = s \in S$ and $\delta(C, D) = w$, then $\delta(C, D') = ws$ or w. If, in addition, $l(ws) = l(w) + 1$, then $\delta(C, D') = ws$.*

(WD3′) *If $\delta(C, D) = w$, then for any $s \in S$ there is a chamber $D' \in \mathcal{C}$ such that $\delta(D, D') = s$ and $\delta(C, D') = ws$.*

We also have, as in Lemma 5.5, that the D' in (WD3′) is uniquely determined if $l(ws) = l(w) - 1$. And, as in Remark 5.6, one can interpret this unique D' as the chamber closest to C in the s-panel containing D.

Exercises

5.19. Prove Remark 5.18.

5.20. (a) If Γ is a gallery of type s in a building of type (W, S), show that Γ is minimal if and *only if* s is reduced.
(b) Given $C, D \in \mathcal{C}$, show that there is a 1–1 correspondence between minimal galleries from C to D and reduced decompositions of $w := \delta(C, D)$.

5.2 Buildings as Chamber Systems

The central idea of our new approach to buildings is to encode properties of galleries in a Weyl-group-valued distance function δ. In Definition 5.1 we did this by requiring certain algebraic properties of δ, which enabled us to define adjacency and galleries. There is a slightly different but closely related way of achieving the same goal using "chamber systems." This approach was introduced by Tits [255], and a slight variant of it was taken as the definition of "building" in the books by Ronan [200] and Weiss [281]. In this section we will show that the definition of Ronan and Weiss is equivalent to the one we gave in Section 5.1.

First of all we have to define the notion of a chamber system. This is done in Section A.1.4 in the context of chamber complexes. Here we do not presuppose any simplicial structure, so we just consider a chamber system as a set together with a family of equivalence relations.

Definition 5.21. A *chamber system* over a set S is a nonempty set $\mathcal{C}$ (whose elements are called *chambers*) together with a family of equivalence relations $(\sim_s)_{s \in S}$ on $\mathcal{C}$ indexed by S. The equivalence classes with respect to $\sim_s$ are called *s-panels*. A *panel* is an *s*-panel for some $s \in S$. Two distinct chambers C and D are called *s-adjacent* if they are contained in the same *s*-panel and *adjacent* if they are *s*-adjacent for some $s \in S$. A *gallery* of length n connecting C_0 and C_n is a sequence $\Gamma \colon C_0, \ldots, C_n$ of $n+1$ chambers such that C_{i-1} and C_i are adjacent for all $1 \le i \le n$. If C_{i-1} and C_i are s_i-adjacent with $s_i \in S$ for all i, then we say that Γ is a gallery of type $(s_1, \ldots, s_n)$.

A chamber system can be viewed as a graph with colored edges; the vertices are the chambers, and *s*-adjacent chambers are connected by an edge of color s. See Exercise 5.25 for more details concerning this point of view.

Example 5.22. Let $(\mathcal{C}, \delta)$ be a building of type (W, S). Then $\mathcal{C}$, with the *s*-equivalence relations $\sim_s$ defined in Section 5.1.1, is a chamber system over S. In this situation the notions of panel, adjacency, gallery and so on as introduced in Definition 5.21 agree with the notions defined in the previous section for buildings of type (W, S).

We are not interested in the notion of chamber system for its own sake, so we immediately introduce the property that turns chamber systems into buildings. To this end, S is not an arbitrary set but is the set of distinguished generators of a Coxeter group W as in the example. Thus we fix a *Coxeter system* (W, S) for the rest of this section. Then types of galleries are words in S, and we can talk as before of *reduced* types. The main building axiom in this setup requires the existence of a map $\delta \colon \mathcal{C} \times \mathcal{C} \to W$ that is related to galleries by the following property:

(G) *For any $C, D \in \mathcal{C}$ and any reduced S-word $\mathbf{s} = (s_1, \ldots, s_n)$ there exists a gallery of type $\mathbf{s}$ connecting C and D if and only if $\delta(C, D) = s_1 \cdots s_n$.*

Here by definition the empty word is considered to be reduced, and the corresponding product is the identity element $1 \in W$. The link with buildings of type (W, S) is now provided by the following result.

Proposition 5.23. *Let $\big(\mathcal{C}, (\sim_s)_{s \in S}\big)$ be a chamber system over S such that each panel contains at least two chambers. Then a map $\delta \colon \mathcal{C} \times \mathcal{C} \to W$ satisfies (G) if and only if it satisfies* (WD1)–(WD3).

Proof. It is shown in Lemma 5.16 that the WD axioms imply (G). So we assume that δ satisfies (G) and deduce the WD axioms, which we may take in the form (WD1), (WD2a), (WD2b), and (WD3) (see Exercise 5.10).

(WD1) By (G) applied to the empty word, two chambers C, D satisfy $\delta(C, D) = 1$ if and only if there is gallery of empty type connecting them, i.e., if and only if $C = D$.

(WD2a) Note first that (G) applied to words of length 1 says that two chambers C' and C are s-adjacent in the sense of Definition 5.21 if and only if $\delta(C', C) = s$. Hence the relation of s-equivalence in Definition 5.2 coincides with the given relation $\sim_s$, which proves (WD2a).

(WD2b) Assume that $C', C, D \in \mathcal{C}$ satisfy $\delta(C', C) = s \in S$ and $\delta(C, D) = w \in W$. We must show that $\delta(C', D) = sw$ if $l(sw) = l(w) + 1$. Choose a reduced decomposition $w = s_1 \cdots s_n$ of w (so $n = l(w)$) and, by (G), a gallery $\Gamma: C = C_0, \ldots, C_n = D$ of type $\mathbf{s} = (s_1, \ldots, s_n)$ connecting C and D. Then $(s, s_1, \ldots, s_n)$ is reduced and is the type of the gallery

$$C' \xrightarrow{\ s\ } C \xrightarrow{\ \Gamma\ } D \ .$$

Hence (G) implies that $\delta(C', D) = sw$.

(WD3) Given $C, D \in \mathcal{C}$ and $s \in S$, let $w = \delta(C, D)$. If $l(sw) = l(w) + 1$, we choose an arbitrary chamber C' with $\delta(C', C) = s$ (here we use the assumption that all panels contain at least two chambers). We can then apply (WD2b), which we have already proved, to obtain $\delta(C', D) = sw$, and we are done. If $l(sw) = l(w) - 1$, we choose a reduced decomposition $w = s_1 \cdots s_n$ of w with $s_1 = s$ and a gallery $C = C_0, \ldots, C_n = D$ of reduced type $(s_1, \ldots, s_n)$ connecting C and D. Then C_1 satisfies $\delta(C_1, C) = s_1 = s$ and (by (G)) $\delta(C_1, D) = s_2 \cdots s_n = s_1 w = sw$. So we may take $C' = C_1$ in order to satisfy (WD3). $\qquad\square$

In view of the proposition, it is a matter of taste whether one introduces buildings of type (W, S) by requiring the properties (WD1)–(WD3) for δ or by first introducing chamber systems and then assuming (G) (provided all panels contain at least two chambers). In this book we will stick with the approach via (WD1)–(WD3) except in two optional sections (5.12 and 8.7) where it is more convenient to use chamber systems in the sense of Definition 5.21.

Exercises

5.24. Give examples of chamber systems over S that satisfy (G) but also have panels with only one chamber.

5.25. Interpret Definition 5.21 in the language of graph theory.

5.3 Residues and Projections

In this section, we fix a Coxeter system (W, S) and a building $(\mathcal{C}, \delta)$ of type (W, S). We first want to generalize the notion of panel and introduce *residues*. The latter substitute, in the context of W-metric spaces, for the links that occurred in the theory of simplicial buildings. Residues will later be identified with the simplices of a simplicial building that we will construct from $(\mathcal{C}, \delta)$. They generalize panels in the sense that we now allow J-equivalences for any $J \subseteq S$ instead of just s-equivalences. In the following, we use the notation $W_J := \langle J \rangle \leq W$ for any $J \subseteq S$.

5.3.1 J-Residues

Definition 5.26. Given $J \subseteq S$, we say that two chambers $C, D \in \mathcal{C}$ are *J-equivalent*, and we write $C \sim_J D$, if $\delta(C, D) \in W_J$. By Lemma 5.16 we have $C \sim_J D$ if and only if there is a gallery of type $(s_1, \ldots, s_n)$ connecting C and D with $s_i \in J$ for all $1 \leq i \leq n$. This implies that J-equivalence is an equivalence relation. The equivalence classes are called *J-residues*, and the J-residue containing a given chamber C is denoted by $R_J(C)$. Thus

$$R_J(C) := \{C' \in \mathcal{C} \mid \delta(C, C') \in W_J\} \ .$$

A subset $\mathcal{R} \subseteq \mathcal{C}$ is called a *residue* if it is a J-residue for some $J \subseteq S$. The set J is called the *type* of the residue, and the cardinality $|J|$ is called its *rank*.

Note, for example, that panels are residues (of rank 1 and of type $\{s\}$ for some $s \in S$), and single chambers are residues (of rank 0 and type $\emptyset$). At the other extreme, $\mathcal{C}$ itself is the unique residue of type S.

Remark 5.27. Given a J-residue $\mathcal{R}$, we can recover the standard subgroup W_J from $\mathcal{R}$ by

$$W_J = \delta(\mathcal{R}, \mathcal{R}) := \{\delta(D, E) \mid D, E \in \mathcal{R}\} \ ;$$

we will prove this in Lemma 5.29 below. Since J is uniquely determined by W_J, it follows that each residue $\mathcal{R}$ has a well-defined type J (and hence also a well-defined rank).

Before proceeding with the study of residues, we state and prove an appropriate version of the "triangle inequality" for our W-metric space $\mathcal{C}$.

Lemma 5.28. *Given $C, D, E \in \mathcal{C}$, set $u := \delta(C, D)$ and $v := \delta(D, E)$. Then the following hold:*

(1) *If $u = s_1 \cdots s_m$ with $s_i \in S$ for all i, then $\delta(C, E) = s_{i_1} \cdots s_{i_k} v$ for some $0 \leq k \leq m$ and $1 \leq i_1 < \cdots < i_k \leq m$.*
(2) *If $v = t_1 \cdots t_n$ with $t_j \in S$ for all j, then $\delta(C, E) = u t_{j_1} \cdots t_{j_l}$ for some $0 \leq l \leq n$ and $1 \leq j_1 < \cdots < j_l \leq n$.*

(3) *If $l(uv) = l(u) + l(v)$, then $\delta(C, E) = uv$.*

Proof. (1) The argument is essentially the same as in Lemma 5.16(1): Choose a gallery $\Gamma\colon C = C_0, \dots, C_m = D$ of type $\mathbf{s} = (s_1, \dots, s_m)$ connecting C and D (this is possible by Lemma 5.16(2)). Then $\delta(C_{m-1}, E) = s_m v$ or v by (WD2). Proceeding inductively we see that $\delta(C, E) = u'v$, where u' is the product of the elements of a subword of $\mathbf{s}$.

(2) The proof is symmetric to that of part (1) (recall that we can use Corollary 5.17(2) and Remark 5.18).

(3) Here we choose $u = s_1 \cdots s_m$ to be a reduced decomposition of u, i.e., $m = l(u)$. Then $l(uv) = l(u) + l(v)$ implies that $l(s_{i+1}s_{i+2} \cdots s_m v) = m - i + l(v)$ for all $0 \le i \le m$ and so in particular $l(s_i \cdots s_m v) = l(s_{i+1} \cdots s_m v) + 1$. Hence if we again choose a gallery $\Gamma\colon C = C_0, \dots, C_m = D$ of type $\mathbf{s} = (s_1, \dots, s_m)$ connecting C and D, then by induction and (WD2) we obtain $\delta(C, D) = s_1 \cdots s_m v = uv$. [Alternatively, concatenate minimal galleries and apply the results of Section 5.1.1.] $\qquad\square$

We will use a diagram of the form

$$
\begin{array}{ccc}
 & C & \\
u \downarrow & \diagdown{}^{u'v} & \\
 & & \\
D & \xrightarrow{\;v\;} & E
\end{array}
\tag{5.1}
$$

as an aid in remembering the statement of part (1) of the lemma, and similarly for the other parts. Here u', as in the proof, is obtained from a decomposition of u by deleting zero or more letters.

As a first application of our triangle inequality we can evaluate the function δ on pairs of residues. In this context we set

$$
\delta(\mathcal{M}, \mathcal{N}) := \{\delta(X, Y) \mid X \in \mathcal{M} \text{ and } Y \in \mathcal{N}\}
$$

for any two subsets $\mathcal{M}, \mathcal{N} \subseteq \mathcal{C}$. If one of the two subsets, say $\mathcal{N}$, is a singleton $\{D\}$, then we will sometimes write $\delta(\mathcal{M}, D)$ instead of $\delta(\mathcal{M}, \{D\})$.

Lemma 5.29. *Let $\mathcal{R}$ be a J-residue and $\mathcal{S}$ a K-residue with $J, K \subseteq S$. Then $\delta(\mathcal{R}, \mathcal{S})$ is a double coset of the form $W_J w W_K$. In particular, $\delta(\mathcal{R}, \mathcal{R}) = W_J$.*

Proof. The second assertion follows from the first since $1 \in \delta(\mathcal{R}, \mathcal{R})$. To prove the first assertion, choose $C \in \mathcal{R}$ and $D \in \mathcal{S}$ and set $w = \delta(C, D)$. Given $C' \in \mathcal{R}$ and $D' \in \mathcal{S}$, set $u = \delta(C', C) \in W_J$ and $v = \delta(D, D') \in W_K$. Then Lemma 5.28 first implies $\delta(C', D) = u'w$ with $u' \in W_J$ and then $\delta(C', D') = u'wv'$ with $v' \in W_K$, as illustrated in the following diagrams:

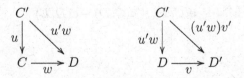

Hence $\delta(\mathcal{R}, \mathcal{S}) \subseteq W_J w W_K$.

In order to establish equality, it suffices to show that $\delta(\mathcal{R}, \mathcal{S})$ is closed under left multiplication by elements of J and right multiplication by elements of K. But this is immediate from (WD3) and (WD3′).   □

Note that if K is empty, then $\mathcal{S}$ consists of a single chamber D, and the assertion is that $\delta(\mathcal{R}, D)$ is a right coset $W_J w$. A similar remark applies if J is empty.

Corollary 5.30. *If $J \subseteq S$ and $\mathcal{R}$ is a J-residue, then $(\mathcal{R}, \delta|_{\mathcal{R} \times \mathcal{R}})$ is a building of type (W_J, J).*

(This is the analogue of the statement in the simplicial theory that the link of a simplex in a building is again a building.)

Proof. We have $\delta(\mathcal{R}, \mathcal{R}) = W_J$ by the lemma, so the restricted function $\delta|_{\mathcal{R} \times \mathcal{R}}$ does indeed take values in W_J. The verification of (WD1)–(WD3) for $\mathcal{R}$ is now straightforward and is left to the reader. [In verifying the second sentence in (WD2), recall that the restriction of the length function l to W_J is equal to the length function of W_J with respect to J by Proposition 2.14. In verifying (WD3), note that if C' is s-adjacent to $C \in \mathcal{R}$ with $s \in J$, then $C' \sim_J C$ and so $C' \in \mathcal{R}$.]   □

Exercises

5.31. Let $\mathcal{R}$ be a residue of type J and $\mathcal{S}$ a residue of type K $(J, K \subseteq S)$. Show that $\mathcal{R} \subseteq \mathcal{S}$ if and only if $\mathcal{R} \cap \mathcal{S}$ is nonempty and $J \subseteq K$.

5.32. Show that the set of all residues, together with the empty set, is closed under intersection. More precisely, suppose $\mathcal{R}_i$ is a J_i-residue for all i in some index set I. Then $\mathcal{R} := \bigcap_{i \in I} \mathcal{R}_i$ is a residue of type $J := \bigcap_{i \in I} J_i$ if it is nonempty.

5.33. This exercise assumes familiarity with the Bruhat order.

(a) For any three chambers $C, D, E \in \mathcal{C}$, deduce from Lemma 5.28 that

$$\delta(C, D) \geq \delta(C, E)\delta(E, D) . \tag{5.2}$$

Thus the lemma really is a triangle inequality, though the direction of the inequality sign may seem surprising at first.

(b) Take lengths on both sides of (5.2) to deduce the ordinary triangle inequality

$$d(C, E) \leq d(C, D) + d(D, E) .$$

5.3.2 Projections and the Gate Property

Next we want to define projections onto residues and deduce, in our present setup, their main properties (cf. Section 4.9). The only tools required here are Lemmas 5.28 and 5.29 together with some basic properties of Coxeter groups that are collected in Section 2.3.2. There is no need in the W-metric approach to first define projections in apartments as we did in the simplicial theory.

Recall that a double coset $W_J w W_K$ $(J, K \subseteq S,\ w \in W)$ has a unique element of minimal length, which we denote by $\min(W_J w W_K)$; thus

$$w_1 = \min(W_J w W_K) : \iff w_1 \text{ has minimal length in } W_J w W_K .$$

In particular, we have a well-defined minimal element $w_1 = \min(\delta(\mathcal{R}, \mathcal{S})) \in W$ for any two residues $\mathcal{R}, \mathcal{S}$. Note that in view of Corollary 5.17(1), we can characterize this element $w_1 \in \delta(\mathcal{R}, \mathcal{S})$ by the property that

$$w_1 = \delta(C, D)$$

for any $C \in \mathcal{R}$ and $D \in \mathcal{S}$ such that

$$d(C, D) = d(\mathcal{R}, \mathcal{S}) ,$$

where

$$d(\mathcal{R}, \mathcal{S}) := \min \{d(C, D) \mid C \in \mathcal{R},\ D \in \mathcal{S}\} .$$

We start by studying projections of chambers onto residues. The "gate property" that we studied earlier in the simplicial setting reads as follows in the present context:

Proposition 5.34. *Let $\mathcal{R}$ be a residue and D a chamber. Then there exists a unique $C_1 \in \mathcal{R}$ such that $d(C_1, D) = d(\mathcal{R}, D)$. This chamber C_1 has the following properties:*

(1) $\delta(C_1, D) = \min(\delta(\mathcal{R}, D))$.
(2) $\delta(C, D) = \delta(C, C_1)\delta(C_1, D)$ *for all $C \in \mathcal{R}$.*
(3) $d(C, D) = d(C, C_1) + d(C_1, D)$ *for all $C \in \mathcal{R}$.*

Proof. Choose $C_1 \in \mathcal{R}$ at minimal distance from D. Then C_1 satisfies (i) by the discussion above, with $\mathcal{S} = \{D\}$. Now let J be the type of $\mathcal{R}$, and let $w_1 = \min(\delta(\mathcal{R}, D)) = \delta(C_1, D)$. We then have (see Proposition 2.20)

$$l(w_J w_1) = l(w_J) + l(w_1)$$

for all $w_J \in W_J$. Since $\delta(C, C_1) \in W_J$ for any $C \in \mathcal{R}$, it follows that $l(\delta(C, C_1)w_1) = l(\delta(C, C_1)) + l(w_1)$, i.e.,

$$l(\delta(C, C_1)\delta(C_1, D)) = l(\delta(C, C_1)) + l(\delta(C_1, D))$$

for all $C \in \mathcal{R}$. So by Lemma 5.28(3), $\delta(C, D) = \delta(C, C_1)\delta(C_1, D)$, proving that C_1 satisfies (ii). Applying Corollary 5.17(1) once more, we also obtain

$$d(C, D) = l\big(\delta(C, C_1)\delta(C_1, D)\big)$$
$$= l\big(\delta(C, C_1)\big) + l\big(\delta(C_1, D)\big)$$
$$= d(C, C_1) + d(C_1, D)$$

for all $C \in \mathcal{R}$, which is statement (iii). It follows from (iii) that $d(C, D) > d(C_1, D)$ for all $C \in \mathcal{R} \smallsetminus \{C_1\}$; this proves the uniqueness of C_1. $\qquad \square$

Definition 5.35. Let $\mathcal{R}$ be a residue in $\mathcal{C}$.

(a) Given $D \in \mathcal{C}$, the unique chamber $C_1 \in \mathcal{R}$ at minimal distance from D is called the *projection of D onto $\mathcal{R}$* and is denoted by $\mathrm{proj}_{\mathcal{R}} D$. It can also be described as the unique chamber in $\mathcal{R}$ satisfying

$$\delta(C_1, D) = \min\big(\delta(\mathcal{R}, D)\big) .$$

(b) If $\mathcal{S}$ is another residue, we set

$$\mathrm{proj}_{\mathcal{R}} \mathcal{S} := \{\mathrm{proj}_{\mathcal{R}} D \mid D \in \mathcal{S}\}$$

and call it the *projection of $\mathcal{S}$ onto $\mathcal{R}$*. Thus $\mathrm{proj}_{\mathcal{R}} \mathcal{S}$ is a subset of $\mathcal{R}$; we will prove in the next lemma that it is a residue in its own right.

We now derive some important properties of projections, which are similar to those formulated in Section 4.9 for projections in simplicial buildings.

Lemma 5.36. *Let $\mathcal{R}$ and $\mathcal{S}$ be residues of types J and K, respectively, where $J, K \subseteq S$. Let $w_1 := \min\big(\delta(\mathcal{R}, \mathcal{S})\big)$.*

(1) *The projection $\mathcal{P} := \mathrm{proj}_{\mathcal{R}} \mathcal{S}$ is given by*

$$\mathcal{P} = \{C \in \mathcal{R} \mid w_1 \in \delta(C, \mathcal{S})\} .$$

In other words, if $C \in \mathcal{R}$, then

$$C \in \mathcal{P} \iff \text{there exists } D \in \mathcal{S} \text{ with } \delta(C, D) = w_1$$
$$\iff \text{there exists } D \in \mathcal{S} \text{ with } d(C, D) = d(\mathcal{R}, \mathcal{S}) .$$

Moreover, if $C \in \mathcal{R}$ and $D \in \mathcal{S}$ satisfy $\delta(C, D) = w_1$, then $C = \mathrm{proj}_{\mathcal{R}} D$.
(2) *$\mathcal{P} = \mathrm{proj}_{\mathcal{R}} \mathcal{S}$ is a residue of type $J_1 := J \cap w_1 K w_1^{-1}$.*
(3) *Given $C, C' \in \mathcal{R}$ and $D, D' \in \mathcal{S}$ with $\delta(C, D) = \delta(C', D') = w_1$, we have*

$$\delta(C, C') = w_1 \delta(D, D') w_1^{-1} \tag{5.3}$$

and $d(C, C') = d(D, D')$.

Proof. (1) Set $\mathcal{P}' = \{C \in \mathcal{R} \mid w_1 \in \delta(C, \mathcal{S})\}$. Suppose $C \in \mathcal{R}$ and $D \in \mathcal{S}$ satisfy $\delta(C, D) = w_1$ or, equivalently, $d(C, D) = d(\mathcal{R}, \mathcal{S})$. Then certainly $d(C, D) = d(\mathcal{R}, D)$, so $C = \mathrm{proj}_{\mathcal{R}} D \in \mathcal{P}$ by definition of the projection. This proves the inclusion $\mathcal{P}' \subseteq \mathcal{P}$ and the last assertion of (1).

To prove $\mathcal{P} \subseteq \mathcal{P}'$, let $C = \mathrm{proj}_{\mathcal{R}} D'$ for some $D' \in \mathcal{S}$, and set $w :=$ $\delta(C, D') = \min(\delta(\mathcal{R}, D'))$. Since

$$\delta(\mathcal{R}, D') \subseteq \delta(\mathcal{R}, \mathcal{S}) = W_J w_1 W_K \ ,$$

Proposition 2.23 implies that w can be written in the form $w = w_J w_1 w_K$ with $w_J \in W_J$, $w_K \in W_K$, and $l(w_J w_1 w_K) = l(w_J) + l(w_1) + l(w_K)$. And since w is minimal in $\delta(\mathcal{R}, D') = W_J w$, we must have $w_J = 1$, i.e., $w = w_1 w_K$. Hence $w_1 \in w W_K = \delta(C, \mathcal{S})$, so $C \in \mathcal{P}'$.

(2) Fix $C_1 \in \mathcal{P}$ and (using (1)) $D_1 \in \mathcal{S}$ with $\delta(C_1, D_1) = w_1$. We will reinterpret (1) and see that that $\mathcal{P}$ is the residue $R_{J_1}(C_1)$. Let $C \in \mathcal{C}$ be arbitrary, and set $u := \delta(C, C_1)$. If $C \in \mathcal{R}$, then, since $C_1 = \mathrm{proj}_{\mathcal{R}} D_1$ by (1), Proposition 5.34 implies that $\delta(C, D_1) = \delta(C, C_1)\delta(C_1, D_1) = uw_1$:

$$
\begin{array}{ccc}
C & & \\
\ \downarrow u & \searrow uw_1 & \\
C_1 & \xrightarrow{\ w_1\ } & D_1
\end{array}
\tag{5.4}
$$

[Note also, for future reference, that $d(C, D_1) = d(C, C_1) + l(w_1)$, again by Proposition 5.34.] In order for our arbitrary chamber C to be in $\mathcal{P}$, then, we need first that $u \in W_J$ (so that $C \in \mathcal{R}$) and then that

$$w_1 \in \delta(C, \mathcal{S}) = \delta(C, D_1) W_K = u w_1 W_K \ .$$

This last condition is equivalent to $u \in w_1 W_K w_1^{-1}$, so we have proven

$$C \in \mathcal{P} \iff u \in W_J \cap w_1 W_K w_1^{-1} \ .$$

Now by Lemma 2.25, $W_J \cap w_1 W_K w_1^{-1} = W_{J_1}$, so $C \in \mathcal{P}$ if and only if $\delta(C, C_1) \in W_{J_1}$. But this just means that $\mathcal{P} = R_{J_1}(C_1)$.

(3) Consider the diagram

$$
\begin{array}{ccc}
C & \xrightarrow{\ w_1\ } & D \\
\ \downarrow u & & \downarrow v \\
C' & \xrightarrow{\ w_1\ } & D'
\end{array}
\tag{5.5}
$$

where $u = \delta(C, C') \in W_J$ and $v = \delta(D, D') \in W_K$. We will use this square to compute $\delta(C, D')$ and $d(C, D')$ in two ways. First, as we noted above (see (5.4)), we have $\delta(C, D') = uw_1$ and $d(C, D') = d(C, C') + l(w_1)$. Similarly, reversing the roles of $\mathcal{R}$ and $\mathcal{S}$, and noting that $w_1^{-1} = \min(\delta(\mathcal{S}, \mathcal{R}))$, we find that $\delta(C, D') = w_1 v$ and $d(C, D') = l(w_1) + d(D, D')$. Hence $uw_1 = w_1 v$ and $d(C, C') + l(w_1) = l(w_1) + d(D, D')$. Part (3) of the lemma follows at once. $\square$

Lemma 5.36 leads quickly to a basic result that can be viewed as a refined version of Proposition 4.96(2).

Proposition 5.37. *Let $\mathcal{R}$ be a residue of type J and $\mathcal{S}$ a residue of type K in $\mathcal{C}$ ($J, K \subseteq S$). Set $\mathcal{R}_1 = \mathrm{proj}_{\mathcal{R}} \mathcal{S}$ and $\mathcal{S}_1 = \mathrm{proj}_{\mathcal{S}} \mathcal{R}$.*

(1) *$\mathcal{R}_1$ is a residue of type $J_1 := J \cap w_1 K w_1^{-1}$, and $\mathcal{S}_1$ is a residue of type $K_1 := K \cap w_1^{-1} J w_1$.*

(2) *Define maps $f \colon \mathcal{R}_1 \to \mathcal{S}_1$ and $g \colon \mathcal{S}_1 \to \mathcal{R}_1$ by $f(C) = \mathrm{proj}_{\mathcal{S}} C$ and $g(D) = \mathrm{proj}_{\mathcal{R}} D$. Then f and g are mutually inverse bijections.*

(3) *Two chambers $C \in \mathcal{R}_1$ and $D \in \mathcal{S}_1$ correspond under the bijections in (2) if and only if $\delta(C, D) = w_1 := \min(\delta(\mathcal{R}, \mathcal{S}))$ or, equivalently, if and only if $d(C, D) = d(\mathcal{R}, \mathcal{S})$.*

(4) *We have*

$$\delta(f(C), f(C')) = w_1^{-1} \delta(C, C') w_1$$

for all $C, C' \in \mathcal{R}_1$ and

$$\delta(g(D), g(D')) = w_1 \delta(D, D') w_1^{-1}$$

for all $D, D' \in \mathcal{S}_1$. The maps f and g preserve adjacency and gallery distance (but not necessarily types of adjacency).

Proof. The first assertion of (1) is already contained in Lemma 5.36, and the second follows by reversing the roles of $\mathcal{R}$ and $\mathcal{S}$. Now consider any $C \in \mathcal{R}_1$. Then, by the lemma, there exists $D \in \mathcal{S}$ with $d(C, D) = d(\mathcal{R}, \mathcal{S})$, and we have $C = \mathrm{proj}_{\mathcal{R}} D$. Reversing the roles of $\mathcal{R}$ and $\mathcal{S}$, we can also conclude from the lemma that $D = \mathrm{proj}_{\mathcal{S}} C \in \mathcal{S}_1$. Thus $g(f(C)) = C$ for all $C \in \mathcal{R}_1$ and, by symmetry, $f(g(D)) = D$ for all $D \in \mathcal{S}_1$. This proves (2), and (3) follows as a byproduct of the proof. Finally, (4) has already been proved as part (3) of the lemma.   □

Remark 5.38. Recall that $\mathcal{R}_1$ and $\mathcal{S}_1$, being residues, are buildings in their own right (Corollary 5.30). What the proposition essentially says is that these two buildings are isomorphic, provided we take account of the type-change given by conjugation by w_1. We will introduce the appropriate language for expressing this in Section 5.5; see Example 5.60(b). In the simplicial setting of Section 4.9, the corresponding result is that AB and BA have isomorphic links (Exercise 4.100).

By way of illustration, we consider Proposition 5.37 in the simplest non-trivial case, in which $K = \{s\}$ for some $s \in S$, i.e., where $\mathcal{S}$ is a panel. [If $K = \emptyset$, then $\mathcal{S}$ is a single chamber and the proposition is vacuous.] Set $w := \min(\delta(\mathcal{R}, \mathcal{S}))$. Then $\mathcal{R}_1 := \mathrm{proj}_{\mathcal{R}} \mathcal{S}$ is a J_1-residue, where $J_1 := J \cap wKw^{-1}$. Now J_1 is either empty or a singleton. If $J_1 = \emptyset$ the conclusion is that the projection $\mathcal{R}_1$ is a single chamber; in other words, $\mathrm{proj}_{\mathcal{R}} D_1 = \mathrm{proj}_{\mathcal{R}} D_2$ for any two chambers $D_1, D_2 \in \mathcal{S}$. Otherwise, $\mathcal{R}_1$ is an s'-panel, where $s' = wsw^{-1} \in J \subseteq S$. We then have $K_1 = K = \{s\}$ and $\mathcal{S}_1 = \mathcal{S}$. Consider now two distinct chambers $D_1, D_2 \in \mathcal{S}$, and set $C_i = \mathrm{proj}_{\mathcal{R}} D_i \in \mathcal{P}$ for $i = 1, 2$. Then the proposition implies that $\delta(C_i, D_i) = w$ for $i = 1, 2$ and that C_1 and C_2 are s'-adjacent. In particular, we have the following situation:

$$C_1 \xrightarrow{\ w\ } D_1$$

$$s' \Big\downarrow \qquad \Big\downarrow s$$

$$C_2 \xrightarrow[\ w\ ]{} D_2$$

We summarize our discussion of this case in the following corollary, which can be stated without reference to the panel S. Indeed, we can just start with two adjacent chambers and apply the results above with S equal to the (unique) panel containing them.

Corollary 5.39. *Let $\mathcal{R}$ be a residue in $\mathcal{C}$, let D_1, D_2 be adjacent chambers in $\mathcal{C}$, and let $C_i = \mathrm{proj}_{\mathcal{R}} D_i$ for $i = 1, 2$. Then C_1 and C_2 are either equal or adjacent. In the latter case, the adjacency type is given by $\delta(C_1, C_2) = s' = wsw^{-1}$, where $w = \delta(C_1, D_1) = \delta(C_2, D_2)$ and $s = \delta(D_1, D_2)$.* □

Exercises

5.40. (a) Let C_1 and C_2 be adjacent chambers, and let D be a chamber such that $d(C_1, D) < d(C_2, D)$. If $\mathcal{P}$ is the panel containing C_1 and C_2, show that $C_1 = \mathrm{proj}_{\mathcal{P}} D$.

(b) Use (a) or some other argument to show that the unique C' in Lemma 5.5 is the projection of D onto the s-panel containing C.

5.41. Let $\mathcal{R}$ be a residue and D a chamber. Suppose $C_1 \in \mathcal{R}$ locally minimizes distance to D in the sense that $d(C_1, D) \leq d(C, D)$ for every $C \in \mathcal{R}$ adjacent to C_1. Show that $C_1 = \mathrm{proj}_{\mathcal{R}} D$.

5.42. If $\mathcal{R}$ and $\mathcal{S}$ are any two residues in $\mathcal{C}$, show that the residues $\mathrm{proj}_{\mathcal{R}} \mathcal{S}$ and $\mathrm{proj}_{\mathcal{S}} \mathcal{R}$ have the same rank.

5.4 Convexity and Subbuildings

As in the chamber complex approach to buildings, the existence of galleries enables one to introduce a notion of *convexity*. It will turn out to be closely related to projections on the one hand and to subbuildings (which we will introduce here) on the other hand. Throughout this section, $(\mathcal{C}, \delta)$ denotes a building of type (W, S).

5.4.1 Convex Sets

Definition 5.43. A nonempty subset $\mathcal{M} \subseteq \mathcal{C}$ is called *(gallery) connected* if for any two chambers $C, D \in \mathcal{M}$, there is a gallery $C = C_0, \ldots, C_n = D$ such that $C_i \in \mathcal{M}$ for all $0 \leq i \leq n$. And $\mathcal{M}$ is called *convex* if for any two chambers $C, D \in \mathcal{M}$, every minimal gallery connecting C and D in $\mathcal{C}$ is contained in $\mathcal{M}$.

Examples 5.44. (a) $\mathcal{C}$ is connected by Lemma 5.16. This implies that every convex subset of $\mathcal{C}$ is connected. Of course $\mathcal{C}$ is also convex.

(b) Every *residue* $\mathcal{R} \subseteq \mathcal{C}$ is convex. To see this, denote by J the type of $\mathcal{R}$, consider two chambers $C, D \in \mathcal{R}$, and set $w := \delta(C, D) \in W_J$. If Γ is a minimal gallery from C to D, its type $(s_1, \ldots, s_n)$ is a reduced decomposition of w (see Lemma 5.16 and Exercise 5.20). We must therefore have $s_i \in J$ for all $1 \leq i \leq n$ by Propositions 2.14 and 2.16, so Γ is contained in $\mathcal{R}$.

(c) If $(\mathcal{M}_i)_{i \in I}$ is a family of convex subsets of $\mathcal{C}$, then $\bigcap_{i \in I} \mathcal{M}_i$, if nonempty, is also convex. There is therefore a smallest convex set containing a given nonempty subset $\mathcal{M} \subseteq \mathcal{C}$, called the *convex hull* of $\mathcal{M}$. It is the intersection of all convex subsets of $\mathcal{C}$ containing $\mathcal{M}$.

We remark in passing that the intersection of connected subsets $\mathcal{M}_i \subseteq \mathcal{C}$ may well be nonempty and nonconnected. This is the case, for instance, if $C, D \in \mathcal{C}$ can be connected by more than one minimal gallery, and we set $\mathcal{M}_i := \{C \mid C \text{ occurs in } \Gamma_i\}$, where $(\Gamma_i)_{i \in I}$ is the family of all these minimal galleries.

It turns out that convex subsets of $\mathcal{C}$ are closed under taking projections and that they are characterized by this property (cf. Proposition 4.115).

Lemma 5.45. *Let $\mathcal{M}$ be a convex subset of $\mathcal{C}$, and let $\mathcal{R}$ be a residue in $\mathcal{C}$ that meets $\mathcal{M}$. Then for any chamber $D \in \mathcal{M}$, we have $\mathrm{proj}_{\mathcal{R}} D \in \mathcal{M}$.*

Proof. Choose $C \in \mathcal{R} \cap \mathcal{M}$ and set $C' = \mathrm{proj}_{\mathcal{R}} D$. By the gate property (Proposition 5.34), $d(C, D) = d(C, C') + d(C', D)$. This says precisely that there is a minimal gallery from C to D passing through C'; hence $C' \in \mathcal{M}$ by convexity. $\qquad\square$

Proposition 5.46. *The following conditions on a nonempty subset $\mathcal{M} \subseteq \mathcal{C}$ are equivalent.*

(i) *$\mathcal{M}$ is convex.*
(ii) *$\mathrm{proj}_{\mathcal{P}} D \in \mathcal{M}$ for every chamber $D \in \mathcal{M}$ and every panel $\mathcal{P}$ in $\mathcal{C}$ that meets $\mathcal{M}$.*

Proof. (i) $\implies$ (ii) by Lemma 5.45. For the converse, suppose (ii) holds. Given a minimal gallery $\Gamma: C_0, \ldots, C_n$ with $C_0, C_n \in \mathcal{M}$, we have to show that all C_i are in $\mathcal{M}$. Since Γ is minimal, $d(C_i, C_n) = n - i$ for all i. This implies that $C_i = \mathrm{proj}_{\mathcal{P}_i} C_n$ for $1 \leq i \leq n$, where $\mathcal{P}_i$ is the panel containing C_{i-1} and C_i (see Exercise 5.40). Since $C_0, C_n \in \mathcal{M}$, it follows inductively that $C_i \in \mathcal{M}$ for all i. $\qquad\square$

Our next observation is that convexity for a subset $\mathcal{M} \subset \mathcal{C}$ is an *intrinsic* property of the pair $(\mathcal{M}, \delta_{\mathcal{M}})$, where $\delta_{\mathcal{M}}$ is the restriction $\delta|_{\mathcal{M} \times \mathcal{M}}$. This may seem surprising, initially, but it follows easily from the close connection between galleries and reduced words (Lemma 5.16 and Exercise 5.20):

Proposition 5.47. *A nonempty subset $M \subseteq C$ is convex if and only if (M, δ_M) has the following property: Given $C, D \in M$ and a reduced decomposition $\mathbf{s}$ of $\delta(C, D)$, there is a gallery from C to D in M of type $\mathbf{s}$.*

Proof. Suppose M is convex. Given C, D in M and a reduced decomposition $\mathbf{s}$ of $w := \delta(C, D)$, there is a minimal gallery from C to D in C of type $\mathbf{s}$ by Lemma 5.16. This gallery is contained in M by convexity, so the condition of the proposition is satisfied. Conversely, suppose this condition holds, and consider an arbitrary minimal gallery $\Gamma: C = C_0, \ldots, C_n = D$ in C with $C, D \in M$. The type $\mathbf{s}$ of this gallery is reduced by Exercise 5.20, so by assumption there is a gallery Γ' from C to D in M of type $\mathbf{s}$. But there is a *unique* gallery from C to D of this type in C (by Lemma 5.16 again), so $\Gamma' = \Gamma$ and M is convex. $\qquad\square$

Exercise 5.48. Generalize Lemma 5.45 as follows. If M is a convex subset of C, and if $\mathcal{R}, \mathcal{S}$ are residues in C satisfying $\mathcal{R} \cap M \neq \emptyset$ and $\mathcal{S} \cap M \neq \emptyset$, then $\mathrm{proj}_{\mathcal{R}} \mathcal{S} \cap M \neq \emptyset$.

5.4.2 Subbuildings

We now specialize to the case that M is a *subbuilding* in the sense of the following definition.

Definition 5.49. Let M be a nonempty subset M of C, and let δ_M be the restriction of δ to $M \times M$. If (M, δ_M) is a building of type (W, S), then it is called a *subbuilding* of (C, δ). We will also say, more briefly, that M is a subbuilding of C.

The notion of a subbuilding is not treated uniformly in the literature. Some authors allow M and C to have different types. In that case residues would also be considered subbuildings, whereas in our definition the only residue that is a subbuilding is $\mathcal{R} = C$. Our convention is chosen so that embeddings of subbuildings will be isometries (these are the natural morphisms in the category of buildings of type (W, S) and will be discussed in the next section). Our convention is also chosen so that it is consistent with the one given in the simplicial context (Section 4.6); this is the content of Exercise 5.97 below, but a preliminary indication can already be found in Exercise 4.89.

By (WD3), each chamber C in a subbuilding M of C is, for any given $s \in S$, s-adjacent to at least one other element of M. We will show below that this, together with convexity, characterizes subbuildings. Let us first introduce some terminology.

Definition 5.50. A nonempty subset M of C is called *thin* (resp. *thick*) if $\mathcal{P} \cap M$ has cardinality 2 (resp. > 2) for every panel $\mathcal{P}$ of C with $\mathcal{P} \cap M \neq \emptyset$. There is an intermediate notion for which the terminology is less standard: M is called *weak* if $\mathcal{P} \cap M$ has cardinality ≥ 2 for every $\mathcal{P}$ as above.

Remarks 5.51. (a) Although these properties are stated in terms of panels $\mathcal{P}$ of $\mathcal{C}$, they are actually intrinsic properties of $(\mathcal{M}, \delta_{\mathcal{M}})$ since a nonempty intersection $\mathcal{P} \cap \mathcal{M}$, where $\mathcal{P}$ is an s-panel, can also be described as an s-equivalence class in $\mathcal{M}$.

(b) The term "weak" is motivated by the fact that buildings as we have defined them are sometimes called "weak buildings" in the literature, the term "building" being reserved for thick buildings.

In view of Remark 5.6, Exercise 5.9, and Lemma 5.45, the reader will not be surprised to see the following characterization of subbuildings. This shows that the property of being a subbuilding, like convexity, is an intrinsic property of $(\mathcal{M}, \delta_{\mathcal{M}})$.

Proposition 5.52. *A nonempty subset* $\mathcal{M} \subseteq \mathcal{C}$ *is a subbuilding if and only if it is weak and convex.*

Proof. Method 1: If $\mathcal{M}$ is a subbuilding, then it is weak by (WD3) and it is convex by Proposition 5.47 (and Lemma 5.16). Conversely, if $\mathcal{M}$ is weak and convex, we must show that it satisfies (WD3). Given $C, D \in \mathcal{M}$ and $s \in S$, set $w := \delta(C, D)$. If $l(sw) = l(w) + 1$, we choose (using the fact that $\mathcal{M}$ is weak) a chamber $C' \in \mathcal{M}$ with $\delta(C', C) = s$. Then $\delta(C', D) = sw$ by (WD2). If $l(sw) = l(w) - 1$, we know there is a (unique) $C' \in \mathcal{C}$ such that $\delta(C', C) = s$ and $\delta(C', D) = sw$. We have to show that $C' \in \mathcal{M}$. Now $d(C', D) = d(C, D) - 1$, so there is a minimal gallery in $\mathcal{C}$ of the form $C, C', \dots, D$. Hence $C' \in \mathcal{M}$ by convexity.

Method 2 (sketch): $\mathcal{M}$ is a subbuilding if and only if it satisfies conditions (WD3a) and (WD3b) of Exercise 5.9. The first condition says that $\mathcal{M}$ is weak, and the second says that it satisfies the convexity criterion of Proposition 5.46. $\square$

We now introduce one of the central notions of building theory in our present setup.

Definition 5.53. A thin subbuilding of $\mathcal{C}$ is called an *apartment* of $\mathcal{C}$.

We note the following important consequence of Proposition 5.52, which should be compared with Theorem 4.86 in Section 4.8.

Corollary 5.54. *A nonempty subset* $\mathcal{A} \subseteq \mathcal{C}$ *is an apartment of* $\mathcal{C}$ *if and only if it is thin and convex.* $\square$

We now show that the "triangle inequality" of Lemma 5.28 always becomes an equality in a thin building and hence, in particular, in an apartment. We will use this in the next section to show that, up to isomorphism, the standard thin building of type (W, S) is the *only* thin building of type (W, S).

Lemma 5.55. *If* $\mathcal{A}$ *is a thin building, then for all* $C, D, E \in \mathcal{A}$, *we have* $\delta(C, E) = \delta(C, D)\delta(D, E)$.

Proof. Consider first the case $\delta(C, D) = s \in S$, as in axiom (WD2). Since $\mathcal{A}$ is thin, C is the only chamber in $\mathcal{A}$ that is s-adjacent to D. Now (WD3) implies that some chamber $D' \in \mathcal{A}$ must satisfy $\delta(D', D) = s$ and $\delta(D', E) = s\delta(D, E)$, so this D' can only be C. Hence $\delta(C, E) = s\delta(D, E) = \delta(C, D)\delta(D, E)$.

In the general case, choose a minimal gallery $\Gamma \colon C = C_0, \ldots, C_n = D$, and let $(s_1, \ldots, s_n)$ be its type. Then we have $\delta(C_{i-1}, E) = s_i\delta(C_i, E)$ for all $n \geq i \geq 1$ by the previous paragraph. An obvious induction now yields $\delta(C, E) = s_1 \cdots s_n\delta(D, E) = \delta(C, D)\delta(D, E)$. $\qquad\square$

Exercise 5.56. Give an example of two buildings $\mathcal{C}', \mathcal{C}$ of type (W, S), with connected Coxeter diagram of rank $|S| \geq 3$, such that $\mathcal{C}'$ is a proper *thick* subbuilding of $\mathcal{C}$. (For $|S| = 2$, trees provide trivial examples of proper thick subbuildings; see Exercise 5.14.)

*5.4.3 2-Convexity

The importance of the notion of convexity should be clear from the previous subsection. It is therefore of interest to minimize what one has to check in order to establish that a set is convex. We present here a result of Abramenko and Van Maldeghem [17], showing that surprisingly little has to be checked. Our discussion will use the concepts of *homotopy* and *elementary homotopy* of galleries, which the reader may need to review before proceeding. These concepts were introduced in Definition 3.23 for Coxeter complexes, and they generalize to buildings in the obvious way.

Definition 5.57. A set of chambers $\mathcal{M} \subseteq \mathcal{C}$ is called 2-*convex* if it has the following property: Suppose $\mathcal{M}$ contains a gallery of alternating type $(s, t, s, \ldots)$ and of length $m(s, t)$ for some $s \neq t$ in S with $m(s, t) < \infty$; then $\mathcal{M}$ also contains the gallery of type $(t, s, t, \ldots)$ with the same length and extremities.

Note that this property says precisely that $\mathcal{M}$ is closed under homotopy of galleries, i.e., if two galleries are homotopic and one is contained in $\mathcal{M}$, then so is the other. Note also that $\mathcal{M}$ is 2-convex if the intersection $\mathcal{M} \cap \mathcal{R}$ is convex (if nonempty) for every spherical rank-2 residue $\mathcal{R}$. Here a residue of type J is called *spherical* if W_J is finite. This explains the terminology "2-convex."

Proposition 5.58. *A set of chambers is convex if and only if it is 2-convex and gallery connected.*

Proof. The "only if" part is trivial. To prove the "if" part, suppose that $\mathcal{M}$ is gallery connected and 2-convex (hence closed under homotopy). Suppose we are given two chambers $C, D \in \mathcal{M}$. To prove that $\mathcal{M}$ is convex, it suffices to show that $\mathcal{M}$ contains *one* minimal gallery from C to D. For then it will contain all such minimal galleries, since they all lie in one apartment and hence are all homotopic by Proposition 3.24.

To find a minimal gallery from C to D in $\mathcal{M}$, start with an arbitrary gallery Γ in $\mathcal{M}$ from C to D. If it is not minimal, then its type is not reduced. The solution to the word problem (Section 2.3.3) therefore implies that Γ is homotopic to a gallery (still in $\mathcal{M}$) whose type has a repetition. This means that there are three consecutive chambers that are all s-equivalent to one another for some $s \in S$. If the first and third of these are equal, we can delete the second and third; otherwise, we can delete the second. Continuing in this way, we reach a minimal gallery in $\mathcal{M}$ after finitely many steps. $\square$

5.5 Isometries and Apartments

Corollary 5.54 provides us with one important characterization of apartments in buildings. In this section we will give a second characterization, similar to Proposition 4.59 in Section 4.5. We first need to introduce the concept of *isometry*, which is the appropriate notion of "isomorphism" in the present setup. We will allow our isometries to involve a change of Coxeter system, just as isomorphisms of simplicial buildings are not required to be type-preserving.

5.5.1 Isometries and σ-Isometries

By an *isomorphism of Coxeter systems* $\sigma\colon (W, S) \to (W', S')$ we mean a group isomorphism $\sigma\colon W \xrightarrow{\sim} W'$ satisfying $\sigma(S) = S'$. Equivalently, we have a bijection $\sigma\colon S \to S'$ such that the order of $\sigma(s)\sigma(t)$ is equal to the order of st for all $s, t \in S$, in which case this map σ uniquely extends to a group isomorphism between W and W' in view of the standard presentation of Coxeter groups (see Section 2.4). So σ may be considered as a relabeling of types. Isomorphisms between Coxeter systems can be identified with isomorphisms between their Coxeter diagrams.

Definition 5.59. Let $(\mathcal{C}, \delta)$ be a building of type (W, S), let $(\mathcal{C}', \delta')$ be a building of type (W', S'), and let $\sigma\colon (W, S) \to (W', S')$ be an isomorphism of Coxeter systems. A *σ-isometry* from $\mathcal{C}$ to $\mathcal{C}'$ is a map $\phi\colon \mathcal{C} \to \mathcal{C}'$ satisfying

$$\delta'(\phi(C), \phi(D)) = \sigma(\delta(C, D))$$

for all $C, D \in \mathcal{C}$. If $(W, S) = (W', S')$ and $\sigma = \mathrm{id}_W$, we simply call ϕ an *isometry*. For general σ, we will sometimes call ϕ an *almost isometry* if we do not need to specify σ. We can also talk about isometries, σ-isometries, and almost isometries in case ϕ is defined only on a subset $\mathcal{M} \subseteq \mathcal{C}$. The same definitions apply without change.

Some of the most important properties of buildings can be phrased in terms of extensions of isometries. For example, we will prove below (Theorem 5.73) that isometries defined on subsets of the standard thin building can always be extended, and this result turns out to encode the fundamental properties of apartments. And a fundamental classification theorem for spherical buildings (to be discussed in Chapter 9) is based on an extension theorem for isometries between spherical buildings that we will describe in Section 5.10.

In view of axiom (WD1), an almost isometry $\phi\colon \mathcal{C} \to \mathcal{C}'$ is obviously injective. However, ϕ need not be surjective. If there exists a surjective almost isometry $\phi\colon \mathcal{C} \to \mathcal{C}'$, the buildings $\mathcal{C}$ and $\mathcal{C}'$ are called *almost isometric*. They are called *isometric* if in addition, $\sigma = \mathrm{id}_W$. Similar notions apply to subsets of buildings. Note that the inverse of a surjective σ-isometry is a σ^{-1}-isometry, so the relation of being "almost isometric" is symmetric. It will become clear as we proceed that "almost isometry" is often the most appropriate notion of "isomorphism" in the W-metric category. Example 5.60(c) below gives a first indication of this. We denote by $\mathrm{Aut}(\mathcal{C}, \delta)$ (or simply $\mathrm{Aut}\,\mathcal{C}$) the group of almost isometries from $(\mathcal{C}, \delta)$ to itself, and we denote by $\mathrm{Aut}_0(\mathcal{C}, \delta)$ (or simply $\mathrm{Aut}_0\,\mathcal{C}$) the group of isometries from $(\mathcal{C}, \delta)$ to itself.

Examples 5.60. (a) If $\mathcal{M}$ is a subbuilding of $\mathcal{C}$, then the inclusion $\mathcal{M} \hookrightarrow \mathcal{C}$ is an isometry.

(b) If $\mathcal{R}$ and $\mathcal{S}$ are residues in $\mathcal{C}$, then by Proposition 5.37, the residues $\mathrm{proj}_{\mathcal{R}}\,\mathcal{S}$ and $\mathrm{proj}_{\mathcal{S}}\,\mathcal{R}$ are almost-isometric buildings. Indeed, using the notation of that proposition, the map $f\colon \mathrm{proj}_{\mathcal{R}}\,\mathcal{S} \to \mathrm{proj}_{\mathcal{S}}\,\mathcal{R}$ is a surjective σ-isometry, where $\sigma\colon (W_{J_1}, J_1) \to (W_{K_1}, K_1)$ is given by $\sigma(w) = w_1^{-1} w w_1$ for all $w \in W_{J_1}$.

(c) Suppose $\mathcal{C}$ and $\mathcal{C}'$ arise as in Section 4.8 from simplicial buildings Δ and Δ' of types (W, S) and (W', S'), respectively. Then any simplicial isomorphism $\phi\colon \Delta \xrightarrow{\sim} \Delta'$ induces a surjective σ-isometry $\mathcal{C} \to \mathcal{C}'$, where $\sigma\colon S \to S'$ is the type-change map ϕ_* (Proposition A.14). Thus isomorphic simplicial buildings yield almost-isometric W-metric buildings. We can delete the word "almost" if $(W, S) = (W', S')$ and ϕ is type-preserving. In particular, there are homomorphisms $\mathrm{Aut}\,\Delta \to \mathrm{Aut}\,\mathcal{C}$ and $\mathrm{Aut}_0\,\Delta \to \mathrm{Aut}_0\,\mathcal{C}$; here $\mathrm{Aut}\,\Delta$ is the group of simplicial automorphisms of Δ, and $\mathrm{Aut}_0\,\Delta$ is the subgroup consisting of type-preserving automorphisms. After we have proved that the W-metric approach to buildings is equivalent to the simplicial approach, it will be easy to prove that these homomorphisms are isomorphisms (see Exercise 5.101).

The following lemma can be viewed as reinterpreting the notion of σ-isometry from the point of view of buildings as chamber systems (Section 5.2).

Lemma 5.61. *Let $(\mathcal{C}, \delta)$ be a building of type (W, S), let $(\mathcal{C}', \delta')$ be a building of type (W', S'), and let $\sigma\colon (W, S) \to (W', S')$ be an isomorphism of Coxeter systems. Then a map $\phi\colon \mathcal{C} \to \mathcal{C}'$ is a σ-isometry if and only if it takes s-adjacent chambers to $\sigma(s)$-adjacent chambers for all $s \in S$.*

Proof. The "only if" part is trivial. To prove the "if" part, suppose ϕ takes s-adjacent chambers to $\sigma(s)$-adjacent chambers for all $s \in S$. Then ϕ takes any gallery in $\mathcal{C}$ of reduced type $\mathbf{s}$ to a gallery in $\mathcal{C}'$ of reduced type $\sigma(\mathbf{s})$. So $\delta'(\phi(C), \phi(D)) = \sigma(\delta(C, D))$ for all $C, D \in \mathcal{C}$ by Lemma 5.16. $\qquad\square$

The next lemma will be extremely useful as we proceed.

Lemma 5.62. *Let $\mathcal{C}$ be a building of type (W, S), $\mathcal{C}'$ a building of type (W', S'), and $\sigma \colon (W, S) \to (W', S')$ an isomorphism of Coxeter systems.*

(1) *If $\mathcal{M}$ is a convex subset of $\mathcal{C}$ and $\phi \colon \mathcal{M} \to \mathcal{C}'$ is a σ-isometry, then $\phi(\mathcal{M})$ is convex in $\mathcal{C}'$.*

(2) *If $\phi \colon \mathcal{C} \to \mathcal{C}'$ is a σ-isometry, then $\phi(\mathcal{C})$ is a subbuilding of $\mathcal{C}'$.*

Proof. (1) This is an easy consequence of the convexity criterion given in Proposition 5.47. In detail, suppose we are given two chambers $C, D \in \mathcal{M}$ and a reduced decomposition $\mathbf{s}'$ of $w' := \delta(\phi(C), \phi(D))$. We must show that there is a gallery from $\phi(C)$ to $\phi(D)$ of type $\mathbf{s}'$ in $\phi(\mathcal{M})$. By the definition of "σ-isometry," we have $w' = \phi(w)$, where $w := \delta(C, D)$. So $\mathbf{s}' = \sigma(\mathbf{s})$ for some reduced decomposition $\mathbf{s}$ of w. There is therefore a gallery of type $\mathbf{s}$ from C to D in $\mathcal{M}$, and we can apply ϕ to get the desired gallery from $\phi(C)$ to $\phi(D)$ in $\phi(\mathcal{M})$.

(2) By (1), $\phi(\mathcal{C})$ is convex, and it is easy to see that $\phi(\mathcal{C})$ is weak since $\mathcal{C}$ is weak. So $\phi(\mathcal{C})$ is a subbuilding of $\mathcal{C}'$ by Proposition 5.52.   □

Exercises

5.63. Let (W, δ_W) be the standard thin building of type (W, S). Show that the set $\mathrm{Iso}(W)$ of isometries $W \to W$ is a group (under composition), which is isomorphic to W and acts simply transitively on W. (Part of what is to be proved is that isometries are automatically surjective in this case.)

5.64. Let $\sigma \colon (W, S) \xrightarrow{\sim} (W', S')$ be an isomorphism of Coxeter systems.

(a) Let $(\mathcal{C}, \delta)$ be a building of type (W, S), and let $\delta^\sigma \colon \mathcal{C} \times \mathcal{C} \to W'$ be the composite $\sigma \circ \delta$. Show that $(\mathcal{C}, \delta^\sigma)$ is a building of type (W', S') that is almost isometric to $(\mathcal{C}, \delta)$.

(b) Interpret the construction in (a) from the simplicial point of view.

5.5.2 Characterizations of Apartments

We begin by proving our earlier claim that the standard thin building (Example 5.7) is essentially the only thin building.

Proposition 5.65. *If $(\mathcal{C}, \delta)$ is a thin building of type (W, S), then $\mathcal{C}$ is isometric to W, where the latter is viewed as the set of chambers of the standard thin building (W, δ_W).*

Proof. Fix a chamber $C_0 \in \mathcal{C}$ and define $\psi \colon \mathcal{A} \to W$ by $\psi(C) := \delta(C_0, C)$ for $C \in \mathcal{C}$. Applying Corollary 5.17(2) and Lemma 5.55, we obtain, for all $C, D \in \mathcal{C}$,

$$\delta_W(\psi(C), \psi(D)) = \delta(C_0, C)^{-1}\delta(C_0, D)$$
$$= \delta(C, C_0)\delta(C_0, D)$$
$$= \delta(C, D) .$$

Hence ψ is an isometry. Using (WD3) (or rather (WD3')) for the building $\mathcal{C}$, one immediately checks that ψ is surjective. Thus $\mathcal{A}$ is isometric to W.   □

The fact that the map ψ in the proof is bijective can be restated as follows:

Corollary 5.66. *Let $(\mathcal{C}, \delta)$ be a thin building of type (W, S), and fix a chamber $C_0 \in \mathcal{C}$. Then for any $w \in W$, there is a unique chamber $C \in \mathcal{C}$ such that $\delta(C_0, C) = w$.*   □

The corollary is trivially true in the standard thin building (most obviously with $C_0 = 1$, since $\delta_W(1, w) = w$), and this is what motivated the definition of ψ in the proof of the proposition.

We can now give the following important characterization of apartments:

Corollary 5.67. *Let $(\mathcal{C}, \delta)$ be a building of type (W, S) and let (W, δ_W) be the standard thin building. Then a subset $\mathcal{A} \subseteq \mathcal{C}$ is an apartment of $\mathcal{C}$ if and only if it is isometric to W.*

Proof. If $\mathcal{A}$ is an apartment, then it is a thin building and hence is isometric to W by Proposition 5.65. Conversely, if there exists an isometry $\phi\colon W \to \mathcal{C}$ with $\phi(W) = \mathcal{A}$, then $\mathcal{A}$ is a subbuilding of $\mathcal{C}$ by Lemma 5.62, and it is obviously thin. Hence, according to Definition 5.53, $\mathcal{A}$ is an apartment.   □

Corollary 5.68. *Any two apartments $\mathcal{A}_1, \mathcal{A}_2$ of $\mathcal{C}$ are isometric. Furthermore, we can always find a surjective isometry $\phi\colon \mathcal{A}_1 \to \mathcal{A}_2$ with $\phi(C) = C$ for all $C \in \mathcal{A}_1 \cap \mathcal{A}_2$.*

Proof. By Corollary 5.67, there exist surjective isometries $\phi_1\colon W \to \mathcal{A}_1$ and $\phi_2\colon W \to \mathcal{A}_2$, so $\phi := \phi_2 \circ \phi_1^{-1}\colon \mathcal{A}_1 \to \mathcal{A}_2$ is a surjective isometry. If $\mathcal{A}_1 \cap \mathcal{A}_2 = \emptyset$, there is nothing more to show. If $\mathcal{A}_1 \cap \mathcal{A}_2$ contains a chamber C_0, we can choose ϕ_1 and ϕ_2 such that $\phi_1(1) = \phi_2(1) = C_0$. (This follows from the fact that the group of all isometries of W acts transitively on W by Exercise 5.63. Alternatively, it follows from the proof of Proposition 5.65.) So we have $\phi(C_0) = C_0$. If $C \in \mathcal{A}_1 \cap \mathcal{A}_2$ is now arbitrary, then we have $\delta(C_0, C) = \delta(\phi(C_0), \phi(C)) = \delta(C_0, \phi(C))$. Since C_0, C, and $\phi(C)$ are all in $\mathcal{A}_2$, Corollary 5.66 implies that $\phi(C) = C$.   □

We remark in passing that the last part of the proof above can be viewed as a version of the *standard uniqueness argument* that we used several times in Chapter 3. [The earlier uses involved galleries, but here Weyl distances play the same role.]

We now have three characterizations of apartments: Given $\mathcal{A} \subseteq \mathcal{C}$,

$$\mathcal{A} \text{ is an apartment} \iff \mathcal{A} \text{ is a thin subbuilding}$$
$$\iff \mathcal{A} \text{ is thin and convex}$$
$$\iff \mathcal{A} \text{ is isometric to } W.$$

Exercises

5.69. If $\mathcal{A}_1$ and $\mathcal{A}_2$ are thin buildings, show that every isometry $\mathcal{A}_1 \to \mathcal{A}_2$ is surjective.

5.70. In this exercise we introduce *retractions* in our present setup (cf. Section 4.4). Let $\mathcal{C}$ be a building of type (W, S). Given an apartment $\mathcal{A}$ and a chamber $C \in \mathcal{A}$, there is a unique isometry $\phi \colon W \to \mathcal{A}$ with $\phi(1) = C$ (check this). Now define $\rho = \rho_{\mathcal{A},C} \colon \mathcal{C} \to \mathcal{A}$ by $\rho(D) = \phi(\delta(C, D))$ for all $D \in \mathcal{C}$. Prove that ρ has the following properties:

(a) ρ preserves s-equivalence for all $s \in S$.
(b) $\delta(C, \rho(D)) = \delta(C, D)$ for all $D \in \mathcal{C}$.
(c) If $\mathcal{A}'$ is an apartment of $\mathcal{C}$ with $C \in \mathcal{A}'$, then ρ maps $\mathcal{A}'$ isometrically onto $\mathcal{A}$.

5.5.3 Existence of Apartments

A serious gap in our knowledge about apartments is that we do not yet know that buildings in the sense of the present chapter contain even a single apartment. We now fill this gap by proving the main result of this section (Theorem 5.73 below), which says that isometries defined on subsets of W can always be extended to isometries defined on all of W. The key step in the argument is provided by the following lemma.

Lemma 5.71. *Let $\mathcal{C}$ be a building of type (W, S), let C, D_1, D_2 be chambers in $\mathcal{C}$, and let $s \in S$ satisfy $l\bigl(s\delta(C, D_i)\bigr) < l\bigl(\delta(C, D_i)\bigr)$ for $i = 1, 2$. Suppose that*

$$\delta(D_1, D_2) = \delta(D_1, C)\delta(C, D_2) \, . \tag{5.6}$$

Then $\mathrm{proj}_\mathcal{P} D_1 = \mathrm{proj}_\mathcal{P} D_2$, where $\mathcal{P}$ denotes the s-panel containing C.

Remark 5.72. Before beginning the proof, we recall that the hypothesis (5.6) always holds in the standard thin building (W, δ_W). Consequently, it holds if the set $\{C, D_1, D_2\}$ is isometric to a subset of the standard thin building. This is the context in which we will apply the lemma. Note also, for motivation, that the lemma gives us the following plausible diagram:

$$
\begin{array}{ccc}
 & & D_1 \\
 & \overset{v_1}{\nearrow} & \\
C \overset{s}{\rule{1.2em}{0.5pt}} C' & & \Big| v_1^{-1}v_2 \\
 & \underset{v_2}{\searrow} & \\
 & & D_2
\end{array}
$$

Here $C' = \mathrm{proj}_\mathcal{P} D_1 = \mathrm{proj}_\mathcal{P} D_2$, $v_i = s\delta(C, D_i)$ (for $i = 1, 2$), and the equation $\delta(D_1, D_2) = v_1^{-1}v_2$ is a restatement of (5.6).

Proof of the lemma. Set $w_i = \delta(C, D_i)$, $v_i = sw_i$, and $C_i = \text{proj}_P D_i$ for $i = 1, 2$. Since $l(v_i) < l(w_i)$, we have $v_i = \delta(C_i, D_i)$, $C_i \neq C$, and $\delta(C_i, C) = s$. We must show that $C_1 = C_2$. If this is false, then $\delta(C_1, C_2) = s$ and hence, by (WD2), $\delta(C_2, D_1) = s\delta(C_1, D_1) = sv_1$, as illustrated in the following diagram:

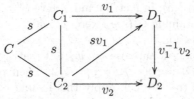

We will now apply the triangle inequality (Lemma 5.28) to the lower triangle and get a contradiction.

Choose a reduced decomposition $v_2 = s_1 \cdots s_m$. Then Lemma 5.28 gives us an element $v'_2 = s_{i_1} \cdots s_{i_k}$ (for some indices $1 \leq i_1 < \cdots < i_k \leq m$) such that

$$\delta(D_1, D_2) = \delta(D_1, C_2)v'_2 = v_1^{-1}sv'_2 \ .$$

Since $\delta(D_1, D_2) = v_1^{-1}v_2$, we can cancel v_1^{-1} to get $sv'_2 = v_2$; hence $v'_2 = sv_2$, contradicting the assumption that $l(sv_2) > l(v_2) \geq l(v'_2)$. $\square$

Theorem 5.73. *Let (C, δ) be a building of type (W, S), and let $V \subseteq W$ be an arbitrary subset. Then any isometry $\phi\colon V \to C$ can be extended to an isometry $\tilde{\phi}\colon W \to C$. Consequently, any subset of C that is isometric to a subset of W is contained in an apartment.*

Proof. The second assertion follows from the first, since the image of W under an isometry is an apartment by Corollary 5.67. To prove the first assertion, we may assume that $V \neq \emptyset$ and $V \neq W$. Then there exist $v_0 \in V$ and $s \in S$ with $v_0s \notin V$. We will show first that ϕ can be extended to an isometry $\phi'\colon V' \to C$, where $V' = V \cup \{v_0s\}$. Since the group of isometries of W acts transitively on W (see Exercise 5.63), we may assume (by changing (V, ϕ) appropriately) that $v_0 = 1$. Set $C = \phi(1)$. We want to find a chamber $C' \in C$ that satisfies

$$\delta(C', \phi(w)) = \delta_W(s, w) = sw \tag{5.7}$$

for all $w \in V$. For it is then clear that the map $\phi'\colon V' \to C$ defined by $\phi'(s) = C'$ and $\phi'(w) = \phi(w)$ for all $w \in V$ is an isometry. Note that the desired C' will have to be s-adjacent to C [set $w = 1$ in (5.7)] and hence must fit into a diagram like the following for each $w \in V$:

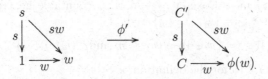

Here $\delta(C, \phi(w)) = w$, as indicated in the diagram, since ϕ is an isometry.

For any $w \in V$ such that $l(sw) > l(w)$, equation (5.7) will hold for *any* C' that is s-adjacent to C by (WD2). In particular, we can choose an arbitrary such C' if $l(sw) > l(w)$ for all $w \in V$. If, on the other hand, there is an element $w \in V$ with $l(sw) < l(w)$, then there is a unique C' that will work for that w (see Lemma 5.5); in fact, we must have $C' = \text{proj}_{\mathcal{P}} \, \phi(w)$, where $\mathcal{P}$ is the s-panel in C containing C (see Exercise 5.40).

The crucial observation, now, is that $C' = \text{proj}_{\mathcal{P}} \, \phi(w)$ *does not depend on the choice of* w. Indeed, suppose $w_1, w_2 \in V$ satisfy $l(sw_1) < l(w_1)$ and $l(sw_2) < l(w_2)$. Set $D_i := \phi(w_i)$ for $i = 1, 2$. Then $\text{proj}_{\mathcal{P}} \, D_2 = \text{proj}_{\mathcal{P}} \, D_1$ by Lemma 5.71. Note that we can apply this lemma by Remark 5.72, since $\{C, D_1, D_2\}$ is isometric via ϕ to the subset $\{1, w_1, w_2\}$ of W:

Thus we have found the desired C' satisfying (5.7) for all $w \in V$.

If S is countable, we complete the proof as follows. We may still assume that $1 \in V$. Then we arrange the elements of $W \smallsetminus V$ in a finite or infinite sequence $(w_1, w_2, \ldots)$ such that for each $n \geq 0$, w_{n+1} is adjacent to an element of $V_n := V \cup \{w_1, \ldots, w_n\}$. Since S is countable, it is easy to achieve this (see Exercise 5.75 below). Using the first part of our proof and a straightforward induction, we obtain a sequence of isometries $\phi_n \colon V_n \to C$, each extending the previous one, with $\phi_0 = \phi$. These fit together to give the desired isometry $\tilde{\phi} \colon W \to C$.

Since we often even assume that S is finite, this induction covers the most interesting cases. However, it is easy to complete the proof in the general case using Zorn's lemma (see Exercise 5.76 below). □

Theorem 5.73 has many applications, including the following:

Corollary 5.74. *For any two chambers $C, D \in \mathcal{C}$, there exists an apartment $\mathcal{A}$ of $\mathcal{C}$ with $C, D \in \mathcal{A}$.*

Proof. This follows from the second assertion of the theorem, since $\{C, D\}$ is isometric to $\{1, w\}$, where $w = \delta(C, D)$. □

Exercises

5.75. Let (W, S) be a Coxeter system with countable generating set $S = \{s_1, s_2, \ldots\}$. Set $S_m := \{s_1, \ldots, s_m\}$ and

$$W_m := \{w \in W \mid l(w) \leq m \text{ and } S(w) \subseteq S_m\} \ .$$

(See Proposition 2.16 for the definition of $S(w)$.)

(a) Show that there is a total order on W such that $w' < w$ if $w' \in W_m$ and $w \notin W_m$ or if w and w' are both in $W_m \smallsetminus W_{m-1}$ and $l(w') < l(w)$.

(b) Show that for each $w \in W \smallsetminus \{1\}$, there exist $w' \in W$ and $s \in S$ with $w = w's$ and $w' < w$.

(c) Given $V \subset W$, modify the total order so that $v < w$ for all $v \in V$ and $w \in W \smallsetminus V$, and deduce that there exists a sequence $(w_1, w_2, \ldots)$ as in the last part of the proof of Theorem 5.73.

5.76. Let V, W, $\mathcal{C}$, and ϕ be as in Theorem 5.73. Consider the set P of all pairs (V', ϕ'), where $V \subseteq V' \subseteq W$ and $\phi' \colon V' \to \mathcal{C}$ is an isometry extending ϕ. Define a partial order on this set by declaring $(V', \phi') \leq (V'', \phi'')$ if and only if $V' \subseteq V''$ and $\phi''|_{V'} = \phi'$. Show that P has a maximal element $(\tilde{V}, \tilde{\phi})$ and that necessarily $\tilde{V} = W$.

5.77. (a) Let C, D, E be chambers in a building $\mathcal{C}$. Show that they are contained in a common apartment if and only if $\delta(C, E) = \delta(C, D)\delta(D, E)$.

(b) More generally, if $\mathcal{B}$ is an arbitrary set of chambers, show that $\mathcal{B}$ is contained in an apartment if and only if $\delta(C, E) = \delta(C, D)\delta(D, E)$ for all $C, D, E \in \mathcal{B}$.

5.78. We finished the proof of Lemma 5.71 by appealing to the "triangle inequality." Give an alternative proof based on the version of the triangle inequality given in Exercise 5.33 (see equation (5.2)).

5.5.4 Roots

Isometries provide a way of defining *roots* in a building of type (W, S). We begin by adapting the notion of a root in a Coxeter complex, as given informally in Chapter 2 and formally in Chapter 3, to our present way of thinking about thin buildings.

Definition 5.79. Let (W, δ_W) be the standard thin building of type (W, S). For $s \in S$, we set

$$\alpha_s := \{w \in W \mid l(sw) > l(w)\}$$

and call it the *simple root* of W corresponding to s.

The motivation for this definition can be found in Sections 3.3.3 and 3.4. To review this briefly, recall that $l(sw) = d(s, w)$ and $l(w) = d(1, w)$. So α_s is the set of all chambers in W that are closer to 1 than to s. By Lemma 3.45, this is the set of chambers of $\phi_s(\Sigma(W, S))$, where ϕ_s denotes the folding of the standard Coxeter complex $\Sigma(W, S)$ with $\phi_s(s) = 1$. Thus α_s is indeed the set of chambers of a root in the sense of our earlier treatment. We now use isometries to define roots in arbitrary buildings.

Definition 5.80. Let $\mathcal{C}$ be a building of type (W, S). A subset $\alpha \subseteq \mathcal{C}$ is called a *root* of $\mathcal{C}$ if it is isometric to a simple root $\alpha_s \subseteq W$ for some $s \in S$.

Proposition 5.81. *Let α be a root of $\mathcal{C}$. Then we have:*

(1) α *is a convex subset of $\mathcal{C}$.*
(2) α *is contained in an apartment of $\mathcal{C}$.*
(3) *If $\mathcal{C}$ is the standard thin building of type (W, S), then $\alpha = v\alpha_s$ for some $s \in S$ and $v \in W$.*

Proof. By definition, there is an isometry $\phi\colon \alpha_s \to \mathcal{C}$ for some $s \in S$, with image α. Since α_s is a convex subset of W by Lemma 3.44, it follows from Lemma 5.62(1) that α is a convex subset of $\mathcal{C}$. This proves (1). Now use Theorem 5.73 to extend ϕ to an isometry (still called ϕ) from W to $\mathcal{C}$. The image is an apartment by Corollary 5.67, whence (2). In case $\mathcal{C}$ is the standard thin building, $\phi\colon W \to W$ is given by left-translation by some $v \in W$ (see Exercise 5.63 and its solution); so $\alpha = v\alpha_s$, proving (3). $\square$

Remark 5.82. It follows from part (3) of the proposition that Definition 5.80 is consistent with the definition of a root in a Coxeter complex given in Chapter 3 if one identifies a Coxeter complex with its set of chambers. Indeed, if ϕ is an arbitrary folding of $\Sigma(W, S)$, then there exist adjacent chambers v, vs ($v \in W$, $s \in S$) such that $\phi(vs) = v$. In view of the uniqueness of foldings (see Lemma 3.46), we must have $\phi = \lambda_v \circ \phi_s \circ \lambda_{v^{-1}}$, where λ_u (for $u \in W$) denotes the automorphism of $\Sigma(W, S)$ given by left multiplication by u, and ϕ_s, as before, is the folding such that $\phi_s(s) = 1$. Hence the set of chambers of $\phi(\Sigma(W, S))$ is $\lambda_v(\alpha_s) = v\alpha_s$.

Exercises

5.83. (a) Let α be a root of a building $\mathcal{C}$ of type (W, S). Let $\mathcal{P}$ be a panel of $\mathcal{C}$ that intersects α in a single chamber C. Given $D \in \mathcal{P} \smallsetminus \{C\}$, show that there exists an apartment $\mathcal{A}$ of $\mathcal{C}$ containing α and D.
  (b) If, furthermore, $D' \in \mathcal{P} \smallsetminus \{C, D\}$ and $\mathcal{A}'$ is an apartment containing $\alpha \cup \{D'\}$, prove that $\mathcal{A} \cap \mathcal{A}' = \alpha$.
  (c) Deduce that if $\mathcal{C}$ is thick, any convex subset of $\mathcal{C}$ that is contained in some apartment of $\mathcal{C}$ is an intersection of apartments of $\mathcal{C}$.

5.84. Part (c) of the previous exercise can be sharpened as follows: If $\mathcal{C}$ is a thick building of type (W, S), $\mathcal{A}$ is an apartment of $\mathcal{C}$, and $\mathcal{M} \subseteq \mathcal{A}$ is convex, then there exists an apartment $\mathcal{A}'$ of $\mathcal{C}$ with $\mathcal{A} \cap \mathcal{A}' = \mathcal{M}$. Prove this stronger statement. To get started, look at Exercise 3.97.

5.6 W-Metric Spaces Versus Chamber Complexes

Now that we have apartments at our disposal, we can show that our W-metric spaces give rise to buildings in the sense of Chapter 4. The first step is to associate a *simplicial complex* to a W-metric building. As we already remarked, it is the *residues* that will be identified with the simplices of this complex. We begin by collecting a few easy facts about residues.

In this section, $\mathcal{C}$ will again denote a building of type (W, S). Recall from Proposition 2.16 that for each $w \in W$, there exists a subset $S(w) \subseteq S$ such that $S(w) = \{s_1, \ldots, s_n\}$ for any reduced decomposition $w = s_1 \cdots s_n$.

Lemma 5.85. *Let $\mathcal{R}$ and $\mathcal{S}$ be two residues in $\mathcal{C}$ of respective types J and K.*

(1) *There is an inclusion-preserving bijection from the set of residues of $\mathcal{C}$ containing $\mathcal{R}$ to the set of subsets of S containing J. It associates to any residue $\mathcal{R}' \supseteq \mathcal{R}$ its type J'.*

(2) *There is a smallest element $\mathcal{M}$ in the set of all residues of $\mathcal{C}$ that contain $\mathcal{R}$ and $\mathcal{S}$. The type of $\mathcal{M}$ is $J \cup K \cup S(w)$, where $w = \delta(C, D)$ for some arbitrarily chosen $C \in \mathcal{R}$ and $D \in \mathcal{S}$.*

Proof. (1) If we fix $C \in \mathcal{R}$, then $J' \mapsto R_{J'}(C)$ defines an inclusion-preserving map from $\{J' \mid J \subseteq J' \subseteq S\}$ to the set of residues of $\mathcal{C}$ containing $\mathcal{R}$. It is easy to check that this is inverse to the (inclusion-preserving) map that associates to any residue $\mathcal{R}' \supseteq \mathcal{R}$ its type J'.

(2) Choose $C \in \mathcal{R}$ and $D \in \mathcal{S}$ and set $w := \delta(C, D)$ and $L := J \cup K \cup S(w)$. Note first that the definition of L does not depend on the choices of C and D. In fact, since $\delta(\mathcal{R}, \mathcal{S}) = W_J w W_K$ by Lemma 5.29, we have $S(w') \subseteq L$ for any $w' \in \delta(\mathcal{R}, \mathcal{S})$ and also, by symmetry, $S(w) \subseteq J \cup K \cup S(w')$, so that $L = J \cup K \cup S(w')$ for all $w' \in \delta(\mathcal{R}, \mathcal{S})$. Now let $\mathcal{R}'$ be any residue containing $\mathcal{R}$ and $\mathcal{S}$, and let L' be its type. By (1), we have $J \cup K \subseteq L'$. But also $w \in W_{L'}$ (since $C, D \in \mathcal{R}'$) and hence $S(w) \subseteq L'$. This implies $L \subseteq L'$. On the other hand, we clearly have $\mathcal{R} \cup \mathcal{S} \subseteq R_L(C) = R_L(D)$. So L is the minimal type of any residue containing $\mathcal{R}$ and $\mathcal{S}$. By (1) again, this implies that $R_L(C)$ is the smallest residue containing $\mathcal{R}$ and $\mathcal{S}$. $\square$

Lemma 5.86. *If $\mathcal{C}'$ is a subbuilding of $\mathcal{C}$, then each residue $\mathcal{R}'$ of $\mathcal{C}'$ is contained in a smallest residue $\mathcal{R}$ of $\mathcal{C}$. The map $\mathcal{R}' \mapsto \mathcal{R}$ is an inclusion-preserving bijection from the set of all residues of $\mathcal{C}'$ to the set of all residues of $\mathcal{C}$ meeting $\mathcal{C}'$.*

Proof. Denote the type of $\mathcal{R}'$ by J and choose $C \in \mathcal{R}'$. Then each residue of $\mathcal{C}$ containing $\mathcal{R}'$ must also contain $\mathcal{R} := R_J(C)$, which is therefore the smallest residue of $\mathcal{C}$ containing $\mathcal{R}'$. Define a map $f \colon \{\text{residues of } \mathcal{C}'\} \to \{\text{residues of } \mathcal{C} \text{ meeting } \mathcal{C}'\}$ by $f(\mathcal{R}') := \mathcal{R}$. Then one checks immediately that the inverse of f is given by $\mathcal{R} \mapsto \mathcal{R} \cap \mathcal{C}'$. $\square$

For the rest of this section we will assume that S *is finite* in order to be consistent with the convention that simplices in simplicial complexes are always assumed to have finite rank. The simplicial complex that we will construct is essentially the set of residues of $\mathcal{C}$, ordered by reverse inclusion. But, for reasons to be explained below, we will make a notational distinction between a residue $\mathcal{R}$ and the corresponding simplex, which will be denoted by $F_{\mathcal{R}}$. Formally, $F_{\mathcal{R}}$ is just another name for $\mathcal{R}$.

Definition 5.87. We associate a poset to $\mathcal{C}$ by setting

$$\Delta(\mathcal{C}) := \{F_{\mathcal{R}} \mid \mathcal{R} \text{ is a residue of } \mathcal{C}\} ,$$

with partial order

$$F_{\mathcal{R}} \leq F_{\mathcal{S}} : \Longleftrightarrow \mathcal{R} \supseteq \mathcal{S} .$$

We refer to the elements of $\Delta(\mathcal{C})$ as *simplices* or *faces*. If J is the type of $\mathcal{R}$, we set

$$\tau(F_{\mathcal{R}}) = S \smallsetminus J \tag{5.8}$$

and say that $S \smallsetminus J$ is the *type* of $F_{\mathcal{R}}$ or that $F_{\mathcal{R}}$ is of *cotype* J. If $\mathcal{R}$ is a singleton $\{C\}$, we identify $F_{\{C\}}$ with C. The simplices of this form are again called the *chambers* of $\Delta(\mathcal{C})$; thus the set of chambers of $\Delta(\mathcal{C})$ is identified with $\mathcal{C}$. If $F_{\mathcal{R}} \leq F_{\mathcal{S}}$, we also say that $F_{\mathcal{R}}$ is a *face* of $F_{\mathcal{S}}$ or that $F_{\mathcal{S}}$ *contains* $F_{\mathcal{R}}$ (though $\mathcal{S} \subseteq \mathcal{R}$ as a subset of $\mathcal{C}$).

There are two reasons why we introduce the new notation $F_{\mathcal{R}}$ instead of just talking about the poset of all residues of $\mathcal{C}$. First of all, it makes it clearer that we think of $F_{\mathcal{R}}$ as a new object now, namely a simplex in a simplicial complex. Secondly, we still want to consider sets of chambers in $\Delta(\mathcal{C})$, and a statement like "$\mathcal{R}$ is the set of all chambers in $\Delta(\mathcal{C})$ containing $\mathcal{R}$" is more confusing than "$\mathcal{R}$ is the set of all chambers in $\Delta(\mathcal{C})$ containing $F_{\mathcal{R}}$."

Let us now formally establish that $\Delta(\mathcal{C})$ is indeed a simplicial complex. See Appendix A.1 for the relevant definitions concerning simplicial and chamber complexes.

Lemma 5.88. $\Delta(\mathcal{C})$ *is a colorable chamber complex of rank equal to* $|S|$. *Its set of chambers* $\mathcal{C}(\Delta(\mathcal{C}))$ *is equal to* $\mathcal{C}$, *and the function* τ *defined in* (5.8) *is a type function. If* $\mathcal{C}'$ *is a subbuilding of* $\mathcal{C}$, *then* $\Delta(\mathcal{C}')$ *is canonically isomorphic to a convex chamber subcomplex of* $\Delta(\mathcal{C})$.

Proof. We first check that the poset $\Delta := \Delta(\mathcal{C})$ has the properties (a) and (b) that characterize simplicial complexes (Definition A.1). By definition of Δ, two simplices $A = F_{\mathcal{R}}$ and $B = F_{\mathcal{S}}$ have a greatest lower bound $A \cap B$ in Δ if and only if the residues $\mathcal{R}$ and $\mathcal{S}$ have a least upper bound in the poset of all residues of $\mathcal{C}$ ordered by inclusion. So (a) follows from Lemma 5.85(2). Likewise, Lemma 5.85(1) implies that $\Delta_{\leq A}$ is isomorphic to the poset $\{J' \mid J \subseteq J' \subseteq S\}$, ordered by reverse inclusion (where J is the type of $\mathcal{R}$), and hence is also isomorphic to the set of subsets of $S \smallsetminus J$. Therefore, condition (b) is also satisfied. At the same time we have shown that the rank of a simplex $A \in \Delta$ of cotype J is equal to $|S \smallsetminus J|$.

The maximal simplices correspond to the residues of type $\emptyset$, i.e., the elements of $\mathcal{C}$. (And the empty simplex is given by the unique residue of type S, which is $\mathcal{C}$ itself.) Codimension-1 simplices correspond to rank-1 residues, i.e., to panels. Hence galleries in $\mathcal{C}$ in the sense of Definition 5.15 also yield galleries in Δ, which is thus shown to be a chamber complex. It is immediate from the first paragraph of the proof that the function τ defined in (5.8) is a chamber map to the complex consisting of subsets of S, so it is a type function (see Section A.1.3). Thus Δ is colorable.

Now assume that $\mathcal{C}'$ is a subbuilding of $\mathcal{C}$. By Lemma 5.86, the poset $\Delta(\mathcal{C}')$ is isomorphic to the poset $P := \{\mathcal{R} \mid \mathcal{R}$ is a residue of $\mathcal{C}$ with $\mathcal{R} \cap \mathcal{C}' \neq \emptyset\}$, ordered by reverse inclusion. The set of maximal elements of P can be identified with $\mathcal{C}'$, and P is canonically isomorphic to the simplicial subcomplex Δ' of Δ given by

$$\Delta' := \{F_{\mathcal{R}} \mid \mathcal{R} \in P\} \ .$$

This is a chamber complex in its own right (being isomorphic to $\Delta(\mathcal{C}')$), and its set of chambers is $\mathcal{C}'$, which is a convex subset of $\mathcal{C}$ by Proposition 5.52. Hence Δ' is a convex chamber subcomplex of Δ. $\square$

In the following, we will identify $\Delta(\mathcal{C}')$ with the chamber subcomplex Δ' of $\Delta(\mathcal{C})$ that occurred in the proof, i.e., with

$$\{F_{\mathcal{R}} \mid \mathcal{R} \text{ is a residue of } \mathcal{C} \text{ with } \mathcal{R} \cap \mathcal{C}' \neq \emptyset\} \ .$$

Example 5.89. If (W, δ_W) is the standard thin building of type (W, S), then $\Delta(W)$ is equal to the standard Coxeter complex $\Sigma(W, S)$ introduced in Chapter 3. Indeed, the J-residues in W are the standard cosets wW_J; hence, up to notation, $\Delta(W)$ is the set of standard cosets, ordered by reverse inclusion, and that was precisely our definition of $\Sigma(W, S)$.

So the standard thin buildings correspond to Coxeter complexes. In order to verify that $\Delta(\mathcal{C})$ satisfies the building axioms of Chapter 4, the only preparation still needed is to see how isometries translate into chamber maps. This is in fact easy and requires only the following remark.

Remark 5.90. Let $(\mathcal{C}', \delta')$ be of type (W', S'), let $\sigma \colon (W', S') \overset{\sim}{\longrightarrow} (W, S)$ be an isomorphism of Coxeter systems, and let $\phi \colon \mathcal{C}' \to \mathcal{C}$ be a σ-isometry. Assume first that ϕ is *surjective*. Then the inverse $\phi^{-1} \colon \mathcal{C} \to \mathcal{C}'$ is a σ^{-1}-isometry. It follows easily that J'-equivalence in $\mathcal{C}'$ corresponds under ϕ to $\sigma(J')$-equivalence in $\mathcal{C}$ for all subsets $J' \subseteq S'$. So ϕ takes residues to residues and induces a simplicial isomorphism $\Delta(\phi) \colon \Delta(\mathcal{C}') \overset{\sim}{\longrightarrow} \Delta(\mathcal{C})$, defined by $\Delta(\phi)(F_{\mathcal{R}'}) = F_{\phi(\mathcal{R}')}$ for any residue $\mathcal{R}'$ of $\mathcal{C}'$. The inverse of this isomorphism is $\Delta(\phi^{-1}) \colon \Delta(\mathcal{C}) \overset{\sim}{\longrightarrow} \Delta(\mathcal{C}')$.

In the general case, where ϕ is not necessarily surjective, recall that $\mathcal{B} := \phi(\mathcal{C}')$ is a subbuilding of $\mathcal{C}$ by Lemma 5.62. In view of Lemma 5.88, we can identify $\Delta(\mathcal{B})$ with a chamber subcomplex of $\Delta(\mathcal{C})$. In this case, we define the chamber map $\Delta(\phi) \colon \Delta(\mathcal{C}') \to \Delta(\mathcal{C})$ as the composite of the isomorphism $\Delta(\mathcal{C}') \overset{\sim}{\longrightarrow} \Delta(\mathcal{B})$ with the canonical embedding $\Delta(\mathcal{B}) \hookrightarrow \Delta(\mathcal{C})$. Concretely, we have $\Delta(\phi)(F_{\mathcal{R}'}) = F_{\mathcal{R}}$, where $\mathcal{R}$ is the smallest residue of $\mathcal{C}$ containing $\phi(\mathcal{R}')$. Note that the type-change map associated to $\Delta(\phi)$ is σ.

We can now easily prove the main result of this section.

Theorem 5.91. *If $(\mathcal{C}, \delta)$ is a building of type (W, S), then $\Delta(\mathcal{C})$ is a simplicial building of type (W, S) in the sense of Definition 4.37.*

Proof. We already verified in Lemma 5.88 that $\Delta(\mathcal{C})$ is a chamber complex, and we have given it a type function with values in S. Set

$$\Omega := \{\Delta(\mathcal{A}) \mid \mathcal{A} \text{ is an apartment of } \mathcal{C}\} \ .$$

Recall that we can consider $\Delta(\mathcal{A})$ as a chamber subcomplex of $\Delta(\mathcal{C})$ by Lemma 5.88. We now check the conditions (B0), (B1), and (B2″) for $\Delta :=$ $\Delta(\mathcal{C})$ with respect to the family Ω of subcomplexes (which will turn out to be the complete system of apartments).

(B0) By Corollary 5.67, any apartment $\mathcal{A}$ of $\mathcal{C}$ is an isometric image of the standard thin building (W, δ_W). Hence, by Remark 5.90 and Example 5.89, $\Delta(\mathcal{A})$ is isomorphic, as a chamber complex, to $\Delta(W) = \Sigma(W, S)$. Note that one in fact gets a type-preserving isomorphism $\Delta(\mathcal{A}) \xrightarrow{\sim} \Sigma(W, S)$, so the fact that Δ is a building of type (W, S) will follow as soon as we finish proving that Δ is a building.

(B1) Since we already know that Δ is a chamber complex, it suffices to show that any two chambers are contained in an element of Ω. This is precisely the content of Corollary 5.74.

(B2″) Consider two elements $\Sigma = \Delta(\mathcal{A})$ and $\Sigma' = \Delta(\mathcal{A}')$ of Ω, and assume that $\mathcal{A} \cap \mathcal{A}'$ contains a chamber C. By Corollary 5.68, there exists a surjective isometry $\phi \colon \mathcal{A} \to \mathcal{A}'$ with $\phi(D) = D$ for all $D \in \mathcal{A} \cap \mathcal{A}'$. Remark 5.90 yields a type-preserving simplicial isomorphism $\Delta(\phi) \colon \Sigma \xrightarrow{\sim} \Sigma'$. We have to show that $\Delta(\phi)(A) = A$ for any $A = F_{\mathcal{R}} \in \Sigma \cap \Sigma'$, where $\mathcal{R}$ is a residue of $\mathcal{C}$ that meets $\mathcal{A}$ and $\mathcal{A}'$. Since $\Delta(\phi)$ is type-preserving and fixes the chambers in $\Sigma \cap \Sigma'$, it suffices to show that A is a face of such a chamber. In other words, we have to show that $\mathcal{R}$ contains a chamber $D \in \mathcal{A} \cap \mathcal{A}'$. To this end we can take $D = \mathrm{proj}_{\mathcal{R}} C$, which is in $\mathcal{A} \cap \mathcal{A}'$ by Lemma 5.45.   □

Remarks 5.92. (a) As indicated, $\Omega := \{\Delta(\mathcal{A}) \mid \mathcal{A} \text{ is an apartment of } \mathcal{C}\}$ is the *complete* system of apartments of the building $\Delta = \Delta(\mathcal{C})$. To see this, let Σ be any apartment of Δ and set $\mathcal{A} := \mathcal{C}(\Sigma)$. By Proposition 4.40, Σ is a thin convex chamber subcomplex of Δ, so $\mathcal{A}$ is a thin and convex subset of $\mathcal{C}$. It is therefore an apartment of $\mathcal{C}$ in the sense of Section 5.4 by Corollary 5.54; hence $\Sigma = \Delta(\mathcal{A}) \in \Omega$.

In the W-metric approach to buildings, it is quite natural to work with the complete system of apartments from the beginning. However, there are situations in which incomplete systems arise naturally, e.g., if $\mathcal{C}$ is part of a twin building (see Section 5.8). And we will see in Chapter 6 that incomplete systems also arise in connection with group actions on buildings

(b) If $\mathcal{R}$ is a J-residue of $\mathcal{C}$ then it can be considered a building of type (W_J, J) as we saw in Corollary 5.30. By definition, the associated simplicial complex $\Delta(\mathcal{R})$ is equal to the poset $\Delta(\mathcal{C})_{\geq F_{\mathcal{R}}}$; it is, up to notation, just the poset of all residues contained in $\mathcal{R}$, ordered by reverse inclusion. So $\Delta(\mathcal{R})$ is canonically isomorphic to the *link* $\mathrm{lk}_{\Delta(C)} F_{\mathcal{R}}$.

(c) If $\mathcal{C}'$ is a subbuilding of $\mathcal{C}$, then $\Delta' := \Delta(\mathcal{C}')$ can be identified with a chamber subcomplex of $\Delta := \Delta(\mathcal{C})$ by Lemma 5.88. This subcomplex is a building in its own right and has the same Coxeter matrix as Δ (both Coxeter matrices being the same as the Coxeter matrix of (W, S)). Hence Δ' is a subbuilding of Δ in the sense of Definition 4.62.

We have now almost completed the proof of the equivalence of the chamber complex approach and the W-metric approach to buildings. We started this with Proposition 4.81 in Chapter 4, where we showed that a simplicial building gives rise to a W-metric building. We have now shown how, conversely, one can construct a simplicial building from a W-metric building. It remains to show that the transitions $\Delta \mapsto (\mathcal{C}(\Delta), \delta)$ and $(\mathcal{C}, \delta) \mapsto \Delta(\mathcal{C})$ are inverse to one another.

Corollary 5.93.

(1) *Let Δ be a simplicial building of type (W, S), and let $(\mathcal{C}(\Delta), \delta)$ be the W-metric building associated to Δ as in Section 4.8. Then the chamber complex $\Delta(\mathcal{C}(\Delta))$ is canonically isomorphic to Δ.*
(2) *Let $(\mathcal{C}, \delta)$ be a (W-metric) building of type (W, S), and let $\Delta = \Delta(\mathcal{C})$ be the corresponding simplicial building of type (W, S). Then the W-metric building associated to $\Delta(\mathcal{C})$ is equal to the original building $(\mathcal{C}, \delta)$.*

Proof. (1) This statement is an immediate application of Proposition A.20 in Appendix A.

(2) By Theorem 5.91, the simplicial building $\Delta(\mathcal{C})$ is of type (W, S), and its set of chambers can be identified with $\mathcal{C}$. We have to show that its Weyl distance function is equal to the function δ that we started with. Consider two chambers $C, D \in \mathcal{C}$, and let $\Gamma \colon C = C_0, \dots, C_n = D$ be a minimal gallery in $\Delta(\mathcal{C})$ connecting them. Then the type $\mathbf{s} = (s_1, \dots, s_n)$ of Γ is the same whether Γ is considered as a gallery in the simplicial building $\Delta(\mathcal{C})$ or in the W-metric building $\mathcal{C}$. And, of course, Γ is also a minimal gallery in $\mathcal{C}$. By Proposition 4.81 and Lemma 5.16, the Weyl distance $\delta_{\Delta(\mathcal{C})}(C, D)$ as defined in Section 4.8 is equal to $s_1 \cdots s_n = \delta(C, D)$. Hence $\delta_{\Delta(\mathcal{C})} = \delta$. $\qquad\square$

Before proceeding further, we digress to complete the proof of Theorem 4.66. For ease of reference, we restate the part whose proof we deferred:

Proposition 5.94. *Let Δ be a simplicial building, and let Δ' be a chamber subcomplex that is weak and convex. Then Δ' is a subbuilding of Δ in the sense of Definition 4.62.*

Proof. We may assume that Δ comes equipped with a type function, so that it is a building of type (W, S) for some Coxeter system (W, S). Then $\mathcal{C}' := \mathcal{C}(\Delta')$ is a weak and convex subset of $\mathcal{C} := \mathcal{C}(\Delta)$, so it is a subbuilding of $\mathcal{C}$ in the W-metric sense. The simplicial building associated to $\mathcal{C}$ can be identified with Δ, and the simplicial building associated to $\mathcal{C}'$ can be identified with a subbuilding of Δ by Remark 5.92(c). This subbuilding and Δ' both have $\mathcal{C}'$ as their set of chambers; since they are both chamber subcomplexes of Δ, they coincide.   □

In Remark 5.90 we described how σ-isometries between W-metric buildings give rise to injective chamber maps between the corresponding simplicial buildings. Recall that a *chamber map* $\psi\colon \Delta' \to \Delta$ between two chamber complexes Δ and Δ' of the same dimension is a simplicial map that takes chambers to chambers (see Section A.1.3).

We conclude this section by investigating how, conversely, injective chamber maps between simplicial buildings induce σ-isometries. In order to do this, we need to make further assumptions, since one can for instance often embed a tree into the flag complex of a generalized m-gon (see Exercise 5.99 below), and these injective chamber maps certainly do not correspond to σ-isometries. But if the simplicial buildings are of the same type, or if the chamber map is bijective, then we will show that there exists an associated σ-isometry between the corresponding W-metric buildings:

Proposition 5.95. *Let Δ' and Δ be simplicial buildings of respective types (W',S') and (W,S). Let $\psi\colon \Delta' \to \Delta$ be an injective chamber map. If ψ is bijective, or if $\Sigma(W',S')$ and $\Sigma(W,S)$ are isomorphic chamber complexes, then ψ induces a σ-isometry $\phi\colon \mathcal{C}(\Delta') \to \mathcal{C}(\Delta)$ for some isomorphism $\sigma\colon (W',S') \overset{\sim}{\longrightarrow} (W,S)$ of Coxeter systems.*

Proof. Recall that Δ comes equipped with a type function having values in S, such that the diameter of $\mathrm{lk}_\Delta A$ is equal to the order $m(s,t)$ of st for all simplices A of cotype $\{s,t\} \subseteq S$, and similarly for Δ'. We also know by Proposition A.14 that the chamber map ψ induces a bijection $f\colon S' \to S$ such that $\tau(\psi(A)) = f(\tau(A))$ for any $A \in \Delta'$. Here τ (resp. τ') is the type function on Δ (resp. Δ'). We claim that $m(f(s'),f(t')) = m'(s',t')$ for all $s',t' \in S'$, so that f induces an isomorphism $\sigma\colon (W',S') \overset{\sim}{\longrightarrow} (W,S)$. It then follows from Lemma 5.61 that the restriction $\phi\colon \mathcal{C}(\Delta') \to \mathcal{C}(\Delta)$ of ψ is a σ-isometry. It remains to prove the claim.

In the first case, when ψ is a simplicial isomorphism, our claim is obvious. In fact, ψ induces isomorphisms between $\mathrm{lk}_{\Delta'} A$ and $\mathrm{lk}_\Delta \psi(A)$ for all $A \in \Delta'$; specializing to simplices A of cotype $\{s',t'\}$, we obtain $m'(s',t') = m(f(s'),f(t'))$ for all $s',t' \in S'$.

The second case is more subtle, but we have done the necessary work in Section 4.5. Choose an apartment Σ' of Δ', and note that it is isomorphic, as a chamber complex, to any apartment of Δ. Proposition 4.59 therefore implies that $\Sigma := \psi(\Sigma')$ is an apartment of Δ (in the complete system of apartments). In particular, the diameter of $\mathrm{lk}_\Delta B$ is equal to the diameter of $\mathrm{lk}_\Sigma B$ for any $B \in \Sigma$ (see Corollary 4.34). We now apply the first case to the simplicial isomorphism $\psi|_{\Sigma'}\colon \Sigma' \to \Sigma$, and we again obtain $m'(s',t') = m(f(s'),f(t'))$ for all $s',t' \in S'$. This proves the claim and hence the proposition.   □

The following consequence of Proposition 5.95 and Lemma 5.62 is by no means obvious a priori.

Corollary 5.96. *If* $\psi \colon \Delta' \to \Delta$ *is an injective chamber map between simplicial buildings that have isomorphic apartments, then* $\psi(\Delta')$ *is a convex chamber subcomplex of* Δ. □

Exercises

5.97. We defined the notion of *subbuilding* in both the simplicial setting (Section 4.6) and the W-metric setting (Section 5.4.2). Show that the two definitions are consistent under the equivalence of theories established in the present section.

5.98. Let $(\mathcal{C}, \delta)$ be a building of type (W, S) with S finite. Show that the simplicial complex $\Delta = \Delta(\mathcal{C})$ admits the following alternative description: There is one vertex for each residue of rank $|S| - 1$, and a collection of such vertices forms a simplex if and only if the corresponding residues have nonempty intersection.

5.99. Let Δ be a building of type $I_2(m)$ such that each vertex of Δ is contained in infinitely many chambers. (One can for instance take $m = 3$ and let Δ be the incidence graph of the projective plane associated to a 3-dimensional vector space over an infinite field.) Show that Δ contains chamber subcomplexes that are thick trees.

5.100. Make the following statements precise and prove them:

(a) The category of simplicial buildings, with bijective chamber maps as morphisms, is equivalent to the category of W-metric buildings, with surjective σ-isometries as morphisms.

(b) For a fixed Coxeter system (W, S), the category of simplicial buildings of type (W, S), with injective chamber maps as morphisms, is equivalent to the category of buildings of type (W, S), with σ-isometries as morphisms. (Here σ ranges over all automorphisms of (W, S).)

5.101. Let Δ be a simplicial building of type (W, S), and let $(\mathcal{C}(\Delta), \delta_{\mathcal{C}(\Delta)})$ be the associated W-metric building.

(a) Show that the group $\mathrm{Aut}_0\,\Delta$ of type-preserving simplicial automorphisms of Δ is isomorphic to the group of surjective isometries of $\mathcal{C}(\Delta)$ onto itself.

(b) Show that the group $\mathrm{Aut}\,\Delta$ of simplicial automorphisms of Δ is isomorphic to the group of surjective σ-isometries of $\mathcal{C}(\Delta)$ onto itself, where σ ranges over all automorphisms of (W, S).

5.102. Let Δ and Δ' be simplicial buildings of type (W, S). Show that any type-preserving chamber map $\phi \colon \Delta \to \Delta'$ induces a map $\phi \colon \mathcal{C}(\Delta) \to \mathcal{C}(\Delta')$ that is "distance-decreasing" with respect to the Bruhat order on W, i.e.,

$$\delta(\phi(C), \phi(D)) \leq \delta(C, D)$$

for all $C, D \in \mathcal{C}(\Delta)$. Generalize to the case that the map is not necessarily type-preserving.

5.7 Spherical Buildings

In Section 4.7 we introduced spherical buildings in the context of simplicial buildings. We now take a fresh look at this subject from our new W-metric point of view. Let us start with the obvious definition.

Definition 5.103. A building $(\mathcal{C}, \delta)$ of type (W, S) is called *spherical* if W is finite.

In this section, (W, S) will always be a Coxeter system with W finite, and $(\mathcal{C}, \delta)$ will be a (spherical) building of type (W, S).

5.7.1 Opposition

We start by recalling a few standard facts about finite Coxeter groups, which all follow from Corollary 2.19 in Section 2.3.2. Recall first that W has a unique element of maximal length. This element w_0 is equal to its inverse w_0^{-1} and satisfies $l(ww_0) = l(w_0) - l(w) = l(w_0 w)$ for all $w \in W$. In particular, $l(w_0 s w_0) = l(w_0) - l(w_0 s) = l(w_0) - (l(w_0) - 1) = 1$ for all $s \in S$, so w_0 normalizes S.

We now fix some notation.

Definition 5.104. We denote by w_0 the unique element of maximal length in W, and we set $d_0 := l(w_0)$. For $J \subseteq S$, $w_0(J)$ denotes the longest element of W_J, and $d_0(J) := l(w_0(J))$. In particular, $w_0(S) = w_0$ and $d_0(S) = d_0$.

The existence of a longest element w_0 in W has the important consequence that the *diameter* of the building $\mathcal{C}$ is finite. Here the diameter of $\mathcal{C}$ is defined as usual in a metric space by

$$\operatorname{diam} \mathcal{C} := \sup \{d(C, D) \mid C, D \in \mathcal{C}\} = l(w_0) = d_0 \,,$$

where w_0 and d_0 are as in Definition 5.104. This leads to the fundamental concept that distinguishes spherical buildings from general buildings:

Definition 5.105. Two chambers $C, D \in \mathcal{C}$ are called *opposite* if $d(C, D) = d_0$ or, equivalently, $\delta(C, D) = w_0$, in which case we also write C op D. Two residues $\mathcal{R}$ and $\mathcal{S}$ of $\mathcal{C}$ are called *opposite* if for each $C \in \mathcal{R}$, there exists a $D \in \mathcal{S}$ with D op C and for each $D' \in \mathcal{S}$, there exists a $C' \in \mathcal{R}$ with C' op D'. If this is the case, we also use the notation $\mathcal{R}$ op $\mathcal{S}$.

We noted above that w_0 normalizes S. Hence conjugation by w_0 induces an automorphism of (W, S).

Definition 5.106. We denote by σ_0 the automorphism of (W, S) given by $\sigma_0(w) = w_0 w w_0$ for all $w \in W$. If $J \subseteq S$, we set $J^0 := \sigma_0(J)$. We call two subsets J and K of S *opposite*, and we write J op K, if $K = J^0$.

We have already encountered σ_0 in Chapter 1, where it occurred in several exercises as well as in Proposition 1.130 and Corollary 1.131. We will not use those results here except in our discussion of examples in Section 5.7.4.

Here is an alternative characterization of opposite residues, which also justifies the definition of opposite types in Definition 5.106.

Lemma 5.107. *A J-residue $\mathcal{R}$ and a K-residue $\mathcal{S}$ of C are opposite if and only if $J = K^0$ and there is a chamber in $\mathcal{R}$ that is opposite to some chamber in $\mathcal{S}$. If this is the case, then $\delta(\mathcal{R}, \mathcal{S}) = W_J w_0 = w_0 W_K$, and $w_0(J) w_0 = w_0 w_0(K)$ is the unique element of minimal length in $\delta(\mathcal{R}, \mathcal{S})$.*

Proof. Suppose there are chambers $C \in \mathcal{R}$ and $D \in \mathcal{S}$ with C op D, i.e., $\delta(C, D) = w_0$. Applying Lemma 5.29, we obtain $\delta(\mathcal{R}, \mathcal{S}) = W_J w_0 W_K$. If $\mathcal{R}$ and $\mathcal{S}$ are opposite, then for each $D' \in \mathcal{S}$ there exists a $C' \in \mathcal{R}$ with $\delta(C', D') = w_0$. This implies that $\delta(\mathcal{R}, D') = W_J w_0$. Since this is true for every $D' \in \mathcal{S}$, it follows that $\delta(\mathcal{R}, \mathcal{S}) = W_J w_0$. Similarly, $\delta(\mathcal{R}, \mathcal{S}) = w_0 W_K$. Thus $W_J w_0 = w_0 W_K$, which implies that $W_J = w_0 W_K w_0 = W_{K^0}$ and hence that $J = K^0$ by the basic properties of standard parabolic subgroups.

If, conversely, $J = K^0$, then $W_J = w_0 W_K w_0$; hence

$$\delta(\mathcal{R}, \mathcal{S}) = W_J w_0 W_K = (w_0 W_K w_0) w_0 W_K = w_0 W_K \ ,$$

and similarly $\delta(\mathcal{R}, \mathcal{S}) = W_J w_0$. This implies $\delta(C', \mathcal{S}) = w_0 W_K$ for any $C' \in \mathcal{R}$ (since $\delta(C', \mathcal{S})$ is a left coset of W_K and is contained in $\delta(\mathcal{R}, \mathcal{S})$), and similarly $\delta(\mathcal{R}, D') = W_J w_0$ for any $D' \in \mathcal{S}$. But this just means that each $C' \in \mathcal{R}$ is opposite a chamber in $\mathcal{S}$ and each $D' \in \mathcal{S}$ is opposite a chamber in $\mathcal{R}$. Hence the residues $\mathcal{R}$ and $\mathcal{S}$ are opposite, and the first part of the lemma is proved.

Now assume that $\mathcal{R}$ op $\mathcal{S}$, so that $\delta(\mathcal{R}, \mathcal{S}) = W_J w_0 = w_0 W_K$ by the argument above. Consider an arbitrary element $w_J w_0 \in W_J w_0$ ($w_J \in W_J$). We have $l(w_J w_0) = l(w_0) - l(w_J)$, so we minimize $l(w_J w_0)$ by maximizing $l(w_J)$, i.e., by taking $w_J = w_0(J)$. Thus $w_0(J) w_0$ is the unique element of minimal length in $W_J w_0$. Similarly, $w_0 w_0(K)$ is the element of minimal length in $w_0 W_K$. Finally, we must have $w_0(J) w_0 = w_0 w_0(K)$, since $W_J w_0 = w_0 W_K$. $\square$

*5.7.2 A Metric Characterization of Opposition

It is natural to ask whether the opposition relation on residues admits a direct characterization in terms of distances between residues, as in the definition of opposition for chambers. In this optional subsection we will give such a characterization.

Note first that the distance between residues makes sense, as a special case of the distance between subsets of a metric space. Namely, if $\mathcal{R}$ and $\mathcal{S}$ are residues, then

$$d(\mathcal{R}, \mathcal{S}) := \min\{d(C, D) \mid C \in \mathcal{R},\ D \in \mathcal{S}\}\ .$$

If $(\mathcal{C}, \delta)$ is the W-metric building associated to a simplicial building Δ, then this is a familiar concept. Indeed, we have $\mathcal{R} = \mathcal{C}_{\geq A}$ and $\mathcal{S} = \mathcal{C}_{\geq B}$ for some simplices $A, B \in \Delta$, and $d(\mathcal{R}, \mathcal{S})$ is the same as the gallery distance $d(A, B)$ that we have worked with in earlier chapters. As usual, we will write $d(\mathcal{R}, D)$ instead of $d(\mathcal{R}, \{D\})$ in case $\mathcal{S}$ is a singleton. From the theory of projections, we know that

$$d(\mathcal{R}, D) = d(\mathrm{proj}_{\mathcal{R}} D, D)\ .$$

The key to our metric characterization of opposition is the following lemma:

Lemma 5.108. *Let $\mathcal{R}$ be a J-residue and D a chamber of $\mathcal{C}$.*

(1) $\max\{d(C, D) \mid C \in \mathcal{R}\} = d_0(J) + d(\mathcal{R}, D)$.
(2) $d(\mathcal{R}, D) \leq d_0 - d_0(J)$, *with equality if and only if $\mathcal{R}$ contains a chamber opposite D.*

Proof. Let $C' := \mathrm{proj}_{\mathcal{R}} D$. Then by the gate property (Proposition 5.34) we have

$$d(C, D) = d(C, C') + d(C', D)$$

for all $C \in \mathcal{R}$. This is maximal when $d(C, C')$ is maximal. Since $\delta(\mathcal{R}, C') = W_J$, it follows that we maximize $d(C, D)$ by taking $C \in \mathcal{R}$ such that $\delta(C, C') = w_0(J)$, in which case $d(C, D) = d_0(J) + d(C', D)$. This proves (1), and (2) follows immediately. $\square$

We can now give our promised characterization of opposition. See Exercise 4.80 for the same result stated in the language of simplicial buildings.

Proposition 5.109. *Let $\mathcal{R}$ be a J-residue, and let $\mathcal{S}$ be a residue with rank $\mathcal{S} \geq$ rank $\mathcal{R}$. Then the following conditions are equivalent.*

(i) $\mathcal{R}$ op $\mathcal{S}$.
(ii) $d(\mathcal{R}, \mathcal{S}) = d_0 - d_0(J)$.
(iii) $d(\mathcal{R}, \mathcal{S}) = \max\{d(\mathcal{R}, \mathcal{S}') \mid \mathcal{S}'$ *is a residue of* $\mathcal{C}\}$.

Proof. It is immediate from the definitions that the maximum on the right side of (iii) is equal to

$$\max\{d(\mathcal{R}, D) \mid D \in \mathcal{C}\}\ .$$

By Lemma 5.108, this maximum is equal to $d_0 - d_0(J)$. So (ii) and (iii) are equivalent. The last assertion of Lemma 5.107 shows that (i) implies (ii). To complete the proof, we assume that (ii) and (iii) hold, and we show that $\mathcal{R}$ op $\mathcal{S}$.

It follows from (ii) and (iii) that *every* chamber $D \in \mathcal{S}$ satisfies $d(\mathcal{R}, D) = d_0 - d_0(J)$ and hence, by the lemma again, $\mathcal{R}$ contains a chamber C op D. As in the proof of Lemma 5.107, we conclude that $\delta(\mathcal{R}, \mathcal{S}) = W_J w_0 = w_0 W_{J^0}$. On the other hand, if K is the type of $\mathcal{S}$, then $\delta(\mathcal{R}, \mathcal{S})$ contains $w_0 W_K$, so $K \subseteq J^0$. But $|K| = \operatorname{rank} \mathcal{S} \geq |J|$, so we must have $K = J^0$, and then $\mathcal{R}$ op $\mathcal{S}$ by Lemma 5.107. $\qquad\square$

It is worth explicitly mentioning the following corollary of the proof:

Corollary 5.110. *Let $\mathcal{R}$ be a J-residue and $\mathcal{S}$ a K-residue. If $\mathcal{R}$ and $\mathcal{S}$ satisfy condition (iii) of the proposition, then $K \subseteq J^0$.* $\qquad\square$

5.7.3 The Thin Case

Next we want to investigate the opposition relation in an apartment, which we can identify with the standard thin building. Some basic properties are collected in the following lemma.

Lemma 5.111. *Let (W, δ) be the standard thin building of type (W, S) (where W is finite). Then the following hold:*

(1) *Any $w \in W$ is opposite precisely one $w' \in W$, namely $w' = w w_0$.*
(2) *The map $\operatorname{op}_W \colon W \to W$ defined by $\operatorname{op}_W(w) := w w_0$ for all $w \in W$ is a surjective σ_0-isometry.*
(3) *Any residue $w W_J$ in W (with $w \in W$ and $J \subseteq S$) is opposite precisely one residue in W, namely $\operatorname{op}_W(w W_J) = w w_0 W_{J^0}$.*

Proof. (1) Given $w, w' \in W$, we have w op $w' \iff \delta(w, w') = w_0 \iff w^{-1} w' = w_0 \iff w' = w w_0$.

(2) Just observe that

$$\delta(\operatorname{op}_W(v), \operatorname{op}_W(w)) = (v w_0)^{-1}(w w_0) = w_0 \delta(v, w) w_0 = \sigma_0(\delta(v, w))$$

for all $v, w \in W$ and that op_W is obviously surjective.

(3) Set $\mathcal{R} := w W_J$, and let $\mathcal{S}$ be any residue in W. By Definition 5.105 and part (1) of this lemma, $\mathcal{R}$ and $\mathcal{S}$ are opposite if and only if $\mathcal{S} = \mathcal{R} w_0$. But this is equivalent to $\mathcal{S} = \operatorname{op}_W(\mathcal{R}) = \operatorname{op}_W(w W_J) = w W_J w_0 = w w_0(w_0 W_J w_0) = w w_0 W_{J^0}$. $\qquad\square$

Remarks 5.112. (a) Since every thin building is isometric to the standard one, Lemma 5.111 implies that every thin spherical building $\mathcal{A}$ admits an *opposition involution* $\operatorname{op}_{\mathcal{A}}$; it is an almost isometry characterized by the property that it maps every chamber of $\mathcal{A}$ to its opposite.

(b) Recall that the standard Coxeter complex $\Sigma = \Sigma(W, S)$ is the simplicial building associated to (W, δ_W). By Remark 5.90, op_W induces a simplicial automorphism ϕ of Σ, again characterized by the property that it maps every chamber of Σ to its opposite. If W is given as a finite reflection group acting on a Euclidean space V and we identify Σ with the cell complex $\Sigma(W, V)$ introduced in Section 1.5, then ϕ coincides with the opposition involution op_Σ introduced in Section 1.6.2 and given by $\mathrm{op}_\Sigma(A) = -A$ for each cell $A \in \Sigma(W, V)$. We remark in passing that right multiplication by an element $w \in W$ usually does not map residues of W onto residues of W; the special property of w_0 that makes this work is that w_0 normalizes S.

(c) The fact that op_W is a σ_0-isometry translates to the statement that σ_0 is the type-change map associated to op_Σ (see the last sentence of Remark 5.90). We already noted this from the point of view of finite reflection groups in Proposition 1.130.

For future reference, we record an easy fact about the opposition involution. Recall that every root α in a thin building $\mathcal{A}$ has an opposite root $-\alpha$ (cf. Section 3.4.1); from the W-metric point of view, $-\alpha$ is simply the complement $\mathcal{A} \smallsetminus \alpha$.

Lemma 5.113. *Let $\mathcal{A}$ be a thin spherical building. Then the opposition involution maps every root α of $\mathcal{A}$ to the opposite root $-\alpha$.*

Proof. This can be proved in many ways. For example, one can identify $\mathcal{A}$ with the set of chambers of a finite reflection group and use the fact that the opposition involution is given by multiplication by -1. Alternatively, one can use the fact that opposite chambers in a spherical Coxeter complex are separated by every wall. Details are left to the reader. $\square$

5.7.4 Computation of σ_0

In this subsection we describe σ_0 explicitly for all finite Coxeter groups. Note first that we may assume that (W, S) is irreducible. For if (W, S) and (W', S') are given with finite W and W' and respective longest elements w_0 and w_0', then $w_0 w_0'$ is obviously the longest element in $W \times W'$ with respect to $S \cup S'$. Next, recall that σ_0, being an automorphism of (W, S), can be identified with a diagram automorphism. This greatly simplifies the analysis. In fact, it will turn out in all cases that we can determine σ_0 as soon as we can decide whether it is the identity. Recall, finally, that if W is given to us as a finite reflection group, then Corollary 1.131 gives us several equivalent ways of deciding whether σ_0 is trivial. We now proceed case by case, using the classification of irreducible finite Coxeter systems given in Sections 1.3 and 1.5.6.

(a) If (W, S) is of type A_1, then σ_0 is the identity.

(b) Suppose $|S| = 2$, so that W is a dihedral group

$$D_{2m} = \left\langle\, s_1, s_2 \;;\; s_1^2 = s_2^2 = (s_1 s_2)^m = 1 \,\right\rangle \;.$$

Then one can write down w_0 explicitly and check that σ_0 is the identity if m is even and that it interchanges s_1 and s_2 if m is odd. [Alternatively, consider a $2m$-gon with colored vertices, and observe that opposite vertices have the same color if m is even and different colors if m is odd.]

(c) Suppose (W, S) is of type A_n with $n \geq 2$. Then σ_0 is nontrivial, and hence it is the unique nontrivial diagram automorphism. There are many ways to see this. For example, we can use the fact that $W = S_{n+1}$ has trivial center, so conjugation by w_0 cannot be the identity. Or, using Corollary 1.131, we can examine the permutation action of W on the n-dimensional space $V := \left\{ x \in \mathbb{R}^{n+1} \mid \sum_i x_i = 0 \right\}$ and observe that $-\mathrm{id}_V \notin W$ for $n \geq 2$. Or we can compute w_0 explicitly (Exercise 1.86). When $n = 2$, the nontriviality of σ_0 also follows from case (b) with $m = 3$.

(d) If (W, S) is of type C_n with $n \geq 2$, then the description of W as the group of signed permutation matrices shows that $-\mathrm{id}_V \in W$; hence σ_0 is trivial. If $n = 2$, the result is also included in (b) with $m = 4$. And if $n > 2$, the result could have been deduced from the fact that the Coxeter diagram has no nontrivial automorphisms.

(e) If (W, S) is of type D_n $(n \geq 4)$, then σ_0 is trivial if n is even and is the unique nontrivial diagram automorphism if n is odd. To see this, recall that W is the group of signed permutation matrices with an even number of minus signs. Hence $-\mathrm{id}_V \in W$ if and only if n is even.

(f) The Coxeter diagrams of type E_7 and E_8 have no nontrivial automorphisms, so σ_0 is trivial in those cases. For W of type E_6, on the other hand, there is a unique nontrivial diagram automorphism, and this is σ_0. The proof, which uses ideas that we have not treated in this book, will be omitted; see Bourbaki [44, Chapter VI, Section 4.12].

(g) The Coxeter diagram of type F_4 has a unique nontrivial automorphism, but σ_0 is trivial in this case. One can see this by using the Dynkin diagram instead of the Coxeter diagram and noting that it does not have any nontrivial automorphisms. For readers not familiar with Dynkin diagrams, we can explain this as follows. Let σ_1 be the nontrivial automorphism of the Coxeter diagram. If we had $\sigma_0 = \sigma_1$, then Exercise 1.126, which describes the action of w_0 on the simple roots, would imply that $w_0 \alpha_s = -\alpha_{\sigma_1(s)}$ for all $s \in S$. But this is impossible, because the roots α_s and $\alpha_{\sigma_1(s)}$ do not have the same length [44, Chapter VI, Section 4.9].

(h) The Coxeter diagrams of type H_3 and H_4 do not have any nontrivial automorphisms, so σ_0 is trivial in those cases.

5.7.5 Projections

We return to the general theory of spherical buildings. The characterization of opposite residues in Lemma 5.107 has important consequences for projections.

Proposition 5.114. *If $\mathcal{R}$ and $\mathcal{S}$ are opposite residues of $\mathcal{C}$, then $\mathrm{proj}_\mathcal{R} \mathcal{S} = \mathcal{R}$ and $\mathrm{proj}_\mathcal{S} \mathcal{R} = \mathcal{S}$.*

Proof. Let J be the type of $\mathcal{R}$ and let K be the type of $\mathcal{S}$. By Lemma 5.107, $\delta(\mathcal{R}, \mathcal{S}) = W_J w_0 = w_0 W_K$. This implies that $\delta(C, \mathcal{S}) = w_0 W_K = \delta(\mathcal{R}, \mathcal{S})$ for any $C \in \mathcal{R}$. In particular, there exists a chamber $D \in \mathcal{S}$ such that $\delta(C, D) = \min(\delta(\mathcal{R}, \mathcal{S}))$. Using Lemma 5.36, we conclude that $C \in \mathrm{proj}_\mathcal{R} \mathcal{S}$. Since $C \in \mathcal{R}$ was arbitrary, we have shown that $\mathcal{R} = \mathrm{proj}_\mathcal{R} \mathcal{S}$. Interchanging the roles of $\mathcal{R}$ and $\mathcal{S}$, we also obtain $\mathrm{proj}_\mathcal{S} \mathcal{R} = \mathcal{S}$. $\square$

Remark 5.115. It is easy to see that the necessary conditions $\mathrm{proj}_\mathcal{R} \mathcal{S} = \mathcal{R}$ and $\mathrm{proj}_\mathcal{S} \mathcal{R} = \mathcal{S}$ are not sufficient to imply that $\mathcal{R}$ and $\mathcal{S}$ are opposite, even if one excludes the trivial counterexamples in which $\mathcal{R}$ and $\mathcal{S}$ are chambers or $\mathcal{R} = \mathcal{S}$. For instance, if $\mathcal{R}$ and $\mathcal{S}$ correspond to two codimension-1 simplices that lie on a common wall in an apartment containing them (here we refer to the associated simplicial building), then one easily checks that $\mathrm{proj}_\mathcal{R} \mathcal{S} = \mathcal{R}$ and $\mathrm{proj}_\mathcal{S} \mathcal{R} = \mathcal{S}$. If we identify the apartment with the set of cells associated to a finite reflection group, the explanation for this is that two cells A and B have the same support if and only if $AB = A$ and $BA = B$; see Proposition 1.41.

However, there is an interesting special case in which these conditions are in fact also sufficient to imply $\mathcal{R}$ op $\mathcal{S}$. Namely, if $\mathcal{R}$ and $\mathcal{S}$ are two distinct residues of rank equal to $|S| - 1$ (and hence corresponding to distinct vertices in $\Delta(\mathcal{C})$) such that $\mathrm{proj}_\mathcal{R} \mathcal{S} = \mathcal{R}$ or $\mathrm{proj}_\mathcal{S} \mathcal{R} = \mathcal{S}$, then these residues are already opposite. To see this, we can work in an apartment as above. Then two distinct vertices correspond to two cells A and B that are rays. If $AB = A$, then $\mathrm{supp}\, B \subseteq \mathrm{supp}\, A$, and hence $\mathrm{supp}\, B = \mathrm{supp}\, A$. This means that A and B are opposite rays in a line. See Sections 3.6.6 and 3.6.7 for further remarks about this, as well as a discussion of the nonspherical case.

Proposition 5.114 has some interesting consequences, which we give in the following corollaries.

Corollary 5.116. *Let $\mathcal{R}$ and $\mathcal{S}$ be opposite residues in $\mathcal{C}$, and let J be the type of $\mathcal{R}$. Then the projection map $\mathrm{proj}_\mathcal{S}$ induces a surjective σ-isometry from $\mathcal{R}$ onto $\mathcal{S}$, where σ is given by conjugation by $w_0 w_0(J)$. The inverse of this σ-isometry is the σ^{-1}-isometry that we obtain by restricting $\mathrm{proj}_\mathcal{R}$ to $\mathcal{S}$.*

Proof. Since $\mathrm{proj}_\mathcal{R} \mathcal{S} = \mathcal{R}$ and $\mathrm{proj}_\mathcal{S} \mathcal{R} = \mathcal{S}$, Proposition 5.37 implies that the projection map from $\mathcal{R}$ onto $\mathcal{S}$ is a σ-isometry, with the projection from $\mathcal{S}$ onto $\mathcal{R}$ as its inverse, where σ is given by conjugation by $w_1^{-1} = \min(\delta(\mathcal{R}, \mathcal{S}))^{-1}$ (see Example 5.60(b)). By Lemma 5.107, $w_1^{-1} = (w_0(J) w_0)^{-1} = w_0 w_0(J)$. $\square$

Here is an important special case, for which we have already stated the simplicial analogue in Proposition 4.103:

Corollary 5.117. *If $\mathcal{P}$ and $\mathcal{Q}$ are opposite panels, then the relation of non-opposition induces a bijection between $\mathcal{P}$ and $\mathcal{Q}$.*

Proof. Given $C \in \mathcal{P}$ and $D \in \mathcal{Q}$, we have C op D if and only if $d(C,D) = d_0$. Otherwise $d(C,D) = d_0 - 1$, and C and D correspond under the bijection of Corollary 5.116. In other words, C and D correspond under that bijection if and only if they are *not* opposite. $\qquad\square$

Note that σ in Corollary 5.116 depends only on the type J of $\mathcal{R}$. This leads to the next corollary.

Corollary 5.118. *If the spherical building $\mathcal{C}$ is thick, then residues of $\mathcal{C}$ of the same type are isometric.*

Proof. Let $\mathcal{R}$ and $\mathcal{T}$ be two J-residues of $\mathcal{C}$. Choose arbitrary chambers $C \in \mathcal{R}$ and $D \in \mathcal{T}$. By Proposition 4.104, there exists a chamber $E \in \mathcal{C}$ that is opposite both C and D. [This is where we use thickness.] Set $K := J^0$ and $\mathcal{S} := R_K(E)$. Then Lemma 5.107 implies $\mathcal{R}$ op $\mathcal{S}$ and $\mathcal{S}$ op $\mathcal{T}$. By Corollary 5.116, there exists a surjective σ-isometry $\phi\colon \mathcal{R} \to \mathcal{S}$, where σ is induced by conjugation by $w_0 w_0(J)$. Similarly, there is a surjective σ'-isometry $\phi'\colon \mathcal{S} \to \mathcal{T}$, where σ' is induced by conjugation by $w_0 w_0(K) = (w_0 w_0(K) w_0) w_0 = w_0(J) w_0$, since $K = J^0$. Hence $\sigma' = \sigma^{-1}$, and $\phi' \circ \phi$ is an isometry of $\mathcal{R}$ onto $\mathcal{T}$. $\qquad\square$

In view of Section 5.6, the last two corollaries have simplicial interpretations that are worth stating explicitly. We first introduce, from our present point of view, an opposition relation for simplices in a simplicial spherical building. Let Δ be a simplicial spherical building of type (W,S). Thus Δ has a type function with values in S such that the link of any simplex of cotype $\{s,t\}$ $(s,t \in S)$ has diameter $m(s,t) =$ the order of st. We still call two subsets J and K of S opposite if $K = J^0$, and we say that two simplices A and B in Δ are *opposite* if they are of opposite types and are faces of opposite chambers. Equivalently, A and B are opposite if the residues $\mathcal{C}(\Delta)_{\geq A}$ and $\mathcal{C}(\Delta)_{\geq B}$ are opposite in the sense of Definition 5.105.

Remark 5.119. We already defined the concept of opposition for simplices in a simplicial spherical building in Section 4.7, based on the theory of finite reflection groups. The interested reader can refer to Exercise 4.79 in that section for a proof that the definition there is equivalent to the one we have just given.

The following result should be compared with Exercise 4.100:

Corollary 5.120. *Let A and B be simplices of a simplicial spherical building Δ of type (W,S).*

(1) *If A and B are opposite, then $\mathrm{lk}_\Delta A$ and $\mathrm{lk}_\Delta B$ are isomorphic simplicial complexes.*

(2) *If Δ is thick and A and B are of the same type, then there is a type-preserving isomorphism between $\mathrm{lk}_\Delta A$ and $\mathrm{lk}_\Delta B$.*

Proof. Let $(\mathcal{C}, \delta)$ be the W-metric building of type (W, S) associated with Δ (so $\mathcal{C} = \mathcal{C}(\Delta)$).

(1) By assumption, the residues $\mathcal{C}_{\geq A}$ and $\mathcal{C}_{\geq B}$ are opposite. So by Corollary 5.116 they are almost isometric. In view of Remark 5.90 this implies that the associated simplicial complexes are isomorphic. These associated simplicial buildings are $\mathrm{lk}_\Delta A$ and $\mathrm{lk}_\Delta B$ by Remark 5.92(b).

(2) Corollary 5.118 implies that the residues $\mathcal{C}_{\geq A}$ and $\mathcal{C}_{\geq B}$ are isometric. By the remarks from Section 5.6 that we already quoted, this isometry induces a type-preserving simplicial isomorphism from $\mathrm{lk}_\Delta A$ onto $\mathrm{lk}_\Delta B$. $\square$

5.7.6 Apartments

The next result is the analogue of Lemma 4.69 in the W-metric setup. (See Example 5.44(c) for the definition of *convex hull* in the present context.)

Lemma 5.121. *If w and w' are opposite chambers of the standard thin building (W, δ_W), then W is the convex hull of $\{w, w'\}$.*

Proof. This is essentially a reformulation of the equation

$$l(w_0) = l(x) + l(x^{-1}w_0) \tag{5.9}$$

for all $x \in W$, which follows from Corollary 2.19. Here are the details. Given $v \in W$, we want to show that there is a minimal gallery from w to w' that contains v or, equivalently, that

$$d(w, v) + d(v, w') = d(w, w') \,,$$

where $d(-, -)$ denotes gallery distance. Recalling from Lemma 5.111 that $w' = ww_0$, we can rewrite this as

$$l(w^{-1}v) + l(v^{-1}ww_0) = l(w_0) \,.$$

This is precisely (5.9), with $x = w^{-1}v$. $\square$

We can use this to re-prove the important Theorem 4.70 from the W-metric point of view:

Theorem 5.122. *Let $(\mathcal{C}, \delta)$ be a spherical building.*

(1) *If C and C' are opposite chambers of $\mathcal{C}$, then there is precisely one apartment $\mathcal{A}$ of $\mathcal{C}$ containing them both, and $\mathcal{A}$ is the convex hull of $\{C, C'\}$.*

(2) *If $\mathcal{A}$ is an apartment of $\mathcal{C}$ and $C \in \mathcal{A}$, then there is precisely one chamber $C' \in \mathcal{A}$ that is opposite C. Therefore, $\mathcal{A}$ is the convex hull of $\{C, C'\}$.*

(3) *The set of apartments of $\mathcal{C}$ is equal to the set of all convex hulls of pairs of opposite chambers.*

Proof. Recall that apartments are convex subsets of $\mathcal{C}$ by Corollary 5.54 and are isometric images of (W, δ_W) by Corollary 5.67, where $(\mathcal{C}, \delta)$ is of type (W, S).

(1) There exists an apartment $\mathcal{A}$ containing C and C' by Corollary 5.74. Since $\mathcal{A}$ is convex in $\mathcal{C}$, the convex hull of $\{C, C'\}$ in $\mathcal{C}$ is the same as the convex hull of $\{C, C'\}$ in $\mathcal{A}$, which is the entire apartment $\mathcal{A}$ by Lemma 5.121. This proves the last assertion of (1), and the uniqueness of $\mathcal{A}$ follows.

(2) The first statement follows from Lemma 5.111, and the second statement is then a consequence of (1).

(3) This is immediate from (1) and (2). □

Remark 5.123. One can define as in Chapter 4 the notion of *system of apart- ments* for a building $(\mathcal{C}, \delta)$. By definition, this is a collection Ω of apartments in the sense of Definition 5.53 such that any two chambers in $\mathcal{C}$ are contained in some $\mathcal{A} \in \Omega$. As we observed in the simplicial setting, it follows from Theorem 5.122 that a spherical building admits a unique system of apartments.

5.7.7 The Dual of a Spherical Building

We complete this section by introducing the "dual" of a spherical building. One can define this notion using either the simplicial approach or the W-metric approach, but it seems more natural from the W-metric point of view.

Definition 5.124. Let $(\mathcal{C}, \delta)$ be a spherical building of type (W, S), and let w_0 continue to denote the longest element of W. Define a new function $\delta_- : \mathcal{C} \times \mathcal{C} \to W$ by $\delta(C, D) := w_0 \delta(C, D) w_0$ for $C, D \in \mathcal{C}$. Then $(\mathcal{C}, \delta_-)$ is called the *dual* of $(\mathcal{C}, \delta)$. We write $\mathcal{C}_-$ instead of $\mathcal{C}$ when we want to emphasize that we are thinking of $\mathcal{C}$ as the set of chambers of the dual building.

It is very easy, as we will see below, to verify that $(\mathcal{C}_-, \delta_-)$ is indeed a building of type (W, S). It is equal to the original building $(\mathcal{C}, \delta)$ if and only if w_0 is central in W, i.e., if and only if the function σ_0 introduced in Definition 5.104 is the identity. We use the notation δ_- instead of δ^* for the dual Weyl distance function in order to be consistent with standard conventions in the theory of twin buildings (Section 5.8 below).

Lemma 5.125. *The dual $(\mathcal{C}, \delta_-)$ of $(\mathcal{C}, \delta)$ is also a building of type (W, S), and it is σ_0-isometric to $(\mathcal{C}, \delta)$. The associated simplicial buildings $\Delta(\mathcal{C})$ and $\Delta(\mathcal{C}_-)$ are identical. If one considers $\Delta(\mathcal{C})$ and $\Delta(\mathcal{C}_-)$ as buildings of type (W, S) with their natural colorings, then the identity map between $\Delta(\mathcal{C})$ and $\Delta(\mathcal{C}')$ has σ_0 (or more precisely $\sigma_0|_S$) as the associated type-change map.*

Proof. Observe that $\delta_- = \delta^{\sigma_0}$ with the notation introduced in Exercise 5.64. So it follows from this exercise (which was a straightforward verification) that $(\mathcal{C}, \delta)$ is a building of type (W, S). By the very definition of the dual, the identity function $\mathrm{id}_\mathcal{C} : \mathcal{C} \to \mathcal{C}$ defines a σ_0-isometry from $\mathcal{C}$ onto $\mathcal{C}_-$. Since $\mathcal{C}$ and $\mathcal{C}_-$ have the same residues, Definition 5.87 implies $\Delta(\mathcal{C}) = \Delta(\mathcal{C}_-)$. If a residue $\mathcal{R}$ has type J in $\mathcal{C}$, then it has type $w_0 J w_0 = \sigma_0(J)$ in $\mathcal{C}_-$. Hence σ_0 is the type-change map of the identity isomorphism from $\Delta(\mathcal{C})$ onto $\Delta(\mathcal{C}_-)$. $\square$

From the simplicial point of view, then, the distinction between a spherical building and its dual appears only when one considers *colored* buildings.

Lemma 5.126. *If $(\mathcal{C}', \delta')$ is another building of type (W, S), the following are equivalent:*

(i) *$\mathcal{C}'$ and $\mathcal{C}_-$ are isometric.*
(ii) *There is a type-preserving simplicial isomorphism $\Delta(\mathcal{C}') \xrightarrow{\sim} \Delta(\mathcal{C}_-)$.*
(iii) *There is a simplicial isomorphism $\Delta(\mathcal{C}') \xrightarrow{\sim} \Delta(\mathcal{C})$ with σ_0 as the associated type-change map.*

Proof. By Remark 5.90, an isometry between $\mathcal{C}'$ and $\mathcal{C}_-$ induces a type-preserving simplicial automorphism between $\Delta(\mathcal{C}')$ and $\Delta(\mathcal{C}_-)$. Conversely, a type-preserving automorphism between $\Delta(\mathcal{C}')$ and $\Delta(\mathcal{C}_-)$ induces an isometry between $\mathcal{C}'$ and $\mathcal{C}_-$ by Proposition 5.95. This proves the equivalence of (i) and (ii). By Lemma 5.125, there is an isomorphism between the simplicial buildings $\Delta(\mathcal{C})$ and $\Delta(\mathcal{C}_-)$ of type (W, S) with associated type-change map σ_0. This immediately implies the equivalence of (ii) and (iii). $\square$

Specializing to $\mathcal{C}' = \mathcal{C}$, we obtain the following:

Corollary 5.127. *The spherical building $(\mathcal{C}, \delta)$ is isometric to its dual $(\mathcal{C}, \delta_-)$ if and only if the simplicial building $\Delta(\mathcal{C})$ of type (W, S) admits an automorphism with associated type-change map σ_0.* $\square$

Remark 5.128. Lemmas 5.125 and 5.126 show how the dual building should be defined in the category of simplicial spherical buildings of type (W, S). Namely, let Δ be a building of type (W, S) with type function τ having values in S. Then the dual of Δ is the same simplicial complex Δ, endowed with the type function $\tau_- := \sigma_0 \circ \tau$.

Example 5.129. The motivation for the notion of "dual building" comes from the case that Δ is the building $\Delta(V)$ that we introduced in Section 4.3. The dual building in that case was essentially computed in Exercise 4.31. We will repeat some of the details.

Let us first recall the setup. We are given a division ring k and a left vector space V over k of finite dimension n. We assume $n \geq 3$ to avoid trivial cases. Then $\Delta(V)$ is the flag complex of the poset of proper nontrivial subspaces of V. We have seen in Section 4.3 that $\Delta(V)$ is a building of type A_{n-1}.

It has a canonical coloring having values in $\{1, \ldots, n-1\}$, where the type of a vertex U of $\Delta(V)$ is its dimension as a subspace of V.

As we saw in Section 5.7.4, σ_0 is the unique nontrivial automorphism of the Coxeter diagram of type A_{n-1}, i.e., $\sigma_0(i) = n-i$ for all $i \in \{1, \ldots, n-1\}$. So the dual building $\Delta(V)_-$, according to Remark 5.128, is the same building $\Delta(V)$, where now a vertex U is declared to have type equal to its *codimension* in V instead of its dimension. Equivalently, we can identify $\Delta(V)_-$ with $\Delta(V^*)$, where V^* is the vector space dual to V, and the (type-preserving) isomorphism $\Delta(V)_- \to \Delta(V^*)$ sends U to its annihilator in V^*.

There is one issue that deserves some attention in case the division ring k is not commutative. (This was already pointed out in the solution to Exercise 4.32.) In that case, the dual V^* of V has to be considered either as a right k-vector space or as a left vector space over the *opposite* skew field k^{op}. Therefore, even though we "only" changed the coloring, $\Delta(V)_-$ should really be viewed as different from $\Delta(V)$, since it has a different coordinatizing division ring k^{op} instead of k.

This brings us to the question of when $\Delta(V)_-$ is isomorphic, as a *colored* simplicial building, to $\Delta(V)$. Equivalently, we ask when there is a type-preserving isomorphism between $\Delta(V^*)$ and $\Delta(V)$. From the point of view of projective geometry, this is the question of when the projective space $P(V)$ associated with V admits a *correlation*. The answer is classical and well known: $P(V)$ admits a correlation if and only if the division rings k and k^{op} are isomorphic (or, in other words, if and only if k admits an antiautomorphism). In view of Corollary 5.127, we can restate this result as follows: The building $\Delta(V)$ admits an automorphism that is *not* type-preserving (and so has σ_0 as its type-change map) if and only if the underlying field k admits an antiautomorphism.

Exercises

In the exercises below, (W, S) always denotes a Coxeter system with finite W, and $(\mathcal{C}, \delta)$ denotes a building of type (W, S).

5.130. Let $\mathcal{C}'$ be a subbuilding of $\mathcal{C}$ (e.g., an apartment), and let $\mathcal{R}$ and $\mathcal{S}$ be two residues of $\mathcal{C}$ that meet $\mathcal{C}'$. Show that $\mathcal{R}$ and $\mathcal{S}$ are opposite in $\mathcal{C}$ if and only if $\mathcal{R} \cap \mathcal{C}'$ and $\mathcal{S} \cap \mathcal{C}'$ are opposite in $\mathcal{C}'$.

5.131. Let $\mathcal{R}$ be a residue and D a chamber of $\mathcal{C}$ that is opposite (at least) one chamber of $\mathcal{R}$. Show that the set of chambers in $\mathcal{R}$ opposite D in $\mathcal{C}$ is precisely the set of chambers opposite $\mathrm{proj}_{\mathcal{R}} D$ in $\mathcal{R}$ (note that $\mathcal{R}$ can also be considered as a spherical building).

5.132. Generalize the previous exercise as follows. Let $(\mathcal{C}', \delta')$ be an arbitrary building of type (W', S') (W' need not be finite), and let $\mathcal{R}$ be a residue in $\mathcal{C}'$ of type J, with $J \subseteq S'$. Assume that $\mathcal{R}$ is spherical, i.e., W'_J is finite. Let D be any chamber of $\mathcal{C}'$. Show that the chambers E opposite $\mathrm{proj}_{\mathcal{R}} D$ in $\mathcal{R}$ are precisely the chambers E in $\mathcal{R}$ satisfying $d(D, E) = \max \{d(D, E') \mid E' \in \mathcal{R}\}$.

*5.8 Twin Buildings

This optional section is long and treats an advanced topic. Most readers will want to omit it on first reading. We are presenting this topic in the present chapter because, historically, the theory of twin buildings provided much of the impetus for the development of the W-metric approach to buildings.

Motivated by Tits's fundamental paper [260] on Kac–Moody groups over fields, Ronan and Tits introduced twin buildings in the late 1980s. Twin buildings generalize spherical buildings, and they are naturally associated to "groups of Kac–Moody type" in the same way that spherical buildings are associated to algebraic groups. We will treat the group theory later (Sections 6.3 and 8.6–8.11). Here we simply collect the basic facts about twin buildings.

The central idea is that a twin building consists of a pair $(\mathcal{C}_+, \mathcal{C}_-)$ consisting of two buildings of the same type together with an *opposition relation* between the chambers of $\mathcal{C}_+$ and those of $\mathcal{C}_-$, with properties similar to those of the opposition relation on the chambers of a spherical building. In this way, the *pair* $(\mathcal{C}_+, \mathcal{C}_-)$ behaves in many respects like a spherical building, whereas the individual buildings $\mathcal{C}_+$ and $\mathcal{C}_-$ are generally not spherical (e.g., they may be Euclidean or hyperbolic).

5.8.1 Definition and First Examples

There are various ways of axiomatizing the opposition relation in a twin building. We will follow Tits [261] and introduce a Weyl-group-valued "codistance" between the chambers of $\mathcal{C}_+$ and those of $\mathcal{C}_-$, with two chambers declared to be opposite if their codistance is 1. A different approach due to Abramenko and Van Maldeghem [16], which axiomatizes the opposition relation directly, is described in Remark 5.154 below.

We again fix an arbitrary Coxeter system (W, S).

Definition 5.133. A *twin building* of type (W, S) is a triple $(\mathcal{C}_+, \mathcal{C}_-, \delta^*)$ consisting of two buildings $(\mathcal{C}_+, \delta_+)$ and $(\mathcal{C}_-, \delta_-)$ of type (W, S) together with a *codistance* function

$$\delta^* : (\mathcal{C}_+ \times \mathcal{C}_-) \cup (\mathcal{C}_- \times \mathcal{C}_+) \to W$$

satisfying the following conditions for each $\epsilon \in \{+, -\}$, any $C \in \mathcal{C}_\epsilon$, and any $D \in \mathcal{C}_{-\epsilon}$, where $w := \delta^*(C, D)$:

(Tw1) $\delta^*(C, D) = \delta^*(D, C)^{-1}$.

(Tw2) *If* $C' \in \mathcal{C}_\epsilon$ *satisfies* $\delta_\epsilon(C', C) = s$ *with* $s \in S$ *and* $l(sw) < l(w)$, *then* $\delta^*(C', D) = sw$.

(Tw3) *For any* $s \in S$, *there exists a chamber* $C' \in \mathcal{C}_\epsilon$ *with* $\delta_\epsilon(C', C) = s$ *and* $\delta^*(C', D) = sw$.

As a minor technical issue, we assume that $\mathcal{C}_+$ and $\mathcal{C}_-$ are disjoint sets, so that δ^* is never defined on pairs of chambers that are contained in the same building.

Definition 5.134. We define the *numerical codistance* between chambers $C \in \mathcal{C}_\epsilon$ and $D \in \mathcal{C}_{-\epsilon}$ ($\epsilon = \pm$) by

$$d^*(C, D) := l\big(\delta^*(C, D)\big) .$$

We say that C and D are *opposite*, and we write C op D, if $d^*(C, D) = 0$ or, equivalently, if $\delta^*(C, D) = 1$. Two residues $\mathcal{R}$ in $\mathcal{C}_+$ and $\mathcal{S}$ in $\mathcal{C}_-$ are called *opposite* if they have the same type and contain opposite chambers.

Remarks 5.135. (a) Although there are no connecting galleries between a chamber $C \in \mathcal{C}_+$ and a chamber $D \in \mathcal{C}_-$, one should think of $d^*(C, D)$ as a measure of how far away C and D are from each other. Intuitively, the bigger $d^*(C, D)$ is, the closer C and D are; they are "at maximal distance" if they are opposite. This explains why we have $l(sw) < l(w)$ in (Tw2): Decreasing codistance should be thought of as increasing distance. We will see soon that any s-panel $\mathcal{P}$ in $\mathcal{C}_\epsilon$ (or, more generally, any spherical residue in $\mathcal{C}_\epsilon$) contains precisely one chamber that is "closest" to a given chamber $D \in \mathcal{C}_{-\epsilon}$ in the sense that it has maximal codistance from D among all the chambers in $\mathcal{P}$.

(b) At first sight our convention that opposite residues have the same type seems to be inconsistent with what we have observed about spherical buildings. Note, however, that we have more leeway in the context of twin buildings since we can always adjust one of the two type functions, either for $\mathcal{C}_+$ or for $\mathcal{C}_-$. This is precisely what happens in Example 5.136 below when we associate a twin building to a spherical building. See also Exercise 5.161 below for further justification.

Before deriving some elementary consequences of the axioms, we briefly discuss two examples. The first shows that twin buildings do in fact generalize spherical buildings. The second shows that, just as for buildings, W provides us with the prototype of a thin twin building (and hence, as we will see, of a twin apartment).

Examples 5.136. (a) Assume that W is finite with w_0 as its element of maximal length, and let $(\mathcal{C}, \delta)$ be a building of type (W, S). Denote by $\mathcal{C}_+$ and $\mathcal{C}_-$ two disjoint copies of $\mathcal{C}$, and, for any $C \in \mathcal{C}$, denote by C_ϵ the corresponding chamber in $\mathcal{C}_\epsilon$. We then define $\delta_\epsilon \colon \mathcal{C}_\epsilon \times \mathcal{C}_\epsilon \hookrightarrow W$ by

$$\delta_+(C_+, D_+) := \delta(C, D) \quad \text{and} \quad \delta_-(C_-, D_-) := w_0 \delta(C, D) w_0 .$$

We now obtain a twin building $(\mathcal{C}_+, \mathcal{C}_-, \delta^*)$ by setting

$$\delta^*(C_+, D_-) := \delta(C, D) w_0 \quad \text{and} \quad \delta^*(D_-, C_+) := w_0 \delta(D, C)$$

for $C, D \in \mathcal{C}$. The reader should check as an exercise that $(\mathcal{C}_+, \mathcal{C}_-, \delta^*)$ is in fact a twin building of type (W, S). Note that $\mathcal{C}_-$ is the dual of $\mathcal{C}_+$ as defined in Section 5.7.7. Moreover, opposition as defined above coincides with the usual opposition relation in a spherical building. See Exercise 5.138 below for the interpretation of numerical codistance.

(b) Take two disjoint copies W_+ and W_- of W. For $w \in W$, we denote by w_+ and w_- the corresponding elements of W_+ and W_-, respectively. We now define two distance functions $\delta_\pm$ and a codistance function δ^* by setting

$$\delta_+(v_+, w_+) = \delta_-(v_-, w_-) = \delta^*(v_+, w_-) = \delta^*(v_-, w_+) = v^{-1}w$$

for all $v, w \in W$. It is even easier than in the first example to check that (W_+, W_-, δ^*) is in fact a twin building of type (W, S). The codistance δ^* introduced here will be denoted by δ_W^* in the following, and we will refer to (W_+, W_-, δ_W^*) as the *standard thin twin building* of type (W, S).

Of course, these two examples are not sufficient motivation for introducing twin buildings. As we already mentioned, however, we have to postpone further examples until we have provided the necessary group-theoretic background. The first interesting example will be given in Section 6.12, and a source of many additional examples will be described in Section 8.11.

Remark 5.137. In view of (Tw1), the other two axioms have "right" analogues. For each $\epsilon \in \{+, -\}$, any $C \in \mathcal{C}_\epsilon$, and any $D \in \mathcal{C}_{-\epsilon}$, let $w := \delta^*(C, D)$; then we have:

(Tw2′) If $D' \in \mathcal{C}_{-\epsilon}$ satisfies $\delta_{-\epsilon}(D, D') = s$ with $s \in S$ and $l(ws) < l(w)$, then $\delta^*(C, D') = ws$.

(Tw3′) For any $s \in S$, there exists a chamber $D' \in \mathcal{C}_{-\epsilon}$ with $\delta_{-\epsilon}(D, D') = s$ and $\delta^*(C, D') = ws$.

So every statement referring to "left" properties of δ^* has a companion referring to "right" properties of δ^*. We will tacitly use this in the following.

Exercise 5.138. Let $(\mathcal{C}_+, \mathcal{C}_-, \delta^*)$ be the twin building associated to a spherical building $(\mathcal{C}, \delta)$ as in Example 5.136(a). Show that

$$d^*(C_+, D_-) = d_0 - d(C, D)$$

for all $C, D \in \mathcal{C}$. (This explains the term "codistance.") Show further that

$$d^*(C_+, D_-) = \min\{d(C, C') \mid C' \text{ op } D\}$$

and that the minimum is achieved if and only if C' is the chamber opposite D in an apartment of $(\mathcal{C}, \delta)$ containing C and D.

5.8.2 Easy Consequences

Assume that $(\mathcal{C}_+, \mathcal{C}_-, \delta^*)$ is a twin building of type (W, S).

Lemma 5.139. *Given $\epsilon \in \{+, -\}$, $C \in \mathcal{C}_\epsilon$, $D \in \mathcal{C}_{-\epsilon}$, and $s \in S$, let $w :=$
$\delta^*(C, D)$. Then we have:*

(1) $\delta^*(C', D) \in \{w, sw\}$ *for any $C' \in \mathcal{C}_\epsilon$ with $\delta_\epsilon(C', C) = s$.*
(2) *If $l(sw) > l(w)$, there exists precisely one chamber $C' \in \mathcal{C}_\epsilon$ satisfying
$\delta_\epsilon(C', C) = s$ and $\delta^*(C', D) = sw$.*

Proof. (1) follows from (Tw2) if $l(sw) < l(w)$, so assume $l(sw) > l(w)$. Then,
by (Tw3), there exists $C'' \in \mathcal{C}_\epsilon$ with $\delta_\epsilon(C'', C) = s$ and $\delta^*(C'', D) = sw$. If
$C' = C''$, (1) is again satisfied. If $C' \neq C''$, then C' and C'' are both contained
in the s-panel through C, and hence $\delta_\epsilon(C', C'') = s$. Since $l(s(sw)) < l(sw)$
by assumption, (Tw2) implies that $\delta^*(C', D) = s(sw) = w$. So statement (1)
is proved. At the same time, our argument has shown that $\delta^*(C', D) = w$ for
all chambers $C' \neq C''$ in the s-panel of C if $l(sw) > l(w)$. This proves (2). $\square$

Lemma 5.140. *Given $\epsilon \in \{+, -\}$, $C, D \in \mathcal{C}_\epsilon$, and $E \in \mathcal{C}_{-\epsilon}$, let $w :=$
$\delta^*(D, E)$.*

(1) *If Γ is a gallery of type $\mathbf{s} = (s_1, \ldots, s_n)$ from C to D in $\mathcal{C}_\epsilon$, then there
exists a subword $(s_{i_1}, \ldots, s_{i_m})$ of $\mathbf{s}$ such that $\delta^*(C, E) = s_{i_1} \cdots s_{i_m} w$.*
(2) *If $v := \delta_\epsilon(C, D)$ satisfies $l(vw) = l(w) - l(v)$, then $\delta^*(C, E) = vw$.*

Proof. (1) follows immediately from Lemma 5.139(1) by an obvious induction
on n.

(2) The hypothesis can also be written as $l(w) = l(v^{-1}) + l(vw)$, so there is
a reduced decomposition $w = s_1 \cdots s_n$ such that some initial segment $s_1 \cdots s_l$
is a reduced decomposition of v^{-1}. Choose a minimal gallery from C to D
of type $(s_l, \ldots, s_1)$. Then repeated application of (Tw2) yields $\delta^*(C, E) =$
$s_l \cdots s_1 w = vw$. $\square$

Corollary 5.141.

(1) *If $C, D \in \mathcal{C}_\epsilon$ and $C' \in \mathcal{C}_{-\epsilon}$ satisfy $\delta_\epsilon(D, C) = \delta^*(D, C')$, then C op C'.*
(2) *Given any chamber $C' \in \mathcal{C}_{-\epsilon}$, any apartment of $\mathcal{C}_\epsilon$ contains at least one
chamber opposite C'.*

Proof. (1) If $w := \delta^*(D, C')$, then the hypothesis says that $\delta_\epsilon(C, D) = w^{-1}$.
Now apply Lemma 5.140(2) to get $\delta^*(C, C') = 1$.

(2) Let Σ be an apartment of $\mathcal{C}_\epsilon$. Choose an arbitrary chamber $D \in \Sigma$,
and let $C \in \Sigma$ be the chamber such that $\delta_\epsilon(D, C) = \delta^*(D, C')$. Then C op C'
by (1). $\square$

Remark 5.142. Note that we have reverted to using the letter Σ for a typical
apartment, as in Chapter 4. The reason is that in dealing with twin buildings,
it is useful to have the symbol $\mathcal{A}$ available for the system of "admissible
apartments" that will arise in Section 5.8.4. From the present W-metric point
of view, of course, Σ is a set of chambers, not a simplicial complex.

The final result of this subsection shows that the codistance δ^* is uniquely determined by the opposition relation. It should be compared with Exercise 5.138 above.

Lemma 5.143. *For any $C \in \mathcal{C}_\epsilon$ and any $D \in \mathcal{C}_{-\epsilon}$, $\delta^*(C, D)$ is the unique element of minimal length in the set $W(C, D) \subseteq W$ defined by*

$$W(C, D) := \{\delta_\epsilon(C, D') \mid D' \in \mathcal{C}_\epsilon \text{ and } D' \text{ op } D\} \ .$$

In particular,

$$d^*(C, D) = \min \{d(C, D') \mid D' \in \mathcal{C}_\epsilon \text{ and } D' \text{ op } D\} \ ,$$

where $d(-, -)$ denotes gallery distance.

Proof. Set $w := \delta^*(C, D)$. Then we have $w \in W(C, D)$ by Corollary 5.141(1). Now consider any $D' \in \mathcal{C}_\epsilon$, and set $v := \delta_\epsilon(D', C)$. By Lemma 5.140(1), we have $\delta^*(D', D) = v'w$, where either $l(v') < l(v)$ or $v' = v$. So if $\delta^*(D', D) = 1$, i.e., $v' = w^{-1}$, then either $l(v) > l(v') = l(w)$ or $v = v' = w^{-1}$. This implies that all elements $v^{-1} \in W(C, D)$ different from w satisfy $l(v^{-1}) > l(w)$.   □

Remark 5.144. Since the codistance is determined by the opposition relation, we can, in principle, view a twin building as a triple $(\mathcal{C}_+, \mathcal{C}_-, \text{op})$, where op is a relation between $\mathcal{C}_+$ and $\mathcal{C}_-$ having suitable properties. We will elaborate on this in Remark 5.154 below.

Exercises

5.145. In our treatment of ordinary buildings, the statement analogous to Lemma 5.139(1) was taken as part of axiom (WD2). Could we have omitted it from the axioms and then deduced it as we have done for twin buildings?

5.146. (For readers familiar with the Bruhat order.) With the notation of Lemma 5.143, show that $\delta^*(C, D)$ is the smallest element of $W(C, D)$ with respect to the Bruhat order.

5.147. Given $C, D \in \mathcal{C}_\epsilon$ and $E \in \mathcal{C}_{-\epsilon}$ such that $d^*(C, E) = d(C, D) + d^*(D, E)$, show that $\delta^*(C, E) = \delta_\epsilon(C, D)\delta^*(D, E)$.

5.8.3 Projections and Convexity

We continue to assume that $(\mathcal{C}_+, \mathcal{C}_-, \delta^*)$ is a twin building of type (W, S).

Given $\mathcal{M} \subseteq \mathcal{C}_\epsilon$ and $\mathcal{N} \subseteq \mathcal{C}_{-\epsilon}$ (where $\epsilon = +$ or $-$), we set

$$\delta^*(\mathcal{M}, \mathcal{N}) := \{\delta^*(C, D) \mid C \in \mathcal{M}, D \in \mathcal{N}\} \ .$$

We will also write $\delta^*(\mathcal{M}, D)$ for $\delta^*(\mathcal{M}, \{D\})$ and $\delta^*(C, \mathcal{N})$ for $\delta^*(\{C\}, \mathcal{N})$.

Lemma 5.148. *If $\mathcal{R}$ is a residue of $\mathcal{C}_\epsilon$ of type J and $\mathcal{S}$ is a residue of $\mathcal{C}_{-\epsilon}$ of type K, then $\delta^*(\mathcal{R}, \mathcal{S}) = W_J \delta^*(C, D) W_K$ for any $C \in \mathcal{R}$ and $D \in \mathcal{S}$.*

Proof. This follows immediately from (Tw3) and Lemma 5.139(1) (and their "right" analogues). □

In the following, we will mainly work with ordinary cosets rather than double cosets. As before, we write $w_1 = \min(W_J w)$ if w_1 is the element of minimal length in $W_J w$ and hence $l(v w_1) = l(v) + l(w_1)$ for all $v \in W_J$. If J is *spherical*, which means that W_J is finite, there is also a unique element of maximal length in $W_J w$, namely $w_1^* := w_0(J) w_1$. Here, as in Section 5.7, $w_0(J)$ is the longest element in W_J. Recall that it satisfies $l(v w_0(J)) = l(w_0(J)) - l(v)$ for all $v \in W_J$. It follows easily that

$$l(v w_1^*) = l(w_1^*) - l(v) \tag{5.10}$$

for all $v \in W_J$ if J is spherical.

Lemma 5.149. *If $\mathcal{R}$ is a residue of $\mathcal{C}_\epsilon$ of spherical type and D is a chamber in $\mathcal{C}_{-\epsilon}$, then there is a unique $C_1 \in \mathcal{R}$ such that $\delta^*(C_1, D)$ is of maximal length in $\delta^*(\mathcal{R}, D)$. This chamber C_1 satisfies*

$$\delta^*(C, D) = \delta_\epsilon(C, C_1) \delta^*(C_1, D)$$

for all $C \in \mathcal{R}$.

Proof. By Lemma 5.148, $\delta^*(\mathcal{R}, D)$ is a coset of the form $W_J w$, where J is the type of $\mathcal{R}$. Choose $C_1 \in \mathcal{R}$ such that $\delta^*(C_1, D)$ is the element w_1^* of maximal length in this coset. Then, by Lemma 5.140(2),

$$\delta^*(C, D) = \delta_\epsilon(C, C_1) \delta^*(C_1, D)$$

for any $C \in \mathcal{R}$, since

$$l\big(\delta_\epsilon(C, C_1) \delta^*(C_1, D)\big) = l(\delta^*(C_1, D)) - l(\delta_\epsilon(C, C_1))$$

by (5.10) (which applies because $\delta_\epsilon(C, C_1) \in W_J$). This also shows that $l(\delta^*(C, D)) < l(\delta^*(C_1, D))$ for all $C \in \mathcal{R} \setminus \{C_1\}$. □

Definition 5.150. If $\mathcal{R}$, D, and C_1 are as in Lemma 5.149, then C_1 is called the *projection* of D onto $\mathcal{R}$ and is denoted by $\mathrm{proj}_\mathcal{R} D$.

Remarks 5.151. (a) As is common in the theory of twin buildings, we do not distinguish in notation and terminology between the projections in one building and the projections between the two "halves" of a twin building. Note, however, that the latter exist only for *spherical* residues. It will always be clear from the context which type of projection we mean.

(b) Projections in the sense of Definition 5.150 are a powerful tool, provided there exist "enough" spherical residues. In particular, if all residues of rank 2 are spherical (this is the important 2-*spherical* case), then twin buildings, like spherical buildings, admit a nice classification. We will discuss this briefly in Section 9.12.

(c) Note that $C_1 = \mathrm{proj}_{\mathcal{R}}\, D$ is the unique chamber in $\mathcal{R}$ that maximizes the numerical codistance $d^*(C_1, D)$. Moreover, as a byproduct of the proof of Lemma 5.149, we have the following analogue of the gate property:

$$d^*(C, D) = d^*(C_1, D) - d(C, C_1) \tag{5.11}$$

for all $C \in \mathcal{R}$, where $d(-, -)$ is the gallery distance in $\mathcal{C}_\epsilon$. To understand this intuitively, recall that decreasing codistance should be thought of as increasing distance.

As a first application of projections, we will prove an analogue of Corollary 5.116. If $J \subseteq S$ is spherical, let σ_J be the automorphism of (W_J, J) defined by $\sigma_J(v) := w_0(J) v w_0(J)$ for $v \in W_J$.

Proposition 5.152. *Let $\mathcal{R}$ and $\mathcal{S}$ be opposite residues of spherical type J in the twin building $(\mathcal{C}_+, \mathcal{C}_-, \delta^*)$. Then the projection maps $\mathrm{proj}_{\mathcal{R}}$ and $\mathrm{proj}_{\mathcal{S}}$ induce mutually inverse σ_J-isometries between $\mathcal{R}$ and $\mathcal{S}$.*

Proof. By assumption, we have $1 \in \delta^*(\mathcal{R}, \mathcal{S})$; hence $\delta^*(\mathcal{R}, \mathcal{S}) = W_J$ by Lemma 5.148. The same lemma also implies that $\delta^*(\mathcal{R}, D) = \delta^*(C, \mathcal{S}) = W_J$ for any $C \in \mathcal{R}$ and $D \in \mathcal{S}$. In view of the definition of the projection, it follows that

$$C = \mathrm{proj}_{\mathcal{R}}\, D \iff \delta^*(C, D) = w_0(J) \iff D = \mathrm{proj}_{\mathcal{S}}\, C\,.$$

This shows that $\mathrm{proj}_{\mathcal{R}}$ and $\mathrm{proj}_{\mathcal{S}}$ induce mutually inverse bijections between $\mathcal{R}$ and $\mathcal{S}$. The proposition will now follow from Lemma 5.61 if we show that these projections transform s-adjacency to s'-adjacency for any $s \in J$, where $s' := \sigma_J(s) = w_0(J) s w_0(J)$. Consider, for instance, $C, C' \in \mathcal{R}$ with $\delta_\epsilon(C', C) = s$ (where we assume that $\mathcal{R} \subseteq \mathcal{C}_\epsilon$). If we set $D := \mathrm{proj}_{\mathcal{S}}\, C$, then $\delta^*(C, D) = w_0(J)$ and $\delta^*(C', D) = s w_0(J) = w_0(J) s'$. By (Tw3'), there exists $D' \in \mathcal{S}$ with $\delta_{-\epsilon}(D, D') = s'$ and $\delta^*(C', D') = \delta^*(C', D) s' = w_0(J)$. Hence $\mathrm{proj}_{\mathcal{S}}\, C' = D'$, which is s'-adjacent to $\mathrm{proj}_{\mathcal{S}}\, C = D$. $\qquad\square$

Here is an important special case, which could also have been deduced directly from Definition 5.133:

Corollary 5.153. *If $\mathcal{P}$ and $\mathcal{Q}$ are opposite panels, then the relation of non-opposition induces a bijection between $\mathcal{P}$ and $\mathcal{Q}$. In particular, the cardinalities of $\mathcal{P}$ and $\mathcal{Q}$ are equal.* $\qquad\square$

Remark 5.154. Given two buildings $\mathcal{C}_\pm$ of the same type, a symmetric opposition relation between them is called a 1-*twinning* if it has the property described in Corollary 5.153. This notion was introduced by Mühlherr [173]. The main result of Abramenko and Van Maldeghem [16] is that $(\mathcal{C}_+, \mathcal{C}_-, \mathrm{op})$ defines a twin building if and only if the following two conditions are satisfied:

(1) op is a 1-twinning.
(2) There is a chamber $C \in \mathcal{C}_+$ with the following property: Any chamber $C' \in \mathcal{C}_-$ opposite C is contained in an apartment Σ_- of $\mathcal{C}_-$ such that C' is the only chamber of Σ_- opposite C.

The second condition is motivated by the theory of twin apartments, which we will discuss in the next subsection.

Definition 5.155. The twin building $(\mathcal{C}_+, \mathcal{C}_-, \delta^*)$ is called *thick* (resp. *thin*) if each of the buildings $\mathcal{C}_+$ and $\mathcal{C}_-$ is thick (resp. thin). Corollary 5.153 shows that it suffices to require only one of the two buildings to be thick (resp. thin).

The following lemma generalizes Proposition 4.104 to twin buildings.

Lemma 5.156. *Let $(\mathcal{C}_+, \mathcal{C}_-, \delta^*)$ be a thick twin building of type (W, S). Then for each $\epsilon \in \{+, -\}$ and any two chambers $C, D \in \mathcal{C}_\epsilon$, there is a chamber $E \in \mathcal{C}_{-\epsilon}$ that is opposite to both C and D.*

Proof. Choose, by applying Corollary 5.141 for instance, a chamber $E' \in \mathcal{C}_{-\epsilon}$ that is opposite C. Set $w := \delta^*(E', D)$, and choose a reduced decomposition $w = s_1 \cdots s_n$. By Corollary 5.153 and thickness, there is a gallery $E' = E_0, E_1, \ldots, E_n =: E$ of type $(s_1, \ldots, s_n)$ in $\mathcal{C}_{-\epsilon}$ such that each E_i is opposite C. Then $\delta_{-\epsilon}(E', E) = w$ by Lemma 5.16, and E op D by Corollary 5.141(1). We also have E op C by construction. $\square$

We can now also generalize Corollary 5.118 to twin buildings.

Corollary 5.157. *Let $(\mathcal{C}_+, \mathcal{C}_-, \delta^*)$ be a thick twin building of type (W, S). Then for each $\epsilon \in \{+, -\}$ and any spherical subset $J \subseteq S$, any two residues of $\mathcal{C}_\epsilon$ of type J are isometric.*

Proof. Let $\mathcal{R}$ and $\mathcal{T}$ be two residues of $\mathcal{C}_\epsilon$ of type J. Choose $C \in \mathcal{R}$, $D \in \mathcal{T}$, and (by Lemma 5.156) $E \in \mathcal{C}_{-\epsilon}$ such that E op C and E op D. Denote by $\mathcal{S}$ the J-residue of $\mathcal{C}_{-\epsilon}$ containing E. Then $\mathcal{S}$ is opposite $\mathcal{R}$ and $\mathcal{T}$. By Proposition 5.152, there are surjective σ_J-isometries $f \colon \mathcal{R} \to \mathcal{S}$ and $g \colon \mathcal{S} \to \mathcal{T}$. Hence $g \circ f \colon \mathcal{R} \to \mathcal{T}$ is a surjective isometry. $\square$

We close this section by using projections to define a notion of convexity for twin buildings.

Definition 5.158. A pair $(\mathcal{M}_+, \mathcal{M}_-)$ of nonempty subsets $\mathcal{M}_+ \subseteq \mathcal{C}_+$ and $\mathcal{M}_- \subseteq \mathcal{C}_-$ is called *convex* if $\operatorname{proj}_{\mathcal{P}} C \in \mathcal{M}_+ \cup \mathcal{M}_-$ for any $C \in \mathcal{M}_+ \cup \mathcal{M}_-$ and any panel $\mathcal{P} \subseteq \mathcal{C}_+ \cup \mathcal{C}_-$ that meets $\mathcal{M}_+ \cup \mathcal{M}_-$.

Remark 5.159. Note that $\operatorname{proj}_{\mathcal{P}} C$ could be either the usual projection in one of the buildings $\mathcal{C}_\epsilon$ (if $\mathcal{P}$ and C are contained in $\mathcal{C}_\epsilon$) or the projection in the sense of Definition 5.150 (if $\mathcal{P} \subseteq \mathcal{C}_\epsilon$ and $C \in \mathcal{C}_{-\epsilon}$). The first case implies that for a convex pair $(\mathcal{M}_+, \mathcal{M}_-)$, each $\mathcal{M}_\epsilon$ is convex in $\mathcal{C}_\epsilon$; see Proposition 5.46.

Hence, by Lemma 5.45, $\text{proj}_{\mathcal{R}} C \in \mathcal{M}_\epsilon$ for *any* residue $\mathcal{R}$ of $\mathcal{C}_\epsilon$ that meets $\mathcal{M}_\epsilon$ and any $C \in \mathcal{M}_\epsilon$. It is not difficult to verify that a similar statement holds for the second type of projection (see Exercise 5.169 below): If the pair $(\mathcal{M}_+, \mathcal{M}_-)$ is convex and $\mathcal{R} \subseteq \mathcal{C}_\epsilon$ is a residue of spherical type that meets $\mathcal{M}_\epsilon$, then $\text{proj}_{\mathcal{R}} C \in \mathcal{M}_\epsilon$ for any $C \in \mathcal{M}_{-\epsilon}$. Hence Definition 5.158 can be formulated more simply as follows:

A pair $(\mathcal{M}_+, \mathcal{M}_-)$ of nonempty subsets $\mathcal{M}_+ \subseteq \mathcal{C}_+$ and $\mathcal{M}_- \subseteq \mathcal{C}_-$ is convex if and only if it is closed under projections.

Given a family of convex pairs $(\mathcal{M}_+, \mathcal{M}_-)$, it is immediate that their intersection (computed componentwise) is again convex if both components are nonempty. We can therefore form convex hulls in the usual way:

Definition 5.160. The *convex hull* of a pair $(\mathcal{M}_+, \mathcal{M}_-)$ of nonempty subsets $\mathcal{M}_+ \subseteq \mathcal{C}_+$ and $\mathcal{M}_- \subseteq \mathcal{C}_-$ is the smallest convex pair containing $(\mathcal{M}_+, \mathcal{M}_-)$. [Here pairs $(\mathcal{M}_+, \mathcal{M}_-)$ are ordered componentwise.] Equivalently, the convex hull of $(\mathcal{M}_+, \mathcal{M}_-)$ is the intersection of all convex pairs containing it.

In the following, an important role will be played by convex hulls of pairs of opposite chambers. These turn out to be *twin apartments*, which are as fundamental for twin buildings as apartments are for ordinary buildings.

Exercises

5.161. Let $\mathcal{R}$ be a residue of $\mathcal{C}_+$ and $\mathcal{S}$ a residue of $\mathcal{C}_-$. Show that $\mathcal{R}$ and $\mathcal{S}$ are opposite in the sense of Definition 5.134 if and only if every chamber in $\mathcal{R}$ is opposite some chamber in $\mathcal{S}$ and vice versa.

5.162. (a) Give an example of a convex pair $(\mathcal{M}_+, \mathcal{M}_-)$ such that each $\mathcal{M}_\epsilon$ contains exactly one chamber.
 (b) If (W, S) is infinite and $(\mathcal{M}_+, \mathcal{M}_-)$ is a convex pair, show that each $\mathcal{M}_\epsilon$ contains more than one chamber.

5.163. If W is finite, show that every twin building of type (W, S) is isometric to one of the form described in Example 5.136(a). Here an *isometry* between two twin buildings of the same type is a pair of isometries preserving codistance.

5.164. Show that every thin twin building of type (W, S) is isometric to the standard thin twin building described in Example 5.136(b).

5.165. Define a notion of "almost isometry" for twin buildings of possibly different types. Show that an almost isometry is the same thing as a pair of simplicial isomorphisms preserving the opposition relation.

5.166. Let $\mathcal{R}$ and $\mathcal{S}$ be opposite residues. Show that $(\mathcal{R}, \mathcal{S}, \delta^*|)$ is again a twin building, where $\delta^*|$ denotes the restriction of δ^* to $(\mathcal{R} \times \mathcal{S}) \cup (\mathcal{S} \times \mathcal{R})$.

5.167. Let $\mathcal{R}$ be a residue of $\mathcal{C}_\epsilon$, and let D be a chamber of $\mathcal{C}_{-\epsilon}$. Set

$$\mathcal{M}(\mathcal{R}, D) := \{C \in \mathcal{R} \mid \delta^*(C, D) = \min(\delta^*(\mathcal{R}, D))\} .$$

If $\mathcal{R}$ is spherical, show that $\mathcal{M}(\mathcal{R}, D)$ is the set of all chambers that are opposite $\mathrm{proj}_\mathcal{R} D$ in the spherical building $\mathcal{R}$. (We will give another description of $\mathcal{M}(\mathcal{R}, D)$, valid for arbitrary residues $\mathcal{R}$, in Exercise 5.189.)

5.168. Let $\mathcal{R}$ be a spherical residue in $\mathcal{C}_\epsilon$ and $\mathcal{S}$ an arbitrary residue in $\mathcal{C}_{-\epsilon}$. Set

$$\mathrm{proj}_\mathcal{R} \mathcal{S} := \{\mathrm{proj}_\mathcal{R} D \mid D \in \mathcal{S}\} .$$

Try to give a characterization of $\mathrm{proj}_\mathcal{R} \mathcal{S}$ similar to the one in Lemma 5.36(1) and to prove that $\mathrm{proj}_\mathcal{R} \mathcal{S}$ is again a residue. Do you need to assume that $\mathcal{S}$ is spherical?

5.169. Suppose $(\mathcal{M}_+, \mathcal{M}_-)$ is a convex pair in the sense of Definition 5.158. If $\mathcal{R}$ is a spherical residue in $\mathcal{C}_\epsilon$ that meets $\mathcal{M}_\epsilon$, show that $\mathrm{proj}_\mathcal{R} D \in \mathcal{M}_\epsilon$ for all $D \in \mathcal{M}_{-\epsilon}$.

5.170. Let $(\mathcal{M}_+, \mathcal{M}_-)$ be a convex pair that is *weak* in the sense that each panel $\mathcal{P}$ that meets $\mathcal{M}_+ \cup \mathcal{M}_-$ contains at least two chambers of $\mathcal{M}_+ \cup \mathcal{M}_-$. Show that $(\mathcal{M}_+, \mathcal{M}_-, \delta^*|)$ is a twin building, where $\delta^*|$ denotes the restriction of δ^* to $(\mathcal{M}_+ \times \mathcal{M}_-) \cup (\mathcal{M}_- \times \mathcal{M}_+)$.

5.8.4 Twin Apartments

For brevity, we will start suppressing δ^* from the notation and simply saying that $(\mathcal{C}_+, \mathcal{C}_-)$ is a twin building of type (W, S). We may also write $\mathcal{C} = (\mathcal{C}_+, \mathcal{C}_-)$ when there is no ordinary building $(\mathcal{C}, \delta)$ under discussion. Twin apartments will be defined as certain subpairs $\Sigma = (\Sigma_+, \Sigma_-)$ of $\mathcal{C}$. There are various equivalent ways of characterizing twin apartments, and it is a matter of taste which one is taken as the definition. Here we choose one that stresses the opposition relation.

Definition 5.171. A *twin apartment* of a twin building $\mathcal{C}$ is a pair $\Sigma = (\Sigma_+, \Sigma_-)$ such that Σ_+ is an apartment of $\mathcal{C}_+$, Σ_- is an apartment of $\mathcal{C}_-$, and every chamber in $\Sigma_+ \cup \Sigma_-$ is opposite precisely one other chamber in $\Sigma_+ \cup \Sigma_-$.

Recall that for each $\epsilon = \pm$, there is always at least one chamber in $\Sigma_{-\epsilon}$ opposite any given chamber in $\mathcal{C}_\epsilon$ (see Corollary 5.141(2)). So "precisely one" can be replaced by "at most one" in the definition. Figure 5.1 shows a schematic representation of a twin apartment and two pairs of opposite chambers. The two "halves" Σ_+, Σ_- of Σ have been drawn so as to suggest the analogy with apartments in spherical buildings. Definition 5.171 leads immediately to the following generalization of the opposition involution of a spherical Coxeter complex (see Sections 1.6.2 and 5.7.3).

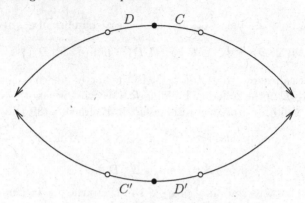

Fig. 5.1. A twin apartment.

Definition 5.172. If $\Sigma = (\Sigma_+, \Sigma_-)$ is a twin apartment, then the *opposition involution*, denoted by op_Σ, associates to each chamber $C \in \Sigma_+ \cup \Sigma_-$ the unique chamber $C' = \mathrm{op}_\Sigma(C) \in \Sigma_+ \cup \Sigma_-$ such that C' op C.

Figure 5.1 suggests that op_Σ preserves s-adjacency for all $s \in S$ and hence induces an isometry $\Sigma_\epsilon \to \Sigma_{-\epsilon}$ for each $\epsilon = \pm$. Let's prove this, along with some other useful facts about twin apartments:

Lemma 5.173. *Let $\Sigma = (\Sigma_+, \Sigma_-)$ be a twin apartment and let $\epsilon = +$ or $-$.*

(1) $\mathrm{op}_\Sigma \colon \Sigma_\epsilon \to \Sigma_{-\epsilon}$ *is an isometry.*
(2) *Given $C \in \Sigma_\epsilon$ and $D' \in \Sigma_{-\epsilon}$, set $D := \mathrm{op}_\Sigma D' \in \Sigma_\epsilon$. Then*

$$\delta^*(C, D') = \delta_\epsilon(C, D) . \tag{5.12}$$

(3) *Given $C \in \Sigma_\epsilon$ and $w \in W$, there is a unique chamber $D' \in \Sigma_{-\epsilon}$ such that $\delta^*(C, D') = w$.*
(4) *For any three chambers $C, D, E \in \Sigma_+ \cup \Sigma_-$,*

$$\delta(C, E) = \delta(C, D)\delta(D, E) ,$$

where each δ is to be interpreted as δ_+, δ_-, or δ^*, whichever one makes sense.
(5) *Σ is a thin twin building in its own right, isomorphic to the standard thin twin building of type (W, S).*
(6) *Σ is convex in $\mathcal{C}$.*

Proof. (1) Let C and D be s-adjacent chambers of Σ_ϵ, and set $C' := \mathrm{op}_\Sigma(C)$. Then $\delta^*(C, C') = 1$ and hence, by Lemma 5.139(1), $\delta^*(D, C') = s$, since D and C' are not opposite. If we denote by D' the chamber of $\Sigma_{-\epsilon}$ that is s-adjacent to C', then (Tw2') now yields $\delta^*(D, D') = ss = 1$, so $D' = \mathrm{op}_\Sigma(C)$. Thus op_Σ preserves s-adajacency and hence is an isometry by Lemma 5.61.

(2) Let D_1 be the unique chamber in Σ_ϵ such that $\delta_\epsilon(C, D_1) = \delta^*(C, D')$. Then D_1 op D' by Corollary 5.141(1); hence $D_1 = D$ and (5.12) holds.

(3) This is immediate from (2) and a standard property of apartments.

(4) This is a standard property of ordinary apartments if all three chambers are in a single Σ_ϵ. In general, one can use (2) to reduce to the case that all chambers are in one Σ_ϵ. Suppose, for instance, that $C, D \in \Sigma_+$ and $E \in \Sigma_-$. Then we obtain, using (2) twice,

$$\delta^*(C, E) = \delta_+(C, \mathrm{op}_\Sigma E)$$
$$= \delta_+(C, D)\delta_+(D, \mathrm{op}_\Sigma E)$$
$$= \delta_+(C, D)\delta^*(D, E) .$$

(5) This follows easily from the previous results.

(6) Since Σ_+ and Σ_- are apartments and hence convex, it suffices to prove the following: If $\mathcal{R}$ is a spherical residue in $\mathcal{C}_\epsilon$ that meets Σ_ϵ and D is a chamber in $\Sigma_{-\epsilon}$, then $\mathrm{proj}_\mathcal{R} D \in \Sigma_\epsilon$. Set $\mathcal{R}_0 := \mathcal{R} \cap \Sigma_\epsilon$, and choose $C \in \mathcal{R}_0$. Since $\delta_\epsilon(\mathcal{R}_0, C) = W_J$, it follows from (4) that $\delta^*(\mathcal{R}_0, D) = W_J\delta^*(C, D) = \delta^*(\mathcal{R}, D)$. By Lemma 5.149 and Definition 5.150, this implies that $\mathrm{proj}_\mathcal{R} D \in \mathcal{R}_0 \subseteq \Sigma_\epsilon$, as required.   □

Examples 5.174. (a) Let $(\mathcal{C}_0, \delta)$ be a spherical building, and let $\mathcal{C} = (\mathcal{C}_+, \mathcal{C}_-)$ be the associated twin building as in Example 5.136(a). Then every apartment Σ_0 of $\mathcal{C}_0$ gives rise to a twin apartment $\Sigma = (\Sigma_+, \Sigma_-)$, where Σ_ϵ is the copy of Σ_0 in $\mathcal{C}_\epsilon$ for $\epsilon = \pm$. The opposition involution op_Σ is essentially the same as the opposition involution of Σ_0 previously defined, except that it is now viewed as interchanging two disjoint copies $\Sigma_\pm$ of Σ_0. In other words, we have a commutative square

$$
\begin{array}{ccc}
\Sigma_\epsilon & \xrightarrow{\ \mathrm{op}_\Sigma\ } & \Sigma_{-\epsilon} \\
\| & & \| \\
\Sigma_0 & \xrightarrow[\mathrm{op}_{\Sigma_0}]{} & \Sigma_0
\end{array}
$$

for each ϵ. Note that when viewed in this way, op_Σ becomes an isometry rather than a σ_0-isometry because of the way the Weyl distance in $\mathcal{C}_-$ is defined.

(b) In the standard thin twin building (W_+, W_-, δ_W^*), the pair $\Sigma = (W_+, W_-)$ is a twin apartment, and op_Σ maps each copy W_ϵ of W to the other copy $W_{-\epsilon}$ by the identity map.

Our goal for the rest of this subsection is to show that every twin building has "sufficiently many" twin apartments. Let C and C' be opposite chambers, with $C \in \mathcal{C}_\epsilon$. If there exists a twin apartment Σ containing C and C', then Σ_ϵ contains, for each $w \in W$, exactly one chamber C_w such that $\delta_\epsilon(C, C_w) = w$, and this chamber also satisfies $\delta^*(C', C_w) = w$ by Lemma 5.173(2). The following lemma will allow us to conclude that Σ is completely determined by C and C'.

Lemma 5.175. *Let C and C' be opposite chambers with $C \in C_\epsilon$.*

(1) *For any $w \in W$ there is a unique chamber $C_w \in C_\epsilon$ such that*

$$\delta_\epsilon(C, C_w) = w = \delta^*(C', C_w) .$$

(2) *The chamber C_w is contained in the convex hull of the pair (C, C').*
(3) *The map $f_{C,C'} \colon W \to C_\epsilon$ given by $w \mapsto C_w$ is an isometry.*

Proof. (1) and (2) are proved by induction on $l(w)$. We may assume $l(w) > 0$, so that $w = w's$ with $s \in S$ and $l(w') = l(w) - 1$. By the induction hypothesis, there is a unique chamber $C_{w'} \in C_\epsilon$ with $\delta_\epsilon(C, C_{w'}) = w' = \delta^*(C', C_{w'})$, and this chamber $C_{w'}$ is in the convex hull K of (C, C').

To prove the existence of C_w, let $\mathcal{P}$ be the s-panel containing $C_{w'}$, and set $C_w := \mathrm{proj}_\mathcal{P} C'$. Note that $C_w \in K$ because $C_{w'}$ and C' are in K. Since $l(w's) > l(w')$, we have $\delta^*(C', C_w) = w's = w$. In particular, $C_w \neq C'_{w'}$, and hence $\delta_\epsilon(C_w, C_{w'}) = s$. Using (WD2'), we now get $\delta_\epsilon(C, C_w) = \delta_\epsilon(C, C_{w'})s = w's = w$. So C_w has the desired properties.

To prove the uniqueness of C_w, assume that $D \in C_\epsilon$ also satisfies $\delta_\epsilon(C, D) = w = \delta^*(C', D)$. Let $\mathcal{P}_1$ be the s-panel containing D, and set $D_1 := \mathrm{proj}_{\mathcal{P}_1} C$. (This is the ordinary projection in C_ϵ.) Then $\delta_\epsilon(C, D_1) = w'$ since $l(w') = l(ws) < l(w)$. Now $\delta_\epsilon(D_1, D) = s$, so (Tw2') implies that $\delta^*(C', D_1) = \delta^*(C', D)s = w'$. By uniqueness of $C_{w'}$, we must have $D_1 = C_{w'}$. Hence $\mathcal{P}_1 = \mathcal{P}$. And since D is s-adjacent to $C_{w'}$ and satisfies $\delta^*(C', D) = w$, we also obtain $D = \mathrm{proj}_\mathcal{P} C' = C_w$. This proves the uniqueness of C_w.

(3) The map $f_{C,C'}$ is obviously injective, so it suffices to show that it preserves s-adajcency for all $s \in S$. Let w and w' be s-adjacent in W, i.e., $w = w's$. We may assume that $l(w') < l(w)$, in which case $C_{w'}$ and C_w are s-adjacent in C_ϵ by the existence proof above. $\square$

Definition 5.176. With C and C' as in Lemma 5.175, we define

$$\Sigma(C, C') := f_{C,C'}(W) = \{D \in C_\epsilon \mid \delta_\epsilon(C, D) = \delta^*(C', D)\} .$$

We also define

$$\Sigma\{C, C'\} := \begin{cases} (\Sigma(C, C'), \Sigma(C', C)) & \text{if } \epsilon = + , \\ (\Sigma(C', C), \Sigma(C, C')) & \text{if } \epsilon = - . \end{cases}$$

Note that $\Sigma(C, C')$ is an apartment of C_ϵ and $\Sigma(C', C)$ is an apartment of $C_{-\epsilon}$ by Lemma 5.175 and Corollary 5.67. We will show soon that $\Sigma\{C, C'\}$ is a twin apartment. But first we need another uniqueness statement, which is also of independent interest. In this context, the following notation will be useful:

Definition 5.177. For $\epsilon \in \{+, -\}$ and $C \in C_\epsilon$, we set

$$C^{\mathrm{op}} := \{C' \in C_{-\epsilon} \mid C' \text{ op } C\} .$$

Lemma 5.178. *Let C and C' be opposite chambers with $C \in \mathcal{C}_\epsilon$, and let Σ_ϵ be an apartment of $\mathcal{C}_\epsilon$ such that $(C')^{\mathrm{op}} \cap \Sigma_\epsilon = \{C\}$.*

(1) $\Sigma_\epsilon = \Sigma(C, C')$.
(2) *If C'' is a second chamber of $\mathcal{C}_{-\epsilon}$ such that $(C'')^{\mathrm{op}} \cap \Sigma_\epsilon = \{C\}$, then $C'' = C'$.*

Proof. (1) For each $w \in W$, let D_w be the chamber of Σ_ϵ such that $\delta_\epsilon(C, D_w) = w$. We have to show that $D_w = C_w$, where C_w is defined as in Lemma 5.175. In other words, we have to show that $\delta^*(C', D_w) = w$. Let D be the chamber in Σ_ϵ such that $\delta_\epsilon(D, D_w) = \delta^*(C', D_w)$. Then D op C' by Corollary 5.141(1). Our hypothesis now implies that $D = C$ and hence that $\delta^*(C', D_w) = \delta_\epsilon(C, D_w) = w$, as required.

(2) Set $w := \delta_{-\epsilon}(C', C'')$, and consider the chamber $C_w \in \Sigma_\epsilon = \Sigma(C, C')$. Since $\delta^*(C', C_w) = w = \delta_{-\epsilon}(C', C'')$, we have C'' op C_w by Corollary 5.141(1). Our hypothesis now implies that $C_w = C$; hence $w = 1$ and $C'' = C'$. $\qquad\square$

It is now easy to derive the main properties of twin apartments.

Proposition 5.179.

(1) *If C and C' are opposite chambers, then $\Sigma\{C, C'\}$ is a twin apartment and is the unique twin apartment containing C and C'. Moreover, $\Sigma\{C, C'\}$ is the convex hull of the pair (C, C').*
(2) *For any apartment Σ_ϵ of $\mathcal{C}_\epsilon$, there is at most one apartment $\Sigma_{-\epsilon}$ of $\mathcal{C}_{-\epsilon}$ such that (Σ_+, Σ_-) is a twin apartment.*
(3) *Given any two chambers $C, D \in \mathcal{C}_+ \cup \mathcal{C}_-$, there is a twin apartment containing them.*

Proof. (1) We may assume that $C \in \mathcal{C}_+$. The main thing to show here is that $\Sigma\{C, C'\}$ is a twin apartment. For $w \in W$, denote by C_w the chamber of $\Sigma(C, C')$ satisfying $\delta_+(C, C_w) = w = \delta^*(C', C_w)$ and by C'_w the chamber of $\Sigma(C', C)$ satisfying $\delta_-(C', C'_w) = w = \delta^*(C, C'_w)$. By Corollary 5.141(1), C_w op C'_w for all $w \in W$. Assume now that C_w op C'_v for some $w, v \in W$, and consider the set $W(C, C'_v)$ defined in Lemma 5.143. According to that lemma, $v = \delta^*(C, C'_v)$ is the unique element of minimal length in $W(C, C'_v)$. But $w = \delta_+(C, C_w)$ is in that set, so we have

$$w = v \text{ or } l(w) > l(v). \tag{5.13}$$

Similarly, we have $v \in W(C', C_w)$ and $w = \delta^*(C', C_w)$, so

$$v = w \text{ or } l(v) > l(w). \tag{5.14}$$

Assertions (5.13) and (5.14) can be consistent with one another only if $v = w$. So for each chamber $C_w \in \Sigma(C, C')$, C'_w is the unique chamber of $\Sigma(C', C)$ that is opposite C_w, and similarly with the roles of C and C' interchanged.

Thus $\Sigma\{C, C'\}$ is indeed a twin apartment. It is then the unique twin apartment containing C and C' by Lemma 5.175 and the remarks preceding it. It is convex by Lemma 5.173(6) and is contained in the convex hull of (C, C') by Lemma 5.175(2), so it is equal to that convex hull.

(2) We may assume that $\epsilon = +$. Suppose there is an apartment Σ_- of $\mathcal{C}_-$ such that $\Sigma := (\Sigma_+, \Sigma_-)$ is a twin apartment. Fix a chamber $C \in \Sigma_+$, and denote by C' the unique chamber in Σ_- opposite C. Then Lemma 5.178(1) (or part (1) of the present proposition) implies that $\Sigma = \Sigma\{C, C'\}$. If $\Sigma' = (\Sigma_+, \Sigma'_-)$ is another twin apartment with first component Σ_+, then the unique chamber of Σ'_- opposite C is the same C' by Lemma 5.178(2). Hence Σ' is also equal to $\Sigma\{C, C'\}$.

(3) Let C be in $\mathcal{C}_\epsilon$. If D is also in $\mathcal{C}_\epsilon$, choose (using (Tw3)) a chamber $C' \in \mathcal{C}_{-\epsilon}$ with $\delta^*(C', D) = \delta_\epsilon(C, D)$. Then C' op C by Corollary 5.141(1), and C and D are both contained in $\Sigma\{C, C'\}$. Similarly, if D is in $\mathcal{C}_{-\epsilon}$, choose a chamber $C' \in \mathcal{C}_{-\epsilon}$ such that $\delta_{-\epsilon}(C', D) = \delta^*(C, D)$. Then C op C', and C and D are contained in $\Sigma\{C, C'\}$. $\qquad\square$

Proposition 5.179 shows that the twinning of $\mathcal{C}_+$ and $\mathcal{C}_-$ distinguishes certain apartment systems in these buildings. We introduce some notation for these systems.

Definition 5.180. We set

$$\mathcal{A}_+ := \{\Sigma(C, C') \mid C \in \mathcal{C}_+, \, C' \in \mathcal{C}_-, \, C \text{ op } C'\} \,,$$
$$\mathcal{A}_- := \{\Sigma(C', C) \mid C' \in \mathcal{C}_-, \, C \in \mathcal{C}_+, \, C \text{ op } C'\} \,.$$

The elements of $\mathcal{A}_\epsilon$ are called *admissible* apartments of $\mathcal{C}_\epsilon$ ($\epsilon = \pm$) with respect to the given twinning. We also denote by $\mathcal{A}$ the set of all twin apartments $\Sigma = (\Sigma_+, \Sigma_-)$. Equivalently,

$$\mathcal{A} = \{\Sigma\{C, C'\} \mid C \in \mathcal{C}_+, \, C' \in \mathcal{C}_-, \, C \text{ op } C'\} \,.$$

Remark 5.181. The apartment systems $\mathcal{A}_+$ and $\mathcal{A}_-$ are usually far from being complete. If the buildings $\mathcal{C}_+$ and $\mathcal{C}_-$ are countable, for example, then $\mathcal{A}_+$ and $\mathcal{A}_-$ are also countable. But a thick, infinite, nonspherical building typically has uncountably many apartments (think for instance of a tree). When we treat concrete examples of twin buildings, we will also be able to describe $\mathcal{A}_+$ and $\mathcal{A}_-$ as orbits of a "small" group; see Sections 6.12 and 8.11.

We close this subsection with a result that will be needed in Section 5.11. Recall from Corollary 5.30 that a J-residue $\mathcal{R}$ in a building is itself a building of type (W_J, J). In particular, there is an opposition relation on $\mathcal{R}$ if the latter is spherical.

Proposition 5.182. *Let* $\Sigma = \Sigma\{C, C'\}$ *be a twin apartment, with* $C \in \mathcal{C}_+$, $C' \in \mathcal{C}_-$, *and* C *op* C'. *Let* $\mathcal{R}$ *be a spherical residue of* $\mathcal{C}_+$ *containing* C, *and let* $J \subseteq S$ *be its type.*

(1) *There is a unique chamber $C_1 \in \mathcal{R} \cap \Sigma_+$ that is opposite C in the spherical building $\mathcal{R}$, and*

$$\mathrm{proj}_{\mathcal{R}} C' = C_1 . \tag{5.15}$$

(2) *For any chamber $D \in \mathcal{R}$,*

$$\delta^*(D, C') = \delta_+(D, C_1) w_0(J) . \tag{5.16}$$

Proof. The first assertion of (1) follows from the fact that $\mathcal{R} \cap \Sigma_+$ is an apartment of $\mathcal{R}$. We then have $\delta^*(C', C_1) = \delta_+(C, C_1) = w_0(J)$, which is the longest element of $W_J = \delta^*(C', \mathcal{R})$. Equation (5.15) now follows from the definition of the projection. For (2), we use (5.15) and Lemma 5.149 to obtain

$$\begin{aligned}
\delta^*(D, C') &= \delta_+(D, C_1) \delta^*(C_1, C') \\
&= \delta_+(D, C_1) \delta_+(C_1, C) \\
&= \delta_+(D, C_1) w_0(J) . \qquad \square
\end{aligned}$$

Remarks 5.183. (a) Equation (5.16) looks more natural if we interpret the right side as a codistance in the spherical building $\mathcal{R}$ (see Example 5.136(a)).

(b) The significance of the proposition for us will be that it describes the restriction of $\delta^*(-, C')$ to $\mathcal{R}$ entirely in terms of $\mathcal{R}$ and Σ_+.

Exercises

5.184. Prove the following analogue of condition (B2'') in Section 4.1: If Σ and Σ' are twin apartments, then there is a surjective isometry $\phi \colon \Sigma \to \Sigma'$ fixing every chamber in $\Sigma \cap \Sigma'$. [See Exercise 5.163 for the definition of "isometry" in this context.]

5.185. (a) Given a twin apartment $\Sigma = \Sigma\{C, C'\}$ of the twin building $(\mathcal{C}_+, \mathcal{C}_-)$, construct a retraction ρ of $(\mathcal{C}_+, \mathcal{C}_-)$ onto (Σ_+, Σ_-) that preserves codistances from C and C'.

(b) Interpret this retraction in terms of the usual retractions in the spherical case.

(c) Let $\alpha \colon (\mathcal{C}_+, \mathcal{C}_-) \to (\Sigma_-, \Sigma_+)$ be the composite $\mathrm{op}_\Sigma \circ \rho$. Show that $\alpha(E)$ op E for all chambers $E \in \mathcal{C}_+ \cup \mathcal{C}_-$.

5.186. Let C, D, E be chambers with $C, D \in \mathcal{C}_+$ and $E \in \mathcal{C}_-$. If $\delta^*(C, E) = \delta_+(C, D) \delta^*(D, E)$, show that there is a twin apartment containing C, D, E.

5.187. Given opposite chambers $C \in \mathcal{C}_\epsilon$ and $C' \in \mathcal{C}_{-\epsilon}$, show that

$$\Sigma(C, C') = \{ D \in \mathcal{C}_\epsilon \mid d(C, D) = d^*(C', D) \} .$$

5.188. Recall from Lemma 5.143 that for any $C \in \mathcal{C}_\epsilon$ and $D \in \mathcal{C}_{-\epsilon}$,

$$d^*(C, D) = \min \{ d(C, D') \mid D' \text{ op } D \} .$$

Show that the minimum is achieved precisely when D' is the chamber opposite D in a twin apartment containing C and D. This generalizes Exercise 5.138.

5.189. Recall from Exercise 5.167 that

$$\mathcal{M}(\mathcal{R}, D) := \big\{ C \in \mathcal{R} \mid \delta^*(C, D) = \min\big(\delta^*(\mathcal{R}, D)\big) \big\}$$

for any residue $\mathcal{R}$ of $\mathcal{C}_+$ and any chamber $D \in \mathcal{C}_-$. Choose a chamber $C \in \mathcal{R}$ and a twin apartment $\Sigma = (\Sigma_+, \Sigma_-)$ containing (C, D). Set $C' := \mathrm{op}_\Sigma(C)$, and denote by $\mathcal{S}$ the residue in $\mathcal{C}_-$ of the same type as $\mathcal{R}$ that contains C'.

(a) Recall that $(\mathcal{R}, \mathcal{S}, \delta^*|)$ is a twin building, and prove that $\mathcal{M}(\mathcal{R}, D) = (\mathrm{proj}_\mathcal{S} D)^{\mathrm{op}}$ with respect to opposition in this restricted twin building.

(b) If $\mathcal{R}$ is spherical, verify that $\mathrm{proj}_\mathcal{R}(\mathrm{proj}_\mathcal{S} D) = \mathrm{proj}_\mathcal{R} D$. Explain how part (a) translates into the statement of Exercise 5.167 in this case.

5.8.5 Twin Roots

Twin roots can be thought of as "half twin apartments," just as roots are "half apartments." We will continue to use the W-metric approach, so that roots are considered to be certain sets of chambers (see Section 5.5.4). But one basic fact concerning roots that was discussed in Chapter 3 in the simplicial setup will be used repeatedly in the following. Namely, if C and D are adjacent chambers in some apartment Σ, then the unique root α in Σ containing C but not D is given (as a set of chambers) by

$$\alpha = \alpha(C, D) := \{ X \in \Sigma \mid d(C, X) < d(D, X) \} \; ;$$

see Lemma 3.45. Here $d(-, -)$, as usual, denotes gallery distance. We also recall our convention that given a fixed apartment Σ, the root opposite a root α is denoted by $-\alpha$.

In the following, $\mathcal{C} = (\mathcal{C}_+, \mathcal{C}_-)$ will continue to denote a twin building of type (W, S). Fix a twin apartment $\Sigma = (\Sigma_+, \Sigma_-)$. Given a pair of roots $\alpha = (\alpha_+, \alpha_-)$ in Σ, we set $-(\alpha_+, \alpha_-) := (-\alpha_+, -\alpha_-)$. Recall from Definition 5.172 that Σ has an opposition involution op_Σ. For a pair of roots (α_+, α_-) as above, we set $\mathrm{op}_\Sigma(\alpha) := \big(\mathrm{op}_\Sigma(\alpha_-), \mathrm{op}_\Sigma(\alpha_+)\big)$. The following definition is motivated by Lemma 5.113.

Definition 5.190. Let $\alpha = (\alpha_+, \alpha_-)$ be a pair of roots in the twin apartment Σ. We call α a *twin root* of Σ if $\mathrm{op}_\Sigma(\alpha) = -\alpha$.

Note that the equation $\mathrm{op}_\Sigma(\alpha) = -\alpha$ can be written as

$$\mathrm{op}_\Sigma(\alpha_-) = -\alpha_+ \quad \text{and} \quad \mathrm{op}_\Sigma(\alpha_+) = -\alpha_- \; . \tag{5.17}$$

These two equations are in fact equivalent to one another, since op_Σ is an involution. Moreover, we could start with one α_ϵ and use (5.17) to define $\alpha_{-\epsilon}$; the latter will be a root in $\Sigma_{-\epsilon}$, since op_Σ is an isometry. This leads to the following concrete description of a typical twin root α. Start with a pair of adjacent chambers C, D in one Σ_ϵ, say Σ_+, and let α_+ be the root containing C but not D, i.e.,

$$\alpha_+ = \alpha_+(C, D) = \{X \in \Sigma_+ \mid d(C, X) < d(D, X)\} . \tag{5.18}$$

Define α_- by (5.17), and set $C' := \mathrm{op}_\Sigma(C)$ and $D' = \mathrm{op}_\Sigma(D)$. Then α_- contains D' but not C', so

$$\alpha_- = \alpha_-(D', C') = \{Y \in \Sigma_- \mid d(D', Y) < d(C', Y)\} . \tag{5.19}$$

See Figure 5.2.

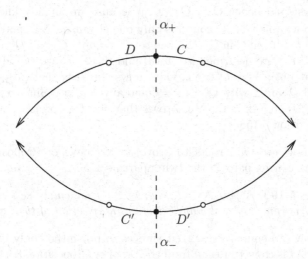

Fig. 5.2. A twin root.

Lemma 5.191. *Let Σ be a twin apartment.*

(1) *Let α be a twin root in Σ, described as in (5.18) and (5.19). Let $\mathcal{P}$ be the panel of C_+ containing C and D, and let $\mathcal{P}'$ be the panel of C_- containing C' and D'. Then*

$$\alpha_+ = \{X \in \Sigma_+ \mid \mathrm{proj}_{\mathcal{P}'} X = D'\}$$

and

$$\alpha_- = \{Y \in \Sigma_- \mid \mathrm{proj}_{\mathcal{P}} Y = C\} .$$

(2) *Every twin root α in Σ is convex in $\mathcal{C}$.*

Remark 5.192. By parts (5) and (6) of Lemma 5.173, Σ is a (thin) twin building in its own right and is convex in $\mathcal{C}$. So an equivalent formulation of (2) is that α is convex in Σ.

Proof of Lemma 5.191. (1) For any chamber $X \in \Sigma_+$, Lemma 5.173(4) implies that $\delta^*(D', X) = s\delta^*(C', X)$, where $s = \delta_-(C', D') = \delta_+(C, D)$. Hence, by Definition 5.150, $D' = \text{proj}_{\mathcal{P}'} X$ if and only if $d^*(D', X) > d^*(C', X)$. By Lemma 5.173(2), this holds if and only if $d(D, X) > d(C, X)$, i.e., if and only if $X \in \alpha_+$. This proves the first equation in (1), and the second is proved similarly.

(2) Let $\mathcal{Q}$ be any panel in $\mathcal{C}_+$ with $\tilde{\mathcal{Q}} := \mathcal{Q} \cap \alpha_+ \neq \emptyset$, and let Y be a chamber in α_-. We want to show that $\text{proj}_{\mathcal{Q}} Y \in \alpha_+$. This is clear if $\tilde{\mathcal{Q}}$ has two chambers (and hence $\tilde{\mathcal{Q}} = \mathcal{Q} \cap \Sigma_+$) because we already know that Σ is convex. So assume that $\tilde{\mathcal{Q}}$ contains only one chamber C. Denote by D the chamber different from C in $\mathcal{Q} \cap \Sigma_+$, and set $C' := \text{op}_\Sigma(C)$ and $D' := \text{op}_\Sigma(D)$. Then α_+ and α_- are described by (5.18) and (5.19), so we can apply (1) to conclude that $\text{proj}_{\mathcal{Q}} Y = C \in \alpha_+$. One shows similarly that $\text{proj}_{\mathcal{Q}'} X \in \alpha_-$ for any panel $\mathcal{Q}'$ in $\mathcal{C}_-$ with $\mathcal{Q}' \cap \alpha_- \neq \emptyset$ and any $X \in \alpha_+$. Since α_+ is convex in $\mathcal{C}_+$ and α_- is convex in $\mathcal{C}_-$, this proves that $\alpha = (\alpha_+, \alpha_-)$ is convex in the sense of Definition 5.158. $\qquad\square$

We can now use twin roots to prove an analogue of Proposition 3.94, characterizing convex pairs in the twin apartment Σ.

Proposition 5.193. *A pair $\mathcal{M} = (\mathcal{M}_+, \mathcal{M}_-)$ of nonempty sets in the twin apartment Σ is convex if and only if it is an intersection of twin roots of Σ.*

Proof. In view of Lemma 5.191(2), it suffices to prove the "only if" part. So assume that $\mathcal{M}$ is convex. Recall from the proof of Proposition 3.94 that $\mathcal{M}_+$ is the intersection of the roots of the form $\alpha_+ = \alpha_+(C, D)$, where C and D are adjacent chambers of Σ_+ with $C \in \mathcal{M}_+$ and $D \notin \mathcal{M}_+$. For each such pair C, D, let α be the corresponding twin root, with α_- defined by (5.17) or, equivalently, by (5.19). We claim that $\mathcal{M}_- \subseteq \alpha_-$, and hence $\mathcal{M} \subseteq \alpha$. Indeed, if $\mathcal{P}$ is the panel containing C and D, then $\text{proj}_{\mathcal{P}} \mathcal{M}_- \subseteq \mathcal{P} \cap \mathcal{M}_+ = \{C\}$ by convexity, so our claim follows from the characterization of α_- in Lemma 5.191(1). In view of the claim, the intersection of the twin roots containing $\mathcal{M}$ has $\mathcal{M}_+$ as its first component. Similarly, that intersection has $\mathcal{M}_-$ as its second component. $\qquad\square$

As an easy but important special case, we get the following characterization of twin roots, which should be compared with Exercise 3.146.

Corollary 5.194. *Let Σ be a twin apartment.*

(1) *If α is a twin root of Σ described as in (5.18) and (5.19), then α is the convex hull of the pair (C, D').*

(2) *The twin roots of Σ are precisely the convex hulls of pairs (C, D') with $C \in \Sigma_+$, $D' \in \Sigma_-$, and $d^*(C, D') = 1$ (i.e., $\delta^*(C, D') \in S$).*

Proof. Both parts of the corollary will follow if we show that for any pair (C, D') as in (2), there is a unique twin root of Σ containing it. Set $C' :=$

$\mathrm{op}_{\Sigma}(C)$ and $D := \mathrm{op}_{\Sigma} D'$. Then a twin root α containing (C, D') cannot contain D (since then it would contain the convex hull of (D, D'), which is the whole twin apartment Σ). So necessarily α is the twin root given by (5.18) and (5.19). □

Definition 5.195. A *twin root* of the twin building $\mathcal{C}$ is a pair $\alpha = (\alpha_{+}, \alpha_{-})$ such that α is a twin root in some twin apartment Σ.

(It then follows easily that α is a twin root in *every* twin apartment that contains it; see Exercise 5.199 below.)

Since any pair of chambers is contained in a twin apartment, Corollary 5.194 has the following immediate consequence:

Corollary 5.196. *The twin roots of $\mathcal{C}$ are precisely the convex hulls of pairs (C, D') with $C \in \mathcal{C}_{+}$, $D' \in \mathcal{C}_{-}$, and $d^{*}(C, D') = 1$.* □

This corollary is useful, for instance, when one wants to describe all twin apartments containing a given twin root α.

Definition 5.197. If α is a twin root, we denote by $\mathcal{A}(\alpha)$ the set of all twin apartments containing α.

The following lemma, which generalizes Lemma 4.118, will be applied in Chapter 8.

Lemma 5.198. *Let $\alpha = (\alpha_{+}, \alpha_{-})$ be a twin root, and for $\epsilon \in \{+, -\}$, let $\mathcal{P}$ be a panel in $\mathcal{C}_{\epsilon}$ that contains precisely one chamber $C \in \alpha_{\epsilon}$. Then there is a bijection $\mathcal{P} \smallsetminus \{C\} \to \mathcal{A}(\alpha)$ that assigns to each $D \in \mathcal{P} \smallsetminus \{C\}$ the convex hull of $\{D\} \cup \alpha$.*

Proof. We may assume that $\epsilon = +$. Observe first that there is a panel $\mathcal{P}'$ of $\mathcal{C}_{-}$ that is opposite $\mathcal{P}$ and contains precisely one chamber $D' \in \alpha_{-}$. (Work in a fixed twin apartment, and use the description of twin roots given in (5.18) and (5.19).) We have $C = \mathrm{proj}_{\mathcal{P}} D'$ by Lemma 5.191, and α is the convex hull of C and D' by Corollary 5.194. Moreover, $\delta^{*}(C, D') = s$, where s is the type of $\mathcal{P}$ and $\mathcal{P}'$, as one again sees by working in a twin apartment.

Note next that by Corollary 5.153, C is the unique chamber in $\mathcal{P}$ that is *not* opposite D'. So every chamber $D \in \mathcal{P} \smallsetminus \{C\}$ is opposite D' and hence gives rise to a twin apartment $\Sigma\{D, D'\}$, which is the convex hull of D and D'. This convex hull contains C and hence α, so it coincides with the convex hull of $\{D\} \cup \alpha$. There is therefore a map $f: \mathcal{P} \smallsetminus \{C\} \to \mathcal{A}(\alpha)$ such that $f(D) = \Sigma\{D, D'\}$ for $D \in \mathcal{P} \smallsetminus \{C\}$, and also $f(D)$ is the convex hull of $\{D\} \cup \alpha$.

To see that f is bijective, we define a map $g: \mathcal{A}(\alpha) \to \mathcal{P} \smallsetminus \{C\}$ by $g(\Sigma) := \mathrm{op}_{\Sigma}(D')$ for $\Sigma \in \mathcal{A}(\alpha)$. If $D = g(\Sigma)$, then $\delta_{+}(C, D) = \delta^{*}(C, D') = s$ by Lemma 5.173(2), so D is indeed in $\mathcal{P} \smallsetminus \{C\}$. Since $\Sigma = \Sigma\{D, D'\}$, we have $f(g(\Sigma)) = \Sigma$. And if we start with any $D \in \mathcal{P} \smallsetminus \{C\}$ and set $\Sigma := \Sigma\{D, D'\}$, then $D = \mathrm{op}_{\Sigma}(D')$, so $g(f(D)) = D$. Thus f and g are mutually inverse bijections. □

Exercises

5.199. Let α be a twin root of $\mathcal{C}$, and let Σ be any twin apartment containing α. Show that α is a twin root of Σ.

5.200. Let α and β be twin roots with $\alpha \subseteq \beta$. Show that $\alpha = \beta$.

5.201. Let α be a twin root of a twin apartment Σ. Show that α is a maximal (proper) convex subpair of Σ.

5.202. Let α be a twin root, and let Σ and Σ' be two distinct twin apartments containing α. Show that $\Sigma \cap \Sigma' = \alpha$.

5.203. Let $\Sigma = \Sigma\{C, C'\}$ be a twin apartment, where C op C' with $C \in \mathcal{C}_+$ and $C' \in \mathcal{C}_-$. Let D be a chamber of $\mathcal{C}_+$ that is adjacent to C but not in Σ_+.

(a) Show that D op C', so that there is a twin apartment $\Sigma' = \Sigma\{D, C'\}$.
(b) Show that $\Sigma \cap \Sigma'$ is a twin root.

5.9 A Rigidity Theorem

In this section and the next we discuss two fundamental theorems of Tits [247]. Taken together, they say, roughly speaking, that a thick spherical building is completely determined by a very small portion of it. The precise statements involve a family of "neighborhoods" of a given chamber C.

Definition 5.204. Let $(\mathcal{C}, \delta)$ be a building of type (W, S). Given a chamber C and a natural number k, we define

$$E_k(C) := \bigcup_{\substack{J \subseteq S \\ |J| \leq k}} R_J(C) .$$

Thus D is in $E_k(C)$ if and only if there is a residue of rank $\leq k$ containing both C and D. From the simplicial point of view, residues of rank r correspond to simplices of codimension r. So if $\mathcal{C} = \mathcal{C}(\Delta)$ for a simplicial building Δ, then there is a residue of rank r containing two given chambers if and only if they have a common face of codimension r. Hence

$$E_k(C) = \{D \in \mathcal{C} \mid \mathrm{codim}(C \cap D) \leq k\} .$$

Note that the sets $E_k(C)$ increase with k:

$$E_0(C) \subseteq E_1(C) \subseteq E_2(C) \subseteq \cdots . \tag{5.20}$$

Note also that $E_0(C) = \{C\}$ and that $E_1(C)$ consists of C together with all chambers adjacent to C. Figure 5.3 shows the intersection of $E_2(C)$ with an apartment in a rank-3 example; it consists of all chambers that have at least

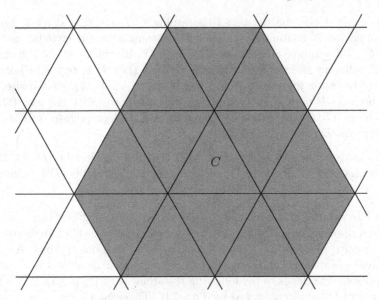

Fig. 5.3. $E_2(C)$ in an apartment.

one vertex in common with C. In general, if $|S| = n < \infty$, then the sequence in (5.20) stabilizes with $E_n(C) = C$.

By an *automorphism* of $(\mathcal{C}, \delta)$ we mean a surjective almost isometry of $\mathcal{C}$ onto $\mathcal{C}$. If $\mathcal{C} = \mathcal{C}(\Delta)$, this is the same thing as a simplicial automorphism of Δ.

Theorem 5.205. *Let $(\mathcal{C}, \delta)$ be a thick spherical building, and let C, C' be opposite chambers in $\mathcal{C}$. If an automorphism ϕ of $\mathcal{C}$ fixes $E_1(C) \cup \{C'\}$ pointwise, then ϕ is the identity.*

Proof. Note first that ϕ is actually an isometry, since it fixes every chamber adjacent to C. (From the simplicial point of view, ϕ is type-preserving.) The proof now consists of two steps. The first step is to remove the apparent asymmetry in the hypothesis.

(a) ϕ fixes every chamber in $E_1(C')$.

Let D' be a chamber adjacent to C', and let $\mathcal{P}'$ be the panel containing C' and D'. Let $\mathcal{P}$ be the panel containing C and having type opposite to that of $\mathcal{P}'$. By Lemma 5.107 the panels $\mathcal{P}$ and $\mathcal{P}'$ are opposite. We therefore have a bijection between $\mathcal{P}$ and $\mathcal{P}'$ given by nonopposition (Corollary 5.117). Since $\phi(C') = C'$ and ϕ is a surjective isometry, we have $\phi(\mathcal{P}') = \mathcal{P}'$. Now ϕ is compatible with the nonopposition bijection and fixes $\mathcal{P}$ pointwise, so it follows at once that ϕ fixes $\mathcal{P}'$ pointwise. In particular, $\phi(D') = D'$.

(b) For every chamber D' adjacent to C', there is a chamber $D \in E_1(C)$ such that D op D' and ϕ fixes $E_1(D) \cup \{D'\}$ pointwise.

As in (a), let $\mathcal{P}'$ be the panel containing C' and D', and let $\mathcal{P}$ be the panel opposite $\mathcal{P}'$ containing C. Then $\mathcal{P}$ contains a unique chamber not opposite C' and a unique chamber not opposite D'. By thickness, we can choose $D \in \mathcal{P}$ different from these two chambers, so that D is opposite both C' and D'. By (a), ϕ fixes $E_1(C')$ and hence also $E_1(C') \cup \{D\}$ pointwise, the latter because D is in $E_1(C)$. Applying (a) once more, with the pair (C, C') replaced by (C', D), we conclude that ϕ fixes $E_1(D)$ pointwise. And we also have $\phi(D') = D'$, since $D' \in E_1(C')$.

An easy induction using (b) now shows that for any chamber $D' \in \mathcal{C}$ there exists a chamber D opposite D' such that ϕ fixes $E_1(D) \cup \{D'\}$ pointwise. Hence $\phi = \mathrm{id}$. $\qquad\square$

By Theorem 5.122, there exists precisely one apartment of $\mathcal{C}$ that contains two given opposite chambers C and C', and each chamber C of an apartment $\mathcal{A}$ is opposite precisely one chamber of $\mathcal{A}$. Now an isometry of $\mathcal{C}$ fixes an apartment pointwise if and only it stabilizes this apartment and fixes one of its chambers. This leads to the following restatement of Theorem 5.205, which is the rigidity theorem as stated by Tits [247, Theorem 4.1.1]:

Corollary 5.206. *Let $(\mathcal{C}, \delta)$ be a thick spherical building, let $\mathcal{A}$ be an apartment of $\mathcal{C}$, and let C be a chamber of $\mathcal{A}$. If an automorphism ϕ of $(\mathcal{C}, \delta)$ fixes $E_1(C) \cup \mathcal{A}$ pointwise, then $\phi = \mathrm{id}$.* $\qquad\square$

We will see many applications of the rigidity theorem in Chapter 7. The following corollary of the proof of the rigidity theorem will also be useful in that chapter:

Corollary 5.207. *Let $(\mathcal{C}, \delta)$ be a thick spherical building, and let C, C' be opposite chambers in $\mathcal{C}$. Then $\mathcal{C}$ is the convex hull of $E_1(C) \cup \{C'\}$.*

Proof. Let $\mathcal{D}$ be the convex hull of $E_1(C) \cup \{C'\}$.

(a) $\mathcal{D}$ contains $E_1(C')$.

This follows from step (a) of the proof of Theorem 5.205. [Recall that the nonopposition bijection from $\mathcal{P}$ to $\mathcal{P}'$ in that proof is given by projection onto $\mathcal{P}'$.]

(b) For every chamber D' adjacent to C' (and hence in $\mathcal{D}$ by (a)), there is a chamber $D \in E_1(C)$ such that D op D' and $\mathcal{D}$ contains $E_1(D)$.

Choose D as in step (b) of the proof of Theorem 5.205. By (a), $\mathcal{D}$ contains $E_1(C')$ and hence $E_1(C') \cup \{D\}$, so another application of (a) shows that $\mathcal{D}$ contains $E_1(D)$.

An easy induction using (b) now gives us $\mathcal{D} = \mathcal{C}$. $\qquad\square$

Remark 5.208. Readers familiar with twin buildings (Section 5.8) will note that the rigidity theorem and its proof remain valid, essentially verbatim, for automorphisms of thick twin buildings. The key result that is needed is the nonopposition bijection between opposite panels, which for twin buildings is given by Corollary 5.153. Similarly, Corollary 5.207 remains valid for thick twin buildings.

5.10 An Extension Theorem

The rigidity theorem implies that an isomorphism between thick spherical buildings is uniquely determined by what it does on a small part of the domain. There is also an existence theorem, which says that an isomorphism can be arbitrarily prescribed in the neighborhood of a given chamber, provided the buildings are irreducible and of rank at least 3. Here, as in the previous section, an *isomorphism* is a surjective almost isometry (or, equivalently, a simplicial isomorphism).

Theorem 5.209. *Let $(\mathcal{C}, \delta)$ and $(\mathcal{C}', \delta')$ be thick, irreducible, spherical buildings of rank at least 3. Let $\phi\colon E_2(C) \to E_2(C')$ be an adjacency-preserving bijection for some $C \in \mathcal{C}$ and $C' \in \mathcal{C}'$. Then ϕ extends to an isomorphism $\mathcal{C} \to \mathcal{C}'$.*

This is Theorem 4.1.2 of Tits [247]. The original proof was long and technical, but there is a simplified proof (with a slightly different hypothesis on ϕ) in Weiss [281, Chapter 10], based on ideas of Ronan. This proof makes systematic use of the W-metric point of view, which was not available when Tits wrote [247]. We will not give a proof here, but we remark only that it consists of the following two steps. First, one extends ϕ to a map $E_2(C) \cup \mathcal{A} \to E_2(C') \cup \mathcal{A}'$ that maps $\mathcal{A}$ isomorphically onto $\mathcal{A}'$, where $\mathcal{A}$ is an apartment of $\mathcal{C}$ containing C and $\mathcal{A}'$ is an apartment of $\mathcal{C}'$ containing C'. Second, one shows that any such map extends to an isomorphism $\mathcal{C} \to \mathcal{C}'$. This second step is of interest in its own right, and we state it explicitly for future reference.

Theorem 5.210. *Let $(\mathcal{C}, \delta)$ and $(\mathcal{C}', \delta')$ be thick, irreducible, spherical buildings of rank at least 3, and let $\mathcal{A}$ (resp. $\mathcal{A}'$) be an apartment of $\mathcal{C}$ (resp. $\mathcal{C}'$). Let $\phi\colon E_2(C) \to E_2(C')$ be an adjacency-preserving bijection for some $C \in \mathcal{A}$ and $C' \in \mathcal{A}'$, and let $\psi\colon \mathcal{A} \to \mathcal{A}'$ be an isomorphism that coincides with ϕ on $\mathcal{A} \cap E_2(C)$. Then there is an isomorphism $\mathcal{C} \xrightarrow{\sim} \mathcal{C}'$ extending ϕ and ψ.*

For the proof, see Tits [247, Proposition 4.16] or Weiss [281, Theorem 10.1]. The following special case will be needed in Chapter 7. We state it in simplicial language, since that is the context in which it will arise.

Corollary 5.211. *Let Δ be a thick, irreducible, spherical building of rank at least 3. Let α be a root of Δ, and let Σ and Σ' be apartments containing α. Then there is an automorphism of Δ that fixes α pointwise and maps*

Σ onto Σ'. Moreover, the automorphism can be chosen to fix $E_2(C)$ pointwise, where C is a chamber in α that is disjoint from $\partial\alpha$.

Proof. By axiom (B2″) for buildings, there is an isomorphism $\psi\colon \Sigma \to \Sigma'$ that fixes α pointwise. Now apply Proposition 3.125 to get a chamber $C \in \alpha$ that is disjoint from $\partial\alpha$, and let $\phi\colon E_2(C) \to E_2(C)$ be the identity map. It remains to verify that ϕ and ψ agree on $E_2(C) \cap \Sigma$. To this end we need only note that (since rank $\Delta \geq 3$) every chamber in $E_2(C)$ has a vertex in common with C. This implies that $E_2(C) \cap \Sigma = E_2(C) \cap \alpha$ and hence that ψ is the identity on $E_2(C) \cap \Sigma$.   □

*5.11 An Extension Theorem for Twin Buildings

We mentioned in Remark 5.208 that the rigidity theorem for spherical buildings generalizes to twin buildings. The situation for the extension theorem is more complicated, but there does exist an extension theorem for a class of twin buildings, due to work of Mühlherr–Ronan [177] that is based on earlier results of Tits [261]. The statement involves two supplementary hypotheses. The first is 2-sphericity, already mentioned briefly in Remark 5.151(b). [An equivalent formulation is that every entry $m(s,t)$ of the Coxeter matrix is finite.] The second is the following connectivity condition:

(co) *For every chamber $C \in \mathcal{C}_\epsilon$ ($\epsilon = \pm$), the set C^{op} of chambers opposite C is a gallery-connected subset of $\mathcal{C}_{-\epsilon}$.*

Remark 5.212. "Almost all" thick, irreducible, 2-spherical twin buildings $\mathcal{C}$ of rank at least 3 satisfy (co). More precisely:

- (a) $\mathcal{C}$ satisfies (co) if all of its rank-2 residues, viewed as spherical buildings, satisfy (co); see Mühlherr and Ronan [177, Theorem 1.5].
- (b) The rank-2 residues of $\mathcal{C}$ necessarily have the Moufang property (which will be defined in Chapter 7); see Tits [261, Section 5.6] and Ronan [203, Section 8].
- (c) There are only four Moufang spherical buildings of rank 2 that do *not* satisfy (co), namely, the buildings associated to the finite Chevalley groups $\mathrm{Sp}_4(\mathbb{F}_2)$, $\mathrm{G}_2(\mathbb{F}_2)$, $\mathrm{G}_2(\mathbb{F}_3)$, and $^2\mathrm{F}_4(\mathbb{F}_2)$. (The latter is a "twisted" Chevalley group; see Section 9.6.) This is due to Abramenko. A proof is sketched in [9, Section II.2], and a detailed proof can be found in [14].

Combining (a)–(c), we see that a thick, irreducible, 2-spherical twin building of rank at least 3 satisfies (co) unless it has a rank-2 residue isomorphic to one of the four buildings listed in (c).

We now quote the extension theorem from [177].

Theorem 5.213. *Let $\mathcal{C} = (\mathcal{C}_+, \mathcal{C}_-)$ and $\mathcal{C}' = (\mathcal{C}'_+, \mathcal{C}'_-)$ be thick 2-spherical twin buildings of the same type. Assume that $\mathcal{C}$ and $\mathcal{C}'$ satisfy condition* (co). *Given two pairs of opposite chambers $(C_+, C_-) \in \mathcal{C}_+ \times \mathcal{C}_-$ and $(C'_+, C'_-) \in \mathcal{C}'_+ \times \mathcal{C}'_-$ and a surjective isometry $\phi \colon E_2(C_+) \cup \{C_-\} \to E_2(C'_+) \cup \{C'_-\}$, there is a unique extension of ϕ to an isomorphism $\mathcal{C} \xrightarrow{\sim} \mathcal{C}'$ of twin buildings.*

(The hypothesis on ϕ means that it preserves Weyl distances *and* codistances.)

Remark 5.214. The extension theorem is not valid for arbitrary thick twin buildings; there are counterexamples involving twin trees [18, 102, 206]. This explains the 2-sphericity assumption. Moreover, it is not expected that 2-sphericity alone is sufficient. Work in progress by Abramenko and Mühlherr suggests that the theorem can fail in some cases in which $\mathcal{C}$ is 2-spherical but does not satisfy (co).

As in the spherical case, we will be interested in the following consequence of the extension theorem:

Corollary 5.215. *Let $\mathcal{C}$ be a thick, irreducible, 2-spherical twin building of rank at least 3 satisfying* (co). *Let α be a twin root of $\mathcal{C}$, and let Σ and Σ' be twin apartments containing α. Then there is an automorphism of $\mathcal{C}$ that fixes α pointwise and maps Σ onto Σ'. This automorphism can be chosen to fix $E_2(C_+)$ pointwise, where C_+ is a chamber in α_+ that is disjoint from $\partial \alpha_+$.*

("Disjoint" in the last sentence should be interpreted from the simplicial point of view.)

Proof. As in the proof of Corollary 5.211, we can find a chamber $C_+ \in \alpha_+$ that is disjoint from $\partial \alpha_+$, and we then have $E_2(C_+) \cap \Sigma_+ = E_2(C_+) \cap \alpha_+ = E_2(C_+) \cap \Sigma'_+$. Let C_- and C'_- be the chambers opposite C_+ in Σ and Σ', respectively, and define $\phi \colon E_2(C_+) \cup \{C_-\} \to E_2(C_+) \cup \{C'_-\}$ by $\phi(C_-) = C'_-$ and $\phi|_{E_2(C_+)} = \mathrm{id}$. Then ϕ is an isometry by Proposition 5.182, so we can apply Theorem 5.213 to extend ϕ to an automorphism ψ of $\mathcal{C}$. By construction, ψ maps Σ onto Σ'. We must show that it fixes α pointwise. Since ψ fixes C_+, it also fixes (by the standard uniqueness argument) the intersection $\Sigma_+ \cap \Sigma'_+$, and hence α_+, pointwise. Now let D_- be any chamber in α_-, and note that $\delta^*(C_+, D_-) = \delta^*(C_+, \psi(D_-))$. This implies that $\psi(D_-) = D_-$, since Σ'_- contains only one chamber at any given codistance from C_+ (see Lemma 5.173(3)). $\qquad\square$

*5.12 Covering Maps

This optional section gives an introduction to the notion of "covering map." This is important for some aspects of the theory of buildings, but it will play a minimal role in the present book. We will refer to it only in Section 8.7.

It will be convenient to view buildings of type (W, S) as chamber systems over S as in Section 5.2. Here (W, S) is an arbitrary Coxeter system. We will also need to deal with chamber systems $(\mathcal{C}, (\sim_s)_{s \in S})$ over S that are not known, a priori, to be buildings. But we will always think of S as a set of generators of W, and this will play a role in some of our definitions. We begin by extending the concept of "residue" to general chamber systems.

Definition 5.216. Given $J \subseteq S$, we say that two chambers $C, D \in \mathcal{C}$ are *J-equivalent* if there is a gallery of type $(s_1, \dots, s_n)$ connecting C and D with $s_i \in J$ for all $1 \leq i \leq n$. The equivalence classes are called *J-residues*, or residues of *type* J, and the J-residue containing a given chamber C is denoted by $R_J(C)$. A subset $\mathcal{R} \subseteq \mathcal{C}$ is called a *residue* if it is a J-residue for some $J \subseteq S$.

In the present generality of arbitrary chamber systems, it is *not* true that a residue has a well-defined type J, or even a well-defined rank $|J|$. Nevertheless, we will allow ourselves to say "$\mathcal{R}$ is a residue of rank m" as shorthand for "$\mathcal{R} = R_J(C)$ for some $C \in \mathcal{C}$ and some $J \subseteq S$ such that $|J| = m$." We will also say that a residue $\mathcal{R}$ is *spherical* if it is a J-residue for some spherical subset $J \subseteq S$, where J is said to be spherical if W_J is finite. [Note that this notion makes sense only because we have a fixed Coxeter system in mind.] For example, panels are spherical residues of rank 1. The chamber system $\mathcal{C}$ is called *connected* if $\mathcal{C} = R_S(C)$ for some (and hence any) $C \in \mathcal{C}$. Finally, note that a residue $R_J(C)$ can be viewed as a chamber system in its own right, over the set J.

Definition 5.217. A *morphism* between two chamber systems $\mathcal{C}'$ and $\mathcal{C}$ over S is a map $\kappa \colon \mathcal{C}' \to \mathcal{C}$ such that

$$C' \sim_s D' \implies \kappa(C') \sim_s \kappa(D')$$

for any $C', D' \in \mathcal{C}'$ and $s \in S$. A morphism κ is called an *isomorphism* if it is bijective and the inverse κ^{-1} is also a morphism; in other words,

$$\kappa(C') \sim_s \kappa(D') \implies C' \sim_s D' \tag{5.21}$$

for any $C', D' \in \mathcal{C}'$ and $s \in S$.

The implication (5.21) says that the bijection κ maps every s-panel in $\mathcal{C}'$ *onto* an s-panel in $\mathcal{C}$. This leads naturally to our next definition.

Definition 5.218. Given two chamber systems $\mathcal{C}'$ and $\mathcal{C}$ over S and a natural number m, we call a morphism $\kappa \colon \mathcal{C}' \to \mathcal{C}$ an *m-covering* if for every spherical subset $J \subseteq S$ of cardinality $|J| \leq m$, every J-residue of $\mathcal{C}'$ is mapped bijectively onto a J-residue of $\mathcal{C}$.

More briefly, the definition says that κ maps every spherical residue of rank at most m bijectively onto a (spherical) residue of the same rank. But we have spelled this out carefully in order to avoid ambiguity.

Remarks 5.219. (a) Note that any m-covering maps s-panels of $\mathcal{C}'$ bijectively onto s-panels of $\mathcal{C}$, since the natural number m, by convention, is at least 1. It follows that the bijections between J-residues in the definition are actually isomorphisms of chamber systems over J.

(b) Some of the literature uses a more restrictive notion of m-covering, in which J is allowed to be an *arbitrary* subset of S with $|J| \leq m$. We have chosen the definition that will be useful for us in what follows. Note that it makes use of our standing assumption that S is the set of distinguished generators of a Coxeter group W.

The first observation about covering maps is that they have a "path-lifting" property that should look familiar to anyone who has studied covering maps in topology:

Lemma 5.220. *Let* $\kappa \colon \mathcal{C}' \to \mathcal{C}$ *be a morphism of chamber systems over S such that for all $s \in S$, κ maps every s-panel of $\mathcal{C}'$ onto an s-panel of $\mathcal{C}$. Let C' be a chamber in $\mathcal{C}'$, let $\mathbf{s}$ be an S-word, and let Γ be a gallery in $\mathcal{C}$ of type $\mathbf{s}$ starting at $\kappa(C')$. Then there is a gallery Γ' in $\mathcal{C}'$ of type $\mathbf{s}$ starting at C' such that $\kappa(\Gamma') = \Gamma$. Consequently, κ maps every J-residue of $\mathcal{C}'$ onto a J-residue of $\mathcal{C}$ for all $J \subseteq S$.*

Proof. This is immediate from the definitions. $\square$

Lemma 5.221. *Let* $\kappa \colon \mathcal{C}' \to \mathcal{C}$ *be a 1-covering, where $\mathcal{C}'$ is a chamber system over S and $\mathcal{C}$ is a building of type (W, S). Assume that any two chambers $C', D' \in \mathcal{C}'$ can be connected by a gallery of reduced type. Then $\mathcal{C}'$ is a building of type (W, S) and κ is an isomorphism.*

Proof. (a) κ is surjective.

Choose an arbitrary $C' \in \mathcal{C}'$. Then κ maps $R_S(C')$ onto $R_S(\kappa(C'))$ by the last assertion of Lemma 5.220. Since $\mathcal{C}$, being a building, is connected, $R_S(\kappa(C')) = \mathcal{C}$.

(b) κ is an isomorphism.

Let C' and D' be distinct chambers in $\mathcal{C}'$. We have to show that $\kappa(C') \neq \kappa(D')$ and that
$$\kappa(C') \sim_s \kappa(D') \implies C' \sim_s D'$$
for any $C', D' \in \mathcal{C}'$ and $s \in S$. By assumption, there exists a gallery Γ' of reduced type $\mathbf{s} = (s_1, \ldots, s_n)$ in $\mathcal{C}'$ connecting C' and D'. Since κ sends s-panels in $\mathcal{C}'$ bijectively onto s-panels in $\mathcal{C}$, $\kappa(\Gamma')$ is a gallery in $\mathcal{C}$ of the same reduced type $\mathbf{s}$; hence $\delta(\kappa(C'), \kappa(D')) = s_1 \cdots s_n$ by condition (G) in Section 5.2. In particular, $\delta(\kappa(C'), \kappa(D')) \neq 1$, so $\kappa(C') \neq \kappa(D')$. If we now assume that $\kappa(C') \sim_s \kappa(D')$ for some $s \in S$, then $\delta(\kappa(C'), \kappa(D')) = s$ and $\mathbf{s}$ is a reduced decomposition of s. Hence $n = 1$ and $\mathbf{s} = (s)$, so $C' \sim_s D'$.

(c) $\big(\mathcal{C}', (\sim_s)_{s \in S}\big)$ is a building of type (W, S).

Since we are working in the context of buildings as chamber systems, what we mean here is that there exists a function $\delta' \colon \mathcal{C}' \times \mathcal{C}' \to W$ such that the conditions of Proposition 5.23 are satisfied. This follows at once from (b) if we set $\delta'(C', D') := \delta(\kappa(C'), \kappa(D'))$ for $C', D' \in \mathcal{C}'$. $\square$

Remark 5.222. Suppose the building $\mathcal{C}$ in Lemma 5.221 is spherical and of rank 2. Viewing it as a graph with colored edges, we can form its universal cover $\kappa \colon \mathcal{C}' \to \mathcal{C}$, which is a 1-covering (see Exercise 5.224 below) but not an isomorphism. This shows that the assumption we made in Lemma 5.221 is not vacuous. However, it is a characteristic feature of buildings that we *only* have to check spherical rank-2 residues in order to make sure that a covering is an isomorphism:

Proposition 5.223. *Let $\kappa \colon \mathcal{C}' \to \mathcal{C}$ be a 2-covering, where $\mathcal{C}'$ is a connected chamber system over S and $\mathcal{C}$ is a building of type (W, S). Then $\mathcal{C}'$ is a building of type (W, S), and κ is an isomorphism.*

Proof. We will verify the assumption of Lemma 5.221. Since $\mathcal{C}'$ is connected, it suffices to show that any minimal gallery in $\mathcal{C}'$ has reduced type. For this we first observe the following:

(∗) *Let Γ' be a gallery of type $\mathbf{s}$ in $\mathcal{C}'$ connecting C' and D', and let $\mathbf{t}$ be an S-word homotopic to $\mathbf{s}$. Then there exists a gallery of type $\mathbf{t}$ in $\mathcal{C}'$ from C' to D'.*

It suffices to prove this when there is an *elementary* homotopy between $\mathbf{s}$ and $\mathbf{t}$. But then we only have to change a part of the gallery Γ' in some spherical rank-2 residue of $\mathcal{C}'$, and this is possible because $\mathcal{C}$ is a building and κ is a 2-covering.

Now if the type $\mathbf{s}$ of Γ' is not reduced, then by Tits's solution of the word problem for Coxeter groups (Section 2.3.3), there is an S-word $\mathbf{t}$ homotopic to $\mathbf{s}$ that contains a subword of the form (s, s) for some $s \in S$. By (∗), there exists a gallery Γ'' of type $\mathbf{t}$ with the same extremities as Γ'. But then Γ'' contains three consecutive chambers that are s-equivalent to one another, so Γ'' is not minimal and hence Γ' is not minimal. $\square$

Exercise 5.224. In this exercise, continuing Exercise 5.25, we interpret some of the concepts of the present section in terms of graphs with colored edges.

(a) Show that a morphism κ of chamber systems is the same thing as a color-preserving graph homomorphism. [Here graph homomorphisms are allowed to be degenerate, i.e., an edge might be collapsed to a vertex.]
(b) Show that κ is an isomorphism in the sense of Definition 5.217 if and only if it is an isomorphism of graphs.
(c) Show that κ is a 1-covering if and only if it is a covering map of graphs in the usual topological sense.

Buildings and Groups

In this chapter we will develop the group theory that goes along with the theory of buildings, in much the same way that the theory of Coxeter groups goes along with the theory of Coxeter complexes. In particular, we will discover a class of groups G for which we can construct an associated building Δ on which G acts as a group of automorphisms.

We begin by assuming that we have a group G that acts in a nice way on a building Δ. This imposes some conditions on G, and we will then take those conditions as axioms for the class of groups we are seeking. Recall from Section 2.5 our convention that *the generating set S of a Coxeter group is always assumed finite.* It will be obvious to the reader that this assumption is largely unnecessary in the present chapter, especially if one adopts the W-metric point of view (Chapter 5), but it simplifies some terminology.

6.1 Group Actions on Buildings

Assume throughout this section that Δ is a simplicial building of type (W, S) with a type-preserving action of a group G. We will certainly want to assume that the action is *chamber transitive*, i.e., that it is transitive on the set of chambers $\mathcal{C} := \mathcal{C}(\Delta)$. But in contrast to the situation with Coxeter groups and Coxeter complexes, chamber transitivity is not strong enough to lead to a useful theory. In this section we will discuss two ways of strengthening this condition. The first, called *strong transitivity*, will be introduced in Section 6.1.1. It arises naturally when one thinks of buildings in terms of apartment systems, as in Chapter 4. After making some easy observations, we will introduce the second strengthening of chamber transitivity, called *Weyl transitivity*, in Section 6.1.3. It is the more natural condition if one approaches buildings from the W-metric point of view of Chapter 5.

6.1.1 Strong Transitivity

Suppose $\mathcal{A}$ is a system of apartments that is G-invariant. Thus if Σ is an apartment in $\mathcal{A}$, then so is its image $g\Sigma$. This is automatic, of course, if $\mathcal{A}$ is the complete system of apartments, but we want to allow $\mathcal{A}$ to be arbitrary.

Definition 6.1. We say that the G-action is *strongly transitive* (with respect to $\mathcal{A}$) if G acts transitively on the set of pairs (Σ, C) consisting of an apartment $\Sigma \in \mathcal{A}$ and a chamber $C \in \Sigma$.

This is equivalent to saying that the action is chamber transitive and that the stabilizer of a given chamber C is transitive on the set of apartments in $\mathcal{A}$ containing C. Alternatively, it is equivalent to saying that G is transitive on $\mathcal{A}$ and that the stabilizer of a given apartment $\Sigma \in \mathcal{A}$ is transitive on $\mathcal{C}(\Sigma)$.

Assume that the G-action is strongly transitive, and choose an arbitrary pair (Σ, C) as in the definition. We will often refer to C as the *fundamental chamber* and to Σ as the *fundamental apartment*. Whenever it is convenient, we will identify Σ with $\Sigma(W, S)$ via the unique type-preserving isomorphism $\Sigma \xrightarrow{\sim} \Sigma(W, S)$ taking C to the fundamental chamber of $\Sigma(W, S)$. In particular, we can view W as the group of type-preserving automorphisms of Σ and S as the set of reflections with respect to the panels of C.

We now introduce three subgroups of G:

$$B := \{g \in G \mid gC = C\} \,,$$
$$N := \{g \in G \mid g\Sigma = \Sigma\} \,,$$
$$T := \{g \in G \mid g \text{ fixes } \Sigma \text{ pointwise}\} \,.$$

Note that T is a normal subgroup of N, being the kernel of the homomorphism $f \colon N \to W$ induced by the action of N on Σ. Note also that f is surjective, so that $W \cong N/T$. For if we are given $w \in W$, then we can find $n \in N$ such that $nC = wC$; since n and w are both type-preserving, they agree pointwise on C; hence $f(n) = w$ by the standard uniqueness argument. In what follows we will identify W with N/T. Note, finally, that $T = B \cap N$; for if $n \in B \cap N$, then n fixes C pointwise and hence acts trivially on Σ. Figure 6.1 summarizes the notation.

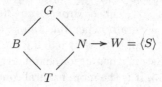

Fig. 6.1. BN data.

Remark 6.2. Everything we are going to do in what follows would go through with no essential change if we replaced N by a subgroup $N' \leq N$ such that N' acts transitively on $\mathcal{C}(\Sigma)$ or, equivalently, such that N' surjects onto W. We would then have to replace T by $T' := N' \cap T$.

We will see that strong transitivity has many consequences for the structure of G, especially if we assume further that Δ is thick. For the moment, we confine ourselves to one simple observation:

Proposition 6.3. *Suppose a group G acts strongly transitively on a building Δ with respect to a system of apartments $\mathcal{A}$. Then, with the notation above,*

$$G = BNB .$$

In particular, G is generated by B and N.

This is a special case of a more precise result, involving a set of apartments $\mathcal{A}$ that is not necessarily a *system* of apartments. Suppose $\mathcal{A}$ is a nonempty G-invariant subset of the complete apartment system. We say that the action of G is *strongly transitive* on $\mathcal{A}$ if G acts transitively on the set of pairs (Σ, C) with $\Sigma \in \mathcal{A}$ and C a chamber in Σ. If $\mathcal{A}$ is in fact a system of apartments, this reduces to strong transitivity as defined in Definition 6.1. In the more general setting, we can still choose a fundamental pair (Σ, C) and define subgroups B, N, T as above.

Lemma 6.4. *Suppose a group G acts strongly transitively on a set $\mathcal{A}$ of apartments in a building Δ. Then, with the notation above, the following conditions are equivalent:*

(i) *The subcomplex $\Delta' := \bigcup_{\Sigma' \in \mathcal{A}}$ is a subbuilding of Δ with $\mathcal{A}$ as a system of apartments.*

(ii) *$G = BNB$.*

(Note that the content of Proposition 6.3 is the implication (i) $\implies$ (ii) in the special case $\Delta' = \Delta$.)

Proof. (i) holds if and only if any two chambers of Δ' are contained in an apartment in $\mathcal{A}$. We may assume that one of the two chambers is the fundamental chamber C, and the other chamber is then gC for some $g \in G$. So a restatement of (i) is that for any $g \in G$, there is an apartment in $\mathcal{A}$ containing C and gC. By strong transitivity, B acts transitively on the apartments in $\mathcal{A}$ containing C. Thus these apartments are the transforms $b\Sigma$ ($b \in B$), and we obtain the following formulation of (i): For any $g \in G$, there is an element $b \in B$ such that $gC \in b\Sigma$. Now gC is in $b\Sigma$ if and only if $gC = bnC$ for some $n \in N$. So a final reformulation of (i) is that for any $g \in G$ there are elements $b \in B$ and $n \in N$ such that $gC = bnC$. This is clearly equivalent to (ii). $\square$

We record, for future reference, the following corollary of the proof (in the case $\Delta' = \Delta$):

Corollary 6.5. *Suppose G acts strongly transitively on Δ with respect to an apartment system $\mathcal{A}$. If Σ is any apartment in $\mathcal{A}$ and B is the stabilizer of a chamber in Σ, then*

$$\Delta = \bigcup_{b\in B} b\Sigma .$$ □

We close this subsection by recording a consequence of strong transitivity that will be useful in Chapter 7.

Proposition 6.6. *Suppose a group G acts strongly transitively on a building Δ with respect to an apartment system $\mathcal{A}$.*

(1) *Given two apartments $\Sigma, \Sigma' \in \mathcal{A}$ and a type-preserving isomorphism $\phi \colon \Sigma \to \Sigma'$, there is an element $g \in G$ such that $gA = \phi(A)$ for all simplices $A \in \Sigma$.*

(2) *Given two apartments $\Sigma, \Sigma' \in \mathcal{A}$, there is an element $g \in G$ such that $g\Sigma = \Sigma'$ and g fixes $\Sigma \cap \Sigma'$ pointwise.*

Proof. (1) Choose an arbitrary chamber $C \in \Sigma$. By strong transitivity, there is an element $g \in G$ such that $g\Sigma = \Sigma'$ and $gC = \phi(C)$. Then g agrees with ϕ on Σ by the standard uniqueness argument.

(2) This follows from (1) and Proposition 4.101.   □

Statement (2) has the following immediate consequence:

Corollary 6.7. *Suppose a group G acts strongly transitively on a building Δ with respect to an apartment system $\mathcal{A}$. Let K be an arbitrary set of simplices in Δ, and let $\mathrm{Fix}_G(K)$ be the pointwise fixer of K, i.e.,*

$$\mathrm{Fix}_G(K) := \{g \in G \mid gA = A \text{ for all } A \in K\} .$$

Then $\mathrm{Fix}_G(K)$ acts transitively on the set of apartments in $\mathcal{A}$ containing K.
     □

6.1.2 Example

Let P be the projective plane over a field k, as defined in Section 4.2, and let Δ be its flag complex. It is a rank-2 building, with one vertex for every proper nonzero subspace of k^3 and one edge for each pair consisting of a 1-dimensional subspace contained in a 2-dimensional subspace. The unique apartment system for Δ has an apartment for every triple $\{L_1, L_2, L_3\}$ of 1-dimensional subspaces such that $k^3 = L_1 \oplus L_2 \oplus L_3$. Given any subset $X \subseteq k^3$, we denote by $[X]$ the subspace spanned by X. Then a triple as above has the form $\{[e_1], [e_2], [e_3]\}$ for some basis e_1, e_2, e_3 of k^3, and the corresponding apartment is the subcomplex of Δ shown in Figure 6.2. As fundamental apartment Σ we take the apartment associated to the standard basis of k^3. And as fundamental chamber C we take the edge joining $[e_1]$ to $[e_1, e_2]$.

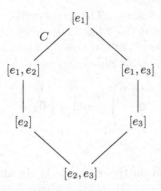

Fig. 6.2. An apartment Σ.

Let G be the group $\mathrm{GL}_3(k)$ of linear automorphisms of k^3. Then any $g \in G$ takes subspaces to subspaces and induces a type-preserving automorphism of Δ. It is easy to check that this action of G is strongly transitive (with respect to the unique apartment system). Let's compute B, N, T, W, and S.

The stabilizer B of C consists of all automorphisms of k^3 that leave the subspaces $[e_1]$ and $[e_1, e_2]$ invariant; hence B is the *upper-triangular group*

$$\begin{pmatrix} * & * & * \\ 0 & * & * \\ 0 & 0 & * \end{pmatrix}.$$

The stabilizer N of Σ consists of all automorphisms that permute the three subspaces $[e_1], [e_2], [e_3]$. Hence N is the *monomial group*, consisting of all matrices with exactly one nonzero element in every row and every column. Given $n \in N$, the action of n on Σ is determined by the permutation of $\{[e_1], [e_2], [e_3]\}$ induced by n; so T consists of the diagonal matrices (which induce the trivial permutation), and $W = N/T$ can be identified with the symmetric group on 3 letters, or, equivalently with the group of 3×3 permutation matrices. [Thus we have a splitting $N = T \rtimes W$.] Finally, it is easy to check that the set S of fundamental reflections consists of the permutations represented by

$$\begin{pmatrix} 0 & 1 & 0 \\ 1 & 0 & 0 \\ 0 & 0 & 1 \end{pmatrix} \quad \text{and} \quad \begin{pmatrix} 1 & 0 & 0 \\ 0 & 0 & 1 \\ 0 & 1 & 0 \end{pmatrix}.$$

As a mnemonic aid, we remark that B is what is called a Borel subgroup of G in the theory of algebraic groups, T is a maximal torus, N is the normalizer of T, and W is the Weyl group.

Remarks 6.8. (a) Instead of taking $G = \mathrm{GL}_3(k)$ above, we could equally well have taken G to be the subgroup $\mathrm{SL}_3(k)$ consisting of matrices of determinant 1. The groups B, N, and T would then be the intersections with $\mathrm{SL}_3(k)$

of the groups B, N, and T above. The quotient $W = N/T$ would still be the symmetric group on 3 letters (as it has to be, since $W = \mathrm{Aut}_0 \Sigma$, independent of G). The set $S \subset W$ consists of the same two permutations as above, which can be represented by the monomial matrices

$$\begin{pmatrix} 0 & -1 & 0 \\ 1 & 0 & 0 \\ 0 & 0 & 1 \end{pmatrix} \quad \text{and} \quad \begin{pmatrix} 1 & 0 & 0 \\ 0 & 0 & -1 \\ 0 & 1 & 0 \end{pmatrix}$$

of determinant 1.

(b) Still another variation on this example is obtained by replacing $\mathrm{GL}_3(k)$ by its quotient $\mathrm{PGL}_3(k) := \mathrm{GL}_3(k)/Z$, where $Z < \mathrm{GL}_3(k)$ is the central subgroup consisting of scalar multiples of the identity matrix. The subgroup Z acts trivially on Δ, so we obtain an action of $\mathrm{PGL}_3(k)$ on Δ; and clearly this action is still strongly transitive. Similarly, $\mathrm{PSL}_3(k) := \mathrm{SL}_3(k)/(\mathrm{SL}_3(k) \cap Z)$ acts strongly transitively on Δ.

(c) As we indicated in Remark 6.2, we do not have to take N to be the full stabilizer of Σ; any subgroup of the monomial group that surjects onto the symmetric group would work just as well. For example, we could use the group of permutation matrices. Or, in the case of $\mathrm{SL}_3(k)$, we could use the monomial matrices of determinant 1 whose nonzero entries are ± 1.

(d) In view of Section 4.3, everything we have done generalizes from GL_3 to GL_n (or SL_n, or PGL_n, or PSL_n).

Exercise 6.9. With G as in this section, prove the equation $G = BNB$ of Proposition 6.3 by direct matrix calculations (row and column operations).

6.1.3 Weyl Transitivity

We return now to an arbitrary building Δ of type (W, S) with a type-preserving G-action. Let $(\mathcal{C}, \delta)$ be the corresponding W-metric building. Thus $\mathcal{C} := \mathcal{C}(\Delta)$ is the set of chambers, δ is the Weyl distance function (Section 4.8), and G acts on $\mathcal{C}$ as a group of *isometries*:

$$\delta(gC, gD) = \delta(C, D)$$

for all chambers $C, D \in \mathcal{C}$ and all $g \in G$.

Definition 6.10. We say that the action of G on Δ is *Weyl transitive* if for each $w \in W$, the action is transitive on the set of ordered pairs (C, D) of chambers with $\delta(C, D) = w$.

This is equivalent to saying that the action is chamber transitive and that the stabilizer of a given chamber C is transitive on the "w-sphere"

$$\{D \in \mathcal{C} \mid \delta(C, D) = w\}$$

for every $w \in W$. Weyl transitivity is analogous to a condition called "two-point transitivity" or "distance-transitivity" in the setting of group actions on ordinary metric spaces.

There is a convenient characterization of Weyl transitivity that does not explicitly refer to Weyl distance:

Proposition 6.11. *Assume that the action of G is chamber transitive. Let C be an arbitrary chamber, and let Σ an arbitrary apartment (in the complete apartment system) containing C. Let B be the stabilizer of C in G. Then the action of G on Δ is Weyl transitive if and only if*

$$\Delta = \bigcup_{b \in B} b\Sigma . \tag{6.1}$$

Proof. As we noted above, Weyl transitivity holds if and only if the action of B is transitive on the w-sphere centered at C for each $w \in W$. Given w, there is a unique chamber $C_w \in \mathcal{C}(\Sigma)$ with $\delta(C, C_w) = w$. (If we identify Σ with $\Sigma(W, S)$ in such a way that C corresponds to the fundamental chamber, then C_w is simply wC. Or, if we identify $\mathcal{C}(\Sigma)$ with the standard thin building W in such a way that C corresponds to $1 \in W$, then C_w is simply w.) So Weyl transitivity says that the B-orbit of C_w is the entire w-sphere for each $w \in W$, and this is precisely equation (6.1). $\square$

Combining the proposition with Corollary 6.5, we see that, as the name suggests, strong transitivity is indeed stronger than Weyl transitivity:

Corollary 6.12. *Strong transitivity (with respect to some apartment system) implies Weyl transitivity.* $\square$

If one wants to try to show, conversely, that a given Weyl-transitive action is strongly transitive, one needs a way to use Weyl transitivity to construct an apartment system. This is easy:

Lemma 6.13. *Suppose the action of G on Δ is Weyl transitive, and let Σ be an arbitrary apartment (in the complete system of apartments). Then the set $G\Sigma := \{g\Sigma \mid g \in G\}$ is a system of apartments.*

Proof. Since $G\Sigma$ is given to us as a subset of the complete system of apartments, it suffices to show that any two chambers C, D are contained in some apartment in $G\Sigma$. By chamber transitivity we may assume that $C \in \mathcal{C}(\Sigma)$. Then equation (6.1) gives us an apartment $b\Sigma \in G\Sigma$ that contains C and D. $\square$

We can now clarify the relationship between our two notions of transitivity:

Proposition 6.14. *The following conditions are equivalent:*

(i) *The G-action on Δ is strongly transitive with respect to some apartment system.*

(ii) *The G-action on Δ is Weyl transitive, and there is an apartment Σ (in the complete system of apartments) such that the stabilizer of Σ acts transitively on $\mathcal{C}(\Sigma)$.*

Proof. The implication (i) $\implies$ (ii) is immediate from the definitions and Corollary 6.12. Conversely, if (ii) holds, then the action is strongly transitive with respect to $G\Sigma$, which is an apartment system by Lemma 6.13. $\square$

We will see examples in Sections 6.10 and 6.11 of actions that are Weyl transitive but are not strongly transitive with respect to any apartment system. In the spherical case, however, such examples cannot exist. Recall first that there is a unique system of apartments in a spherical building, so we can talk about strong transitivity without specifying an apartment system. And, as we will now prove, strong transitivity then turns out to be equivalent to Weyl transitivity.

Proposition 6.15. *The following conditions are equivalent for a (type-preserving) action of a group on a spherical building:*

(i) *The action is strongly transitive.*
(ii) *The action is Weyl transitive.*
(iii) *The action is transitive on ordered pairs (C, C') of opposite chambers.*

Proof. We already know that (i) $\implies$ (ii). To prove (ii) $\implies$ (iii), just note that C and C' are opposite if and only if $\delta(C, C') = w_0$, where w_0 is the longest element of W (Section 1.5.2). Finally, (iii) is equivalent to the assertion that G is transitive on $\mathcal{C}$ and the stabilizer of a given chamber C is transitive on the chambers opposite C. Using the correspondence between chambers opposite C and apartments containing C (Section 4.7), we conclude that (iii) $\implies$ (i). $\square$

Exercise 6.16. Characterize Weyl transitivity without referring to Weyl distances or apartments in case Δ is a tree.

6.1.4 The Bruhat Decomposition

Assume now that we have a *Weyl transitive* action of G on Δ. Choose a fundamental chamber C, and let B be its stabilizer in G. We can then identify the set $\mathcal{C}$ of chambers with the set G/B of left cosets gB via $gC \leftrightarrow gB$ for $g \in G$. Now Weyl transitivity implies that the B-orbits in $\mathcal{C}$ are in 1–1 correspondence with the elements of W, with the orbit of a chamber D corresponding to $w = \delta(C, D)$. But the B-orbits in $\mathcal{C}$ correspond to the B-orbits in G/B and hence to double cosets BgB. Thus we have a bijection $B\backslash G/B \to W$ and hence a set-theoretic decomposition

$$G = \coprod_{w \in W} C(w), \tag{6.2}$$

where $C(w)$ is the double coset corresponding to w. This decomposition is known as the *Bruhat decomposition* for historical reasons that we will explain in Section 6.4.

It is easy to chase through the definitions and see that the bijection $B\backslash G/B \to W$ is given by $BgB \mapsto \delta(C, gC)$. Thus the relation "$g \in C(w)$" can be represented schematically by the diagram

$$C \xrightarrow{\ w\ } gC \ .$$

The following theorem summarizes the discussion:

Theorem 6.17. *Assume that the action of G on Δ is Weyl transitive, and let B be the stabilizer of a chamber C. Then there is a bijection $B\backslash G/B \to W$ given by $BgB \mapsto \delta(C, gC)$. Hence (6.2) holds, where $w \mapsto C(w)$ is the inverse bijection.* $\square$

Exercise 6.18. Choose an apartment Σ containing C, and identify it with $\Sigma(W, S)$ as above. Let $\rho\colon \Delta \to \Sigma$ be the retraction onto Σ centered at C. Show that the double coset $C(w)$ containing g can be characterized by the equation $\rho(gC) = wC$.

With the aid of the Bruhat decomposition, we can completely reconstruct the Weyl distance function δ, and hence the building Δ, from group-theoretic data. Namely, if we identify C with G/B, then δ becomes a function

$$G/B \times G/B \to W \ ,$$

still denoted by δ, and we claim that it is a "difference" function, much like those that occurred in Section 3.5. More precisely, given cosets gB, hB, the set $(gB)^{-1}(hB)$ of differences is the double coset $Bg^{-1}hB$ and hence is $C(w)$ for a unique $w \in W$. The claim is that $\delta(gB, hB)$ is this element w. In other words, δ is the composite

$$G/B \times G/B \to B\backslash G/B \to W \ , \tag{6.3}$$

where the first arrow is given by $(gB, hB) \mapsto Bg^{-1}hB$ and the second is the bijection in Theorem 6.17. This claim follows immediately from the fact that by G-invariance, the Weyl distance function on C satisfies $\delta(gC, hC) = \delta(C, g^{-1}hC)$.

6.1.5 The Strongly Transitive Case

If we assume further that the action is strongly transitive, then there is a slight simplification, in that we can easily describe $C(w)$ in terms of the "BN data" (Figure 6.1). Namely, $C(w)$ is simply the double coset $B\tilde{w}B$, where $\tilde{w}$ is any lift of w to N. This follows from the fact that in the Coxeter complex Σ

with fundamental chamber C, we have $\delta(C, wC) = w$; see equation (3.8) in Section 3.5. Thus the Bruhat decomposition becomes

$$G = \coprod_{w \in W} B\tilde{w}B \, .$$

This is a considerable strengthening of Proposition 6.3. Taking $G = \mathrm{GL}_n(k)$ as in Section 6.1.2, for example, we can choose the elements $\tilde{w}$ to be permutation matrices; the Bruhat decomposition then becomes

$$\mathrm{GL}_n(k) = \coprod_w BwB \, ,$$

where B is the upper-triangular group and w ranges over the permutation matrices.

Remark 6.19. To simplify notation, we will often write BwB instead of $B\tilde{w}B$. This should cause no problem since the double coset is independent of the choice of the lift $\tilde{w}$. Similar remarks apply to expressions like wB, wBw', and so on.

Exercise 6.20. Suppose $G = \mathrm{GL}_n(k)$ acting on $\Delta(k^n)$ as in Section 6.1.2. Deduce from the Bruhat decomposition the following result, which has already occurred in Exercise 4.93: The Weyl distance between two chambers is given by the Jordan–Hölder permutation.

6.1.6 Group-Theoretic Consequences

We return now to the setup of Section 6.1.4, where G acts Weyl transitively on a building Δ of type (W, S). The arguments in that subsection were purely set-theoretic and made no use of the fact that we were working with a *building*. We now wish to use the properties of δ, as expressed by the WD axioms in Section 5.1.1, to derive properties of double cosets. Axiom (WD1) leads to the unsurprising result that

$$C(w) = B \iff w = 1 \, , \tag{6.4}$$

which is immediate from the definitions anyway. Axioms (WD2) and (WD3), on the other hand, lead to much more interesting results, involving products of double cosets.

Note first that a product $BgB \cdot Bg'B$ of two double cosets is a set that contains gg' and is closed under left and right multiplication by B; hence it is a union of double cosets, one of which is $Bgg'B$. This is all that can be said in general. In the present situation, however, we can say much more:

Theorem 6.21. *Given $s \in S$ and $w \in W$, we have*

$$C(sw) \subseteq C(s)C(w) \subseteq C(sw) \cup C(w) \, . \tag{6.5}$$

In particular, $C(s)C(w)$ is either the double coset $C(sw)$ or else the union of two double cosets. If $l(sw) = l(w) + 1$, then $C(s)C(w) = C(sw)$.

Proof. Given $h \in C(s)$ and $g \in C(w)$, we want to know which double coset contains the product hg. In other words, we are given that $\delta(C, hC) = s$ and $\delta(C, gC) = w$, and we want to compute $\delta(C, hgC)$. To this end we apply axiom (WD2) to the situation

$$
\begin{array}{ccc}
C & & \\
{\scriptstyle s}\downarrow & \searrow & \\
hC & \xrightarrow[w]{} & hgC
\end{array}
\tag{6.6}
$$

and we conclude that $\delta(C, hgC) = sw$ or w; moreover, $\delta(C, hgC) = sw$ if $l(sw) = l(w) + 1$. Hence $hg \in C(sw) \cup C(w)$, and $hg \in C(sw)$ if $l(sw) = l(w) + 1$. This proves the second inclusion in (6.5) and the last assertion of the proposition. Finally, consider the diagram

$$
\begin{array}{ccc}
h^{-1}C & & \\
{\scriptstyle s}\downarrow & \searrow & \\
C & \xrightarrow[w]{} & gC
\end{array}
\tag{6.7}
$$

obtained by applying the action of h^{-1} to (6.6). Axiom (WD3) implies that the representative $h \in C(s)$ can be chosen such that $\delta(h^{-1}C, gC) = sw$. Thus $hg \in C(sw)$ for this h, which proves that $C(s)C(w)$ meets $C(sw)$ and hence contains it. This proves the first inclusion in (6.5).   □

Remark 6.22. There is another property of δ that has a simple group-theoretic interpretation, namely the symmetry property (Corollary 5.17(2)). We will show that this implies

$$
C(w)^{-1} = C(w^{-1})
\tag{6.8}
$$

and hence lets us remove the apparent asymmetry in Theorem 6.21. In other words, we could state an analogue of the theorem for $C(w)C(s)$ instead of $C(s)C(w)$. To prove (6.8), note that it is equivalent to the assertion that the bijection $B\backslash G/B \to W$ given by $BgB \mapsto \delta(C, gC)$ preserves inverses, i.e.,

$$
\delta(C, g^{-1}C) = \delta(C, gC)^{-1} .
\tag{6.9}
$$

Now $\delta(C, g^{-1}C) = \delta(gC, C)$ by G-invariance of δ, so (6.9) follows from the fact that $\delta(gC, C) = \delta(C, gC)^{-1}$.

Exercise 6.23. Assume, in addition to the hypotheses of this subsection, that Δ is thick and spherical. Let $w_0 \in W$ be the longest element. Show that $C(w_0)C(w_0) = G$.

6.1.7 The Thick Case

We continue to assume that we have a Weyl-transitive action of a group G on a building Δ of type (W, S). In the most important applications, Δ is *thick*. This again has a group-theoretic interpretation, and it leads to a sharpening of Theorem 6.21.

First of all, Δ is thick if and only if for each $s \in S$ there are at least two chambers C' such that $\delta(C, C') = s$. Writing $C' = hC$ with $h \in G$, we have $\delta(C, hC) = s$ if and only if $h \in C(s)$. So the condition for thickness is that

$$[C(s) : B] \geq 2 \text{ for all } s \in S , \qquad (6.10)$$

where $[C(s) : B]$ is the number of left cosets hB in the double coset $C(s)$. Since $C(s)^{-1} = C(s)$ by (6.8), we could equally well use right cosets Bh.

Let's return now to the situation of (6.6) (or, equivalently, (6.7)), where we appealed to axiom (WD3) to find $h \in C(s)$ such that $\delta(C, hgC) = sw$. In case $l(sw) = l(w) - 1$, we know from Lemma 5.5 (or Proposition 4.84) that the chamber hC, and hence the coset hB, is unique; all other cosets $h'B \subseteq C(s)$ would give $\delta(C, h'gC) = w$, and hence $h'g \in C(w)$. Now thickness says precisely that there exists such an h', so $C(s)C(w)$ meets $C(w)$ and hence contains it. This proves the following result, which makes Theorem 6.21 more precise in the thick case:

Proposition 6.24. *Suppose Δ is thick. For any $s \in S$ and $w \in W$ with $l(sw) = l(w) - 1$,*

$$C(s)C(w) = C(sw) \cup C(w) . \qquad \square$$

Exercise 6.25. Give an example to show that the assumption of thickness cannot be removed from the proposition.

Remark 6.26. We can rewrite the criterion (6.10) for thickness in a way that might look more natural from the point of view of group theory. Namely, if we choose any element $h \in C(s)$, then $[C(s) : B] \geq 2$ if and only if $C(s) \neq hB$, i.e., $BhB \neq hB$. This holds if and only if $Bh \nsubseteq hB$, so the criterion can be stated in terms of conjugation:

$$B \nsubseteq hBh^{-1} .$$

Note that we could have used right cosets instead of left cosets, so another equivalent formulation is

$$hBh^{-1} \nsubseteq B .$$

Finally, since these last two conditions are equivalent to one another (in the present context of a Weyl-transitive action on a building), they are also equivalent to the more symmetric condition

$$hBh^{-1} \neq B .$$

6.1.8 Stabilizers

We return to the general case, where we have a Weyl-transitive action on a building Δ that is not necessarily thick.

So far we have concentrated on chambers. From the simplicial viewpoint, however, there are simplices other than chambers. In particular, if we want to reconstruct Δ as a simplicial complex directly from the group G, then we need to know the stabilizers of the faces of the fundamental chamber. These are quite easy to work out:

Proposition 6.27. *Given $J \subseteq S$, let A be the face of C of cotype J. Then the stabilizer of A in G is*

$$P_J := \bigcup_{w \in W_J} C(w) . \tag{6.11}$$

In particular, the union of double cosets in (6.11) *is a subgroup of G.*

Proof. Recall that simplices are in 1–1 correspondence with residues (Corollary 4.11). Given $g \in G$, it follows that $gA = A$ if and only if gC and C are in the same J-residue. As we noted in Section 5.3.1 (or Exercise 4.92), this happens if and only if $\delta(C, gC) \in W'$, i.e., if and only if $g \in C(w)$ for some $w \in W_J$. $\square$

Definition 6.28. We will call the subgroups of the form P_J *standard parabolic subgroups*, and we will call the left cosets gP_J *standard parabolic cosets*.

As in the case of Coxeter complexes, the stabilizer calculation leads immediately to the following result:

Corollary 6.29. *The building Δ is isomorphic, as a poset, to the set of standard parabolic cosets, ordered by reverse inclusion.* $\square$

Remark 6.30. Let $Z := \bigcap_{g \in G} gBg^{-1}$; this is the normal subgroup of G consisting of the elements that act trivially on Δ. Let $\bar{G} := G/Z$. By analogy with the situation for Coxeter groups and their associated complexes, one might expect to be able to recover $\bar{G}$ from Δ as the group $\mathrm{Aut}_0 \, \Delta$ of type-preserving automorphisms. This turns out to be false in general; counterexamples will be given in Section 6.9 below (see Remark 6.112(c)).

6.2 Bruhat Decompositions, Tits Subgroups, and BN-Pairs

6.2.1 Bruhat Decompositions

We have seen that a Weyl-transitive action of a group G on a building leads to a subgroup B and a bijection $C \colon W \to B \backslash G / B$ with certain properties. Conversely, we will show that such a bijection leads easily to a Weyl-transitive action of G on a building.

Definition 6.31. Suppose we are given a group G, a subgroup B, a Coxeter system (W, S), and a bijection $C \colon W \to B\backslash G/B$ satisfying the following condition:

(B) *For all $s \in S$ and $w \in W$,*

$$C(sw) \subseteq C(s)C(w) \subseteq C(sw) \cup C(w) .$$

If $l(sw) = l(w) + 1$, then $C(s)C(w) = C(sw)$.

Then the bijection C is said to provide a *Bruhat decomposition of type (W, S)* for (G, B).

Note that we necessarily have $C(1) = B$. For if $w_1 \in W$ is the element such that $C(w_1) = B$, then we can take any $s \in S$ and deduce that

$$C(s) = C(s)C(w_1) \supseteq C(sw_1) ;$$

hence $s = sw_1$ and $w_1 = 1$.

It is now completely routine to reverse the arguments given in Section 6.1.6 and construct a building Δ, provided we use the W-metric approach to buildings. Namely, set $\mathcal{C} := G/B$ and define $\delta \colon \mathcal{C} \times \mathcal{C} \to W$ to be the composite

$$G/B \times G/B \to B\backslash G/B \to W ,$$

where the first map is $(gB, hB) \mapsto Bg^{-1}hB$, and the second is C^{-1}. Thus

$$\delta(gB, hB) = w \iff g^{-1}h \in C(w) .$$

One easily verifies the axioms (WD1), (WD2), and (WD3) of Section 5.1.1, so, by Section 5.6, we have a building Δ with $\mathcal{C}(\Delta) = \mathcal{C}$. Moreover, the natural action of G on G/B induces an action of G on Δ, which we claim is Weyl transitive: Take the coset B as fundamental chamber, and consider two chambers $gB, g'B$ with $\delta(B, gB) = \delta(B, g'B)$. Then g and g' are in the same double coset $C(w)$, so $g' = bgb'$ for some $b, b' \in B$; hence $g'B = bgB$. Thus B, the stabilizer of the fundamental chamber, is transitive on the chambers at given Weyl distance from the fundamental chamber. This proves the claim.

Note that Corollary 6.29 gives an explicit description of the building Δ as a simplicial complex: It can be identified with the poset of standard parabolic cosets, ordered by reverse inclusion. [As a byproduct, we obtain the fact that $P_J := \bigcup_{w \in W_J} C(w)$ is in fact a subgroup of G, which we will also verify algebraically in the next subsection.] Alternatively, one can derive this description from the W-metric theory, which says that Δ is the poset of residues, ordered by reverse inclusion. Indeed, given $J \subseteq S$, one checks directly from the definitions that two chambers gB, hB are in the same J-residue if and only if $g^{-1}h \in P_J$, so the J-residues are in 1–1 correspondence with the left P_J-cosets.

Definition 6.32. Given a Bruhat decomposition for (G, B), we denote by $\Delta(G, B)$ the poset of standard parabolic cosets, ordered by reverse inclusion.

Remark 6.33. The notation $\Delta(G, B)$ is somewhat misleading, since one needs the bijection $C \colon W \to B\backslash G/B$ in order to define the standard parabolic subgroups and hence the poset $\Delta(G, B)$. This abuse of notation is not serious, however, because it turns out that $\Delta(G, B)$, if it is thick, depends only on the pair (G, B). See Corollary 6.44.

Combining the discussion above with Theorems 6.17 and 6.21, we obtain the following:

Proposition 6.34. *Given a Bruhat decomposition for* (G, B)*, the poset* $\Delta = \Delta(G, B)$ *is a building, and the natural action of* G *on* Δ *by left translation is Weyl transitive and has* B *as the stabilizer of a fundamental chamber. Conversely, if a group* G *admits a Weyl-transitive action on a building* Δ *and* B *is the stabilizer of a fundamental chamber, then* (G, B) *admits a Bruhat decomposition and* Δ *is canonically isomorphic to* $\Delta(G, B)$*.* □

Thus there is essentially a 1–1 correspondence between Bruhat decompositions and Weyl-transitive actions.

Remark 6.35. In this subsection we have used the W-metric approach to buildings since it meshes so perfectly with the algebraic theory of Bruhat decompositions. Moreover, we do not know any way to prove Proposition 6.34 from the simplicial point of view, since there is no obvious way to construct apartments in $\Delta(G, B)$ from the data given in Definition 6.31. In Section 6.2.5, however, when we develop the algebraic theory corresponding to strongly transitive actions, it will be possible to give an alternative treatment that is purely simplicial. This will be outlined in Exercise 6.54.

Our next goal is to get a better algebraic understanding of Bruhat decompositions.

6.2.2 Axioms for Bruhat Decompositions

If one wants to construct a building from group-theoretic data, it is of interest to minimize what has to be verified. It turns out that we can get by with axioms that appear to be weaker than the requirements in Section 6.2.1.

Let G be a group, B a subgroup, (W, S) a Coxeter system, and

$$C \colon W \to B\backslash G/B$$

a function. Consider the following three axioms:

(Bru1) $C(w) = B$ *if and only if* $w = 1$.

(Bru2) $C \colon W \to B\backslash G/B$ *is surjective, i.e.,*

$$G = \bigcup_{w \in W} C(w).$$

(Bru3) *For any $s \in S$ and $w \in W$,*

$$C(sw) \subseteq C(s)C(w) \subseteq C(sw) \cup C(w) .$$

There appears to be asymmetry in (Bru3), which involves *left* multiplication by elements of S. But we will see in the next proposition that the axioms (Bru1)–(Bru3) imply the following "right" analogue of (Bru3):

(Bru3') *For any $s \in S$ and $w \in W$,*

$$C(ws) \subseteq C(w)C(s) \subseteq C(ws) \cup C(w) .$$

We now show that (Bru1), (Bru2), and (Bru3) suffice for a Bruhat decomposition. In other words, they imply that C is bijective and that the second assertion of (B) holds. For reasons that will become obvious in Section 6.2.3, we will take care to prove this without using the assumption that (W, S) is a Coxeter system.

Proposition 6.36. *Let G be a group and B a subgroup. Suppose we are given a group W, a generating set S consisting of elements of order 2, and a function $C \colon W \to B \backslash G / B$ satisfying (Bru1), (Bru2), and (Bru3). Then the six conditions below are satisfied. In particular, C provides a Bruhat decomposition for (G, B) if (W, S) is a Coxeter system.*

(1) *C is a bijection, i.e.,*

$$G = \coprod_{w \in W} C(w) .$$

(2) *$C(w)^{-1} = C(w^{-1})$ for all $w \in W$. Consequently, (Bru3') holds.*
(3) *If $l(sw) \geq l(w)$ with $s \in S$ and $w \in W$, then $C(s)C(w) = C(sw)$.*
(4) *Given a reduced decomposition $w = s_1 \cdots s_l$ of an element $w \in W$, we have $C(w) = C(s_1) \cdots C(s_l)$.*
(5) *If $l(sw) \leq l(w)$ with $s \in S$ and $w \in W$, and if $[C(s) : B] \geq 2$, then $C(s)C(w) = C(sw) \cup C(w)$.*
(6) *Let $J \subseteq S$ be an arbitrary subset. Then $P_J := \bigcup_{w \in W_J} C(w)$ is a subgroup of G. It is generated by the cosets $C(s)$ with $s \in J$.*

Proof. For (1), we must show that $C(v) = C(w) \implies v = w$ for $v, w \in W$. We argue by induction on $\min\{l(v), l(w)\}$, which we may assume is equal to $l(v)$. The case $l(v) = 0$ is covered by (Bru1), so suppose $l(v) > 0$ and choose $s \in S$ such that $l(sv) < l(v)$. If $C(v) = C(w)$, then we can multiply by $C(s)$ to get $C(s)C(v) = C(s)C(w)$. If these equal products consist of one double coset, then the equation becomes $C(sv) = C(sw)$. Otherwise, there are two double cosets on each side, and we have $C(sv) \cup C(v) = C(sw) \cup C(w)$; subtracting $C(v) = C(w)$, we again obtain $C(sv) = C(sw)$. In either case, the induction hypothesis implies $sv = sw$, and hence $v = w$.

Next we prove (2), (3), and (4) simultaneously, by induction on $l(w)$. We may assume $l(w) > 0$, since the assertions are all trivial if $w = 1$. We will carry out the induction by means of the following steps:

(a) (2) is true if $l(w) = 1$.
(b) Given $l > 0$, if (3) and (4) hold when $l(w) < l$ then (4) holds when $l(w) = l$.
(c) Given $l > 0$, if (4) holds when $l(w) = l$, then (2) holds when $l(w) = l$.
(d) Given $l > 0$, if (2) holds when $l(w) \leq l$ and (3) holds when $l(w) < l$, then (3) holds when $l(w) = l$.

For (a) we must show that $C(s)^{-1} = C(s)$ for $s \in S$. To prove this, note that $1 \in B = C(1) \subseteq C(s)C(s)$, so $C(s)^{-1}$ meets $C(s)$ and hence is equal to it. (b) is easy and is left to the reader. For (c), take a reduced decomposition $w = s_1 \cdots s_l$. Then $C(w) = C(s_1) \cdots C(s_l)$ and $C(w^{-1}) = C(s_l) \cdots C(s_1)$; now apply the result of (a). To prove (d), finally, write $w = w_1 t$ with $t \in S$ and $l(w_1) = l-1$. Then the hypothesis of (d) implies that $C(w) = C(w_1)C(t)$, as one sees by taking inverses. Hence

$$C(s)C(w) = C(s)C(w_1)C(t) . \tag{6.12}$$

Our assumption $l(sw) \geq l(w)$ implies that $l(sw_1) \geq l(w_1)$; for otherwise we would have

$$l(sw) = l(sw_1 t) \leq l(sw_1) + 1 < l(w_1) + 1 = l(w) .$$

So we may apply the hypothesis of (d) again to rewrite (6.12) as

$$C(s)C(w) = C(sw_1)C(t) . \tag{6.13}$$

We wish to show that $C(s)C(w)$ is a single double coset, in which case it is necessarily $C(sw)$. Suppose this is false. Then we have

$$C(s)C(w) = C(sw) \cup C(w) \tag{6.14}$$

by (Bru3). On the other hand, we get a different expression for $C(s)C(w)$ as a union of two double cosets by applying (Bru3) to the inverse of the right side of (6.13):

$$\begin{aligned}
C(s)C(w) &= C(sw_1)C(t) \\
&= \left(C(t)C(w_1^{-1}s) \right)^{-1} \\
&= \left(C(tw_1^{-1}s) \cup C(w_1^{-1}s) \right)^{-1} \\
&= C(w^{-1}s)^{-1} \cup C(sw_1) .
\end{aligned} \tag{6.15}$$

Note that our hypothesis does not allow us to go one step further and claim that $C(w^{-1}s)^{-1} = C(sw)$, but fortunately we do not need this. Indeed, (6.15) and (6.14) imply that $C(sw_1) \subseteq C(sw) \cup C(w)$ and hence that $sw_1 = sw$ or $sw_1 = w$. The first possibility would imply $w_1 = w$, contradicting the fact that $l(w_1) < l(w)$, while the second would contradict the assumption that $l(sw) \geq l(w)$. This completes the proof of (d) and hence the inductive proof of (2)–(4).

To prove (5), note first that as in Remark 6.26, the assumption that $[C(s){:}B] \geq 2$ can be rewritten as $hBh^{-1} \nleq B$, where $C(s) = BhB = Bh^{-1}B$. Now $C(s)C(s) = BhBh^{-1}B$, so $C(s)C(s) \nsubseteq B$. In view of (Bru3), we must have $C(s)C(s) = B \cup C(s)$. This is a special case of (5), and this special case, together with (3), easily yields the general case. Indeed, if $l(sw) \leq l(w)$, then $l(s \cdot sw) \geq l(sw)$. So (3) implies

$$C(w) = C(s \cdot sw) = C(s)C(sw) \, ,$$

hence

$$\begin{aligned} C(s)C(w) &= C(s)C(s)C(sw) \\ &= (B \cup C(s))C(sw) \\ &= C(sw) \cup C(s)C(sw) \\ &= C(sw) \cup C(w) \, . \end{aligned}$$

To prove (6), finally, let $G_J := \langle C(s) \rangle_{s \in J}$. Then G_J contains P_J by (4), and it follows from (Bru3) (and the fact that $C(s)^{-1} = C(s)$ for $s \in S$) that P_J is closed under left multiplication by G_J. So $P_J = G_J$.   □

Exercises

6.37. Let the hypotheses be as in Proposition 6.36. Given any two elements $w, w' \in W$, write $w = s_1 \cdots s_d$ with $s_i \in S$, and show that

$$C(ww') \subseteq C(w)C(w') \subseteq \bigcup_{w''} C(w''w') \, ,$$

where w'' ranges over the elements of W obtained from the word $s_1 \cdots s_d$ by deleting zero or more letters. This shows again that P_J is a subgroup of G for every $J \subseteq S$.

6.38. Prove the following generalization of the Bruhat decomposition: For any standard parabolic subgroup P_J ($J \subseteq S$), there is a bijection $B \backslash G / P_J \to W/W_J$. Still more generally, show for any two standard parabolics P_J and P_K ($J, K \subseteq S$) that there is a bijection $P_J \backslash G / P_K \to W_J \backslash W / W_K$.

6.2.3 The Thick Case

Definition 6.39. We say that a Bruhat decomposition for (G, B) is *thick* if the building $\Delta(G, B)$ is thick.

As we saw in Section 6.1.7, this is equivalent to the following condition:

(Th) $[C(s) : B] \geq 2$ *for all* $s \in S$.

We can also express thickness by saying that B has index ≥ 3 in the group $P_s := B \cup C(s)$. Yet another formulation (cf. Remark 6.26) is that for every $s \in S$ and $h \in C(s)$,

$$hBh^{-1} \not\subseteq B .\tag{6.16}$$

Thickness has some remarkable consequences. First of all, in the presence of (Th), we can further weaken the conditions that need to be verified in order to construct a Bruhat decomposition. Namely, we do not need to assume that (W, S) is a Coxeter system or even that the elements of S have order 2. This is the group-theoretic analogue of the fact that in defining thick buildings, apartments do not have to be assumed to be Coxeter complexes (Theorem 4.131). Moreover, we do not even have to specify S, since as we will show, it is uniquely determined by the rest of the data.

In stating the precise result, we have to be careful about whether we interpret $[C(s) : B]$ in (Th) using left cosets or right cosets, since it might make a difference if we do not know that $C(s)^{-1} = C(s)$. It turns out that we need to take $[C(s) : B]$ to be the number of *right* cosets Bh in $C(s)$. With this convention, (Th) is equivalent to the condition that (6.16) holds for every $s \in S$ and for some (or every) $h \in C(s)$.

Proposition 6.40. *Let G be a group and B a subgroup. Suppose we are given a group W, a generating set S of W, and a function $C \colon W \to B \backslash G / B$ satisfying (Bru1), (Bru2), (Bru3), and (Th). Then (W, S) is a Coxeter system, and hence the given data constitute a thick Bruhat decomposition for (G, B). Moreover, the generating set S is uniquely determined; it consists of all nontrivial elements $w \in W$ such that $B \cup C(w)$ is a subgroup of G.*

Proof. We show first that elements of S have order 2. Take $w = s^{-1}$ in (Bru3) to get

$$B \subseteq C(s)C(s^{-1}) \subseteq B \cup C(s^{-1}) .\tag{6.17}$$

This implies that $C(s^{-1})$ meets $C(s)^{-1}$ and hence is equal to it. Since $C(s)C(s^{-1}) = C(s)C(s)^{-1} \not\subseteq B$ by (Th) (and our convention about how to interpret it), we conclude that the two double cosets on the right of (6.17) are distinct and that

$$C(s)C(s^{-1}) = B \amalg C(s^{-1}) .$$

Taking inverses, we obtain $C(s)C(s^{-1})$ again on the left but $B \amalg C(s)$ on the right. This implies that $C(s^{-1}) = C(s)$, and the equality above becomes

$$C(s)C(s) = B \amalg C(s) .$$

On the other hand, if we take $w = s$ in (Bru3) and use the fact that $C(s)C(s)$ is known to consist of two double cosets, then we obtain

$$C(s)C(s) = C(s^2) \amalg C(s) .$$

Hence $C(s^2) = B$ and $C(s) \neq B$, so s has order 2 by (Bru1).

Next, we show that (W, S) is a Coxeter system by verifying the folding condition (F) of Section 2.3.1. Given $w \in W$ and $s, t \in S$ such that $l(sw) = l(w) + 1 = l(wt)$ but $l(swt) < l(w) + 2$, we must show that $sw = wt$. The proof is very similar to the proof of (3) in Proposition 6.36, with the present w playing the role of the element w_1 in that proof. Namely, we compute $C(s)C(w)C(t)$ in two different ways. (Note that we can freely use the results of Proposition 6.36 when we do these computations.) First, we have

$$C(s)C(w)C(t) = C(s)C(wt) = C(swt) \cup C(wt) \,,$$

where the first equality comes from $l(wt) = l(w) + 1$ and the second from $l(s(wt)) \leq l(wt)$. Similarly,

$$C(s)C(w)C(t) = C(sw)C(t) = C(swt) \cup C(sw).$$

Hence $C(sw) = C(wt)$, and so $sw = wt$.

Finally, the last assertion of the proposition is an immediate consequence of the lemma that follows.   □

Lemma 6.41. *Suppose we are given a thick Bruhat decomposition for (G, B). If $w \in W$ admits a reduced decomposition $w = s_1 \cdots s_l$, then the subgroup of G generated by $C(w)$ contains the double cosets $C(s_i)$ for $i = 1, \ldots, l$ and is generated by them. Moreover, this subgroup is generated by B and gBg^{-1} for any $g \in C(w)$.*

Proof. The subgroup generated by $C(w) = BgB$ contains g and B. We therefore have

$$\langle B, gBg^{-1} \rangle \leq \langle C(w) \rangle \leq \langle C(s_1), \ldots, C(s_l) \rangle \,,$$

where the second inclusion follows from Proposition 6.36(4). The lemma will follow if we can show that the subgroup $P := \langle B, gBg^{-1} \rangle$ contains $C(s_i)$ for each i. We argue by induction on $l = l(w)$. Since $l(s_1w) < l(w)$, we know that $C(s_1)C(w) = C(s_1w) \cup C(w)$. Writing $C(s_1) = Bh_1B$, this becomes $Bh_1BgB = C(s_1w) \cup C(w)$, so h_1Bg meets $C(w) = BgB$. Hence h_1B meets $BgBg^{-1}$, which implies that $C(s_1) \subseteq P$. Let $w' = s_1w = s_2 \cdots s_l$. Then $C(w) = C(s_1)C(w')$, so we can assume that $g = h_1g'$ with $g' \in C(w')$. Since P contains h_1 and $gBg^{-1} = h_1g'Bg'^{-1}h_1^{-1}$, we also have $g'Bg'^{-1} \leq P$. We can therefore apply the induction hypothesis to w' to conclude that P contains $C(s_i)$ for $i = 2, \ldots, l$, whence the lemma.   □

To summarize the results so far, a thick Bruhat decomposition (and hence a Weyl-transitive action on a thick building) is determined by a group G, a subgroup B, a group W, and a function $C \colon W \to B\backslash G/B$ such that W admits a set of generators S for which (Bru1), (Bru2), (Bru3), and (Th) are satisfied. Since S is unique, it does not have to be specified as part of the structure.

We will see in the next subsection that in fact, a thick Bruhat decomposition is uniquely determined by the pair (G, B) alone.

Remark 6.42. The fact that the elements of S do not have to be assumed to be of order 2 in Proposition 6.40 is interesting but of no practical importance. For in all examples that we know of, it is trivial to verify that S consists of elements of order 2. If we simply add this as a hypothesis, then we have $C(s)^{-1} = C(s)$ by Proposition 6.36, so we can forget about the annoying distinction between left cosets and right cosets in interpreting (Th).

6.2.4 Parabolic Subgroups

We introduced Lemma 6.41 in order to finish off the proof of Proposition 6.40. But the lemma has much more striking consequences. As in Section 6.1.8, a subgroup of the form $P_J := \bigcup_{w \in W_J} C(w)$, where $J \subseteq S$, will be called a *standard parabolic subgroup*.

Theorem 6.43. *Suppose (G, B) admits a thick Bruhat decomposition.*

(1) *The standard parabolic subgroups are precisely the subgroups of G containing B.*

(2) *If P is a standard parabolic subgroup and $gBg^{-1} \leq P$ for some $g \in G$, then $g \in P$.*

(3) *Every standard parabolic subgroup is equal to its own normalizer, and no two of them are conjugate.*

Proof. The standard parabolics certainly contain $B = C(1)$. Conversely, suppose P is a subgroup containing B. Then P is a union of double cosets $C(w)$; hence it is generated by certain double cosets $C(s)$ with $s \in S$ by Lemma 6.41. But any such subgroup is a standard parabolic by Proposition 6.36(6). This proves (1)

(2) is an immediate consequence of the last assertion of Lemma 6.41, which says that the subgroup generated by B and gBg^{-1} contains BgB.

Finally, suppose P and P' are standard parabolics and $gP'g^{-1} = P$ for some $g \in G$. Then $gBg^{-1} \leq P$, so (2) implies that $g \in P$. Thus $P = P'$ and is its own normalizer, whence (3). $\square$

It follows from the theorem that the building $\Delta(G, B)$ associated with a thick Bruhat decomposition can be described entirely in terms of (G, B), as we claimed in Remark 6.33:

Corollary 6.44. *Suppose (G, B) admits a thick Bruhat decomposition. Then the building $\Delta(G, B)$ of Definition 6.32 is the poset of cosets gP, where P ranges over the subgroups of G containing B and the cosets are ordered by reverse inclusion.* $\square$

Following Bourbaki [44, Section IV.2, Exercise 3], we introduce the following terminology:

Definition 6.45. A subgroup B of a group G is called a *Tits subgroup* if (G, B) admits a thick Bruhat decomposition.

Thus giving a group G and a Tits subgroup B is essentially the same as giving a Weyl-transitive action of G on a thick building. We have used the first assertion of Theorem 6.43 to give one description of this building in terms of (G, B). We now use the second assertion to give a different description.

Definition 6.46. Given a group G and a Tits subgroup B, a subgroup $Q \leq G$ is called *parabolic* if Q contains a conjugate of B, or, equivalently, if Q is conjugate to a standard parabolic subgroup.

It follows from statement (3) of Theorem 6.43 that there is a bijection from the set of standard parabolic cosets to the set of parabolic subgroups, given by $gP \mapsto gPg^{-1}$. This bijection is compatible with the inclusion relation. Consequently:

Corollary 6.47. *Let B be a Tits subgroup of a group G. Then the building $\Delta(G, B)$ is isomorphic to the set of parabolic subgroups of G, ordered by reverse inclusion, with G acting by conjugation.* $\square$

Remarks 6.48. (a) Since the simplices of $\Delta = \Delta(G, B)$ correspond to the parabolic subgroups of G, one can use geometric language to express properties of the parabolic subgroups. Consider, for example, the minimal parabolics (conjugates of B). These correspond to the chambers of Δ, so we can talk about the distance between two minimal parabolics. In the spherical case, we can ask whether two minimal parabolics are opposite to one another. In $GL_n(k)$, for example, the upper-triangular group and the lower-triangular group are opposite to one another (Exercise 4.78).

(b) We have seen that a Tits subgroup B gives rise to a building $\Delta(G, B)$, which can be described without reference to the group W and the function C. The building then gives us the Weyl group W and the function $w \mapsto C(w)$ providing the Bruhat decomposition. In principle, then, it ought to be possible to get this structure directly from (G, B). It is indeed quite easy to do this:

Index the double cosets in $B\backslash G/B$ by a *set* W, so that we are given a bijection $w \mapsto C(w)$ from W to $B\backslash G/B$. Denote by 1 the element of W such that $C(1) = B$, and let S be the set of elements $s \neq 1$ in W such that $B \cup C(s)$ is a subgroup of G. For $s \in S$ and $w \in W$, define sw to be the element of W such that $C(s)C(w) \subseteq C(sw) \cup C(w)$. This uniquely determines a product on W such that it is a group with S as a set of generators of order 2. The resulting pair (W, S) is in fact a Coxeter system, and the bijection C that we started with gives the essentially unique thick Bruhat decomposition for (G, B).

The verification of these assertions, which the reader is asked to carry out in Exercise 6.51 below, is based on the fact that one knows a priori that (G, B) admits a thick Bruhat decomposition. It is natural to go further and try to axiomatize Tits subgroups entirely in terms of the pair (G, B). Giving the details would take us too far afield, but an outline can be found in Bourbaki [44, Section IV.2, Exercise 3].

Exercises

6.49. If B is a Tits subgroup of G, show that B contains the center $Z(G)$. Deduce that $Z(G)$ acts trivially on $\Delta = \Delta(G, B)$. Show further that the image of G in Aut Δ has trivial center.

6.50. If a group G contains a Tits subgroup, show that G cannot be nilpotent.

6.51. Verify the assertions made in Remark 6.48(b).

6.2.5 Strongly Transitive Actions

We know that Weyl-transitive actions correspond to Bruhat decompositions. We now develop the algebraic theory corresponding to *strongly transitive* actions.

Suppose we are given a quadruple (G, B, N, S), where G is a group; B and N are subgroups that generate G; the intersection $T := B \cap N$ is normal in N; and S is a finite set of generators of the quotient group $W := N/T$. Thus we have the setup described in Section 6.1.1:

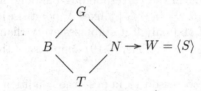

As in Remark 6.19, we will find it convenient to write expressions like BwB, wB, wBw', and so on, where $w, w' \in W$. These are defined by using representatives in N for elements of W, and they are independent of the choice of representative.

Proposition 6.52. *Assume that (W, S) is a Coxeter system and that*

$$sBw \subseteq BswB \cup BwB \tag{6.18}$$

for all $s \in S$ and $w \in W$. Then the function $w \mapsto C(w) := BwB$ provides a Bruhat decomposition for (G, B). Let $\Delta = \Delta(G, B)$ be the associated building. Then Δ contains an apartment Σ stabilized by N such that N is transitive on $\mathcal{C}(\Sigma)$. Hence the action of G on Δ is strongly transitive with respect to the apartment system $\mathcal{A} := G\Sigma$.

Proof. The inclusion (6.18) is a restatement of the second inclusion in (Bru3), since $C(s)C(w) = BsBwB$. The first inclusion in (Bru3) holds trivially in the present setup, as does (Bru1). To get a Bruhat decomposition we must show that the conditions of the proposition imply (Bru2). To this end, note that $\bigcup_{w \in W} BwB$ is a subgroup; the proof is the same as that of statement (6) in Proposition 6.36. This subgroup contains B and N; hence it is G.

Let $\Delta = \Delta(G, B)$ be the associated building, with $\mathcal{C}(\Delta) = G/B$. Then the chambers wB ($w \in W$) form a subset W-isometric to W. This gives the desired apartment Σ. The last assertion of the proposition follows from the proof of Proposition 6.14.   □

Roughly speaking, then, strongly transitive actions with a chosen fundamental pair (Σ, C) correspond to quadruples (G, B, N, S) satisfying (6.18). This statement is slightly misleading, however, because the subgroup N in the quadruple might not be the full stabilizer of Σ. We have already seen examples of this in connection with $G = \mathrm{GL}_n(k)$; see Remark 6.8(c). So if we want the group theory to precisely reflect the theory of strongly transitive actions, we need to add a new axiom that guarantees that N is big enough. The appropriate axiom turns out to be

$$T = \bigcap_{w \in W} wBw^{-1} . \tag{6.19}$$

To see why this is the right axiom, consider an arbitrary (G, B, N, S) satisfying the conditions of Proposition 6.52, and let $\widetilde{N}$ be the stabilizer of the fundamental apartment Σ in the corresponding building. Then it is easy to see that $\widetilde{N} = N\widetilde{T}$, where $\widetilde{T} := \{g \in G \mid g \text{ fixes } \Sigma \text{ pointwise}\}$. Now an element of G fixes Σ pointwise if and only if it stabilizes every chamber of Σ [why?]; so we have $\widetilde{T} = \bigcap_{w \in W} wBw^{-1}$. Thus (6.19) simply says that $\widetilde{T} = T$, which implies that $\widetilde{N} = N$.

In practice, there is no reason to impose the condition (6.19). For if we want to apply geometry to group theory, the important thing is to be able to construct a building associated to a given group.

Remark 6.53. We have based our proof of Proposition 6.52 on the W-metric approach to buildings. In the (G, B, N, S) context, it is possible to carry out the construction of the building entirely in the simplicial world, without reference to Weyl distances, but one has to work a little harder. See the following exercise.

Exercise 6.54. Let (G, B, N, S) satisfy the hypotheses of Proposition 6.52. Let $\Delta = \Delta(G, B)$ be as in Definition 6.32. Show that Δ is a building with a strongly transitive type-preserving G-action by proving the following:

(a) The poset Δ is a simplicial complex.
(b) Let Σ be the subcomplex of Δ consisting of the cosets of the form wP with $w \in W$ and P a standard parabolic subgroup of G. Then Σ is a Coxeter complex.
(c) Let $\mathcal{A} = G\Sigma$. Then $\mathcal{A}$ is a system of apartments for Δ, which is therefore a building. The G-action is type-preserving and strongly transitive.

6.2.6 BN-Pairs

We have saved for last the algebraic theory that goes with the most common geometric situation: strongly transitive actions on thick buildings. All the work has been done already, and we just need to combine the results of Sections 6.2.3 and 6.2.5.

Definition 6.55. We say that a pair of subgroups B and N of a group G is a *BN-pair* if B and N generate G, the intersection $T := B \cap N$ is normal in N, and the quotient $W := N/T$ admits a set of generators S such that the following two conditions hold:

(BN1) *For $s \in S$ and $w \in W$,*

$$sBw \subseteq BswB \cup BwB .$$

(BN2) *For $s \in S$,*

$$sBs^{-1} \nsubseteq B .$$

The group W will be called the *Weyl group* associated to the BN-pair. One also says in this situation that the quadruple (G, B, N, S) is a *Tits system*.

The following theorem summarizes Propositions 6.40 and 6.52, together with some of the results of Section 6.1:

Theorem 6.56.

(1) *Given a BN-pair in G, the generating set S is uniquely determined, and (W, S) is a Coxeter system. There is a thick building $\Delta = \Delta(G, B)$ that admits a strongly transitive G-action such that B is the stabilizer of a fundamental chamber and N stabilizes a fundamental apartment and is transitive on its chambers.*

(2) *Conversely, suppose a group G acts strongly transitively on a thick building Δ with fundamental apartment Σ and fundamental chamber C. Let B be the stabilizer of C, and let N be a subgroup of G that stabilizes Σ and is transitive on the chambers of Σ. Then (B, N) is a BN-pair in G, and Δ is canonically isomorphic to $\Delta(G, B)$.* $\qquad\square$

We emphasize, once again, that the building $\Delta(G, B)$ associated to a BN-pair depends only on (G, B). But the subgroup N lets us exhibit an apartment system with respect to which the G-action is strongly transitive. We will say more about the role of N below.

Since we presented the theory of BN-pairs as a special case of the theory of Tits subgroups, which was developed over several subsections, we review here some of the important features of the theory. Let G be a group with a BN-pair. Then the axioms (Bru1)–(Bru3) of Section 6.2.2 and (Th) of Section 6.2.3 are satisfied, so all of the results derived from those axioms are true for (G, B). In particular:

(1) G has a Bruhat decomposition, which is a bijection

$$C \colon W \xrightarrow{\sim} B \backslash G / B \ .$$

given by

$$C(w) = BwB := B\tilde{w}B$$

for $w \in W$, where $\tilde{w}$ is a representative of w in N. More briefly,

$$G = BWB = \coprod_{w \in W} BwB \ .$$

(2) There is a more precise version of axiom (BN1): Given $s \in S$ and $w \in W$, we have

$$C(s)C(w) = \begin{cases} C(sw) & \text{if } l(sw) > l(w), \\ C(sw) \cup C(w) & \text{if } l(sw) < l(w). \end{cases}$$

(3) The subgroups of G containing B are precisely the standard parabolic subgroups $P_J = BW_J B$ ($J \subseteq S$). These are self-normalizing, and no two of them are conjugate.

(4) The building $\Delta = \Delta(G, B)$ has several equivalent descriptions. From the W-metric point of view, $\Delta = (\mathcal{C}, \delta)$, where $\mathcal{C} = G/B$ and $\delta \colon G/B \times G/B \to W$ is characterized by

$$\delta(gB, hB) = w \iff Bg^{-1}hB = C(w) \ .$$

The action of G is by left translation. The fundamental apartment is

$$\Sigma = \{wB \mid w \in W\} \ .$$

From the simplicial point of view, Δ is the set of standard parabolic cosets gP ($g \in G$, $P \geq B$), ordered by reverse inclusion. Again, G acts by left translation. The fundamental apartment is

$$\Sigma = \{wP \mid w \in W, \, P \geq B\} \ .$$

A second simplicial version takes Δ to be the poset of parabolic subgroups, ordered by reverse inclusion. The action of G is by conjugation. The fundamental apartment is

$$\Sigma = \{wPw^{-1} \mid w \in W, \, P \geq B\} \ .$$

Recall, finally, that N is not necessarily the full stabilizer of Σ. We analyzed this situation near the end of Section 6.2.5, and our discussion there leads to the following definition:

Definition 6.57. The BN-pair (B, N) is said to be *saturated* if it satisfies

$$T = \bigcap_{w \in W} wBw^{-1}$$

or, equivalently, if N is the full stabilizer of the fundamental apartment of $\Delta(G, B)$.

Thus saturated BN-pairs are essentially in 1–1 correspondence with strongly transitive group actions on thick buildings.

Remarks 6.58. (a) Although the building $\Delta = \Delta(G, B)$ does not depend on N, there appears to be a dependence on N in our description of Δ from the W-metric point of view. Here is one way to get rid of the apparent dependence on N. To begin with, suppose we have a Tits subgroup $B \leq G$ that is not necessarily part of a BN-pair. Let $(P_s)_{s \in S}$ be the family of minimal parabolic subgroups properly containing B. Here S is just an index set. Then we get a type function τ on $\Delta := \Delta(G, B)$ with values in S such that the panel P_s has cotype s. There is therefore a well-defined Coxeter group W generated by S, which is simply the Weyl group of Δ with respect to τ. Explicitly, the entry $m(s, t)$ of the Coxeter matrix of (W, S) is obtained by considering the type of the rank-2 building $\Delta(G_{s,t}, B)$, where $G_{s,t} := \langle P_s, P_t \rangle$ for $s \neq t$ in S. Consequently, there is a well-defined W-metric building depending only on (G, B).

Suppose now that the Tits subgroup B happens to be part of a Tits system (G, B, N, S'). Then there is a unique bijection $S \to S'$, denoted by $s \mapsto s'$, such that $P_s = B \cup Bs'B$ for all $s \in S$. Thus we can identify S' with S, and we can identify the Weyl group $N/(N \cap B)$ of the Tits system with W. [For this last assertion, consider the Coxeter matrix of $N/(N \cap B)$ with respect to S'; it can be described in the same way we described the Coxeter matrix of (W, S).] It is not hard to deduce now that the W-metric building associated to a BN-pair is independent of N up to canonical isomorphism. Further details are left to the interested reader.

(b) A second question that naturally arises is the extent to which N itself is unique. The following lemma and exercise shed some light on this.

Lemma 6.59. *Let (B, N) be a saturated BN-pair in G, and let N' be a subgroup of N. Then (B, N') is a BN-pair if and only if $N'T = N$ or, equivalently, N' surjects onto $W = N/T$.*

Proof. If N' surjects onto W, then it is trivial to verify that the BN-pair axioms are still satisfied by (B, N'). [Alternatively, note that N' stabilizes the fundamental apartment and is transitive on its chambers, so (B, N') is a BN-pair by Section 6.1.] Conversely, suppose (B, N') is a BN-pair, and let W' be the image of N' in W. Then we have $G = BW'B$, so $W' = W$ by the Bruhat decomposition. [Alternatively, note that N' acts transitively on the chambers of an apartment Σ' that is contained in the fundamental apartment Σ. By the thinness of apartments, we must have $\Sigma' = \Sigma$; hence N' surjects onto W.]   □

Exercise 6.60.

(a) Let (B, N) be a saturated BN-pair in G of spherical type (i.e., $\Delta(G, B)$ is spherical). If N' is a subgroup of G such that (B, N') is also a saturated BN-pair, show that $N' = bNb^{-1}$ for some $b \in B$.
(b) Continuing with the hypotheses and notation of (a), set $T := B \cap N$ and $T' := B \cap N'$. Show that the isomorphism $N/T \xrightarrow{\sim} N'/T'$ induced by conjugation by b is the same isomorphism that one gets via Remark 6.58(a).
(c) What happens in (a) if $\Delta(G, B)$ is not spherical?

6.2.7 Simplicity Results

In this section we present results of Tits [244, Section 2], except that we formulate these results in terms of Tits subgroups, whereas Tits worked with BN-pairs. This requires minor modifications, but otherwise our treatment follows his very closely. Throughout this subsection, G denotes a group with a Tits subgroup B of *irreducible* type (W, S), and $\Delta := \Delta(G, B)$ is the associated thick building. Thus we have a Weyl-transitive action of G on Δ, and B is the stabilizer of a fundamental chamber C. We wish to give criteria under which we can prove that G (or a closely related subquotient) is a simple group. We set

$$Z := \bigcap_{g \in G} gBg^{-1} \, ;$$

it is the kernel of the action of G on Δ. The letter Z is chosen because Z turns out to be the center of G in many examples. For instance if $G = \mathrm{GL}_n(k)$ as in our standard example (see Section 6.1.2 and Remark 6.8(d)), then Z is the set of scalar multiples of the identity.

Everything we do in this subsection will be based on the following lemma:

Lemma 6.61. *If H is a normal subgroup of G, then either $H \leq Z$ or else $HB = G$. In other words, either H acts trivially on Δ or else H acts chamber transitively on Δ.*

Proof. Since H is normal, HB is a subgroup of G containing B; hence it is a standard parabolic subgroup (Theorem 6.43). Thus

$$HB = \bigcup_{w \in W_J} C(w)$$

for some subset $J \subseteq S$. We claim that if $s \in J$ and $t \in S \setminus J$, then $st = ts$. Since (W, S) is irreducible, this implies that $J = \emptyset$ or $J = S$. If $J = \emptyset$, then $HB = B$, i.e., $H \leq B$, and hence $H \leq Z$ by normality. If $J = S$, then $HB = G$. It remains to prove the claim.

Since $s \in J$, we have $C(s) \subseteq HB$ and hence $C(s)$ meets H. Choose $h \in C(s) \cap H$ and $g \in C(t)$. Then $ghg^{-1} \in H \leq HB$, so $ghg^{-1} \in C(w)$ for some $w \in W_J$. On the other hand, basic facts about products of double cosets (Section 6.2.2) imply that

$$ghg^{-1} \in C(t)C(s)C(t) = C(t)C(st) \subseteq C(tst) \cup C(st) .$$

So either $tst \in W_J$ or $st \in W_J$. Since $s \in J$ and $t \notin J$, we cannot have $st \in W_J$, so the only possibility is that $tst \in W_J$. But also $tst \in W_{\{s,t\}}$, so Proposition 2.16 implies that $tst \in W_K$ with $K = J \cap \{s,t\} = \{s\}$, i.e., $tst = s$. This proves the claim. $\square$

We can immediately deduce the following simplicity theorem:

Theorem 6.62. *Assume that G is perfect and B is solvable. Then every proper normal subgroup of G is contained in Z. Hence G/Z is a simple group.*

Proof. Let H be a proper normal subgroup of G. By the lemma, either $H \leq Z$ or $HB = G$. But the second case is impossible, because it would imply that the (nontrivial) perfect group G/H is isomorphic to the solvable group $B/(B \cap H)$. $\square$

This applies, for example, to $G = \mathrm{SL}_n(k)$ unless $n = 2$ and $k = \mathbb{F}_2$ or $\mathbb{F}_3$. Indeed, the upper-triangular subgroup B is solvable, and easy computations show that G is perfect. [We will have occasion to do those computations in Chapter 7; see Example 7.133 and Exercise 7.139.] So one concludes from the theorem that $\mathrm{PSL}_n(k)$ is simple.

With a little more work, we can improve Theorem 6.62 to a result that is slightly more technical to state but has wider applicability. It is based on the following lemma:

Lemma 6.63. *Let U be a normal subgroup of B, and let G_1 be the normal closure of U in G. Assume that G_1 is perfect and that U is solvable. Then every normal subgroup of G either is contained in Z or contains G_1.*

Proof. Let H be a normal subgroup of G that is not contained in Z. Then $HB = G$ by Lemma 6.61. Since U is normal in B, HU is normal in $HB = G$, so $HU \geq G_1$. We now have

$$U/(U \cap H) \cong HU/H = HG_1/H \cong G_1/(G_1 \cap H) .$$

The group on the left is solvable and the one on the right is perfect, so these groups are trivial. Hence $G_1 \cap H = G_1$, i.e., $G_1 \leq H$. $\square$

We are ready now for the main result. In Tits's paper [244], he assumed that B was part of a BN-pair, and he made use of the group $T = B \cap N$. In our more general setup, we will work instead with the (thick) building $\Delta = \Delta(G, B)$. Choose an apartment Σ (in the complete apartment system) containing the fundamental chamber C, and suppose we are given a subgroup $T \leq B$ that fixes Σ pointwise.

Theorem 6.64. *Let U be a normal subgroup of B such that $B = UT$, and let G_1 be the normal closure of U in G. Assume that G_1 is perfect and that U is solvable. Then every subgroup of G that is normalized by G_1 either is contained in Z or contains G_1. In particular, $G_1/(G_1 \cap Z)$ is a simple group.*

Proof. Recall that by Weyl transitivity, $\Delta = \bigcup_{b \in B} b\Sigma$ (Proposition 6.11). Since $B = UT$, this can be rewritten as

$$\Delta = \bigcup_{u \in U} u\Sigma. \tag{6.20}$$

In particular, U does not act trivially on Δ, since that would imply $\Delta = \Sigma$, contradicting thickness. We now apply Lemma 6.61 with $H = G_1$ to conclude that $G = G_1 B$, i.e., that G_1 acts chamber transitively on Δ. It then follows from (6.20) and another application of Proposition 6.11 that G_1 acts Weyl transitively on Δ. Another consequence of the equation $G = G_1 B$ is that G_1, which is generated by the G-conjugates of U, is actually generated by the G_1-conjugates of U.

Suppose now that H is a subgroup of G that is normalized by G_1, and consider the group $G' := G_1 H$. It acts Weyl transitively on Δ, since it contains G_1, and G_1 is the normal closure of U in G' by the last sentence of the previous paragraph. Setting $B' := G' \cap B$, we can now apply Lemma 6.63 with (G, B) replaced by (G', B') (but with the same U and G_1). Since H is normal in G', the result is that either H acts trivially on Δ (hence $H \leq Z$) or else $H \geq G_1$. □

The following further observations are taken directly from Tits [244, 2.9].

Remarks 6.65. (a) The assumption that G_1 is perfect is equivalent to saying that U is contained in the commutator subgroup of G_1.

(b) Suppose $Z \cap U = \{1\}$. Then, since Z and U normalize each other, it follows that Z centralizes U and hence G_1. The last sentence of the theorem therefore implies that G_1 modulo its center is simple.

(c) It is clear from the proof that we could weaken the assumption that U is solvable; it suffices to assume that U has no nontrivial perfect quotients. In classical examples, U is in fact solvable, but applications with nonsolvable U have arisen recently in connection with Kac–Moody groups; see [74]. See also [71, Theorem 19] for a related simplicity theorem.

Finally, it is instructive to see what Theorem 6.64 says about our standard example. We noted above that we could apply Theorem 6.62 with $G = SL_n(k)$ and deduce (with two exceptions) that $PSL_n(k)$ is simple. We get a stronger result if we instead apply Theorem 6.64 with $G = GL_n(k)$, B equal to the upper-triangular group, U equal to the strict upper-triangular group, and T equal to the group of diagonal matrices. The normal closure G_1 of U is $SL_n(k)$, since the latter is generated by elementary matrices. So the conclusion is that (if we exclude the two exceptional cases) every subgroup of $GL_n(k)$ that is normalized by $SL_n(k)$ either is contained in the center of $GL_n(k)$ or contains $SL_n(k)$.

We will see further applications of the theorem later in this chapter, as well as in the next chapter.

Exercise 6.66. Did the proof of Theorem 6.64 use the assumption that T fixes Σ pointwise, or could we have just assumed that T stabilizes Σ?

*6.3 Twin BN-Pairs and Twin Buildings

Just as a strongly transitive group action on a building corresponds to a BN-pair, we will see that a strongly transitive group action on a twin building corresponds to a "twin BN-pair." The first step is to figure out what we should mean by a strongly transitive action on a twin building. We assume that the reader is familiar with the material of Section 5.8 on twin buildings.

6.3.1 Group Actions on Twin Buildings

In the following, $\mathcal{C} = (\mathcal{C}_+, \mathcal{C}_-)$ denotes a twin building of type (W, S), and G is a group. (As usual, we have suppressed the codistance δ^* from the notation.)

Definition 6.67. We say that G *acts* on $\mathcal{C}$ if it acts simultaneously on the two sets $\mathcal{C}_+$ and $\mathcal{C}_-$ and preserves Weyl distances and the codistance. This means that for all $g \in G$, all $C, C' \in \mathcal{C}_+$, and all $D, D' \in \mathcal{C}_-$, we have

(1) $\delta_+(gC, gC') = \delta_+(C, C')$.
(2) $\delta_-(gD, gD') = \delta_-(D, D')$.
(3) $\delta^*(gC, gD) = \delta^*(C, D)$.

Remark 6.68. In view of Lemma 5.143, (3) can be replaced by

(3') $C \text{ op } D \implies gC \text{ op } gD$.

Assume that G acts on $\mathcal{C}$, and fix a "fundamental" pair of chambers (C_+, C_-), where $C_+ \in \mathcal{C}_+$, $C_- \in \mathcal{C}_-$, and $C_+ \text{ op } C_-$. Let $B_\pm$ be the stabilizer of $C_\pm$:

$$B_+ := \{g \in G \mid gC_+ = C_+\},$$
$$B_- := \{g \in G \mid gC_- = C_-\}.$$

Next, the choice of (C_+, C_-) yields a "fundamental twin apartment" $\Sigma = (\Sigma_+, \Sigma_-) = \Sigma\{C_+, C_-\}$, which is the unique twin apartment containing (C_+, C_-) by Proposition 5.179. This leads to two more subgroups of G:

$$N := \{g \in G \mid g\Sigma = \Sigma\},$$
$$T := \{g \in G \mid gC = C \text{ for all } C \in \Sigma_+ \cup \Sigma_-\}.$$

Note, finally, that G acts on the apartment systems $\mathcal{A}_+$, $\mathcal{A}_-$, and $\mathcal{A}$ introduced in Definition 5.180, since it preserves distances and codistances.

The first results concerning the groups B_+, B_-, and N do not require any transitivity assumptions.

Lemma 6.69. *With the notation above, we have:*

(1) N *is the stabilizer of* Σ_+ *as well as the stabilizer of* Σ_-.
(2) T *is the pointwise fixer of each* Σ_ϵ, *i.e.,*

$$T = \{g \in G \mid gC = C \text{ for all } C \in \Sigma_+\} ,$$
$$= \{g \in G \mid gC = C \text{ for all } C \in \Sigma_-\} .$$

(3) $B_+ \cap B_- = B_+ \cap N = B_- \cap N = T$.

Proof. (1) follows immediately from the fact that Σ_ϵ, for $\epsilon = +$ or $-$, has only one twin partner with which it can form a twin apartment, namely $\Sigma_{-\epsilon}$ (see Proposition 5.179(2)). And if an element of N fixes Σ_ϵ pointwise, then it also has to fix $\Sigma_{-\epsilon}$ pointwise, since the action preserves opposition. This yields (2). Now the stabilizer in N of C_ϵ fixes Σ_ϵ pointwise by the standard uniqueness argument. Therefore, $B_+ \cap N = T = B_- \cap N$. Finally, $B_+ \cap B_-$ stabilizes $\Sigma = \Sigma\{C_+, C_-\}$, and hence $B_+ \cap B_- \subseteq T$ by the previous argument; the opposite inclusion is trivial.   □

We remark in passing that T is the kernel of the canonical homomorphism from N to W, where the latter is identified with the group of isometries of Σ_ϵ for $\epsilon = \pm$. Moreover, we get the same homomorphism $N \to W$ for each choice of ϵ. Indeed, consider an element $n \in N$, and suppose that its action on Σ_+ is given by $w \in W$. Then $\delta_+(C_+, nC_+) = w$; hence

$$\delta_-(C_-, nC_-) = \delta^*(C_+, nC_-) = \delta_+(C_+, nC_+) = w ,$$

since all the chambers that occur are in $\Sigma\{C_+, C_-\} = \Sigma\{nC_+, nC_-\}$. Thus n also acts as w on Σ_-, as claimed. If the action of G is strongly transitive in the sense to be defined below, then N acts transitively on Σ_ϵ, so the homomorphism $N \to W$ is surjective and $W \cong N/T$.

By analogy with the theory of Section 6.1, one might imagine several reasonable transitivity conditions to impose on the action of G on $\mathcal{C}$. It turns out that they are all equivalent:

Lemma 6.70. *The following conditions are equivalent:*

(i) *For any* $w \in W$, G *acts transitively on*

$$\{(C, D) \in \mathcal{C}_+ \times \mathcal{C}_- \mid \delta^*(C, D) = w\} .$$

(ii) *For* $\epsilon = +$ *or* $-$, G *acts transitively on* $\mathcal{C}_\epsilon$, *and* B_ϵ *acts transitively on* $\{D \in \mathcal{C}_{-\epsilon} \mid \delta^*(C_\epsilon, D) = w\}$ *for each* $w \in W$.

(iii) G *acts transitively on* $\{(C, C') \in \mathcal{C}_+ \times \mathcal{C}_- \mid C \text{ op } C'\}$.
(iv) G *acts transitively on* $\mathcal{A}$, *and* N *acts transitively on*

$$\{(C, C') \in \Sigma_+ \times \Sigma_- \mid C \text{ op } C'\} \ .$$

(v) *For* $\epsilon = + \ or \ -$, G *acts transitively on* $\mathcal{A}_\epsilon$, *and* N *acts transitively on* Σ_ϵ.

(Note that we are taking the W-metric point of view here, so Σ_ϵ in (v) is a set of chambers.)

Proof. The equivalence of (i) and (ii) is routine, and trivially (i) $\implies$ (iii). Suppose now that (iii) is satisfied. Then (by Proposition 5.179) G acts transitively on $\mathcal{A}$. And if $g \in G$ satisfies $g(C_+, C_-) = (C, C')$ for some $(C, C') \in \Sigma_+ \times \Sigma_-$ with C op C', then g is automatically in N since $\Sigma \{C, C'\} = \Sigma$. Hence (iii) $\implies$ (iv). Now if G acts transitively on $\mathcal{A}$, then by the definition of $\mathcal{A}$, $\mathcal{A}_+$, and $\mathcal{A}_-$, it also acts transitively on $\mathcal{A}_+$ and $\mathcal{A}_-$. And the transitive action of N on $\{(C, C') \in \Sigma_+ \times \Sigma_- \mid C \text{ op } C'\}$ is, in view of the definition of twin apartments, equivalent to the transitive action of N on Σ_+ or Σ_-. So (v) follows from (iv). We verify, finally, that (v) implies (i), where we may assume that $\epsilon = +$ in (v).

Suppose we are given $C, C_1 \in \mathcal{C}_+$, $D, D_1 \in \mathcal{C}_-$, and $w \in W$ with $\delta^*(C, D) = w = \delta^*(C_1, D_1)$. We have to find a $g \in G$ such that $g(C_1, D_1) = (C, D)$. Choose a twin apartment containing (C, D) and a twin apartment containing (C_1, D_1); this is possible by Proposition 5.179(3). Note that G, acting transitively on $\mathcal{A}_+$, also acts transitively on $\mathcal{A}$, since each element of $\mathcal{A}_+$ is part of a unique twin apartment. So we may assume that (C, D) and (C_1, D_1) are both contained in the fundamental twin apartment Σ. Applying the transitive action of N on Σ_+, we can also achieve $C_1 = C$. But in the twin apartment Σ, there is only one chamber D satisfying $\delta^*(C, D) = w$ (see Lemma 5.173(3)). Hence also $D_1 = D$. $\qquad\square$

Definition 6.71. We say that the action of G on $\mathcal{C}$ is *strongly transitive* if it satisfies the five equivalent conditions of Lemma 6.70.

In view of condition (i), it would also be reasonable to call the action Weyl transitive. However, since all these conditions are equivalent in the present situation, we will stick to the more common term "strongly transitive."

Corollary 6.72. *Suppose that* G *acts strongly transitively on* $\mathcal{C}$.

(1) *For* $\epsilon \in \{+, -\}$, G *acts strongly transitively on the building* $\mathcal{C}_\epsilon$ *with respect to the apartment system* $\mathcal{A}_\epsilon$. *In particular,* G *acts Weyl transitively on* $\mathcal{C}_+$.
(2) *If* $\mathcal{C}$ *is thick, then* (B_+, N) *and* (B_-, N) *are BN-pairs in* G *with common Weyl group* $N/T \cong W$.

Proof. The first assertion of (1) is immediate from the definition of strong transitivity via condition (v) of Lemma 6.70, and the second assertion follows from the first by Corollary 6.12. We already remarked after Lemma 6.69 that W can be identified with N/T if N acts transitively on Σ_ϵ. Assertion (2) now follows from Theorem 6.56(2). $\qquad\square$

We close this subsection by recording an analogue of Proposition 6.6(2).

Lemma 6.73. *Suppose a group G acts strongly transitively on a twin building. Given two twin apartments Σ, Σ', there is an element $g \in G$ such that $g\Sigma = \Sigma'$ and g fixes every chamber in $\Sigma \cap \Sigma'$.*

Proof. This is immediate if $\Sigma \cap \Sigma' = \emptyset$, so we may assume without loss of generality that $\Sigma_+ \cap \Sigma'_+$ contains a chamber C. By strong transitivity, there exists $g \in G$ such that $g\Sigma_+ = \Sigma'_+$ and $gC = C$. Then necessarily $g\Sigma_- = \Sigma'_-$, and it follows by a variant of the standard uniqueness argument that g fixes $\Sigma \cap \Sigma'$ pointwise. [For chambers in $\Sigma_+ \cap \Sigma'_+$, one can use the usual standard uniqueness argument. To see that g fixes every chamber $D \in \Sigma_- \cap \Sigma'_-$, observe that $\delta^*(C, gD) = \delta^*(C, D)$; since gD and D are both in Σ'_- and C is in Σ'_+, it follows from Lemma 5.173(3) that $gD = D$.] $\square$

The following immediate consequence will be needed in Chapter 8.

Corollary 6.74. *Suppose a group G acts strongly transitively on a twin building, let α be a twin root, and let $\mathrm{Fix}_G(\alpha)$ be the pointwise fixer of α in G. Then $\mathrm{Fix}_G(\alpha)$ acts transitively on the set $\mathcal{A}(\alpha)$ of twin apartments containing α.* $\square$

6.3.2 Group-Theoretic Consequences

We now proceed as in Sections 6.1.4–6.1.7 to interpret strong transitivity in group-theoretic terms, involving double cosets. Our discussion can be briefer here, since the arguments are identical to those we have already seen. First of all, strong transitivity in the form of condition (ii) of Lemma 6.70 implies that the B_+-orbits in $\mathcal{C}_-$ are in 1–1 correspondence with the elements of W, with the orbit of a chamber $D \in \mathcal{C}_-$ corresponding to $w = \delta^*(C_+, D)$. But the B_+-orbits in $\mathcal{C}_-$ correspond to the B_+-orbits in G/B_-, so we get a bijection

$$B_+ \backslash G / B_- \overset{\sim}{\longrightarrow} W ,$$

given by

$$B_+ g B_- \mapsto \delta^*(C_+, gC_-) \tag{6.21}$$

for $g \in G$. We claim that the inverse of this bijection is given by

$$w \mapsto B_+ w B_-$$

for $w \in W$. To see this, recall that by the paragraph following the proof of Lemma 6.69, we have $\delta^*(C_+, nC_-) = w$ if $n \in N$ represents $w \in W$. Hence the double coset $B_+ w B_- := B_+ n B_-$ maps to w under the map in (6.21).

Obviously, we could have interchanged the roles of $\mathcal{C}_+$ and $\mathcal{C}_-$ above. We therefore obtain the following analogue of the Bruhat decomposition:

Proposition 6.75. *If a group G acts strongly transitively on a twin building, then, with the notation introduced above,*

$$G = \coprod_{w \in W} B_\epsilon w B_{-\epsilon} \tag{6.22}$$

for each $\epsilon = \pm$. Given $g \in G$ and $w \in W$, we have

$$g \in B_\epsilon w B_{-\epsilon} \iff \delta^*(C_\epsilon, gC_{-\epsilon}) = w \ . \qquad \square$$

Equation (6.22) is called the *Birkhoff decomposition*. For reasons that will become obvious in the next subsection, we call attention to the following consequence of it:

Corollary 6.76. *If G acts strongly transitively on a twin building, then*

$$B_+ s \cap B_- = \emptyset$$

for all $s \in S$.

Proof. This just says that the double cosets $B_+ s B_-$ and $B_+ B_-$ are different.
$\square$

Our discussion so far has made very little use of the fact that δ^* is the codistance function of a twin building. As before (see Theorem 6.21 and Proposition 6.24), we get results about products of double cosets from the properties of δ^*:

Proposition 6.77. *Assume that G acts strongly transitively on $\mathcal{C}$. For $\epsilon = \pm$, $w \in W$, and $s \in S$, we have:*

(1) $B_\epsilon s B_\epsilon w B_{-\epsilon} \subseteq B_\epsilon \{sw, w\} B_{-\epsilon}$.
(2) *If $l(sw) < l(w)$, then $B_\epsilon s B_\epsilon w B_{-\epsilon} = B_\epsilon sw B_{-\epsilon}$.*
(3) *If $\mathcal{C}$ is thick and $l(sw) > l(w)$, then*

$$B_\epsilon s B_\epsilon w B_{-\epsilon} = B_\epsilon \{sw, w\} B_{-\epsilon} \ .$$

Proof. We may assume $\epsilon = +$. Given $h \in B_+ s B_+$ and $g \in B_+ w B_-$, we have $\delta_+(C_+, hC_+) = s$ and $\delta^*(hC_+, hgC_-) = w$. Hence $\delta^*(C_+, hgC_-) \in \{sw, w\}$ by Lemma 5.139(1), so $hg \in B_+ \{sw, w\} B_-$; this proves (1). If $l(sw) < l(w)$, then $\delta^*(C_+, hgC_-) = sw$ by (Tw2), whence (2). If $l(sw) > l(w)$, then both sw and w are possible by Lemma 5.139(2) and thickness, so (3) holds. $\square$

We close this subsection by noting that we can reconstruct the twin building $\mathcal{C}$ from group-theoretic data, as in the discussion at the end of Section 6.1.4. That discussion already showed how to reconstruct $(\mathcal{C}_\epsilon, \delta_\epsilon)$ for $\epsilon = \pm$, so it remains only to describe the codistance δ^*. If we make the identifications $\mathcal{C}_\epsilon = G/B_\epsilon$, then the result is that

$$\delta^*(gB_\epsilon, hB_{-\epsilon}) = w \iff g^{-1}h \in B_\epsilon w B_{-\epsilon} \tag{6.23}$$

for $g, h \in G$, $w \in W$, and $\epsilon \in \{+, -\}$.

6.3.3 Twin BN-Pairs

We have seen that a strongly transitive action of a group G on a twin building imposes certain conditions on G. We now wish to reverse the procedure and give axioms on a group G from which we can construct a strongly transitive action on a twin building. For simplicity, we will treat only the thick case; see Exercise 6.92 for the general case.

Definition 6.78. Let B_+, B_-, and N be subgroups of a group G such that $B_+ \cap N = B_- \cap N =: T$. Assume that $T \trianglelefteq N$, and set $W := N/T$. The triple (B_+, B_-, N) is called a *twin BN-pair* with Weyl group W if W admits a set S of generators such that the following conditions hold for all $w \in W$ and $s \in S$ and each $\epsilon \in \{+, -\}$:

(TBN0) (G, B_ϵ, N, S) *is a Tits system.*

(TBN1) *If* $l(sw) < l(w)$, *then* $B_\epsilon s B_\epsilon w B_{-\epsilon} = B_\epsilon sw B_{-\epsilon}$.

(TBN2) $B_+ s \cap B_- = \emptyset$.

In this situation we also say that the quintuple (G, B_+, B_-, N, S) is a *twin Tits system*.

Note that the definition is symmetric with respect to $+$ and $-$. In the case of (TBN2), we see this by right-multiplying by a representative of s in N.

Our discussion so far has shown the following:

Corollary 6.79. *If G acts strongly transitively on a thick twin building of type (W, S), then, with the notation introduced in Section 6.3.1, the quintuple (G, B_+, B_-, N, S) is a twin Tits system.* □

Here, of course, we identify the given W with N/T so that S can be viewed as a set of generators of the latter. Note that as in the theory of ordinary BN-pairs, the corollary is still valid if N is replaced by any subgroup that surjects onto N/T.

We now proceed to show that conversely, a twin BN-pair in a group G gives rise to a thick twin building on which G acts strongly transitively. The crucial step is to derive a Birkhoff decomposition from the TBN axioms.

Lemma 6.80. *Let (G, B_+, B_-, N, S) be a twin Tits system. For each $\epsilon = \pm$ and all $s \in S$ and $w \in W$,*

$$B_\epsilon s B_\epsilon w B_{-\epsilon} \subseteq B_\epsilon \{s, w\} B_{-\epsilon} . \tag{6.24}$$

Equality holds if $l(sw) > l(w)$.

Proof. We may assume that $\epsilon = +$ and that $l(sw) > l(w)$. Since (B_+, N) is a BN-pair, we have $B_+sB_+sB_+ = B_+ \cup B_+sB_+$. Combining this with the assumption $l(sw) > l(w)$ and using (TBN1), we obtain

$$
\begin{aligned}
B_+sB_+wB_- &= B_+s(B_+wB_-) \\
&= B_+s(B_+sB_+swB_-) \\
&= (B_+sB_+sB_+)swB_- \\
&= (B_+ \cup B_+sB_+)swB_- \\
&= B_+swB_- \cup B_+sB_+swB_- \\
&= B_+swB_- \cup B_+wB_- \ .
\end{aligned}
$$
$\square$

We can now derive the Birkhoff decomposition.

Proposition 6.81. *Let (G, B_+, B_-, N, S) be a twin Tits system with Weyl group W. Then*
$$
G = \coprod_{w \in W} B_\epsilon w B_{-\epsilon}
$$
for each $\epsilon \in \{+, -\}$.

Proof. We may assume that $\epsilon = +$. The proof is similar to the proof of the Bruhat decomposition in Section 6.2.2, except that we do not have an assumption analogous to (Bru1). So we begin by proving that $B_+wB_- \neq B_+B_-$ for $w \neq 1$ in W. Choose $s \in S$ such that $l(sw) < l(w)$. Assume, to get a contradiction, that $B_+wB_- = B_+B_-$. Multiplying by B_+s and using (TBN1), we obtain
$$
B_+swB_- = B_+sB_+B_- = B_+\{s, 1\}B_- \ ,
$$
where the second equation follows from Lemma 6.80 (applied to $w = 1$). The left side is a single double coset, but the right side consists of two distinct double cosets by (TBN2), so we have a contradiction.

Next, we show that $B_+vB_- = B_+wB_- \implies v = w$ for $v, w \in W$. We argue by induction on $\min\{l(v), l(w)\}$, which we may assume is equal to $l(v)$. The case $l(v) = 0$ is provided by the previous paragraph, so assume $l(v) > 0$, and choose $s \in S$ such that $l(sv) < l(v)$. If $B_+vB_- = B_+wB_-$, then we can multiply by B_+s and apply (TBN1) to obtain $B_+svB_- = B_+sB_+wB_-$. Since there is only one double coset on the left side of this equation, the right side must be the single double coset B_+swB_-. We therefore have $sv = sw$ by the induction hypothesis, and hence $v = w$.

Finally, we show that $G = \bigcup_{w \in W} B_+wB_-$. To this end, note that (by Lemma 6.80) the union is closed under left multiplication by N and B_+. So it is closed under left multiplication by $G = B_+NB_+$ and is therefore the entire group G.
$\square$

Given a twin Tits system (G, B_+, B_-, N, S) with Weyl group $W = N/T = \langle S \rangle$, we can now construct a (thick) twin building $\mathcal{C}$ of type (W, S) in the obvious way (see the end of Section 6.3.2): First, we take $(\mathcal{C}_\epsilon, \delta_\epsilon)$, for $\epsilon = \pm$, to be the building associated to the BN-pair (B_ϵ, N). Thus $\mathcal{C}_\epsilon := G/B_\epsilon$, with $\delta_\epsilon(gB_\epsilon, hB_\epsilon) = w$ (for $g, h \in G$) if and only if $g^{-1}h \in B_\epsilon w B_\epsilon$. We now define the codistance

$$\delta^* : \mathcal{C}_+ \times \mathcal{C}_- \cup \mathcal{C}_- \times \mathcal{C}_+ \to W$$

by equation (6.23). It is easy to verify that $\mathcal{C} = (\mathcal{C}_+, \mathcal{C}_-)$ satisfies the axioms (Tw1)–(Tw3); see Exercise 6.88. Note, for future reference, the following characterization of the opposition relation. Given $g, h \in G$, we have

$$gB_+ \text{ op } hB_- \iff \delta^*(gB_+, hB_-) = 1 \iff g^{-1}h \in B_+B_- . \qquad (6.25)$$

There is a natural action of G on $\mathcal{C}$ by left translation. It is obvious that this is indeed an action in the sense of Definition 6.67. As fundamental chambers we take the cosets $C_+ := B_+$ and $C_- := B_-$. [These are opposite by (6.25) with $g = h = 1$.] If (gB_+, hB_-) (with $g, h \in G$) is any other pair of opposite chambers, then $g^{-1}h \in B_+B_-$, so we can find $b \in B_+$ and $b' \in B_-$ such that $gb = hb' =: x$. Then $(gB_+, hB_-) = (xB_+, xB_-) = x(C_+, C_-)$, so G acts strongly transitively on $\mathcal{C}(G, B_+, B_-)$.

Definition 6.82. If (G, B_+, B_-, N, S) is a twin Tits system with associated Coxeter system (W, S), the (thick) twin building of type (W, S) constructed above with $\mathcal{C}_+ = G/B_+$ and $\mathcal{C}_- = G/B_-$ will be called the *twin building associated to* (G, B_+, B_-, N, S) and denoted by $\mathcal{C}(G, B_+, B_-)$.

Remark 6.83. We will see below that the notation is legitimate, i.e., that $\mathcal{C}(G, B_+, B_-)$ does not depend on N. To get started, note that the following are independent of N:

- the simplicial buildings $\Delta_\pm$ associated to $(\mathcal{C}_\pm, \delta_\pm)$ [since these are the buildings $\Delta(G, B_\pm)$; see Section 6.2.6];
- the opposition relation [see (6.25)];
- the numerical codistance d^* [see Lemma 5.143];
- the fundamental chambers C_+ and C_-;
- the apartment systems $\mathcal{A}$, $\mathcal{A}_+$, and $\mathcal{A}_-$ [since these depend only on the complete apartment systems in $\Delta_\pm$ and on the opposition relation between $\mathcal{C}_+$ and $\mathcal{C}_-$; see Remark 5.92(a) and Definition 5.171];
- the fundamental twin apartment $\Sigma = \Sigma\{C_+, C_-\}$ [since this is the unique element of $\mathcal{A}$ containing C_+ and C_-].

On the other hand, the following appear to depend on the choice of N:

- the Coxeter system (W, S);
- the Weyl distances $\delta_\pm$ or, equivalently, the canonical type functions on $\Delta_\pm$;
- the (Weyl) codistance δ^*.

One could deal with these last three items using the ideas of Remark 6.58(a). But it turns out that there is a much easier way. In fact, we will see below that N itself is almost determined by B_+ and B_-.

Although the fundamental twin apartment Σ does not depend on N, we can describe it using N, as in the theory of BN-pairs. Namely,

$$\Sigma_\epsilon = \{nC_\epsilon \mid n \in N\}$$

for $\epsilon \in \{+, -\}$. To see this, we need only note that by the definition of δ_ϵ and δ^* we have

$$\delta_\epsilon(C_\epsilon, nC_\epsilon) = w = \delta^*(C_{-\epsilon}, nC_\epsilon)$$

for all $n \in N$, where $w = nT \in W$. Our claim now follows from the definition of $\Sigma\{C_+, C_-\}$. This shows, in particular, that N stabilizes Σ and acts transitively on Σ_+ and Σ_-. But there is no reason to expect N to be the full stabilizer of Σ (which is the same as the stabilizer of Σ_+ as well as the stabilizer of Σ_-). As in the discussion near the end of Section 6.2.5, the full stabilizer of Σ (or Σ_+ or Σ_-) is the group

$$\widetilde{N} := N\widetilde{T},$$

where $\widetilde{T} = B_+ \cap B_-$ is the set of elements of G that fix Σ_+ (or Σ_-) pointwise. Obviously we could replace N by $\widetilde{N}$ and still have a twin BN-pair with the same associated twin building. So to get a 1–1 correspondence between twin BN-pairs and strongly transitive actions on thick twin buildings, we should impose a further condition:

Definition 6.84. The twin BN-pair (B_+, B_-, N) is said to be *saturated* if it satisfies

$$T = B_+ \cap B_-$$

or, equivalently, if N is the full stabilizer of the fundamental twin apartment of $\mathcal{C}(G, B_+, B_-)$.

The discussion leading up to this definition shows that (B_+, B_-, N) is saturated if and only if one (or both) of the BN-pairs (B_ϵ, N) is saturated in the sense of Definition 6.57.

It is now easy to understand how much freedom there is in the choice of N for a given B_+ and B_-. The first statement of the following lemma should be compared with Lemma 6.59. The second statement has no counterpart in the theory of ordinary BN-pairs.

Lemma 6.85.

(1) Let (B_+, B_-, N) be a saturated twin BN-pair in a group G. If N' is an arbitrary subgroup of G, then (B_+, B_-, N') is a twin BN-pair if and only if $N'T = N$.

(2) If a pair of subgroups B_+, B_- of G is part of a twin BN-pair, then there is a unique subgroup $N \leq G$ such that (B_+, B_-, N) is a saturated twin BN-pair.

Proof. (1) If $N'T = N$, it is trivial that (B_+, B_-, N') satisfies the TBN axioms. Conversely, suppose that (B_+, B_-, N') is a second twin BN-pair with the same groups $B_\pm$. By Remark 6.83, our two twin BN-pairs give rise to the same (simplicial) buildings $\varDelta_\pm$ and the same fundamental twin apartment $\varSigma$. Since N' stabilizes $\varSigma$ and N is the full stabilizer, it follows that $N' \le N$. We can now apply Lemma 6.59 to conclude that $N'T = N$.

(2) The existence of N was discussed prior to Definition 6.84, and uniqueness follows from (1).   □

Remark 6.86. It is now clear that the twin building $\mathcal{C}(G, B_+, B_-)$ associated to a twin BN-pair is independent of N, since it does not change if we replace the given N by the full stabilizer $\widetilde{N}$ as above, and this is independent of N by Lemma 6.85(2).

We summarize the main results of this section in the following theorem.

Theorem 6.87.

(1) *Let G be a group that acts strongly transitively on a thick twin building $\mathcal{C} = (\mathcal{C}_+, \mathcal{C}_-)$ of type (W, S). Let $(C_+, C_-) \in \mathcal{C}_+ \times \mathcal{C}_-$ be a pair of opposite chambers, and let $\varSigma = \varSigma\{C_+, C_-\}$ be the associated twin apartment. Then, if we denote by B_+, B_-, and N the stabilizers of C_+, C_-, and $\varSigma$ in G, the triple (B_+, B_-, N) is a saturated BN-pair in G with Weyl group W. The twin building $\mathcal{C}(G, B_+, B_-)$ associated to this twin BN-pair is canonically isomorphic to $\mathcal{C}$.*

(2) *Let (G, B_+, B_-, N, S) be a twin Tits system with Weyl group W. Then $\mathcal{C}(G, B_+, B_-)$ is a thick twin building of type (W, S) on which G acts strongly transitively. If we set $C_+ := B_+$, $C_- := B_-$, and $\varSigma := \varSigma\{C_+, C_-\}$, then we recover $B_\pm$ as the stabilizer of $C_\pm$ in G. Furthermore, $N(B_+ \cap B_-)$ is the stabilizer of $\varSigma$ in G, and hence we recover N as this stabilizer if the twin BN-pair (B_+, B_-, N) is saturated.*   □

Exercises

6.88. Verify the axioms (Tw1)–(Tw3) for $\mathcal{C}(G, B+, B-)$.

6.89. Let (W, S) be a Coxeter system with W finite, and let w_0 be the longest element of W. Prove the following:

(a) If (B, N) is a BN-pair with Weyl group W, then $(B, w_0 B w_0, N)$ is a twin BN-pair
(b) If (B_+, B_-, N) is a twin BN-pair with Weyl group W, then $B_- = w_0 B_+ w_0$.

6.90. Let (G, B_+, B_-, N, S) be a twin Tits system with infinite Weyl group.

(a) Show that B_+ and B_- are not conjugate in G.

(b) Suppose (W, S) is irreducible, and let J be a proper subset of S. Show that B_+ is not conjugate to a subgroup of $B_- W_J B_-$. Equivalently, no proper subgroup of G can be parabolic with respect to both B_+ and B_-.

6.91. If (B_+, B_-, N) is a twin BN-pair in G, show that G is generated by B_+ and B_-.

6.92. How should we modify the TBN axioms if we want an analogue of Theorem 6.87 that does not require thickness?

6.4 Historical Remarks

We began Chapter 4 by writing down, with no motivation, the strange-looking axioms for buildings. These were reformulated in Chapter 5 in terms of W-metric spaces. We then showed in the present chapter how the axioms lead in a fairly natural way to equally strange-looking axioms for groups. In this brief section we will attempt to put all of these axiom systems in their historical context. They may still seem strange when we are done, but at least you will have some idea of where they came from.

Our starting point is a 1954 paper of F. Bruhat [58] on the representation theory of complex Lie groups. Bruhat was especially interested in the four classical families A_n, B_n, C_n, D_n of simple groups G. Readers who are not familiar with these can just think about the group $G = \mathrm{SL}_n(\mathbb{C})$ for now; this is the group of type A_{n-1}. [The group $\mathrm{SL}_n(\mathbb{C})$ is not really simple, but it is "almost simple." More precisely, its center Z is finite, and the quotient $\mathrm{PSL}_n(\mathbb{C}) := \mathrm{SL}_n(\mathbb{C})/Z$ is simple, as we saw in Section 6.2.7.]

At the time of Bruhat's work, it had been known for a long time how to associate to G a finite reflection group W, called the *Weyl group* of G. It is given by $W := N/T$, where T is a "maximal torus" and N is its normalizer. And people were becoming aware of the importance of a certain subgroup $B \leq G$ (which eventually became known as the "Borel subgroup" of G as a result of the fundamental work of Borel [38]). What was not yet known, however, was the connection between B and W provided by the *Bruhat decomposition* $G = \coprod_{w \in W} BwB$.

Bruhat discovered this while studying so-called induced representations. Questions about these led him to ask whether the set $B \backslash G / B$ of double cosets was finite. He was apparently surprised to discover, by a separate analysis for each of the four families of classical simple groups, that the set of double cosets was not only finite but was in 1–1 correspondence with W.

The Bruhat decomposition was a fundamental fact that had previously gone unnoticed. Chevalley [81, 83] picked up on it immediately, and it became a basic tool in his work on the construction and classification of simple algebraic groups. He replaced Bruhat's case-by-case proof by a unified proof that applied not only to the classical groups (types A–D) but also to the five exceptional groups (types E_6, E_7, E_8, F_4, and G_2). Moreover, he worked over an arbitrary field k, not just $k = \mathbb{C}$. In particular, since k could be finite, one now had for each of the types A–G examples of *finite* simple groups that admitted a Bruhat decomposition, with the Weyl group W being the finite reflection group of the given type. Finally, Chevalley's work included a study of the basic properties of the parabolic subgroups of the groups he constructed, now known as *Chevalley groups*.

Meanwhile, Tits had been trying since the mid 1950s to find geometric interpretations for the exceptional simple Lie groups. "Geometric" here refers to incidence geometry, as exemplified by the close relationship between projective geometry and the group SL_n. See [254] for Tits's own account of this project. One of his motivations was the hope that, armed with geometric interpretations of the exceptional simple Lie groups, he would be able to·define analogues of them over arbitrary fields, and in particular, he would have new families of finite simple groups. (The cases G_2 and E_6 had previously been done by Dickson.) But then Chevalley, as we noted above, constructed the groups directly. So it was natural to turn the question around: Now that we have the groups, can we construct the geometries? Although Tits had not yet come up with the definitive notion of "building," he knew that he needed retractions to make the geometric theory work. So he studied Chevalley's work and, in the course of trying to construct geometries and retractions, arrived at the axioms for a BN-pair.

This axiomatization can be found in a short 1962 paper [241], which contains most of the results about BN-pairs that we have presented. [The only serious omission from this paper is the proof that the Weyl group W associated to a BN-pair is necessarily a Coxeter group; this fact was discovered a year or two later by Tits [243] and, independently, by Matsumoto [160].] A second paper [242] contains axioms for a class of incidence geometries, together with an indication of how a group with a BN-pair leads to such a geometry.

The axioms given in that paper look very much like the three axioms for buildings with which we began Chapter 4, except that they are stated in terms of incidence geometries instead of simplicial complexes. One also finds in this paper some of the fundamental ideas in the theory of buildings, such as retractions onto apartments (phrased in the language of incidence geometry). The paper makes extensive use of flags but does not explicitly talk about flag complexes. It was only a matter of time before Tits focused on the flag complexes themselves and restated his axioms in terms of simplicial complexes. The first published account of this was given in a 1965 Bourbaki Seminar exposé [243], where buildings were called "complexes with Weyl structure." This paper contains, among other things, an outline of much of the basic theory of Coxeter complexes and buildings that we gave in Chapters 3 and 4. It also contains the correspondence between BN-pairs and strongly transitive actions on thick buildings.

Our story so far has described the development of the simplicial approach to buildings, as in Chapter 4. Tits first introduced the newer approach in a

1981 paper [255]. This definition also evolved over a period of years, and it reached a mature form in the late 1980s. The final version, in which buildings are thought of as W-metric spaces as in Chapter 5, was first published in [261, Section 2.1]. Since apartments were no longer part of the definition, this made it natural to replace strongly transitive actions by Weyl-transitive actions and to replace BN-pairs by what we have called Tits subgroups. This is sketched in [261, Section 3.1], but the basic ideas, formulated without the aid of δ, can already be found in Bourbaki [44, Chapter IV, Section 2, Exercises 3, 10, and 11]. Bourbaki attributes these exercises to Tits in a footnote on p. 39.

This completes our highly condensed account of the origin of buildings. We hope it gives you some idea, admittedly vague, as to how Tits discovered buildings by combining (a) years of work on incidence geometries associated with groups and (b) ideas inspired by Chevalley's theory.

Remark 6.93. Even though Tits did not succeed in his original goal of constructing exceptional algebraic groups via geometry, his geometric approach to group theory did in fact yield some new groups. For example, in a 1959 paper [239] Tits gave a geometric construction of the group now known as the twisted Chevalley group of type $^3D_4$. Moreover, generalized m-gons (Definition 4.20), which are essentially the rank-2 spherical buildings, made their first appearance in this paper.

There is an interesting footnote to this story. We mentioned that the types A–G of finite reflection groups all arose in Chevalley's work as the Weyl group W of a finite simple group with a BN-pair. What about the remaining types H_3, H_4, and $I_2(m)$ ($m = 5$ or $m \geq 7$)? The type $I_2(8)$ (dihedral group of order 16) was observed fairly early; it arises, for instance, from a BN-pair in a finite simple group called the *Tits group*, of order $17,971,200 = 2^{11} \cdot 3^3 \cdot 5^2 \cdot 13$. [See also Section 9.9 below for more examples of this type.] But it turns out that this is the only "unusual" Weyl group that can arise from a finite group (simple or not) with a BN-pair. This is a consequence of a theorem of Feit and Higman [101], which can be restated as follows in the language of buildings:

Theorem 6.94. *If Δ is a finite thick building, then every connected component of its Coxeter diagram is of type A_n, C_n, D_n, E_n, F_4, G_2, or $I_2(8)$.*

Here are a few words about the proof. First, it suffices to consider the case that Δ is irreducible, by which we mean that its Coxeter diagram is connected. For in the general case, Δ can be decomposed as a join of irreducible buildings, one for each component of the diagram. Next, it suffices to consider the case that Δ is of rank 2. For the only other cases to worry about are H_3 and H_4; and if Δ had either of these types, then a suitable link in Δ would be a finite thick building of the prohibited type $I_2(5)$.

We are therefore reduced to the following question: For which $m \geq 3$ do there exist finite generalized m-gons in which every point is on at least three lines and every line contains at least three points? The bulk of the Feit–Higman paper is devoted to answering this question, and what they show is that the only possibilities for m are 3, 4, 6, and 8. These correspond, respectively, to the types A_2, C_2, G_2, and $I_2(8)$, whence the theorem.

Finally, if we drop the finiteness assumption but instead impose a symmetry condition called the Moufang property, then it turns out that the conclusion of the Feit–Higman theorem remains true. It follows that there are no thick buildings (finite or infinite) of type H_3 or H_4. We will explain this in the next chapter (see Remark 7.60 and Corollary 7.61).

6.5 Example: The General Linear Group

Let $G = \mathrm{GL}_n(k)$ ($n \geq 2$), where k is an arbitrary field. [Everything we are about to do goes through if we instead take G to be $\mathrm{SL}_n(k)$, or $\mathrm{PGL}_n(k)$, or $\mathrm{PSL}_n(k)$.] We have already obtained a BN-pair in G by using the action of G on the complex Δ of flags of proper nonzero subspaces of k^n (Section 6.1.2). This approach relies, of course, on the fact that Δ is a building, which we proved in Section 4.3.

In the present section we will give an alternative approach, which does not depend on that result. Namely, we will simply verify the BN-pair axioms by direct matrix computations. As a byproduct, we will obtain a new proof that the flag complex Δ is indeed a building.

Let $B \leq G$ be the upper-triangular group, i.e., the stabilizer of the standard flag

$$[e_1] < [e_1, e_2] < \cdots < [e_1, \ldots, e_{n-1}] ,$$

where $e_1, \ldots, e_n$ is the standard basis of k^n. Let $N \leq G$ be the monomial group, i.e., the stabilizer of the set of lines $\{[e_1], \ldots, [e_n]\}$. Then N acts as a group of permutations of this set, and we obtain a surjection from N onto the symmetric group on n letters. The kernel of this homomorphism is $T := B \cap N$, which is the diagonal subgroup of G. So N normalizes T, and $W := N/T$ can be identified with the symmetric group on n letters.

To see that B and N generate G, we need only note that the subgroup $\langle B, N \rangle$ contains the lower-triangular group, which is wBw^{-1} for a suitable $w \in W$; hence it contains all elementary matrices. [Recall that an elementary matrix is one that has 1's on the diagonal and exactly one nonzero off-diagonal entry; left multiplication (resp. right multiplication) by such a matrix corresponds to an elementary row (resp. column) operation.] It now follows from elementary linear algebra that $\langle B, N \rangle = G$.

Let $S \subset W$ be the standard set of generators $\{s_1, \ldots, s_{n-1}\}$, where s_i is the transposition that interchanges i and $i + 1$. To simplify the notation, we will verify the axioms (BN1) and (BN2) only for $s = s_1$; the other elements of S are treated similarly. Our s, then, is represented by any monomial matrix of the form

$$\begin{pmatrix} 0 & * & & & \\ * & 0 & & & \\ & & * & & \\ & & & \ddots & \\ & & & & * \end{pmatrix},$$

where the blank regions are understood to be filled with zeros.

Axiom (BN1) says that $sBw \subseteq BswB \cup BwB$. Multiplying on the right by w^{-1}, we can rewrite this as

$$sB \subseteq BsB' \cup BB',$$

where $B' = wBw^{-1}$. In other words, we must show that any matrix in sB is reducible to either s or 1 via left multiplication by B and right multiplication by B'. It turns out that we will need to use only the elementary matrices in B and B', so that we will simply be doing some elementary row and column operations (also known as "pivoting").

Note first that left multiplication by upper-triangular elementary matrices allows us to pivot upward, i.e., to add a multiple of a row to any higher row. Now a typical element of sB has the form

$$\begin{pmatrix} 0 & * & * & \cdots & * \\ * & * & * & \cdots & * \\ & & * & \cdots & * \\ & & & \ddots & \vdots \\ & & & & * \end{pmatrix},$$

which is easily reduced to

$$\begin{pmatrix} 0 & * & & & \\ * & * & & & \\ & & * & & \\ & & & \ddots & \\ & & & & * \end{pmatrix}$$

by pivoting upward. If the $(2,2)$-entry is zero, then we have already reduced the matrix to s. So we may assume that all three $*$'s in the upper left 2×2 block are nonzero.

Now let's use right multiplication by B'. To avoid messing up what we have already achieved, we will use only $B' \cap \mathrm{GL}_2$; here GL_2 is identified with the subgroup

$$\{g \in \mathrm{GL}_n \mid g[e_1, e_2] = [e_1, e_2] \text{ and } ge_i = e_i \text{ for } i > 2\} .$$

Note that $B' = wBw^{-1}$ is the stabilizer of the flag

$$[e_{w(1)}] < [e_{w(1)}, e_{w(2)}] < \cdots < [e_{w(1)}, \ldots, e_{w(n-1)}] .$$

It follows easily that $B' \cap \mathrm{GL}_2$ is the stabilizer in GL_2 of the line spanned by either e_1 or e_2, whichever occurs first in the list $e_{w(1)}, \ldots, e_{w(n)}$. In other words, $B' \cap \mathrm{GL}_2$ is the upper-triangular subgroup of GL_2 if $w^{-1}(1) < w^{-1}(2)$ and the lower-triangular subgroup otherwise.

Looking at the elementary matrices in $B' \cap \mathrm{GL}_2$, we see that we now have a column operation available: If $w^{-1}(1) < w^{-1}(2)$, then we can add a multiple of column 1 to column 2, and otherwise we can add a multiple of column 2 to column 1. In the first case, we pivot on the $(2,1)$-entry of our matrix above in order to clear out the $(2,2)$-entry; this reduces the matrix to s. In the second case, we pivot on the $(2,2)$-entry in order to clear out the $(2,1)$-entry; the resulting matrix is in B, and we are done. (Note that the proof of (BN1) actually showed that $sBw \subseteq BswB$ for half of the elements $w \in W$. This should not be surprising.)

Finally, it is trivial to check that (BN2) holds; for we have

$$
sBs = \begin{pmatrix} * & 0 & * & \cdots & * \\ * & * & * & \cdots & * \\ & & * & \cdots & * \\ & & & \ddots & \vdots \\ & & & & * \end{pmatrix} \not\subseteq B \,.
$$

Having verified the BN-pair axioms [with remarkably little effort], we obtain a building $\Delta(G, B)$. Let's show that this building is isomorphic to the complex Δ of flags of proper nonzero subspaces of k^n. Consider the action of G on Δ. If C is the standard flag, it is immediate that $\overline{C} := \Delta_{\leq C}$ is a simplicial fundamental domain for the action. Moreover, the stabilizers of the faces of C are precisely the standard parabolic subgroups of G. Indeed, they are standard parabolics, since they contain B, and it is trivial to verify that they are all distinct. Since C has 2^{n-1} faces and G contains only 2^{n-1} standard parabolics (one for each subset of S), the stabilizers must exhaust the standard parabolics. The desired isomorphism now follows easily. In particular, we have obtained a group-theoretic proof, independent of Section 4.3, that the flag complex Δ is a building.

Finally, we recall from Section 6.2.7 that the existence of this BN-pair leads to results about the normal subgroups of G and, in particular, to the result that $\mathrm{PSL}_n(k)$ is simple (unless $n = 2$ and $k = \mathbb{F}_2$ or $\mathbb{F}_3$).

6.6 Example: The Symplectic Group

Let k continue to be an arbitrary field. Let $\langle -, - \rangle$ be the bilinear form on k^{2n} ($n \geq 1$) defined as follows on the standard basis vectors:

$$
\langle e_i, e_j \rangle = \begin{cases} 0 & \text{if } i + j \neq 2n + 1, \\ 1 & \text{if } i + j = 2n + 1 \text{ and } i < j, \\ -1 & \text{if } i + j = 2n + 1 \text{ and } i > j. \end{cases}
$$

If we denote the standard basis vectors by $e_1, e_2, \ldots, e_n, f_n, f_{n-1}, \ldots, f_1$, then the nonzero "inner products" above can be written more simply as

$$\langle e_i, f_i \rangle = 1 = -\langle f_i, e_i \rangle .$$

The bilinear form $\langle -, - \rangle$ is *alternating*, by which we mean that $\langle v, v \rangle = 0$ for all v. [This implies skew-symmetry: $\langle v, v' \rangle = -\langle v', v \rangle$. Conversely, skew-symmetry of a bilinear form implies that the form is alternating, provided char $k \neq 2$.]

It is easy to explicitly compute $\langle -, - \rangle$ in terms of coordinates: If we write a typical element of k^{2n} as a pair (X, Y) with $X, Y \in k^n$, then we have

$$\langle (X, Y), (Z, W) \rangle = X \cdot W' - Y \cdot Z',$$

where the prime means "reverse the coordinates" and the dot denotes the ordinary dot product of vectors in k^n:

$$(x_1, \ldots, x_n) \cdot (y_1, \ldots, y_n) := \sum_{i=1}^{n} x_i y_i .$$

When $n = 1$, for instance, $\langle v, w \rangle$ is simply the determinant of the 2×2 matrix with v and w as columns.

Definition 6.95. The *symplectic group* $\mathrm{Sp}_{2n}(k)$ is the group of linear automorphisms g of k^{2n} that preserve $\langle -, - \rangle$, i.e., that satisfy $\langle gv, gw \rangle = \langle v, w \rangle$ for all $v, w \in k^{2n}$. It is enough to check this equation when v and w are basis vectors. So an element of $\mathrm{Sp}_{2n}(k)$ is a $2n \times 2n$ matrix g whose columns $v_1, \ldots, v_{2n}$ satisfy the same inner product relations as the standard basis vectors $e_1, \ldots, e_{2n}$. When $n = 1$, this simply says that $\det g = 1$; thus $\mathrm{Sp}_2 = \mathrm{SL}_2$.

For each $i = 1, \ldots, n$ there is a copy of Sp_2 [$= \mathrm{SL}_2$] in Sp_{2n}, which stabilizes the plane $[e_i, f_i]$ and fixes all basis vectors other than e_i and f_i. Taking $n = 2$ and $i = 1$, for instance, we obtain a copy of SL_2 in Sp_4 that looks like this:

$$\begin{pmatrix} * & & & * \\ & 1 & 0 & \\ & 0 & 1 & \\ * & & & * \end{pmatrix} .$$

In addition, there are various ways to embed GL_2 in Sp_{2n}. Namely, given $1 \leq i < j \leq n$, there is a copy of GL_2 that stabilizes $[e_i, e_j]$ and $[f_i, f_j]$ and fixes all basis vectors other than these four; an automorphism g of this type can do anything at all on $[e_i, e_j]$, but its effect on $[f_i, f_j]$ is then forced by the requirement that g be symplectic. Suppose, for example, that we take $n = 2$ again and try to construct an element $g \in \mathrm{Sp}_4$ that is given by an elementary matrix $\left(\begin{smallmatrix} 1 & a \\ 0 & 1 \end{smallmatrix} \right)$ on $[e_1, e_2]$ and that stabilizes $[f_1, f_2] = [e_4, e_3]$. Then g must have the form

$$\begin{pmatrix} 1 & a & & \\ 0 & 1 & & \\ & & * & * \\ & & * & * \end{pmatrix},$$

and a simple computation shows that this will be symplectic if and only if the lower right 2×2 block is $\left(\begin{smallmatrix} 1 & -a \\ 0 & 1 \end{smallmatrix} \right)$. Finally, for each $i < j$ as above there is also a copy of GL_2 in Sp_{2n} that stabilizes $[e_i, f_j]$ and $[e_j, f_i]$.

Call a symplectic matrix *elementary* if it is the image of a 2×2 elementary matrix under one of the embeddings described in the three previous paragraphs. The reader might find it a useful exercise to explicitly write down all the types of elementary matrices in Sp_4. (There are two copies of SL_2 and two copies of GL_2, hence 8 types of elementary matrices.) One can also check as an exercise that Sp_{2n} is generated by elementary matrices. The idea is to interpret multiplication by elementary matrices in terms of row and column operations. It is then easy to use these operations to reduce any symplectic matrix to the form

$$\begin{pmatrix} I & \\ & A \end{pmatrix},$$

where I and A are $n \times n$ matrices and I is the identity. But then A is forced to be the identity also, since the matrix is symplectic.

We need one last bit of terminology before constructing a BN-pair.

Definition 6.96. A subspace $V \leq k^{2n}$ is called *totally isotropic* if $\langle v, v' \rangle = 0$ for all $v, v' \in V$. This is equivalent to saying that $V \leq V^{\perp}$, where $V^{\perp}$ is the orthogonal subspace, defined in the usual way:

$$V^{\perp} := \{ u \in V \mid \langle u, v \rangle = 0 \text{ for all } v \in V \} .$$

For example, a subset of the standard basis spans a totally isotropic subspace provided it contains no pair $\{e_i, f_i\}$. The chain of totally isotropic subspaces

$$[e_1] < [e_1, e_2] < \cdots < [e_1, \ldots, e_n]$$

will be called the *standard isotropic flag* in k^{2n}. Note that the subspaces orthogonal to these totally isotropic subspaces form a descending chain

$$[e_1, \ldots, e_{2n-1}] > [e_1, \ldots, e_{2n-2}] > \cdots > [e_1, \ldots, e_n] ;$$

so if we take the standard isotropic flag together with the orthogonal subspaces, we get the standard ordinary flag in k^{2n} (with the subspace $[e_1, \ldots, e_n]$ counted twice). Incidentally, the set of nonzero totally isotropic subspaces, with inclusion as the incidence relation, is an example of what is called an n-dimensional *polar space*.

Now let B be the group of upper-triangular symplectic matrices, i.e., the stabilizer in $G = Sp_{2n}(k)$ of the standard flag in k^{2n}. If a symplectic matrix stabilizes a subspace V, then it stabilizes $V^{\perp}$ too; hence B can also be described as the stabilizer in G of the standard isotropic flag. Let $N \leq G$ be the group of symplectic monomial matrices, i.e., the stabilizer in G of the set of lines $\{L_1, \ldots, L_n, L'_n, \ldots, L'_1\}$, where $L_i = [e_i]$ and $L'_i = [f_i]$. Then $T := B \cap N$ is the group of diagonal symplectic matrices, i.e., the group of matrices of the form $\operatorname{diag}(\lambda_1, \ldots, \lambda_n, \lambda_n^{-1}, \ldots, \lambda_1^{-1})$. In particular, N normalizes T. Note that T is isomorphic to the product of n copies of k^*; in the language of the theory of algebraic groups, T is a torus of rank n. Although we will not need that theory, the interested reader can refer to Appendix C for an explanation of the terminology.

The quotient $W := N/T$ can be identified with a group of permutations of the set of $2n$ lines above. We will show that W is equal to the group W' consisting of all permutations that map each pair $\{L_i, L'_i\}$ to another such pair. The inclusion $W \leq W'$ is immediate, since W preserves the orthogonality relations among the given lines. To prove the opposite inclusion, note first that W' is generated by the following set $S := \{s_1, \ldots, s_n\}$ of permutations: s_i for $i < n$ is the product of the two transpositions $L_i \leftrightarrow L_{i+1}$ and $L'_i \leftrightarrow L'_{i+1}$; and s_n is the transposition $L_n \leftrightarrow L'_n$. So the inclusion $W' \leq W$ follows from the easy observation that each $s \in S$ can be represented by a symplectic monomial matrix in one of our embedded SL_2's or GL_2's.

One can now use elementary row and column operations, exactly as in the case of GL_n, to complete the proof that we have a BN-pair. Details are left to the reader. [You might want to check (BN1) for $n = 2$ and $s = s_1$, for instance, just to convince yourself that the same method really does work.] The crucial thing that keeps the proof from becoming unpleasant is that each s_i is in a GL_2 or SL_2; one is thereby able to reduce (BN1) to a 2×2 computation before ever having to think about what the group $B' := wBw^{-1}$ looks like.

One can now check that the associated building $\Delta = \Delta(G, B)$ is isomorphic to the flag complex of the set of nonzero totally isotropic subspaces. The proof is essentially the same as the proof of the analogous statement for GL_n, provided one knows the basic linear algebra of alternating forms. This can be found in many places, such as Artin [22, Chapter III] and Aschbacher [24, Chapter 7].

Exercise 6.97. Draw a picture of the fundamental apartment in Δ when $n = 2$; it is a barycentrically subdivided quadrilateral, whose 8 vertices consist of the totally isotropic subspaces spanned by subsets of the standard basis.

Remark 6.98. Note that the Weyl group W is isomorphic to the "signed permutation group on n letters," which is the finite reflection group of type C_n (see Section 1.3). [Strictly speaking, type C_n was defined only for $n \geq 2$ in Section 1.3; but we make the convention that $C_1 = A_1$.] This calculation of W is consistent with the fact that Sp_{2n} is a classical group of type C_n.

Remark 6.99. The "inner product" $\langle -, - \rangle$ that we worked with may have seemed arbitrary. But in fact, one can show that it is the typical nondegenerate alternating bilinear form, in the following sense: If V is a finite-dimensional vector space with a nondegenerate alternating bilinear form $\langle -, - \rangle$, then dim V is even and V has a basis $e_1, \ldots, e_n, f_n, \ldots, f_1$ whose inner products look like those of our example. Proofs can be found in the books of Artin and Aschbacher cited above.

Finally, we mention that the simplicity results of Section 6.2.7 imply that the group $\mathrm{PSp}_{2n}(k) := \mathrm{Sp}_{2n}(k)/\{\pm 1\}$ is simple for $n \geq 2$, except for $\mathrm{PSp}_4(\mathbb{F}_2)$. The latter has a simple subgroup of index 2. [In fact, $\mathrm{PSp}_4(\mathbb{F}_2)$ happens to be isomorphic to the symmetric group S_6, and its simple subgroup of index 2 is the alternating group.] We have excluded the case $n = 1$ only because $\mathrm{Sp}_2 = \mathrm{SL}_2$, which has already been treated. One needs to check, of course, that $\mathrm{Sp}_{2n}(k)$ [or its subgroup of index 2 in the exceptional case] is perfect for $n \geq 2$; see, for example, [22, Chapter V; 24, Section 22; 123, Chapter 3].

Remark 6.100. We will see in Chapter 7 (see especially Sections 7.7.2 and 7.9) that there is a conceptual explanation for the fact that groups associated with buildings often turn out to be perfect.

6.7 Example: Orthogonal Groups

Throughout this section we assume for simplicity that k is a field of characteristic $\neq 2$, although the theory extends to characteristic 2 with suitable modifications.

6.7.1 The Standard Quadratic Form

The "standard" quadratic form, for us, is the one associated to the symmetric bilinear form $\langle -, - \rangle$ on k^m ($m \geq 2$) defined as follows on the standard basis vectors $e_1, \ldots, e_m$:

$$\langle e_i, e_j \rangle = \begin{cases} 1 & \text{if } i + j = m + 1, \\ 0 & \text{otherwise.} \end{cases}$$

We will write $m = 2n$ (resp. $m = 2n + 1$) if m is even (resp. odd), and we will denote the last n basis vectors by $f_n, \ldots, f_1$. The nonzero inner products, then, are

$$\langle e_i, f_i \rangle = 1 = \langle f_i, e_i \rangle$$

and, if $m = 2n + 1$,

$$\langle e_{n+1}, e_{n+1} \rangle = 1 .$$

It is easy to explicitly compute $\langle -, - \rangle$ in terms of coordinates: Write a typical vector $v \in k^m$ as (X, Y) if m is even and as (X, λ, Y) if m is odd, with $X, Y \in k^n$ and $\lambda \in k$; then we have

$$\langle (X,Y),(Z,W)\rangle = X \cdot W' + Y \cdot Z' \,,$$
$$\langle (X,\lambda,Y),(Z,\mu,W)\rangle = X \cdot W' + Y \cdot Z' + \lambda\mu \,,$$

where the prime and the dot product have the same meaning as in Section 6.6. In particular, the associated quadratic form Q is given by

$$Q(X,Y) = 2X \cdot Y' \quad \text{or} \quad Q(X,\lambda,Y) = 2X \cdot Y' + \lambda^2 \,.$$

This form is equivalent, under change of coordinates, to the form Q' given by

$$Q'(z_1,\ldots,z_m) = -z_1^2 - \cdots - z_n^2 + z_{n+1}^2 + \cdots + z_m^2 \,.$$

And if k contains $\sqrt{-1}$, then it is equivalent to the familiar quadratic form $\sum_{i=1}^m z_i^2$.

Definition 6.101. The *orthogonal group* $O_m(k,Q)$ is the group of linear automorphisms g of k^m that preserve $\langle -,-\rangle$ or, equivalently, Q. The *special orthogonal group* $SO_m(k,Q)$ is defined to be the group of orthogonal matrices of determinant 1. We will suppress Q from the notation and simply write $O_m(k)$ and $SO_m(k)$ whenever it is clear from the context that we are considering the standard quadratic form.

Note that SO_2 is the "rank-1 torus" consisting of diagonal matrices $\mathrm{diag}(\lambda,\lambda^{-1})$. It has index 2 in O_2, the nontrivial coset being the set of matrices

$$\begin{pmatrix} 0 & \lambda^{-1} \\ \lambda & 0 \end{pmatrix} ,$$

which have determinant -1.

Let's focus now on $G = SO_m$, returning to the case of O_m afterward. For each $i = 1,\ldots,n$ we have a copy of SO_2 in G that stabilizes $[e_i, f_i]$ and fixes all the other basis vectors. In case $m = 2n+1$, we can extend this to an embedding $O_2 \hookrightarrow G$ using the "extra" basis vector $v = e_{n+1}$: For if $g \in O_2$ has determinant -1, then we can copy g on $[e_i, f_i]$ and then send v to $-v$ in order to make the determinant 1.

Next, given $1 \le i < j \le n$, there are two ways of embedding GL_2 in $G = SO_m$, exactly analogous to the two embeddings used for Sp_{2n}. In particular, this gives us many elementary matrices to work with.

We now construct the BN-pair in the usual way: B is the upper-triangular subgroup of G, and N is the monomial subgroup. Then $T := B \cap N$ consists of the diagonal elements of G. If $m = 2n$, these elements necessarily have the form

$$\mathrm{diag}(\lambda_1,\ldots,\lambda_n,\lambda_n^{-1},\ldots,\lambda_1^{-1}) \,,$$

and if $m = 2n+1$ they have the form

$$\mathrm{diag}(\lambda_1,\ldots,\lambda_n,1,\lambda_n^{-1},\ldots,\lambda_1^{-1}) \,.$$

In both cases, T is a "rank-n torus." The Weyl group $W := N/T$ can be identified with a group of permutations of the $2n$ lines $L_i := [e_i]$ and $L_i' := [f_i]$, $i = 1, \ldots, n$. This is clear if $m = 2n$, but it is also true if $m = 2n + 1$; for e_{n+1} is the only nonisotropic basis vector, so W necessarily fixes the line it spans.

If $m = 2n + 1$, then W is the permutation group called W' in our treatment of Sp_{2n}. One proves this by exhibiting elements of N that represent the generators s_i of W' ($i = 1, \ldots, n$) constructed in Section 6.6. For $i < n$, the required element of N can be found in the embedded GL_2 acting on $[e_i, e_{i+1}]$ and $[f_i, f_{i+1}]$. And for $i = n$, the required element can be found in the embedded O_2 acting on $[e_n, e_{n+1}, f_n]$. Hence W is the reflection group of type C_n, also said to be of type B_n (see Section 1.3). This calculation of W is consistent with the fact that SO_{2n+1} is a classical group of type B_n.

If $m = 2n$, however, it is impossible to represent s_n by an orthogonal monomial matrix of determinant 1. The group W in this case turns out to be a subgroup of index 2 in W'. This subgroup is generated by the s_i for $i < n$ together with one additional element t, which is the product of the transpositions $L_{n-1} \leftrightarrow L_n'$ and $L_n \leftrightarrow L_{n-1}'$. Note that t is in the embedded GL_2 acting on $[e_{n-1}, f_n]$ and $[e_n, f_{n-1}]$. (We are assuming here that $n \geq 2$; if $n = 1$, then $G = \mathrm{SO}_2$, and we have already said everything there is to say about that group.)

In both cases, we now have a set of n generators for W, and it is a routine (although somewhat tedious) matter to verify the BN-pair axioms, using the same methods as in Sections 6.5 and 6.6. We have already identified the Weyl group in case $m = 2n + 1$. In case $m = 2n$, one can check, by computing orders of products of generators, that W is of type D_n. [Strictly speaking, D_n was defined only for $n \geq 4$; but the appropriate convention is that $\mathrm{D}_3 = \mathrm{A}_3$ and that the diagram of type D_2 is the union of two copies of the diagram of type A_1.] This calculation of W is consistent with the fact that SO_{2n} is a classical group of type D_n.

Remark 6.102. Did the definition of t above seem ad hoc? Was there a different choice of t that seemed more natural to you? If so, you would have struggled in vain to verify the BN-pair axioms for your choice. Indeed, we know from the general theory that given B and N, there can be only one set S for which the BN-pair axioms hold (Theorem 6.56).

Let's try now to figure out what the building $\Delta(G, B)$ is. The naïve guess is that it is the flag complex Δ of nonzero totally isotropic subspaces of k^m. This guess is correct if m is odd, and the proof is the same as the proofs of the analogous assertions in Sections 6.5 and 6.6. But it is wrong if m is even, as one can see in a variety of ways. For one thing, the flag complex Δ has the wrong type. [It is the flag complex of a polar space and hence has type C_n; but we've already seen that the Weyl group of SO_{2n} has type D_n.] For another thing, one can show that the action of SO_{2n} on $\mathcal{C}(\Delta)$ is not transitive. [There are precisely two orbits.] Yet a third thing that goes wrong is that Δ is not thick. [A simple computation shows that there are only two ways of extending the flag

$$[e_1] < \cdots < [e_1, \ldots, e_{n-1}]$$

to a maximal flag.]

The correct answer, when m is even, turns out to be that $\Delta(G, B)$ is the flag complex of the following so-called oriflamme geometry P: The elements of P are the nonzero totally isotropic subspaces of k^{2n} of dimension $\neq n - 1$; two such subspaces are called incident if one is contained in the other or if both have dimension n and their intersection has dimension $n - 1$. As an example of a flag in P we have the following chamber C:

$$[e_1, \ldots, e_{n-1}, e_n]$$

$$[e_1] < [e_1, e_2] < \cdots < [e_1, \ldots, e_{n-2}]$$

$$[e_1, \ldots, e_{n-1}, f_n]$$

We can get some feeling for this by thinking about the case $n = 2$ and describing P in the language of projective geometry: An isotropic line in k^4 is simply a point in the 3-dimensional projective space $\mathbb{P}^3$ over k that lies on the quadric surface X defined by $Q = 0$. And an isotropic plane in k^4 is simply a line in $\mathbb{P}^3$ that is contained in X. Our geometry P, then, consists of lines in the surface X, two such being called incident if they intersect. If you believe, as asserted above, that the flag complex of P is a building of type $D_2 = A_1 \times A_1$, then it must be true that there are two types of lines in X, and that every line of one type intersects every line of the other type (see Example 4.16). One can actually see this directly, by exhibiting an isomorphism between X and the direct product $\mathbb{P}^1 \times \mathbb{P}^1$ of two copies of the projective line; the two types of lines in X, then, are simply the two types of slices of the product. Details can be found in van der Waerden [263, Section I.7], which contains an interesting discussion of the groups SO_m for $3 \leq m \leq 6$.

Next, what happens if we look at the full orthogonal group O_m instead of SO_m? For m odd, everything goes through with no essential change. In particular, we again get a BN-pair, with the associated building being the same as the building for SO_m. This is not surprising, since $O_m = SO_m \times \{\pm 1\}$ if m is odd, so there is virtually no difference between the two groups.

When m is even, on the other hand, the situation is more complicated. One still has an action of $G = O_{2n}$ on the oriflamme complex Δ, but the action is not type-preserving. For it is easy to give examples of orthogonal matrices that stabilize the flag C constructed above but do not fix it pointwise. So the action of G on this building does not yield a BN-pair in G. We obtain, instead, something called a *generalized BN-pair*. See Bourbaki [44, Section IV.2, Exercise 8] to find out precisely what this means.

One further comment about $G = O_{2n}$: In addition to the building Δ of type D_n, one still has the nonthick building of type C_n, consisting of flags of totally isotropic subspaces. The action of G on this building is type-preserving and strongly transitive, so all of the results of Sections 6.1 and 6.2 are applicable except those that used thickness. In particular, we get a Bruhat decomposition, with B equal to the upper-triangular subgroup of O_{2n} [which happens to be the same as the upper-triangular subgroup of SO_{2n}] and with the Weyl group being of type C_n. One thus has a choice of two geometries associated to G, one of type D_n [which yields a generalized BN-pair] and one of type C_n [which yields what could be called a "weak" BN-pair]. The former might seem more natural, since G is the classical group of type D_n, but the latter has had applications also; see, for instance, Vogtmann [276].

Finally, we remark that the results of Section 6.2.7 can be used to give a simple subquotient of $SO_m(k)$, whose description is more complicated than in the cases of GL_n and Sp_{2n} treated in the two previous sections. We will be brief in our statement of the results, referring to [22, Chapter V; 24, Section 22; 123, Chapters 6 and 9] for more details. (See also Remark 6.100.) We assume $m \geq 5$ to avoid uninteresting cases or cases in which the orthogonal groups can be described in terms of other classical groups. Recall also that we have excluded fields of characteristic 2 from our discussion.

It turns out that $SO_m(k)$ is quite often not perfect, but its commutator subgroup $\Omega_m(k)$ is perfect, and the theory of Section 6.2.7 then shows that $\Omega_m(k)/\Omega_m(k) \cap \{\pm 1\}$ is simple. Moreover, $SO_m(k)/\Omega_m(k) \cong k^*/(k^*)^2$. Thus, for example, $\Omega_m(k) = SO_m(k)$ if k is algebraically closed. On the other hand, $\Omega_m(k)$ has index 2 in $SO_m(k)$ if k is finite. To complete the picture, we need to know when $-1 \in \Omega_m(k)$. This obviously cannot happen if m is odd, since $-1 \notin SO_m(k)$, so assume $m = 2n$. Then if $\sqrt{-1} \in k$, we always have $-1 \in \Omega_{2n}(k)$, and otherwise $-1 \in \Omega_{2n}(k)$ if and only if n is even. If k is a finite field $\mathbb{F}_q$, for example, then we can combine the two cases and say that $-1 \in \Omega_{2n}(k)$ if and only if $q^n \equiv 1 \bmod 4$.

6.7.2 More General Quadratic Forms

Everything in the previous subsection goes through with very little change if we replace our standard quadratic form by an arbitrary nondegenerate quadratic form Q on a finite-dimensional vector space V (still assuming char $k \neq 2$). We describe the situation briefly, assuming that the reader has some familiarity with quadratic forms. See, for instance [150, 181, 208, 215]; see also [24] for a treatment emphasizing the case that k is a finite field.

Choose a maximal totally isotropic subspace $[e_1, \ldots, e_n]$, and choose vectors $f_1, \ldots, f_n$ that are paired with the e_i as in the standard form. In other words, if $\langle -, - \rangle$ is the symmetric bilinear form associated with Q, then $\langle e_i, f_i \rangle = 1$ and all other inner products among the e_i and f_i are zero. It may happen that $n = 0$, i.e., that $Q(v) = 0$ only if $v = 0$; in this case Q is said to be *anisotropic* and the building is empty. Otherwise Q is said to *represent* 0 and is called *isotropic*. The integer n is independent of the choices and is called the *Witt index* of Q.

Let V_0 be the orthogonal complement of $[e_1, \ldots, e_n, f_1, \ldots, f_n]$. If $V_0 = 0$, then Q is the standard form on k^{2n}. If $V_0 \neq 0$, the theory is similar to the case of the standard form on k^{2n+1}, with V_0 playing the role of the line spanned by e_{n+1}, and one gets a building of type C_n. Details are left to the interested reader.

Remarks 6.103. (a) The theory even goes through, with minor modifications, if V is infinite-dimensional, provided the Witt index n is still finite. Thus V_0 is infinite-dimensional, but this is harmless.

(b) As in the case of the standard quadratic form, there is a (somewhat complicated) simplicity theorem, whose statement we omit.

We close by spelling out in more detail what happens in case the field k is finite. In this case a nondegenerate quadratic form on k^m is completely determined up to equivalence by its *discriminant*, i.e., by the determinant of any representing symmetric matrix. The discriminant is well defined as an element of $k^*/(k^*)^2$, which is a group of order 2. Concretely, suppose we diagonalize the form and write it in suitable coordinates as

$$Q(x_1, \ldots, x_m) = \sum_{i=1}^{m} \lambda_i x_i^2$$

with $\lambda_i \in k^*$; then the discriminant is the product $\lambda_1 \cdots \lambda_m$, viewed as an element of $k^*/(k^*)^2$. Convention: It will be convenient in what follows to use angle bracket notation $\langle \lambda_1, \ldots, \lambda_m \rangle$ for a diagonal form as above.

In particular, there are precisely two equivalence classes of quadratic forms for any fixed $m \geq 1$. They can be represented, for instance, as $\langle 1, \ldots, 1 \rangle$ and $\langle 1, \ldots, 1, a \rangle$, where a is any nonsquare. If the dimension m is odd, these two forms are equivalent to scalar multiples of one another, as one can see by checking the discriminant. (If one multiplies a form by any nonsquare, then the discriminant changes if m is odd.) So there is only one orthogonal group in this case, which is therefore the standard one discussed in Section 6.7.1. When m is even, however, the two forms are fundamentally different. In fact, one of them has Witt index $n := m/2$, as we already know, and this is generally denoted by $O_m^+(k)$, while the other has Witt index $m/2 - 1$ and is denoted by $O_m^-(k)$. The latter, then, gives rise to a thick building of type C_n, where $n = m/2 - 1$. (We assume $n > 0$, i.e., $m > 2$, so that the form is isotropic.)

6.8 Example: Unitary Groups

Let K be a field with an automorphism of order 2 called *conjugation* and denoted by $\lambda \mapsto \bar{\lambda}$. The motivating example is $K = \mathbb{C}$ with ordinary complex conjugation. By a *Hermitian form* on a K-vector space V we mean a function $V \times V \to K$, denoted by $v, w \mapsto \langle v, w \rangle$, which is linear in the first variable and satisfies $\langle w, v \rangle = \overline{\langle v, w \rangle}$ for $v, w \in V$. This implies that the form is conjugate-linear in the second variable, i.e., it is additive and satisfies $\langle v, \lambda w \rangle = \bar{\lambda}\langle v, w \rangle$ for $v, w \in V$ and $\lambda \in K$. We assume that our form is nondegenerate.

There is a *standard Hermitian form* on K^m, which should look familiar by now. It is defined as follows on the standard basis vectors $e_1, \ldots, e_m$:

$$\langle e_i, e_j \rangle = \begin{cases} 1 & \text{if } i + j = m + 1, \\ 0 & \text{otherwise.} \end{cases}$$

We will write $m = 2n$ (resp. $m = 2n + 1$) if m is even (resp. odd), and we will denote the last n basis vectors by $f_n, \ldots, f_1$. The nonzero inner products, then, are

$$\langle e_i, f_i \rangle = 1 = \langle f_i, e_i \rangle$$

and, if $m = 2n + 1$,

$$\langle e_{n+1}, e_{n+1} \rangle = 1 .$$

Remark 6.104. Let k be the fixed field of the conjugation, so that K/k is a Galois extension of degree 2. If the norm map $N: K^* \to k^*$ is surjective, then every nondegenerate Hermitian form is equivalent to the standard one. This is *not* the case for the canonical example with $K = \mathbb{C}$, but it does hold if K is a finite field. In this case we can write $k = \mathbb{F}_q$, $K = \mathbb{F}_{q^2}$, and $\bar{\lambda} = \lambda^q$. The norm is then given by $N(\lambda) = \lambda^{q+1}$, and the surjectivity is easily verified. [Use the fact that the multiplicative groups are cyclic.]

Definition 6.105. The *unitary group* $U(V)$ is the group of K-linear automorphisms of V that preserve $\langle -, - \rangle$. The *special unitary group* $SU(V) \le U(V)$ is the subgroup consisting of automorphisms of determinant 1. The groups depend on the particular conjugation on K, which we have omitted from the notation. If $V = K^m$ with the standard form, we will write $U_m(K)$ and $SU_m(K)$.

The construction of a BN-pair in $U(V)$ (or $SU(V)$) now proceeds as in the case of the orthogonal groups, except that it is easier. In particular, $U(V)$ works as well as $SU(V)$, and they both yield the same thick building of type C_n, where n is the Witt index, again defined to be the common dimension of the maximal totally isotropic subspaces. (In the case of the standard form, the Witt index is the same number n that occurred in the definition of the form.) The building can be identified with the flag complex of totally isotropic subspaces of V. Details are left to the interested reader.

Finally, we remark that one can use the simplicity theorems of Section 6.2.7 to deduce that $PSU(V)$, which is defined to be $SU(V)$ modulo its center [consisting of multiples of the identity by scalars of norm 1], is simple if $\dim V \ge 2$ and the Witt index is > 0, except for $PSU_2(\mathbb{F}_4)$, $PSU_2(\mathbb{F}_9)$, and $PSU_3(\mathbb{F}_4)$. As usual, one must first show that $PSU(V)$ is perfect; see [24, Section 22; 123, Chapter 11]. See also Remark 6.100.

6.9 Example: SL$_n$ over a Field with Discrete Valuation

Up to now, all of our examples of BN-pairs have had finite Weyl groups (and hence spherical buildings). It turns out that many of the same groups that occurred in those examples admit a second BN-pair structure whenever the ground field comes equipped with a discrete valuation. This was first noticed by Iwahori and Matsumoto [137] and was later generalized to a much larger class of groups by Bruhat and Tits [59]. The Weyl group for this second BN-pair is an infinite Euclidean reflection group, and the associated building has apartments that are Euclidean spaces. We will illustrate this by treating the groups SL$_n$. But first we must review discrete valuations.

6.9.1 Discrete Valuations

Let K be a field and K^* its multiplicative group of nonzero elements.

Definition 6.106. A *discrete valuation* on K is a surjective homomorphism $v\colon K^* \twoheadrightarrow \mathbb{Z}$ satisfying the following inequality:

$$v(x + y) \geq \min\{v(x), v(y)\}$$

for all $x, y \in K^*$ with $x + y \neq 0$. It is convenient to extend v to a function defined on all of K by setting $v(0) = +\infty$; the inequality then remains valid for all $x, y \in K$. Note that we necessarily have $v(-1) = 0$, since $\mathbb{Z}$ is torsion-free; hence $v(-x) = v(x)$. It follows from this and the inequality above that the set $A := \{x \in K \mid v(x) \geq 0\}$ is a subring of K; it is called the *valuation ring* associated to K. And any ring A that arises in this way from a discrete valuation is called a *discrete valuation ring*.

The group A^* of units of A is precisely the kernel $v^{-1}(0)$ of v. So if we pick an element $\pi \in K$ with $v(\pi) = 1$, then every element $x \in K^*$ is uniquely expressible in the form $x = \pi^n u$ with $n \in \mathbb{Z}$ and $u \in A^*$. In particular, K is the field of fractions of A.

The principal ideal πA generated by π can be described in terms of v as $\{x \in K \mid v(x) > 0\}$. It is a maximal ideal, since every element of A not in πA is a unit. The quotient ring $k := A/\pi A$ is therefore a field, called the *residue field* associated to the valuation v.

Example 6.107. Let K be the field $\mathbb{Q}$ of rational numbers, and let p be a prime number. The *p-adic valuation* on $\mathbb{Q}$ is defined by setting $v(x)$ equal to the exponent of p in the prime factorization of x. More precisely, given $x \in \mathbb{Q}^*$, write $x = p^n u$, where n is a (possibly negative) integer and u is a rational number whose numerator and denominator are not divisible by p; then $v(x) = n$. The valuation ring A is the ring of fractions a/b with $a, b \in \mathbb{Z}$ and b not divisible by p. [The ring A happens to be the localization of $\mathbb{Z}$ at p, but we will not make any use of this.] The residue field k is the field $\mathbb{F}_p$ of integers mod p; one sees this by using the homomorphism $A \twoheadrightarrow \mathbb{F}_p$ given by $a/b \mapsto (a \bmod p)(b \bmod p)^{-1}$, where a and b are as above.

The valuation ring A in this example can be described informally as the largest subring of $\mathbb{Q}$ on which reduction mod p makes sense. It is thus the natural ring to introduce if one wants to relate the field $\mathbb{Q}$ to the field $\mathbb{F}_p$. This illustrates our point of view toward valuations: We will be interested in studying objects (such as matrix groups) defined over a field K, and we wish to "reduce" to a simpler field k as an aid in this study; a discrete valuation makes this possible by providing us with a ring A to serve as intermediary between K and k:

Examples 6.108. Let $K = k(t)$, the field of rational functions in one variable over a field k.

(a) Let $v = v_0$ be the valuation on K that gives the order of vanishing at 0 of a rational function. In other words, if $f(t) = t^n g(t)$, where t does not divide the numerator or denominator of $g(t)$, then $v(f) = n$. This is the analogue of the p-adic valuation, with the polynomial ring $k[t]$ playing the role of $\mathbb{Z}$ and t playing the role of p. The valuation ring A is the set of rational functions such that $f(0)$ is defined, the maximal ideal is generated by $\pi := t$, and the residue field $A/\pi A$ can be identified with the original field k via $f \mapsto f(0)$ for $f \in A$.

(b) Now take $v = v_\infty$, the order of vanishing of a rational function at infinity. In other words, if $f(t) = g(t)/h(t)$ with g and h polynomials, then

$$v_\infty(f) := \deg h - \deg g \,.$$

Note that $v_\infty(f(t)) = v_0(f(1/t))$. The valuation ring A is the set of rational functions such that $f(\infty)$ is defined, i.e., the degree of the numerator is less than or equal to the degree of the denominator. We can take $\pi = 1/t$, and we can identify the residue field with k via $f \mapsto f(\infty)$.

Returning now to the general theory, we note that the study of the arithmetic of A (e.g., ideals and prime factorization) is fairly trivial:

Proposition 6.109. *A discrete valuation ring A is a principal ideal domain, and every nonzero ideal is generated by π^n for some $n \geq 0$. In particular, πA is the unique nonzero prime ideal of A.*

Proof. Let I be a nonzero ideal and let $n := \min \{v(a) \mid a \in I\}$. Then I contains π^n, and every element of I is divisible by π^n; hence $I = \pi^n A$. $\square$

One consequence of this is that we can apply the basic facts about modules over a principal ideal domain (e.g., a submodule of a free module is free). Let's recall some of these facts, in the form in which we'll need them later.

Definition 6.110. Let V be the vector space K^n. By a *lattice* (or *A-lattice*) in V we will mean an A-submodule $L < V$ of the form $L = Ae_1 \oplus \cdots \oplus Ae_n$ for some basis $e_1, \ldots, e_n$ of V. In particular, L is a free A-module of rank n. If we take $e_1, \ldots, e_n$ to be the standard basis of V, then the resulting lattice is A^n, which we call the *standard lattice.*

If L' is a second lattice in V, then we can choose our basis $e_1, \ldots, e_n$ for L in such a way that L' admits a basis of the form $\lambda_1 e_1, \ldots, \lambda_n e_n$ for some scalars $\lambda_i \in K^*$. This fact should be familiar for the case $L' \leq L$, and the general case follows easily. [Choose a large integer M such that $\pi^M L' \leq L$, and apply the usual theory to $L'' := \pi^M L'$.] The scalars λ_i can be taken to be powers of π, and they are then unique up to order. They are called the *elementary divisors* of L' with respect to L.

All of this follows from well-known results about modules over principal ideal domains. But we will sketch the proof of part of it (namely, the existence of the e_i and λ_i) in the case at hand, where A is a discrete valuation ring; for the proof involves ideas that will be needed later anyway.

Start with arbitrary bases of L and L', and express the basis elements of L' as linear combinations of those of L; this yields an element of $\mathrm{GL}_n(K)$. It is easy to see that this matrix can be reduced to a monomial matrix by means of integral row and column operations, where "integral" means that the operation is given by multiplication by an elementary matrix in $\mathrm{SL}_n(A)$. [In other words, when we add a scalar multiple of one row or column to another, the scalar is required to be in A.] To see that this is possible, choose a matrix entry a_{ij} with $v(a_{ij})$ minimal. Then pivot to clear out everything other than a_{ij} in the ith row and jth column, noting that this pivoting requires only integral row and column operations. Now ignore the ith row and jth column and repeat the process, using an element of minimal valuation in the rest of the matrix. It is clear that we will eventually obtain a monomial matrix by continuing in this way.

The row and column operations above correspond to changes of basis in L and L'. So what we have just done is to replace the given bases of L and L' by new ones, such that the new basis elements of L' are scalar multiples of the new basis elements of L. This completes the proof.

Note that if the matrix in $\mathrm{GL}_n(K)$ that we started with above happened to be in $\mathrm{SL}_n(K)$, then the same would be true of the monomial matrix that we ended with. So we obtain, as a byproduct of the proof, the following:

Proposition 6.111. $\mathrm{SL}_n(K)$ *is generated by its monomial subgroup together with the elementary matrices in* $\mathrm{SL}_n(A)$. $\qquad\square$

We end this review of discrete valuations by commenting briefly on the notion of *completeness*. A discrete valuation v induces a real-valued absolute value on K, defined by

$$|x| := e^{-v(x)}.$$

We then have

$$|xy| = |x| \cdot |y| \quad \text{and} \quad |x + y| \le \max\{|x|, |y|\} \ .$$

This inequality is a very strong form of the triangle inequality. In particular, we get a metric on K by setting $d(x, y) := |x - y|$. It therefore makes sense to ask whether K is complete, in the sense that every Cauchy sequence converges. If not, then one can form the *completion* $\hat{K}$ of K by formally adjoining limits of Cauchy sequences, in exactly the same way that one constructs $\mathbb{R}$ in elementary analysis by completing $\mathbb{Q}$. The only difference is that the construction is actually easier in the present context, because of the strong form of the triangle inequality. In fact, one can build the completion purely algebraically using inverse limits; see, for instance, Atiyah–MacDonald [27, Chapter 10].

The field operations and the function v extend to $\hat{K}$ by continuity, and $\hat{K}$ is again a field with a discrete valuation. Its valuation ring is the completion $\hat{A}$ of A (i.e., the closure of A in $\hat{K}$), and its residue field is the same as that of K. In case the residue field k is finite, one can show that $\hat{A}$ is compact; since $\hat{A}$ is the closed unit ball in $\hat{K}$, the latter is locally compact in this case.

The canonical example for all this is the p-adic valuation on $\mathbb{Q}$ discussed above. The completion is the field $\mathbb{Q}_p$ of *p-adic numbers*. It is a complete, locally compact, discretely valued field, with residue field $\mathbb{F}_p$. Its valuation ring is called the ring of *p-adic integers*. Another familiar example is $K = k(t)$ · with $v = v_0$ as in Example 6.108(a). The completion can be identified with the field $k((t))$ of formal Laurent series $\sum_{n \in \mathbb{Z}} a_n t^n$ with $a_n \in k$ and $a_n = 0$ for $n \ll 0$. [The point here is that the partial sums of such a series form a Cauchy sequence because $v\left(\sum_{i=n}^{m} a_i t_i\right) \ge n$.] Similarly, the completion with respect to v_∞ (Example 6.108(b)) is $k((1/t))$.

6.9.2 The Group $\mathrm{SL}_n(K)$

Let K continue to denote a field with a discrete valuation v, and let A, π, and k be as in Section 6.9.1. We then have a diagram of matrix groups

$$\begin{array}{ccc} \mathrm{SL}_n(A) & \lhook\joinrel\longrightarrow & \mathrm{SL}_n(K) \\ \downarrow & & \\ \mathrm{SL}_n(k) & & \end{array}$$

which we will use to construct a BN-pair in $\mathrm{SL}_n(K)$ by "lifting" the BN-pair in $\mathrm{SL}_n(k)$ that we studied in Section 6.5. More precisely, we will take B to be the inverse image in $\mathrm{SL}_n(A)$ of the upper-triangular subgroup of $\mathrm{SL}_n(k)$, but we will take N, as before, to be the monomial subgroup of $\mathrm{SL}_n(K)$. [It would not make sense to also construct N as an inverse image, since B and N would then both be subgroups of $\mathrm{SL}_n(A)$ and hence could not possibly generate $\mathrm{SL}_n(K)$.]

Note that B contains the upper-triangular subgroup of $\mathrm{SL}_n(A)$; the subgroup generated by B and N therefore contains both the upper-triangular and lower-triangular subgroups of $\mathrm{SL}_n(A)$ and hence all elementary matrices in $\mathrm{SL}_n(A)$. So this subgroup is the whole group $\mathrm{SL}_n(K)$ by Proposition 6.111. The intersection $T := B \cap N$ is the diagonal subgroup of $\mathrm{SL}_n(A)$, which is easily checked to be normalized by N; in fact, the conjugation action of N on T simply permutes the diagonal entries of a matrix in T.

We need some notation in order to describe the group $W := N/T$. For any commutative ring R, let $N(R)$ (resp. $T(R)$) denote the monomial (resp. diagonal) subgroup of $\mathrm{SL}_n(R)$. Then our N and T above are $N(K)$ and $T(A)$, so $W = N(K)/T(A)$. Let $\overline{W} = N(K)/T(K)$, identified as usual with the symmetric group on n letters. Then $\overline{W}$ is a quotient of W, and we have a short exact sequence

$$1 \to T(K)/T(A) \to W \to \overline{W} \to 1 .$$

This sequence splits, since the subgroup $N(A)/T(A)$ of W maps isomorphically to the quotient $\overline{W}$; so we have

$$W \cong F \rtimes \overline{W} ,$$

where $F := T(K)/T(A) \cong (K^*/A^*)^{n-1}$. Note that the valuation v induces an isomorphism $K^*/A^* \overset{\sim}{\longrightarrow} \mathbb{Z}$, so the normal subgroup F above is free abelian of rank $n - 1$. In order to understand the action of $\overline{W}$ on this free abelian group, identify F with $\{(x_1, \ldots, x_n) \in \mathbb{Z}^n \mid \sum_{i=1}^n x_i = 0\}$. [The isomorphism of F with this group is obtained by applying v to the n diagonal entries of an element of $T(K)$.] The action of $\overline{W}$ on F, then, simply permutes the n coordinates. We now need to find a suitable set of generators for W and verify the BN-pair axioms. Let's start with the case $n = 2$, deferring the general case to Section 6.9.3.

When $n = 2$, we have $W \cong \mathbb{Z} \rtimes \{\pm 1\}$, with the nontrivial action of $\{\pm 1\}$ on $\mathbb{Z}$; this is the infinite dihedral group. It is generated by s_1 and u, where s_1 is the nontrivial element of the $\{\pm 1\}$ factor and u is a generator of the infinite cyclic normal subgroup $F := T(K)/T(A) < W$. So W is generated by the set $S := \{s_1, s_2\}$ of elements of order 2, where $s_2 := s_1 u$. Let's take u to be represented by the element $\mathrm{diag}(\pi, \pi^{-1}) \in T(K)$; then s_1 is represented by

$$\begin{pmatrix} 0 & -1 \\ 1 & 0 \end{pmatrix}$$

and s_2 is represented by

$$\begin{pmatrix} 0 & -1 \\ 1 & 0 \end{pmatrix} \begin{pmatrix} \pi & 0 \\ 0 & \pi^{-1} \end{pmatrix} = \begin{pmatrix} 0 & -\pi^{-1} \\ \pi & 0 \end{pmatrix} .$$

Note that it is by no means clear, a priori, that we have made the right choice of u—we could replace u by u^{-1} and still get a set of two generators of W of order 2. But, as we noted when discussing SO_{2n} in Section 6.7, at most one of these choices can be "right" (in the sense that the BN-pair axioms hold). We have, in fact, made the right choice, and one can easily verify the axioms.

The verification of (BN1) is slightly tedious, since it requires consideration of several cases. As an example, let's take $s = s_2$ and $w = (s_1 s_2)^r$, where $r > 0$. In this case we will show that $sBw \subseteq BswB$, which is what should be true, since $l(sw) = l(w) + 1$.

Note first that the subgroup B of $\mathrm{SL}_2(K)$ can be described by the following conditions on the valuations of the matrix entries:

$$\begin{pmatrix} v = 0 & v \geq 0 \\ v \geq 1 & v = 0 \end{pmatrix}.$$

Computing sBw, we find that its elements satisfy

$$\begin{pmatrix} v \geq r & v = -r - 1 \\ v = r + 1 & v \geq -r + 1 \end{pmatrix}.$$

The unique entry with minimal valuation is in the upper right-hand corner, so we pivot at this position to clear out the diagonal entries. This requires multiplication by elementary matrices of the form

$$\begin{pmatrix} 1 & 0 \\ \pi^2 a & 1 \end{pmatrix}$$

with $a \in A$. Elementary matrices of this form are in B, so the pivoting operations are legal and we have reduced our matrix to a monomial matrix in $\mathrm{SL}_2(K)$ whose upper right-hand corner has valuation $-r - 1$. This monomial matrix is equivalent mod T to the matrix

$$\begin{pmatrix} 0 & -\pi^{-r-1} \\ \pi^{r+1} & 0 \end{pmatrix},$$

which represents sw. This completes the verification of our special case of (BN1). The other cases are equally easy.

Now let's try to describe the building $\Delta(G, B)$. Note first that one of the standard parabolics is $\mathrm{SL}_2(A)$, since this is a subgroup containing B. Since $\mathrm{SL}_2(A)$ is the stabilizer in $\mathrm{SL}_2(K)$ of the standard A-lattice A^2 in K^2, this suggests that the vertices of $\Delta(G, B)$ should correspond to lattices in K^2. [There is, of course, an obvious action of $\mathrm{SL}_2(K)$, and even $\mathrm{GL}_2(K)$, on the set of lattices.] On the other hand, our experience in Section 6.5 suggests that Δ should admit an action of $\mathrm{PGL}_2(K) := \mathrm{GL}_2(K)/Z$, where Z is the group of scalar multiples of the identity. So it is more reasonable to expect the vertices to correspond to Z-orbits of lattices. With this as motivation, we proceed to describe the building.

Call two A-lattices L, L' in K^2 *equivalent* if $L = \lambda L'$ for some $\lambda \in K^*$. Note that the scalar λ can then be taken to be a power of π. Let $[L]$ denote the equivalence class of a lattice L. If L is given as $Af_1 \oplus Af_2$ for some basis f_1, f_2 of K^2, then we will also write $[[f_1, f_2]]$ for the class $[L]$.

We want to assign a *type* to a lattice class. To this end, consider the obvious action of $GL_2(K)$ on the set of lattice classes. This action is transitive, and the stabilizer of $[A^2]$ is $Z \cdot GL_2(A)$, where Z is as above. It follows that $v(\det g)$ is an even integer for every g in this stabilizer. We can now say that a lattice class Λ is of type 0 (resp. type 1) if $v(\det g)$ is even (resp. odd) for every $g \in GL_2(K)$ such that $g[A^2] = \Lambda$. In other words, the type of $[[f_1, f_2]]$ is $v(\det(f_1, f_2))$ mod 2, where $\det(f_1, f_2)$ is the determinant of the matrix with f_1 and f_2 as columns.

Call two distinct lattice classes Λ, Λ' *incident* if they have representatives L, L' that satisfy

$$\pi L < L' < L .$$

Note that the representatives $\pi L, L'$ then satisfy $\pi L' < \pi L < L'$, so the incidence relation is symmetric. Note also that in this situation, the elementary divisors of L' with respect to L are necessarily 1 and π, so we have $\Lambda = [[f_1, f_2]]$ and $\Lambda' = [[f_1, \pi f_2]]$ for some basis f_1, f_2 of K^2. It follows that Λ and Λ' are of different types, so we have a plane incidence geometry.

Let's show now that the flag complex Δ of this geometry is isomorphic to $\Delta(G, B)$ for $G = SL_2(K)$ as above. We will prove this, as usual, by finding a fundamental domain and computing stabilizers. Let C be the edge with vertices $[[e_1, e_2]]$ and $[[e_1, \pi e_2]]$, where e_1, e_2 is the standard basis of K^2. Let $C' = \{\Lambda, \Lambda'\}$ be an arbitrary edge, with Λ of type 0. Then there is a basis f_1, f_2 such that $\Lambda = [[f_1, f_2]]$, $\Lambda' = [[f_1, \pi f_2]]$, and $\det(f_1, f_2) = \pi^{2r} u$ for some $r \in \mathbb{Z}$ and $u \in A^*$. Replacing f_1 by $\pi^{-r} u^{-1} f_1$ and f_2 by $\pi^{-r} f_2$, we still have $\Lambda = [[f_1, f_2]]$ and $\Lambda' = [[f_1, \pi f_2]]$, but now $\det(f_1, f_2) = 1$. So the matrix g with f_1 and f_2 as columns is an element of G such that $gC = C'$. Since the action of G is type-preserving, it follows easily that $\overline{C}$ is a simplicial fundamental domain.

The stabilizer of $[A^2]$ in G is $SL_2(K) \cap (Z \cdot GL_2(A)) = SL_2(A)$. And the stabilizer of $[[e_1, \pi e_2]]$ is the conjugate $g\, SL_2(A) g^{-1}$, where $g := \mathrm{diag}(1, \pi)$. This conjugate is the subgroup of G defined by

$$\begin{pmatrix} v \geq 0 & v \geq -1 \\ v \geq 1 & v \geq 0 \end{pmatrix} .$$

The stabilizer of C, then, which is the intersection of the stabilizers of its two vertices, is precisely B. The desired isomorphism $\Delta \cong \Delta(G, B)$ follows easily.

The fundamental apartment Σ is obtained by applying the elements of the monomial group N to C; it is a line, with vertices $[[\pi^a e_1, \pi^b e_2]]$, $a, b \in \mathbb{Z}$. An arbitrary apartment $g\Sigma$ is the same sort of line, but with e_1, e_2 replaced by an arbitrary basis of K^2.

Remarks 6.112. (a) Since $\Delta(G, B)$ is a building of type $\circ\!\!\!\overset{\infty}{-\!\!-\!\!-}\!\!\!\circ$, we know from Section 4.4 that the flag complex Δ is a tree. We could have simply proven this directly (see Serre [217, Chapter II]) and deduced the BN-pair structure in G without any matrix computations. (We would then have had to check strong transitivity, but this is easy.)

(b) This example shows that incomplete apartment systems arise naturally. In fact, it is not hard to see that the apartment system $G\Sigma := \{g\Sigma \mid g \in G\}$ is complete if and only if the field K is complete with respect to the valuation v. One can show further that if an arbitrary K is replaced by its completion $\hat{K}$, then the building Δ remains the same—all that changes is the apartment system, which gets replaced by the complete system of apartments. These assertions, and their analogues for SL_n with $n > 2$, will be proved in Section 11.8.6. See also Serre [217, Section II.1]. Consider, for example, the case that K is the field $\mathbb{Q}_2$ of 2-adic numbers. Then one can show that Δ is the regular tree of degree 3 (Figure 6.3), with uncountably many apartments.[*] If we instead take $K = \mathbb{Q}$ (with the 2-adic valuation), then we get the same

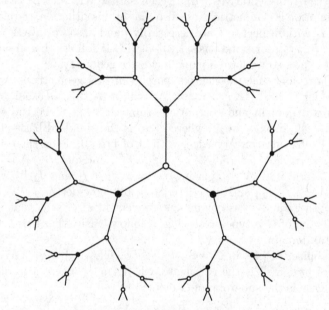

Fig. 6.3. The building for $SL_2(\mathbb{Q}_2)$.

tree, equipped with a certain countable apartment system.

(c) The isomorphism type of the tree Δ associated to $SL_2(K)$ depends only on the cardinality of the residue field $k = A/\pi A$. If $k = \mathbb{F}_2$, for instance, then every vertex is on exactly 3 edges, so the tree is necessarily the one in Figure 6.3. This shows that many different choices of (K, v) can yield the same tree Δ, even if we stick to the case that K is complete. Thus there is no hope of recovering the group G, or even the quotient $\bar{G}$ defined in Remark 6.30,

---

[*] Figure 6.3 was drawn by Kai-Uwe Bux; we are grateful to him for permission to reproduce it here.

from the tree Δ. In particular, one cannot expect that $\bar{G}$ is the group of type-preserving automorphisms of Δ; see Exercise 6.114.

This discussion might seem to suggest that there is a very poor correspondence between groups with a BN-pair and buildings, unlike the situation for Coxeter groups and Coxeter complexes. But the tree case is atypical in this regard, and the correspondence is much better for some other classes of buildings; see Chapter 9 and Section 11.9.

Before moving on to SL$_n$ for $n > 2$, we look briefly at what happens if we replace SL$_2(K)$ by GL$_2(K)$. It is clear that GL$_2(K)$ acts on the flag complex Δ; but the action does not preserve types. This is the same situation that we saw at the end of Section 6.7, so we obtain some kind of generalized BN-pair structure in GL$_2(K)$. We illustrate this by proving one result about double cosets in GL$_2(K)$.

Let (G, B, N, S) continue to have the same meaning as above, with $G = $ SL$_2(K)$, etc. The result to be proved is that

$$sBm \subseteq BsmB \cup BmB \tag{6.26}$$

for any $s \in S$ and any monomial matrix $m \in$ GL$_2(K)$. Note first that m stabilizes the fundamental apartment, so $mC = wC$ for some $w \in W$. This implies that $mB' = wB'$, where B' is the stabilizer of C in GL$_2(K)$. Now we know that

$$sBw \subseteq BswB \cup BwB ,$$

so we can right-multiply by a suitable element of B' to get

$$sBm \subseteq BswB' \cup BwB' = BsmB' \cup BmB' .$$

Since all elements of $Bsm \cup Bm$ have the same determinant as the elements of sBm, we can replace B' by $B' \cap$ SL$_2(K) = B$ to get (6.26).

Exercises

6.113. Show that the canonical map SL$_n(A) \to$ SL$_n(k)$ is surjective.

6.114. The purpose of this exercise is to show, by three different methods, that the action of PSL$_2(\mathbb{Q}_p)$ on the tree Δ constructed above is very far from giving all automorphisms of Δ. Let $G =$ PSL$_2(\mathbb{Q}_p)$ and let H be the full group of type-preserving automorphisms of Δ.

(a) For any vertex $u \in \Delta$, show that the stabilizer H_u induces all possible permutations of the neighbors of u. Show that this is not true of G_u, on the other hand, except for some small values of p.

(b) If $g \in G$ is any element that stabilizes an apartment and acts as a reflection on it, show that g^2 is the identity on the entire tree Δ. But it is easy to construct elements $h \in H$ of order $\neq 2$ such that h acts as a reflection on an apartment.

(c) Show that the set of ends of Δ can be identified with the projective line over K, with its natural action of $G = \mathrm{PSL}_2(K)$. On the other hand, H induces many more automorphisms of the set of ends than those given by elements of G.

6.115. Let Λ be the vertex $[A^2]$. Recall that its stabilizer in $\mathrm{SL}_2(K)$ is $\mathrm{SL}_2(A)$.

(a) Show that the set of vertices of the link of Λ can be identified with the projective line over k, on which $\mathrm{SL}_2(A)$ acts via the canonical homomorphism $\mathrm{SL}_2(A) \twoheadrightarrow \mathrm{SL}_2(k)$.

(b) Let $K = k(t)$ with the valuation v_0 of Example 6.108(a). Show that $\mathrm{SL}_2(A)$ contains a subgroup U that fixes the vertex $[[e_1, \pi e_2]]$ and permutes the remaining vertices of the link of Λ simply transitively.

6.9.3 The Group $\mathrm{SL}_n(K)$, Concluded

We continue with the notation of Section 6.9.2, but we now assume that $n \geq 3$. Recall that $W = F \rtimes \overline{W}$, where $F = T(K)/T(A) \cong \mathbb{Z}^{n-1}$ and $\overline{W} = N(A)/T(A)$. We need a set $S = \{s_1, \ldots, s_n\}$ of generators of W. For the first $n - 1$ of these we use the standard generators of the symmetric group $\overline{W}$; thus s_i for $i < n$ is represented by a monomial matrix in the embedded $\mathrm{SL}_2(A) < \mathrm{SL}_2(K) \hookrightarrow \mathrm{SL}_n(K)$ acting on $[e_i, e_{i+1}]$. And for s_n we take the element of W represented by

$$
\tilde{s}_n = \begin{pmatrix}
0 & & & & -\pi^{-1} \\
& 1 & & & \\
& & \ddots & & \\
& & & 1 & \\
\pi & & & & 0
\end{pmatrix}.
$$

This monomial matrix is in the embedded $\mathrm{SL}_2(K)$ acting on $[e_n, e_1]$. To see that S generates W, note first that the subgroup $W' := \langle S \rangle$ contains $\overline{W}$. Multiplying s_n by a suitable element of $\overline{W}$, we conclude that W' also contains the element of F represented by $\mathrm{diag}(\pi, 1, \ldots, 1, \pi^{-1})$. Conjugating this element by $\overline{W}$, we obtain a set of generators for F, so $W' = W$.

The verification of (BN2) presents no problem. To check (BN1), one proceeds as in Sections 6.5–6.7, the idea being to use row operations to reduce to a 2×2 matrix computation. As an illustration, here are the details for the case $n = 3$ and $s = s_3$.

The statement to be proved is that $sB \subseteq BsB' \cup BB'$, where $B' := mBm^{-1}$ for some monomial matrix m of determinant 1. Motivated by what we just did at the end of Section 6.9.2, let's prove this inclusion more generally for an arbitrary monomial matrix. Now an element of sB has the form

$$
\begin{pmatrix}
v \geq 0 & v \geq 0 & v = -1 \\
v \geq 1 & v = 0 & v \geq 0 \\
v = 1 & v \geq 1 & v \geq 1
\end{pmatrix},
$$

so we start by pivoting at the upper right-hand corner to clear out the two entries below it. (The row operations required for this are given by left multiplication by lower-triangular elements of B.) This leaves us with

$$\begin{pmatrix} v \geq 0 & v \geq 0 & v = -1 \\ v \geq 1 & v = 0 & 0 \\ v = 1 & v \geq 1 & 0 \end{pmatrix}.$$

We can also make the middle entry (the one with $v = 0$) equal to 1; for this can be achieved by multiplication by an element of T.

Now pivot on this middle entry to clear out the entries above and below it. This yields

$$\begin{pmatrix} v \geq 0 & 0 & v = -1 \\ v \geq 1 & 1 & 0 \\ v = 1 & 0 & 0 \end{pmatrix}.$$

Pivoting at the lower left-hand corner, finally, reduces us to a matrix in the copy of SL_2 that contains $s = s_3$:

$$\begin{pmatrix} v \geq 0 & 0 & v = -1 \\ 0 & 1 & 0 \\ v = 1 & 0 & 0 \end{pmatrix}.$$

For the rest of the proof we ignore the middle row and middle column and work entirely in the SL_2 that remains. We need some notation. Let $G_3 := SL_3(K)$ and let G_2 be the embedded $SL_2(K)$ that we have just reduced to, i.e., $G_2 = \{g \in G_3 \mid ge_2 = e_2, \ g[e_1, e_3] = [e_1, e_3]\}$. Let B_3 and B_2 be the corresponding B's. The matrix above, then, is an element of $sB_3 \cap G_2$, and we wish to reduce it to 1 or s by multiplying on the left by $B_3 \cap G_2$ and on the right by $B' \cap G_2$.

It is easy to check that $B_3 \cap G_2 = B_2$ and that $sB_3 \cap G_2 = \tilde{s}B_2$, where $\tilde{s}$ is the monomial matrix $\tilde{s}_3$ that we wrote down above. So we will have reduced our problem about G_3 to the same sort of problem for G_2 [which we solved in Section 6.9.2], provided $B' \cap G_2$ is a "B'-type" subgroup of G_2. To finish the proof, then, we need to understand what kind of subgroup of G_3 can arise as a B', and we need to show that $B' \cap G_2$ is the "same kind" of subgroup of G_2.

Let L be the standard lattice in K^3, with basis e_1, e_2, e_3. Then we can identify $L/\pi L$ with the vector space k^3 over the residue field k. Note that any $g \in SL_3(A)$ stabilizes L and hence acts on $L/\pi L = k^3$; indeed, this is one way of describing the homomorphism $SL_3(A) \to SL_3(k)$ that we wrote down at the beginning of Section 6.9.2. We can now describe $B = B_3$ as follows: Given $g \in G_3$, we have

$$g \in B_3 \iff gL = L \text{ and } gC = C,$$

where C is the standard flag in k^3. Consequently, $B' = mBm^{-1}$ admits a similar description:

$$g \in B' \iff gL' = L' \text{ and } gC' = C' \, ,$$

where $L' = mL$ and C' is the flag in $L'/\pi L'$ corresponding to C under the isomorphism $L/\pi L \to L'/\pi L'$ induced by m.

Now the lattice L' has a basis $\pi^a e_1, \pi^b e_2, \pi^c e_3$ for some $a, b, c \in \mathbb{Z}$. So the k-vector space $L'/\pi L'$ comes equipped with a "standard basis," and C' is simply the "permuted standard flag" obtained from a permutation of that standard basis. [The permutation that arises is the one corresponding to the monomial matrix m.] It is now easy to describe $B' \cap G_2$: Let L'_2 be the lattice in $[e_1, e_3]$ with basis $\pi^a e_1, \pi^c e_3$; given $g \in G_2$, we have

$$g \in B' \cap G_2 \iff gL'_2 = L'_2 \text{ and } gC'_2 = C'_2 \, ,$$

where C'_2 is a certain permuted standard flag in $L'_2/\pi L'_2$. This characterization of $B' \cap G_2$ as a subgroup of G_2 is the exact analogue of the characterization of B' as a subgroup of G_3, so we are done.

Remark 6.116. If we had simply tried to prove (BN1) instead of a generalization of it, then we would have assumed $\det m = 1$. The only difference this would have made is that we would have had $a + b + c = 0$ in the description of L'. But the analogous sum $a + c$ for L'_2 would not necessarily have been 0, so we would still have needed the generalized (BN1) for the 2×2 case. In other words, we needed to understand GL_2, and not just SL_2, in order to deal with SL_n for $n \geq 3$.

The building Δ associated to $\mathrm{SL}_n(K)$ is a flag complex as in the case $n = 2$: One considers classes of lattices in K^n, one assigns a type to any such class by taking the valuation of a determinant and reducing mod n, and one defines incidence exactly as before. The fundamental chamber is the simplex with vertices $[[e_1, \ldots, e_i, \pi e_{i+1}, \ldots, \pi e_n]]$, $i = 1, \ldots, n$. Further details are left to the interested reader.

There are a number of interesting things to say about this example, and they motivate much of what we will do in Chapter 11 when we develop the theory of *Euclidean buildings*. Consider first the Weyl group $W = \mathbb{Z}^{n-1} \rtimes \overline{W}$. [Recall from Section 6.9.2 that $\mathbb{Z}^{n-1}$ is identified with the subgroup of $\mathbb{Z}^n$ defined by $x_1 + \cdots + x_n = 0$, and the symmetric group $\overline{W}$ acts by permuting the n coordinates.] This Coxeter group is not one that we have seen before. Computing the orders of the products $s_i s_j$, one finds that its Coxeter diagram is

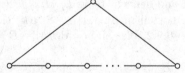

where there are n vertices altogether. [This is the diagram when $n \geq 3$; the diagram for $n = 2$ is of course different.] We will see in Section 10.1.7 that

W is a Euclidean reflection group acting on $\mathbb{R}^{n-1}$. More precisely, it is the *affine Weyl group* associated to the root system of type A_{n-1}, and W itself is said to have type $\tilde{A}_{n-1}$. If $n = 3$, for instance, W is the group of isometries of the plane generated by the reflections in the sides of an equilateral triangle. The apartments in this case are planes tiled by equilateral triangles; see Example 3.7 and Figure 3.1. These planes are glued together to form the building Δ, which can then be viewed as a 2-dimensional analogue of a tree. (A tree is constructed by gluing lines together.)

If we delete a vertex from the Coxeter diagram above, we obtain the diagram of type A_{n-1}. So the link of a vertex in our building Δ is a spherical building of type A_{n-1}. (This is true if $n = 2$ also.) If we take, for instance, the vertex $[[e_1, \ldots, e_n]]$ whose stabilizer is $\mathrm{SL}_n(A)$, this link is some building of type A_{n-1} that comes equipped with an action of $\mathrm{SL}_n(A)$. The obvious guess is that the link is the spherical building associated to $\mathrm{SL}_n(k)$ as in Section 6.5, and it is easy to check that this guess is correct.

There is, of course, another building of type A_{n-1} that one would naturally think of, namely, the building Δ' obtained by forgetting that K has a valuation and applying Section 6.5 to $\mathrm{SL}_n(K)$. This building admits an action of the full group $\mathrm{SL}_n(K)$. Is it somehow related to our building Δ also? The answer is that Δ' is the spherical building "at infinity" in Δ, obtained by attaching a sphere at infinity to each apartment of Δ. Details will be given in Chapter 11.

This connection between Δ and Δ' provides a nice geometric explanation of the fact that both of our BN-pairs in $\mathrm{SL}_n(K)$ used the same N. For N is the stabilizer of the fundamental apartment in both Δ and Δ'; the use of the same N therefore yields a 1–1 correspondence between the apartments in Δ and those in Δ'. The previous paragraph explains this correspondence geometrically.

All of this is easy to understand when $n = 2$ in terms of ends of trees. The building Δ' in this case is 0-dimensional, so it is simply the discrete set G/B [with B equal to the upper-triangular group], in which the two-element subsets have been called apartments. This set can be identified with a set of ends of the tree Δ [217, Section II.1.3], so Δ' is clearly a building "at infinity" associated to Δ. And the 1–1 correspondence between apartment systems simply reflects the fact that a line in a tree gives rise to a pair of ends and that, conversely, a pair of ends determines a unique line.

Remark 6.117. The construction of a Euclidean building associated to the group SL_n over a field with discrete valuation is a special case of the theory of Bruhat and Tits [59], and the corresponding buildings are often called *Bruhat–Tits buildings*. The method of Bruhat and Tits involves constructing something called a "root datum with valuation" and showing that this gives rise to a BN-pair. But it is possible to construct the buildings associated to the classical groups explicitly in terms of lattices, as we have done for SL_n; see Abramenko–Nebe [13] and Garrett [106, Chapters 19 and 20].

*6.10 Example: Weyl-Transitive Actions

We are now in a position to give examples of Weyl-transitive actions that are not strongly transitive with respect to any apartment system. In group-theoretic language, the examples exhibit Tits subgroups B that do not come from BN-pairs. A method for constructing such examples was outlined by Tits [261, Section 3.1, Example (b)]. Namely, start with a strongly transitive action of a group G on a building Δ. Assume that G is a topological group and that the stabilizer of some (and hence every) chamber is an open subgroup. (This is the case, for instance, if $G = \mathrm{SL}_n(\mathbb{Q}_p)$ as in Section 6.9.) Then the action of any dense subgroup $G' \leq G$ will still be Weyl transitive but not necessarily strongly transitive. The details follow.

6.10.1 Dense Subgroups

Lemma 6.118. *Let G be a topological group and G' a dense subgroup.*

(1) *If U is an open subgroup of G, then G' maps onto G/U under the quotient map $G \to G/U$.*

(2) *If G acts transitively on a set X and the stabilizer of some $x \in X$ is an open subgroup, then the action of G' on X is transitive.*

(3) *If G acts on an arbitrary set X and the stabilizers are open subgroups, then the G'-orbits in X are the same as the G-orbits.*

Proof. For (1), observe that every coset gU is a nonempty open set, so it meets G'. (2) is a restatement of (1), and (3) follows from (2).   □

Proposition 6.119. *Suppose a group G acts Weyl transitively on a building Δ. Assume that G is a topological group and that the stabilizer B of some chamber is an open subgroup. If G' is a dense subgroup of G, then the action of G' on Δ is also Weyl transitive.*

Proof. Consider the diagonal action of G on $\mathcal{C} \times \mathcal{C}$, where $\mathcal{C} = \mathcal{C}(\Delta)$. Note that every stabilizer is an open subgroup of G, being an intersection of two conjugates of B. To show that the action of G' is Weyl transitive, we must show that the G'-orbits in $\mathcal{C} \times \mathcal{C}$ are the sets of the form $\{(C, C') \mid \delta(C, C') = w\}$, one for each $w \in W$. But these are precisely the G-orbits by assumption, so the result follows from Lemma 6.118(3).   □

Notice that the action of G might well be strongly transitive, but there is no reason to think that the same is true of the action of G'.

6.10.2 Dense Subgroups of $\mathrm{SL}_2(\mathbb{Q}_p)$

Start with the ring $\mathbb{Z}[1/p]$, consisting of all rational numbers of the form a/p^n $(a, n \in \mathbb{Z})$. This is dense in $\mathbb{Q}_p$ because its closure is a subring containing $\mathbb{Z}_p$

and $1/p$. It follows that the group $\mathrm{SL}_2(\mathbb{Z}[1/p])$ is dense in $G := \mathrm{SL}_2(\mathbb{Q}_p)$, since its closure contains the elementary subgroups $\left(\begin{smallmatrix} 1 & * \\ 0 & 1 \end{smallmatrix}\right)$ and $\left(\begin{smallmatrix} 1 & 0 \\ * & 1 \end{smallmatrix}\right)$.

Now fix an integer $q \geq 3$ and relatively prime to p, and consider the subgroup $G' < \mathrm{SL}_2(\mathbb{Z}[1/p])$ consisting of matrices that are congruent to the identity mod q. In other words, G' is the kernel of the canonical homomorphism $\mathrm{SL}_2(\mathbb{Z}[1/p]) \to \mathrm{SL}_2(\mathbb{Z}/q\mathbb{Z})$. Then G' is still dense in G by the same argument as above, since it contains all elementary matrices whose off-diagonal entry is in $q\mathbb{Z}[1/p]$, and $q\mathbb{Q}_p = \mathbb{Q}_p$. (Note: G' is an example of a *congruence subgroup*.)

Consider the action of G on the tree constructed in Section 6.9.2. It is strongly transitive with respect to the complete apartment system (see Remark 6.112(b)). The action of G' on this tree is therefore Weyl transitive by Proposition 6.119. But a very simple argument will show that this action is not strongly transitive with respect to any apartment system.

Recall that in the action of $G = \mathrm{SL}_2(\mathbb{Q}_p)$ on the tree, the stabilizer of the fundamental apartment Σ is the monomial group, with the diagonal matrices acting as translations and the nondiagonal matrices acting as reflections. In particular, any $g \in G$ that acts as a reflection on Σ satisfies $g^2 = -1$. Since G acts transitively on the complete apartment system, it follows that any $g \in G$ that stabilizes *any* apartment and acts as a reflection on it satisfies $g^2 = -1$. But our group G' does not contain the matrix -1; hence, a fortiori, it does not contain any such g. Thus the stabilizer of an apartment in G' cannot be chamber transitive on that apartment, and hence, by Proposition 6.14, the action of G' on the tree is not strongly transitive with respect to any apartment system.

*6.11 Example: Norm-1 Groups of Quaternion Algebras

The examples in the previous section may have seemed artificial, though they should at least provide convincing evidence that Weyl-transitive actions are much more abundant than strongly transitive actions. In this section we study a family of examples that "arises in nature" and so may seem less ad hoc. These examples were first worked out in Abramenko–Brown [10]. A surprising feature is that although we are in a setting where we might expect strong transitivity to be rare, it turns out to hold roughly half of the time.

6.11.1 Quaternion Algebras

In this subsection we will use standard facts about quadratic forms that can be found in many places, such as [150, 181, 208, 215]. The first three of these books also contain treatments of quaternion algebras. All results that we state without proof can be found in at least one of these references. See also Section 6.7.2 above for some of the terminology and notation that we use in connection with quadratic forms.

The well-known quaternion algebra constructed by Hamilton in the nineteenth century is a 4-dimensional division algebra D over $\mathbb{R}$ with basis $1, i, j, k$, where 1 is the identity element, $i^2 = j^2 = -1$, and $ij = -ji = k$. (The rules for multiplying other pairs of basis elements then follow. For example, $k^2 = -1$, $jk = -kj = i$.) Given a quaternion $x = x_1 + x_2 i + x_3 j + x_4 k$, its *conjugate* is $\bar{x} := x_1 - x_2 i - x_3 j - x_4 k$, and its *norm* is $N(x) := x\bar{x} = x_1^2 + x_2^2 + x_3^2 + x_4^2 \in \mathbb{R}$. Note that D is indeed a division algebra, as claimed above, because the quadratic form N is anisotropic: The inverse of an element $x \neq 0$ is $\bar{x}/N(x)$. Note further that the norm is multiplicative $[N(xy) = N(x)N(y)$ for $x, y \in D]$, so the set of elements of norm 1 is a multiplicative group, called the *norm*-1 *group* of D. It is a Lie group, whose underlying topological space is the 3-sphere.

One can identify D with a real subalgebra of $M_2(\mathbb{C})$, the algebra of 2×2 matrices over $\mathbb{C}$, via

$$x_1 + x_2 i + x_3 j + x_4 k \mapsto \begin{pmatrix} z & -\bar{w} \\ w & \bar{z} \end{pmatrix} ; \tag{6.27}$$

here $z = x_1 + x_2 i$, $w = x_3 - x_4 i$, and the bar denotes complex conjugation. To derive this, view D as a 2-dimensional right vector space over the subfield $\mathbb{C} = \mathbb{R} + \mathbb{R}i < D$, with basis $1, j$, and let D act on itself by left multiplication; see [150, pp. 53–54] for more details. One can check that the map in (6.27) induces an isomorphism from the norm-1 group of D to the special unitary group $\mathrm{SU}_2(\mathbb{C})$.

Hamilton's construction generalizes as follows: Let F be an arbitrary field of characteristic $\neq 2$, and let α, β be nonzero elements of F. There is a 4-dimensional associative F-algebra D with basis e_1, e_2, e_3, e_4, where e_1 is the identity element, $e_2^2 = \alpha$, $e_3^2 = \beta$, and $e_2 e_3 = -e_3 e_2 = e_4$. [Here α and β are identified with αe_1 and βe_1.] As in the classical case, these relations yield rules for multiplying all pairs of basis elements. For example, $e_4^2 = e_2 e_3 e_2 e_3 = -e_2^2 e_3^2 = -\alpha\beta$, and $e_3 e_4 = e_3 e_2 e_3 = -\beta e_2$. The algebra D is called the *quaternion algebra* determined by F, α, β, and we will denote it by $(\alpha, \beta)_F$. Thus the classical quaternion algebra of Hamilton is $(-1, -1)_{\mathbb{R}}$.

If α does not have a square root in F, then we can form the quadratic extension $F(\sqrt{\alpha})$ of F and identify D with an F-subalgebra of $M_2(F(\sqrt{\alpha}))$. This is done as in the derivation of (6.27): View D as a 2-dimensional right vector space over the subfield $F + Fe_2$, which is isomorphic to $F(\sqrt{\alpha})$, and let D act on itself by left multiplication. The resulting map is given by

$$x_1 + x_2 e_2 + x_3 e_3 + x_4 e_4 \mapsto \begin{pmatrix} x_1 + x_2\sqrt{\alpha} & \beta(x_3 + x_4\sqrt{\alpha}) \\ x_3 - x_4\sqrt{\alpha} & x_1 - x_2\sqrt{\alpha} \end{pmatrix} . \tag{6.28}$$

In spite of our use of the letter D, quaternion algebras are not always division algebras. In fact, the argument given above for the classical case can be used to show that D is a division algebra if and only if its norm form N is anisotropic. Here the norm of $x = x_1 + x_2 e_2 + x_3 e_3 + x_4 e_4$ is

$$N(x) := x\bar{x} = x_1^2 - \alpha x_2^2 - \beta x_3^2 + \alpha\beta x_4^2 \in F \,,$$

where $\bar{x} := x_1 - x_2 e_2 - x_3 e_3 - x_4 e_4$ as before. Moreover, if D is not a division algebra, then it is isomorphic to the algebra $M_2(F)$ of 2×2 matrices, with the norm form corresponding to the determinant; D is said to *split* in this case.

For example, D splits if α has a square root in F. In this case there is an isomorphism of D with $M_2(F)$ given by the same formula as in (6.28), where $\sqrt{\alpha}$ now denotes a square root of α in F rather than in an extension field.

There are several useful criteria for a quaternion algebra to split [150, Theorem III.2.7; 181, 57:9; 208, Corollary 2.11.10]. We have already mentioned one: $D = (\alpha, \beta)_F$ splits if and only if the norm form N is isotropic. Here is a second criterion that we will have occasion to use, which we call *Hilbert's criterion* following Lam [150, p. 59]: D splits if and only if the binary quadratic form $\langle \alpha, \beta \rangle$ over F represents 1. (Here, as in Section 6.7.2, $\langle \alpha, \beta \rangle$ denotes the quadratic form $Q(x, y) = \alpha x^2 + \beta y^2$; it represents 1 if and only if the equation $\alpha x^2 + \beta y^2 = 1$ has a solution x, y over F.)

Next, we show that quaternion division algebras are less special than they might seem at first glance:

Proposition 6.120. *Let D be a noncommutative 4-dimensional division algebra over a field F of characteristic $\neq 2$. Then D is isomorphic to a quaternion algebra.*

This is proved in Bourbaki [43, Section 11.2, Proposition 1] and Lam [150, Theorem III.5.1]. Since we will need to refer to the proof in Exercise 6.121 below, we sketch it here. We follow Bourbaki's proof, which uses standard facts from Wedderburn theory. Lam's proof is slightly more elementary.

Sketch of proof. The hypotheses easily imply that F is the center of D. Let K be a maximal commutative subfield of D. Then K is a quadratic extension of F and hence has the form $K = F(e_2)$, where $e_2^2 = \alpha$ for some $\alpha \in F^*$. By the Skolem–Noether theorem, the nontrivial Galois automorphism of K over F, taking e_2 to $-e_2$, extends to an inner automorphism of D. In other words, there is an element $e_3 \in D^*$ such that $e_3 e_2 e_3^{-1} = -e_2$, or $e_3 e_2 = -e_2 e_3$. Then $e_3 \notin K$ because conjugation by e_3 is nontrivial on K, so D is a 2-dimensional left K-vector space with basis $1, e_3$, i.e., $D = K \oplus K e_3$. It follows that D has an F-basis $1, e_2, e_3, e_4$, where $e_4 = e_2 e_3$, and the proof will be complete if we show that $e_3^2 \in F$. To this end we need only note that conjugation by e_3^2 induces the trivial automorphism of K. Thus e_3^2 commutes with K and e_3, so e_3^2 is central and hence is in F. $\qquad\square$

Finally, we collect, in the form of exercises, some facts that will be used in the next subsection.

Exercises

6.121. Show that the proof of Proposition 6.120 yields the following more precise conclusion: There is an element $\alpha \in F^*$ such that α has a square root in D but not in F. For any such α, there is a $\beta \in F^*$ such that $D \cong (\alpha, \beta)_F$.

6.122. (Cf. [150, Exercise III.5]) Let D be an arbitrary quaternion algebra. If -1 has a square root in D but not in F, show that D is isomorphic to a quaternion algebra $(\gamma, -1)_F$ for some $\gamma \in F$.

6.123. Let F be the field $\mathbb{Q}_p$ of p-adic numbers, where p is an odd prime. If α and β are p-adic units, i.e., $v_p(\alpha) = v_p(\beta) = 0$, show that $(\alpha, \beta)_{\mathbb{Q}_p}$ splits.

6.124. Let $D := (\alpha, \beta)_{\mathbb{Q}}$ be a quaternion algebra over $\mathbb{Q}$, and, for any prime p, let D_p be the $\mathbb{Q}_p$-algebra obtained from D by extension of scalars; equivalently, $D_p = (\alpha, \beta)_{\mathbb{Q}_p}$. Assume that $p \neq 2$ and that α and β are p-adic units, so that D_p splits by Exercise 6.123. Show that $D \cong (\alpha', \beta)_{\mathbb{Q}}$ for some $\alpha' \in \mathbb{Q}^*$ such that α' is a p-adic unit and has a square root in $\mathbb{Q}_p$.

6.11.2 Density Lemmas

We now specialize to the case that the ground field F is $\mathbb{Q}$. Choose $\alpha, \beta \in \mathbb{Q}^*$ such that $D := (\alpha, \beta)_{\mathbb{Q}}$ is a division algebra. We can ensure this, for example, by taking $\alpha, \beta < 0$. If -1 has a square root in D, then we will assume, without loss of generality, that $\beta = -1$ (see Exercise 6.122). Choose a prime $p \neq 2$ such that $v_p(\alpha) = v_p(\beta) = 0$. (Note that all but finitely many primes satisfy these conditions.) Then $D_p := (\alpha, \beta)_{\mathbb{Q}_p}$ splits by Exercise 6.123. Finally, we assume, again without loss of generality, that α has a square root in $\mathbb{Q}_p$ (Exercise 6.124).

It is obvious that D is dense in D_p, where the latter is topologized as a 4-dimensional $\mathbb{Q}_p$-vector space. It is also true, but not obvious, that the density persists when one passes to elements of norm 1. These elements form a multiplicative group, sometimes denoted by $\mathrm{SL}_1(D)$.

Lemma 6.125. $G := \mathrm{SL}_1(D)$ is dense in $G_p := \mathrm{SL}_1(D_p)$.

Proof. This is a special case of the weak approximation theorem [188, Chapter 7], but we will give a direct proof. The main point is to construct many elements of G, which we do by the following "normalization": Given $x \in D^*$, let $x' = x\bar{x}^{-1} = x^2/N(x)$; then x' has norm 1. Using this construction, we see that the closure of G in G_p contains all elements of the form $y^2/N(y)$ with $y \in D_p^*$. In particular, it contains all squares of elements of G_p, so the proof will be complete if we show that G_p is generated by squares. This follows, for instance, from the fact that $G_p \cong \mathrm{SL}_2(\mathbb{Q}_p)$; the latter is generated by strictly triangular matrices, all of which are squares. □

Since α has a square root in $\mathbb{Q}_p$, we have a specific isomorphism $G_p \xrightarrow{\sim}$ $\mathrm{SL}_2(\mathbb{Q}_p)$ given by (6.28), as we noted in Section 6.11.1; we use this to identify G_p with $\mathrm{SL}_2(\mathbb{Q}_p)$. Let T be the "torus" in $G = \mathrm{SL}_1(D)$ consisting of quaternions of the form $x = x_1 + x_2 e_2$ with $N(x) = 1$, and let T_p be the similarly defined subgroup of G_p. Under our identification of G_p with $\mathrm{SL}_2(\mathbb{Q}_p)$, the subgroup T_p is simply the standard torus that we have considered before, consisting of the diagonal matrices of determinant 1.

Lemma 6.126. T *is dense in* T_p.

Proof. We use the same normalization trick as in the proof of Lemma 6.125. Namely, we construct elements of T by starting with an arbitrary $x = x_1 + x_2 e_2 \in D^*$ and forming $x' := x^2/N(x)$. Computing the images of such elements x' in $\mathrm{SL}_2(\mathbb{Q}_p)$, we find that they are the diagonal matrices with diagonal entries λ, λ^{-1}, where

$$\lambda = \frac{x_1 + x_2\sqrt{\alpha}}{x_1 - x_2\sqrt{\alpha}} \tag{6.29}$$

for some $x_1, x_2 \in \mathbb{Q}$ that are not both zero. [Note that the denominator is not zero, since in view of our assumption that D is a division algebra, α does not have a square root in $\mathbb{Q}$.] The closure of T in G_p therefore contains all diagonal matrices of the same form, where now $x_1, x_2 \in \mathbb{Q}_p$ and the numerator and denominator are assumed to be nonzero. To complete the proof, we will show that every $\lambda \in \mathbb{Q}_p^*$ can be expressed in this way. Given $\lambda \in \mathbb{Q}_p^*$, let's first try to achieve this with $x_2 = 1$, i.e., we try to solve

$$\lambda = \frac{x + \sqrt{\alpha}}{x - \sqrt{\alpha}} \tag{6.30}$$

for $x \in \mathbb{Q}_p$ with $x \neq \pm\sqrt{\alpha}$. Formally solving (6.30) for x, we obtain

$$x = \sqrt{\alpha}\,\frac{\lambda + 1}{\lambda - 1}\,,$$

so we are done if $\lambda \neq 1$. But we can take care of $\lambda = 1$ by putting $x_2 = 0$ in (6.29). $\square$

6.11.3 Norm-1 Groups over $\mathbb{Q}$ and Buildings

We continue with the hypotheses and notation of the previous subsection. In particular, $D = (\alpha, \beta)_{\mathbb{Q}}$ is a quaternion division algebra, p is an odd prime such that $v_p(\alpha) = v_p(\beta) = 0$, $G = \mathrm{SL}_1(D)$, and $G_p = \mathrm{SL}_1(D_p)$. Moreover, $\beta = -1$ if -1 has a square root in D. Let Δ_p be the tree associated to the BN-pair in G_p discussed in Section 6.9. The action of G_p on Δ_p is strongly transitive with respect to the complete apartment system (see Remark 6.112(b)).

Proposition 6.127. *The action of* G *on* Δ_p *is Weyl transitive.*

Proof. Since G is dense in G_p by Lemma 6.125, and since the B of the BN-pair in G_p is an open subgroup, this follows from Proposition 6.119.

Remark 6.128. Note that just from the fact that G is transitive on the chambers, the theory of groups acting on trees gives a decomposition of G as an amalgamated free product [217], as in the better-known case of $SL_2(\mathbb{Q})$ (or a p-adically dense subgroup, such as $SL_2(\mathbb{Z}[1/p])$).

As we recalled in Section 6.10.2, the stabilizer of the fundamental apartment Σ in G_p is the monomial group, with the diagonal matrices acting as translations and the nondiagonal matrices acting as reflections. The translation action of the diagonal group T_p is given by a surjective homomorphism $T_p \twoheadrightarrow \mathbb{Z}$ whose kernel is $T_p \cap B$, which is an open subgroup of T_p. In view of Lemma 6.126 and Lemma 6.118(1), it follows that all of the translations can be achieved by elements of T. But as we are about to see, one cannot in general realize the reflections by elements of G.

Proposition 6.129. *The following conditions are equivalent:*

(i) -1 *has a square root in* D.
(ii) G *contains an element that stabilizes the fundamental apartment* Σ *and acts as a reflection on it.*
(iii) *The action of* G *on* Δ_p *is strongly transitive with respect to some apartment system.*

Proof. If (i) holds, then $\beta = -1$ by our choices at the beginning of this subsection. The quaternion e_3 is therefore in $G = SL_1(D)$ and maps to $\left(\begin{smallmatrix} 0 & -1 \\ 1 & 0 \end{smallmatrix}\right) \in SL_2(\mathbb{Q}_p)$. This proves (ii). The latter implies (iii) by Proposition 6.14, since the dihedral group of type-preserving automorphisms of an apartment is generated by the translations and any one reflection. Finally, suppose (iii) holds. Then G contains an element g that stabilizes an apartment and acts as a reflection on it. But we already noted in Section 6.10.2 that any such g satisfies $g^2 = -1$, so (i) holds. $\square$

To get specific examples of actions that are Weyl transitive but not strongly transitive, we need to choose α, β in such a way that $-1 \notin D^2$. Now direct calculation shows that $-1 \in D^2$ if and only if the ternary quadratic form $\langle \alpha, \beta, -\alpha\beta \rangle$ represents -1. Hence

$$-1 \notin D^2 \iff \langle 1, \alpha, \beta, -\alpha\beta \rangle \text{ is anisotropic.} \tag{6.31}$$

Let l be a prime such that $l \equiv 1 \mod 4$, so that $-1 \in \mathbb{Q}_l^2$. Set $\beta = -l$, and let α be any negative integer such that α is not a square mod l. Then the quaternary form in (6.31) is equivalent over $\mathbb{Q}_l$ to the form $\langle 1, \alpha, l, \alpha l \rangle$, which is anisotropic over $\mathbb{Q}_l$. In fact, it is the essentially unique anisotropic quaternary form over $\mathbb{Q}_l$ [150, Theorem VI.2.2(3); 181, 63:17; 215, Section IV.2.3, Corollary to Theorem 7]. The form is therefore anisotropic over $\mathbb{Q}$, so -1 does not have a square root in $D := (\alpha, -l)_\mathbb{Q}$. For a concrete example, take $l = 5$ and $\alpha = -2$.

Corollary 6.130. *Let* $D = (-2, -5)_{\mathbb{Q}}$ *and let* $G = \mathrm{SL}_1(D)$. *Then for all primes* $p \neq 2, 5$, *there is a Weyl-transitive action of* G *on* Δ_p *that is not strongly transitive with respect to any apartment system, where* Δ_p *is the regular tree of degree* $p + 1$. □

Readers familiar with the language of algebraic groups (see Appendix C) can check that $\mathrm{SL}_1(D)$ is the group $\mathcal{G}(\mathbb{Q})$ of $\mathbb{Q}$-points of a linear algebraic group $\mathcal{G}$ defined over $\mathbb{Q}$ (still called the norm-1 group), and $G_p = \mathcal{G}(\mathbb{Q}_p)$. So Proposition 6.129 yields the following dichotomy:

Corollary 6.131. *Let* D *be a quaternion division algebra over* $\mathbb{Q}$, *and let* $\mathcal{G}$ *be its norm-1 group, viewed as a linear algebraic group over* $\mathbb{Q}$. *For each prime* p *such that* D_p *splits, let* Δ_p *be the tree associated to* $\mathcal{G}(\mathbb{Q}_p)$. *Then one of the following conditions holds:*

(a) *For almost all primes* p, *the action of* $\mathcal{G}(\mathbb{Q})$ *on* Δ_p *is strongly transitive with respect to some apartment system.*

(b) *For almost all primes* p, *the action of* $\mathcal{G}(\mathbb{Q})$ *on* Δ_p *is not strongly transitive with respect to any apartment system.* □

The conclusion in case (b) can also be phrased in group-theoretic terms: For almost all primes p, we have a Tits subgroup B in $\mathcal{G}(\mathbb{Q})$ that does not come from a BN-pair.

Remark 6.132. What we have done in this section, following [10], is to work out the simplest possible case of a suggestion of Tits [261, Section 3.1, Example (b)] involving anisotropic algebraic groups. We were surprised that such groups act strongly transitively as often as they do. Norm-1 groups are not the only examples in which this happens, but we have some evidence to suggest that the phenomenon is still relatively rare, as Tits suggested.

Exercise 6.133. With D as in Corollary 6.130, verify directly that $-1 \notin D^2$.

*6.12 Example: A Twin BN-Pair

In this section we assume that the reader is familiar with twin buildings and twin BN-pairs (Sections 5.8 and 6.3). Let K be the rational function field $k(t)$ over a field k, and let v_+ (resp. v_-) be the valuation that gives the order of a function at 0 (resp. ∞); see Examples 6.108(a) and (b). Let $A_\pm$ be the corresponding valuation rings, and let R be the ring $k[t, t^{-1}]$ of Laurent polynomials. Informally, A_+ consists of the functions that are defined at 0, A_- consists of the functions that are defined at ∞, and R consists of the functions that are defined *except* possibly at 0 and ∞. In this section we will sketch the construction of a twin BN-pair in the group

$$G := \mathrm{SL}_n(R) .$$

Let $\Delta_\pm$ be the Bruhat–Tits buildings associated to $SL_n(K)$ and the valuations $v_\pm$ (see Section 6.9). We have a strongly transitive action of $SL_n(K)$ on each Δ_ϵ ($\epsilon = \pm$), so we get actions of the subgroup G. We claim that these actions are also strongly transitive. To see this, one checks first that R is dense in K with respect to the topology induced by the valuation v_ϵ for each ϵ. It follows that G is dense in $SL_n(K)$ and hence that the action of G on Δ_ϵ is Weyl transitive (see Proposition 6.119). Now let N be the group of monomial matrices in G, i.e., N is the stabilizer in G of the fundamental apartment Σ_ϵ of Δ_ϵ. Then it is easily checked that N contains representatives of all elements of the Weyl group W discussed in Section 6.9. Hence N is transitive on the chambers of Σ_ϵ. Our claim that G acts strongly transitively on Δ_ϵ therefore follows from Proposition 6.14.

Recall from Section 6.9 that the buildings $\Delta_\pm$ can be described in terms of lattice classes. We will use the same notation as in that section. Let C_+ be the chamber in Δ_+ whose vertices are the A_+-lattice classes

$$[[e_1, \ldots, e_i, te_{i+1}, \ldots, te_n]] ,$$

$i = 1, \ldots, n$. This is the fundamental chamber that we used before; its stabilizer in $SL_n(K)$ is the set of matrices in $SL_n(A_+)$ that are upper triangular mod t. As our fundamental chamber C_- in Δ_- we take the chamber whose vertices admit the same formal description as those of C_+, i.e., they are the A_--lattice classes $[[e_1, \ldots, e_i, te_{i+1}, \ldots, te_n]]$, $i = 1, \ldots, n$. This is *not* the fundamental chamber that we used in Section 6.9. The latter has vertices $[[e_1, \ldots, e_i, t^{-1}e_{i+1}, \ldots, t^{-1}e_n]]$, since the "$\pi$" for A_- is t^{-1}, whereas the present vertices are

$$[[e_1, \ldots, e_i, te_{i+1}, \ldots, te_n]] = [[t^{-1}e_1, \ldots, t^{-1}e_i, e_{i+1}, \ldots, e_n]] .$$

One can check that the stabilizer of C_- in $SL_n(K)$ is the set of matrices in $SL_n(A_-)$ that are *lower* triangular mod t^{-1}.

We now get our candidate (B_+, B_-, N) for a twin BN-pair by letting B_ϵ be the stabilizer of C_ϵ in G. Since $R \cap A_+ = k[t]$ and $R \cap A_- = k[t^{-1}]$, B_+ is the set of matrices in $SL_n(k[t])$ that are upper triangular mod t, and B_- is the set of matrices in $SL_n(k[t^{-1}])$ that are lower triangular mod t^{-1}.

In view of the way we got (B_+, B_-, N) from strongly transitive actions, we know that (TBN0) is satisfied. The interested reader can now verify (TBN1) and (TBN2) by direct matrix calculations. This yields a twinning of the two buildings $\Delta_\pm$ via Section 6.3.3. Alternatively, one could construct the twinning directly and verify that the G-action on the resulting twin building is strongly transitive; the TBN axioms then follow from Sections 6.3.1 and 6.3.2. This second approach is carried out in detail in [16, Section 4], based on the characterization of twin buildings that we mentioned in Remark 5.154. We will confine ourselves here to describing the twin apartments and the opposition relation.

Let M be the free R-module in K^n generated by the standard basis vectors $e_1, \ldots, e_n$. For each R-basis $f_1, \ldots, f_n$ of M, there is a twin apartment

$\Sigma = (\Sigma_+, \Sigma_-)$ whose vertices are the lattice classes $[[t^{a_1} f_1, \ldots, t^{a_n} f_n]]$ with $a_1, \ldots, a_n \in \mathbb{Z}$. Here, of course, we use A_+-lattices for Σ_+ and A_--lattices for Σ_-. The opposition involution $\mathrm{op}_\Sigma : \Sigma_\epsilon \to \Sigma_{-\epsilon}$ is the obvious map on vertices: It takes the A_+-lattice class $[[t^{a_1} f_1, \ldots, t^{a_n} f_n]]$ to the A_--lattice class described by the same symbol. Finally, we declare two chambers (or, more generally, two simplices) to be *opposite* if they are contained in a twin apartment and are related by the opposition involution. See [16] for further details.

Remark 6.134. Note that our example involved function fields rather than number fields. This is no accident. In fact, we will show in Section 11.10 that the Bruhat–Tits building associated to $\mathrm{SL}_n(\mathbb{Q}_p)$ cannot be part of a twin building if $n \geq 3$.

Root Groups and the Moufang Property

In this chapter we will be concerned mostly with thick spherical buildings Δ. The Moufang property, very roughly, says that Δ has a great deal of symmetry; in particular, Δ arises from a group with a BN-pair. A remarkable theorem of Tits says that *every* thick, irreducible, spherical building of rank at least 3 (dimension at least 2) has the Moufang property. Moreover, thick spherical buildings with the Moufang property have turned out to be classifiable [262]. This explains the importance of the concept. The Moufang property has also become useful in connection with certain classes of nonspherical buildings that arise in the theory of Kac–Moody groups. We will treat this general (nonspherical) theory in the next chapter.

We begin slowly, by describing a situation in which one can construct a BN-pair from a building. When we impose the Moufang property on the building, we will get a refinement of the BN-pair, involving a system of "root groups." The properties of these root groups motivate the algebraic theory of "RGD systems," to be developed in Section 7.8. By the end of the chapter we will have, essentially, a 1–1 correspondence between groups with an RGD system and buildings with the Moufang property.

7.1 Pre-Moufang Buildings and BN-Pairs

In this section Δ denotes a (simplicial) building, not necessarily spherical. We assume for simplicity that Δ is *thick*, though some of what we do is valid without this assumption. Choose a "fundamental apartment" Σ and a "fundamental chamber" $C_0 \in \Sigma$, and let Φ be the set of roots of Σ (see Sections 3.4 and 5.5.4). We denote by Φ_+ (resp. Φ_-) the set of positive (resp. negative) roots. Here a root is *positive* if it contains the fundamental chamber and negative otherwise. Given a root α and a panel $P \in \partial\alpha$, the set of chambers of Δ that are "attached to α along P" will play an important role in this chapter. We recall some notation that was introduced in Section 4.11.1. Let $\mathcal{C} := \mathcal{C}(\Delta)$, and let $\mathcal{C}_P$ be the set $\mathcal{C}_{\geq P}$ of chambers having P as a face.

Definition 7.1. Given a root α of Δ and a panel $P \in \partial\alpha$, we set

$$\mathcal{C}(P,\alpha) := \mathcal{C}_P \smallsetminus \{C\} \ ,$$

where C is the unique chamber in α having P as a face.

Note that any automorphism of Δ that fixes α pointwise induces a permutation of $\mathcal{C}(P,\alpha)$ for every panel $P \in \partial\alpha$. These permutations will be used in the next definition. Recall that $\mathrm{Aut}_0\,\Delta$ denotes the group of type-preserving automorphisms of Δ.

Definition 7.2. A family $(X_\alpha)_{\alpha \in \Phi}$ of subgroups of $\mathrm{Aut}_0\,\Delta$ is called a system of *pre–root groups* if it satisfies the following three conditions:

(1) X_α fixes α pointwise for each $\alpha \in \Phi$.
(2) For each $\alpha \in \Phi$ and each panel $P \in \partial\alpha$, the action of X_α on $\mathcal{C}(P,\alpha)$ is transitive.
(3) For each $\alpha \in \Phi$ there is an element n_α in the subgroup $\langle X_\alpha, X_{-\alpha}\rangle$ such that $n_\alpha(\alpha) = -\alpha$ and $n_\alpha(-\alpha) = \alpha$. In other words, n_α stabilizes Σ and acts on Σ as the reflection s_α.

We say that Δ is a *pre-Moufang* building if it admits a system of pre–root groups.

Remark 7.3. Pre-Moufang buildings have some (but not all) of the features of Moufang buildings, which will be defined in Section 7.3 along with root groups. The root groups in a (spherical) Moufang building satisfy conditions (1)–(3) above; this explains the terminology "pre–root groups."

Conditions (2) and (3) take a little time to digest. We begin by reformulating (2) in the spherical case. For any root α of Δ, we denote by $\mathcal{A}(\alpha)$ the set of apartments of Δ containing α. Recall from Lemma 4.118 that for any panel $P \in \partial\alpha$, there is a canonical bijection between $\mathcal{C}(P,\alpha)$ and $\mathcal{A}(\alpha)$. This leads immediately to the following result:

Lemma 7.4. *Assume that Δ is spherical and that $(X_\alpha)_{\alpha \in \Phi}$ is a system of subgroups of $\mathrm{Aut}_0\,\Delta$ satisfying condition (1) of Definition 7.2. Then the following conditions are equivalent:*

(i) *The system $(X_\alpha)_{\alpha \in \Phi}$ satisfies condition (2).*
(ii) *For each $\alpha \in \Phi$ there exists a panel $P \in \partial\alpha$ such that the action of X_α on $\mathcal{C}(P,\alpha)$ is transitive.*
(iii) *For each $\alpha \in \Phi$, the action of X_α on $\mathcal{A}(\alpha)$ is transitive.*

Proof. Let α be a root and let P be a panel in $\partial\alpha$. Then Lemma 4.118 implies that the action of X_α is transitive on $\mathcal{C}(P,\alpha)$ if and only if it is transitive on $\mathcal{A}(\alpha)$. Thus (ii) $\Longrightarrow$ (iii) $\Longrightarrow$ (i), and the implication (i) $\Longrightarrow$ (ii) is trivial. $\qquad\square$

Next, we attempt to demystify condition (3) in Definition 7.2 by showing that in the spherical case, it actually follows from (1) and (2). Moreover, we will see that it has some geometric content. Our standing assumption that Δ is thick is crucial here.

Lemma 7.5. *If Δ is spherical and $(X_\alpha)_{\alpha \in \Phi}$ is a system of subgroups of $\mathrm{Aut}_0\,\Delta$ satisfying (1) and (2), then it also satisfies (3).*

Proof. Choose a panel $P \in \partial\alpha$, and let C and D be the chambers of Σ having P as a face, with $C \in \alpha$ and $D \in -\alpha$. By thickness and condition (2), we can find an element $x \in X_\alpha$ such that $xD \neq D$; see Figure 7.1. We

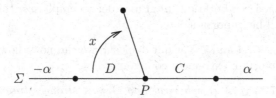

Fig. 7.1. Constructing n_α; step 1.

claim now that there are elements $x', x'' \in X_{-\alpha}$ such that the composite $m(x) := x'xx''$ interchanges C and D. Indeed, it suffices to choose, by (2), an element $x'' \in X_{-\alpha}$ such that $x''C = x^{-1}D$ [so that $xx''C = D$] and an element $x' \in X_{-\alpha}$ such that $x'(xD) = C$. The claim now follows from the fact that x' and x'' fix D. See Figure 7.2.

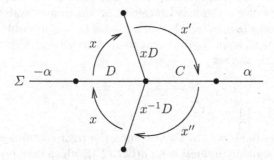

Fig. 7.2. Constructing n_α; step 2.

Note that we have not yet used the assumption that Δ is spherical, and indeed, without that assumption there is no reason to expect $m(x)$ to interchange $\pm\alpha$ just because it interchanges C and D. In the presence of sphericity, however, we know that α is the convex hull of C and $\partial\alpha$ (see Example 3.133(d)), and similarly $-\alpha$ is the convex hull of D and $\partial\alpha$. Since

$m(x)$ fixes $\partial\alpha$, it must therefore interchange $\pm\alpha$, so it is the desired element $n_\alpha \in \langle X_\alpha, X_{-\alpha} \rangle$. $\qquad\square$

For future reference, we record the following corollary of the proof:

Corollary 7.6. *Under the hypotheses of Lemma 7.5, let α be a root of Σ, let $D \in -\alpha$ be a chamber with a panel in $\partial\alpha$, and let $x \in X_\alpha$ be an element such that $xD \neq D$. Then there is an element $m(x) \in X_{-\alpha} x X_{-\alpha}$ such that $m(x)$ interchanges $\pm\alpha$.* $\qquad\square$

Remark 7.7. It often happens that the action of X_α on $\mathcal{C}(P, \alpha)$ not only is transitive, but is in fact *simply* transitive for each α. In this case the elements x', x'' in the proof of the lemma are uniquely determined by x; hence so is $x'xx''$. We used the notation $m(x)$ in order to emphasize this dependence on x, which will be important later.

With Lemmas 7.4 and 7.5 at our disposal, we can now show that systems of pre–root groups are extremely common:

Proposition 7.8. *If Δ is spherical and G is a strongly transitive group of type-preserving automorphisms of Δ, then G contains a system of pre–root groups.*

Proof. For each $\alpha \in \Phi$, let X_α be the pointwise fixer of α in G, i.e.,

$$X_\alpha = \operatorname{Fix}_G(\alpha) := \{g \in G \mid gA = A \text{ for all } A \in \alpha\} .$$

Then X_α acts transitively on $\mathcal{A}(\alpha)$ by Corollary 6.7, so the proposition follows from Lemmas 7.4 and 7.5. $\qquad\square$

The main goal of the rest of the section is to prove a converse of Proposition 7.8, i.e., to construct a strongly transitive action from a system of pre–root groups. This works in general; Δ does not have to be spherical. The following lemma is the crucial step. Assume that we have a system $(X_\alpha)_{\alpha \in \Phi}$ of pre–root groups in $\operatorname{Aut}_0 \Delta$, and set

$$U := \langle X_\alpha \mid \alpha \in \Phi_+ \rangle . \tag{7.1}$$

Lemma 7.9. $\Delta = \bigcup_{u \in U} u\Sigma.$

Proof. It suffices to show that the union on the right contains every chamber $C \in \mathcal{C}(\Delta)$. We argue by induction on $d(C, \mathcal{C}(\Sigma))$, which may be assumed > 0. Choose a gallery $D_0, \dots, D_l = C$ of minimal length $l := d(C, \mathcal{C}(\Sigma))$ with $D_0 \in \Sigma$. Let $P := D_0 \cap D_1$, let D_0' be the chamber of Σ adjacent to D_0 along P, and let α be the root of Σ containing D_0 but not D_0'; see Figure 7.3. If α is a positive root, choose $x \in X_\alpha$ such that $xD_1 = D_0'$. Otherwise choose $x \in X_{-\alpha}$ such that $xD_1 = D_0$. In either case we have $x \in U$ with $xD_1 \in \Sigma$, and the gallery $xD_1, \dots, xD_l = xC$ shows that $d(xC, \mathcal{C}(\Sigma)) < l$. So we may apply the induction hypothesis to find $u' \in U$ with $u'xC \in \Sigma$; hence C is in $\bigcup_{u \in U} u\Sigma$. $\qquad\square$

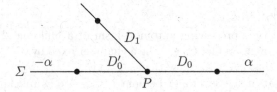

Fig. 7.3. A gallery leaving Σ.

We can now prove the main result of this section. Suppose we have a subgroup $G \leq \mathrm{Aut}_0\, \Delta$ containing a system of pre–root groups $(X_\alpha)_{\alpha \in \Phi}$. Let B be the stabilizer in G of the fundamental chamber C_0, and set

$$N := \langle n_\alpha \mid \alpha \in \Phi \rangle$$

for some choice of elements n_α as in Definition 7.2.

Theorem 7.10.

(1) *The action of G on Δ is strongly transitive with respect to the apartment system $G\Sigma := \{g\Sigma \mid g \in G\}$.*

(2) *The subgroups B and N defined above form a BN-pair in G, and $\Delta \cong \Delta(G, B)$.*

Proof. Note that N stabilizes Σ and acts transitively on the chambers of Σ. Note further that $U \leq B$, where U is the group defined in (7.1). Using Lemma 7.9, we conclude, first, that the action of G on Δ is chamber transitive, and then that the action is Weyl transitive (Proposition 6.11) and hence strongly transitive (Proposition 6.14). This proves (1). Assertion (2) follows from (1) and the fact that a strongly transitive action on a thick building yields a BN-pair (see Theorem 6.56). □

We close this section by proving the following result of Tits [247], which shows that (certain) buildings have much more symmetry than one would expect from the definition. (Tits actually proved a stronger theorem of this type, which we will treat in Section 7.6 below.)

Proposition 7.11. *If Δ is a thick, irreducible, spherical building of rank at least 3, then Δ is pre-Moufang. In particular, $\Delta \cong \Delta(G, B)$ for some group G with a BN-pair.*

Proof. Let $G = \mathrm{Aut}_0\, \Delta$. As in the proof of Proposition 7.8, it suffices to show, for each root α of Σ, that $X_\alpha := \mathrm{Fix}_G(\alpha)$ acts transitively on the set of apartments containing α. This is precisely what we proved in Corollary 5.211, as an application of Tits's extension theorem. □

Exercise 7.12.

(a) Let g be a type-preserving automorphism of a building Δ, let Σ be an apartment of Δ, and let $C_0 \in \Sigma$ be a chamber fixed by g. If A is a simplex of Σ such that $gA \in \Sigma$, show that $gA = A$.

(b) Deduce that in the setting of Lemma 7.9, Σ is a simplicial fundamental domain for the action of U on Δ.

7.2 Calculation of Fixers

Suppose Δ is a spherical building and G is a strongly transitive group of type-preserving automorphisms. Our purpose in this section is to present some results of Tits [247, Chapter 13], calculating the fixer $\mathrm{Fix}_G(K)$ for suitable subsets $K \subseteq \Delta$. We will phrase the results in terms of a system of pre–root groups X_α. (Tits considered the special case that X_α is the full fixer $\mathrm{Fix}_G(\alpha)$.) The calculation of fixers will be useful in our study of root groups later in this chapter. For example, it leads to a simple proof of some important commutator relations among the root groups (Section 7.7.2).

7.2.1 Preliminaries: Convex Sets of Roots

Let Σ be a spherical Coxeter complex and Φ its set of roots. If K is any set of simplices in Σ, we define a set of roots $\Psi(K)$ by

$$\Psi(K) := \{\alpha \in \Phi \mid \alpha \supseteq K\} . \tag{7.2}$$

Definition 7.13. A set of roots $\Psi \subseteq \Phi$ is said to be *convex* if it has the form $\Psi = \Psi(K)$ for some set of simplices $K \subseteq \Sigma$ containing at least one chamber.

(Tits [247] calls such sets Ψ *strictly* convex.)

The following lemma explains the terminology:

Lemma 7.14. *There is an order-reversing 1–1 correspondence between convex chamber subcomplexes of Σ and convex sets of roots. It associates to a convex chamber subcomplex $K \subseteq \Sigma$ the set of roots $\Psi(K)$. Its inverse is given by $\Psi \mapsto \bigcap_{\alpha \in \Psi} \alpha$ for any convex set of roots Ψ.*

Proof. For any subset $K \subseteq \Sigma$, the set $\Psi(K)$ does not change if we replace K by its convex hull, which is a convex chamber subcomplex of Σ (see Section 3.6.6). So the function $K \mapsto \Psi(K)$ is a surjection from the set of convex chamber subcomplexes of Σ to the set of convex subsets of Φ. This function is injective and has the inverse described in the lemma, since every convex chamber subcomplex of Σ is an intersection of roots by Theorem 3.131. $\square$

Note that the empty set of roots is convex (corresponding to $K = \Sigma$), but the set of all roots Φ is *not* convex. Indeed, a convex set of roots Ψ contains at most half of the roots. (Write $\Psi = \Psi(K)$, and take the fundamental chamber to be in K; then $\Psi \subseteq \Phi_+$.)

Example 7.15. Given two chambers $C, D \in \Sigma$, set

$$\Phi(C, D) := \{\alpha \in \Phi \mid C \in \alpha,\ D \notin \alpha\}\ .$$

Then $\Phi(C, D) = \Psi(\{C, D'\})$, where $D' := \mathrm{op}_\Sigma D$ is the chamber of Σ opposite D. [Recall that a spherical Coxeter complex Σ has an *opposition involution* op_Σ, which is just multiplication by -1 if we think of Σ as the complex associated to a finite reflection group; see Sections 1.6.2 and 5.7.3.] Hence $\Phi(C, D)$ is a convex set of roots. It contains precisely $d = d(C, D)$ roots, one for each wall of Σ that separates C from D. We can enumerate these roots by choosing a minimal gallery $C = C_0, \ldots, C_d = D$ from C to D. If α_i is the root containing C_{i-1} but not C_i $(1 \le i \le d)$, then $\Phi(C, D) = \{\alpha_1, \ldots, \alpha_d\}$.

Note that the ordering of the roots $\alpha_1, \ldots, \alpha_d$ in Example 7.15 has the property that the set $\{\alpha_i, \ldots, \alpha_d\}$ is convex for each $1 \le i \le d$. Indeed, this set is $\Phi(C_{i-1}, D)$. This motivates the following definition:

Definition 7.16. Let Ψ be a convex set of roots. An ordering $\alpha_1, \ldots, \alpha_m$ of Ψ will be called *admissible* if the set $\{\alpha_i, \ldots, \alpha_m\}$ is convex for each $1 \le i \le m$.

Lemma 7.17. *Every convex set of roots admits an admissible ordering.*

Proof. Let Ψ be convex, and write $\Psi = \Psi(K)$ with K a convex chamber subcomplex. We may assume that $\Psi \neq \emptyset$, so that $K \neq \Sigma$. Choose a pair of adjacent chambers D, D' with $D \in K$ and $D' \notin K$, and let α_1 be the root containing D but not D'. Then we have $\alpha_1 \in \Psi$ (see the proof of Proposition 3.94). Moreover, $\Psi \smallsetminus \{\alpha_1\} = \Psi(K \cup \{D'\})$, since α_1 is the only root containing D but not D'. Hence $\Psi \smallsetminus \{\alpha_1\}$ is convex, and an obvious induction on $|\Psi|$ completes the proof.   □

We close this subsection by mentioning that the theory of convex sets of roots has a useful generalization to Coxeter complexes that are not necessarily spherical.

Definition 7.18. Let Σ be an arbitrary Coxeter complex, and let Φ be its set of roots. A subset $\Psi \subseteq \Phi$ is said to be *convex* if it has the form

$$\Psi = \{\alpha \in \Phi \mid K \subseteq \alpha \text{ and } K' \subseteq (-\alpha)\}\ , \tag{7.3}$$

where K and K' are subsets of Σ that each contain at least one chamber.

Note that the set Ψ in (7.3) contains one element for each wall of Σ that separates K from K'. In particular, a convex set Ψ is always finite. A simple example of a convex set is $\Phi(C, D)$, defined exactly as in Example 7.15, for two chambers $C, D \in \Sigma$.

Remark 7.19. In the spherical case, Definition 7.18 is consistent with Definition 7.13. Indeed, if Σ is spherical and $\Psi = \Psi(K)$ (where K is a subset of Σ containing at least one chamber), then Ψ has the form (7.3) with $K' := \mathrm{op}_\Sigma(K)$. Conversely, if Ψ is defined as in (7.3), then $\Psi = \Psi(K \cup \mathrm{op}_\Sigma(K'))$.

The theory of convex sets in the nonspherical case is best understood from the point of view of twin buildings. We will return to it in Section 8.2.

7.2.2 Fixers

Assume now that Δ is a spherical building and that G is a group of type-preserving automorphisms of Δ that contains a system $(X_\alpha)_{\alpha \in \Phi}$ of pre–root groups. As usual, Φ is the set of roots of a chosen fundamental apartment Σ. For any subset $\Psi \subseteq \Phi$ we set

$$X_\Psi := \langle X_\alpha \mid \alpha \in \Psi \rangle .$$

We also set

$$T := \mathrm{Fix}_G(\Sigma) .$$

The following proposition is the main result of this section.

Proposition 7.20. *Let Ψ be a convex set of roots in the fundamental apartment Σ, let $\alpha_1, \ldots, \alpha_m$ be an admissible ordering of Ψ, and set $X_i := X_{\alpha_i}$ for $i = 1, \ldots, m$.*

(1) *$X_\Psi T$ is a subgroup of G, and*

$$X_\Psi T = X_1 \cdots X_m T . \tag{7.4}$$

(2) *If $x_1 \cdots x_m t = x_1' \cdots x_m' t'$ with $x_i, x_i' \in X_i$ for $i = 1, \ldots, m$ and $t, t' \in T$, then there are elements $t_1, \ldots, t_m \in T$ such that*

$$x_1' = x_1 t_1 ,$$
$$x_2' = t_1^{-1} x_2 t_2 ,$$
$$\vdots$$
$$x_m' = t_{m-1}^{-1} x_m t_m ,$$
$$t' = t_m^{-1} t .$$

(3) *If $\Psi = \Psi(K)$ for some subset $K \subseteq \Sigma$ containing at least one chamber, then the pointwise fixer of K is given by*

$$\mathrm{Fix}_G(K) = X_\Psi T .$$

Remarks 7.21. (a) In examples that arise in practice, each group X_α is normalized by T, and hence $X_\Psi T$ is a subgroup for *any* set of roots Ψ. Thus the important part of (1) is equation (7.4).

(b) If X_α is the full fixer of α for each root α, then X_α contains T and we can simplify the statement of the proposition. In particular, T can be deleted from the statements of (1) and (3), except in the trivial case $\Psi = \emptyset$. At the other extreme, suppose $X_\alpha \cap T = \{1\}$ for each α. (This situation will arise when we consider root groups; see Section 7.7.) Then the statement of (2) can be simplified, since we necessarily have $t_i = 1$ for each $i = 1, \ldots, m$. So the content of (2) in this case is that every element of $X_\Psi T$ is uniquely expressible as $x_1 \cdots x_m t$ with $x_i \in X_i$ for $i = 1, \ldots, m$ and $t \in T$.

Proof of Proposition 7.20. We prove (1) and (3) simultaneously, by induction on $m = |\Psi|$. We may assume that the subset K in (3) is a convex chamber subcomplex. To see this, let L be the convex hull of K in Δ (Definition 4.116). Then L is a convex chamber subcomplex of Σ, and we claim that $\mathrm{Fix}_G(L) = \mathrm{Fix}_G(K)$. Indeed, if g is an element of G that fixes K pointwise, then g stabilizes L and fixes at least one chamber of L, so it fixes L pointwise by the standard uniqueness argument. This proves the claim and justifies our assumption that K is a convex chamber complex. We may further assume that $m > 0$. (If $m = 0$, then $K = \Sigma$ and $\mathrm{Fix}_G(K) = T$.)

Since $\Psi' := \Psi \smallsetminus \{\alpha_1\} = \{\alpha_2, \ldots, \alpha_m\}$ is a convex set of roots, we have $\Psi' = \Psi(K')$ for a convex chamber subcomplex $K' \supsetneq K$. Choose adjacent chambers D, D' with $D \in K$ and $D' \in K' \smallsetminus K$, and let α be the root containing D but not D'. Then

$$\Psi \smallsetminus \{\alpha_1\} = \Psi' = \Psi(K') \subseteq \Psi(K \cup \{D'\}) = \Psi \smallsetminus \{\alpha\} \ ,$$

where the last inequality follows from the proof of Lemma 7.17. So $\alpha = \alpha_1$, and $\Psi' = \Psi(K \cup \{D'\})$. We now obtain, by the induction hypothesis,

$$\mathrm{Fix}_G(K \cup \{D'\}) = X_{\Psi'} T = X_2 \cdots X_m T \ . \tag{7.5}$$

Now consider an arbitrary element $g \in \mathrm{Fix}_G(K)$. Setting $P := D \cap D'$, we have $gD' \in \mathcal{C}(P, \alpha_1)$; so $xgD' = D'$ for some $x \in X_1$, and hence $xg \in \mathrm{Fix}(K \cup \{D'\})$. In view of (7.5), this proves that

$$\mathrm{Fix}_G(K) \subseteq X_1 X_2 \cdots X_m T \subseteq X_\Psi T \ . \tag{7.6}$$

On the other hand, it is trivial that $X_\Psi T \subseteq \mathrm{Fix}_G(K)$, so both inclusions in (7.6) are equalities, and the proof of (1) and (3) is complete.

To prove (2), note that $X_\alpha \le \mathrm{Fix}_G(\alpha)$ for any root α, so it suffices to prove (2) in the case that $X_i = \mathrm{Fix}_G(\alpha_i)$ for all i. In particular, $X_i \ge T$ for all i. We again argue by induction on m, which may be assumed ≥ 2. Let Ψ' and D' be as in the previous paragraph, and recall that $X_{\Psi'} = \mathrm{Fix}_G(K \cup \{D'\})$. The equation

$$x_1 \cdots x_m t = x_1' \cdots x_m' t'$$

therefore implies that $x_1^{-1} x_1' \in X_1 \cap X_{\Psi'} = \mathrm{Fix}_G(\alpha_1 \cup \{D'\}) = T$, since the convex hull of $\alpha_1 \cup \{D'\}$ is the entire apartment Σ. [α_1 contains the chamber of Σ opposite D'.] So we may take $t_1 := x_1^{-1}x_1'$. We now have $x_2x_3 \cdots x_m = (t_1x_2')x_3 \cdots x_m$, and an application of the induction hypothesis completes the proof. $\square$

We close this section by interpreting part (1) of the proposition in terms of galleries in case $\Psi = \Phi(C, D)$ as in Example 7.15. The result in this case will be useful to us when Δ is not necessarily spherical, so we give an independent proof under weaker assumptions. Assume for the rest of this section that Δ is an *arbitrary* building with a chosen fundamental apartment Σ. We can still talk about the set of roots Φ of Σ, as well as the sequence of roots $\alpha_1, \ldots, \alpha_d$ associated to a minimal gallery as in Example 7.15. Assume that we are given a family $(X_\alpha)_{\alpha \in \Phi}$ of subgroups of $\operatorname{Aut} \Delta$ satisfying conditions (1) and (2) of Definition 7.2, which we recall here:

(1) X_α fixes α pointwise for each $\alpha \in \Phi$.
(2)   For each $\alpha \in \Phi$ and each panel $P \in \partial\alpha$, the action of X_α on $\mathcal{C}(P, \alpha)$ is transitive.

We will also consider the following strengthening of (2):

(2+)   For each $\alpha \in \Phi$ and each panel $P \in \partial\alpha$, the action of X_α on $\mathcal{C}(P, \alpha)$ is simply transitive.

Lemma 7.22. *Assume that the X_α satisfy (1) and (2). Let $\Gamma \colon C_0, \ldots, C_l$ be a minimal gallery in Σ. For $1 \le i \le l$ let α_i be the root of Σ such that $C_{i-1} \in \alpha_i$ and $C_i \notin \alpha_i$, and set $X_i := X_{\alpha_i}$. Suppose $\Gamma' \colon C_0', \ldots, C_l'$ is a gallery in Δ of the same type as Γ and having $C_0' = C_0$. Then there are elements $x_i \in X_i$ $(1 \le i \le l)$ such that $x := x_k \cdots x_1$ satisfies $xC_l' = C_l$. We then have $x\Gamma' = \Gamma$. If, in addition, the X_α satisfy (2+), then the x_i are unique.*

Proof. Since Γ is minimal, it crosses each wall separating C_0 from C_l exactly once. Hence $C_0, \ldots, C_{i-1} \in \alpha_i$ and $C_i, \ldots, C_l \notin \alpha_i$ for each $1 \le i \le l$. In particular, $C_0 \in \alpha_i$ for all i, so C_0 is stabilized by each X_i. Similarly, C_1 is stabilized by X_i for $i \ge 2$, and so on.

We now prove existence of the x_i by induction on l, which may be assumed > 0. This part of the proof is similar to the proof of Lemma 7.9. Let $P := C_0 \cap C_1 = C_0 \cap C_1'$, where the equality follows from the fact that Γ and Γ' have the same type. We then have $C_1, C_1' \in \mathcal{C}(P, \alpha_1)$, so condition (2) implies that there is an element $x_1 \in X_1$ such that $x_1C_1' = C_1$; see Figure 7.4. Now consider the galleries $\Gamma_1 \colon C_1, \ldots, C_l$ and $\Gamma_1' \colon x_1C_1', \ldots, x_1C_l'$. We can apply the induction hypothesis to these galleries to obtain elements $x_i \in X_i$ $(2 \le i \le l)$ such that $x_k \cdots x_2(x_1C_l') = C_l$. This completes the existence proof.

Note that we could have proved, as part of the induction, that $x\Gamma' = \Gamma$, where $x := x_k \cdots x_1$. But we did not need to do this because the equations $xC_0' = C_0$ and $xC_l' = C_l$ imply that $x\Gamma' = \Gamma$. Indeed, $x\Gamma'$ and $x\Gamma$ are galleries of the same reduced type and the same extremities, so they are equal by Corollary 4.42.

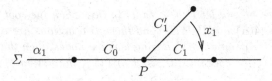

Fig. 7.4. Constructing x_1.

We now assume (2+) and prove uniqueness of the x_i by induction on l, which again may be assumed > 0. Suppose we have a second sequence $y_1, \ldots, y_l$ satisfying the conditions of the lemma. Then, as we just observed, $y\Gamma' = \Gamma = x\Gamma'$, where $y := y_k \cdots y_1$. In particular, $yC_1' = C_1 = xC_1'$. Since the x_i and y_i with $i \geq 2$ fix C_1, it follows that $y_1 C_1' = x_1 C_1'$ and hence that $y_1 = x_1$ by (2+). (Recall that $C_1' \in \mathcal{C}(P, \alpha_1)$ with P as in the existence proof.) Now consider the galleries $\Gamma_1 \colon C_1, \ldots, C_l$ and $\Gamma_1' \colon x_1 C_1', \ldots, x_1 C_l'$ and the sequences $x_2, \ldots, x_l$ and $y_2, \ldots, y_l$. By the induction hypothesis, $x_i = y_i$ for $i \geq 2$. □

Exercise 7.23. Returning to the hypotheses stated at the beginning of this subsection, show that $X_\Psi \cap X_{\Psi'} \leq X_{\Psi \cap \Psi'} T$ for all convex sets $\Psi, \Psi' \subseteq \Phi$.

7.3 Root Groups and Moufang Buildings

Throughout this section Δ denotes a thick spherical building unless the contrary is explicitly stated.

7.3.1 Definitions and Simple Consequences

Recall that the *star* of a simplex $A \in \Delta$, denoted by $\mathrm{st}_\Delta A$, is the set of simplices joinable to A.

Definition 7.24. For any root α of Δ, the *root group* U_α is defined to be the set of automorphisms g of Δ such that (a) g fixes α pointwise and (b) g fixes $\mathrm{st}_\Delta P$ pointwise for every panel $P \in \alpha \smallsetminus \partial\alpha$.

Note that $U_\alpha \leq \mathrm{Aut}_0 \Delta$. Note also that (a) follows from (b), and hence could be omitted, if Δ has rank at least 2. See Exercise 7.30 below for the rank-1 case. The reader anxious to see examples of root groups can look ahead at Section 7.3.4. Here are some simple properties of these groups:

Lemma 7.25.

(1) *For any root α and any $g \in \mathrm{Aut}\, \Delta$,*

$$gU_\alpha g^{-1} = U_{g\alpha} .$$

(2) *Let α be a root and let P be a panel in $\partial\alpha$. Then the root group U_α acts on the sets $\mathcal{A}(\alpha)$ and $\mathcal{C}(P,\alpha)$, and these two actions are equivalent.*

(3) *If the Coxeter diagram of Δ has no isolated nodes, then the actions in (2) are free, i.e., for every $u \in U_\alpha^* := U_\alpha \smallsetminus \{1\}$, the action of u has no fixed points.*

Proof. Assertion (1) is straightforward and is left to the reader, and (2) is an immediate consequence of Lemma 4.118. To prove (3), recall from Example 3.128 that α contains a chamber C having no panel in $\partial\alpha$. So every element $u \in U_\alpha^*$ fixes every chamber adjacent to C. The rigidity theorem (Corollary 5.206) now implies that u cannot fix any apartment containing α. $\qquad\square$

Remark 7.26. The assumption that the Coxeter diagram has no isolated nodes cannot be removed. See Exercise 7.30.

Definition 7.27. We say that Δ is *Moufang*, or is a *Moufang building*, if the actions in Lemma 7.25(2) are transitive for every root α of Δ. If, in addition, these actions are simply transitive, then we say that Δ is *strictly Moufang*, or is a *strictly Moufang building*.

Note that by part (3) of the lemma, a Moufang building whose Coxeter diagram has no isolated nodes is strictly Moufang.

The following proposition is an immediate consequence of the results of Section 7.1 (see Lemma 7.5 and Theorem 7.10). It also justifies the terms "pre–root group" and "pre-Moufang."

Proposition 7.28. *If Δ is Moufang, then it is pre-Moufang. More precisely, if we choose an apartment Σ and let Φ be its set of roots, then $(U_\alpha)_{\alpha\in\Phi}$ is a system of pre–root groups. Hence $G := \langle U_\alpha \mid \alpha \in \Phi \rangle$ acts strongly transitively on Δ, and $\Delta \cong \Delta(G, B)$, where B is the stabilizer in G of any chamber.* $\qquad\square$

Remarks 7.29. (a) If we want to check that a spherical building is Moufang, it suffices to choose a fundamental apartment as in the proposition and check that U_α is transitive on $\mathcal{A}(\alpha)$ for every $\alpha \in \Phi$. Indeed, that implies that G is strongly transitive, and then we can use the G-action to transport what we know about the U_α for $\alpha \in \Phi$ to the case of an arbitrary root of Δ. [The point here is that G is transitive on the apartments of Δ, so every root of Δ is G-equivalent to a root of Σ.]

(b) The same argument, along with Lemma 7.25(1), implies that the group G in the proposition contains U_α for every root of Δ, so we have

$$G = \langle U_\alpha \mid \alpha \text{ is a root of } \Delta \rangle .$$

In particular, G does not depend on the choice of Σ.

(c) Let the hypotheses and notation be as in the proposition. Since Δ is pre-Moufang, we know that for each $\alpha \in \Phi$, there is an element $n_\alpha \in \langle U_\alpha, U_{-\alpha} \rangle$ that stabilizes Σ and induces the reflection s_α on it. In view of Lemma 7.25(1), we have the following additional property of n_α in the Moufang case:

$$n_\alpha U_\beta n_\alpha^{-1} = U_{s_\alpha(\beta)}$$

for all $\beta \in \Phi$. If Δ is *strictly* Moufang, we can say still more about n_α (see Remark 7.7 and Lemma 7.25(3)): For any $u \in U_\alpha^*$, there are unique elements $u', u'' \in U_{-\alpha}$ such that $m(u) := u'uu''$ interchanges $\pm\alpha$ and hence can serve as n_α.

(d) One needs to be careful in reading the literature, since there are various definitions of "Moufang" that differ slightly from one another. In particular, some authors require from the start that the Coxeter diagram have no isolated nodes (in which case there is no need for the concept of "strictly Moufang building"), and some even require irreducibility. The definition that we have chosen, which is Tits's original definition in [247, p. 274], has the advantage that links in Moufang buildings are again Moufang (see Proposition 7.32 below). The disadvantage, however, is that some results require the explicit assumption that the Coxeter diagram has no isolated nodes.

Exercises

7.30. If Δ is a building of type A_1 or $A_1 \times A_1$, show that the actions in Lemma 7.25(2) are not necessarily free.

7.31. Suppose Δ is reducible, so that it is a join $\Delta' * \Delta''$ (see Exercise 4.17).

(a) Describe the root groups of Δ in terms of those of Δ' and Δ''.
(b) Show that Δ is Moufang (resp. strictly Moufang) if and only if Δ' and Δ'' are both Moufang (resp. strictly Moufang).

7.3.2 Links

For any $A \in \Delta$, the link $\Delta' := \mathrm{lk}_\Delta A$ is again a thick spherical building. Its apartments are the intersections $\Sigma' := \Sigma \cap \Delta'$, where Σ ranges over the apartments of Δ containing A (see Proposition 4.9 and its proof). Combining this fact with Proposition 3.79, we see that the roots of Δ' are the intersections $\alpha' := \alpha \cap \Delta'$, where α ranges over the roots of Δ with $A \in \partial\alpha$. The bounding wall $\partial\alpha'$ is $\partial\alpha \cap \Delta'$.

With α and α' as above, note that the root group U_α fixes A; hence it stabilizes Δ'. Moreover, it fixes $\mathrm{st}_{\Delta'} P'$ pointwise for every Δ'-panel P' in $\alpha' \smallsetminus \partial\alpha'$, since $P := P' \cup A$ is a panel of Δ in $\alpha \smallsetminus \partial\alpha$. Hence we have a restriction homomorphism

$$\rho \colon U_\alpha \to U_{\alpha'} \, .$$

Now consider a *boundary* panel P' of α' and let $P := P' \cup A$ be the corresponding boundary panel of α. Then there is a bijection

$$\mathcal{C}(P', \alpha') \overset{\sim}{\longrightarrow} \mathcal{C}(P, \alpha)$$

given by $C' \mapsto C' \cup A$ for $C' \in \mathcal{C}(P', \alpha')$, so the action of U_α on $\mathcal{C}(P, \alpha)$ is equivalent to an action of U_α on $\mathcal{C}(P', \alpha')$. And this is the same action one would get by composing ρ with the natural action of $U_{\alpha'}$ on $\mathcal{C}(P', \alpha')$. These considerations lead to the following result:

Proposition 7.32. *Let A be a simplex of Δ and let $\Delta' := \mathrm{lk}_\Delta A$.*

(1) *If Δ is Moufang, then so is Δ'.*
(2) *If Δ is strictly Moufang, then the restriction map $\rho \colon U_\alpha \to U_{\alpha'}$ is injective for any root α of Δ with $A \in \partial\alpha$.*
(3) *If Δ is strictly Moufang and the Coxeter diagram of Δ' has no isolated nodes, then Δ' is strictly Moufang, and $\rho \colon U_\alpha \to U_{\alpha'}$ is an isomorphism.*

Proof. (1) Consider a root α' of Δ' and a panel $P' \in \partial\alpha'$, and set $\mathcal{D} := \mathcal{C}(P', \alpha')$. With α as above, we have a transitive action of U_α on $\mathcal{D}$ that factors as

$$U_\alpha \overset{\rho}{\longrightarrow} U_{\alpha'} \longrightarrow \mathrm{Sym}\,\mathcal{D} \ .$$

Here $\mathrm{Sym}\,\mathcal{D}$ is the symmetric group, consisting of all permutations of $\mathcal{D}$. Hence the action of $U_{\alpha'}$ on $\mathcal{D}$ is also transitive.

(2) Since the action of U_α is free, the composite above is injective. The first map ρ is therefore also injective.

(3) Our hypothesis implies that the action of $U_{\alpha'}$ on $\mathcal{D}$ is simply transitive, so no proper subgroup of $U_{\alpha'}$ can act transitively. Hence $\rho(U_\alpha) = U_{\alpha'}$. □

Finally, while we are on the subject of links, we make one further observation about root groups (Corollary 7.34 below), in which we use the rigidity theorem of Section 5.9 to deduce that U_α often fixes more simplices than those explicitly required by Definition 7.24. We give this application of the rigidity theorem in the following lemma, in which we allow Δ to be nonspherical. Recall that $E_1(C)$, for $C \in \mathcal{C}(\Delta)$, is the set of chambers having a panel in common with C.

Lemma 7.33. *Let Δ be an arbitrary thick building, let α be a root of Δ, let C be a chamber of α, and let A be a face of C such that $A \notin \partial\alpha$. Assume that $\Delta' := \mathrm{lk}_\Delta A$ is spherical. If g is an automorphism of Δ that fixes α and $E_1(C) \cap \mathrm{st}_\Delta A$ pointwise, then g fixes Δ' pointwise, hence also $\mathrm{st}_\Delta A$.*

Proof. Note that $\alpha \cap \Delta'$ is an apartment of Δ', since $\alpha \cap \Delta' = \Sigma \cap \Delta'$ for any apartment Σ of Δ containing α. So g fixes an apartment of Δ' pointwise as well as $E_1(C')$, where C' is the chamber $C \smallsetminus A$ of Δ'. The lemma therefore follows from the rigidity theorem (see Corollary 5.206) applied to the action of g on Δ'. □

Returning to the spherical case and root groups, we deduce the following:

Corollary 7.34. *Suppose the Coxeter diagram of Δ has no isolated nodes. Then for every root α of Δ and every vertex $v \in \alpha \smallsetminus \partial\alpha$, the root group U_α fixes $\mathrm{lk}_\Delta\, v$ pointwise. In other words, U_α fixes every vertex that is joinable to an interior vertex of α.*

Proof. The hypothesis on the Coxeter diagram implies that there is a chamber $C \in \alpha$ such that v is a vertex of C and C has no panel in $\partial\alpha$ (Example 3.128). Now apply the lemma, with A being the vertex v. $\qquad\qquad\square$

Remark 7.35. The technique of applying the rigidity theorem to links is very powerful, and we will give other applications of it in Section 7.5.

Exercise 7.36. Let Δ be a Moufang building whose Coxeter diagram has no isolated nodes. For every root α of Δ, show that there is a simplex $A \in \partial\alpha$ whose link Δ' is a rank-2 irreducible Moufang building. Hence $U_\alpha \cong U_{\alpha'}$, where α' is the root $\alpha \cap \Delta'$ of Δ'.

7.3.3 Subbuildings

Our goal in this subsection is to prove that under mild hypotheses, subbuildings of Moufang buildings are again Moufang. Readers may wish to review Section 4.6 for the basic facts about subbuildings before proceeding.

Proposition 7.37. *Let Δ be a Moufang building whose Coxeter diagram has no isolated nodes. If Δ' is a thick subbuilding of Δ, then Δ' is also Moufang. If, moreover, Σ is an apartment of Δ', α is a root of Σ, and U_α is the corresponding root group of Δ, then $U'_\alpha := \{u \in U_\alpha \mid u(\Delta') = \Delta'\}$ is the root group of Δ' associated to α.*

Our proof is inspired by the proof of [264, Lemma 5.2.2], which treats the rank-2 case. We will use the following lemma:

Lemma 7.38. *Let Δ be a thick spherical building, and let Δ' be a thick subbuilding. If C and C' are opposite chambers of Δ', then Δ' is the convex hull in Δ of $E_1(C) \cap \Delta'$ and C'.*

Proof. This follows from Corollary 5.207 applied to Δ', together with the fact that Δ' is convex in Δ (see Theorem 4.66). $\qquad\qquad\square$

Proof of Proposition 7.37. Let V_α be the root group of Δ' associated to α (in the sense of Definition 7.24). We have a homomorphism $U'_\alpha \to V_\alpha$ induced by restriction. This homomorphism is injective in view of the simply transitive action of U_α on $\mathcal{A}(\alpha)$. So all claims will be proved if we show that U'_α acts transitively on the set $\mathcal{A}'(\alpha)$ of apartments of Δ' containing α. Let $\Sigma' \in \mathcal{A}'(\alpha)$ be given. Since Δ is Moufang, there exists $u \in U_\alpha$ with $u(\Sigma) = \Sigma'$. Choose a chamber C of α having no panel in $\partial\alpha$. (This is possible by Example 3.128.) Then u fixes $E_1(C)$ pointwise, so u maps the convex hull of $(E_1(C) \cap \Delta') \cup \Sigma$ onto the convex hull of $(E_1(C) \cap \Delta') \cup \Sigma'$. By Lemma 7.38, both these convex hulls are equal to Δ'. Hence u is in U'_α. $\hfill\square$

Corollary 7.39. *Suppose a Moufang building Δ' can be embedded as a subcomplex of a Moufang building Δ of the same type. If the Coxeter diagram has no isolated nodes, then the root groups of Δ' are isomorphic to subgroups of the root groups of Δ.*

Proof. Δ' is a subbuilding of Δ by Proposition 4.63, so we may apply Proposition 7.37. $\hfill\square$

Remark 7.40. We will see in Chapter 9 that "most" Moufang buildings have an associated field (possibly noncommutative). A simple example is the building associated to a vector space. (We will show in the next subsection that this building is Moufang.) Moreover, one can recover the characteristic of the field from the structure of the root groups. In the situation of Corollary 7.39, it follows that the fields associated to Δ and Δ' have the same characteristic. Using results of Tits–Weiss [262], one can even show that the field associated to Δ' has to be a subfield of the field associated to Δ; but we will not need this stronger result.

7.3.4 The Building Associated to a Vector Space

Let $\Delta = \Delta(V)$ be the building associated to an n-dimensional vector space ($n \geq 2$) over a possibly noncommutative division ring D. (See Section 4.3 and Exercise 4.32.) For definiteness, we take $V = D^n$, which we view as a *right* vector space. Thus we have a natural *left* action of $\mathrm{GL}_n(D)$ on D^n by matrix multiplication (where vectors are viewed as column vectors). This induces an action of $\mathrm{GL}_n(D)$ on Δ, which we will use as an aid in constructing enough automorphisms of Δ to prove the Moufang property.

Recall that the vertices of Δ are the nonzero proper subspaces of V and that the simplices are the chains of such subspaces. Recall further that Δ has one apartment $\Sigma = \Sigma(\mathcal{F})$ for every frame $\mathcal{F} = \{L_1, \ldots, L_n\}$. The standard basis $e_1, \ldots, e_n$ for V yields a standard frame, with $L_i := e_i D$ for all i, and hence a standard apartment Σ. It has one vertex for each nonempty proper subset $I \subset \{1, \ldots, n\}$, that vertex being the subspace $Y := \bigoplus_{i \in I} L_i$. Using Exercise 3.58, we get the following description of the roots of Σ: There is one root α_{ij} for each ordered pair of indices $1 \leq i, j \leq n$ with $i \neq j$. The vertices of α_{ij} are the vertices Y of Σ such that

$$L_j \leq Y \implies L_i \leq Y .$$

The vertices of $\partial\alpha_{ij}$ are those Y such that

$$L_j \leq Y \iff L_i \leq Y ,$$

and the interior vertices of α_{ij} are those such that

$$L_i \leq Y \text{ and } L_j \not\leq Y .$$

To verify the Moufang property, we will use the group

$$U_{ij} := \{E_{ij}(\lambda) \mid \lambda \in D\} ,$$

where $E_{ij}(\lambda)$ is the elementary matrix with 1's on the diagonal, λ in the (i, j) position, and 0's elsewhere. The explanation of why this works will be conceptually clearer if we describe the automorphism of V given by $E_{ij}(\lambda)$ in a coordinate-free way.

Let τ be a linear automorphism of V. Then τ is called a *transvection* if there exist $e \in V$ and $f \in V^*$ such that $f(e) = 0$ and

$$\tau(v) = v + ef(v)$$

for all $v \in V$. Here V^* is the dual of V. [Recall, in order to make sense of the expression for τ, that scalars operate on the right of vectors in V.] We then write $\tau = \tau_{e,f}$. For example, the automorphism given by an elementary matrix $E_{ij}(\lambda)$ is the transvection $\tau_{e_i\lambda, e_j^*}$, where $e_1^*, \ldots, e_n^*$ is the basis of V^* dual to $e_1, \ldots, e_n$. Given a transvection $\tau = \tau_{e,f}$, set $L := eD \leq V$ and $H := \ker f \leq V$. Then one checks immediately that a subspace $Y \leq V$ is invariant under τ if and only if either $L \leq Y$ or $Y \leq H$. [Observe that Y is invariant under τ if and only if Y is mapped into itself by the rank-1 operator $v \mapsto ef(v)$.]

Returning now to our root α_{ij}, we will prove the following lemma, which implies the Moufang property:

Lemma 7.41.

(1) U_{ij} *fixes every vertex of* Δ *that is joinable to an interior vertex of* α_{ij}. *In particular,* U_{ij} *fixes* α_{ij} *pointwise.*

(2) *Let* P *be a panel in* $\partial\alpha_{ij}$. *Then* U_{ij} *acts simply transitively on* $\mathcal{C}(P, \alpha_{ij})$.

Proof. (1) Let Y be a vertex of Δ that is joinable to an interior vertex Y' of α_{ij}, and consider a nontrivial element $\tau \in U_{ij}$. Thus $\tau = \tau_{e,f}$ with $eD = L_i$ and $\ker f = \bigoplus_{k \neq j} L_k$. According to our description of the vertices of α_{ij} above, we have $L_i \leq Y'$ and $L_j \not\leq Y'$; hence $eD \leq Y' \leq \ker f$. [To rewrite $L_j \not\leq Y'$ as $Y' \leq \ker f$, use the fact that Y' is a vertex of Σ.] Since Y is joinable to Y', we therefore have either $Y \leq Y' \leq \ker f$ or $Y \geq Y' \geq eD$. In either case, $\tau(Y) = Y$ by our discussion of subspaces invariant under a transvection.

(2) Recall that a simplex of Δ is a chain $Y_1 < \cdots < Y_q$ of nonzero proper subspaces of V. As in Section 4.3, it is convenient to enlarge the chain by setting $Y_0 = 0$ and $Y_{q+1} = V$. In particular, a panel P is a chain of the form

$$0 = Y_0 < \cdots < Y_{k-1} < Y_{k+1} < \cdots < Y_n = V$$

for some $0 < k < n$, where $\dim Y_l = l$ for all l. If $P \in \partial\alpha_{ij}$, then Y_{k+1} contains both L_i and L_j, while Y_{k-1} contains neither. The chambers of Δ containing P are obtained by inserting a k-dimensional subspace Y_k into the chain, and the chamber C of this form that is in α_{ij} is the one with $Y_k = Y_{k-1} + L_i$. For all the others, we have $Y_k = Y_{k-1} + vD$, where $v = e_i\lambda + e_j$ for a uniquely determined $\lambda \in D$. This says precisely that U_{ij} acts simply transitively on the $Y_k \neq Y_{k-1} + L_i$ and hence on the corresponding chambers. $\qquad\square$

Remark 7.42. It follows from the lemma that U_{ij} acts faithfully on Δ and so can be viewed as a subgroup of $\operatorname{Aut}\Delta$. Part (1) then says that $U_{ij} \leq U_{\alpha_{ij}}$. [In fact, we proved that U_{ij} fixes more simplices than Definition 7.24 requires; but this should not be surprising in view of Corollary 7.34.] It then follows from part (2) that U_{ij} is the full root group if $n \geq 3$; indeed, the full group acts freely on $\mathcal{C}(P, \alpha_{ij})$ by Lemma 7.25(3), so a proper subgroup cannot act transitively. If $n = 2$, however, U_{ij} is *not* the full group unless $k = \mathbb{F}_2$; see Exercise 7.30.

In case $n = 3$, we can give a concrete geometric interpretation of the lemma. Recall that Δ in this case is the incidence graph of the projective plane over D. Suppose we are given a line L in this plane and a point $p \in L$; these will play the role of the interior vertices of a root α. Choose a point $q \neq p$ on L and a line $M \neq L$ containing p; these will be the boundary vertices of α. See Figure 7.5, where the root α is on the left and the corresponding picture in the projective plane is on the right. [Note in passing that if we wanted to complete α to an apartment, as indicated by the dashed line on the left, we would have to pick another line through q or, equivalently, another point on M.] What Lemma 7.41 says about the root group U_α is the following:

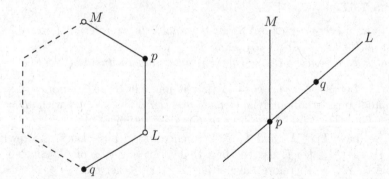

Fig. 7.5. A root in the incidence graph of a projective plane.

(1) U_α stabilizes every point on L and every line through p.
(2) U_α acts simply transitively on the points of M different from p and on the lines through q different from L.

This is a very strong symmetry property of a projective plane, and planes satisfying it are called *Moufang planes* in honor of Ruth Moufang [168, 169, 225]. This is the origin of the term "Moufang building." See Tits–Weiss [262, p. 176] for more details about Moufang's contributions.

We remark, finally, that the Moufang planes we have studied here, coming from division rings, are those that are *Desarguesian*. A general Moufang plane arises from an "alternative division ring," which need not be associative. The nonassociative alternative division rings are called *Cayley division algebras*. For example, many readers will have heard of the Cayley numbers, also called the octonions, which form an 8-dimensional nonassociative division algebra over the field $\mathbb{R}$ of real numbers. There is then a corresponding Moufang projective plane.

7.4 k-Interiors of Roots

Our next major goal is to prove that thick, irreducible, spherical buildings of rank at least 3 are always Moufang. We will do this in Section 7.6. The proof will use a result of Tits [251] about Coxeter complexes, which we prove in the present section (Theorem 7.51). This result could have been proved in Chapter 3, but it is quite special, and we know of no use for it except in connection with the theory of Moufang buildings. The proof is somewhat technical. Readers anxious to get back to the theory of Moufang buildings may wish to just take Theorem 7.51 on faith and move ahead to the next section.

Throughout this section Σ denotes a fixed but arbitrary Coxeter complex. We will view Σ as a collection of finite subsets of a vertex set, as in the standard definition of "simplicial complex," rather than as an abstract poset with certain properties. As usual, we identify a vertex v with the corresponding simplex $\{v\}$. For any simplex $A \in \Sigma$ and any wall H of Σ, we abuse notation slightly and write $A \cap H$ for the maximal face of A in H. Thus the vertices of $A \cap H$ are the vertices of A fixed by the reflection s_H.

Let $\alpha \subseteq \Sigma$ be a root with bounding wall $\partial\alpha$, and let $\mathcal{C}(\alpha)$ be the set of chambers of α.

Definition 7.43. For any integer $k \geq 0$, the *k-interior* of α, denoted by $\alpha^{(k)}$, is defined by

$$\alpha^{(k)} := \{C \in \mathcal{C}(\alpha) \mid \operatorname{codim}(C \cap \partial\alpha) > k\}$$
$$= \{C \in \mathcal{C}(\alpha) \mid C \text{ has at least } k+1 \text{ vertices not in } \partial\alpha\} \ .$$

We can also describe $\alpha^{(k)}$ in terms of the "neighborhoods" $E_k(C)$ defined in Section 5.9. Namely, one can check that

$$\alpha^{(k)} = \{C \in \mathcal{C}(\alpha) \mid E_k(C) \subseteq \alpha\} \ . \tag{7.7}$$

If we set $n := \operatorname{rank} \Sigma$, then we have

$$\mathcal{C}(\alpha) = \alpha^{(0)} \supseteq \alpha^{(1)} \supseteq \cdots \supseteq \alpha^{(n-1)} \supseteq \alpha^{(n)} = \emptyset \,.$$

Note that $\alpha^{(n-1)}$ is simply the set of chambers of α disjoint from $\partial \alpha$.

We have already proven some results about the sets $\alpha^{(k)}$ in Section 3.6.5, without using that notation. For example, we showed that $\alpha^{(n-1)} \neq \emptyset$ if Σ is irreducible (Proposition 3.125). And we showed that $\alpha^{(1)} \neq \emptyset$ if the Coxeter diagram of Σ has no isolated nodes (Example 3.128), a result that we have already used in this chapter. Our main goal in this section is to prove that $\alpha^{(k)}$ is always gallery connected if Σ is 2-spherical in the sense of the following definition.

Definition 7.44. A Coxeter system is said to be *2-spherical* if every rank-2 standard subgroup is finite, or, equivalently, if every entry $m(s,t)$ of the Coxeter matrix is finite. A Coxeter complex Σ or a building Δ is said to be *2-spherical* if its Weyl group is 2-spherical or equivalently, if the link of every codimension-2 simplex is spherical, or, equivalently, if every rank-2 residue is spherical.

We need a few preparatory lemmas, whose proofs will make extensive use of some of the basic results about convex subcomplexes of Coxeter complexes proved in Chapter 3. We begin by restating, for ease of reference, two such results.

Lemma 7.45.

(1) Let K be a convex subcomplex of Σ and let A be a maximal simplex of K. Then $K \subseteq \operatorname{supp} A$.
(2) If A and B are simplices of Σ such that $\dim A = \dim B$ and $A \in \operatorname{supp} B$, then $\operatorname{supp} A = \operatorname{supp} B$. $\qquad\square$

Both parts of this lemma have easy proofs using the Tits cone, and they also have easy combinatorial proofs that we gave in Section 3.6.6 (Lemma 3.140(2) and Corollary 3.141(2)).

Corollary 7.46. *Suppose K is a convex subcomplex of Σ, and let A and B be simplices of K such that $\dim A = \dim B = \dim K$. Then $\operatorname{supp} A = \operatorname{supp} B$.* $\qquad\square$

Next, we need some observations about links of simplices. For any simplex $A \in \Sigma$, we set

$$L_A := \operatorname{lk}_\Sigma A \,.$$

It is a Coxeter complex of rank equal to $\operatorname{codim}_\Sigma A$. Recall from Proposition 3.79 that the walls of L_A are the intersections $H' := H \cap L_A$, where H ranges over the walls of Σ containing A. For any such H, if α is a root of Σ bounded by H, then $\alpha' := \alpha \cap L_A$ is a root of L_A bounded by H'.

We will also have occasion to consider intersections with L_A of walls H that do *not* contain A. The following simple observation will be crucial in this connection:

Lemma 7.47. *If H is a wall of Σ and A is a simplex disjoint from H, then $L_A \cap H$ is a convex subcomplex of Σ.*

Proof. Consider the star of A, as defined in Example 3.133(b). According to that example, $\operatorname{st} A$ is a convex subcomplex of Σ, hence so is $\operatorname{st} A \cap H$. Now note that $\operatorname{st} A \cap H = L_A \cap H$, since A is disjoint from H.   □

We now specialize to links of codimension-2 simplices. If A is such a simplex, then L_A is a 1-dimensional Coxeter complex; hence either L_A is a line or else it is a circle decomposed into an even number of arcs. In the former case, it has one wall for each vertex, and the corresponding roots are half-lines. In the latter case, it has one wall for each pair of opposite vertices, and the corresponding roots are semicircles.

Our next two lemmas will be used only in order to avoid appealing to the optional Section 2.6 on the Tits cone. Readers familiar with the Tits cone can skip ahead to Lemma 7.50.

Lemma 7.48. *Let A be a simplex in Σ of codimension 2. Suppose K is a 0-dimensional subcomplex of L_A such that K is convex as a subcomplex of Σ. If K has more than one vertex, then we have:*

(1) *L_A is finite, and K contains exactly two vertices x, y, which are opposite in L_A.*
(2) *The two roots of L_A bounded by x and y are convex subcomplexes of Σ.*

Proof. (1) Let x be a vertex of K, and consider the wall H of Σ containing the panel $A \cup \{x\}$. Then $H \supseteq \operatorname{supp} x \supseteq K$, so $H \cap L_A$ is a wall of L_A containing K; see Figure 7.6. Since K has more than one vertex, the same is true of this wall. Hence L_A is a circle, $H \cap L_A$ has exactly two vertices (which are opposite), and $K = H \cap L_A$.

(2) Let β' be one of the roots of L_A bounded by x and y, and let E be the edge of β' containing x. We will show that β' is the convex hull of E and y in Σ. Let γ be this convex hull. Since γ is connected (by Proposition 3.136) and contains E and y, it suffices to show that $\gamma \subseteq \beta'$.

Note first that $\operatorname{supp} E$ contains $\operatorname{supp} x$; hence it contains y. Since $\operatorname{supp} E$ is convex in Σ, it follows that $\operatorname{supp} E \supseteq \gamma$; in particular, γ is 1-dimensional. Similarly, since $\operatorname{st} A$ is a convex subcomplex of Σ containing E and y, we have $\operatorname{st} A \supseteq \gamma$. Observe next that no vertex v of A can be in γ. For if v were in γ, then we would have $E \cup \{v\} \in \gamma$. [This follows, for instance, from Remark 3.138.] But this would contradict the fact that γ is 1-dimensional. Hence $\gamma \subseteq L_A$. Finally, if we write $\beta' = \beta \cap L_A$ with β a root of Σ, then we have $\gamma \subseteq \beta$ and hence $\gamma \subseteq \beta'$, as required.   □

Lemma 7.49. *Let Σ' be a Coxeter complex, and let u, v be vertices of Σ' such that $\operatorname{supp} u = \operatorname{supp} v$ and $d(u, v) = 1$, where $d(-, -)$ denotes gallery distance in Σ'. Then Σ' is the join $L * M$, where $L = \operatorname{lk}_{\Sigma'} u = \operatorname{lk}_{\Sigma'} v$ and $M = \operatorname{supp} u = \operatorname{supp} v = \{u, v, \emptyset\}$. In particular, L is the only wall of Σ' not containing u and v, and this wall strictly separates u and v.*

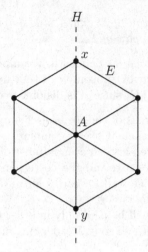

Fig. 7.6. Roots in L_A.

Proof. To deduce the second statement from the first, one need only check that in a join $\Sigma_1 * \Sigma_2$ of Coxeter complexes, the walls are the subcomplexes of the form $H_1 * \Sigma_2$ and $\Sigma_1 * H_2$, where H_i is a wall of Σ_i for $i = 1, 2$. We now prove the first statement. We have $uv = u$, so any chamber C containing u can start a minimal gallery from u to v. Since $d(u, v) = 1$, this means that any chamber containing u is adjacent to a chamber containing v. A similar statement holds with the roles of u and v reversed. It follows easily that $\mathrm{st}_{\Sigma'}\, u \cup \mathrm{st}_{\Sigma'}\, v$ is a thin chamber subcomplex of Σ' and hence is equal to Σ'. The remaining details are routine and are left to the reader.   □

We come, finally, to the key lemma.

Lemma 7.50. *Let $A \in \Sigma$ be a simplex of codimension 2, and let α be a root of Σ such that $A \in \alpha \smallsetminus \partial\alpha$. Then for the intersection $I := L_A \cap \partial\alpha$ we have precisely the following four possibilities:*

(1) *I has at most one vertex, so that $I = \{\emptyset\}$ or $I = \{\emptyset, x\}$ for some vertex x.*
(2) *$\mathrm{diam}\, L_A = 2$, and $I = \{\emptyset, x, y\}$ for two vertices x, y that are opposite in L_A.*
(3) *I is a path of length 1 or 2 (i.e., having 2 or 3 vertices).*
(4) *$I = L_A$.*

Proof. The first step is to reduce to the case that A is disjoint from $\partial\alpha$. Let A' be the face of A given by the vertices not in $\partial\alpha$, and let A'' be the complementary face $A \cap \partial\alpha$. Set $\Sigma' := \mathrm{lk}_\Sigma A''$, and let $\alpha' := \alpha \cap \Sigma'$. Then A' is a codimension-2 simplex of Σ', and $L_A = \mathrm{lk}_{\Sigma'} A'$. Moreover, α' is a root of Σ' whose bounding wall $\partial\alpha' = \partial\alpha \cap \Sigma'$ is disjoint from A', and we have $I = (\mathrm{lk}_{\Sigma'} A') \cap \partial\alpha'$. Thus we may replace Σ, α, and A by Σ', α', and A', thereby reducing to the case that A is disjoint from $\partial\alpha$.

Assuming now that A is disjoint from $\partial\alpha$, our intersection I is a convex subcomplex of Σ by Lemma 7.47. In particular, I is connected if it has dimension 1 (Proposition 3.136). The lemma now reduces to the following two claims:

Claim I. If I contains more than two edges, then $I = L_A$.

Claim II. If $\dim I = 0$ and I has more than one vertex, then the conditions of case (2) are satisfied.

To prove Claim I, consider a path of length 3 in I, given by consecutive edges E_1, E_2, E_3. Let E_4 be the L_A-edge adjacent to E_3 and different from E_2. It suffices to show that $E_4 \in I$. Let $C_i := E_i \cup A$ for $i = 1, 2, 3, 4$, so that we have a gallery C_1, C_2, C_3, C_4 in Σ; see Figure 7.7, in which we have also indicated the wall of Σ separating C_2 and C_3. Consider the reflection r of Σ

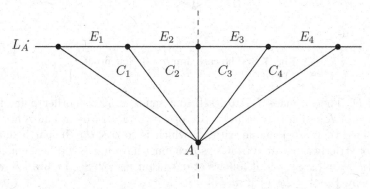

Fig. 7.7. A path in L_A.

with respect to this wall. Then r interchanges C_2 and C_3 as well as C_1 and C_4. It follows that r interchanges E_2 and E_3 as well as E_1 and E_4. Now the edges E_1, E_2, E_3 are maximal simplices of the convex subcomplex I of Σ, so they all have the same support by Corollary 7.46. Since the automorphism r of Σ maps $\{E_1, E_2, E_3\}$ to $\{E_2, E_3, E_4\}$, it follows that E_2, E_3, and E_4 all have the same support; hence the four edges E_1, E_2, E_3, E_4 have the same support. Since some of these edges are contained in the wall $\partial\alpha$, we conclude from the definition of "support" that all of them are contained in $\partial\alpha$. In particular, E_4 is in $\partial\alpha$; hence E_4 is in I. This completes the proof of Claim I.

Turning now to Claim II, we give two proofs. The first, which is essentially the proof given by Tits [251, Lemma 1], makes use of the Tits cone, while the second is purely combinatorial and is based on Lemmas 7.48 and 7.49.

Proof 1: By Lemma 3.143 and Remark 3.144, I has exactly two vertices x, y, which are *opposite in* Σ in the sense that they are strictly separated by every wall of Σ not containing them. Moreover, L_A is a $2m$-gon with $2 \le m < \infty$. Our task is to show that $m = 2$. Consider a path of length 2 in L_A starting at x. Let the edges of this path be E and F, and set $C := E \cup A$ and $D := F \cup A$; see Figure 7.8. Let r be the reflection of Σ that interchanges

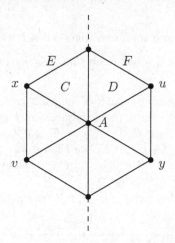

Fig. 7.8. The case dim $I = 0$, first proof.

C and D. Then r maps L_A to itself and sends x, y to another pair of vertices u, v of L_A such that u and v are opposite in Σ. Now u and v are both in α, since they are joinable with A, which is in $\alpha \smallsetminus \partial\alpha$. In particular, $\partial\alpha$ does not strictly separate u from v, so we must have $u, v \in \partial\alpha$; hence $u, v \in I$. Thus $\{u, v\} = \{x, y\}$, and it follows at once that $m = 2$. See Figure 7.8, which shows why $\{u, v\} \neq \{x, y\}$ if $m = 3$.

Proof 2: By Lemma 7.48, L_A is a $2m$-gon with $2 \leq m < \infty$, and $I = \{\emptyset, x, y\}$, where x and y are opposite vertices of L_A. Moreover, x and y cut L_A into two halves β'_1, β'_2 that are convex subcomplexes of Σ; see Figure 7.9. Let E and F be the edges of L_A containing x, with $E \in \beta'_1$ and $F \in \beta'_2$, and let u (resp. v) be the vertex of E (resp. F) different from x. Let r be the reflection of Σ that fixes the panel $A \cup \{v\}$; the corresponding wall is indicated in Figure 7.9.

If $m \geq 3$, then β'_2 contains at least 3 edges, so $r(\beta'_2)$ contains both E and F. By the convexity of $r(\beta'_2)$ and Corollary 7.46, it follows that supp $E = $ supp F and hence that the vertices u, v have the same support in $\Sigma' := \text{lk}_\Sigma x$. Since $d_{\Sigma'}(u, v) = 1$, we can therefore apply Lemma 7.49 to conclude that there is a unique wall of Σ' not containing u and v and that this wall strictly separates them. Now we know that $u, v \notin \partial\alpha$, so the unique wall of Σ' not containing them must be $\partial\alpha \cap \Sigma'$. In particular, u and v cannot both be in α. But A is joinable to both u and v and A is in $\alpha \smallsetminus \partial\alpha$, which implies that u and v are both in α. This contradiction shows that $m < 3$. $\square$

We are ready now for the main result of this section.

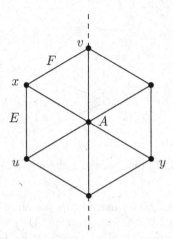

Fig. 7.9. The case $\dim I = 0$, second proof.

Theorem 7.51. *Let Σ be a 2-spherical Coxeter complex and let $\alpha \subseteq \Sigma$ be a root. Then for any integer $k \geq 0$, the k-interior $\alpha^{(k)}$ of α is gallery connected. For $k \geq 1$, the following more precise result holds: Given chambers $C \in \alpha^{(k-1)}$ and $D \in \alpha^{(k)}$, there is a gallery $C = C_0, \ldots, C_n = D$ with $C_i \in \alpha^{(k)}$ for $i \geq 1$.*

Proof. We argue by induction on k. The case $k = 0$ is already known by the convexity of roots, since $\alpha^{(0)} = \mathcal{C}(\alpha)$, so we may assume $k > 0$. Given chambers $C \in \alpha^{(k-1)}$ and $D \in \alpha^{(k)}$, the induction hypothesis implies that we can connect C and D by a gallery $\Gamma \colon C = C_0, \ldots, C_n = D$ in $\alpha^{(k-1)}$. Let $m(\Gamma)$ be the number of indices i with $1 \leq i \leq n$ and $C_i \notin \alpha^{(k)}$. It suffices to show that if $m(\Gamma) > 0$, then there is a new gallery Γ' from C to D in $\alpha^{(k-1)}$ with $m(\Gamma') < m(\Gamma)$.

Since $m(\Gamma) > 0$, we can find an index $0 < i < n$ with $C_i \in \alpha^{(k-1)} \smallsetminus \alpha^{(k)}$ and $C_{i+1} \in \alpha^{(k)}$. We may assume $C_{i-1} \neq C_{i+1}$, so that the simplex $A := C_{i-1} \cap C_i \cap C_{i+1}$ has codimension 2 in Σ. Consider the path in $L_A := \mathrm{lk}_\Sigma A$ corresponding to the gallery C_{i-1}, C_i, C_{i+1}, and let the vertices along this path be denoted by v_1, v_2, v_3, v_4 as in Figure 7.10. (The vertex colors will be justified in the next paragraph.) Since Σ is 2-spherical, L_A is a circle; so there is a second path from v_1 to v_4 in L_A. Let Γ' be the gallery obtained from Γ by removing C_i and inserting the chambers $D_1, \ldots, D_p$ corresponding to the edges along this second path. To see that $m(\Gamma') < m(\Gamma)$, we need to show that $D_j \in \alpha^{(k)}$ for all $1 \leq j \leq p$.

Note first that we must have $v_2 \in \partial\alpha$ and $v_4 \notin \partial\alpha$, since C_{i+1} has more vertices in the interior of α (i.e., in $\alpha \smallsetminus \partial\alpha$) than C_i has. This implies that $A \notin \partial\alpha$; for if we had $A \in \partial\alpha$, then the panel $A \cup \{v_2\}$ would be in $\partial\alpha$, contradicting the fact that C_{i-1} and C_i are both in α. Note next that if $v_3 \notin \partial\alpha$, then also $v_1 \notin \partial\alpha$, because C_{i-1} has at least as many vertices in the interior of α as C_i has.

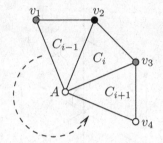

Fig. 7.10. Modifying Γ; black in $\partial\alpha$, white not in $\partial\alpha$, gray unknown.

We now apply Lemma 7.50 to analyze the possibilities for the intersection $I := L_A \cap \partial\alpha$. In view of the information in the previous paragraph, this intersection must have vertex set $\{v_2\}$, $\{v_2, v_3\}$, or $\{v_1, v_2, v_3\}$. (Note that case (2) of the lemma cannot occur, because if $\operatorname{diam} L_A = 2$, then v_4 is the vertex of L_A opposite v_2.) In particular, all the vertices of L_A other than v_1, v_2, v_3 are in the interior of α. We can now conclude that the chambers D_j that we have inserted into Γ' are in $\alpha^{(k)}$. For $j \geq 2$, this follows from the fact that A has at least $k-1$ vertices $\notin \partial\alpha$, and D_j has 2 more such vertices coming from L_A. This argument also works for $j = 1$ unless $v_1 \in \partial\alpha$. But in this case A has k vertices $\notin \partial\alpha$, since $C_{i-1} \in \alpha^{(k-1)}$; and L_A contributes one more such vertex to D_1, so we still obtain $D_1 \in \alpha^{(k)}$. □

Corollary 7.52. *Under the hypotheses of the theorem, suppose C is a chamber in $\alpha^{(k)} \smallsetminus \alpha^{(k+1)}$. If $\alpha^{(k+1)} \neq \emptyset$, then there is a chamber $D \in \alpha^{(k+1)}$ adjacent to C.* □

Exercises

7.53. Give an example to show that we cannot drop the 2-sphericity assumption in Theorem 7.51.

7.54. In the 2-spherical case, is $\alpha^{(k)}$ necessarily convex for $k > 0$?

7.5 Consequences of the Rigidity Theorem

We continue our preparation for the proof that thick, irreducible, spherical buildings of rank at least 3 are Moufang. That proof will be given in the next section. Throughout the present section we assume:

- Δ is a thick, 2-spherical building.
- α is a root of Δ.
- g is an automorphism of Δ that fixes α pointwise.

Our goal is to find minimal hypotheses under which we can deduce that g fixes $\mathrm{st}_\Delta\, P$ pointwise for every panel $P \in \alpha \smallsetminus \partial\alpha$ (and hence, in the spherical case, that g is in the root group U_α). There are two ingredients to the proofs. The first is the rigidity theorem, applied to the action of g on the links of various codimension-2 simplices; we have already recorded this application of the rigidity theorem in Lemma 7.33. The second ingredient is the information about k-interiors of roots that we collected in the previous section.

For the purposes of the applications in the present chapter, the main result of this section is Proposition 7.57 below. In the next chapter, when we consider Moufang twin buildings, we will also need Proposition 7.58.

Lemma 7.55.

(1) *If C is a chamber in $\alpha^{(2)}$ such that g fixes $E_1(C)$ pointwise, then g fixes $E_2(C)$ pointwise.*

(2) *Let C and D be adjacent chambers with $C \in \alpha^{(1)}$ and $D \in \alpha^{(2)}$. Then g fixes $E_1(C)$ pointwise if and only if it fixes $E_1(D)$ pointwise.*

Proof. (1) Given $D \in E_2(C)$, the face $A := D \cap C$ has codimension at most 2, and it is in $\alpha \smallsetminus \partial\alpha$ since $C \in \alpha^{(2)}$. We can now apply Lemma 7.33 to get $gD = D$.

(2) If g fixes $E_1(D)$ pointwise, then it also fixes $E_2(D)$ pointwise by (1). Since $E_1(C) \subseteq E_2(D)$, this implies that g fixes $E_1(C)$ pointwise. Conversely, suppose g fixes $E_1(C)$ pointwise. Given $E \in E_1(D)$, the face $A := E \cap D \cap C$ has codimension at most 2, and it is in $\alpha \smallsetminus \partial A$, since $D \in \alpha^{(2)}$. We can now apply Lemma 7.33 to get $gE = E$. □

Lemma 7.56. *Suppose the Coxeter diagram of Δ has no isolated nodes. If g fixes $E_1(C)$ pointwise for every $C \in \alpha^{(1)}$, then g fixes $\mathrm{st}_\Delta\, P$ pointwise for every panel $P \in \alpha \smallsetminus \partial\alpha$.*

Proof. Let P be a panel in $\alpha \smallsetminus \partial\alpha$, and choose a chamber $C \geq P$ in α. If C is in $\alpha^{(1)}$, then g fixes $\mathrm{st}_\Delta\, P$ pointwise by hypothesis. Otherwise, Corollary 7.52 implies that there is a chamber $C' \in \alpha^{(1)}$ adjacent to C. [Recall that $\alpha^{(1)} \neq \emptyset$ by Example 3.128, so the corollary is applicable.] By hypothesis, g fixes $E_1(C')$ pointwise. Since $C \notin \alpha^{(1)}$, it has exactly one vertex $x \notin \partial\alpha$, and x must also be a vertex of P and C' (because $P \notin \partial\alpha$ and $C' \in \alpha^{(1)}$). So $A := C' \cap P$ is a simplex of $\alpha \smallsetminus \partial\alpha$ having codimension at most 2. We can therefore apply Lemma 7.33 to conclude that g fixes $\mathrm{st}_\Delta\, A$ pointwise; hence it also fixes $\mathrm{st}_\Delta\, P$ pointwise. □

Proposition 7.57. *Suppose Δ is irreducible and of rank at least 3. If g fixes $E_1(C)$ pointwise for some chamber $C \in \alpha^{(1)}$, then g fixes $\mathrm{st}_\Delta\, P$ pointwise for any panel $P \in \alpha \smallsetminus \partial\alpha$.*

Proof. In view of Lemma 7.56, it suffices to show that g fixes $E_1(C')$ pointwise for every chamber $C' \in \alpha^{(1)}$. Recall now that $\alpha^{(2)} \neq \emptyset$ by Proposition 3.125.

So two applications of Theorem 7.51 yield a gallery $C = C_0, \ldots, C_n = C'$ with $C_i \in \alpha^{(2)}$ for $0 < i < n$. Since g fixes $E_1(C)$ pointwise, Lemma 7.55(2) now implies that g fixes $E_1(C')$ pointwise. ☐

Proposition 7.58. *Suppose Δ is irreducible and of rank at least 3. If g fixes* $\mathrm{st}_\Delta P$ *pointwise for every panel $P \in \alpha$ such that* $\mathrm{codim}_\Delta(P \cap \partial\alpha) = 2$, *then g fixes* $\mathrm{st}_\Delta P$ *pointwise for every panel* $P \in \alpha \smallsetminus \partial\alpha$.

(Here, as in Section 7.4, $P \cap \partial\alpha$ denotes the maximal face of P in $\partial\alpha$.)

Proof. Let C_0 be any chamber of α with a panel in $\partial\alpha$, and choose (by Corollary 7.52) a chamber $C_1 \in \alpha^{(1)}$ adjacent to C_0. We will show that g fixes $E_1(C)$ pointwise, so that an application of Proposition 7.57 will complete the proof. Let x be the vertex of C_0 not in $\partial\alpha$, and let y be the vertex of C_1 not in C_0. Then x and y are the two vertices of C_1 not in $\partial\alpha$. Set $A := C_1 \smallsetminus \{x, y\}$; see Figure 7.11. By assumption, g fixes $\mathrm{st}_\Delta P$ pointwise for the two panels

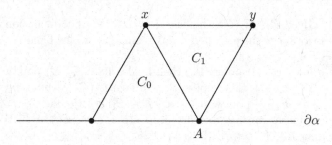

Fig. 7.11. $C_0 \in \alpha \smallsetminus \alpha^{(1)}$, $C_1 \in \alpha^{(1)}$.

$P = A \cup \{x\}$ and $P = A \cup \{y\}$ of C_1. We must show that g fixes $\mathrm{st}_\Delta P$ pointwise for any panel P of C_1 containing both x and y. We will prove the stronger result that g fixes $\mathrm{st}_\Delta B$ pointwise, where $B := P \smallsetminus \{y\} \leq C_0$. (In the 2-dimensional example shown in the picture, P is the edge joining x and y, and B is the vertex x.) In view of Lemma 7.33, it suffices to show that g fixes $E_1(C_0) \cap \mathrm{st}_\Delta B$ pointwise. This follows from our hypothesis, since any panel Q of C_0 containing B also contains x and hence satisfies $\mathrm{codim}_\Delta(Q \cap \partial\alpha) = 2$. ☐

7.6 Spherical Buildings of Rank at Least 3

In Section 7.1 we showed, as an easy consequence of the extension theorem, that if Δ is a thick, irreducible, spherical building of rank ≥ 3, then Δ is pre-Moufang (see Proposition 7.11). Using the results of Section 7.5, we will see that Δ is actually Moufang. This was announced by Tits in [247, p. 274] and proved in [251].

Theorem 7.59. *If Δ is a thick, irreducible, spherical building of rank at least 3, then Δ is Moufang.*

Proof. Let α be a root of Δ, and let Σ and Σ' be apartments containing α. By Corollary 5.211, there is an automorphism g of Δ such that $g\Sigma = \Sigma'$, g fixes α pointwise, and g fixes $E_2(C)$ pointwise for some $C \in \alpha^{(2)}$. Proposition 7.57 implies that g is in the root group U_α, so the latter acts transitively on $\mathcal{A}(\alpha)$. Thus Δ has the Moufang property. $\qquad\qquad\square$

Remark 7.60. This is a continuation of the discussion at the end of Section 6.4. Recall that by the Feit–Higman theorem, finite thick buildings of type $I_2(m)$ ($m \geq 3$) exist only for $m = 3$, 4, 6, and 8. Tits [250, 253] later proved that the same conclusion holds if the building is assumed to be Moufang instead of finite. A slightly more general result was proved almost simultaneously by Weiss [280]. See also Ronan [200, Appendix 1] or Tits–Weiss [262, Chapter 18]. The consequence we need right now is that one cannot have $m = 5$. Since links in Moufang buildings are again Moufang by Proposition 7.32, we can combine this result with Theorem 7.59 to see that there do not exist thick buildings of type H_3 or H_4. We therefore obtain the following result:

Corollary 7.61. *If Δ is a thick, irreducible, spherical building of rank $n \geq 3$, then Δ is of type A_n, C_n, D_n, E_n, or F_4.* $\qquad\qquad\square$

7.7 Group-Theoretic Consequences of the Moufang Property

Throughout this section Δ denotes a strictly Moufang (spherical) building unless the contrary is explicitly stated. Choose a fundamental apartment Σ and a fundamental chamber $C \in \Sigma$. As usual, we denote by Φ the set of roots of Σ, by Φ_+ the set of positive roots (those containing C), and by Φ_- the set of negative roots (those not containing C or, equivalently, those containing the chamber C^0 of Σ opposite C).

Our purpose in this section is to prove some algebraic results about the group

$$G := \langle U_\alpha \mid \alpha \in \Phi \rangle$$

that occurred in Proposition 7.28. These algebraic results will then lead us to axioms for an algebraic theory to be developed in the next section.

7.7.1 The Groups $U_\pm$, $B_\pm$, and U_w

We define two subgroups $U_+, U_- \leq G$ by

$$U_\pm := \langle U_\alpha \mid \alpha \in \Phi_\pm \rangle .$$

Recall that U_+ played a crucial role in the proof that the G-action on Δ is strongly transitive (see Lemma 7.9, where U_+ was simply called U). Next, there are two important stabilizers:

$$B_+ := \{g \in G \mid gC = C\} \, ,$$
$$B_- := \{g \in G \mid gC^0 = C^0\} \, .$$

Finally, we set

$$T := \mathrm{Fix}_G(\Sigma) = \{g \in G \mid g \text{ fixes } \Sigma \text{ pointwise}\} \, ,$$

as is customary in the theory of strongly transitive actions (Section 6.1.1). Note that

$$B_+ \cap B_- = T \, , \tag{7.8}$$

since the convex hull of C and C^0 is Σ.

Our main goal in this subsection is to apply the results of Section 7.2.2 to the root groups U_α. The following lemma will play a crucial role.

Lemma 7.62. $U_+ \cap T = \{1\}$.

Proof. Let P be a panel of the fundamental chamber C, let α be the positive root with $P \in \partial\alpha$, and consider the action of U_+ on the set of chambers $\mathcal{C}(P, \alpha) = \mathcal{C}_P \smallsetminus \{C\}$. We claim that if an element $u \in U_+$ fixes one chamber in $\mathcal{C}(P, \alpha)$, then it fixes all chambers in $\mathcal{C}(P, \alpha)$. In other words, the image of U_+ in the group of permutations of $\mathcal{C}(P, \alpha)$ acts freely on that set. Indeed, the image of U_+ is the same as the image of U_α, since U_β acts trivially on $\mathcal{C}(P, \alpha)$ for $\beta \in \Phi_+ \smallsetminus \{\alpha\}$. [We have $P \in \beta \smallsetminus \partial\beta$ for any such β.] So our claim follows from the fact that U_α acts freely on $\mathcal{C}(P, \alpha)$.

Now suppose $u \in U_+ \cap T$. Then u fixes every chamber of Σ adjacent to C, and hence, by the claim, u fixes every chamber of Δ adjacent to C. The rigidity theorem (see Corollary 5.206) therefore implies that $u = 1$. $\square$

Corollary 7.63. $B_+ \cap U_- = \{1\}$.

Proof. By (7.8) we have $B_+ \cap U_- \leq T \cap U_-$. Now note that $T \cap U_- = \{1\}$ by Lemma 7.62, applied with the roles of C and C^0 reversed. $\square$

We now consider the groups

$$U_\Psi := \langle U_\alpha \mid \alpha \in \Psi \rangle \, ,$$

where Ψ is a convex subset of Φ in the sense of Definition 7.13.

Proposition 7.64. *Let Ψ be a convex set of roots in the fundamental apartment Σ, let $\alpha_1, \ldots, \alpha_m$ be an admissible ordering of Ψ, and set $U_i := U_{\alpha_i}$ for $i = 1, \ldots, m$.*

(1) $U_\Psi = U_1 \cdots U_m$. *More precisely, every element of U_Ψ is uniquely expressible as $u_1 \cdots u_m$ with $u_i \in U_i$ for each i.*
(2) *If $\Psi = \Psi(K)$ for some subset $K \subseteq \Sigma$ containing at least one chamber, then the pointwise fixer of K in G is given by*

$$\mathrm{Fix}_G(K) = U_\Psi T = U_\Psi \rtimes T .$$

Proof. Since T acts trivially on Σ, it follows from Lemma 7.25(1) that T normalizes U_α for each $\alpha \in \Phi$. In particular, T normalizes U_Ψ. Note next that we can take the fundamental chamber C to be in K, so that $\Psi \subseteq \Phi_+$ and $U_\Psi \leq U_+$. In particular, Lemma 7.62 implies that

$$U_\Psi \cap T = \{1\} . \tag{7.9}$$

Statement (2) now follows immediately from Proposition 7.20(3), applied with $X_\alpha = U_\alpha$ for all α. Using parts (1) and (2) of the same proposition, we see that every element of $U_\Psi T$ admits a unique representation as $u_1 \cdots u_m t$ with $u_i \in U_i$ for all i and $t \in T$ (see Remark 7.21(b)). In particular, every element of U_Ψ admits such a representation, and necessarily $t = 1$ in view of (7.9). This proves (1). $\qquad\square$

Taking $K = \{C\}$, we obtain the following important special case:

Corollary 7.65.

(1) *There is an ordering $U_1, \ldots, U_m$ of the positive root groups such that every element of U_+ admits a unique representation as $u_1 \cdots u_m$ with $u_i \in U_i$ for all i.*
(2) $B_+ = U_+ \rtimes T$. $\qquad\square$

Next we consider the case $\Psi = \Phi(C, D)$, as in Example 7.15, for some chamber $D \in \Sigma$. Here C is still the fundamental chamber. If we write $D = wC$ with $w \in W$, then the chamber D' of Σ opposite D is wC^0. Hence $\Phi(C, wC) = \Psi(\{C, wC^0\})$. It has precisely $d(C, wC) = l(w)$ elements, and any reduced decomposition of w yields an admissible ordering of it. We will discuss this set in more detail in Section 7.8.4. For the moment, we simply introduce the notation

$$\Phi(w) := \Phi(C, wC) = \Psi(\{C, wC^0\}) \subseteq \Phi_+ ,$$

and we apply Proposition 7.64 to the corresponding groups

$$U_w := U_{\Phi(w)} \leq U_+ .$$

Corollary 7.66. *Given $w \in W$, choose a reduced decomposition $w = s_1 \cdots s_l$, let Γ be the corresponding gallery from C to wC, and let $\alpha_1, \ldots, \alpha_l$ be the sequence of roots associated to Γ as in Example 7.15. Set $U_i := U_{\alpha_i}$ for $i = 1, \ldots, l$.*

(1) *Every element of U_w admits a unique representation as $u_1 \cdots u_l$ with $u_i \in U_i$ for all i.*

(2) *U_w is the stabilizer of wC^0 in U_+. In other words,*

$$U_w = U_+ \cap wU_- w^{-1} .$$

Proof. (1) is just Proposition 7.64(1), specialized to $\Psi = \Phi(w)$. To prove (2), note first that the stabilizer of wC^0 in B_+ is the fixer of the set $K = \{C, wC^0\}$, so it is $U_w T$ by Proposition 7.64(2). The stabilizer of wC^0 in U_+ is therefore $U_+ \cap U_w T$, which equals U_w by Lemma 7.62. $\square$

Finally, we record the result of applying Lemma 7.22 to the gallery Γ in Corollary 7.66.

Corollary 7.67. *U_w acts simply transitively on the w-sphere*

$$\mathcal{C}_w := \{D \in \mathcal{C}(\Delta) \mid \delta(C, D) = w\} .$$

In particular, U_+ acts simply transitively on

$$C^{\mathrm{op}} := \{D \in \mathcal{C}(\Delta) \mid D \text{ op } C\} .$$

Proof. For any $D \in \mathcal{C}_w$, there is a gallery from C to D having type $(s_1, \ldots, s_l)$, so transitivity follows immediately from Lemma 7.22. Simple transitivity follows from the uniqueness assertion of that lemma combined with Corollary 7.66(1). Alternatively, it follows from the fact that if $u \in U_w$ fixes wC, then u fixes the pair $\{wC, wC^0\}$ of opposite chambers by Corollary 7.66(2); hence $u \in U_w \cap T = \{1\}$. Finally, the last assertion of the corollary is obtained by taking w to be the longest element $w_0 \in W$. $\square$

The transitivity assertion in Corollary 7.67 yields a refinement of the Bruhat decomposition in the setting of Moufang buildings:

Corollary 7.68. *For any $w \in W$, $B_+ w B_+ = U_w w B_+$.*

Proof. For any $b \in B_+$, the chamber bwC is in $\mathcal{C}_w$, so there is an element $u \in U_w$ such that $bwC = uwC$, i.e., $bwB_+ = uwB_+$. Thus $B_+ wB \subseteq U_w w B_+$. $\square$

Remark 7.69. We made the assumption that Δ is *strictly* Moufang because we needed the action of the root group U_α on $\mathcal{C}(P, \alpha)$ to be simply transitive for each $\alpha \in \Phi$. But all the results of this subsection remain valid if we weaken the assumption slightly. Namely, it suffices to assume that we are given a system of groups $(U_\alpha)_{\alpha \in \Phi}$ with the following two properties:

- For each $\alpha \in \Phi$, U_α is a subgroup of the root group associated to α.
- For each $\alpha \in \Phi$ and each panel $P \in \partial\alpha$, the action of U_α on $\mathcal{C}(P, \alpha)$ is simply transitive.

Exercise 7.70.

(a) Show that $U_\Psi \cap U_{\Psi'} = U_{\Psi \cap \Psi'}$ for any two convex subsets Ψ, Ψ' of Φ.

(b) If Ψ is a convex subset of Φ, show that

$$\Psi = \{\alpha \in \Phi \mid U_\alpha \leq U_\Psi\} .$$

7.7.2 Commutator Relations

The elementary matrices in general linear groups satisfy well-known commutator identities. (Readers who have not seen these can look ahead at equation (7.21) in Section 7.9.) These identities imply, for example, that

$$[U_{ij}, U_{jk}] \leq U_{ik} \tag{7.10}$$

if i, j, k are distinct indices ($1 \leq i, j, k \leq n$), where the notation is that of Section 7.3.4. Here $[U, V]$ denotes the subgroup generated by the commutators $[u, v] := uvu^{-1}v^{-1}$ for $u \in U$ and $v \in V$. In the present subsection we give a geometric explanation for the existence of relations like (7.10). Our setting is still that Δ is a strictly Moufang (spherical) building with a fundamental pair (Σ, C), and Φ is the set of roots of Σ. In order to state the commutator relations, we need the following definition.

Definition 7.71. Given $\alpha, \beta \in \Phi$, we define the *closed interval* $[\alpha, \beta]$ by

$$[\alpha, \beta] := \{\gamma \in \Phi \mid \gamma \supseteq \alpha \cap \beta\} \ .$$

We define the *open interval* (α, β) by

$$(\alpha, \beta) := [\alpha, \beta] \smallsetminus \{\alpha, \beta\} \ .$$

It is easy to get a concrete picture of the interval $[\alpha, \beta]$ that explains why it is called an interval. We will do this in Section 7.7.4. But first we establish the importance of intervals by proving the commutator relations.

Proposition 7.72. *Given $\alpha, \beta \in \Phi$ with $\alpha \neq \pm\beta$, we have*

$$[U_\alpha, U_\beta] \leq U_{(\alpha,\beta)} := \langle U_\gamma \mid \gamma \in (\alpha, \beta) \rangle \ .$$

Proof. We first show that $[\alpha, \beta]$ and (α, β) are convex sets of roots, so that the results of the previous subsection about the groups U_Ψ are applicable. The convexity of $[\alpha, \beta]$ is trivial, since by definition, $[\alpha, \beta] = \Psi(\alpha \cap \beta)$. [And $\alpha \cap \beta$ contains at least one chamber, since $\alpha \not\subseteq -\beta$ by Lemma 3.53.] To see that (α, β) is also convex, note that $\alpha \cap \beta \subsetneqq \beta$, so we can find adjacent chambers D', D with $D' \in \alpha \cap \beta$ and $D \in \beta \smallsetminus \alpha$. Similarly, there are adjacent chambers E', E with $E' \in \alpha \cap \beta$ and $E \in \alpha \smallsetminus \beta$. One then checks immediately that

$$(\alpha, \beta) = \Psi\big((\alpha \cap \beta) \cup \{D, E\}\big) \ , \tag{7.11}$$

so (α, β) is indeed convex.

We now claim that $[U_\alpha, U_\beta]$ fixes D. To see this, let P be the panel $D \cap D'$. Then we have $P \in \alpha \cap \beta$, so U_α and U_β stabilize the set of chambers $\mathcal{C}_P = \mathcal{C}(\Delta)_{\geq P}$. And we have $P \notin \partial\beta$, so U_β even fixes $\mathcal{C}_P$ pointwise. A trivial computation now proves the claim. Similarly, $[U_\alpha, U_\beta]$ fixes E. Since $[U_\alpha, U_\beta]$

also fixes $\alpha \cap \beta$ pointwise, we may apply Proposition 7.64 and equation (7.11) to conclude that

$$[U_\alpha, U_\beta] \leq \operatorname{Fix}_G\big((\alpha \cap \beta) \cup \{D, E\}\big) = U_{(\alpha,\beta)} \rtimes T \leq U_{[\alpha,\beta]} \rtimes T .$$

Since $[U_\alpha, U_\beta]$ is contained in the first factor of the rightmost semidirect product, this implies the proposition. $\square$

Corollary 7.73. *Let $\alpha_1, \ldots, \alpha_d$ be the roots associated to a minimal gallery in Σ as in Example 7.15, and set $U_i := U_{\alpha_i}$ for $i = 1, \ldots, d$. If $d \geq 2$, then*

$$[U_1, U_d] \leq U_2 \cdots U_{d-1} .$$

Proof. It is easy to check that $[\alpha_1, \alpha_d] \subseteq \{\alpha_1, \ldots, \alpha_d\}$ (see Exercise 7.75); hence $(\alpha_1, \alpha_d) \subseteq \{\alpha_2, \ldots, \alpha_{d-1}\}$. The corollary now follows from the proposition and the convexity of $\{\alpha_2, \ldots, \alpha_{d-1}\}$. $\square$

Remark 7.74. In some treatments of the commutator relations, Corollary 7.73 is proved directly; see, for instance, [200, 281]. One can then deduce Proposition 7.72 after reducing to the rank-2 case. We have followed Tits's original approach, which seems more straightforward and has the advantage that it applies with no essential change to Moufang twin buildings (Section 8.5.2).

Exercises

7.75. If Ψ is a convex set of roots, show that Ψ contains $[\alpha, \beta]$ for any $\alpha, \beta \in \Psi$.

7.76. Verify that the application of Proposition 7.72 in the proof of Corollary 7.73 was legitimate, i.e., that $\alpha_1 \neq \pm\alpha_d$.

7.77. Suppose that W is the Weyl group of a generalized root system. We can then identify the roots in $\Sigma(W, S)$ with root vectors as in Section 1.5.10. Show that after this identification, the interval $[\alpha, \beta]$ determined by two root vectors α, β is the set of root vectors γ such that γ is a nonnegative linear combination of α and β.

7.78. Let (W, S) be the Coxeter system of type A_{n-1}. As we have seen in Examples 1.119 and 3.52, it has one root α_{ij} for each ordered pair $1 \leq i, j \leq n$ with $i \neq j$. Given three distinct indices i, j, k, show that $(\alpha_{ij}, \alpha_{j,k}) = \{\alpha_{ik}\}$. Thus the commutator relation (7.10) is a special case of Proposition 7.72.

7.7.3 The Role of the Commutator Relations

We show here that in some sense, the commutator relations capture the essence of the Moufang property. Let's start with a thick spherical building Δ (with a fundamental apartment as usual), and suppose we are given a family of subgroups $X_\alpha \leq \operatorname{Aut} \Delta$ ($\alpha \in \Phi$) with the following two properties:

(1) X_α fixes α pointwise for each $\alpha \in \Phi$.
(2) For each $\alpha \in \Phi$ and each panel $P \in \partial\alpha$, the action of X_α is transitive on the set of chambers $\mathcal{C}(P, \alpha)$.

Consider now the following two additional properties:

(3) For each $\alpha \in \Phi$ and each panel $P \in \alpha \smallsetminus \partial\alpha$, X_α fixes $\mathcal{C}_P$ pointwise.
(3') For any $\alpha, \beta \in \Phi$ with $\alpha \neq \pm\beta$,

$$[X_\alpha, X_\beta] \leq X_{(\alpha,\beta)} \,.$$

In the case we have been treating, with the X_α being the root groups U_α, (3) holds by definition, and this was crucial in our proof of (3'). We now show that conversely, (3') implies (3) (cf. [200, Proposition 6.14]).

Proposition 7.79. *Let Δ be a spherical building with a family of subgroups $(X_\alpha)_{\alpha \in \Phi}$ of $\mathrm{Aut}\,\Delta$, where Φ is the set of roots of an apartment Σ. If the X_α satisfy (1), (2), and (3'), then they also satisfy (3). Hence Δ, if it is thick, is a Moufang building, and X_α is a subgroup of the root group U_α for each $\alpha \in \Phi$. If, in addition, the Coxeter diagram of Δ has no isolated nodes, then Δ is strictly Moufang and $X_\alpha = U_\alpha$ for all α.*

Proof. Let α and P be as in (3), and consider any chamber $C \in \mathcal{C}_P$. Let D and E be the chambers of Σ in $\mathcal{C}_P$. They are necessarily both in α. We have to show that X_α fixes C. In view of (1), we may assume that C is different from D and E.

The panel P determines a wall in Σ that bounds two opposite roots $\pm\beta$. These are both different from α, since $P \notin \partial\alpha$. Let's say that $D \in \beta$. By (2) there exists $v \in X_\beta$ such that $vC = E$. So X_α fixes C if and only if $vX_\alpha v^{-1}$ fixes E. Since X_α fixes E (by (1) again) and $v \in X_\beta$, we are reduced to showing that $[X_\alpha, X_\beta]$ fixes E. By (3') we can reduce further to showing that X_γ fixes E for any $\gamma \in (\alpha, \beta)$. Now any such γ contains D and, being different from β, also E. Hence X_γ fixes E, and the proof of (3) is complete. The remaining assertions of the proposition follow easily. (To see that $X_\alpha = U_\alpha$ when the Coxeter diagram has no isolated nodes, use the fact that U_α acts freely on $\mathcal{C}(P, \alpha)$.) $\qquad\square$

Remark 7.80. The proof made no use of the assumption that Δ is spherical. But as we will see, the definition of the interval (α, β) that we have given is suitable only in the spherical case. We will return to nonspherical buildings in the next chapter and prove the appropriate analogue of the proposition (see Lemma 8.58).

7.7.4 The Structure of $[\alpha, \beta]$

This subsection is intended to provide some intuition for the notion of "interval" that occurred in the commutator relations. We start with the case that

Σ is the standard rank-2 Coxeter complex of type $I_2(m)$, which we identify with a $2m$-gon. It has $2m$ roots, each of which can be traced out by starting at a vertex and going halfway around in the clockwise direction. Since there is a cyclic order on the starting vertices, this imposes a cyclic order on the roots, which we number as $\alpha_1, \alpha_2, \ldots, \alpha_{2m}$. The choice of which one to call α_1 is arbitrary, and the subscripts $1, 2, \ldots, 2m$ should be thought of as integers mod $2m$; thus $\alpha_{i+m} = -\alpha_i$ for each i. See Figure 7.12 for the case $m = 3$. Consider now a pair of roots α, β with $\beta \neq \pm\alpha$. Interchanging α and β if

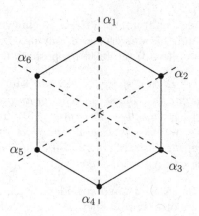

Fig. 7.12. The roots of a rank-2 Coxeter complex.

necessary, we may assume that $\alpha = \alpha_i$ for some i and that $\beta = \alpha_{i+l}$, with $0 < l < m$. It is then easy to check that the interval $[\alpha, \beta]$ consists of the cyclically consecutive roots $\alpha_i, \alpha_{i+1}, \ldots, \alpha_{i+l}$.

In the general case, we will reduce the calculation of $[\alpha, \beta]$ to the rank-2 case, as follows. Recall first that every maximal simplex A of $\partial\alpha \cap \partial\beta$ has codimension 2 in Σ. This is obvious from the point of view of finite reflection groups, and we have also given a combinatorial proof in Lemma 3.164. [The hypothesis of that lemma is automatically satisfied here because spherical Coxeter complexes have no nested roots by Lemma 3.53.]

Choose a maximal simplex $A \in \partial\alpha \cap \partial\beta$, and consider the link $L_A := \mathrm{lk}_\Sigma A$. Recall that the walls H' of L_A are in 1–1 correspondence with the walls H of Σ containing A, via $H' = H \cap L_A$. Similarly, the roots γ' of L_A correspond to the roots γ of Σ such that $A \in \partial\gamma$, via $\gamma' = \gamma \cap L_A$. (See Proposition 3.79.) We can now, as promised, reduce the calculation of $[\alpha, \beta]$ to the rank-2 case:

Lemma 7.81. *Let α and β be roots of Σ such that $\alpha \neq \pm\beta$, let A be a maximal simplex of $\partial\alpha \cap \partial\beta$, and let $\alpha' := \alpha \cap L_A$ and $\beta' := \beta \cap L_A$ be the roots of $L_A := \mathrm{lk}_\Sigma A$ corresponding to α and β. Then there is a bijection*

$$[\alpha, \beta] \xrightarrow{\sim} [\alpha', \beta']$$

given by $\gamma \mapsto \gamma' := \gamma \cap L_A$ *for* $\gamma \in [\alpha, \beta]$.

Proof. Given $\gamma \in [\alpha, \beta]$, we have $\gamma \supseteq \alpha \cap \beta$. Applying the opposition involution, we see that also $-\gamma \supseteq (-\alpha) \cap (-\beta)$; hence $\partial \gamma \supseteq \partial \alpha \cap \partial \beta$. In particular, $A \in \partial \gamma$, so γ corresponds to a root $\gamma' := \gamma \cap L_A$. We have $\gamma' \supseteq \alpha' \cap \beta'$, so $\gamma' \in [\alpha', \beta']$. It remains to show that if we start with an arbitrary $\gamma' \in [\alpha', \beta']$, then the corresponding root γ of Σ is in $[\alpha, \beta]$. Thus we are given that $\gamma' \supseteq \alpha' \cap \beta'$, and we must show that $\gamma \supseteq \alpha \cap \beta$. Since $\alpha \cap \beta$ is a chamber subcomplex of Σ [why?], it suffices to note that γ contains every chamber $D \in \alpha \cap \beta$; this follows at once from the description of γ in terms of γ' given in Lemma 3.162. $\qquad\square$

7.8 RGD Systems of Spherical Type

We have seen that a Moufang building gives rise to an automorphism group $G = \langle U_\alpha \mid \alpha \in \Phi \rangle$ with certain special properties. We now reverse the process and look for group-theoretic axioms that allow one to construct a Moufang building. The resulting structure is called an RGD system. ("RGD" stands for "root group data.") One can think of this as a refinement of a BN-pair. Thus a BN-pair gives us a building Δ on which a group G acts, and an RGD system gives us in addition the root groups, enabling us to prove that Δ is Moufang. Before we can talk about "root groups" in an abstract setting, we need roots. So we start with a Coxeter system.

Throughout this section we denote by (W, S) a Coxeter system with W finite. Set $\Sigma := \Sigma(W, S)$, and identify $\mathcal{C}(\Sigma)$ with W, so that the fundamental chamber is the element $1 \in W$. As before, we denote by Φ the set of roots of Σ and by Φ_+ (resp. Φ_-) the set of positive (resp. negative) roots. Thus

$$\Phi_+ = \{\alpha \in \Phi \mid 1 \in \alpha\}$$

and

$$\Phi_- = \{\alpha \in \Phi \mid 1 \notin \alpha\} = \{\alpha \in \Phi \mid w_0 \in \alpha\} \ ,$$

where w_0 is the longest element of W. Recall, finally, that we have simple roots α_s $(s \in S)$, characterized by

$$\mathcal{C}(\alpha_s) = \{w \in W \mid l(sw) > l(w)\} \ .$$

See equation (3.2) in Section 3.4.1 or Definition 5.79 in Section 5.5.4.

7.8.1 The RGD Axioms

The following definition is taken, with slight modifications, from Tits [261]. Suppose we are given a triple $\big(G, (U_\alpha)_{\alpha \in \Phi}, T\big)$ consisting of a group G, a family of subgroups U_α, and a subgroup T. To avoid complicated subscripts in what follows, we will often write U_s instead of U_{α_s} and U_{-s} instead of $U_{-\alpha_s}$

for $s \in S$. We will also use the notation $U^* := U \smallsetminus \{1\}$ for any group U, and we will use the notion of an interval of roots (Definition 7.71). Finally, we continue to use the notation

$$U_\Psi := \langle U_\alpha \mid \alpha \in \Psi \rangle$$

for any subset $\Psi \subseteq \Phi$. (In practice, we will use this notation only when Ψ is convex in the sense of Definition 7.13.)

Definition 7.82. The triple $\big(G, (U_\alpha)_{\alpha \in \Phi}, T\big)$ is called an *RGD system of type* (W, S) if the following conditions are satisfied:

(RGD0) *For all $\alpha \in \Phi$, $U_\alpha \neq \{1\}$.*

(RGD1) *For all $\alpha, \beta \in \Phi$ with $\beta \neq \pm\alpha$,*

$$[U_\alpha, U_\beta] \leq U_{(\alpha,\beta)} := \langle U_\gamma \mid \gamma \in (\alpha, \beta) \rangle .$$

(RGD2) *For every $s \in S$ there is a function $m \colon U_s^* \to G$ such that for all $u \in U_s^*$ and $\alpha \in \Phi$,*

$$m(u) \in U_{-s} u U_{-s} \quad and \quad m(u) U_\alpha m(u)^{-1} = U_{s\alpha} .$$

Moreover, $m(u)^{-1} m(v) \in T$ for all $u, v \in U_s^*$.

(RGD3) *For all $s \in S$,*
$$U_{-s} \not\leq U_+ ,$$

where $U_\pm := \langle U_\alpha \mid \alpha \in \Phi_\pm \rangle$.

(RGD4) $G = T \langle U_\alpha \mid \alpha \in \Phi \rangle$.

(RGD5) *T normalizes U_α for each $\alpha \in \Phi$, i.e.,*

$$T \leq \bigcap_{\alpha \in \Phi} N_G(U_\alpha) .$$

Example 7.83. Let Δ be a strictly Moufang building with fundamental apartment Σ_0 and fundamental chamber $C_0 \in \Sigma_0$. We identify Σ_0 with $\Sigma(W, S)$, where W is the group of type-preserving automorphisms of Σ_0 and S is the set of reflections with respect to the walls of C_0. The set of roots of Σ_0 is then identified with $\Phi = \Phi(\Sigma(W, S))$. For each $\alpha \in \Phi$ let $U_\alpha \leq \operatorname{Aut} \Delta$ be the corresponding root group. Let $G := \langle U_\alpha \mid \alpha \in \Phi \rangle \leq \operatorname{Aut} \Delta$, and define T, as usual, by
$$T := \operatorname{Fix}_G(\Sigma) .$$

Then $\big(G, (U_\alpha)_{\alpha \in \Phi}, T\big)$ is an RGD system of type (W, S). Axiom (RGD0) follows from the fact that Δ is thick, and (RGD1) is Proposition 7.72. The first part of (RGD2) follows from Corollary 7.6 and Lemma 7.25(1). Since $m(u)$ induces the reflection s_α on Σ_0, the second part of (RGD2) follows from the definition of T. To prove (RGD3), note that U_+ fixes C_0 but U_{-s} does not. Finally, (RGD4) and (RGD5) are true by definition.

Remark 7.84. We will see in Section 7.8.5 that as a consequence of our axioms, one always has

$$T = \bigcap_{\alpha \in \Phi} N_G(U_\alpha) . \tag{7.12}$$

So we could get a simpler but equivalent set of axioms by deleting (RGD5) and the second assertion of (RGD2) and then *defining* T by (7.12). The only appearance of T in the axioms would then be in (RGD4). This is precisely what Tits does in [261]. We have chosen our definition, however, for two reasons. First, we find it interesting that under our axioms, T is uniquely determined by G and the U_α. Second, our axioms seem slightly more natural in connection with examples. One typically has a canonical choice of T (often a torus), and it is usually easy to check (RGD5), but it may not be obvious a priori that (7.12) holds.

Our goal for the rest of this section is to show that the subgroups $B_+ := TU_+$ and $N := \langle T, \{m(u) \mid u \in U_s^*, s \in S\} \rangle$ constitute a BN-pair in G with Weyl group $N/(B_+ \cap N) = N/T \cong W$. We will then get a thick spherical building $\Delta = \Delta(G, B_+)$, which we will prove to be Moufang. Moreover, the U_α will be the root groups of Δ if the Coxeter diagram of (W, S) has no isolated nodes. Roughly speaking, then, all RGD systems arise as in Example 7.83. (This is not quite true, because the given G in an RGD system need not be generated by the U_α in general and need not act faithfully on Δ. We will analyze the kernel of the action in Proposition 7.127. Also, one has to deal with the possibility of isolated nodes in the Coxeter diagram.)

The first step is to work out the algebraic theory in the rank-1 case. This is not interesting from the point of view of buildings, but we will find it useful to apply the rank-1 results to the groups $G_s := T\langle U_s, U_{-s} \rangle$ $(s \in S)$ in the general case.

7.8.2 Rank-1 Groups

In this subsection we derive some consequences of the axioms in the simplest possible case, in which (W, S) has rank 1 (i.e., $|S| = 1$). The treatment that follows uses ideas from Bruhat–Tits [59, Section 6].

A rank-1 Coxeter complex has exactly two roots $\pm\alpha$, and we will write $U_\pm$ instead of $U_{\pm\alpha}$. Thus a rank-1 RGD system consists of a group G and subgroups U_+, U_-, T satisfying the following axioms (where we omit (RGD1), which is vacuous in the rank-1 case):

(RGD0) $U_\pm \neq \{1\}$.

(RGD2) *There is a function* $m: U_+^* \to G$ *such that* $m(u) \in U_- u U_-$ *and*

$$m(u) U_\pm m(u)^{-1} = U_\mp .$$

for all $u \in U_+^*$. *Moreover,* $m(u)^{-1} m(v) \in T$ *for all* $u, v \in U_+^*$.

(RGD3) $U_- \nleq U_+$.

(RGD4) $G = T\langle U_+, U_- \rangle$.

(RGD5) $T \leq N_G(U_+) \cap N_G(U_-)$.

We note for future reference that in the present rank-1 setup, we can replace (RGD3) by the weaker axiom

(RGD3′) $U_- \neq U_+$.

Indeed, if we make this replacement, then (RGD3) is an easy consequence; see Remark 7.86 below.

Example 7.85. Let $G = \mathrm{SL}_2(k)$, where k is a field. Let U_+ be the strict upper-triangular subgroup, let U_- be the strict lower-triangular subgroup, and let T be the diagonal subgroup. It is easy to verify the axioms, the only interesting one being (RGD2). Here one wants $m(u)$ to be a monomial matrix of the form $\left(\begin{smallmatrix} 0 & * \\ * & 0 \end{smallmatrix} \right)$ so that it will interchange $U_\pm$ by conjugation. In order to find such a monomial matrix in $U_- u U_-$, one needs to perform suitable row and column operations on u. The result of the calculation is that if $u = \left(\begin{smallmatrix} 1 & \lambda \\ 0 & 1 \end{smallmatrix} \right)$ with $\lambda \in k^*$, then we can take

$$m(u) := \begin{pmatrix} 0 & \lambda \\ -\lambda^{-1} & 0 \end{pmatrix} = \begin{pmatrix} 1 & 0 \\ -\lambda^{-1} & 1 \end{pmatrix} \begin{pmatrix} 1 & \lambda \\ 0 & 1 \end{pmatrix} \begin{pmatrix} 1 & 0 \\ -\lambda^{-1} & 1 \end{pmatrix} \in U_- u U_- .$$

$$(7.13)$$

We now derive 15 consequences of the axioms. The reader might find it helpful to keep Example 7.85 in mind while working through these consequences.

(1) For any $v \in U_-^*$ there is an element $m(v) \in U_+ v U_+$ such that

$$m(v) U_\pm m(v)^{-1} = U_\mp .$$

Proof. Set $m := m(u_0)$ for some fixed $u_0 \in U_+^*$; such a u_0 exists by (RGD0). Since $mU_+m^{-1} = U_-$, we can write $v = mum^{-1}$ with $u \in U_+^*$. Now set $m(v) := mm(u)m^{-1} \in m(U_- u U_-)m^{-1} = U_+ v U_+$. Then conjugation by $m(v)$ interchanges U_+ and U_-, since $m(v)$ is a product of an odd number of elements that interchange U_+ and U_-. □

(2) $U_+ \cap N_G(U_-) = \{1\}$ and $U_- \cap N_G(U_+) = \{1\}$. In particular, $U_+ \cap U_- = \{1\}$.

Proof. If there were an element $u \in U_+^* \cap N_G(U_-)$, then we would have $m(u) \in U_- u U_- \subseteq N_G(U_-)$. But $m(u)$ conjugates U_- to U_+, and these two subgroups are distinct by (RGD3′). This contradiction proves the first assertion, and the proof of the second is similar (using (1)). □

Remark 7.86. Note that we have used only (RGD0), (RGD2), and (RGD3′) so far, but (RGD3), and even a stronger version of it, follows from (2). This proves our claim above that (RGD3) can be replaced by (RGD3′).

(3) Given $u \in U_+^*$, there exist unique elements $u_1, u_2 \in U_-$ such that $(u_1 u u_2) U_\pm (u_1 u u_2)^{-1} = U_\mp$. In particular, $m(u)$ is the unique element of $U_- u U_-$ that interchanges U_+ and U_- by conjugation. Moreover, $u_1, u_2 \neq 1$.

Proof. Suppose $m = u_1 u u_2$ and $n = v_1 u v_2$ both interchange $U_\pm$, where $u_1, u_2, v_1, v_2 \in U_-$. We have $u = u_1^{-1} m u_2^{-1} = v_1^{-1} n v_2^{-1}$, which implies

$$v_1 u_1^{-1} = n v_2^{-1} u_2 m^{-1} = \left(n v_2^{-1} u_2 n^{-1} \right) \left(n m^{-1} \right).$$

The left side is in U_-, the first factor on the right is in U_+, and the second factor on the right normalizes both U_+ and U_-. So $v_1 u_1^{-1} \in U_- \cap N_G(U_+)$, and hence $v_1 = u_1$ by (2). We now have $n v_2^{-1} u_2 n^{-1} \in U_+ \cap N_G(U_-)$, and we again apply (2) to conclude that $v_2 = u_2$. This completes the uniqueness proof. To show $u_1, u_2 \neq 1$, suppose for instance that $u_1 = 1$. Then $u u_2 U_- u_2^{-1} u^{-1} = U_+$. Conjugating by u^{-1}, we conclude that $U_- = U_+$, contradicting (RGD3). Hence $u_1 \neq 1$. A similar argument shows that $u_2 \neq 1$. □

(4) $m(u^{-1}) = m(u)^{-1}$ for all $u \in U_+^*$.

Proof. We have $m(u)^{-1} \in U_- u^{-1} U_-$, and $m(u)^{-1}$ interchanges U_+ and U_- by conjugation. Now use the uniqueness of $m(u^{-1})$ proved in (3). □

(5) $m(t u t^{-1}) = t m(u) t^{-1}$ for all $u \in U_+^*$, $t \in T$.

Proof. As in the proof of (4), note that $t m(u) t^{-1}$ is in $U_- (t u t^{-1}) U_-$ and interchanges $U_\pm$ by conjugation. □

(6) $m(u) \in N_G(T)$ for all $u \in U_+^*$.

Proof. Using (5), we obtain

$$m(u)^{-1} t m(u) = m(u)^{-1} (t m(u) t^{-1}) t = m(u)^{-1} m(t u t^{-1}) t$$

for all $t \in T$. The right side is in T by the second assertion of (RGD2), so $m(u)^{-1} T m(u) \leq T$. Replacing u by u^{-1} and using (4), we get the opposite inclusion. □

Remark 7.87. At this point we can show that the concept of "rank-1 RGD system" is completely symmetric. In other words, if we set $U_+' := U_-$ and $U_-' := U_+$, then $\left(G, (U_\pm'), T \right)$ is still an RGD system. Indeed, we already noted above that (RGD3) can be replaced by the symmetric axiom (RGD3′). The only other asymmetric axiom is (RGD2), and its symmetric counterpart follows from (1) and its proof, together with (6). Consequently, every result in the rank-1 case has an "opposite" or "dual" result with the roles of $+$ and $-$ interchanged.

(7) $m(u)T = m(v)T = m(u)^{-1}T$ for all $u, v \in U_+^*$.

Proof. The first equation follows from (RGD2). Now replace v by u^{-1} and use (4). □

For the next few statements we fix $u_0 \in U_+^*$ and set $m := m(u_0)$.

(8) $U_+^* \subseteq U_-^* m U_-^* T$.

Proof. Given $u \in U_+^*$, we have $u \in U_-^* m(u) U_-^*$ by (RGD2) and the last assertion of (3). Since $m(u) \in mT$ by (RGD2), this implies $u \in U_-^* mTU_-^* = U_-^* m U_-^* T$ [recall that T normalizes U_-]. □

(9) $U_-^* \subseteq U_+^* m U_+^* T$.

Proof. Conjugate (8) by m and use (6). □

(10) $U_+^* mTU_+ = U_-^* TU_+$.

Proof. We have

$$\begin{aligned}
U_+^* mTU_+ &\subseteq U_-^* m U_-^* TmTU_+ && \text{by (8)} \\
&= U_-^* m U_-^* mTU_+ && \text{because } Tm = mT \text{ by (6)} \\
&= U_-^* m U_-^* m^{-1} TU_+ && \text{by (7)} \\
&= U_-^* U_+^* TU_+ && \\
&= U_-^* TU_+ && \text{because } U_+^* T = TU_+^*.
\end{aligned}$$

This proves one inclusion. For the other, we have

$$\begin{aligned}
U_-^* TU_+ &\subseteq U_+^* m U_+^* TTU_+ && \text{by (9)} \\
&= U_+^* m U_+^* TU_+ && \\
&= U_+^* mTU_+^* U_+ && \\
&= U_+^* mTU_+ .
\end{aligned}$$
□

(11) Set $B_+ := TU_+$. Then

$$G = B_+ \cup U_+ m B_+ .$$

Proof. Since $U_+ = \{1\} \cup U_+^*$, we have

$$U_+ m B_+ = m B_+ \cup U_+^* m B_+ = m B_+ \cup U_-^* B_+ ,$$

where the second equation comes from (10). Taking the union with B_+ and writing $U_- = \{1\} \cup U_-^*$, we obtain

$$B_+ \cup U_+ m B_+ = m B_+ \cup U_- B_+ . \tag{7.14}$$

Now set $G' := B_+ \cup U_+ m B_+$. Note that if we left-multiply the right side of (7.14) by m, we get the left side. (Use (7) to see that $m^2 B_+ = m^2 T U_+ = T U_+ = B_+$. And note that $m U_- = m U_- m^{-1} m = U_+ m$.) Hence G' is closed under left multiplication by m. Moreover, G' is also closed under left multiplication by T and U_+. (To see that $U_+ m B_+$ is closed under left multiplication by T, note that $T U_+ m B_+ = U_+ T m B_+ = U_+ m T B_+ = U_+ m B_+$.) Now $G = \langle T, U_+, U_- \rangle$ by (RGD4); hence also $G = \langle T, U_+, m \rangle$. So G' is closed under left multiplication by G, which implies that $G' = G$. $\qquad\square$

(12) $\{1\} \amalg U_+ m$ is a set of representatives for G/B_+.

Proof. Note first that U_+ is a normal subgroup of B_+ and $m \notin N_G(U_+)$, so $m \notin B_+$. The union in (11) is therefore a disjoint union. It remains to show that distinct elements of $U_+ m$ represent different cosets of B_+. Suppose $u_1 m B_+ = u_2 m B_+$ for some $u_1, u_2 \in U_+$. Then $m^{-1} u_2^{-1} u_1 m \in B_+ \cap U_- \leq N_G(U_+) \cap U_- = \{1\}$; hence $u_1 = u_2$. $\qquad\square$

(13) $B_+ = N_G(U_+)$.

Proof. We know that $B_+ \leq N_G(U_+)$. If this were a proper inclusion, then $N_G(U_+)$ would meet $U_+ m B_+$, contradicting the fact that $m \notin N_G(U_+)$. $\qquad\square$

(14) Set $B_- := T U_-$. Then $B_- = N_G(U_-)$.

Proof. Conjugate (13) by m. Alternatively, use Remark 7.87. $\qquad\square$

(15) $T = N_G(U_+) \cap N_G(U_-)$.

Proof. By (13) and (14), $N_G(U_+) \cap N_G(U_-) = B_+ \cap B_- = T(U_+ \cap B_-)$. Now note that $U_+ \cap B_- = \{1\}$ by (2). $\qquad\square$

Remark 7.88. The most important of these 15 properties in what follows are (11) and (15). But of course, the others (with the exception of (12)) were needed along the way.

7.8.3 The Weyl Group

We return now to an arbitrary RGD system $(G, (U_\alpha)_{\alpha \in \Phi}, T)$ of spherical type (W, S). We define a subgroup $N \leq G$ by

$$N := \langle T, \{ m(u) \mid u \in U_s^*, s \in S \} \rangle .$$

Note that each generator $m(u)$ normalizes T by statement (6) of the previous section applied to the rank-1 group $G_s := T \langle U_s, U_{-s} \rangle$. Hence T is a normal subgroup of N. Our goal in this subsection is to prove that $N/T \cong W$.

The first step is to define an action of N on Φ. We have a conjugation action of N on the set of subgroups $\{ U_\alpha \mid \alpha \in \Phi \}$ by (RGD2) and (RGD5), so we will get an action of N on Φ if we show that this set of subgroups is in 1–1 correspondence with Φ. In other words, we need to show the following:

Lemma 7.89. *If α and β are distinct roots of Σ, then $U_\alpha \neq U_\beta$.*

The proof will use a simple lemma that could have been proved in Chapter 1:

Lemma 7.90. *If α and β are distinct roots of Σ, then there are elements $w \in W$ and $s \in S$ such that $w\alpha = -\alpha_s$ and $w\beta \in \Phi_+$.*

Proof. We claim there are adjacent chambers C', D' with $C' \in \beta \smallsetminus \alpha$ and $D' \in \alpha$. Assuming the claim, let $w \in W$ be the element such that wC' is the fundamental chamber 1. [Recall that $\mathcal{C}(\Sigma) = W$.] Then $w\beta$ contains 1, so $w\beta \in \Phi_+$. And $w\alpha$ does not contain 1 but it contains a chamber adjacent to 1, so $w\alpha = -\alpha_s$ for some $s \in S$. It remains to prove the claim.

If $\alpha = -\beta$, we can take any pair of adjacent chambers C', D' with $C' \in \beta$ and $D' \in -\beta = \alpha$. Otherwise, we have $\alpha \not\subseteq -\beta$ by Lemma 3.53, so $\alpha \cap \beta$ contains at least one chamber. We also have $\beta \not\subseteq \alpha$, so there is a minimal gallery that starts in $\beta \smallsetminus \alpha$ and ends in $\alpha \cap \beta$. This gallery must be entirely contained in β by convexity of the latter, so it contains a pair of adjacent chambers $C', D' \in \beta$ such that $C' \not\subseteq \alpha$ but $D' \in \alpha$. This proves the claim. $\square$

Proof of Lemma 7.89. Choose $w \in W$ and $s \in S$ as in Lemma 7.90, and write $w = s_1 \cdots s_l$ with $s_i \in S$ for $1 \leq i \leq l$. Choose $u_i \in U_{s_i}^*$ for $1 \leq i \leq l$, and set $\tilde{w} := m(u_1) \cdots m(u_l)$. Then $\tilde{w} U_\alpha \tilde{w}^{-1} = U_{-s}$ and $\tilde{w} U_\beta \tilde{w}^{-1} = U_\gamma$, where $\gamma = w\beta \in \Phi_+$. We are therefore reduced to showing $U_{-s} \neq U_\gamma$. But this follows from (RGD3), since $U_\gamma \leq U_+$. $\square$

We now have a well-defined action of N on Φ, given by a homomorphism $\nu \colon N \to \operatorname{Sym}\Phi$, where the latter is the symmetric group of Φ, i.e., the group of permutations of Φ. The action is characterized by

$$nU_\alpha n^{-1} = U_{\nu(n)\alpha} \tag{7.15}$$

for $n \in N$ and $\alpha \in \Phi$. In view of (RGD5), the action of T is trivial. We also have, of course, the canonical action of W on Φ. This action is easily seen to be faithful. [For example, we could identify Φ with a generalized root system in a vector space on which W acts as a finite reflection group, in which case we verified the faithfulness of the action in the proof of Lemma 1.4.] Identifying W with its image in $\operatorname{Sym}\Phi$, we conclude from (RGD2) that the image of ν is precisely W. So we can view ν as a surjection $N \twoheadrightarrow W$. It induces a surjection

$$\bar{\nu} \colon N/T \twoheadrightarrow W .$$

We wish to prove that $\bar{\nu}$ is an isomorphism or, equivalently, that $T = \ker \nu$. A key step in the argument is an application of the rank-1 theory to subgroups of the form $G_s := T\langle U_s, U_{-s} \rangle$ for $s \in S$ to obtain the following result:

Lemma 7.91. $\ker \nu \cap G_s = T$ *for each* $s \in S$.

Proof. It is immediate from (7.15) that $\ker \nu = N \cap \bigcap_{\alpha \in \Phi} N_G(U_\alpha)$. Hence

$$\ker \nu \cap G_s \leq N_{G_s}(U_s) \cap N_{G_s}(U_{-s}) = T$$

by statement (15) in Section 7.8.2. The opposite inclusion is trivial.   □

We can now prove the main result of this subsection, which is a step toward the construction of a BN-pair in G:

Lemma 7.92. $\bar{\nu}$ *is an isomorphism. In other words,* $T = \ker \nu$.

Proof. For each $s \in S$, choose $u \in U_s^*$ and set $\tilde{s} := m(u) \in N$. Using both parts of (RGD2), we see that the coset $\tilde{s}T$ is independent of the choice of u and that $\bar{\nu}(\tilde{s}T) = s$. Moreover, the elements $\tilde{s}T$ for $s \in S$ generate N/T. It therefore suffices to show that the $\tilde{s}T$ satisfy the Coxeter relations defining W; equivalently, we have to show that $(\tilde{s}\tilde{t})^{m(s,t)} \in T$ for all $s, t \in S$, where $\big(m(s,t)\big)_{s,t \in S}$ is the Coxeter matrix of (W, S). Indeed, once we have shown this, the Coxeter presentation for W will give us a well-defined homomorphism $\mu \colon W \to N/T$ such that $\mu(s) = \tilde{s}T$ for all $s \in S$, and it is immediate that μ (once we know it exists) is inverse to $\bar{\nu}$.

We already know that $\tilde{s}^2 \in T$ for all $s \in S$ by statement (7) in Section 7.8.2 applied to G_s. So suppose $s \neq t$, and set $m := m(s,t)$. The trick for proving the Coxeter relation, following Tits [260, 5.4, proof of Lemma 3(iii)], is to rewrite that relation in terms of conjugation so that we can make effective use of equation (7.15).

Consider first the Coxeter relation $(st)^m = 1$ in W. This can be written as

$$st \cdots s = t \, ,$$

where the left side is an alternating product of $2m - 1$ factors. The middle letter is s if m is odd and t if m is even. If we denote by w the alternating product $st \cdots$ of length $m - 1$, the relation becomes

$$wxw^{-1} = t \, ,$$

where $x = s$ or t, depending on whether m is odd or even. This implies

$$w\alpha_x = \pm \alpha_t \, , \qquad \qquad (7.16)$$

and in fact the sign is $+$, although we do not really need this. [xw^{-1} is an alternating product of length m, so $l(xw^{-1}) > l(w^{-1}) \implies w^{-1} \in \alpha_x \implies 1 \in w\alpha_x \implies w\alpha_x \in \Phi_+$.] Letting $n \in N$ be the alternating product $\tilde{s}\tilde{t}\cdots$ of length $m - 1$, the relation to be proved becomes

$$n\tilde{x}n^{-1} \equiv \tilde{t} \bmod T \, . \qquad \qquad (7.17)$$

Since $\nu(n) = w$, equations (7.15) and (7.16) imply that $nU_{\pm x}n^{-1} = U_{\pm t}$; hence $nG_xn^{-1} = G_t$. In particular, $n\tilde{x}n^{-1} \in G_t$. We also know that $\nu(n\tilde{x}n^{-1}) = \nu(\tilde{t})$. We can therefore apply Lemma 7.91 to conclude that (7.17) holds.   □

We close this subsection with two further results that we can now prove and that will be needed later. Recall that $B_\pm := TU_\pm$.

Lemma 7.93. $N \cap N_G(U_+) = T$, and hence $N \cap B_+ = T$.

Proof. It suffices to show that

$$N \cap N_G(U_+) \le T = \ker \nu .$$

Given $n \in N \cap N_G(U_+)$, set $w := \nu(n)$. For any $\alpha \in \Phi_+$, we then have

$$U_{w\alpha} = nU_\alpha n^{-1} \le U_+ .$$

In view of (RGD3), it follows that we cannot have $w\alpha = -\alpha_s$ with $s \in S$. Hence $w = 1$ by Exercise 7.97 below, and we have proved $N \cap N_G(U_+) \le \ker \nu$. $\qquad\square$

Corollary 7.94. $U_{-s} \cap B_+ = \{1\}$ for all $s \in S$.

Proof. Suppose there exists an element $v \in U_{-s}^* \cap B_+$, and consider the element $m(v) \in U_s v U_s$ that interchanges $U_{\pm s}$ by conjugation; see Section 7.8.2, statement (1). The construction of $m(v)$ shows that it is in N; hence it is in $N \cap B_+ = T$. But this contradicts the fact that it does not normalize $U_{\pm s}$. $\qquad\square$

Remark 7.95. This is a good place to point out that the theory of RGD systems is completely symmetric with respect to $+$ and $-$. More precisely, suppose we are given an RGD system $(G, (U_\alpha)_{\alpha \in \Phi}, T)$. We define a new system of groups by setting $U'_\alpha := U_{-\alpha}$ for $\alpha \in \Phi$. The claim, then, is that $(G, (U'_\alpha)_{\alpha \in \Phi}, T)$ is again an RGD system. Axioms (RGD0), (RGD1), (RGD4), and (RGD5) are clear. [One has to use the opposition involution in the case of (RGD1).] And (RGD2) follows from the results in Section 7.8.2; see Remarks 7.86 and 7.87. Finally, we can deduce (RGD3) for the new system from (RGD3) for the original system by conjugating by the longest element w_0 or, more precisely, by any representative of w_0 in N. One needs to note here that the action of w_0 on Σ interchanges the positive and negative roots and permutes the walls of the fundamental chamber, so that $w_0 U_+ w_0^{-1} = U_-$ and $\{w_0 U_{-s} w_0 \mid s \in S\} = \{U_s \mid s \in S\}$.

Exercises

7.96. The purpose of this exercise is to show that Lemma 7.90, which was proved here for spherical Coxeter complexes, is not true for general Coxeter complexes.

(a) Give an example of a Coxeter complex Σ that contains roots α, β such that $\alpha \cap \beta$ contains no chamber but $\alpha \ne -\beta$.
(b) Let $\Sigma = \Sigma(W, S)$ for some Coxeter system (W, S). If α and β are roots of Σ for which the conclusion of Lemma 7.90 holds, show that either $\alpha = -\beta$ or $\alpha \cap \beta$ contains a chamber.

(c) Deduce that Lemma 7.90 does not hold for Σ as in (a).

7.97. Let (W, S) be an arbitrary Coxeter system, let $\Sigma = \Sigma(W, S)$, and let Φ be its set of roots. For any $w \neq 1$ in W, show that there exist $\alpha \in \Phi_+$ and $s \in S$ such that $w\alpha = -\alpha_s$.

7.8.4 The Groups U_w

As in the previous subsection, we choose for each $s \in S$ an element $u \in U_s^*$ and set $\tilde{s} := m(u)$. We omit the tilde in expressions that do not depend on the choice of u, as in the following lemma.

Lemma 7.98. $B_+ s B_+ = U_s s B_+$ for any $s \in S$.

Proof. We have $B_+ s B_+ = U_+ T s B_+ = U_+ s T B_+ = U_+ s B_+$, so we must show that $U_+ s B_+ = U_s s B_+$. In view of the definition of U_+, it suffices to show that $U_\alpha U_s s B_+ \subseteq U_s s B_+$ for all $\alpha \in \Phi_+$. This is clear if $\alpha = \alpha_s$, so assume $\alpha \neq \alpha_s$. We also have $\alpha \neq -\alpha_s$, so we may apply the commutator relation (RGD1) to get

$$[U_\alpha, U_s] \leq \langle U_\gamma \mid \gamma \in (\alpha, \alpha_s) \rangle \leq \langle U_\gamma \mid \gamma \in \Phi_+ \smallsetminus \{\alpha_s\} \rangle .$$

Recall now that α_s is the only positive root that is mapped by s to a negative root. Hence $s\gamma \in \Phi_+$ for each γ that occurs in the commutator relation above. Using this fact, we can complete the proof with the following calculation:

$$\begin{aligned}
U_\alpha U_s s B_+ &\subseteq U_s U_\alpha [U_\alpha, U_s] s B_+ \\
&\subseteq U_s U_\alpha \langle U_\gamma \mid \gamma \in \Phi_+ \smallsetminus \{\alpha_s\} \rangle s B_+ \\
&= U_s \langle U_\gamma \mid \gamma \in \Phi_+ \smallsetminus \{\alpha_s\} \rangle s B_+ \\
&= U_s s \langle U_{s\gamma} \mid \gamma \in \Phi_+ \smallsetminus \{\alpha_s\} \rangle B_+ \\
&= U_s s B_+ .
\end{aligned}$$
$\square$

Corollary 7.99. $B_+ \cup B_+ s B_+$ is a subgroup of G for each $s \in S$.

Proof. It suffices to show that $s B_+ s B_+ \subseteq B_+ \cup B_+ s B_+$. By the lemma we have

$$s B_+ s B_+ = s U_s s B_+ \subseteq G_s B_+ ,$$

where, as usual, $G_s := T \langle U_s, U_{-s} \rangle$. We can now apply statement (11) from Section 7.8.2 to obtain $G_s = T U_s \cup U_s s T U_s$, so that

$$G_s B_+ = (T U_s \cup U_s s T U_s) B_+ = B_+ \cup U_s s B_+ = B_+ \cup B_+ s B_+ .$$
$\square$

Definition 7.100. For $w \in W$ we set

$$\begin{aligned}
\Phi(w) &:= \{\alpha \in \Phi_+ \mid w \notin \alpha\} \\
&= \{\alpha \in \Phi \mid 1 \in \alpha, w \notin \alpha\} \\
&= \{\alpha \in \Phi \mid 1 \in \alpha, w \in -\alpha\} .
\end{aligned}$$

We also set

$$U_w := U_{\Phi(w)} .$$

Note that if we think of 1 and w as chambers of Σ, then $\Phi(w)$ coincides with the set $\Phi(1, w)$ that occurred in Example 7.15. We record for ease of reference some simple properties of $\Phi(-)$.

Lemma 7.101. *Fix $w \in W$.*

(1) *If $\alpha, \beta \in \Phi(w)$, then $[\alpha, \beta] \subseteq \Phi(w)$.*

(2) *Given a reduced decomposition $w = s_1 \cdots s_n$, set $w_i := s_1 \cdots s_i$ for $i = 0, \ldots, n$, and set $\alpha_i := w_{i-1}\alpha_{s_i}$ for $i = 1, \ldots, n$. Then $\Phi(w) = \{\alpha_1, \ldots, \alpha_n\}$.*

(3) *If $s \in S$ and $l(sw) > l(w)$, then $\Phi(sw) = \{\alpha_s\} \cup s\Phi(w)$. In particular, $s\Phi(w) \subseteq \Phi_+$.*

Proof. (1) This follows easily from the convexity of $[\alpha, \beta]$; see Exercise 7.75.

(2) Consider the minimal gallery $\Gamma \colon 1 = w_0, \ldots, w_n = w$. Since α_{s_i} is the root containing 1 but not s_i, we can multiply by w_{i-1} to see that α_i is the root containing w_{i-1} but not w_i for $1 \leq i \leq n$. (2) now follows from the observations made in Example 7.15.

(3) Choose a reduced decomposition $w = s_1 \cdots s_n$ as in (2). Then we have a reduced decomposition $sw = ss_1 \cdots s_n$, and (3) follows at once from (2). Alternatively, check (3) directly from Definition 7.100 (see Exercise 7.109). $\square$

Lemma 7.102. *Fix $w \in W$.*

(1) *Given a reduced decomposition $w = s_1 \cdots s_n$, let $\alpha_1, \ldots, \alpha_n$ be as in part (2) of Lemma 7.101. Then $U_w = U_{\alpha_1} \cdots U_{\alpha_n}$.*

(2) *If w_1 and w_2 are elements of W such that $l(w_1 w_2) = l(w_1) + l(w_2)$, then $U_{w_1 w_2} = U_{w_1} w_1 U_{w_2} w_1^{-1}$. In particular, $U_{sw} = U_s s U_w s^{-1}$ for $s \in S$ and $w \in W$ with $l(sw) > l(w)$.*

Proof. (1) We argue by induction on $n = l(w)$. Assuming, as we may, that $w \neq 1$, set $w' := s_1 \cdots s_{n-1}$ and $s := s_n$. By Lemma 7.101, $\Phi(w) = \Phi(w') \cup \{\alpha_n\}$, with $\alpha_n = w'\alpha_s$, so $U_w = \langle U_{w'}, U_{\alpha_n} \rangle$. By the induction hypothesis, $U_{w'} = U_{\alpha_1} \cdots U_{\alpha_{n-1}}$, so we need to show that $U_w = U_{w'} U_{\alpha_n}$. It suffices to show that $U_{w'} U_{\alpha_n} U_{\alpha_i} \subseteq U_{w'} U_{\alpha_n}$ for $1 \leq i < n$. Since $[\alpha_n, \alpha_i] \subseteq \Phi(w)$ by Lemma 7.101, we have $(\alpha_n, \alpha_i) \subseteq \Phi(w) \smallsetminus \{\alpha_n\} = \Phi(w')$. Hence $[U_{\alpha_n}, U_{\alpha_i}] \leq U_{w'}$ by (RGD1), and

$$U_{w'} U_{\alpha_n} U_{\alpha_i} \subseteq U_{w'} [U_{\alpha_n}, U_{\alpha_i}] U_{\alpha_i} U_{\alpha_n} \subseteq U_{w'} U_{\alpha_n}.$$

(2) Choose reduced decompositions $w_1 = s_1 \cdots s_n$ and $w_2 = t_1 \cdots t_m$ with all $s_i, t_j \in S$, and combine these to get a reduced decomposition of $w_1 w_2$. Let $\alpha_1, \ldots, \alpha_n$ (resp. $\beta_1, \ldots, \beta_m$) be the roots associated to the decomposition of w_1 (resp. w_2). Then the roots associated to the decomposition of $w_1 w_2$ are

$$\alpha_1, \ldots, \alpha_n, \gamma_1, \ldots, \gamma_m$$

with $\gamma_j = w_1 \beta_j$ for $1 \leq j \leq m$. The result now follows at once from (1) and the fact that $U_{w_2(\beta)} = w_2 U_\beta w_2^{-1}$ for any root β. $\square$

Remarks 7.103. (a) In spite of the fact that every $s \in S$ has order 2 in W, it is necessary to write s^{-1} instead of s in (2); see Exercise 7.110 below.

(b) Once we have constructed a building associated to our RGD system, we will quickly get the following addendum to (1): Every $u \in U_w$ admits a unique decomposition $u = u_1 \cdots u_n$ with $u_i \in U_{\alpha_i}$ for $1 \le i \le n$; see Corollary 7.119. The interested reader can give an algebraic proof of this fact now (Exercise 7.111).

We can now prove the following generalization of Lemma 7.98 (cf. Corollary 7.68).

Lemma 7.104. $B_+ w B_+ = U_w w B_+$ *for all* $w \in W$.

Proof. We argue by induction on $l(w)$, which may be assumed > 0. It suffices to show that $B_+ w \subseteq U_w w B_+$. Writing $w = sw'$ with $l(w) = l(w') + 1$, we have

$$
\begin{aligned}
B_+ w &= B_+ s w' \\
&\subseteq U_s s B_+ w' && \text{by Lemma 7.98} \\
&\subseteq U_s s U_{w'} w' B_+ && \text{by the induction hypothesis} \\
&= U_s (s U_{w'} s^{-1}) w B_+ \\
&= U_w w B_+ && \text{by Lemma 7.102(2)} . \qquad \square
\end{aligned}
$$

We close this subsection with some elementary observations about the structure of the groups U_w. We need the following result from group theory:

Lemma 7.105. *If H is a group generated by two nilpotent normal subgroups, then H is nilpotent.*

Proof. Call the two subgroups M, N, so that $H = MN$. Let M and N be nilpotent of class c and d, respectively. We will show by induction on $c + d$ that H is nilpotent of class $\le c + d$. We may assume $c, d > 0$ (i.e., M and N are both nontrivial). The center $Z(M)$ of M is normal in H, so we may apply the induction hypothesis to $H/Z(M)$ to conclude that the latter is nilpotent of class $\le c + d - 1$. Similarly, $H/Z(N)$ is nilpotent of class $\le c + d - 1$, so $K := H/Z(M) \times H/Z(N)$ is nilpotent of class $\le c + d - 1$. Observe now that $Z(M) \cap Z(N) \le Z(H)$, so that we have canonical maps

$$
H/Z(H) \leftarrow H/(Z(M) \cap Z(N)) \hookrightarrow H/Z(M) \times H/Z(N) = K .
$$

Thus $H/Z(H)$ is a subquotient of K and is therefore also nilpotent of class $\le c + d - 1$. Hence H is nilpotent of class $\le c + d$. $\qquad \square$

Proposition 7.106. *If U_α is nilpotent for all $\alpha \in \Phi$, then U_w is nilpotent for all $w \in W$. In particular, U_+ is nilpotent.*

Proof. The last assertion follows from the first because $U_+ = U_{w_0}$. [Every wall separates 1 from w_0.] We prove the first assertion by induction on $l(w)$, which may be assumed ≥ 2. Write $w = sw't$ with $s, t \in S$ and $l(w) = l(w') + 2$. Set $w_1 := sw'$ and $w_2 := w't$. Then

$$\Phi(w) = \{\alpha_s\} \cup s\Phi(w') \cup \{w_1\alpha_t\} = \Phi(w_1) \cup s\Phi(w_2) .$$

Hence U_w is generated by U_{w_1} and $sU_{w_2}s^{-1}$, both of which are nilpotent by the induction hypothesis. In view of the lemma, it suffices to show that they are both normal in U_w. Set $\beta := w_1\alpha_t$. To show that $U_{w_1} \trianglelefteq U_w = U_{w_1}U_\beta$, it suffices to show that U_β normalizes U_{w_1}. Now for any $\gamma \in \Phi(w_1) = \Phi(w) \smallsetminus \{\beta\}$, the open interval (γ, β) is contained in $\Phi(w_1)$, so $[U_\gamma, U_\beta] \leq U_{w_1}$. This implies that U_β normalizes U_{w_1}. Similarly, setting $\alpha := \alpha_s$, (RGD1) shows that U_α normalizes $sU_{w_2}s^{-1}$, so the latter is normal in $U_w = U_\alpha sU_{w_2}s^{-1}$. $\qquad\square$

Remarks 7.107. (a) The hypothesis that the U_α are nilpotent is in fact *always* satisfied if the Coxeter diagram of (W, S) has no isolated nodes. For buildings of rank 2, this is known as a result of the classification [262, Chapter 17]; see also Tent [234] for a direct proof. The general case then follows from Exercise 7.36. In concrete examples, one simply sees this by inspection. In fact, the groups U_α are often abelian. In Section 7.9.3 we will see a conceptual explanation for the fact that U_α is nilpotent for RGD systems associated with algebraic groups.

(b) It is also true that the group $\langle U_s, U_{-s} \rangle$ is *never* nilpotent. This fact, which is not as difficult as the one quoted in (a), will be proved in Exercise 7.129 below. A case that often arises in practice is that $\langle U_s, U_{-s} \rangle$ is isomorphic to $SL_2(k)$ or $PSL_2(k)$ for some field k.

(c) The proof of Proposition 7.106 used only (RGD1) together with the following consequence of (RGD0) and (RGD2): For any $s \in S$ there exists an element $\tilde{s} \in G$ such that $\tilde{s}U_\alpha\tilde{s}^{-1} = U_{s\alpha}$ for all $\alpha \in \Phi$. In particular, the proof made no use of (RGD3). So if we are trying to verify the RGD axioms in an example in which we know that the U_α are nilpotent, and if we also happen to know that $\langle U_s, U_{-s} \rangle$ is nonnilpotent for all $s \in S$, then (RGD3) holds automatically; for if we had $U_{-s} \leq U_+$, then we would have a nilpotent group with a nonnilpotent subgroup.

Finally, we sketch how one can use the results of this subsection (including the uniqueness assertion in Remark 7.103(b)) to get a presentation for U_w. In particular, this gives a presentation for $U_+ = U_{w_0}$. The result, roughly speaking, is that the commutator relations in (RGD1) give a set of defining relations among the groups U_α that generate U_w. [In case $w = w_0$, these are the U_α with $\alpha \in \Phi_+$.] More precisely, suppose we choose generators and relations for each group U_α ($\alpha \in \Phi_+$). For each pair of distinct positive roots α, β, (RGD1) says that there are relations of the form

$$[u_\alpha, u_\beta] = v , \tag{7.18}$$

where u_α ranges over the generators of U_α, u_β ranges over the generators of U_β, and v is a word in the generators of the groups U_γ with $\gamma \in (\alpha, \beta)$.

Proposition 7.108. *Given $w \in W$, one obtains a presentation for U_w by taking generators and relations for the groups U_α with $\alpha \in \Phi(w)$ and adding the commutator relations (7.18) for $\alpha, \beta \in \Phi(w)$, $\alpha \neq \beta$.*

Sketch of proof. Choose a reduced decomposition $w = s_1 \cdots s_n$, let $\alpha_1, \ldots, \alpha_n$ be as in Lemma 7.101(2), and set $U_i := U_{\alpha_i}$ for $i = 1, \ldots, n$. In view of Remark 7.103(b), the proof of Proposition 7.106 gives us a semidirect product decomposition $U_w = U_{w'} \rtimes U_n$, where $w' := s_1 \cdots s_{n-1}$. So we get a presentation of U_w by combining a presentation of $U_{w'} = U_1 \cdots U_{n-1}$ with a presentation of U_n and adding conjugation relations describing the action of U_n on $U_1, \ldots, U_{n-1}$ (see Exercise 7.114 below).

These conjugation relations are required to have the form $u_n u_i u_n^{-1} = v$ with $i < n$, where u_i ranges over the generators of U_i, u_n ranges over the generators of U_n, and v is a word in the generators of $U_1, \ldots, U_{n-1}$. We can equally well use commutator relations of the form $[u_n, u_i] = v'$, where v' satisfies the same conditions as v. Relations of this form are provided by (7.18), since $(\alpha_n, \alpha_i) \subseteq \{\alpha_{i+1}, \ldots, \alpha_{n-1}\}$. To complete the proof, we argue inductively that our presentation for $U_{w'} = U_1 \cdots U_{n-1}$ can be taken to have the desired form. $\square$

Exercises

7.109. Prove part (3) of Lemma 7.101 directly from Definition 7.100.

7.110. In Lemma 7.102 and elsewhere we have used expressions of the form wVw^{-1} with $w \in W$ and V a subgroup of G. Under what conditions is this legitimate, and how should such an expression be interpreted?

7.111. Prove the assertion in Remark 7.103(b).

7.112. Suppose that all U_α are nilpotent. For any two roots α, β with $\alpha \neq -\beta$, show that $\langle U_\alpha, U_\beta \rangle$ is nilpotent.

7.113. Consider the standard RGD system in the group $G = \mathrm{GL}_3(k)$, where k is a field. (See Section 7.3.4, or look ahead at Section 7.9.1.) Give a concrete interpretation of Lemma 7.104 in terms of row and column operations.

7.114. Suppose a group G is a semidirect product $N \rtimes Q$. Show that one can get a presentation of G by combining a presentation of Q, a presentation of N, and relations of the form $qnq^{-1} = n'$ describing the action of Q on N. Here q is a generator of Q, n is a generator of N, and n' is a word in the generators of N.

7.8.5 The BN-Pair and the Associated Moufang Building

We continue to assume that $\big(G, (U_\alpha)_{\alpha \in \Phi}, T\big)$ is an RGD system of type (W, S) with W finite. We are now ready to show that $B_+ := TU_+$ and $N := \langle T, \{m(u) \mid u \in U_s^*,\ s \in S\} \rangle$ constitute a BN-pair in G.

Theorem 7.115. (G, B_+, N, S) *is a Tits system with Weyl group*

$$N/(B_+ \cap N) = N/T \cong W.$$

Proof. We have already shown that $B_+ \cap N = T$ (Lemma 7.93), and we have exhibited a homomorphism $\nu \colon N \twoheadrightarrow W$ with kernel T (Lemma 7.92), so that $N/T \cong W$. The canonical generating set of N/T corresponding to S is $\{\tilde{s}T \mid s \in S\}$, where $\tilde{s} = m(u)$ for any $u \in U_s^*$. Thus we have the familiar situation

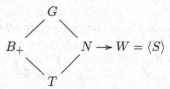

Let's verify now that $G = \langle B_+, N \rangle$. We have $G = T \langle U_\alpha \mid \alpha \in \Phi \rangle$, by (RGD4), so we just need to check that $\langle B_+, N \rangle$ contains U_α for $\alpha \in \Phi_-$. Write $\alpha = w\alpha_s$ for some $w \in W$ and $s \in S$, and choose $n \in \nu^{-1}(w)$. Then $U_\alpha = nU_s n^{-1} \le \langle B_+, N \rangle$, as required.

Next we check (BN1). Given $s \in S$ and $w \in W$, we must show that $sB_+w \subseteq B_+swB_+ \cup B_+wB_+$. Suppose first that $l(sw) > l(w)$, in which case we will show that $sB_+w \subseteq B_+swB_+$. We have

$$
\begin{aligned}
sB_+w &\subseteq sU_w wB_+ && \text{by Lemma 7.104}\\
&= \langle U_{s\alpha} \mid \alpha \in \Phi(w) \rangle swB_+\\
&= \langle U_\beta \mid \beta \in s\Phi(w) \rangle swB_+\\
&\subseteq U_+ swB_+ && \text{by Lemma 7.101(3)}\\
&\subseteq B_+swB_+ .
\end{aligned}
$$

If $l(sw) < l(w)$, set $w' := sw$. Then

$$
\begin{aligned}
sB_+w &= sB_+sw'\\
&\subseteq (B_+ \cup B_+sB_+)w' && \text{by Corollary 7.99}\\
&= B_+w' \cup B_+sB_+w'\\
&\subseteq B_+w' \cup B_+sw'B_+ && \text{by the previous case}\\
&= B_+sw \cup B_+wB_+ .
\end{aligned}
$$

Finally, we check (BN2). Given $s \in S$, we must show that $sB_+s^{-1} \not\le B_+$. For this we need only recall that $U_{-s} = sU_ss^{-1} \le sB_+s^{-1}$ and that $U_{-s} \not\le B_+$ by Corollary 7.94. $\square$

In view of the theorem, we now have a building $\Delta = \Delta(G, B_+)$. Recall from Section 6.2.6 that Δ is the poset of standard cosets, ordered by reverse inclusion, with G acting by left translation. The chambers are the cosets gB_+ ($g \in G$). We have a fundamental chamber $C_0 = B_+$ and a fundamental apartment Σ_0, with chamber set

$$\mathcal{C}(\Sigma_0) = \{wB_+ \mid w \in W\} \ .$$

The apartment Σ_0 is canonically isomorphic to $\Sigma = \Sigma(W, S)$. For each root $\alpha \subseteq \Sigma$, we denote by α_0 the corresponding root in Σ_0, whose set of chambers is $\{wB_+ \mid w \in \alpha\}$.

Theorem 7.116. *The building $\Delta = \Delta(G, B_+)$ is Moufang. For any $\alpha \in \Phi$ and any boundary panel P of α_0, U_α acts simply transitively on $\mathcal{C}(P, \alpha_0)$. In particular, the action of U_α on Δ is faithful, so we may identify U_α with its image in $\operatorname{Aut} \Delta$. It is a subgroup of the root group U_{α_0}. If, in addition, the Coxeter diagram of (W, S) has no isolated nodes, then $U_\alpha = U_{\alpha_0}$, and Δ is strictly Moufang.*

Proof. For each $\alpha \in \Phi$, let $\bar{U}_\alpha$ be the image of U_α in $\operatorname{Aut} \Delta$. We will show that the family of groups $(\bar{U}_\alpha)_{\alpha \in \Phi}$ satisfies the conditions of Proposition 7.79. The third of those conditions consists of the commutator relations (RGD1), so it suffices to check the first two. To simplify the notation, we will identify a root $\alpha \in \Phi$ with the corresponding root α_0 of Σ_0. (But U_α still denotes the group given as part of the RGD system, not the root group U_{α_0}.)

(a) $\bar{U}_\alpha$ fixes α pointwise.

Since the action of G on Δ is type-preserving, it suffices to show that $\bar{U}_\alpha$ stabilizes every chamber of α. In other words, we need to show that $U_\alpha wB_+ = wB_+$ for each $w \in \alpha$. Now

$$U_\alpha wB_+ = wU_{w^{-1}\alpha}B_+ \ ,$$

so the result follows from a familiar calculation: $w \in \alpha \implies 1 \in w^{-1}\alpha \implies w^{-1}\alpha \in \Phi_+ \implies U_{w^{-1}\alpha} \le B_+$.

(b) For any root $\alpha \in \Phi$ and any panel $P \in \partial\alpha$, the action of $\bar{U}_\alpha$ on $\mathcal{C}(P, \alpha)$ is transitive.

We will prove the stronger statement that, as claimed in the theorem, U_α acts simply transitively on $\mathcal{C}(P, \alpha)$ (and hence can be identified with $\bar{U}_\alpha$). Assume first that the chamber $C \ge P$ in α is the fundamental chamber C_0, so that $\alpha = \alpha_s$ for some $s \in S$ and P is the standard parabolic subgroup $P_s := B_+ \cup B_+ sB_+$. We have $P_s = B_+ \cup U_s sB_+$, by Lemma 7.98, so $\mathcal{C}_P$ is the set of cosets gB_+ with $g \in P$, and $\mathcal{C}(P, \alpha) = \mathcal{C}_P \smallsetminus \{C_0\}$ consists of the cosets in $U_s sB_+$. Transitivity of $U_s = U_{\alpha_s}$ on this set is now transparent. To prove simple transitivity, we must show that the stabilizer of sB_+ in U_s is trivial. Now the stabilizer of sB_+ in G is the conjugate sB_+s^{-1} of B_+, so the stabilizer in U_s is $U_s \cap sB_+s^{-1} = s(U_{-s} \cap B_+)s^{-1}$, which is indeed trivial by Corollary 7.94.

We now use the action of W to treat the general case. Given $\alpha \in \Phi$, let P be a boundary panel of α, and let C be the chamber in α with P as a face. Then $C = wC_0$ for some $w \in W$, $P = wP_s$ for some $s \in S$, and $\alpha = w\alpha_s$. Choose $n \in \nu^{-1}(w)$. We have $U_\alpha = nU_sn^{-1}$, so the simple transitivity of the latter on $\mathcal{C}(P, \alpha) = n \cdot \mathcal{C}(P_s, \alpha_s)$ follows from the simple transitivity of U_s on $\mathcal{C}(P_s, \alpha_s)$.

An application of Proposition 7.79 now completes the proof of the theorem.   □

Corollary 7.117. $N_G(U_+) = B_+$ and $N_G(U_-) = B_-$.

Proof. T normalizes U_+, so $B_+ = TU_+ \trianglelefteq N_G(U_+)$. The latter is therefore a standard parabolic subgroup. If it were strictly bigger than B_+, it would contain $\tilde{s}$ for some $s \in S$. But then we would have $U_{-s} = sU_ss^{-1} \leq U_+$, contradicting (RGD3). This proves the first equation, and the second is obtained by conjugating by w_0 or by symmetry (see Remark 7.95).   □

Example 7.118. Suppose $(G, (U_\alpha)_{\alpha \in \Phi}, T)$ is the RGD system associated to a strictly Moufang building Δ as in Example 7.83. Then $\Delta(G, B_+)$ is canonically isomorphic to the building Δ that we started with. This follows from Proposition 7.28 and the fact that B_+ is the stabilizer of the fundamental chamber C of Δ. We proved this fact in Corollary 7.65(2). [Note that B_+, in the context of that corollary, was *defined* to be the stabilizer of C; but the corollary implies that it is equal to what we are now calling B_+.]

Theorem 7.116 implies that the group-theoretic results of Section 7.7 are valid for our groups U_α (see Remark 7.69). We state two of these results explicitly for ease of reference. The first is Corollary 7.66(1), which sharpens Lemma 7.102(1). The second is Corollary 7.67.

Corollary 7.119. *With the notation of Lemma 7.102(1), every $u \in U_w$ admits a unique decomposition $u = u_1 \cdots u_n$ with $u_i \in U_{\alpha_i}$ for $1 \leq i \leq n$.*   □

Corollary 7.120. *For each $w \in W$, U_w acts simply transitively on $\mathcal{C}_w := \{C \in \mathcal{C}(\Delta) \mid \delta(C_0, C) = w\}$.*   □

Note that the stabilizer in U_w of wB_+ is $U_w \cap wB_+w^{-1}$. So the triviality of this stabilizer can be rewritten as

$$U_w \cap wB_+w^{-1} = \{1\} \tag{7.19}$$

for any $w \in W$. Equivalently:

Corollary 7.121. *For any $w \in W$, $w^{-1}U_ww \cap B_+ = \{1\}$.*   □

We now specialize equation (7.19) and Corollary 7.121 to the case $w = w_0$. Recalling that $U_{w_0} = U_+$ and that conjugation by w_0 interchanges $U_\pm$, we obtain the following:

Corollary 7.122. $U_+ \cap B_- = \{1\} = U_- \cap B_+$. *In particular,* $U_+ \cap T = \{1\}$, *so* $B_+ = TU_+$ *is a semidirect product* $T \ltimes U_+$. *Similarly,* $B_- = T \ltimes U_-$. $\square$

Corollary 7.123.

(1) $B_+ \cap B_- = T$.
(2) $T = \bigcap_{\alpha \in \Phi} N_G(U_\alpha)$.

Proof. (1) $B_+ \cap B_- = TU_+ \cap B_- = T(U_+ \cap B_-) = T$ by Corollary 7.122.

(2) $T \leq \bigcap_{\alpha \in \Phi} N_G(U_\alpha) \leq N_G(U_+) \cap N_G(U_-) = B_+ \cap B_-$ by Corollary 7.117. In view of (1), these inequalities must be equalities. $\square$

Remarks 7.124. (a) Since $B_- = w_0 B_+ w_0^{-1}$, B_- is the stabilizer of the chamber $w_0 B_+$ of Σ_0 opposite C_0. It follows that $B_+ \cap B_-$ is the set of elements in G that fix Σ_0 pointwise. A restatement of (1), then, is that the BN-pair (B_+, N) is saturated.

(b) Part (2) of Corollary 7.123 shows that the RGD system is uniquely determined by G and the U_α and that our axioms are equivalent to those of Tits [261]; see Remark 7.84.

The uniqueness of T leads immediately to another description of it in an important special case:

Corollary 7.125. *If* $G = \langle U_\alpha \mid \alpha \in \Phi \rangle$, *then*

$$T = \langle m(u)^{-1} m(v) \mid u, v \in U_s^*, s \in S \rangle .$$

Consequently, $N = \langle m(u) \mid u \in U_s^*, s \in S \rangle$.

Proof. Let $T_1 := \langle m(u)^{-1} m(v) \mid u, v \in U_s^*, s \in S \rangle$. Then, under our assumption that G is generated by the U_α, we still have an RGD system if we replace T by T_1. So $T_1 = T$ by the uniqueness of T (Corollary 7.123(2)). It follows that $\langle m(u) \mid u \in U_s^*, s \in S \rangle$ contains T and hence equals N, since the latter, by definition, is generated by T and the $m(u)$. $\square$

Exercise 7.126. For any subset $J \subseteq S$, let $\Phi_J := \{w\alpha_s \mid w \in W_J, s \in J\}$ and $G_J := T \langle U_\alpha \mid \alpha \in \Phi_J \rangle$.

(a) Show that Φ_J can be identified with the set of roots of $\Sigma(W_J, J)$ and that $\big(G_J, (U_\alpha)_{\alpha \in \Phi_J}, T\big)$ is an RGD system of type (W_J, J).
(b) How is the corresponding building related to $\Delta(G, B_+)$?

7.8.6 The Kernel of the Action

We started this chapter by considering buildings and certain groups of automorphisms of them. We then developed a corresponding algebraic theory in which we constructed a building $\Delta = \Delta(G, B_+)$ from a group G with an RGD system. In general, however, the group G we started with does not act faithfully on Δ. To complete the picture, we wish to describe the kernel of the map $\phi \colon G \to \operatorname{Aut} \Delta$ giving the action of G on Δ.

Proposition 7.127. *Let* $G_1 := \langle U_\alpha \mid \alpha \in \Phi \rangle$.

(1) $\ker \phi = C_G(G_1)$, *where the latter is the centralizer of* G_1 *in* G. *Moreover,*

$$Z(G) \le \ker \phi \le T \,,$$

where $Z(G)$ *is the center of* G.
(2) *If* $G = G_1$ *or if* T *is abelian, then* $\ker \phi = Z(G)$.
(3) $\ker \phi \cap G_1 = Z(G_1)$.

Proof. (1) We have already observed (Remark 7.124(a)) that T is the set of $g \in G$ such that g fixes Σ_0 pointwise. So $\ker \phi \le T$. In particular, $\ker \phi$ normalizes U_α for all $\alpha \in \Phi$. Since $\ker \phi$ is normal in G, it follows that

$$[\ker \phi, U_\alpha] \le \ker \phi \cap U_\alpha \le T \cap U_\alpha = \{1\} \,.$$

So in fact $\ker \phi$ *centralizes* U_α for each $\alpha \in \Phi$, i.e.,

$$\ker \phi \le \bigcap_{\alpha \in \Phi} C_G(U_\alpha) = C_G(G_1) \,.$$

Since $C_G(G_1)$ obviously contains $Z(G)$, all that remains is to show that $C_G(G_1) \le \ker \phi$. Recall that $G = \langle T, G_1 \rangle$ and that T normalizes G_1. Hence $G = TG_1 = G_1 T$, and every chamber of Δ has the form $g_1 B_+$ with $g_1 \in G_1$. So any $z \in C_G(G_1)$ satisfies

$$zg_1 B_+ = g_1 z B_+ = g_1 B_+ \,,$$

where the last equality comes from the fact that $z \in T$ by Corollary 7.123(2). Thus z stabilizes every chamber and so is in $\ker \phi$, since the action of G is type-preserving.

(2) If $G = G_1$, then (1) says that $\ker \phi = C_G(G_1) = Z(G)$. If T is abelian, then (1) gives

$$Z(G) \le \ker \phi \le T \cap C_G(G_1) \le C_G(T) \cap C_G(G_1) = Z(G) \,,$$

so again $\ker \phi = Z(G)$.

(3) Using (1) again, $\ker \phi \cap G_1 = C_G(G_1) \cap G_1 = Z(G_1)$. $\square$

Remark 7.128. The inclusion $Z(G) \le \ker \phi$ holds more generally for any group with a BN-pair; see Exercise 6.49.

Exercises

We continue to assume that $(G, (U_\alpha)_{\alpha \in \Phi}, T)$ is an RGD system.

7.129. For any $\alpha \in \Phi$, show that $\langle U_\alpha, U_{-\alpha} \rangle$ is not nilpotent.

7.130. Set $G_1 = \langle U_\alpha \mid \alpha \in \Phi \rangle$ and $T_1 := \langle m(u)^{-1}m(v) \mid u, v \in U_s^*, s \in S \rangle$. Let T' be a subgroup with $T_1 \leq T' \leq T$, and let $G' := T'G_1$.

(a) Show that $(G', (U_\alpha)_{\alpha \in \Phi}, T')$ is an RGD system.
(b) Show that $T' = G' \cap T$.
(c) Show that the RGD system in (a) gives rise to the same building as the original RGD system.

7.131. Let K be a subgroup of T that is normal in G. Set $\bar{G} := G/K$, $\bar{U}_\alpha := U_\alpha K/K$, and $\bar{T} := TK/K$.

(a) Show that $(\bar{G}, (\bar{U}_\alpha)_{\alpha \in \Phi}, \bar{T})$ is an RGD system and that $U_\alpha \xrightarrow{\sim} \bar{U}_\alpha$ for each $\alpha \in \Phi$.
(b) Show that the RGD system in (a) gives rise to the same building as the original RGD system.

7.8.7 Simplicity Results

We continue to denote by $(G, (U_\alpha)_{\alpha \in \Phi}, T)$ an RGD system of type (W, S) with W finite. We wish to apply Theorem 6.64 to show that G, or a closely related subquotient, is simple if (W, S) is irreducible. All the work has already been done, and we need only put the pieces together.

Set $B := B_+$ and $U = U_+$. Then B is a Tits subgroup of G, and $B = TU = UT$, with $U \trianglelefteq B$. Let $G_1 := \langle U_\alpha \mid \alpha \in \Phi \rangle$, and note that $G_1 \trianglelefteq G = TG_1$, since G_1 is normalized by T. Moreover, G_1 is the normal closure of U_+ in G because every root α is W-equivalent to a positive root (and even to a simple root α_s). Thus our present G_1 coincides with the group called G_1 in Section 6.2.7. Next, we set

$$Z := \bigcap_{g \in G} gBg^{-1} = \ker \phi = C_G(G_1),$$

where we have used Proposition 7.127 for the last equality. As we already noted in that same proposition, $G_1 \cap Z = Z(G_1)$. We can now state the following proposition, as an immediate application of Theorem 6.64 and Proposition 7.106:

Proposition 7.132. *Assume (W, S) is irreducible, G_1 is perfect, and every U_α is nilpotent. Then every subgroup of G that is normalized by G_1 is either contained in $Z = C_G(G_1)$ or contains G_1. In particular, $G_1/Z(G_1)$ is a simple group.* $\qquad\square$

Recall from Remark 7.107(a) that the assumption on the U_α is in fact always satisfied. In any case, it is generally easy to verify in examples. Moreover, the condition that G_1 be perfect is also true and easy to verify in many examples. Indeed, the calculations that verify the RGD axioms often give, as a byproduct, the fact that G_1 is perfect. We will see concrete illustrations of this in the next section.

7.9 Examples of RGD Systems

7.9.1 Classical Groups

Example 7.133. This example is essentially the same as the one in Section 7.3.4, but treated from the algebraic point of view. Let $G = \mathrm{GL}_n(D)$, where $n \geq 2$ and D is a division ring. Let W be the symmetric group S_n on n letters with its standard generating set $S = \{s_1, \ldots, s_{n-1}\}$ consisting of adjacent transpositions. Let $\Sigma = \Sigma(W, S)$. Recall from Example 1.119 or 3.52 that Σ has one root α_{ij} for each ordered pair (i, j) with $1 \leq i, j \leq n$ and $i \neq j$. So we need to exhibit a group T and root groups $U_{\alpha_{ij}}$.

Let T be the group $T_n(D)$ of diagonal matrices, i.e.,

$$T := \{\mathrm{diag}(\lambda_1, \ldots, \lambda_n) \mid \lambda_i \in D^*\} .$$

For $1 \leq i, j \leq n$ with $i \neq j$, let $U_{\alpha_{ij}} = U_{ij} := \{E_{ij}(\lambda) \mid \lambda \in D\}$, where, as in Section 7.3.4, $E_{ij}(\lambda)$ is the elementary matrix with 1's on the diagonal, λ in position (i, j), and 0's elsewhere. Thus U_{ij} is isomorphic to the additive group of D. In particular, $U_{ij} \neq \{1\}$, so we have (RGD0). One can also verify the conjugation formula

$$\mathrm{diag}(\lambda_1, \ldots, \lambda_n) E_{ij}(\lambda) \, \mathrm{diag}(\lambda_1, \ldots, \lambda_n)^{-1} = E_{ij}(\lambda_i \lambda \lambda_j^{-1}) , \qquad (7.20)$$

so T normalizes U_{ij} and (RGD5) holds. Let G_1 be the subgroup generated by all the subgroups U_{ij} or, equivalently, by all elementary matrices $E_{ij}(\lambda)$. Then T normalizes G_1, so $TG_1 = G_1T$ is a subgroup of G. Standard computations with row and/or column operations show that in fact this subgroup is equal to G, which proves (RGD4). It then follows that $G_1 \trianglelefteq G$.

Remark 7.134. The same row/column operations show that every coset $G_1 g$ contains a diagonal matrix of the form $\mathrm{diag}(1, \ldots, 1, \lambda)$. If D is commutative, this implies that $G_1 = \mathrm{SL}_n(D)$. In general, we simply *define* $\mathrm{SL}_n(D)$ to be G_1. Readers familiar with the Dieudonné determinant will note that $\mathrm{SL}_n(D)$ consists of the elements of $\mathrm{GL}_n(D)$ with Dieudonné determinant 1.

Returning to our roots α_{ij} and the corresponding groups U_{ij}, recall (or check) that α_{ij} is positive if and only if $i < j$, that $\alpha_{ji} = -\alpha_{ij}$, and that $w\alpha_{ij} = \alpha_{w(i)w(j)}$ for all $w \in W = S_n$. Note further that U_{ij} consists of strictly upper (resp. lower) triangular matrices if α_{ij} is positive (resp. negative). Thus U_+ is contained in the strict upper-triangular group [and is in fact equal to it, as one again sees by elementary row/column operations]. This implies that $U_\alpha \cap U_+ = \{1\}$ if α is a negative root; hence (RGD3) holds. It remains to verify (RGD1) and (RGD2).

For (RGD1), one checks the following commutator formulas, where our convention is that $[u, v] := uvu^{-1}v^{-1}$:

$$[E_{ij}(\lambda), E_{kl}(\mu)] = \begin{cases} 1 & \text{if } i \neq l \text{ and } j \neq k, \\ E_{il}(\lambda\mu) & \text{if } i \neq l \text{ and } j = k, \\ E_{kj}(-\mu\lambda) & \text{if } i = l \text{ and } j \neq k. \end{cases} \qquad (7.21)$$

Consequently,

$$[U_{ij}, U_{kl}] = \begin{cases} 1 & \text{if } i \neq l \text{ and } j \neq k, \\ U_{il} & \text{if } i \neq l \text{ and } j = k, \\ U_{kj} & \text{if } i = l \text{ and } j \neq k. \end{cases} \qquad (7.22)$$

Recall now from Exercise 7.78 that α_{il} is in the open interval $(\alpha_{ij}, \alpha_{jl})$ if $i \neq l$, and similarly $\alpha_{kj} \in (\alpha_{ij}, \alpha_{ki}) = (\alpha_{ki}, \alpha_{ij})$ if $j \neq k$. Equation (7.22) therefore proves (RGD1).

Remark 7.135. Note: There is no commutator formula analogous to (7.21) if $i = l$ and $j = k$. This is the case corresponding to opposite roots.

We turn, finally, to (RGD2), which we can verify by imitating what we did for SL_2 in Example 7.85. Let's begin by rewriting equation (7.13) in the present notation, starting with the case $n = 2$. Given $\lambda \in D^*$, let

$$m_{12}(\lambda) := \begin{pmatrix} 0 & \lambda \\ -\lambda^{-1} & 0 \end{pmatrix}.$$

If we set $u := E_{12}(\lambda)$, then equation (7.13) says that

$$m_{12}(\lambda) = E_{21}(-\lambda^{-1})uE_{21}(-\lambda^{-1}). \qquad (7.23)$$

To extend this 2×2 calculation to the $n \times n$ case, note that for each fixed $i \neq j$ there is an isomorphic copy of $GL_2(D)$ in $GL_n(D)$ that acts on the 2-dimensional space $e_i D + e_j D$ and fixes e_k for $k \neq i, j$. Here $e_1, \ldots, e_n$ is the standard basis for the right D-vector space D^n. For $\lambda \in D^*$, let $m_{ij}(\lambda)$ be the image of $m_{12}(\lambda)$ under this embedding of $GL_2(D)$ into $GL_n(D)$. We now set $m(u) := m_{ij}(\lambda)$ for $u = E_{ij}(\lambda) \in U_{ij}^*$, and equation (7.23) implies that

$$m_{ij}(\lambda) = E_{ji}(-\lambda^{-1})uE_{ji}(-\lambda^{-1}) \in U_{ji}uU_{ji}. \qquad (7.24)$$

Another 2×2 calculation shows that $m(u)^{-1}m(v) \in T$ for all $u, v \in U_{ij}^*$; this reduces to

$$\begin{pmatrix} 0 & -\lambda \\ \lambda^{-1} & 0 \end{pmatrix} \begin{pmatrix} 0 & \mu \\ -\mu^{-1} & 0 \end{pmatrix} = \begin{pmatrix} \lambda\mu^{-1} & 0 \\ 0 & \lambda^{-1}\mu \end{pmatrix}. \tag{7.25}$$

This proves the second part of (RGD2). For the first part, it is convenient to work with permutation matrices. Given a permutation $\pi \in W = S_n$, let π act on D^n by $\pi e_j = e_{\pi(j)}$. Then the action of π is represented by the matrix in $\mathrm{GL}_n(D)$, still denoted by π, given by $\pi_{ij} = \delta_{i,\pi(j)}$. With this convention, one easily checks that $m_{ij}(\lambda) \in s_{ij}T$, where s_{ij} is the transposition that interchanges i and j. The first part of (RGD2) now follows from the observation that

$$\pi U_{kl} \pi^{-1} = U_{\pi(k)\pi(l)}$$

for all $\pi \in W$ and all $1 \le k, l \le n$ with $k \ne l$.

All of the axioms are now verified, and we have an RGD system in $G = \mathrm{GL}_n(D)$. The associated Moufang building $\Delta(G, B_+)$ is, of course, canonically isomorphic to the building $\Delta(D^n)$ studied in Section 7.3.4. We make one final remark about this example, which also applies to many other examples. Namely, the calculations above show that $G_1 = \mathrm{SL}_n(D)$ is perfect unless $n = 2$ and $D = \mathbb{F}_2$ or $\mathbb{F}_3$. This is especially transparent when $n \ge 3$. In this case, if we are given $i \ne j$ we can choose $k \ne i, j$ and note that

$$U_{ij} = [U_{ik}, U_{kj}] \le [G_1, G_1]$$

by (7.22). If $n = 2$ and D has more than 3 elements, we instead use the conjugation formula (7.20) to deduce that

$$[T_1, U_{12}] = U_{12} \tag{7.26}$$

and similarly for U_{21}, where $T_1 := G_1 \cap T$; see Exercise 7.139 below. Hence Proposition 7.132 is applicable. In particular, $\mathrm{SL}_n(D)/(Z \cap \mathrm{SL}_n(D))$ is simple (unless $n = 2$ and $D = \mathbb{F}_2$ or $\mathbb{F}_3$), where $Z = C_G(G_1) = \bigcap_{i \ne j} C_G(U_{ij})$. A final calculation shows that Z is the set of matrices λI, where I is the identity matrix and λ is in the center of D.

Example 7.136. In this example we describe the RGD systems associated to symplectic, orthogonal, and unitary groups. Our basic reference is Bruhat–Tits [59, Section 10], where all omitted details can be found. See also Abramenko [9, pp. 77–78 and 107–111] for a convenient summary. Our notational conventions will be consistent with those in Chapter 6, which differ slightly from those in the cited references.

Let K be a field with an automorphism σ such that $\sigma^2 = \mathrm{id}$. We denote the action of σ by $\lambda \mapsto \lambda^\sigma$ for $\lambda \in K$. We will always assume $\mathrm{char}\, K \ne 2$, although the theory extends to characteristic 2 with suitable modifications. (And for the unitary groups, the theory also extends to the case that K is a division ring; see Exercise 7.142.)

Definition 7.137. Let V be a K-vector space, possibly infinite-dimensional. Fix $\epsilon = \pm 1$. A function $B: V \times V \to K$ is said to be (σ, ϵ)-*Hermitian* if it is linear in the first variable and satisfies

$$B(y, x) = \epsilon B(x, y)^{\sigma} \tag{7.27}$$

for all $x, y \in V$.

Note that (7.27) implies that B is σ-linear in the second variable, i.e., it is additive and satisfies $B(x, \lambda y) = \lambda^{\sigma} B(x, y)$ for $\lambda \in K$ and $x, y \in V$.

In order to relate this notion to the examples we saw in Chapter 6, note that:

- If $(\sigma, \epsilon) = (\mathrm{id}, 1)$, then B is a symmetric bilinear form; this is the *orthogonal* case.
- If $(\sigma, \epsilon) = (\mathrm{id}, -1)$, then B is a skew-symmetric bilinear form; this is the *symplectic* case.
- If $\sigma \neq \mathrm{id}$ and $\epsilon = 1$, then B is Hermitian in the sense of Section 6.8; this is the *unitary* case.

Remark 7.138. The remaining case, in which $\sigma \neq \mathrm{id}$ and $\epsilon = -1$, is essentially the same as the unitary case. Indeed, if $\sigma \neq \mathrm{id}$, then there is a scalar $a \neq 0$ such that $a^{\sigma} = -a$. [Choose any $b \in K$ with $b^{\sigma} \neq b$, and set $a := b - b^{\sigma}$.] We can then replace B by aB to convert a (σ, ϵ)-Hermitian form to a $(\sigma, -\epsilon)$-Hermitian form. We will therefore also refer to this case as the unitary case.

Assume from now on that we are given a (σ, ϵ)-Hermitian form on V satisfying the following two conditions:

(1) B is *nondegenerate* in the sense that $V^{\perp} = 0$, where

$$V^{\perp} := \{x \in V \mid B(x, -) = 0\} \ .$$

(2) B has finite Witt index $n \geq 1$.

Here, as before, the Witt index is the maximal dimension of a totally isotropic subspace. It follows from these assumptions that we can find vectors $e_1, \ldots, e_n, e_{-n}, \ldots, e_{-1}$ satisfying the same relations as in the examples we treated in Sections 6.6, 6.7, and 6.8, where e_{-i} plays the role of the vector called f_i $(1 \leq i \leq n)$ in those examples. Explicitly, if we set

$$\epsilon(i) = \begin{cases} 1 & \text{if } i > 0, \\ \epsilon & \text{if } i < 0, \end{cases}$$

and $\langle x, y \rangle := B(x, y)$, then the relations are

$$\langle e_i, e_{-i} \rangle = \epsilon(i) \tag{7.28}$$

for all $i \in I := \{\pm 1, \ldots, \pm n\}$ and

$$\langle e_i, e_j \rangle = 0 \tag{7.29}$$

for all $i, j \in I$ with $j \neq -i$. Our assumptions also imply that V splits into a direct sum

$$V = V_1 \oplus \cdots \oplus V_n \oplus V_0 \oplus V_{-n} \oplus \cdots \oplus V_{-1},$$

where $V_i = Ke_i$ for $i \in I$ and $V_0 := \bigcap_{i \in I} e_i^\perp$. We might have $V_0 = 0$; this is necessarily true in the symplectic case. Or at the other extreme, V_0 might be infinite-dimensional.

In what follows we will represent linear maps $V \to V$ by matrices whose rows and columns are labeled by $I \cup \{0\}$, where the (i, j)-entry describes the $(V_j \to V_i)$ component of the map. These components are not scalars, in general, if they involve V_0. For example, the $(0, 1)$-entry is a vector in V_0 (the image of e_1 under a map $V_1 \to V_0$); the $(1, 0)$-entry is an element of V_0^* (representing a map $V_0 \to V_1 \cong K$); and the $(0, 0)$-entry is a linear map $V_0 \to V_0$.

We now consider the *isometry group*

$$G := \{g \in \mathrm{GL}(V) \mid \langle gx, gy \rangle = \langle x, y \rangle \text{ for all } x, y \in V\},$$

and we will exhibit an RGD system. First, we set T equal to the set of all $g \in G$ represented by diagonal matrices (i.e., $gV_i = V_i$ for all $i \in I \cup \{0\}$). Next, we define root groups using "elementary" automorphisms of V. The basic idea for this has already been illustrated in some of the examples in Chapter 6, where we used elementary subgroups in various copies of SL_2 or GL_2 in G as an aid in verifying the BN-pair axioms.

Given any $i, j \in I$ with $i \neq \pm j$ and any $\lambda \in K$, we can perform an "elementary change of basis" in which we replace e_j by $e_j' := e_j + \epsilon(i)\lambda e_i$ and we replace e_{-i} by $e_{-i}' := e_{-i} - \epsilon(j)\lambda^\sigma e_{-j}$. The new basis vectors satisfy the same inner-product relations as the old ones and have the same orthogonal complement V_0. So there is an element $E_{ij}(\lambda) \in G$ given by

$$e_j \mapsto e_j',$$
$$e_{-i} \mapsto e_{-i}',$$
$$e_l \mapsto e_l \quad \text{if } l \in I \setminus \{j, -i\},$$
$$x \mapsto x \quad \text{if } x \in V_0.$$

Thus the restriction of $E_{ij}(\lambda)$ to $V_{ij} := V_i \oplus V_j \oplus V_{-j} \oplus V_{-i}$ is represented by the matrix

	i	j	$-j$	$-i$
i	1	$\epsilon(i)\lambda$	0	0
j	0	1	0	0
$-j$	0	0	1	$-\epsilon(j)\lambda^\sigma$
$-i$	0	0	0	1

and $E_{ij}(\lambda)$ is the identity on $V_{ij}^\perp$. We now set $U_{ij} := \{E_{ij}(\lambda) \mid \lambda \in K\} \le G$. It is isomorphic to the additive group of K. Note that U_{ij} is simply the image of the strict upper-triangular subgroup of $\mathrm{SL}_2(K)$ under an embedding of the latter into G.

In most cases there is a second family of root subgroups U_i $(i \in I)$, where U_i consists of the elements $g \in G$ satisfying the following conditions:

(1) $ge_j = e_j$ for $j \in I \smallsetminus \{\pm i\}$.
(2) g stabilizes $V_i \oplus V_0 \oplus V_{-i}$.
(3) The restriction of g to $V_i \oplus V_0 \oplus V_{-i}$ has a matrix of the form

$$
\begin{array}{c|ccc}
 & i & 0 & -i \\
\hline
i & 1 & f & \lambda \\
0 & 0 & 1 & v \\
-i & 0 & 0 & 1
\end{array}
$$

for some $f \in V_0^*$, $\lambda \in K$, and $v \in V_0$.

In other words, g is given by

$$
\begin{aligned}
e_j &\mapsto e_j & &\text{if } j \in I \smallsetminus \{\pm i\}, \\
e_i &\mapsto e_i, \\
e_{-i} &\mapsto e_{-i} + v + \lambda e_i, \\
x &\mapsto x + f(x)e_i & &\text{if } x \in V_0.
\end{aligned}
$$

It is easy to work out the conditions that f, λ, v must satisfy in order for g to preserve inner products. The crucial relations turn out to be $\langle gx, ge_{-i} \rangle = 0$ (for $x \in V_0$) and $\langle ge_{-i}, ge_{-i} \rangle = 0$, which translate to

$$f(x) = -\epsilon(i)\langle x, v \rangle \tag{7.30}$$

and

$$\lambda + \epsilon \lambda^\sigma = -\epsilon(i)Q(v), \tag{7.31}$$

where $Q(v) := \langle v, v \rangle$. In particular, g is completely determined by the two parameters $v \in V_0$ and $\lambda \in K$. If we write $g = g_i(v, \lambda)$, then we have the multiplication rule

$$g_i(v, \lambda)g_i(v', \lambda') = g_i(v + v', \lambda + \lambda' - \epsilon(i)\langle v', v \rangle), \tag{7.32}$$

which follows from the calculation

$$
\begin{pmatrix} 1 & f & \lambda \\ 0 & 1 & v \\ 0 & 0 & 1 \end{pmatrix}
\begin{pmatrix} 1 & f' & \lambda' \\ 0 & 1 & v' \\ 0 & 0 & 1 \end{pmatrix}
=
\begin{pmatrix} 1 & f + f' & \lambda + \lambda' + f(v') \\ 0 & 1 & v + v' \\ 0 & 0 & 1 \end{pmatrix}
$$

for $f, f' \in V_0^*$, $\lambda, \lambda' \in K$, and $v, v' \in V_0$.

In the symplectic case, where $(\sigma, \epsilon) = (\mathrm{id}, -1)$, we have $V_0 = 0$, and (7.31) holds for all λ. An element $g \in U_i$ is then determined by the parameter λ, and U_i is isomorphic to the additive group of K. It is the image of the strict upper-triangular subgroup of $\mathrm{SL}_2(K)$ under an embedding of the latter into G.

In the orthogonal case, where $(\sigma, \epsilon) = (\mathrm{id}, 1)$, equation (7.31) says that

$$\lambda = -\epsilon(i)Q(v)/2 \, ,$$

so that an element $g = g_i(v, \lambda) \in U_i$ is determined by the parameter v. It then follows from (7.32) that U_i is isomorphic to the additive group of V_0. In particular, U_i is nontrivial (and so is a candidate for a root group) if and only if $V_0 \neq 0$.

The unitary case ($\sigma \neq$ id) turns out to be the most interesting. We will take $\epsilon = -1$ (see Remark 7.138), though we could handle $\epsilon = 1$ with minor modifications. Equation (7.31) then reads

$$\lambda - \lambda^\sigma = -\epsilon(i)Q(v) \, . \tag{7.33}$$

Let k be the fixed field of the automorphism σ, so that K is a quadratic extension of k. For any $v \in V_0$ we have one solution of (7.33), given by $\lambda = \lambda_0(v) := -\epsilon(i)Q(v)/2$. [Use the fact that $Q(v) = -Q(v)^\sigma$.] The general solution is then $\lambda = \lambda_0(v) + \mu$ with $\mu \in k$. If we write $g_i(v, \lambda) =: u_i(v, \mu)$, then we obtain from (7.32) and a short calculation the multiplication rule

$$u_i(v, \mu)u_i(v', \mu') = u_i(v + v', \mu + \mu' + \epsilon(i)\operatorname{tr}\langle v, v'\rangle)$$

for $v, v' \in V_0$ and $\mu, \mu' \in k$, where $\operatorname{tr}: K \to k$ is the trace map ($\operatorname{tr} a = a + a^\sigma$). If $V_0 = 0$, then U_i is isomorphic to the additive group of k. If $V_0 \neq 0$, however, then U_i is a nonabelian nilpotent group of class 2. The noncommutativity shows up in the term $\operatorname{tr}\langle v, v'\rangle$ above, which changes sign if v and v' are interchanged. The center of U_i is $Z := \{u_i(0, \mu) \mid \mu \in k\}$, which is isomorphic to the additive group of k, and U_i/Z is isomorphic to the additive group of V_0.

We are now ready to describe the groups U_α for our RGD system, which will be a system of type C_n in most cases. Let Φ be the set of roots of the standard Coxeter complex of type C_n. It consists of roots α_i for $i \in I$ and α_{ij} for $i, j \in I$ with $i \neq \pm j$. This follows from the descriptions of the root systems of type B_n and C_n in Example 1.11. In more detail, let $b_1, \ldots, b_n$ be the standard basis of $\mathbb{R}^n$, and set $b_{-i} = -b_i$ for $i = 1, \ldots, n$. Then α_{ij} corresponds to the root vector $b_i - b_j$, and α_i corresponds the root vector b_i or $2b_i$, depending on whether we use the root system of type B_n or C_n. Recall now that we chose a specific fundamental chamber in Example 1.82 for the Weyl group, which is the same for both root systems. According to that choice, the set Φ_+ of positive roots consists of the roots α_i with $i > 0$ and the roots α_{ij} such that i precedes j in the ordering

$$1, 2, \ldots, n, -n, \ldots, -2, -1$$

of I.

We now set

$$U_{\alpha_i} := U_i$$

and

$$U_{\alpha_{ij}} := U_{ij} \, ,$$

and one can check that $(G, (U_\alpha)_{\alpha \in \Phi}, T)$ is an RGD system of type C_n unless we are in the orthogonal case with $V_0 = 0$. We refer to the cited references for the details, which are not difficult.

In the orthogonal case with $V_0 = 0$, there are two differences, both of which are to be expected in view of Section 6.7.1. First, the RGD system is of type D_n instead of C_n. [Here we assume $n \geq 2$, since there is no root system of type D_1; the orthogonal group is boring in that case anyway.] Thus Φ consists only of roots α_{ij} with $i, j \in I$ and $i \neq \pm j$ (see Example 1.13), and we define U_α for $\alpha \in \Phi$ as above. Second, the group generated by the U_α and T is not the full orthogonal group $G = O_{2n}(K)$ but only the subgroup $SO_{2n}(K)$ of index 2; see Exercise 7.141. The result, then, is that we have an RGD system $(SO_{2n}(K), (U_\alpha)_{\alpha \in \Phi}, T)$ of type D_n.

In all cases, one gets a Moufang building Δ on which G acts, and, not surprisingly, it is the same building that was constructed in Chapter 6.

Finally, we remark that $G_1 := \langle U_\alpha \mid \alpha \in \Phi \rangle$ is almost always perfect, so that G_1/Z is simple by Proposition 7.132 (where Z now denotes $Z(G_1)$), provided our Coxeter system is irreducible. The unique case in which it is reducible is the orthogonal case with $n = 2$ and $V_0 = 0$, where the system is of type $D_2 = A_1 \times A_1$. In this case $G_1/Z \cong PSL_2(K) \times PSL_2(K)$ and so is not simple but is usually a product of two simple groups. If we exclude this case (as well as SO_2), then we have the following short list of exceptional cases in which G_1/Z is not simple. [Note: Since we have not discussed the theory in characteristic 2, we are not listing exceptions with char $K = 2$.]

- If $n = 1$, then $Sp_{2n} = SL_2$, so it is not perfect when $K = \mathbb{F}_3$, as we noted in connection with Example 7.133.
- In the unitary case with $n = 1$ and $V_0 = 0$, we have $G_1 \cong SL_2(k)$, so there is again an exception when $k = \mathbb{F}_3$ ($K = \mathbb{F}_9$).
- In the orthogonal case with $n = 1$ and $\dim V_0 = 1$, we have $G_1/Z \cong PSL_2(K)$, so there is an exception when $K = \mathbb{F}_3$.

Notice that our three exceptional nonsimple groups G_1/Z are all the same group $PSL_2(\mathbb{F}_3)$, which is isomorphic to the alternating group on 4 letters.

Exercises

7.139. Let D be a division ring.

(a) Show that $h(\lambda) := \text{diag}(\lambda, \lambda^{-1})$ is in $SL_2(D)$ for all $\lambda \in D^*$.
(b) Show that $[h(\lambda), E_{12}(\mu)] = E_{12}((\lambda^2 - 1)\mu)$ if λ and μ commute.
(c) Deduce that equation (7.26) holds, and hence that $SL_2(D)$ is perfect, if $D \neq \mathbb{F}_2$ or $\mathbb{F}_3$.

7.140. For any division ring D, show that the groups $SL_2(D)$ and $PSL_2(D) := SL_2(D)/Z(SL_2(D))$ are not nilpotent.

7.141. (a) In the orthogonal case of Example 7.136 with $V_0 = 0$, show that every element of T has determinant 1.

(b) What can you say about the determinant of an element of T in the symplectic case or in the unitary case with $V_0 = 0$?

7.142. Let D be a division ring with an antiautomorphism σ such that $\sigma^2 = $ id. Thus

$$(\lambda + \mu)^\sigma = \lambda^\sigma + \mu^\sigma \quad \text{and} \quad (\lambda\mu)^\sigma = \mu^\sigma \lambda^\sigma .$$

What happens in Example 7.136 if we replace K by D?

7.9.2 Chevalley Groups

Our basic reference for this section is Steinberg [227]. We follow his approach closely, except for minor changes of notation to conform with the notation we have used elsewhere in this book. The starting point is a semisimple Lie algebra $\mathfrak{g}$ over the field $\mathbb{C}$ of complex numbers. As we indicated briefly in Section 6.4, one can then define a group $G(K)$ for any field K; it can be thought of as the analogue over K of the complex Lie group G with Lie algebra $\mathfrak{g} = \mathcal{L}(G)$. For the benefit of readers familiar with Lie theory, we will state this slightly more precisely. [Warning: The group that we are calling $G(K)$ here, following Steinberg, is in general smaller than the group of K-points of the corresponding algebraic group.]

There is not just one complex Lie group G with Lie algebra $\mathfrak{g}$, but a family of them. The smallest is the *adjoint group*, and all the others are coverings of it. The biggest is the simply connected group, which is the common universal cover of all of them. The family of groups G can be parametrized by a lattice Λ in the real vector space that contains the root system Φ of $\mathfrak{g}$. We have $\Lambda_r \leq \Lambda \leq \Lambda_w$ where Λ_r is the *root lattice* and Λ_w is the *weight lattice*. The lattice Λ_r corresponds to the adjoint group G_a, and the lattice Λ_w corresponds to the simply connected group G_u, also called the *universal group*.

If K is now an arbitrary field, then there is a corresponding family of *Chevalley groups*

$$G(K) = G(\Phi, \Lambda; K)$$

associated to $\mathfrak{g}$. We will not go into the details of the definition but will simply state the facts that enable one to construct an RGD system of type Φ (or type (W, S), where W is the Weyl group W_Φ). The first fact is that $G(K)$ has generators, denoted by $x_\alpha(\lambda)$, with $\alpha \in \Phi$ and $\lambda \in K$. The reader might find it helpful to think of the case $G(K) = \mathrm{SL}_n(K)$, where Φ is the root system of type A_{n-1} and $x_\alpha(\lambda)$ is an elementary matrix. For fixed α, if we set

$$U_\alpha := \{x_\alpha(\lambda) \mid \lambda \in K\} ,$$

then U_α is a subgroup of $G(K)$ and is isomorphic to the additive group of K via $\lambda \mapsto x_\alpha(\lambda)$ for $\lambda \in K$; see [227, Section 3, Corollary 1, p. 26].

We now define elements of $G(K)$ by

$$m_\alpha(\lambda) := x_\alpha(\lambda)x_{-\alpha}(-\lambda^{-1})x_\alpha(\lambda) ,$$
$$m_\alpha := m_\alpha(1) ,$$
$$h_\alpha(\lambda) := m_\alpha(\lambda)m_\alpha^{-1} ,$$

for $\alpha \in \Phi$ and $\lambda \in K^*$. One has $h_\alpha(\lambda\mu) = h_\alpha(\lambda)h_\alpha(\mu)$ [227, Section 3, Lemma 28(a), p. 43], so $\{h_\alpha(\lambda) \mid \lambda \in K^*\}$ is a subgroup of $G(K)$ and is a homomorphic image of the multiplicative group K^*. If $G = \mathrm{SL}_2(K)$ and $\alpha = \alpha_{12}$, for example, then

$$m_\alpha(\lambda) = m_{-\alpha}(-\lambda^{-1}) = \begin{pmatrix} 0 & \lambda \\ -\lambda^{-1} & 0 \end{pmatrix}$$

and $h_\alpha(\lambda) = h_{-\alpha}(\lambda^{-1}) = \mathrm{diag}(\lambda, \lambda^{-1})$.

There are two families of conjugation relations that say how the elements m_α and $h_\alpha(\lambda)$ act by conjugation on the generators of $G(K)$. First,

$$m_\alpha x_\beta(\lambda)m_\alpha^{-1} = x_{s_\alpha(\beta)}(\pm\lambda) \tag{7.34}$$

for $\alpha, \beta \in \Phi$ and $\lambda \in K^*$, where the rule for determining the ambiguous sign need not concern us here; see [227, Section 3, relation (R7) on p. 30]. Second,

$$h_\alpha(\lambda)x_\beta(\mu)h_\alpha(\lambda)^{-1} = x_\beta(\lambda^{\langle\beta,\alpha^\vee\rangle}\mu) \tag{7.35}$$

for $\alpha, \beta \in \Phi$, $\lambda \in K^*$, and $\mu \in K$; see [227, Section 3, relation (R8) on p. 30]. The angle brackets here denote inner products as in Appendix B. [Warning: Steinberg [227] uses angle brackets to denote something different.] Combining (7.34), (7.35), and the definition of $h_\alpha(\lambda)$, one gets

$$m_\alpha(\lambda)x_\beta(\mu)m_\alpha(\lambda)^{-1} = x_{s_\alpha(\beta)}(\pm\lambda^{\langle s_\alpha(\beta),\alpha^\vee\rangle}\mu) . \tag{7.36}$$

In the special case $\beta = \alpha$, (7.35) becomes

$$h_\alpha(\lambda)x_\alpha(\mu)h_\alpha(\lambda)^{-1} = x_\alpha(\lambda^2\mu) ,$$

which implies the commutator formula

$$[h_\alpha(\lambda), x_\alpha(\mu)] = x_\alpha((\lambda^2 - 1)\mu) .$$

Consequently,

$$[h_\alpha(\lambda), U_\alpha] = U_\alpha$$

if $\lambda \in K \smallsetminus \{0, \pm 1\}$. This shows that $G(K)$ is a perfect group if $|K| \geq 4$, i.e., $K \neq \mathbb{F}_2, \mathbb{F}_3$.

Finally, we have commutator formulas involving the generators $x_\alpha(\lambda)$. Given roots $\alpha \neq \pm\beta$ and scalars $\lambda, \mu \in K$, the commutator formulas have the form

$$[x_\alpha(\lambda), x_\beta(\mu)] = \prod_{\substack{i,j\in\mathbb{N} \\ i\alpha+j\beta\in\Phi}} x_{i\alpha+j\beta}(c_{ij}\lambda^i\mu^j) . \tag{7.37}$$

Here one must choose a definite order for the factors in the product (which do not commute with one another), and the c_{ij} are integers that depend on α and β and the chosen order; see [227, Section 3, relation (R2) on p. 30]. Note that U_α is abelian, so we could also write a commutator formula for the case $\beta = \alpha$. But there is no such formula for $\beta = -\alpha$. In fact, it turns out that for all $\alpha \in \Phi$,

$$\langle U_\alpha, U_{-\alpha} \rangle \cong \mathrm{SL}_2(K) \text{ or } \mathrm{PSL}_2(K), \tag{7.38}$$

depending on α and the lattice Λ; see [227, Section 3, Corollary 6, pp. 46–47]. One gets $\mathrm{SL}_2(K)$ for all α in the universal case ($\Lambda = \Lambda_w$), and one gets $\mathrm{PSL}_2(K)$ for all α in the adjoint case ($\Lambda = \Lambda_r$).

We now set

$$T := \langle h_\alpha(\lambda) \mid \alpha \in \Phi, \lambda \in K^* \rangle,$$

and it is straightforward to verify that $(G(K), (U_\alpha)_{\alpha \in \Phi}, T)$ is an RGD system. We run through the axioms briefly. (RGD0) is trivial, since $U_\alpha \cong K$ (additive group). The commutator relations (RGD1) follow from the commutator formulas (7.37), since each root of the form $i\alpha + j\beta$ in (7.37) is in the open interval (α, β) by Exercise 7.77. For (RGD2), consider an element $u = x_\alpha(\lambda) \in U^*$, where $\lambda \in K^*$. If we set

$$m(u) := m_{-\alpha}(-\lambda^{-1}) = x_{-\alpha}(-\lambda^{-1})x_\alpha(\lambda)x_{-\alpha}(-\lambda^{-1}) \in U_{-\alpha}uU_{-\alpha},$$

then we have

$$m(u)U_\beta m(u)^{-1} = U_{s_\alpha(\beta)}$$

by (7.36). Moreover, a short calculation, which we omit, gives $m(u)^{-1}m(v) \in T$ for all $u, v \in U_\alpha^*$. This proves (RGD2). There are two easy ways to prove (RGD3). One is to look at the explicit construction of $G(K)$, which we did not write down. This exhibits $G(K)$ as a matrix group, in such a way that $x_\alpha(\lambda)$ is strictly triangular; it is upper triangular for $\alpha \in \Phi_+$ and lower triangular for $\alpha \in \Phi_-$, so (RGD3) is clear. Alternatively, we could use the method of Remark 7.107(c), which is applicable in view of (7.38). Finally, (RGD4) is trivial, and (RGD5) follows from (7.35).

Remarks 7.143. (a) The group $G(K)$ is almost always perfect, so the simplicity results of Section 7.8.7 are applicable if Φ is irreducible, i.e., if the Lie algebra $\mathfrak{g}$ is simple. The conclusion is that the adjoint group $G(\Phi, \Lambda_r; K)$ is simple except in a few cases involving the fields $\mathbb{F}_2$ and $\mathbb{F}_3$.

(b) There is of course considerable overlap between the Chevalley groups of the present subsection and the classical groups of the previous subsection. In particular, $\mathrm{SL}_n(K)$ and $\mathrm{Sp}_{2n}(K)$ are examples of Chevalley groups $G(K)$. And the special orthogonal group $\mathrm{SO}_m(K)$ associated to the standard quadratic form differs in only a minor way from such a Chevalley group. More precisely, the commutator subgroup $\Omega_m(K) \leq \mathrm{SO}_m(K)$ that we discussed at the end of Section 6.7.1 is a Chevalley group $G(K)$, and we have $\mathrm{SO}_m(K) = T'G(K)$, where T' is the diagonal subgroup of $\mathrm{SO}_m(K)$. The RGD systems that we have described in $G(K)$ and $\mathrm{SO}_m(K)$ give rise to the same building (see Exercise 7.130).

The unitary groups, on the other hand, do *not* fit into the theory of the present subsection, nor do the classical groups involving a noncommutative division ring, nor do orthogonal groups associated to quadratic forms other than the standard one. Some (but not all) of the omitted classical groups will be covered by the next subsection.

*7.9.3 Nonsplit Algebraic Groups

In this subsection we assume some knowledge of the theory of algebraic groups and Lie algebras, although we will try to keep this to a minimum. The terminology involving algebraic groups is summarized in Appendix C, and the standard reference is Borel [39]. See also Abramenko [7] for a survey. We start by considering linear algebraic groups over an *algebraically closed* field K. In this case we will identify a linear algebraic group G with its group $G(K)$ of K-points.

Let G be a (connected) semisimple linear algebraic group over a field K, and let $T \leq G$ be a torus of dimension l. There is an associated *character group*

$$X = X(T) := \mathrm{Hom}(T, \mathbb{G}_m) \cong \mathbb{Z}^l ,$$

where $\mathbb{G}_m$ denotes the multiplicative group. Identifying $\mathbb{G}_m$ with its group K^* of K-points, we can view an element of X as a homomorphism $\alpha \colon T \to K^*$. The centralizer $C_G(T)$ is again a connected K-group, and it is a finite-index normal subgroup of the normalizer $N_G(T)$. We set

$$W(G, T) := N_G(T)/C_G(T) .$$

If T is a maximal torus, then $C_G(T) = T$. In this case we write W instead of $W(G, T)$ and call it the *Weyl group* of G; thus

$$W := N/T ,$$

where $N := N_G(T)$. All maximal tori are conjugate, so W is independent of T up to isomorphism.

Let $\mathfrak{g} := \mathfrak{L}(G)$ be the Lie algebra of G, and consider the adjoint representation

$$\mathrm{Ad} \colon G \to \mathrm{GL}(\mathfrak{g}) .$$

For any torus T in G (not necessarily maximal), the elements of T acting on $\mathfrak{g}$ are simultaneously diagonalizable, so we can decompose $\mathfrak{g}$ into T-eigenspaces:

$$\mathfrak{g} = \bigoplus_{\alpha \in X} \mathfrak{g}_\alpha ,$$

where

$$\mathfrak{g}_\alpha := \{v \in \mathfrak{g} \mid \mathrm{Ad}(t)v = \alpha(t)v \text{ for all } t \in T\} .$$

Here $\alpha(t)v$ makes sense because we are thinking of α as a homomorphism $T \to K^*$. There are, of course, only finitely many $\alpha \in X$ such that $\mathfrak{g}_\alpha \neq 0$. The trivial character $(\alpha(t) \equiv 1)$ is one of them, and the corresponding eigenspace is the Lie algebra $\mathfrak{L}(C_G(T))$. The nontrivial characters that occur are called the *roots* of G relative to T, and we denote the set of all such roots by $\varPhi(G,T)$. The eigenspace (or *root space*) decomposition thus becomes

$$\mathfrak{g} = \mathfrak{L}(C_G(T)) \oplus \bigoplus_{\alpha \in \varPhi(G,T)} \mathfrak{g}_\alpha .$$

We will return to this relative theory shortly, but assume now that T is maximal, in which case we write $\varPhi$ instead of $\varPhi(G,T)$, and we have

$$\mathfrak{g} = \mathfrak{L}(T) \oplus \bigoplus_{\alpha \in \varPhi} \mathfrak{g}_\alpha .$$

Here are some basic facts.

(1) $N = N_G(T)$ acts on T by conjugation. Since T is abelian, this induces an action of the Weyl group $W = N/T$ on T and hence an action of W on the character group X. The set $\varPhi$ of roots is invariant under the W-action. More precisely, given $w \in W$ and a representative n of w in N,

$$\mathrm{Ad}(n)\mathfrak{g}_\alpha = \mathfrak{g}_{w\alpha}$$

for all $\alpha \in \varPhi$.

(2) $\varPhi$ is a (reduced, crystallographic) root system in the l-dimensional real vector space $\mathbb{R} \otimes_\mathbb{Z} X$ endowed with a suitable inner product, and W is its Weyl group $W_\varPhi$.

(3) Each root space $\mathfrak{g}_\alpha$ $(\alpha \in \varPhi)$ is 1-dimensional. There is a unique T-invariant connected algebraic subgroup $U_\alpha \leq G$ such that $\mathfrak{L}(U_\alpha) = \mathfrak{g}_\alpha$. Moreover, U_α is isomorphic to the additive group $\mathbb{G}_a$ via a canonical homomorphism $\lambda \mapsto x_\alpha(\lambda) \in U_\alpha$ for $\lambda \in K$. We have

$$tx_\alpha(\lambda)t^{-1} = x_\alpha(\alpha(t)\lambda)$$

for all $t \in T$ and $\lambda \in K$; in particular, T normalizes U_α. We also have

$$wU_\alpha w^{-1} = U_{w\alpha}$$

for all $w \in W$ and $\alpha \in \varPhi$.

Example 7.144. Let $G = \mathrm{SL}_n$, in which case $\mathfrak{g}$ is the Lie algebra of $n \times n$ matrices of trace 0. Let T be the group of diagonal matrices. For each off-diagonal matrix position (i,j), there is a root subspace of $\mathfrak{g}$ consisting of the matrices that vanish except at that position. The corresponding root α is the character $t_i t_j^{-1}$, where $t_i \in X$ gives the ith diagonal entry of a matrix in T, and U_α is the group U_{ij} discussed in Example 7.133.

Remark 7.145. The theory that we have just sketched is essentially a reformulation of the theory of Chevalley groups. [The character group X plays the role of the lattice Λ that we mentioned in Section 7.9.2.] In particular, we already know that we have an RGD system $\big(G, (U_\alpha)_{\alpha \in \Phi}, T\big)$.

We turn now to the relative theory. Thus we consider k-groups where the field k is not necessarily algebraically closed. Let G be a semisimple k-group, and let K be an algebraic closure of k. Then we can apply the absolute theory described above to the group G_K obtained from G by extension of scalars. Let $\mathfrak{g}$ be the Lie algebra of G_K. There are two tori to which we can apply the theory sketched above. First, we choose a maximal k-split torus $S \leq G$. Second, we choose a maximal torus $T \leq G$ (defined over k) containing S. The centralizer $C_G(S)$ is a connected k-group, and it is a finite-index normal subgroup in the normalizer $N_G(S)$. We define the *relative Weyl group* $_kW$ by

$$_kW := N_G(S)/C_G(S) .$$

Our two tori now give us two character groups $X(T)$ and $X(S)$, which are related by a surjection

$$j \colon X(T) \twoheadrightarrow X(S)$$

obtained by restricting characters from T to the subtorus S. We also get two sets of roots:

$$\Phi := \Phi(G, T) \subseteq X(T)$$

and

$$_k\Phi := \Phi(G, S) \subseteq X(S) .$$

Here one has to interpret $\Phi(G, T)$ and $\Phi(G, S)$ after extension of scalars; we have omitted the subscripts K to simplify notation. It is a fact that the maximal k-torus T remains maximal after extension of scalars to K, so Φ has the same meaning as in our discussion of the absolute case, and $_k\Phi$ is the set of roots of the adjoint representation restricted to S. Thus

$$\mathfrak{g} = \mathfrak{L}\big(C_G(S)\big) \oplus \bigoplus_{\alpha \in {}_k\Phi} \mathfrak{g}_\alpha .$$

Our two sets of roots are related by

$$_k\Phi \subseteq j(\Phi) \subseteq {}_k\Phi \cup \{0\} ,$$

where 0 denotes the trivial character of S. (This convention is consistent with the fact that we often think of $X(S)$ as a lattice in the real vector space $\mathbb{R} \otimes_{\mathbb{Z}} X(S)$ and therefore use additive notation.) We now list the basic facts:

(1) $_k\Phi$ is a (possibly nonreduced) crystallographic root system in $\mathbb{R} \otimes_{\mathbb{Z}} X(S)$ with Weyl group $_kW$. "Possibly nonreduced" means that we might have a root $\alpha \in {}_k\Phi$ such that 2α is also a root; see Section B.5.

(2) By the absolute theory, we have root groups $U_\beta(K) \leq G(K)$ for $\beta \in \Phi$, and each such root group is isomorphic to the additive group $\mathbb{G}_a(K) = K$. It turns out that if we lump these root groups together appropriately, we get algebraic groups that can be defined over k. To make this precise, fix a root $\alpha \in {}_k\Phi$, and set

$$\Psi := \{\beta \in \Phi \mid j(\beta) = \alpha \text{ or } 2\alpha\} .$$

Then one can show that there is an algebraic k-group $U_\alpha \leq G$ such that

$$U_\alpha(K) = \langle U_\beta(K) \mid \beta \in \Psi \rangle = \prod_{\beta \in \Psi} U_\beta(K)$$

for a suitable ordering of the factors. In fact, the multiplication map

$$\prod_{\beta \in \Psi} U_\beta(K) \to U_\alpha(K)$$

defines an isomorphism of algebraic varieties over K. (These assertions are reminiscent of the facts about the groups U_w that we proved in Section 7.7.2.) Moreover, even though the individual groups U_β are not defined over k in general, one can prove that U_α is isomorphic as an algebraic variety over k to m-dimensional affine space, where $m := |\Psi|$.

(3) Continuing with the notation in (2), suppose that $2\alpha \notin {}_k\Phi$. Then U_α is abelian, and in fact, there is an isomorphism

$$U_\alpha \cong \mathbb{G}_a^m$$

of k-groups, where $m = |\Psi|$ as above. If, on the other hand, $2\alpha \in {}_k\Phi$, then U_α is nilpotent but generally nonabelian. What happens is that $U_{2\alpha}(k)$ is a proper central subgroup of $U_\alpha(k)$ (typically the entire center), with

$$[U_\alpha(k), U_\alpha(k)] \leq U_{2\alpha}(k) . \tag{7.39}$$

The classical groups discussed in Section 7.9.1 illustrate this lumping phenomenon. Consider, for example, the group $\mathrm{SL}_n(D)$, where D is a division ring. If D is finite-dimensional over its center k, then $\mathrm{SL}_n(D)$ is the group of k-points of a semisimple k-group G, which becomes isomorphic to SL_N after extension of scalars. [Here $N = nr$ if $\dim_k D = r^2$.] The root groups U_{ij} that we described in Example 7.133 are isomorphic to the additive group of D, and each is obtained by lumping together r^2 root groups of $\mathrm{SL}_N(K)$. The root system ${}_k\Phi$ is of type A_{n-1} in this example. In particular, it is reduced, and all the root groups are abelian.

Consider now the unitary groups of Example 7.136, with $V_0 \neq 0$ and finite-dimensional. With k and n as in that example, the unitary group is a semisimple k-group whose relative root system ${}_k\Phi$ is the root system of type BC_n (see Example B.7 in Appendix B). This root system is nonreduced, so we have a conceptual explanation for the appearance of nonabelian root groups in the unitary group.

Returning now to the general theory, we have almost completed the construction of an RGD system in $G(k)$. Let Σ be the Coxeter complex associated to the relative Weyl group $_kW$. Since $_k\Phi$ is not necessarily reduced, one can have two root vectors $\alpha, 2\alpha \in {}_k\Phi$ corresponding to the same root of Σ. In this case we choose the bigger group $U_\alpha(k)$ as our root group for the RGD system. And the "T" of the RGD system is the group of k-points of $C_G(S)$. We omit the verification of the RGD axioms, which can be found in [59].

Remark 7.146. The theory that we have just sketched actually leads to a "refined" RGD system, with additional root groups $U_{2\alpha}(k)$ whenever α and 2α are both in $_k\Phi$. Such systems were studied by Bruhat and Tits [59]. The general picture is that one has root groups indexed by a possibly nonreduced root system, and the main new ingredient is that there are commutator relations for all pairs of roots α, β with $\alpha \neq -\beta$. The relation in (7.39) is an example. Another commutator relation asserts that U_α is abelian if α is a root such that 2α is not a root.

Moufang Twin Buildings and RGD Systems

In this chapter, which is intended for advanced readers, we generalize to twin buildings the theory developed in Chapter 7 for spherical buildings. The algebraic version is a theory of RGD systems of arbitrary type (W, S). At the time of this writing, it is not very easy to learn this theory from the existing literature. One of our goals in writing this chapter has been to supply some of the missing details in order to make the subject more accessible. In particular, we will provide detailed proofs of results for which only sketches currently appear in the literature. The results of this chapter have important applications to Kac–Moody groups, which we will survey in the final section.

We assume knowledge of the basic facts about twin buildings and twin BN-pairs, which were given in the optional Sections 5.8 and 6.3.

8.1 Pre-Moufang Twin Buildings and Twin BN-Pairs

Throughout this section, $\mathcal{C} = (\mathcal{C}_+, \mathcal{C}_-)$ denotes a thick twin building of type (W, S). [As usual, we have suppressed δ^* from the notation.] We denote by $\mathrm{Aut}_0 \, \mathcal{C}$ the group of type-preserving automorphisms of $\mathcal{C}$; thus elements of $\mathrm{Aut}_0 \, \mathcal{C}$ act on $\mathcal{C}_+$ and $\mathcal{C}_-$, and they preserve Weyl distances and the codistance as in Definition 6.67. Fix a "fundamental pair" (C_+, C_-) of opposite chambers, with $C_\pm \in \mathcal{C}_\pm$, and let $\Sigma = (\Sigma_+, \Sigma_-) = \Sigma \{C_+, C_-\}$ be the corresponding "fundamental twin apartment." Let Φ be the set of twin roots of Σ, and set

$$\Phi_+ := \{\alpha \in \Phi \mid C_+ \in \alpha\} \,,$$
$$\Phi_- := \{\alpha \in \Phi \mid C_- \in \alpha\} \,.$$

It turns out that everything we did in Section 7.1, as specialized to the spherical case, extends with minor modifications to the present setup. We will quickly run through the relevant definitions and results, pointing out the minor modifications that are needed.

Note first that some of the notation and terminology will change in trivial ways because we are now working in the W-metric setting. For example, apartments and roots are now sets of chambers, and a *panel* is now a rank-1 residue. Given a twin root α and a panel $\mathcal{P}$ of $\mathcal{C}_+$ or $\mathcal{C}_-$ that meets α, we say that $\mathcal{P}$ is a *boundary panel* of α if $\mathcal{P} \cap \alpha$ consists of exactly one chamber; otherwise, $\mathcal{P}$ is said to be an *interior panel* of α. If $\mathcal{P}$ is a boundary panel of α, we write $\mathcal{C}(\mathcal{P}, \alpha) := \mathcal{P} \setminus \{C\}$, where C is the chamber in $\mathcal{P} \cap \alpha$.

Definition 8.1. A family $(X_\alpha)_{\alpha \in \Phi}$ of subgroups of $\mathrm{Aut}_0\, \mathcal{C}$ is called a system of *pre–root groups* if it satisfies the following two conditions:

(1) X_α fixes α pointwise for each $\alpha \in \Phi$.
(2) For each $\alpha \in \Phi$ and each boundary panel $\mathcal{P}$ of α, the action of X_α on $\mathcal{C}(\mathcal{P}, \alpha)$ is transitive.

We say that $\mathcal{C}$ is a *pre-Moufang* twin building if it admits a system of pre–root groups.

(Recall that we are generalizing the *spherical* theory, so there is no reason to require an analogue of condition (3) of Definition 7.2; instead we will prove, as in the spherical case, that such an analogue always holds.)

Recall from Lemma 5.198 that for any twin root α and any boundary panel $\mathcal{P}$ of α, there is a canonical bijection between $\mathcal{C}(\mathcal{P}, \alpha)$ and the set $\mathcal{A}(\alpha)$ of twin apartments containing α. We therefore obtain the following generalization of Lemma 7.4:

Lemma 8.2. *Suppose $(X_\alpha)_{\alpha \in \Phi}$ is a system of subgroups of $\mathrm{Aut}_0\, \mathcal{C}$ satisfying condition (1) of Definition 8.1. Then the following conditions are equivalent:*

(i) *The system $(X_\alpha)_{\alpha \in \Phi}$ satisfies condition (2).*
(ii) *For each $\alpha \in \Phi$ there exists a boundary panel $\mathcal{P}$ of α such that the action of X_α on $\mathcal{C}(\mathcal{P}, \alpha)$ is transitive.*
(iii) *For each $\alpha \in \Phi$, the action of X_α on $\mathcal{A}(\alpha)$ is transitive.* $\square$

Next, we have the analogue of Lemma 7.5:

Lemma 8.3. *Let $(X_\alpha)_{\alpha \in \Phi}$ be a system of pre–root groups in $\mathrm{Aut}_0\, \mathcal{C}$. Then for each $\alpha = (\alpha_+, \alpha_-)$ in Φ there is an element $n_\alpha \in \langle X_\alpha, X_{-\alpha} \rangle$ such that $n_\alpha(\alpha) = -\alpha$ and $n_\alpha(-\alpha) = \alpha$. In other words, n_α stabilizes Σ and acts on each Σ_ϵ as the reflection s_{α_ϵ} ($\epsilon = \pm$).*

Proof. Let C and D be the adjacent chambers in Σ_+ with $C \in \alpha_+$ and $D \in -\alpha_+$. Let $C' := \mathrm{op}_\Sigma C$ and $D' := \mathrm{op}_\Sigma D$. Let $\mathcal{P}$ (resp. $\mathcal{P}'$) be the panel containing C and D (resp. C' and D'). See Figure 8.1. The first part of the proof of Lemma 7.5 goes through without change. Thus if we start with an element $x \in X_\alpha$ such that $xD \neq D$, then we can find an element $m(x) \in X_{-\alpha} x X_{-\alpha}$ that interchanges C and D. We claim that $m(x)$ also interchanges C' and D'. Observe first that X_α and $X_{-\alpha}$ stabilize $\mathcal{P}$ and $\mathcal{P}'$,

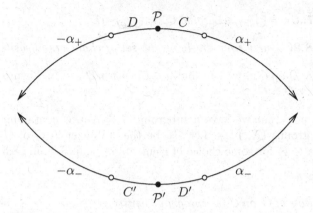

Fig. 8.1. A twin root.

so the same is true of $m(x)$. Our claim now follows from the fact that since $\mathcal{P}$ and $\mathcal{P}'$ are opposite, C' (resp. D') is the unique element of $\mathcal{P}'$ not opposite D (resp. C); see Corollary 5.153. Recalling now that α is the convex hull of C and D' and that $-\alpha$ is the convex hull of D and C' (Corollary 5.194), we conclude that $m(x)$ interchanges $\pm\alpha$.   □

There are also analogues of Corollary 7.6 and Remark 7.7:

Corollary 8.4. *Under the hypotheses of Lemma 8.3, let α be a twin root of Σ, let $D \in -\alpha_+$ be a chamber contained in a boundary panel of α, and let $x \in X_\alpha$ be an element such that $xD \neq D$. Then there is an element $m(x) \in X_{-\alpha}xX_{-\alpha}$ such that $m(x)$ interchanges $\pm\alpha$.*   □

Remark 8.5. Suppose the action of X_α on $\mathcal{C}(\mathcal{P}, \alpha)$ is simply transitive for each $\alpha \in \Phi$. Then for any $x \in X_\alpha$ as in the corollary, there are unique elements $x', x'' \in X_{-\alpha}$ such that $m(x) = x'xx''$ interchanges $\pm\alpha$.

Finally, the connection between the pre-Moufang property and strongly transitive actions generalizes to twin buildings. In one direction, if we start with a strongly transitive automorphism group $G \leq \mathrm{Aut}_0\,\mathcal{C}$, then the point-wise fixers $X_\alpha := \mathrm{Fix}_G(\alpha)$ ($\alpha \in \Phi$) form a system of pre–root groups; the proof is the same as the proof of Proposition 7.8, except that we use Corollary 6.74 instead of Corollary 6.7:

Proposition 8.6. *If G is a strongly transitive group of type-preserving auto-morphisms of $\mathcal{C}$, then G contains a system of pre–root groups.*   □

To go in the other direction, we start with a system of pre–root groups X_α and set
$$U := \langle X_\alpha \mid \alpha \in \Phi_+ \rangle \,.$$
We then have, exactly as in the proof of Lemma 7.9,

Lemma 8.7. $\mathcal{C}_+ = \bigcup_{u \in U} u\Sigma_+$ and $\mathcal{C}_- = \bigcup_{u \in U} u\Sigma_-$. $\qquad\qquad\square$

Corollary 8.8. *U acts transitively on the set of chambers opposite* C_+.

Proof. Given D op C_+, we can find $u \in U$ with $uD \in \Sigma_-$. Then uD op $uC_+ = C_+$; hence $uD = C_-$. $\qquad\qquad\square$

Suppose now that we have a subgroup $G \leq \mathrm{Aut}_0\,\mathcal{C}$ containing a system of pre–root groups $(X_\alpha)_{\alpha \in \Phi}$. Let $B_\pm$ be the stabilizer in G of $C_\pm$, and set $N := \langle n_\alpha \mid \alpha \in \Phi \rangle$ for some choice of elements n_α as in Lemma 8.3.

Theorem 8.9.

(1) *The action of G on $\mathcal{C}$ is strongly transitive.*
(2) *The triple (B_+, B_-, N) is a twin BN-pair in G, and $\mathcal{C} \cong \mathcal{C}(G, B_+, B_-)$.*

Proof. (1) Note that N stabilizes Σ_+ and acts transitively on it. Combining this with Lemma 8.7, we see that G acts transitively on $\mathcal{C}_+$. Since B_+ is transitive on the set of chambers opposite C_+ by Corollary 8.8, strong transitivity follows.

(2) follows from (1) via Corollary 6.79 (and the comments immediately following it). $\qquad\qquad\square$

8.2 Calculation of Fixers

We show here that the results of Section 7.2 all generalize to the setting of twin buildings.

8.2.1 Preliminaries: Convex Sets of Twin Roots

Let $\Sigma = (\Sigma_+, \Sigma_-)$ be a thin twin building. Recall that since we are in a W-metric setting, $\Sigma_\pm$ are sets of chambers, i.e., we are not considering any lower-dimensional simplices. For example, we could just take Σ to be the standard twin building of type (W, S) (Example 5.136(b)), in which case $\Sigma_+ = \Sigma_- = W$.

For any pair $\mathcal{M} = (\mathcal{M}_+, \mathcal{M}_-) \subseteq \Sigma$, we define a set of roots $\Psi(\mathcal{M})$ by

$$\Psi(\mathcal{M}) := \{\alpha \in \Phi \mid \alpha \supseteq \mathcal{M}\} . \tag{8.1}$$

Definition 8.10. A set of twin roots $\Psi \subseteq \Phi$ is said to be *convex* if it has the form $\Psi = \Psi(\mathcal{M})$ for some pair $\mathcal{M}$ with $\mathcal{M}_+$ and $\mathcal{M}_-$ both nonempty.

Remarks 8.11. (a) Recall that a twin root $\alpha = (\alpha_+, \alpha_-)$ is completely determined by its first component α_+, which can be an arbitrary root of Σ_+. We can therefore identify Φ with the set of roots of Σ_+. Moreover, the condition $\alpha \supseteq \mathcal{M}$ in (8.1) is equivalent to the two conditions

$$\alpha_+ \supseteq \mathcal{M}_+ \text{ and } -\alpha_+ \supseteq \text{op}_\Sigma(\mathcal{M}_-) \, .$$

So a set of twin roots is convex if and only if its set of first components is convex in the sense of Definition 7.18. Note, however, how much more natural the notion of "convexity" is from the twin point of view. Moreover, the analogy with the spherical case points the way toward results that would be more cumbersome to state if we worked only with Σ_+.

(b) A convex set of twin roots is always *finite* in view of an observation made after Definition 7.18.

The following lemma is proved in exactly the same way as Lemma 7.14, with the aid of Proposition 5.193. It refers to the concept of "convex pair" that was introduced in Definition 5.158.

Lemma 8.12. *There is an order-reversing* 1–1 *correspondence between convex pairs* $\mathcal{M}$ *and convex subsets of* Φ. *It is given by* $\mathcal{M} \mapsto \Psi(\mathcal{M})$, *and its inverse is given by* $\Psi \mapsto \bigcap_{\alpha \in \Psi} \alpha$. $\qquad\square$

Example 8.13. Given two chambers $C, D \in \Sigma_+$, set

$$\Phi(C, D) := \{\alpha \in \Phi \mid C \in \alpha_+, \, D \notin \alpha_+\} \, .$$

Then $\Phi(C, D) = \Psi(\mathcal{M})$, where $\mathcal{M}_+ = \{C\}$ and $\mathcal{M}_- = \{D'\}$, with $D' := \text{op}_\Sigma D$. Hence $\Phi(C, D)$ is a convex set of roots. Its set of first components is precisely what was called $\Phi(C, D)$ in the setting of ordinary Coxeter complexes (following Definition 7.18). Thus it contains precisely $d = d(C, D)$ roots, one for each wall of Σ_+ that separates C from D, and we can enumerate these roots as in Example 7.15 by choosing a minimal gallery $C = C_0, \ldots, C_d = D$ from C to D.

Observe next that the notion of *admissible ordering* in Definition 7.16 applies verbatim to the present setting. Using Lemma 5.191(1), one can now imitate the proof of Lemma 7.17 to obtain the following:

Lemma 8.14. *Every convex set of twin roots admits an admissible ordering.*
$$\square$$

We are now ready to return to pre–root groups.

8.2.2 Fixers

Assume that $\mathcal{C} = (\mathcal{C}_+, \mathcal{C}_-)$ is a twin building and that G is a subgroup of $\text{Aut}_0\,\mathcal{C}$ that contains a system $(X_\alpha)_{\alpha \in \Phi}$ of pre–root groups. Here Φ is the set of twin roots of a fundamental twin apartment $\Sigma = (\Sigma_+, \Sigma_-)$. For any subset $\Psi \subseteq \Phi$ we set

$$X_\Psi := \langle X_\alpha \mid \alpha \in \Psi \rangle \, .$$

We also set

$$T := \mathrm{Fix}_G(\Sigma) \,.$$

We can now record the analogue of Proposition 7.20, whose proof goes through with minor modifications:

Proposition 8.15. *Let Ψ be a convex set of twin roots in the fundamental twin apartment Σ, let $\alpha_1, \dots, \alpha_m$ be an admissible ordering of Ψ, and set $X_i := X_{\alpha_i}$ for $i = 1, \dots, m$.*

(1) *$X_\Psi T$ is a subgroup of G, and*

$$X_\Psi T = X_1 \cdots X_m T \,.$$

(2) *If $x_1 \cdots x_m t = x_1' \cdots x_m' t'$ with $x_i, x_i' \in X_i$ for $i = 1, \dots, m$ and $t, t' \in T$, then there are elements $t_1, \dots, t_m \in T$ such that*

$$
\begin{aligned}
x_1' &= x_1 t_1 \,, \\
x_2' &= t_1^{-1} x_2 t_2 \,, \\
&\ \ \vdots \\
x_m' &= t_{m-1}^{-1} x_m t_m \,, \\
t' &= t_m^{-1} t \,.
\end{aligned}
$$

(3) *If $\Psi = \Psi(\mathcal{M})$ for some pair $\mathcal{M} = (\mathcal{M}_+, \mathcal{M}_-)$ with $\mathcal{M}_\pm \neq \emptyset$, then the pointwise fixer of $\mathcal{M}$ is given by*

$$\mathrm{Fix}_G(\mathcal{M}) = X_\Psi T \,. \qquad \qquad \square$$

8.3 Root Groups and Moufang Twin Buildings

Throughout this section, $\mathcal{C} = (\mathcal{C}_+, \mathcal{C}_-)$ denotes a thick twin building of type (W, S). We continue to record the (still mostly routine) generalizations of the concepts and results of Chapter 7.

8.3.1 Definitions and Simple Consequences

Definition 8.16. For any twin root α of $\mathcal{C}$, the *root group* U_α is defined to be the set of automorphisms g of $\mathcal{C}$ such that (a) g fixes α pointwise and (b) g fixes $\mathcal{P}$ pointwise for every interior panel $\mathcal{P}$ of α.

Note that $U_\alpha \leq \mathrm{Aut}_0\, \mathcal{C}$ and that as in Definition 7.24, (a) is redundant if the rank is at least 2. Note also that the panel $\mathcal{P}$ might be in either $\mathcal{C}_+$ or $\mathcal{C}_-$.

Lemma 8.17.

(1) *For any twin root α and any $g \in \operatorname{Aut}_0 \mathcal{C}$,*

$$gU_\alpha g^{-1} = U_{g\alpha} \, .$$

(2) *Let α be a twin root and let $\mathcal{P}$ be a boundary panel of α. Then the root group U_α acts on the sets $\mathcal{A}(\alpha)$ and $\mathcal{C}(\mathcal{P}, \alpha)$, and these two actions are equivalent.*

(3) *If the Coxeter diagram of (W, S) has no isolated nodes, then the actions in (2) are free.*

Proof. This is similar to the proof of Lemma 7.25. For (2) one uses Lemma 5.198 instead of Lemma 4.118, and for (3) one needs to recall that the rigidity theorem is valid for twin buildings (see Remark 5.208). □

Definition 8.18. We say that $\mathcal{C}$ is *Moufang*, or is a *Moufang twin building*, if the actions in Lemma 8.17(2) are transitive for every twin root α of $\mathcal{C}$. If, in addition, these actions are simply transitive, then we say that $\mathcal{C}$ is *strictly Moufang*, or is a *strictly Moufang twin building*.

Note that a Moufang twin building whose Coxeter diagram has no isolated nodes is strictly Moufang.

Proposition 8.19. *If $\mathcal{C}$ is Moufang, then it is pre-Moufang. More precisely, if we choose a twin apartment Σ and let Φ be its set of twin roots, then $(U_\alpha)_{\alpha \in \Phi}$ is a system of pre–root groups. Hence $G := \langle U_\alpha \mid \alpha \in \Phi \rangle$ acts strongly transitively on $\mathcal{C}$, and $\mathcal{C} \cong \mathcal{C}(G, B_+, B_-)$, where $B_\pm$ are the stabilizers in G of any pair of opposite chambers.* □

Remarks 8.20. (a) As in Remarks 7.29(a) and (b), a twin building is Moufang if U_α is transitive on $\mathcal{A}(\alpha)$ for every $\alpha \in \Phi$, where Φ is the set of twin roots in a fundamental twin apartment. Moreover, G is then generated by all the root groops U_α with α a twin root of $\mathcal{C}$.

(b) Even more than in the case of Moufang spherical buildings, one needs to be careful in reading the literature. In particular, the original definition of "Moufang twin building" given by Tits [261, p. 261] is based on asymmetrically defined root groups, and we do not know whether these are always the same as our root groups as defined above. But our (symmetric) definition seems to be the "right" one, since, as we will show, it leads to the expected equivalence between Moufang twin buildings and RGD systems; see Example 8.47(a) and Theorem 8.81. Moreover, Ronan and Tits used the symmetric definition of root groups in their paper [205] on twin trees. We suspect, incidentally, that the two definitions do *not* agree for twin trees, but we do not have a counterexample.

(c) As a byproduct of our work in Section 8.4, we will see that the symmetric and asymmetric versions of root groups agree in the 2-spherical case (see Remark 8.26).

We turn next to links (or, rather, residues, since we are now using the W-metric approach). We will not need a systematic study of residues in Moufang twin buildings, so we confine ourselves to recording one result that will be needed later. Its proof is similar to that of Proposition 7.32(1).

Proposition 8.21. *If C is a Moufang twin building, then every spherical residue of C is a Moufang (spherical) building.*   □

Finally, we remark that the results of Section 7.3.3 on subbuildings extend to Moufang twin buildings with no difficulty. In particular, we have the following analogue of Proposition 7.37:

Proposition 8.22. *Let C be a Moufang twin building whose Coxeter diagram has no isolated nodes. If C' is a thick twin subbuilding of C, then C' is also Moufang. If, moreover, Σ is a twin apartment of C', α is a twin root of Σ, and U_α is the corresponding root group of C, then $U'_\alpha := \{u \in U_\alpha \mid u(C') = C'\}$ is the root group of C' associated to α.*   □

8.3.2 Example

In Section 6.12 we briefly described a twin building $\Delta = (\Delta_+, \Delta_-)$ associated to a rational function field $K = k(t)$ and an integer $n \geq 2$. Here k is an arbitrary field. We now sketch a proof that Δ has the Moufang property. In what follows we will work in the simplicial setting and will freely use the notation introduced in Section 6.12, which the reader should review before proceeding.

Recall that we have a fundamental twin apartment $\Sigma = (\Sigma_+, \Sigma_-)$ whose vertices are the $A_\pm$-lattice classes $[[t^{a_1}e_1, \ldots, t^{a_n}e_n]]$, where $e_1, \ldots, e_n$ is the standard basis of $V = K^n$ and $a_1, \ldots, a_n \in \mathbb{Z}$. In order to discuss the Moufang property, we need the following description of the twin roots of Σ: There is one twin root α_{ij}^l for each ordered pair of indices $1 \leq i, j \leq n$ with $i \neq j$ and each integer l; the vertices of its Σ_+-component are the A_+-lattice classes $[[t^{a_1}e_1, \ldots, t^{a_n}e_n]]$ with $a_i - a_j \leq l$, and the vertices of its Σ_--component are the A_--lattice classes $[[t^{a_1}e_1, \ldots, t^{a_n}e_n]]$ with $a_i - a_j \geq l$. To verify this, one needs to understand the Weyl group W of Δ and its Coxeter complex. Recall that W is the Coxeter group of type $\tilde{A}_{n-1}$ that we described in Section 6.9.3, where we claimed that it is a Euclidean reflection group. We will in fact study this group and its Coxeter complex in detail in Section 10.1.7, and the interested reader can look ahead and see that our description of the roots is indeed correct. (Alternatively, the reader may prefer to specialize to the case $n = 2$. Here W is the infinite dihedral group, each Σ_ϵ is a triangulated line, and it is trivial to describe the roots.)

Remark 8.23. For future reference, we note that α_{ij}^l is a positive root (i.e., it contains the fundamental chamber C_+) if and only if either $i < j$ and $l \geq 0$ or $i > j$ and $l \geq 1$.

Recall next that we have an action of the group $G = \mathrm{SL}_n(k[t, t^{-1}])$ on Δ. We will use this action as an aid in proving the Moufang property. For each i, j, l as above, let U_{ij}^l be the subgroup of G defined by

$$U_{ij}^l := \left\{ E_{ij}(ct^l) \mid c \in k \right\}.$$

Here $E_{ij}(-)$ denotes an elementary matrix as in Section 7.3.4. Thus each U_{ij}^l is isomorphic to the additive group of k. The following analogue of Lemma 7.41 implies that Δ has the Moufang property. The proof is straightforward and is left to the interested reader. (See Exercise 6.115 and its solution for some hints related to part (2).)

Lemma 8.24.

(1) U_{ij}^l *fixes every vertex of* Δ *that is joinable to an interior vertex of* α_{ij}^l. *In particular,* U_{ij}^l *fixes* α_{ij}^l *pointwise.*

(2) *Let* P *be a panel in* $\partial \alpha_{ij}^l$. *Then* U_{ij}^l *acts simply transitively on* $\mathcal{C}(P, \alpha_{ij}^l)$.

□

8.4 2-Spherical Twin Buildings of Rank at Least 3

In this section we generalize Theorem 7.59 and derive the Moufang property for a class of twin buildings. Surprisingly, although Theorem 8.27 below has certainly been known to the experts for quite some time, we could not find an explicit statement of it in the literature.

As in the spherical case, the Moufang property for 2-spherical twin buildings is basically (modulo some additional arguments that we will supply) a consequence of the extension theorem that we quoted in Section 5.11. We therefore have to make the same assumptions here as in Section 5.11. Namely, we assume that $\mathcal{C} = (\mathcal{C}_+, \mathcal{C}_-)$ is a thick, irreducible, 2-spherical twin building of rank at least 3 that satisfies condition (co) of Section 5.11. We then want to deduce from Corollary 5.215 that $\mathcal{C}$ is Moufang, just as we deduced from Corollary 5.211 that thick, irreducible, spherical buildings of rank at least 3 are Moufang. In our present context, we will need an additional argument showing that the automorphism provided by Corollary 5.215 fixes pointwise all interior panels of α contained in $\mathcal{C}_-$. (For those contained in $\mathcal{C}_+$, we can apply Proposition 7.57 as in the spherical case.) This additional argument is given in Proposition 8.25 below (which does not require condition (co)). Its proof was kindly suggested to us by Bernhard Mühlherr and is included here with his permission.

Proposition 8.25. *Let* $\mathcal{C} = (\mathcal{C}_+, \mathcal{C}_-)$ *be a thick, irreducible, 2-spherical twin building of rank at least 3, and let* $\alpha = (\alpha_+, \alpha_-)$ *be a twin root of* $\mathcal{C}$. *Assume that* g *is an automorphism of* $\mathcal{C}$ *that fixes* α *pointwise as well as all interior panels of* α *contained in* $\mathcal{C}_+$. *Then* g *fixes all interior panels of* α *pointwise. In other words,* g *is in the root group* U_α *as defined in Definition 8.16.*

Proof. Given an interior panel $\mathcal{P}_-$ of α contained in $\mathcal{C}_-$, we have to show that g fixes all chambers in $\mathcal{P}_-$. In view of Proposition 7.58, it suffices to prove this for $\mathcal{P}_-$ "close" to the boundary of α_-. In order to make this precise in terms of our present W-metric setup, we first generalize the terminology we introduced before Definition 8.1. Namely, we call a residue $\mathcal{R}$ of $\mathcal{C}$ meeting α an *interior residue* of α if $\mathcal{R} \cap \Sigma = \mathcal{R} \cap \alpha$ for any twin apartment Σ containing α, and we call $\mathcal{R}$ a *boundary* residue of α otherwise. So, in view of Proposition 7.58, we may now assume that $\mathcal{P}_-$ is contained in a rank-2 residue $\mathcal{R}_-$ of $\mathcal{C}_-$ that is a boundary residue of α.

Choose a twin apartment $\Sigma = (\Sigma_+, \Sigma_-)$ containing α. Observe that $\mathrm{op}_\Sigma(\mathcal{R}_- \cap \Sigma_-)$ is a (thin) rank-2 residue of Σ_+ having nonempty intersection with α_+ as well as $-\alpha_+$. So $\mathrm{op}_\Sigma(\mathcal{R}_- \cap \Sigma_-)$ is contained in a (unique) rank-2 residue $\mathcal{R}_+$ of $\mathcal{C}_+$ that is a boundary residue of α. Now $\mathcal{R}_+$ and $\mathcal{R}_-$ are opposite spherical residues (recall that $\mathcal{C}$ is 2-spherical). Hence, by Proposition 5.152, $\mathrm{proj}_{\mathcal{R}_+}$ and $\mathrm{proj}_{\mathcal{R}_-}$ induce mutually inverse σ_J-isometries between $\mathcal{R}_-$ and $\mathcal{R}_+$ (where J is the common type of $\mathcal{R}_+$ and $\mathcal{R}_-$). In particular, $\mathcal{P}_+ := \mathrm{proj}_{\mathcal{R}_+} \mathcal{P}_-$ is a panel contained in $\mathcal{R}_+$, and $\mathrm{proj}_{\mathcal{R}_-} \mathcal{P}_+ = \mathcal{P}_-$.

Since α is convex (see Lemma 5.191) and $\mathcal{R}_+$ meets α_+, $\mathrm{proj}_{\mathcal{R}_+}(\mathcal{P}_- \cap \alpha_-)$ is contained in α_+ (see Remark 5.159 and Exercise 5.169). Hence $\mathcal{P}_+$ is an interior panel of α, and g fixes $\mathcal{P}_+$ pointwise by assumption. Since $\mathcal{P}_- = \mathrm{proj}_{\mathcal{R}_-} \mathcal{P}_+$, any chamber $C_- \in \mathcal{P}_-$ is of the form $C_- = \mathrm{proj}_{\mathcal{R}_-} C_+$ for some $C_+ \in \mathcal{P}_+$. Now g stabilizes $\mathcal{R}_-$ (since g is an isometry and stabilizes $\mathcal{P}_-$), and g fixes $C_+ \in \mathcal{P}_+$. Hence g also fixes C_-, and the proof is complete. $\qquad\square$

Remark 8.26. The proposition implies, as we claimed in Remark 8.20(c), that Tits's asymmetric definition of root groups [261, Section 4.3] agrees with our Definition 8.16 for (thick) 2-spherical twin buildings. [One easily reduces to the irreducible case, and then the assertion is trivial unless the rank is at least 3.]

It is now easy to put the pieces together in order to obtain the main result of this section.

Theorem 8.27. *If $\mathcal{C}$ is a thick, irreducible, 2-spherical twin building of rank at least 3 that satisfies condition* (co) *of Section 5.11, then $\mathcal{C}$ is Moufang.*

Proof. Given a twin root α of $\mathcal{C}$ and two twin apartments Σ and Σ' containing α, we need to find an element g in the root group U_α with $g\Sigma = \Sigma'$. By Corollary 5.215, there is an automorphism g of $\mathcal{C}$ that fixes α and $E_2(C_+)$ pointwise for some chamber $C_+ \in \alpha_+^{(2)}$ and that satisfies $g\Sigma = \Sigma'$. By Proposition 7.57, g fixes pointwise all interior panels of α contained in $\mathcal{C}_+$. Proposition 8.25 now implies that g is in U_α. $\qquad\square$

Corollary 8.28. *If $\mathcal{C}$ is as in the theorem, then it is the twin building associated to a twin BN-pair.*

Proof. This follows from the theorem together with Proposition 8.19. $\qquad\square$

Corollary 8.29. *If C is as in the theorem, then all rank-2 residues of C are Moufang. In particular, the Coxeter matrix associated to C has all of its entries $m(s,t)$ in $\{1,2,3,4,6,8\}$.*

Proof. The first assertion follows from the theorem and Proposition 8.21. The second assertion now follows from the result of Tits cited in Remark 7.60. □

Remarks 8.30. (a) Our proof of Corollary 8.29 was based on the extension theorem 5.213. But there are weaker versions of the extension theorem that are sufficient to yield the corollary; see [261, Section 5.6] and [203, Section 8]. And these weaker versions can be proved without using (co). So Corollary 8.29 is in fact true for all thick, irreducible, 2-spherical twin buildings of rank at least 3. This is significant in connection with Remark 5.212 (see assertion (b) of that remark).

(b) Moufang octagons are pretty special (see [256]). So one cannot expect arbitrary Coxeter matrices with all $m(s,t) \in \{1,2,3,4,6,8\}$ to be realizable as the Coxeter matrices of thick twin buildings if 8 actually occurs. But if one excludes the value 8 from this list, i.e., if one restricts to Coxeter matrices with $m(s,t) \in \{1,2,3,4,6\}$, then all of them are in fact realizable. We will see this while discussing Kac–Moody groups (see Remark 8.97), which give rise to twin BN-pairs and hence to twin buildings.

8.5 Group-Theoretic Consequences of the Moufang Property

Throughout this section, $C = (C_+, C_-)$ denotes a strictly Moufang twin building of type (W, S). Choose a fundamental twin apartment $\Sigma = (\Sigma_+, \Sigma_-)$ and a fundamental pair of opposite chambers $C_\pm$, so that $\Sigma = \Sigma\{C_+, C_-\}$. As usual, we denote by Φ the set of twin roots of Σ, by Φ_+ the set of positive twin roots (those containing C_+), and by Φ_- the set of negative twin roots (those containing C_-).

Our purpose in this section is to record the algebraic results about the group

$$G := \langle U_\alpha \mid \alpha \in \Phi \rangle$$

analogous to the results of Section 7.7 for the spherical case. No difficulties arise, and we will be brief.

8.5.1 The Groups $U_\pm$, $B_\pm$, and U_w

We define two subgroups $U_+, U_- \leq G$ by

$$U_\pm := \langle U_\alpha \mid \alpha \in \Phi_\pm \rangle \, ;$$

U_+ is the group that occured in Lemma 8.7, where it was called U. Next, we introduce the two stabilizers

$$B_+ := \{g \in G \mid gC_+ = C_+\} \ ,$$
$$B_- := \{g \in G \mid gC_- = C_-\} \ .$$

Finally, we set

$$T := \mathrm{Fix}_G(\Sigma) = \mathrm{Fix}_G(\Sigma_+) = \mathrm{Fix}_G(\Sigma_-) \ .$$

Note that

$$B_+ \cap B_- = T \ ,$$

since the convex hull of C_+ and C_- is Σ.

Lemma 7.62 and Corollary 7.63 and their proofs go through with no essential change:

Lemma 8.31. $U_+ \cap T = B_+ \cap U_- = \{1\}$.   □

This leads to the following strengthening of Corollary 8.8:

Corollary 8.32. U_+ *acts simply transitively on the set of chambers opposite* C_+.   □

If Ψ is a convex subset of Φ in the sense of Definition 8.10, we set

$$U_\Psi := \langle U_\alpha \mid \alpha \in \Psi \rangle \ ,$$

and one can then prove, as in Proposition 7.64, the following result:

Proposition 8.33. *Let* Ψ *be a convex set of twin roots in the fundamental twin apartment* Σ, *let* $\alpha_1, \ldots, \alpha_m$ *be an admissible ordering of* Ψ, *and set* $U_i := U_{\alpha_i}$ *for* $i = 1, \ldots, m$.

(1) $U_\Psi = U_1 \cdots U_m$. *More precisely, every element of* U_Ψ *is uniquely expressible as* $u_1 \cdots u_m$ *with* $u_i \in U_i$ *for each* i.

(2) *If* $\Psi = \Psi(\mathcal{M})$ *for some pair* $\mathcal{M} = (\mathcal{M}_+, \mathcal{M}_-)$ *with both components nonempty, then*

$$\mathrm{Fix}_G(\mathcal{M}) = U_\Psi T = U_\Psi \rtimes T \ .$$ □

We now specialize to the case $\Psi = \Phi(C_+, wC_+)$ with $w \in W$, where the notation is that of Example 8.13. Equivalently, $\Psi = \Psi(\mathcal{M})$ with $\mathcal{M} = (\{C_+\}, \{wC_-\})$. If we identify Σ_+ with $\Sigma(W, S)$, then the set of first components α_+ of the twin roots $\alpha \in \Psi$ is simply the set $\Phi(w)$ defined exactly as in Definition 7.100. By abuse of notation, we will also write $\Phi(w)$ for the corresponding set of twin roots. We now set

$$U_w := U_{\Phi(w)} \le U_+ \ ,$$

and we have, as in Corollaries 7.66 and 7.67, the following:

Corollary 8.34. *Given $w \in W$, choose a reduced decomposition $w = s_1 \cdots s_l$, let Γ be the corresponding gallery from C_+ to wC_+, and let $\alpha_1, \ldots, \alpha_l$ be the associated sequence of twin roots. Set $U_i := U_{\alpha_i}$ for $i = 1, \ldots, l$.*

(1) *Every element of U_w admits a unique representation as $u_1 \cdots u_l$ with $u_i \in U_i$ for all i.*

(2) *U_w is the stabilizer of wC_- in U_+. In other words,*

$$U_w = U_+ \cap wU_-w^{-1} \, .$$

(3) *U_w acts simply transitively on the w-sphere*

$$\mathcal{C}_w := \{D \in \mathcal{C}_+ \mid \delta_+(C_+, D) = w\} \, . \qquad \square$$

(Note that in the proof of (3), we are allowed to use Lemma 7.22, which we explicitly stated for buildings that are not necessarily spherical.)

Remark 8.35. Part (2) of the corollary enables us to give a complete analysis of the action of U_+ on $\mathcal{C}_-$. Indeed, Lemma 8.7 implies that every U_+-orbit in $\mathcal{C}_-$ is represented by a chamber wC_- for some $w \in W$. And w is uniquely determined by the orbit, since it is equal to $\delta^*(C_+, D)$ for any chamber D in the orbit. Combining this with the stabilizer calculation in the corollary, we obtain a bijection

$$\kappa: \coprod_{w \in W} U_+/U_w \xrightarrow{\sim} \mathcal{C}_- \, , \qquad (8.2)$$

given by $\kappa(xU_w) = xwC_-$ for $x \in U_+$ and $w \in W$. This observation will be useful in Section 8.7.

Exercise 8.36. This is a continuation of Remark 8.35. The goal is to reconstruct the building $(\mathcal{C}_-, \delta_-)$ from the group U_+ and the family of subgroups U_w.

(a) Let $D, D' \in \mathcal{C}_-$ be s-adjacent chambers for some $s \in S$. If $\delta^*(C_+, D) \neq \delta^*(C_+, D')$, show that there is an element $u \in U_+$ such that uD and uD' are both in Σ_-. Equivalently, there are elements $x \in U_+$ and $w \in W$ such that $D = xwC_-$ and $D' = xwsC_-$. If $\delta^*(C_+, D) = \delta^*(C_+, D')$, on the other hand, then we know that D and D' are U_+-equivalent to the same chamber $wC_- \in \Sigma_-$. Show in this case that $l(ws) > l(w)$ and that there are elements $x, y \in U_+$ such that $D = xwC_-$, $D' = ywC_-$, and $xwsC_- = ywsC_-$.

(b) Deduce that $\mathcal{C}_-$, as a chamber system over S (see Section 5.2), is isomorphic to the following chamber system $\mathcal{C}'$. As a set,

$$\mathcal{C}' := \coprod_{w \in W} U_+/U_w \, .$$

Two cosets xU_v and yU_w ($x, y \in U_+$, $v, w \in W$) are declared to be s-equivalent (for $s \in S$) if (i) $v = w$ or ws and (ii) $y^{-1}x$ is in the larger of the two groups U_w, U_{ws}. [Recall or check that $U_w \leq U_{ws}$ if $l(ws) > l(w)$.]

8.5.2 Commutator Relations

Definition 8.37. Given two twin roots $\alpha, \beta \in \Phi$, we say that the pair $\{\alpha, \beta\}$ is *prenilpotent* if both components of $\alpha \cap \beta$ are nonempty. We then define the *closed interval* $[\alpha, \beta]$ by

$$[\alpha, \beta] := \{\gamma \in \Phi \mid \gamma \supseteq \alpha \cap \beta\} \ ,$$

and we define the *open interval* (α, β) by

$$(\alpha, \beta) := [\alpha, \beta] \smallsetminus \{\alpha, \beta\} \ .$$

Remarks 8.38. (a) It is natural to restrict the definition of $[\alpha, \beta]$ to the prenilpotent case, since this guarantees that $[\alpha, \beta]$ is convex in the sense of Definition 8.10. As in the spherical theory, it is then possible to give a very concrete interpretation of the interval. We will do this in the next subsection.

(b) The terminology "prenilpotent" was introduced by Tits [260]. It is motivated by the fact that for a prenilpotent pair $\{\alpha, \beta\}$ of twin roots, the root groups U_α and U_β of our stictly Moufang building $\mathcal{C}$ generate a nilpotent group, provided that all root groups are nilpotent (which is almost always true in concrete examples). This result is stated in Exercise 8.53 below. The spherical case was treated in Exercise 7.112. This group-theoretic background also explains, independently of (a), why one requires the existence of chambers in *both* components of $\alpha \cap \beta$. Indeed, the group $\langle U_\alpha, U_\beta \rangle$ is not nilpotent in general if one component of the intersection is empty, even if all root groups are abelian. Consider, for instance, the example of Section 8.3.2, and take $\alpha = \alpha_{ij}^l$ and $\beta = \alpha_{ji}^m$ for some $i \neq j$ and some $l, m \in \mathbb{Z}$. If $l = -m$, these twin roots are opposite. Otherwise, there are chambers in one (and only one) component of $\alpha \cap \beta$. Regardless of the choice of l, m, the groups U_α and U_β do not generate a nilpotent subgroup, nor do they satisfy any "nice" commutator relations.

The closed interval determined by a prenilpotent pair of roots is convex, as we noted above. The open interval is also convex; one can see this by arguing as in the first part of the proof of Proposition 7.72: We may assume that $\alpha \neq \beta$, which implies that $\beta \not\subseteq \alpha$ (see Exercise 5.200). We can then find adjacent chambers D', D in $\mathcal{C}_+$ or $\mathcal{C}_-$ with $D' \in \alpha \cap \beta$ and $D \in \beta \smallsetminus \alpha$. Similarly, there are adjacent chambers E', E in $\mathcal{C}_+$ or $\mathcal{C}_-$ with $E' \in \alpha \cap \beta$ and $E \in \alpha \smallsetminus \beta$. One then checks immediately that

$$(\alpha, \beta) = \Psi\big((\alpha \cap \beta) \cup \{D, E\}\big) \ , \tag{8.3}$$

so (α, β) is indeed convex.

The rest of the proof of Proposition 7.72 also generalizes with no essential change, as does the proof of Corollary 7.73:

Proposition 8.39. *For all $\alpha \neq \beta$ in Φ such that the pair $\{\alpha, \beta\}$ is prenilpotent,*

$$[U_\alpha, U_\beta] \leq U_{(\alpha, \beta)} .$$ □

Corollary 8.40. *Let $\alpha_1, \ldots, \alpha_d$ be the sequence of twin roots associated to a minimal gallery in Σ_+ as in Example 8.13, and set $U_i := U_{\alpha_i}$ for $i = 1, \ldots, d$. If $d \geq 2$, then*

$$[U_1, U_d] \leq U_2 \cdots U_{d-1} .$$ □

8.5.3 Prenilpotent Pairs

In this final subsection we say a little more about prenilpotent pairs of twin roots, and we describe the interval $[\alpha, \beta]$. For both of these purposes it is convenient to work in just one half of the twin apartment and to take a simplicial approach. Thus Σ now denotes a (simplicial) Coxeter complex of type (W, S), and Φ is its set of roots. If we rewrite Definition 8.37 in terms of first components of twin roots, we are led to the following:

Definition 8.41. Given $\alpha, \beta \in \Phi$, the pair $\{\alpha, \beta\}$ is called *prenilpotent* if $\alpha \cap \beta$ and $(-\alpha) \cap (-\beta)$ each contain at least one chamber. In this case we set

$$[\alpha, \beta] := \{\gamma \in \Phi \mid \gamma \supseteq \alpha \cap \beta \text{ and } -\gamma \supseteq (-\alpha) \cap (-\beta)\} .$$

We record for future reference some immediate consequences of the definition. The proofs are easy and are left to the reader:

Lemma 8.42. *Let α and β be roots.*

(1) *$\{\alpha, \beta\}$ is prenilpotent if and only if $\{-\alpha, -\beta\}$ is prenilpotent. In this case $[-\alpha, -\beta] = -[\alpha, \beta] := \{-\gamma \mid \gamma \in [\alpha, \beta]\}$.*
(2) *If the pair $\{\alpha, \beta\}$ is nested (i.e., $\alpha \subseteq \beta$ or $\beta \subseteq \alpha$), then $\{\alpha, \beta\}$ is prenilpotent.*
(3) *$\{\alpha, \beta\}$ is not prenilpotent if and only if the pair $\{\alpha, -\beta\}$ is nested. In particular, if $\{\alpha, \beta\}$ is not prenilpotent, then $\{\alpha, -\beta\}$ is prenilpotent.* □

Example 8.43. If W is finite, then $\{\alpha, \beta\}$ is prenilpotent if and only if $\beta \neq -\alpha$. (This follows for instance from Lemma 3.53 and part (3) of Lemma 8.42.) And the definition of $[\alpha, \beta]$ in this case coincides with the one given in Definition 7.71, since we can apply the opposition involution to the inclusion $\gamma \supseteq \alpha \cap \beta$.

Example 8.44. Suppose W is a Euclidean reflection group, so that Σ can be identified with a Euclidean space V decomposed into simplices by affine hyperplanes. [Readers who are not familiar with Euclidean reflection groups, which will not be formally introduced until Chapter 10, can think about the case in which V is the plane tiled by equilateral triangles.] The closed half-spaces determined by the hyperplanes can be identified with the roots. For each root α, let e_α be the unit vector orthogonal to $\partial\alpha$ and pointing to the side containing α. Thus α is defined by an inequality $\langle e_\alpha, -\rangle \geq c$ for some $c \in \mathbb{R}$. The result, then, is that two roots α, β form a prenilpotent pair if and only if $e_\alpha \neq -e_\beta$. In more detail, we have the following possibilities for α, β:

(1) $e_\alpha = -e_\beta$ and $\alpha = -\beta$. Then $\alpha \cap \beta = (-\alpha) \cap (-\beta) = \partial\alpha$. Neither intersection contains a chamber, and $\{\alpha, \beta\}$ is not prenilpotent.
(2) $e_\alpha = -e_\beta$ but $\alpha \neq -\beta$. Then $\partial\alpha$ and $\partial\beta$ are distinct and parallel. Of the two intersections $\alpha \cap \beta$ and $(-\alpha) \cap (-\beta)$, one contains a chamber but the other is empty. The pair $\{\alpha, \beta\}$ is not prenilpotent.
(3) $e_\alpha = e_\beta$. Then $\partial\alpha$ and $\partial\beta$ are parallel and the pair $\{\alpha, \beta\}$ is nested. In particular, it is prenilpotent.
(4) $e_\alpha \neq \pm e_\beta$. Then $\partial\alpha$ and $\partial\beta$ intersect and divide V into four quadrants corresponding to the four intersections $\pm\alpha \cap \pm\beta$, each of which contains a chamber. In particular, $\{\alpha, \beta\}$ is prenilpotent.

We now describe the interval $[\alpha, \beta]$ in each of the prenilpotent cases. In case (3), suppose $\alpha \subseteq \beta$. Then $[\alpha, \beta]$ consists of the roots γ such that $\alpha \subseteq \gamma \subseteq \beta$. There are finitely many such γ, one for each wall parallel to $\partial\alpha$ and $\partial\beta$ and between them. See Figure 8.2. In case (4), choose a maximal

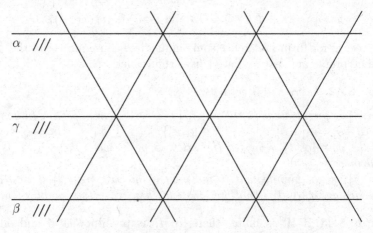

Fig. 8.2. A root $\gamma \in (\alpha, \beta)$; the parallel-walls case.

simplex A in $\partial\alpha \cap \partial\beta$. As in Lemma 7.81, one can then set up a bijection between $[\alpha, \beta]$ and $[\alpha', \beta']$, where α', β' are the roots in the spherical rank-2 Coxeter complex $\Sigma' := \mathrm{lk}_\Sigma A$ obtained by intersecting α and β with Σ'. [See Lemma 8.45 below.] In particular, the bounding wall of every $\gamma \in [\alpha, \beta]$ must contain A, so that γ corresponds to a root γ' of Σ'. See Figure 8.3 for an example; we leave it to the reader to locate α', β', γ' in the picture. Note in both cases that $\{\gamma \in \Phi \mid \gamma \supseteq \alpha \cap \beta\}$ is much bigger than $[\alpha, \beta]$. In fact, it is infinite,

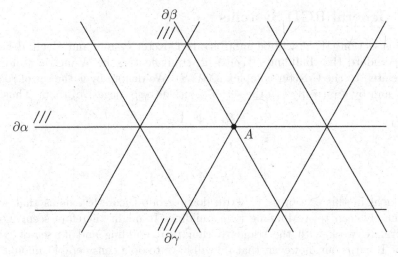

Fig. 8.3. A root $\gamma \in (\alpha, \beta)$; the intersecting-walls case.

whereas $[\alpha, \beta]$ is finite. Thus it is crucial that we require $-\gamma \supseteq (-\alpha) \cap (-\beta)$ in the definition of $[\alpha, \beta]$.

We come now to the main point of this subsection, which is a simple description of the interval $[\alpha, \beta]$ when $\{\alpha, \beta\}$ is prenilpotent (cf. Lemma 7.81).

Lemma 8.45. *Let* $\{\alpha, \beta\}$ *be a prenilpotent pair of roots. Then one of the following holds:*

(a) $\{\alpha, \beta\}$ *is nested, say* $\alpha \subseteq \beta$, *in which case*

$$[\alpha, \beta] = \{\gamma \in \Phi \mid \alpha \subseteq \gamma \subseteq \beta\} .$$

(b) *Every maximal simplex* A *of* $\partial\alpha \cap \partial\beta$ *has codimension 2, and the link* L_A *is spherical. For any such* A, *let* $\alpha' := \alpha \cap L_A$ *and* $\beta' := \beta \cap L_A$ *be the roots of* L_A *corresponding to* α *and* β. *Then there is a bijection*

$$[\alpha, \beta] \xrightarrow{\sim} [\alpha', \beta']$$

given by $\gamma \mapsto \gamma' := \gamma \cap L_A$ *for* $\gamma \in [\alpha, \beta]$.

Proof. If each of the four intersections $(\pm\alpha) \cap (\pm\beta)$ contains a chamber, then we are in case (b). Indeed, the first assertion follows from Lemma 3.164, and the rest is proved exactly as in the proof of Lemma 7.81. [One of course uses the definition of prenilpotence instead of the opposition involution in the proof.] Otherwise, either $\alpha \cap (-\beta)$ or $(-\alpha) \cap \beta$ contains no chamber, which says precisely that $\{\alpha, \beta\}$ is nested. $\qquad\square$

Exercise 8.46. If W is infinite, show that there is always a pair of positive roots $\{\alpha, \beta\}$ that is not prenilpotent.

8.6 General RGD Systems

In this section (W, S) will be an arbitrary Coxeter system and Σ will denote the standard thin building (W, δ_W) or, equivalently, the W-metric building associated to the Coxeter complex $\Sigma(W, S)$. We denote by Φ the set of roots of Σ and by Φ_+ (resp. Φ_-) the set of positive (resp. negative) roots. Thus

$$\Phi_+ = \{\alpha \in \Phi \mid 1 \in \alpha\}$$

and

$$\Phi_- = \{\alpha \in \Phi \mid 1 \notin \alpha\} \ .$$

Our goal in this section is to write down group-theoretic axioms that will ultimately lead to a Moufang twin building. It might therefore seem more natural to work with the standard thin *twin* building and its set of twin roots. It turns out, however, that we will have to do a considerable amount of work with the $\mathcal{C}_+$ half of the desired twin building before we can complete the construction. It is therefore more convenient to work with just the ordinary thin building Σ.

8.6.1 The RGD Axioms

To define an RGD system $\big(G, (U_\alpha)_{\alpha \in \Phi}, T\big)$ of type (W, S), one simply repeats Definition 7.82 verbatim with the exception of (RGD1). This is rewritten as follows:

(RGD1) *For all $\alpha \neq \beta$ in Φ such that $\{\alpha, \beta\}$ is prenilpotent in the sense of Definition 8.41,*

$$[U_\alpha, U_\beta] \leq U_{(\alpha, \beta)} \ .$$

The interval (α, β), of course, now has to be interpreted as in Definition 8.41. In view of Example 8.43, the new (RGD1) is equivalent to the original one if W is finite, so there is no harm in continuing to call it (RGD1). This is the definition given by Tits [261, Section 3.3], except that he *defines* T to be $\bigcap_{\alpha \in \Phi} N_G(U_\alpha)$; our reasons for not doing this were explained in Remark 7.84.

Examples 8.47. (a) Let $\mathcal{C} = (\mathcal{C}_+, \mathcal{C}_-)$ be a strictly Moufang twin building of type (W, S) with fundamental twin apartment (Σ_+, Σ_-) and fundamental chamber $C_+ \in \Sigma_+$. We identify Σ_+ with W. The set of twin roots of (Σ_+, Σ_-) is then identified with the set of roots Φ defined above. For each $\alpha \in \Phi$ let $U_\alpha \leq \operatorname{Aut} \mathcal{C}$ be the corresponding root group. Set

$$G := \langle U_\alpha \mid \alpha \in \Phi \rangle \leq \operatorname{Aut} \mathcal{C} \ ,$$

and define T by

$$T = \operatorname{Fix}_G(\Sigma) \ .$$

Then one checks, as in Example 7.83, that $\big(G, (U_\alpha)_{\alpha \in \Phi}, T\big)$ is an RGD system of type (W, S).

(b) For a specific example, we can apply (a) to the twin building of type $\tilde{A}_{n-1}$ discussed in Section 8.3.2. But it is actually easier to treat this example purely algebraically. Namely, we start with $G = \mathrm{SL}_n\left(k[t, t^{-1}]\right)$, where k is a field, we define T to be the diagonal subgroup of $\mathrm{SL}_n(k)$, and we define the U_α to be the groups $U_{ij}^l \leq G$ introduced in Section 8.3.2. The RGD axioms are easy to verify. [Strictly speaking, we should take $G = \mathrm{PSL}_n\left(k[t, t^{-1}]\right)$ if we really want this to be a special case of (a).]

As we mentioned above, we are heading toward a proof that every RGD system has an associated Moufang twin building. As a first step, one proves as in Theorem 7.115 the existence of a Tits system (G, B_+, N, S), which leads to a (generally nonspherical) building that is Moufang in a suitable sense. This first step is not so different from our Section 7.8, and we will explain it below. But we will have to work considerably harder to get a *twin* Tits system (G, B_+, B_-, N, S). The proof of this will not be completed until Section 8.8.

Remarks 8.48. (a) We could simplify our development substantially if we replaced (RGD3) by the following stronger (and symmetric) axiom:

(RGD3″) *For all* $s \in S$,

$$U_{-s} \not\leq U_+ \quad and \quad U_s \not\leq U_- .$$

This would be harmless for many purposes, since in concrete examples both statements are often easy to verify. But it is interesting and potentially useful that one needs to require only the first of these two conditions, and the second then follows. We will therefore go to some trouble in order to avoid assuming (RGD3″). In particular, our proof of the following assertion will be less straightforward than it was in the spherical case:

$$\alpha \neq \beta \implies U_\alpha \neq U_\beta \quad \text{for } \alpha, \beta \in \Phi. \tag{8.4}$$

If we were willing to assume (RGD3″), we would be able to prove this by a slight variant of the argument that we used in the spherical case; see Remark 8.49 below.

(b) The most serious difficulty in constructing a twin BN-pair is proving the equality

$$U_- \cap B_+ = \{1\} ,$$

to which we will devote the long and technical Section 8.7. This was much easier in the spherical case (see Corollary 7.122), where we were able to use the longest element $w_0 \in W$. To the best of our knowledge, there is no complete proof of this equality in the literature at the time of this writing.

(c) We will later prove, after having done the work referred to in (b), that (RGD3″) is a consequence of our axioms; see Section 8.8. So replacing (RGD3) by (RGD3″) yields an equivalent system of axioms.

Let us now review the program we worked through in Section 7.8 and see how it generalizes. There is no change needed in Section 7.8.2. That dealt with the rank-1 situation and is still applicable to our present setup. It will be important for us that in the rank-1 case, it was actually possible to replace (RGD3) by a *weaker* (and symmetric) assumption; see Remark 7.86. We turn next to the generalization of Section 7.8.3, concerning the Weyl group.

8.6.2 The Weyl Group

Here we have to use a somewhat roundabout approach because the method we used in the spherical case was based on Lemma 7.90. That lemma, as we saw in Exercise 7.96, is not valid for general Coxeter complexes. We proceed in several steps.

(1) By (RGD2) and (RGD5), the group $N := \langle T, \{m(u) \mid u \in U_s^*, s \in S\}\rangle$ acts on $\mathcal{U} := \{U_\alpha \mid \alpha \in \Phi\}$ by conjugation, and the action of T is trivial. By statement (6) in Section 7.8.2 (applied to $G_s := T\langle U_s, U_{-s}\rangle$), we also know that $T \trianglelefteq N$. Hence we have a well-defined action of N/T on $\mathcal{U}$. (However, we cannot at this point identify $\mathcal{U}$ with Φ.)

(2) We wish to define an action of W on $\mathcal{U}$, which we will write as conjugation. The action of an element $w \in W$ will be defined to be conjugation by a suitably chosen $\tilde{w} \in N$. [It will then follow that this action is conjugation by n for any $n \in \tilde{w}T$.] We first define the action of a generator $s \in S$. Choose $u_s \in U_s^*$ and set $\tilde{s} := m(u_s) \in N$. By the last sentence in (RGD2), the coset $\tilde{s}T$ does not depend on the choice of u_s, so we may set

$$sUs^{-1} := \tilde{s}U\tilde{s}^{-1}$$

for $U \in \mathcal{U}$. And by the first part of (RGD2), we then have

$$sU_\alpha s^{-1} = U_{s\alpha} \tag{8.5}$$

for all $\alpha \in \Phi$. Now consider an arbitrary $w \in W$, and choose a decomposition $w = s_1 \cdots s_l$ (not necessarily reduced) with $s_i \in S$ for all i. We then write $\tilde{w} := \tilde{s}_1 \cdots \tilde{s}_l \in N$. This of course depends on the chosen decomposition of w, but by (8.5), the conjugation action of $\tilde{w}$ on $\mathcal{U}$ depends only on w. In fact,

$$\tilde{w}U_\alpha\tilde{w}^{-1} = U_{w\alpha}$$

for all $\alpha \in \Phi$. Hence we can set

$$wUw^{-1} := \tilde{w}U\tilde{w}^{-1}$$

for $U \in \mathcal{U}$, and we then have

$$wU_\alpha w^{-1} = U_{w\alpha} \tag{8.6}$$

for all $w \in W$ and $\alpha \in \Phi$. It follows from (8.6) that we have indeed defined an action of W on $\mathcal{U}$, i.e., a homomorphism $W \to \mathrm{Sym}\,\mathcal{U}$.

We can summarize what we have done so far in the diagram

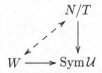

where the dashed arrow is a reminder that we expect $N/T \cong W$. We will show shortly that there is in fact an isomorphism $W \xrightarrow{\sim} N/T$, given by $w \mapsto \tilde{w}T$. At the moment, however, we do not have a well-defined map between W and N/T (in either direction).

(3) For any $w \neq 1$ in W we can find, by Exercise 7.97, a root $\alpha \in \Phi_+$ and an element $s \in S$ such that $w\alpha = -\alpha_s$. Then $wU_\alpha w^{-1} = U_{-s}$ and hence, by (RGD3),

$$wU_\alpha w^{-1} \not\leq U_+ .$$

This shows, in particular, that the action of W on $\mathcal{U}$ is faithful.

(4) Lemma 7.91 can now be modified as follows: Suppose $s_1 \cdots s_l = 1$ in W, and suppose we have an element $n \in \tilde{1}T = \tilde{s}_1 \cdots \tilde{s}_l T$ such that $n \in G_s$ for some $s \in S$. Then $n \in T$.

Proof. By (2), we have $nU_\alpha n^{-1} = U_\alpha$ for all $\alpha \in \Phi$. Since $n \in G_s$, we conclude that $n \in N_{G_s}(U_s) \cap N_{G_s}(U_{-s}) = T$, where the equality follows from statement (15) in Section 7.8.2. □

(5) The main argument in the proof of Lemma 7.92 can now be phrased as follows: There exists a well-defined homomorphism $\mu\colon W \to N/T$ with $\mu(s) = \tilde{s}T$ for all $s \in S$. Moreover, μ is an isomorphism.

Proof. Note first that μ is automatically an isomorphism if it exists. Indeed, it is surjective since, by the definition of N, N/T is generated by the $\tilde{s}T$ for $s \in S$. And it is injective since the composite $W \to N/T \to \mathrm{Sym}\,\mathcal{U}$ is injective by (3). So it suffices to verify that the $\tilde{s}T$ satisfy the Coxeter relations, i.e., that $(\tilde{s}\tilde{t})^{m(s,t)} \in T$ for all $s, t \in S$. We already know that $\tilde{s}^2 \in T$ for all $s \in S$ by statement (7) in Section 7.8.2. So suppose $s \neq t$, and set $m := m(s,t)$. As in the proof of Lemma 7.92, let w be the alternating product $st\cdots$ of length $m-1$, and let $x = s$ or t, depending on whether m is odd or even. We have $w\alpha_x = \alpha_t$ and hence $\tilde{w}G_x\tilde{w}^{-1} = G_t$ by (2). In particular, $\tilde{w}\tilde{x}\tilde{w}^{-1} \in G_t$; hence also $n := (\tilde{s}\tilde{t})^m \in G_t$, since $n \equiv \tilde{w}\tilde{x}\tilde{w}^{-1}\tilde{t}$ mod T (recall that $\tilde{s} \equiv \tilde{s}^{-1}$ and $\tilde{t} \equiv \tilde{t}^{-1}$ mod T). Applying (4) to the relation $(st)^m = 1$, we conclude that $n \in T$, as required. □

(6) We can now define $\nu\colon N \twoheadrightarrow W$ as the composite of the natural projection $N \twoheadrightarrow N/T$ with $\mu^{-1}\colon N/T \to W$. We then have $\ker \nu = T$ as well as

$$nU_\alpha n^{-1} = U_{\nu(n)\alpha}$$

for all $n \in N$ and all $\alpha \in \Phi$.

(7) The proofs of Lemma 7.93 and Corollary 7.94 now go through with no essential change. Thus we have

$$B_+ \cap N = T \tag{8.7}$$

and

$$U_{-s} \cap B_+ = \{1\} \tag{8.8}$$

for all $s \in S$.

So we have managed to prove all the relevant statements in Section 7.8.3 for general (W, S) without knowing that the root groups U_α are distinct. For the sake of completeness, we will finish this section by proving that fact as well. As preparation, we prove the following:

(8) For any root $\alpha \in \Phi_-$, $U_\alpha \nleq U_+$.

Proof. Choose $s \in S$ and $w \in W$ such that $\alpha = w\alpha_s$ and hence $-\alpha = w(-\alpha_s)$. Also choose $n \in N$ with $\nu(n) = w$. Then $nG_s n^{-1} = T\langle U_\alpha, U_{-\alpha}\rangle$ by (6). If we had $U_\alpha \leq U_+$, then we would have $T\langle U_\alpha, U_{-\alpha}\rangle \leq TU_+ = B_+$. This would imply $n\tilde{s}n^{-1} \in B_+ \cap N = T$ and hence $\tilde{s} \in n^{-1}Tn = T$. But then $\tilde{s}$ would normalize U_s and U_{-s}, so we have a contradiction. $\square$

(9) If α and β are distinct roots, then $U_\alpha \neq U_\beta$.

Proof. Suppose, for instance, that $\beta \nsubseteq \alpha$, and choose a chamber $C \in \beta \smallsetminus \alpha$. Let $w \in W$ be the element such that wC is the fundamental chamber. Then $w\beta \in \Phi_+$ and $w\alpha \in \Phi_-$. We have $U_{w\alpha} \nleq U_+$ by (8), so $U_{w\alpha} \neq U_{w\beta}$. Since $U_{w\alpha} = wU_\alpha w^{-1}$ and $U_{w\beta} = wU_\beta w^{-1}$, it follows that $U_\alpha \neq U_\beta$. $\square$

Remarks 8.49. (a) Suppose we replace (RGD3) by the axiom (RGD3″) mentioned in Remark 8.48(a). There is then a straightforward proof of (9), very similar to the proof we gave in the spherical case: First, we may transform α and β by a suitable element of W in order to reduce to the case $\alpha = -\alpha_s$ for some $s \in S$. If $\beta \in \Phi_+$, then we argue as in the proof of Lemma 7.89. If $\beta \in \Phi_-$, then we instead apply the symmetric argument to $s\alpha = \alpha_s$ and $s\beta$, which is still in Φ_-. [Recall that $-\alpha_s$ is the only negative root taken to a positive root by s.] So if we had been willing to adopt (RGD3″) as an axiom, the methods of Section 7.8.3 would have gone through with no change for general (W, S).

(b) At this point in the development of the spherical theory, we were able to use the longest element w_0 to prove that the theory is $\pm$-symmetric. We are not yet at that point for the general theory, but the following observations will be useful in the meantime. Given any RGD system $(G, (U_\alpha)_{\alpha \in \Phi}, T)$, we can try to define a new RGD system by setting $U'_\alpha := U_{-\alpha}$ for $\alpha \in \Phi$. The new system $(G, (U'_\alpha)_{\alpha \in \Phi}, T)$ still satisfies (RGD0), (RGD1), (RGD2), (RGD4), and (RGD5), as well as the following weak form of (RGD3):

(RGD3′) *For all* $s \in S$, $U_{-s} \neq U_s$.

Indeed, it is clear that the new system satisfies (RGD0), (RGD3′), (RGD4), and (RGD5). In view of Lemma 8.42(1), (RGD1) for the old system implies (RGD1) for the new system. Finally, (RGD2) for the new system follows from (RGD2) for the old system together with the results in Section 7.8.2; see Remarks 7.86 and 7.87. As a consequence, everything we have proved about RGD systems using the weaker axioms (i.e., with (RGD3′) in place of (RGD3)) is also true for $\bigl(G, (U'_\alpha)_{\alpha \in \Phi}, T\bigr)$.

8.6.3 The Groups U_w

Almost everything in Section 7.8.4 extends to general RGD systems except, of course, results that use the longest element w_0. But some of the proofs need modification, starting with the very first result, Lemma 7.98, whose proof requires more work than in the spherical case:

Lemma 8.50. $B_+ s B_+ = U_s s B_+$ *for all* $s \in S$.

Proof. As before, it suffices to show that $U_\alpha U_s s B_+ \subseteq U_s s B_+$ for all $\alpha \neq \alpha_s$ in Φ_+. If the pair $\{\alpha, \alpha_s\}$ is prenilpotent, then the proof of Lemma 7.98 remains valid. So suppose $\{\alpha, \alpha_s\}$ is not prenilpotent. Then, by Lemma 8.42(3), $\{\alpha, -\alpha_s\}$ is nested, hence prenilpotent. Since $\alpha \in \Phi_+$ and $-\alpha_s \in \Phi_-$, the only way they can be nested is if $-\alpha_s \subseteq \alpha$; thus

$$[-\alpha_s, \alpha] = \{\gamma \in \Phi \mid -\alpha_s \subseteq \gamma \subseteq \alpha\} \ .$$

It follows that every $\gamma \in (-\alpha_s, \alpha)$ is positive. Indeed, we have $s \in -\alpha_s \implies s \in \gamma \implies 1 \in \gamma$. [Recall that $-\alpha_s$ is the only root containing s but not 1.] It also follows that $\alpha_s \notin [-\alpha_s, \alpha]$. So we may apply (RGD1) to obtain

$$[U_\alpha, U_{-s}] \leq U_{\Phi_+ \smallsetminus \{\alpha_s\}} \ . \tag{8.9}$$

Next, we recall that the rank-1 group $G_s = T\langle U_s, U_{-s}\rangle$ can be written as $G_s = U_s \{1, s\} T U_s$ (see statement (11) in Section 7.8.2); hence $G_s B_+ = U_s \{1, s\} B_+$. Since $G_s = s G_s$, we also have $G_s B_+ = s U_s \{1, s\} B_+ = U_{-s} s \{1, s\} B_+ = U_{-s} \{1, s\} B_+$. Hence

$$G_s B_+ = U_s \{1, s\} B_+ = U_{-s} \{1, s\} B_+ \ . \tag{8.10}$$

Combining (8.9) and (8.10), we obtain

$$\begin{aligned}
U_\alpha U_s s B_+ &\subseteq U_\alpha G_s B_+ \\
&= U_\alpha U_{-s} \{1, s\} B_+ \\
&\subseteq U_{-s} U_\alpha [U_\alpha, U_{-s}] \{1, s\} B_+ \\
&\subseteq U_{-s} U_{\Phi_+ \smallsetminus \{\alpha_s\}} \{1, s\} B_+
\end{aligned}$$

$$= U_{-s} \{1, s\} \, U_{\Phi_+ \smallsetminus \{\alpha_s\}} B_+$$
$$= U_{-s} \{1, s\} \, B_+$$
$$= U_s \{1, s\} \, B_+$$
$$= B_+ \cup U_s s B_+ \, .$$

So either $U_\alpha U_s s B_+ \subseteq U_s s B_+$, which is what we wanted to prove, or else $U_\alpha U_s s B_+$ meets B_+. In the second case s (or rather any representative of s in N) is in B_+, and the lemma is trivial.

Remark 8.51. The second case that just arose is actually impossible by equation (8.7). But the proof of that equation required (RGD3), which we wanted to avoid using (see Remark 8.49(b)).

The rest of Section 7.8.4 now generalizes with only minor changes to arbitrary Coxeter systems (W, S). Specifically:

- Corollary 7.99 and its proof remain valid. Thus $P_s := B_+ \cup B_+ s B_+ = B_+ \cup U_s s B_+ = G_s B_+$ is a subgroup of G. Note that $P_s = s P_s$ as in the proof of Lemma 8.50 (or because P_s is a group), so we also have

$$P_s = s(B_+ \cup U_s s B_+) = s B_+ \cup U_{-s} B_+ \, . \tag{8.11}$$

This alternative description of P_s will be useful later.

- Definition 7.100 requires no modification.

- In the statement of Lemma 7.101(1) we should add the assertion that $\{\alpha, \beta\}$ is prenilpotent for all $\alpha, \beta \in \Phi(w)$; this is immediate from the definition of $\Phi(w)$. And in the proof of (1), we of course use the definition of $[\alpha, \beta]$ for general (W, S) instead of appealing to the opposition involution. The rest of Lemma 7.101 requires no change.

- Lemmas 7.102 and 7.104 and their proofs remain valid. The sharpening of Lemma 7.102 stated in Remark 7.103(b) also remains valid and will be proved after we have constructed a building (see Proposition 8.59).

- The first part of Proposition 7.106 remains valid, with the same proof. But we can no longer conclude that U_+ is nilpotent, and in fact, U_+ is *never* nilpotent if W is infinite. We will explain this further in Section 8.10 (see Remark 8.87).

- Concerning Remark 7.107(a), it is *not* true in general that root groups are always nilpotent. As a practical matter, however, it is true in most "natural" examples. Remarks 7.107(b) and 7.107(c) are still true as stated.

- Proposition 7.108, giving a presentation of U_w, remains valid. But this presentation does not immediately apply to U_+, since the latter is not equal to a U_w in general. With considerably more work, however, one can in fact obtain a similar presentation of U_+. We will do this in Section 8.10.

Note that the results of Section 7.8.4 that we have generalized above do not require (RGD3), so we can apply the observations in Remark 8.49(b). For example, here is the "dualized" version of Lemma 7.104:

Lemma 8.52. *With $B_- := TU_-$ and $U'_w := U_{-\Phi(w)}$, we have*

$$B_- w B_- = U'_w w B_-$$

for all $w \in W$. □

Exercise 8.53. State and prove an analogue of the result of Exercise 7.112 for general RGD systems.

8.6.4 The Building $\mathcal{C}(G, B_+)$

Continuing with our program, we can now construct *one* BN-pair.

Proposition 8.54.

(1) (G, B_+, N, S) *is a Tits system with Weyl group* $N/(B_+ \cap N) = N/T \cong W$.
(2) *For $s \in S$ and $w \in W$, we have $B_- s B_- w B_- = B_- sw B_-$ if $l(sw) > l(w)$ and $B_- s B_- w B_- \subseteq B_- \{w, ws\} B_-$ if $l(sw) < l(w)$.*
(3) *If $s \in S$ and $w \in W$ satisfy $l(sw) < l(w)$, then $B_+ s B_+ w B_- = B_+ sw B_-$ and $B_- s B_- w B_+ = B_- sw B_+$.*

Proof. (1) In view of the preparations we have already made, the proof is the same as that of Theorem 7.115.

(2) This was proved for B_+ instead of B_- in the course of proving (1). Since the proof did not require (RGD3), it also applies to B_- by Remark 8.49(b).

(3) Recall that $B_+ s B_+ = U_s s B_+$ and hence, by taking inverses, $B_+ s B_+ = B_+ s U_s$. Similarly (again by Remark 8.49(b)), $B_- s B_- = B_- s U_{-s}$. If $l(sw) < l(w)$, then $w \notin \alpha_s$, i.e., $\beta := w^{-1}(a_s)$ is a negative root. Hence

$$B_+ s B_+ w B_- = B_+ s U_s w B_- = B_+ sw U_\beta B_- = B_+ sw B_-$$

and

$$B_- s B_- w B_+ = B_- s U_{-s} w B_+ = B_- sw U_{-\beta} B_+ = B_- sw B_+ . \qquad \square$$

As in the proof of Corollary 7.117, the proposition yields the following:

Corollary 8.55. $N_G(U_+) = B_+$. □

(We cannot, however, show that $N_G(U_-) = B_-$ at this point.)

Next, we introduce the building associated to the Tits system of Proposition 8.54. Since we are working in the W-metric framework in this chapter, we will denote this building by $\mathcal{C}(G, B_+)$ instead of $\Delta(G, B_+)$. We will also use the W-metric notions of *boundary panel* and *interior panel* of a root, defined as in Section 8.1. There is no difficulty getting the same consequences for the building $C_+ = C(G, B_+)$ as in Theorem 7.116 (except that we momentarily postpone the discussion of the Moufang property). Let's spell this out in detail. We denote by Σ_+ the fundamental apartment and by C_+ the fundamental chamber B_+.

Proposition 8.56. *The system of groups $(U_\alpha)_{\alpha \in \Phi}$ acting on C_+ satisfies the following properties:*

(1) *For each $\alpha \in \Phi$, U_α fixes the root α of Σ_+ pointwise.*
(2) *For each $\alpha \in \Phi$ and each boundary panel $\mathcal{P}$ of α, the action of U_α is simply transitive on $C(\mathcal{P}, \alpha) := \mathcal{P} \smallsetminus \{C\}$, where C is the chamber in $\mathcal{P} \cap \alpha$.*
(3) *If $\{\alpha, \beta\}$ is a prenilpotent pair of distinct roots, then*

$$[U_\alpha, U_\beta] \leq U_{(\alpha,\beta)} .$$

Proof. Parts (1) and (2) are proved as in the proof of Theorem 7.116, and (3) is just one of our axioms. ◻

Remarks 8.57. (a) Statement (2) shows that for every $\alpha \in \Phi$, the group U_α can be identified with a subgroup of $\mathrm{Aut}(C_+)$.

(b) According to Ronan [200, p. 74], a thick building C (with no isolated nodes in its Coxeter diagram) is called *Moufang* if there exists, for a fixed apartment Σ_0 of C, a system $(U_\alpha)_{\alpha \in \Phi}$ of subgroups of $\mathrm{Aut}_0 C$ satisfying the conditions of Proposition 8.56 together with (RGD2). So what we have shown so far is that a (general) RGD system always gives rise to a (general) Moufang building, provided the Coxeter diagram has no isolated nodes.

(c) Conversely, it is easy to see that a (general) Moufang building C always gives rise to an RGD system $(G, (U_\alpha)_{\alpha \in \Phi}, T)$, where the U_α are as in the definition of a Moufang building, G is the subgroup of $\mathrm{Aut}_0 C$ generated by the U_α, and $T := \langle m(u)^{-1} m(v) \mid u, v \in U_s^*, s \in S \rangle$ (here S is the set of reflections of Σ_0 with respect to the walls of a fixed chamber C_0). Note that we easily get (RGD3) in this context, but it is not at all clear why U_α cannot be contained in U_- for $\alpha \in \Phi_+$.

As Ronan observed, one important consequence of (1)–(3) in Proposition 8.56 is the property of root groups that is the defining property in the spherical case. We already discussed this in the spherical case in Section 7.7.3, and only minor modifications are needed in the present more general setup.

Lemma 8.58. *If a building C admits a system $(U_\alpha)_{\alpha \in \Phi}$ of subgroups of $\mathrm{Aut}_0 C$ satisfying properties (1)–(3) of Proposition 8.56, then for every $\alpha \in \Phi$ and every interior panel $\mathcal{P}$ of α, the group U_α fixes $\mathcal{P}$ pointwise.*

[The proof will show that transitivity suffices in property (2).]

Proof. The proof of Proposition 7.79 goes through with one minor change: In that proof we had a pair of opposite roots determined by an interior panel of α, and we arbitrarily called one of them β and applied the commutator relations. In the present setting we instead choose β so that the pair $\{\alpha, \beta\}$ is prenilpotent. This is possible by Lemma 8.42(3).   □

Finally, we show that Corollaries 7.119 and 7.120 remain valid for general RGD systems (hence Corollary 7.121 also holds). This may not be immediately obvious, since we do not yet have a twin building, so we will write out a proof in detail, using a different method:

Proposition 8.59. *Fix $w \in W$.*

(1) *Given a reduced decomposition $w = s_1 \cdots s_n$, let $\alpha_1, \ldots, \alpha_n$ be as in Lemma 7.101(2). Then every $u \in U_w$ admits a unique decomposition $u = u_1 \cdots u_n$ with $u_i \in U_{\alpha_i}$ for $1 \leq i \leq n$.*

(2) *U_w acts simply transitively on $\mathcal{C}_w := \{C \in \mathcal{C}_+ \mid \delta(C_+, C) = w\}$. Consequently,*

$$w^{-1} U_w w \cap B_+ = \{1\} \tag{8.12}$$

for all $w \in W$.

Proof. We have $U_w = U_{\alpha_1} \cdots U_{\alpha_n}$ by the analogue of Lemma 7.102(1). So (1) and the first assertion of (2) will follow if we can prove the following claim: Given any chamber $C \in \mathcal{C}_w$, there are unique elements $u_i \in U_{\alpha_i}$ $(1 \leq i \leq n)$ such that $C = u_1 \cdots u_n C_w$, where C_w is the chamber wB_+ (i.e., it is the chamber of Σ_+ at Weyl distance w from C_+).

Consider the minimal gallery

$$\Gamma \colon B_+, s_1 B_+, s_1 s_2 B_+, \ldots, s_1 \cdots s_n B_+ = wB_+$$

in Σ_+ corresponding to the given reduced decomposition of w. Then the roots $\alpha_1, \ldots, \alpha_n$ in (1) are the same as the roots associated to Γ as in Lemma 7.22. Our claim now follows at once from that lemma, applied with Γ' equal to the (unique) gallery from C_+ to C having type $(s_1, \ldots, s_n)$.

Finally, the second assertion of (2) follows from the first.   □

Let's look at what we have achieved so far. We have a Tits system (G, B_+, N, S), and we have a Moufang building $\mathcal{C}_+ = \mathcal{C}(G, B_+)$ on which G acts strongly transitively. In order to get a twin Tits system, we still need to prove $N \cap B_- = T$, $sB_-s \neq B_-$ for all $s \in S$, and $B_+ s \cap B_- = \emptyset$ for all $s \in S$. In view of Proposition 8.54, the first two of these conditions would make (G, B_-, N, S) a Tits system with Weyl group W, and the third condition would then make (G, B_+, B_-, N, S) a twin Tits system. It is not at all easy to verify these conditions at this point. Indeed, it is not even clear that $B_- \neq G$.

In the spherical case, all of this was easy, since we could use conjugation by w_0 to interchange the roles of U_+ and U_-. See, for instance, Corollary 7.122

and its proof. In the general case, a completely new idea is needed, involving the covering theory of buildings (Section 5.12). So we interrupt our discussion of the group-theoretic consequences of the RGD axioms in order to construct, in the next section, a covering map that will enable us to prove the fundamental equation

$$U_- \cap B_+ = \{1\} . \tag{8.13}$$

The proof of (8.13) is long and technical and is the only place in this book where covering theory is needed. Some readers may therefore prefer to take (8.13) on faith and move ahead to Section 8.8, where we will complete the construction of a twin BN-pair and give some further algebraic consequences of the RGD axioms.

Exercises

8.60. Show that $U_- W B_+ = G$, and give a geometric interpretation.

8.61. Given $s \in S$ and $w \in W$ with $l(sw) > l(w)$, show that

$$w^{-1} U_{-s} w \nleq B_+ .$$

8.62. Show that B_+ and B_- (or U_+ and U_-) are conjugate in G if and *only if* W is finite. This shows that there is no hope of using conjugacy arguments to interchange $+$ and $-$ in the nonspherical case. [The same result has already occurred in Exercise 6.90 for twin BN-pairs, but we do not yet know that we have a twin BN-pair.]

8.63. Explain the last sentence of Remark 8.57(c).

8.7 A 2-Covering of $\mathcal{C}(G, B_+)$

We continue to assume that we are given an RGD system $(G, (U_\alpha)_{\alpha \in \Phi}, T)$. As we said above, our goal in this section is to prove equation (8.13) using the combinatorial covering theory of buildings. The reader may need to review Section 5.12 before proceeding. Our strategy will be to introduce a chamber system $\mathcal{C}'$, which is simply a disguised version of $\mathcal{C}_+ := \mathcal{C}(G, B_+)$, constructed so that U_- acts on it in the way we expect U_- to act on $\mathcal{C}_+$. We will then exhibit a 2-covering $\kappa: \mathcal{C}' \to \mathcal{C}_+$. The main result of Section 5.12 will then show that κ is an isomorphism, and equation (8.13) will fall out easily. See Rémy [195, Theorem 3.5.2] for a slightly different version of the theory presented here. Both versions are based on hints given by Tits [260, 261].

8.7.1 The Chamber System $\mathcal{C}'$

Recall some notation that was introduced in Remark 8.49(b).

Definition 8.64. For any $\alpha \in \Phi$ we set $U'_\alpha := U_{-\alpha}$, and for any $w \in W$ we set

$$U'_w := \langle U'_\alpha \mid \alpha \in \Phi(w) \rangle = \langle U_\alpha \mid \alpha \in -\Phi(w) \rangle.$$

Here are some simple consequences of the definition.

Lemma 8.65. *For all $w, w_1, w_2 \in W$, we have:*

(1) $U'_w = w U_{w^{-1}} w^{-1}$.

(2) *If $l(w_1 w_2) = l(w_1) + l(w_2)$, then $U'_{w_1 w_2} = U'_{w_1} w_1 U'_{w_2} w_1^{-1}$. In particular,* $U'_{ws} = U'_w w U'_s w^{-1}$ *if $l(ws) > l(w)$ ($s \in S$).*

Proof. (1) follows immediately from the definition of U'_w and the easily checked equation $-\Phi(w) = w\Phi(w^{-1})$.

(2) is already known for the groups U_w (see Lemma 7.102 and the remarks in Section 8.6.3). The proof does not require (RGD3), so it also applies to the groups U'_w. $\qquad\square$

Recall now that we are expecting $\mathcal{C}_+ := \mathcal{C}(G, B_+)$ to be the first component of a Moufang twin building $\mathcal{C} = (\mathcal{C}_+, \mathcal{C}_-)$. We therefore know what $\mathcal{C}_+$ "ought" to look like as a set with U_--action. Namely, the fundamental apartment Σ_+ should be a fundamental domain for this action, and the stabilizers of the chambers in Σ_+ should be the groups U'_w. (See Remark 8.35, with the roles of $+$ and $-$ reversed.) This leads us to define $\mathcal{C}'$ by

$$\mathcal{C}' := \coprod_{w \in W} U_- / U'_w. \tag{8.14}$$

Next, for any $s \in S$ we define a relation of s-equivalence in $\mathcal{C}'$ by

$$x U'_v \sim_s y U'_w \iff \{v, vs\} = \{w, ws\} \text{ and } y^{-1} x \in U'_w \cup U'_{ws} \tag{8.15}$$

for $x, y \in U_-$ and $v, w \in W$. Exercise 8.36 explains the motivation behind this definition.

Remarks 8.66. (a) It follows from Lemma 8.65(2) (or directly from the definitions) that $U'_w \le U'_{ws}$ if $l(ws) > l(w)$ and that the opposite inclusion holds if $l(ws) < l(w)$. Hence the union $U'_w \cup U'_{ws}$ is simply equal to the larger of these two groups. This makes it easy to verify that the relation $\sim_s$ is well defined and is an equivalence relation.

(b) U_- acts on $\mathcal{C}'$ by left translation as a group of chamber-system isomorphisms. If we denote by Σ' the set of chambers U'_w ($w \in W$), then Σ' is a fundamental domain for the action, and the subgroups U'_w are the stabilizers of the chambers of Σ'.

(c) The interested reader can prove, as an exercise, that $U'_v = U'_w$ only if $v = w$ $(v, w \in W)$. Using this, one can identify $\mathcal{C}'$ with the collection of subsets xU'_w of G. We will not use this in what follows, since it will suffice to view $\mathcal{C}'$ as an abstract disjoint union. But we will then have to make the following notational convention: If we write $xU_v = yU_w$ with $x, y \in U_-$ and $v, w \in W$, we want this to be understood as an equation in $\mathcal{C}'$, which means, in particular, that $v = w$.

Lemma 8.67. *The groups U'_w $(w \in W)$ generate U_-.*

Proof. Given $\alpha \in \Phi_+$, choose any $w \in -\alpha$. Then we have $\alpha \in \Phi(w)$ (see Definition 7.100); hence $U'_\alpha \leq U'_w$. Since $U_- = \langle U'_\alpha \mid \alpha \in \Phi_+ \rangle$, this proves the lemma. $\qquad\square$

Lemma 8.68. *The chamber system $\mathcal{C}'$ is connected.*

Proof. In the following, we will use the notation C—D as a schematic representation of a gallery between two chambers $C, D \in \mathcal{C}'$. We will also identify U_- with U_-/U'_w when $w = 1$, i.e., we will write x instead of $x\{1\}$ for $x \in U_-$.

(a) For any $x \in U_-$ and $w \in W$, x is in the same connected component of $\mathcal{C}'$ as xU'_w.

Choose a decomposition $w = s_1 \cdots s_n$ with $s_i \in S$ for all i. Then we have a gallery

$$x \sim_{s_1} xU'_{s_1} \sim_{s_2} xU'_{s_1 s_2} \sim_{s_3} \cdots \sim_{s_n} xU'_w$$

in $\mathcal{C}'$.

(b) For any $w \in W$ and $x \in U'_w$, x is in the same connected component of $\mathcal{C}'$ as 1.

In fact, we have 1 — $U'_w = xU'_w$ — x by (a).

(c) 1 and x are in the same connected component of $\mathcal{C}'$ for any $x \in U_-$.

By Lemma 8.67 we can write $x = x_1 \cdots x_n$ with $x_i \in U'_{w_i}$ for some $w_i \in W$. Using the U_--action and (b), we obtain

$$1 \text{ — } x_1 \text{ — } x_1 x_2 \text{ — } x_1 x_2 x_3 \text{ — } \cdots \text{ — } x_1 \cdots x_n = x \ .$$

(d) Every chamber of $\mathcal{C}'$ is in the connected component of 1.

This follows immediately from (a) and (c). $\qquad\square$

8.7.2 The Morphism $\kappa \colon \mathcal{C}' \to \mathcal{C}_+$

Still motivated by the theory of Moufang twin buildings (see Remark 8.35), we define a function
$$\kappa \colon \mathcal{C}' \to \mathcal{C}_+$$
by
$$\kappa(xU_w') = xwC_+ = xwB_+ \tag{8.16}$$
for $x \in U_-$ and $w \in W$. This is well defined because Lemma 8.65(1) implies that
$$U_w' \leq U_- \cap wB_+ w^{-1} . \tag{8.17}$$
Ultimately we will prove that κ is an isomorphism of chamber systems and hence that equality holds in (8.17). To get started, we prove the following:

Lemma 8.69. κ *is a* U_-*-equivariant morphism of chamber systems over* S.

Proof. U_--equivariance is obvious. To prove that κ preserves s-equivalence for $s \in S$, suppose $xU_v' \sim_s yU_w'$ for some $x, y \in U_-$ and $v, w \in W$. In other words,
$$v \in \{w, ws\} \quad \text{and} \quad y^{-1}x \in U_w' \cup U_{ws}' .$$
We have to show that $xvB_+ \sim_s ywB_+$ in $\mathcal{C}(G, B_+)$. From the definition of the Weyl distance in $\mathcal{C}(G, B_+)$ (see Section 6.2.6), we have $gB_+ \sim_s hB_+$ for $g, h \in G$ if and only if $gP_s = hP_s$, where P_s is the standard parabolic subgroup $B_+ \cup B_+ sB_+$. So the result to be proved is that $xvP_s = ywP_s$ or, equivalently, that $y^{-1}x \in wP_s v^{-1}$. Since $v = w$ or ws, and since (any representative of) s is in P_s, we have $wP_s v^{-1} = wP_s w^{-1}$. It therefore suffices to show that the group $V := U_w' \cup U_{ws}'$ is contained in $wP_s w^{-1}$. If $l(ws) < l(w)$, then $V = U_w' = wU_{w^{-1}}w^{-1}$ by Lemma 8.65(1), and so
$$V \leq wU_+w^{-1} \leq wB_+w^{-1} \leq wP_sw^{-1} .$$
If $l(ws) > l(w)$, then $V = U_{ws}' = U_w'wU_s'w^{-1}$ by Lemma 8.65(2), so $V = w(U_{w^{-1}}U_s')w^{-1}$. This is contained in wP_sw^{-1} because
$$U_s' = U_{-s} = sU_s s^{-1} \leq P_s . \qquad \square$$

Our next goal is to analyze spherical residues in $\mathcal{C}'$. As preparation, we digress to record some properties of the groups U_w'.

8.7.3 The Groups U_w'

We begin with a consequence of the Bruhat decomposition for (G, B_+).

Lemma 8.70. *If* $w, w_1, w_2 \in W$ *are elements such that* $U_w'w_1B_+ \cap w_2B_+ \neq \emptyset$, *then* $w_1 = w_2$.

Proof. By Lemma 8.65(1), we can rewrite the assumption as

$$wU_{w^{-1}}w^{-1}w_1B_+ \cap w_2B_+ \neq \emptyset .$$

This implies that $B_+w^{-1}w_1B_+ = B_+w^{-1}w_2B_+$ and hence, by the Bruhat decomposition, $w^{-1}w_1 = w^{-1}w_2$, i.e., $w_1 = w_2$. $\square$

Next, we derive some consequences of equation (8.12) in Section 8.6.4.

Lemma 8.71. *For all $w, w_1, w_2 \in W$, we have:*

(1) $U'_w \cap B_+ = \{1\}$.
(2) *If $l(w_1w_2) = l(w_1) + l(w_2)$, then $U'_{w_1w_2} \cap w_1B_+w_1^{-1} = U'_{w_1}$.*
(3) *If $l(ww_1) = l(w) + l(w_1)$ and $l(w_1w_2) = l(w_1) + l(w_2)$, then*

$$U'_{ww_1} \cap wU'_{w_1w_2}w^{-1} = wU'_{w_1}w^{-1} . \tag{8.18}$$

Proof. (1) follows from Lemma 8.65(1) and equation (8.12) (with w replaced by w^{-1}).

(2) Lemma 8.65 yields $w_1^{-1}U'_{w_1w_2}w_1 = U_{w_1^{-1}}U'_{w_2}$. Since $U_{w_1^{-1}} \leq U_+$, we obtain

$$w_1^{-1}U'_{w_1w_2}w_1 \cap B_+ = U_{w_1^{-1}}U'_{w_2} \cap B_+$$
$$= U_{w_1^{-1}}(U'_{w_2} \cap B_+)$$
$$= U_{w_1^{-1}} ,$$

since $U'_{w_2} \cap B_+ = \{1\}$ by (1). Conjugating back by w_1 yields

$$U'_{w_1w_2} \cap w_1B_+w_1^{-1} = w_1U_{w_1^{-1}}w_1^{-1} = U'_{w_1} .$$

(3) Applying Lemma 8.65(2) and taking inverses, we obtain

$$U'_{ww_1} = wU'_{w_1}w^{-1}U'_w .$$

We also have

$$U'_w \cap wU'_{w_1w_2}w^{-1} = w(w^{-1}U'_w w \cap U'_{w_1w_2})w^{-1}$$
$$= w(U_{w^{-1}} \cap U'_{w_1w_2})w^{-1}$$
$$\leq w(B_+ \cap U'_{w_1w_2})w^{-1}$$
$$= \{1\} ,$$

where the last equation follows from (1). Using these two facts together with the obvious inclusion $wU'_{w_1}w^{-1} \leq wU'_{w_1w_2}w^{-1}$, we get

$$U'_{ww_1} \cap wU'_{w_1w_2}w^{-1} = wU'_{w_1}w^{-1}(U'_w \cap wU'_{w_1w_2}w^{-1}) = wU'_{w_1}w^{-1} .   \square$$

8.7.4 Spherical Residues

In order to prove that κ is a 2-covering, we will need to analyze spherical residues in $\mathcal{C}'$. Let $J \subseteq S$ be spherical. This means, by definition, that W_J is finite. If we wanted to invoke Exercises 7.126 and the spherical case of Exercise 8.36, we could quickly conclude that κ maps the J-residue of 1 in $\mathcal{C}'$ isomorphically onto the J-residue of the fundamental chamber in $\mathcal{C}_+$. This would not, however, immediately imply that κ maps *every* J-residue isomorphically onto a J-residue in $\mathcal{C}_+$. The problem is that we do not have an obvious chamber-transitive group action on $\mathcal{C}'$, so results about the J-residue of 1 do not immediately yield results about other J-residues. Our purpose in this subsection is to find other ways of relating different J-residues to one another.

Recall from Section 2.3.2 that every left W_J-coset in W ($J \subseteq S$) contains a unique representative w of minimal length, which satisfies $l(wv) = l(w) + l(v)$ for all $v \in W_J$. Recall further that an element $w \in W$ is said to be (right) J-reduced if it is the element of minimal length in its coset wW_J. Finally, recall that if J is spherical then W_J has a unique element of maximal length in W_J, which we will denote by $w_0(J)$. It is of order 2, and $l(w_0(J)) = l(v) + l(v^{-1}w_0(J))$ for all $v \in W_J$. In what follows we will use the abbreviation

$$U'_J := U'_{w_0(J)}$$

if J is spherical. Although we will not use this fact, note that U'_J is simply the "U_-" associated to the spherical RGD system $(G_J, (U_\alpha)_{\alpha \in \Phi_J}, T)$, where $\Phi_J := \{w\alpha_s \mid w \in W_J, s \in J\}$ and $G_J := T \langle U_\alpha \mid \alpha \in \Phi_J \rangle$.

Lemma 8.72. *Let J be a spherical subset of S. Then we have:*

(1) *The U_--orbit of any J-residue in $\mathcal{C}'$ contains a J-residue of the form $R_J(U'_w)$, where w is a J-reduced element of W.*

(2) *If w is J-reduced, then*

$$R_J(U'_w) = \left\{ xU'_{wv} \mid x \in U'_{ww_0(J)} \text{ and } v \in W_J \right\}.$$

In particular,

$$R_J(1) = \{ xU'_v \mid x \in U'_J \text{ and } v \in W_J \}.$$

(3) *If w is J-reduced and n is an element of N such that $\nu(n) = w$, then there is a bijective morphism $f_n \colon R_J(1) \to R_J(U'_w)$ of chamber systems over J, given by $f_n(xU'_v) := nxn^{-1}U'_{wv}$ for $x \in U'_J$ and $v \in W_J$.*

Proof. (1) By the definition of $\mathcal{C}'$ it is clear that any J-residue is U_--equivalent to one of the form $R_J(U'_{w'})$ for some $w' \in W$. Denote by w the J-reduced element of $w'W_J$, and write $w' = ws_1 \cdots s_n$ with all $s_i \in J$. As in part (a) of the proof of Lemma 8.68, one then has a J-gallery connecting U'_w and $U'_{w'}$, which implies that $R_J(U'_{w'}) = R_J(U'_w)$. [Note: This part of the proof does not require J to be spherical.]

(2) Let $Q_J(U'_w) := \{xU'_{wv} \mid x \in U'_{ww_0(J)} \text{ and } v \in W_J\}$. Note that $U'_{wv} \leq U_{ww_0(J)}$ for any $v \in W_J$ by Lemma 8.65(2) and the characteristic properties of w and $w_0(J)$. Hence

$$Q_J(U'_w) = \coprod_{v \in W_J} U'_{ww_0(J)}/U'_{wv} \ . \tag{8.19}$$

In particular,

$$Q_J(1) = \coprod_{v \in W_J} U'_J/U'_v \ . \tag{8.20}$$

Observe first that $Q_J(U'_w) \subseteq R_J(U'_w)$. Indeed, for any $x \in U'_{ww_0(J)}$ and any $v \in W_J$, we have a J-gallery

$$U'_w - U'_{ww_0(J)} = xU_{ww_0(J)} - xU'_{wv}$$

as in part (a) of the proof of Lemma 8.68. So (2) will follow if we show that $Q_J(U'_w)$ is closed under J-equivalence. To see this, consider any chamber $C = xU'_{wv} \in Q_J(U'_w)$ with x and v as above. Given $s \in J$, any chamber D that is s-equivalent to C has the form $D = yU'_{wv'}$ with $v' \in \{v, vs\} \subseteq W_J$ and $x^{-1}y \in U'_{wv} \cup U'_{wvs} \leq U'_{ww_0(J)}$. Hence y is in $U'_{ww_0(J)}$ and D is in $Q_J(U'_w)$.

(3) Define $f_n \colon R_J(1) = Q_J(1) \to Q_J(U'_w) = R_J(U'_w)$ by

$$f_n(xU'_v) := nxn^{-1}U'_{wv}$$

for $x \in U'_J$ and $v \in W_J$. It is well defined because $nU'_v n^{-1} = wU'_v w^{-1} \leq U'_{wv}$ by Lemma 8.65(2), and it takes values in $Q_J(U'_w)$ by this same fact applied to $v = w_0(J)$. Note further that f_n preserves the decompositions in (8.20) and (8.19). To see that it is bijective, then, it suffices to show that it maps U'_J/U'_v bijectively onto $U'_{ww_0(J)}/U'_{wv}$ for each fixed $v \in W_J$.

For surjectivity, recall that $U'_{ww_0(J)} = U'_w w U'_J w^{-1}$ and hence, taking inverses, $U'_{ww_0(J)} = wU'_J w^{-1}U'_w = nU'_J n^{-1}U'_w$. Since $U'_w \leq U'_{wv}$ for all $v \in W_J$, it follows that every coset in $U'_{ww_0(J)}/U'_{wv}$ has a representative in $nU'_J n^{-1}$; this is precisely what is needed for surjectivity.

For injectivity, suppose $nxn^{-1}U'_{wv} = nyn^{-1}U'_{wv}$ with $x, y \in U'_J$. Then

$$ny^{-1}xn^{-1} \in U'_{wv} \cap wU'_J w^{-1} = wU'_v w^{-1} \ ,$$

where the last equation follows from Lemma 8.71(3). Hence $y^{-1}x \in U'_v$, implying $xU'_v = yU'_v$.

Finally, we show that f_n is a morphism of chamber systems over J. Suppose $x_1 U'_{v_1} \sim_s x_2 U'_{v_2}$ with $x_1, x_2 \in U'_J$, $v_1, v_2 \in W_J$, and $s \in J$. Then, first of all, we have $v_2 \in \{v_1, v_1 s\}$, and hence $wv_2 \in \{wv_1, wv_1 s\}$. Furthermore,

$$nx_2^{-1}x_1 n^{-1} \in wU'_{v_1}w^{-1} \cup wU'_{v_1 s}w^{-1} \leq U'_{wv_1} \cup U'_{wv_1 s} \ .$$

Hence $f_n(x_1 U'_{v_1}) = nx_1 n^{-1}U'_{wv_1} \sim_s nx_2 n^{-1}U'_{wv_2} = f_n(x_2 U'_{v_2})$. $\qquad\square$

8.7.5 The Main Result

Proposition 8.73. *The function* $\kappa\colon \mathcal{C}' \to \mathcal{C}_+$ *defined in* (8.16) *is a 2-covering.*

Proof. Before beginning the proof, we remark that we could eliminate steps (a) and (b) below using the exercises cited near the beginning of Section 8.7.4. In order to keep the present discussion self-contained, however, we will avoid using those exercises. We now proceed to the details.

(a) $\kappa(R_s(1)) = R_s(B_+)$ for all $s \in S$, where R_s is an abbreviation for $R_{\{s\}}$.

By the definition of $\mathcal{C}'$ (or by Lemma 8.72(2) with $J = \{s\}$),

$$R_s(1) = \{x \mid x \in U_s'\} \cup \{U_s'\} \ .$$

Hence $\kappa(R_s(1)) = \{xB_+ \mid x \in U_s'\} \cup \{sB_+\}$, which is the set of left B_+-cosets in

$$U_s'B_+ \cup sB_+ = U_{-s}B_+ \cup sB_+ = P_s \ ;$$

here the last equality is equation (8.11) in Section 8.6.3. Now observe that the set of left B_+-cosets in P_s is precisely $R_s(B_+)$ by the definition of $\mathcal{C}(G, B_+)$ (see the proof of Lemma 8.69).

(b) For any spherical $J \subseteq S$, the restriction of κ to $R_J(1)$ is injective.

Let C, D be arbitrary elements of $R_J(1)$. By Lemma 8.72(2), there exist $x, y \in U_J'$ and $v, w \in W_J$ such that $C = xU_v'$ and $D = yU_w'$. Now assume $\kappa(C) = \kappa(D)$, i.e., $xvB_+ = ywB_+$. This implies first that $v = w$ (by Lemma 8.70) and then that $y^{-1}x \in wB_+w^{-1} \cap U_J' = U_w'$, where the last equation follows from Lemma 8.71(2). (Recall that $l(w_0(J)) = l(w)+l(w^{-1}w_0(J))$.) Hence $C = D$.

(c) For any spherical $J \subseteq S$, κ maps every J-residue of $\mathcal{C}'$ bijectively onto a J-residue of $\mathcal{C}_+$.

By Lemma 8.72(1), any J-residue in $\mathcal{C}'$ has the form $R_J(xU_w'')$ for some $x \in U_-$ and some J-reduced $w \in W$. Choose $n \in N$ with $\nu(n) = w$, and consider the following diagram:

$$
\begin{array}{ccccc}
R_J(1) & \xrightarrow{\ f_n\ } & R_J(U_w') & \xrightarrow{\ \lambda_x\ } & R_J(xU_w') \\
\kappa \downarrow & & \kappa \downarrow & & \kappa \downarrow \\
R_J(B_+) & \xrightarrow{\ \lambda_n\ } & R_J(wB_+) & \xrightarrow{\ \lambda_x\ } & R_J(xwB_+)
\end{array}
\qquad (8.21)
$$

where f_n is the bijection given in Lemma 8.72(3), and λ_x (resp. λ_n) is given by left multiplication by x (resp. n). Note that the three vertical arrows labeled κ make sense because κ is a morphism of chamber systems and hence maps any J-residue into a J-residue. One checks immediately that this diagram is commutative. [For the right-hand square, this is just U_--equivariance of κ.] Since all horizontal maps are bijections, all three vertical maps are surjective/injective/bijective if and only if the first is.

We apply this argument first to the special case $J = \{s\}$ with $s \in S$. Then we know by (a) that the first vertical map is surjective; hence so are the other two. This shows that κ maps s-panels *onto* s-panels. As we observed in Section 5.12 (see Lemma 5.220), κ therefore maps all residues of $\mathcal{C}'$ onto residues of $\mathcal{C}_+$ of the same type. By (b), we already know that the first vertical map in (8.21) is injective for any spherical J, so all vertical maps are bijective.

Restricting (c) to spherical residues of rank ≤ 2, we conclude, as desired, that κ is a 2-covering.   □

Combining Proposition 8.73, Lemma 8.68, and Proposition 5.223, we are finally able to conclude that $\mathcal{C}'$ is indeed a disguised version of $\mathcal{C}_+$:

Corollary 8.74. *The map κ is an isomorphism of chamber systems over S.*
                                                                        □

We cooked up $\mathcal{C}'$ so that the chamber U'_w, which maps to wB_+, would have stabilizer U'_w. Since κ is U_--equivariant and the stabilizer of wB_+ in U_- is $U_- \cap wB_+w^{-1}$, this yields the following group-theoretic result:

Corollary 8.75. $U'_w = U_- \cap wB_+w^{-1}$ *for any* $w \in W$.   □

Specializing, finally, to $w = 1$, we obtain the result that was the goal of this entire section:

Proposition 8.76. $U_- \cap B_+ = \{1\}$.   □

8.8 Algebraic Consequences

We continue to assume that we are given an RGD system $\big(G, (U_\alpha)_{\alpha\in\Phi}, T\big)$. Now that we have proved Proposition 8.76, we can continue the line of reasoning that we were pursuing in Section 8.6.4. A trivial consequence of the proposition is that $U_+ \cap U_- = \{1\}$. In particular:

Corollary 8.77. *For all* $s \in S$, $U_s \nleq U_-$.   □

So the stronger axiom (RGD3″) mentioned in Remark 8.48(a) is a consequence of the RGD axioms we started with. Replacing (RGD3) by (RGD3″), we thus get an equivalent set of axioms. Since there is now symmetry with respect to $+$ and $-$, the system $\big(G, (U'_\alpha)_{\alpha\in\Phi}, T\big)$ with $U'_\alpha = U_{-\alpha}$ for all $\alpha \in \Phi$ is also an RGD system (see Remark 8.49(b)). So all the results we have proved concerning $U_\pm$ and $B_\pm$ remain true if we interchange the signs. In particular, we obtain:

- $B_- \cap N = T$ (see (8.7)).

- (G, B_-, N, S) is a Tits system with Weyl group $N/(B_- \cap N) = N/T \cong W$ (see Proposition 8.54).

- $N_G(U_-) = B_-$ (see Corollary 8.55).

- $U_+ \cap B_- = \{1\}$ (see Proposition 8.76).

It follows, in particular, that $U_+ \cap T = \{1\} = U_- \cap T$, so

$$B_+ = T \ltimes U_+ \quad \text{and} \quad B_- = T \ltimes U_- .$$

Proposition 8.76 also has the following easy consequence:

Corollary 8.78. $B_+ \cap B_- = T$.

Proof. $B_- \cap B_+ = TU_- \cap B_+ = T(U_- \cap B_+) = T$. $\qquad\qquad\square$

This implies (as in Corollary 7.123) the following characterization of T:

Corollary 8.79. $T = \bigcap_{\alpha \in \Phi} N_G(U_\alpha) = N_G(U_+) \cap N_G(U_-)$. $\qquad\square$

Now Remark 7.124(b) and Corollary 7.125 are true for general RGD systems as well. Remark 7.124(a), however, has to be modified using twin buildings instead of spherical buildings. We will get back to this in the next section. First we have to complete the proof that we have a twin BN-pair associated to our RGD system. As we will see, all of the work has already been done.

Theorem 8.80. (G, B_+, B_-, N, S) *is a saturated twin Tits system with Weyl group $N/T \cong W$.*

Proof. We have to check the following (see Definitions 6.78 and 6.84):

(a) $N \cap B_+ = N \cap B_- = B_+ \cap B_- \trianglelefteq N$.

We have shown that all intersections are equal to T, and we already know that $T \trianglelefteq N$.

(b) (G, B_+, N, S) is a Tits system.

See Proposition 8.54(1).

(c) (G, B_-, N, S) is a Tits system.

See the discussion above.

(d) For all $w \in W$ and $s \in S$ with $l(sw) < l(w)$, we have

$$B_+ s B_+ w B_- = B_+ s w B_- \quad \text{and} \quad B_- s B_- w B_+ = B_- s w B_+ .$$

See Proposition 8.54(3).

(e) $B_+ s \cap B_- = \emptyset$ for all $s \in S$.

Elements of $B_+ s$ conjugate U_{-s} into U_+, but elements of B_- conjugate U_{-s} into U_-. Since $U_+ \cap U_- = \{1\}$, we must have $B_+ s \cap B_- = \emptyset$. $\qquad\square$

8.9 The Moufang Twin Building

We continue to assume that we are given an RGD system $\left(G, (U_\alpha)_{\alpha \in \Phi}, T\right)$. We have now done all the necessary work to show that this system gives rise to a Moufang twin building. We are therefore in a position to complete the generalization of the theory of spherical RGD systems (Section 7.8). The main result in the previous section (Theorem 8.80) states that (G, B_+, B_-, N, S) is a saturated twin Tits system with Weyl group $W = N/T$, where

$$T = B_+ \cap B_- = B_+ \cap N = B_- \cap N.$$

Let $\mathcal{C}(G, B_+, B_-)$ be the associated thick twin building as in Definition 6.82. Recall that $\mathcal{C}_+ = G/B_+$, $\mathcal{C}_- = G/B_-$, and G acts strongly transitively on the twin building $\mathcal{C} := \mathcal{C}(G, B_+, B_-) = (\mathcal{C}_+, \mathcal{C}_-)$. For $\epsilon \in \{+, -\}$, we denote by C_ϵ the fundamental chamber in $\mathcal{C}_\epsilon$ (so $C_\epsilon = B_\epsilon$ as an element of G/B_ϵ). The chambers C_+ and C_- are opposite, and the associated fundamental twin apartment will be denoted by $\widehat{\Sigma} := \Sigma\{C_+, C_-\} = (\Sigma_+, \Sigma_-)$, where, using the notation introduced in Definition 5.176, $\Sigma_+ := \Sigma(C_+, C_-)$ and $\Sigma_- := \Sigma(C_-, C_+)$. Each Σ_ϵ is canonically isomorphic to $\Sigma = \Sigma(W, S)$.

Recall from Section 6.3.3 that we can recover B_+, B_-, T, and N from the action of G on $\mathcal{C}$ as follows: B_ϵ is the stabilizer of C_ϵ in G, $T = B_+ \cap B_-$ is the (pointwise) fixer of $\widehat{\Sigma}$ in G, and N is the stabilizer of $\widehat{\Sigma}$ in G. We note in passing that this in particular provides the generalization of Remark 7.124(a), which we postponed in the previous section.

For any root α in $\Sigma = \Sigma(W, S)$ and each $\epsilon \in \{+, -\}$, there is an associated root $\alpha_\epsilon := \{wB_\epsilon \mid w \in \alpha\}$ in Σ_ϵ, as in the discussion preceding Theorem 7.116. There is also an associated twin root, namely $\hat{\alpha} := (\alpha_+, -\alpha_-)$. We have $\alpha_- = \mathrm{op}_{\widehat{\Sigma}}(\alpha_+)$ by the definition of the opposition relation in $\mathcal{C}$, so $\hat{\alpha}$ is indeed a twin root in the sense of Definition 5.190. [Warning: Note that α now denotes a root of the Coxeter complex Σ, not a twin root. Note also the minus sign in the definition of $\hat{\alpha}$.]

By Proposition 8.56, U_α acts simply transitively on $\mathcal{C}(\mathcal{P}, \alpha_+)$ for each boundary panel $\mathcal{P}$ of α_+. In particular, we can identify U_α with its image in $\mathrm{Aut}\,\mathcal{C}$. Furthermore, by Lemma 8.58, U_α fixes every interior panel of α_+ pointwise. We now apply the symmetry argument introduced in Section 8.8 in order to obtain the analogous statements for $\mathcal{C}_-$. That is, we consider the RGD system $\left(G, (U_\alpha'), T\right)$ with $U_\alpha' := U_{-\alpha}$, and we apply the results of Section 8.6.4 to the building $\mathcal{C}(G, B_+') = \mathcal{C}_-$, where $B_+' := TU_+' = B_-$ and $U_+' := \langle U_\alpha' \mid \alpha \in \Phi_+ \rangle = U_-$. In this context, Proposition 8.56 yields that $U_\alpha = U_{-\alpha}'$ acts simply transitively on $\mathcal{C}(\mathcal{P}, -\alpha_-)$ for each boundary panel $\mathcal{P}$ of $-\alpha_-$, and Lemma 8.58 yields that U_α fixes every interior panel of $-\alpha_-$ pointwise.

In particular, U_α is a subgroup of the root group $U_{\hat{\alpha}}$ (in the sense of Definition 8.16) associated to the twin root $\hat{\alpha}$, and U_α acts simply transitively on the set $\mathcal{A}(\hat{\alpha})$ of all twin apartments containing $\hat{\alpha}$ (see Lemma 8.17(2)).

Furthermore, U_α is equal to $U_{\hat\alpha}$ if the latter acts freely on $\mathcal{A}(\hat\alpha)$, e.g., if the Coxeter diagram of (W, S) has no isolated nodes. So we have proved the following generalization of Theorem 7.116:

Theorem 8.81. *Let $\mathcal{C} = \mathcal{C}(G, B_+, B_-)$ be the twin building associated to the twin Tits system (G, B_+, B_-, N, S) of Theorem 8.80. Then for any $\alpha \in \Phi$, U_α acts simply transitively on $\mathcal{A}(\hat\alpha)$ and is a subgroup of the root group $U_{\hat\alpha}$. In particular, $\mathcal{C}$ is a Moufang twin building. If, in addition, the Coxeter diagram of (W, S) has no isolated nodes, then $U_\alpha = U_{\hat\alpha}$, and $\mathcal{C}$ is strictly Moufang.* □

We conclude this section by stating the analogue of Proposition 7.127. Given the results of Section 8.8, the proof of Proposition 7.127 works in our present context as well.

Proposition 8.82. *Let $\phi\colon G \to \operatorname{Aut}\mathcal{C}$ be the map giving the action of G on $\mathcal{C} = \mathcal{C}(G, B_+, B_-)$, and set $G_1 := \langle U_\alpha \mid \alpha \in \Phi \rangle$.*

(1) $\ker\phi = C_G(G_1)$, and $Z(G) \le \ker\phi \le T$.
(2) If $G = G_1$ or if T is abelian, then $\ker\phi = Z(G)$.
(3) $\ker\phi \cap G_1 = Z(G_1)$. □

Corollary 8.83. *If $\mathcal{C}$ is a strictly Moufang twin building, then the subgroup of $\operatorname{Aut}\mathcal{C}$ generated by the root groups has trivial center.* □

8.10 A Presentation of U_+

We continue to assume that we are given an RGD system $\left(G, (U_\alpha)_{\alpha \in \Phi}, T\right)$. The main result of this section is Theorem 8.84 below, which generalizes the case $w = w_0$ of Proposition 7.108. As in that proposition, choose generators and relations for each group U_α ($\alpha \in \Phi_+$). For each prenilpotent pair of positive roots $\{\alpha, \beta\}$, choose relations of the form

$$[u_\alpha, u_\beta] = v\,, \tag{8.22}$$

where u_α ranges over the generators of U_α, u_β ranges over the generators of U_β, and v is a word in the generators of the groups U_γ with $\gamma \in (\alpha, \beta)$.

Theorem 8.84. *One obtains a presentation of U_+ by taking generators and relations for each U_α ($\alpha \in \Phi_+$) and adding the commutator relations (8.22).*

We will deduce this from Theorem 8.85 below, which exhibits U_+ as a direct limit of the subgroups U_w ($w \in W$) defined in Section 8.6.3. We assume that the reader is familiar with the notion of *direct limit*, as defined for instance in [217, Section I.1.1]. Consider the system consisting of the groups U_w and the inclusion maps $U_w \hookrightarrow U_{ws}$ for all $w \in W$ and $s \in S$ such that $l(ws) > l(w)$. We write $\varinjlim_w U_w$ for the direct limit of this system. There is a canonical map

$$\varinjlim_w U_w \longrightarrow U_+$$

induced by the inclusions $U_w^{\cdot} \hookrightarrow U_+$.

Theorem 8.85. *The canonical map* $\varinjlim_w U_w \longrightarrow U_+$ *is an isomorphism.*

Remarks 8.86. (a) We sometimes express the conclusion of the theorem saying that U_+ is the *amalgamated sum*, or simply the *amalgam*, of the subgroups U_w, relative to the family of inclusions $U_w \hookrightarrow U_{ws}$.

(b) There is a partial order on W, called the *weak Bruhat order*, defined by

$$w \leq w' \iff w' = wv \text{ with } l(w') = l(w) + l(v) .$$

(As the name suggests, it is weaker than the Bruhat order defined in Exercise 3.59: If $w \leq w'$ in the weak Bruhat order, then $w \leq w'$ in the Bruhat order, but the converse is false in general.) The direct limit in the theorem does not change if we use the inclusions $U_w \hookrightarrow U_{w'}$ for all $w \leq w'$ in W.

(c) Note that Theorem 8.85 is vacuous in the spherical case, since w_0 is the largest element of W with respect to the weak Bruhat order.

It is quite easy to deduce Theorem 8.84 from Theorem 8.85, using the fact that we already have presentations for the groups U_w (Section 8.6.3):

Proof of Theorem 8.84. Let V be the group defined by the presentation in the statement of the theorem. It comes equipped with maps $i_\alpha \colon U_\alpha \to V$ ($\alpha \in \Phi_+$), and there is a surjection $p \colon V \twoheadrightarrow U_+$ such that the composite $p \circ i_\alpha$ is the inclusion $U_\alpha \hookrightarrow U_+$ for each α. We wish to show that p is an isomorphism. For each $w \in W$, the presentation of U_w that we just cited gives us a map $q_w \colon U_w \to V$ such that $q_w|_{U_\alpha} = i_\alpha$ for each $\alpha \in \Phi(w)$. Given $w \in W$ and $s \in S$ with $l(ws) > l(w)$, we have $\Phi(w) \subseteq \Phi(ws)$; hence $q_{ws}|_{U_w} = q_w$. The maps q_w therefore induce a map $q \colon \varinjlim U_w \longrightarrow V$, which is surjective since every positive root α occurs in $\Phi(w)$ for some $w \in W$. (See the proof of Lemma 8.67.) The composite pq of our two surjections is the isomorphism of Theorem 8.85, so p is an isomorphism. $\qquad\square$

Remark 8.87. We can now explain our claim near the end of Section 8.6.3 that U_+ is not nilpotent if W is infinite. Note first that our presentation of U_+ does not involve any relations between U_α and U_β if the pair $\{\alpha, \beta\}$ is not prenilpotent. In fact, it can be shown that no such relations exist, i.e., that the group generated by U_α and U_β is their free product $U_\alpha * U_\beta$ (see [73]). Now a nonprenilpotent pair of positive roots always exists if W is infinite; this was proved in Exercise 8.46, all the work having been done in Corollary 3.166. So our claim follows from the fact that a free product of nontrivial groups is never nilpotent.

We turn now to the proof of Theorem 8.85. It is possible to deduce this theorem from a general result of Tits, as we will explain briefly in Section 14.1.3. But since we do not prove that general result in this book, we will give a different proof of the theorem, in the spirit of [260, proof of Theorem 2]. The basic idea is easy to state: By symmetry we may consider U_- instead of U_+ and U'_w instead of U_w, where the notation is that of Section 8.7. We will then examine the arguments in that section and show that everything goes through if we replace U_- by $\widetilde{U} := \varinjlim_w U'_w$, which admits a canonical surjection

$$p \colon \widetilde{U} \twoheadrightarrow U_- \,.$$

The fact that p is an isomorphism will fall out at the end. Here are the details.

The definition of $\widetilde{U}$ gives us maps $i_w \colon U'_w \to \widetilde{U}$ such that $p \circ i_w$ is the inclusion $U'_w \hookrightarrow U_-$ for each $w \in W$. It follows that i_w is injective, and we denote by $\widetilde{U}_w$ its image. Note that $\widetilde{U}_w \le \widetilde{U}_{ws}$ whenever $l(ws) > l(w)$. More generally,

$$\widetilde{U}_w \le \widetilde{U}_{w'} \text{ if } w \le w'$$

in the weak Bruhat order. Note further that we have canonical isomorphisms

$$U'_w \xrightarrow{\sim} \widetilde{U}_w \xrightarrow{\sim} U'_w$$

whose composite is the identity, where the first isomorphism is given by i_w and the second is the restriction of p. Thus we could identify $\widetilde{U}_w$ with U'_w, but it will help avoid confusion in what follows if we maintain the distinction between them.

Next, we introduce some conjugation operations that will be needed below. Suppose $w \le w'$ as above, and write $w' = wv$. Recall from Lemma 8.65(2) that $U'_{w'} = U'_w(wU'_vw^{-1})$. We now set $w\widetilde{U}_vw^{-1} := i_{w'}(wU'_vw^{-1})$, so that $\widetilde{U}_{w'} = \widetilde{U}_w(w\widetilde{U}_vw^{-1})$. For a fixed representative n of w in N and any $x \in \widetilde{U}_v$, we define $nxn^{-1} := i_{w'}(np(x)n^{-1})$; this makes sense, since $p(x) \in U'_v$ and $np(x)n^{-1} \in wU'_vw^{-1}$. This also shows that $n\widetilde{U}_vn^{-1} = w\widetilde{U}_vw^{-1}$.

We now work our way through Section 8.7. Define a chamber system $\widetilde{\mathcal{C}}$ by

$$\widetilde{\mathcal{C}} := \coprod_{w \in W} \widetilde{U}/\widetilde{U}_w \,,$$

with

$$x\widetilde{U}_v \sim_s y\widetilde{U}_w \iff \{v, vs\} = \{w, ws\} \text{ and } y^{-1}x \in \widetilde{U}_w \cup \widetilde{U}_{ws}$$

for $s \in S$, $x, y \in \widetilde{U}$, and $v, w \in W$. This is an equivalence relation because the union $\widetilde{U}_w \cup \widetilde{U}_{ws}$ is equal to the larger of these two groups. As in the proof of Lemma 8.68, we have the following:

Lemma 8.88. *The chamber system $\widetilde{\mathcal{C}}$ is connected.* □

Now define $\tilde{\kappa} \colon \tilde{\mathcal{C}} \to \mathcal{C}_+$ by $\tilde{\kappa}(x\tilde{U}_w) := p(x)wB_+$ for $x \in \tilde{U}$ and $w \in W$. This is well defined because $p(\tilde{U}_w) = U'_w \leq wB_+w^{-1}$.

Lemma 8.89. $\tilde{\kappa}$ *is a* $\tilde{U}$*-equivariant morphism of chamber systems over* S.

Proof. Note that $y^{-1}x \in \tilde{U}_w \cup \tilde{U}_{ws} \implies p(y)^{-1}p(x) \in U'_w \cup U'_{ws}$. The proof of Lemma 8.69 now goes through with no change. $\qquad\square$

Lemmas 8.70 and 8.71 record some facts about the groups U'_w, and we do not require $\tilde{U}$-analogues of most of them. But we will need the following consequence of 8.71(3):

Corollary 8.90. *Given* $w, w_1, w_2 \in W$ *satisfying*

$$l(ww_1w_2) = l(w) + l(w_1) + l(w_2) \, ,$$

we have

$$\tilde{U}_{ww_1} \cap w\tilde{U}_{w_1w_2}w^{-1} = w\tilde{U}_{w_1}w^{-1} \, . \tag{8.23}$$

(Here the conjugates by w are defined as explained above.)

Proof. Under the hypothesis of the corollary, all the groups occurring in (8.18) are subgroups of $U'_{ww_1w_2}$. Apply $i_{ww_1w_2}$ to both sides to obtain (8.23). $\qquad\square$

Note that it would not be clear how to deal with the intersection in (8.23) under the weaker hypotheses of Lemma 8.71(3). Fortunately, the corollary will suffice. We turn now to the analogue of Lemma 8.72. As in that setting, we set

$$\tilde{U}_J := \tilde{U}_{w_0(J)}$$

if $J \subseteq S$ is spherical.

Lemma 8.91. *Let* J *be a spherical subset of* S. *Then we have:*

(1) *The* $\tilde{U}$*-orbit of any* J*-residue in* $\tilde{\mathcal{C}}$ *contains a* J*-residue of the form* $R_J(\tilde{U}_w)$, *where* w *is a* J*-reduced element of* W.
(2) *If* w *is* J*-reduced, then*

$$R_J(\tilde{U}_w) = \left\{ x\tilde{U}_{wv} \mid x \in \tilde{U}_{ww_0(J)} \text{ and } v \in W_J \right\} \, .$$

In particular,

$$R_J(1) = \left\{ x\tilde{U}_v \mid x \in \tilde{U}_J \text{ and } v \in W_J \right\} \, .$$

(3) *If* w *is* J*-reduced and* n *is an element of* N *such that* $\nu(n) = w$, *then there is a bijective morphism* $f_n \colon R_J(1) \to R_J(\tilde{U}_w)$ *of chamber systems over* J, *given by* $f_n(x\tilde{U}_v) := nxn^{-1}\tilde{U}_{wv}$ *for* $x \in \tilde{U}_J$ *and* $v \in W_J$.

Proof. The proof of Lemma 8.72 goes through with the following minor changes. First, the definition of f_n in (3) uses the remarks we made about conjugation above. It is well defined and surjective as in the proof of Lemma 8.72. The proof of injectivity is also the same, except that we apply Corollary 8.90 instead of Lemma 8.71(3). [To see that this is possible, note that for any $v \in W_J$, there exists $v' \in W_J$ such that $w_0(J) = vv'$ and $l(wvv') = l(w) + l(v) + l(v')$. Note also that for $x, y \in \widetilde{U}_J$,

$$n(y^{-1}x)n^{-1} \in w\widetilde{U}_v w^{-1} \implies y^{-1}x \in \widetilde{U}_v ,$$

since the map $\widetilde{U}_J \to \widetilde{U}_{ww_0(J)}$ given by $z \mapsto nzn^{-1}$ is injective and $n\widetilde{U}_v n^{-1} = w\widetilde{U}_v w^{-1}$.] Finally, the proof that f_n is a morphism is unchanged. $\qquad\square$

Proposition 8.92. $\tilde{\kappa} \colon \widetilde{C} \to C_+$ *is a 2-covering.*

Proof. Step (a) is the same as in the proof of Proposition 8.73 (since $p(\widetilde{U}_s) = U'_s$). Step (b) is also basically the same: We just start with $x, y \in \widetilde{U}_J$ but then derive the same conclusions from $p(x)vB_+ = p(y)wB_+$ as before. Note that $p(y^{-1}x) \in U'_w \implies y^{-1}x \in \widetilde{U}_w$ since $\widetilde{U}_w \le \widetilde{U}_J$.

Finally, only minor changes are needed in step (c), which uses a diagram as in (8.21). Here C'-residues are replaced by $\widetilde{C}$-residues in the top row. And λ_x becomes $\lambda_{p(x)}$ in the second row, since x is now in $\widetilde{U}$. The left-hand square is commutative because $p(nyn^{-1}) = p(i_{ww_0(J)}(np(y)n^{-1})) = np(y)n^{-1}$ for $y \in \widetilde{U}_J$. $\qquad\square$

As before, the proposition has the following consequence:

Corollary 8.93. $\tilde{\kappa}$ *is an isomorphism of chamber systems over* S. $\qquad\square$

This quickly yields what we have been aiming for:

Proof of Theorem 8.85. It suffices to show that $p \colon \widetilde{U} \twoheadrightarrow U_-$ is injective. Given $x \in \ker p$, we have $\tilde{\kappa}(x) = p(x)B_+ = B_+ = \tilde{\kappa}(1)$. So the injectivity of $\tilde{\kappa}$ implies $x = 1$. $\qquad\square$

Exercise 8.94. The proof of Theorem 8.84 may have seemed unnecessarily roundabout. Why couldn't we have just manipulated the direct limit $\varinjlim U_w$ (using the known presentations of the groups U_w) to arrive at the presentation given in Theorem 8.84?

8.11 Groups of Kac–Moody Type

Kac–Moody groups can be viewed as infinite-dimensional generalizations of semisimple linear algebraic groups. The theory of twin buildings was developed to provide a geometric framework for studying these groups, in the same way that spherical buildings provide a geometric framework for studying ordinary (finite-dimensional) semisimple groups. We give here a very brief introduction to Kac–Moody theory, referring to [73] for a detailed survey and many references. We assume that the reader is familiar with the theory of root systems and Lie algebras. We will freely use the notation of Appendix B, which the reader may need to review before proceeding.

8.11.1 Cartan Matrices

Let Φ be a crystallographic root system in a Euclidean vector space V. Choose a fundamental chamber for the Weyl group $W = W_\Phi$, and let $\Pi = \{\alpha_1, \ldots, \alpha_n\}$ be the corresponding system of simple roots. The coroots $\alpha_i^\vee$ of the α_i correspond to elements $h_i \in V^*$, also called coroots, under the canonical isomorphism between V and V^* ($i = 1, \ldots, n$). Thus the fundamental reflections s_i that generate W are given by

$$s_i(x) = x - \langle h_i, x \rangle \alpha_i$$

for $x \in V$, where the angle brackets denote the canonical pairing between V and V^*. In particular,

$$s_i(\alpha_j) = \alpha_j - a_{ij}\alpha_i \,,$$

where $a_{ij} = \langle h_i, \alpha_j \rangle$. The matrix $A = (a_{ij})$ is the *Cartan matrix* of Φ relative to Π; it consists of nonpositive integers except on the diagonal, where we have $a_{ii} = 2$. It is related to the Coxeter matrix $M = (m_{ij})$ by

$$a_{ij} = -2\frac{\|\alpha_j\|}{\|\alpha_i\|} \cos \frac{\pi}{m_{ij}} \,,$$

which implies

$$a_{ij}a_{ji} = 4\cos^2 \frac{\pi}{m_{ij}} \,; \tag{8.24}$$

see Exercise B.3. The left side of (8.24) is an integer, and the right side is in the interval $[0, 4)$ if $i \neq j$. Thus the only possible values for $a_{ij}a_{ji}$, if $i \neq j$, are $0, 1, 2, 3$, and we have

$$
\begin{aligned}
m_{ij} &= \quad 2 \quad 3 \quad 4 \quad 6 \\
\text{if} \quad a_{ij}a_{ji} &= \quad 0 \quad 1 \quad 2 \quad 3 \,.
\end{aligned}
\tag{8.25}
$$

More briefly, $a_{ij}a_{ji}$ is the number of bonds between i and j if one uses the convention for Coxeter diagrams described in Remark 1.98(c).

Remarks 8.95. (a) One can also prove (8.24), and hence (8.25), by direct computation. To do this, write out the 2×2 matrices that represent the restrictions of s_i and s_j to the plane spanned by e_i and e_j. Multiply them together, compute the eigenvalues of the product, and deduce (8.24).

(b) The discussion above shows that the Coxeter matrix is determined by the Cartan matrix. Conversely, the Cartan matrix is determined by the Coxeter matrix in the *simply laced* case, i.e., the case in which the Coxeter diagram contains no double or triple bonds. This is equivalent to saying that $m_{ij} = 2$ or 3 for $i \neq j$, or that $a_{ij} = 0$ or -1 for $i \neq j$.

8.11.2 Finite-Dimensional Lie Algebras

Root systems arose historically from the study of finite-dimensional complex semisimple Lie algebras. Every such algebra $\mathfrak{g}$ gives rise to a crystallographic root system Φ, and one can recover $\mathfrak{g}$ from the Cartan matrix of Φ. We review this briefly here.

Choose a maximal subspace $\mathfrak{h}$ of $\mathfrak{g}$ consisting of commuting elements h such that $\operatorname{ad} h$ is diagonalizable. Here "ad" refers to the adjoint representation of $\mathfrak{g}$. Then $\mathfrak{h}$ is an abelian subalgebra of $\mathfrak{g}$, called a *Cartan subalgebra*. Elementary linear algebra now implies that the elements $h \in \mathfrak{h}$ are simultaneously diagonalizable in the adjoint representation, so we can decompose $\mathfrak{g}$ into eigenspaces for $\mathfrak{h}$. On a given eigenspace, the adjoint action of any $h \in \mathcal{H}$ is given by

$$[h, x] = \alpha(h)x .$$

Here the eigenvalue $\alpha(h)$ is a scalar that depends linearly on h, so α is an element of the dual space $\mathfrak{h}^*$. The nonzero linear functionals α that arise in this way are called *roots*, because the numbers $\alpha(h)$ are roots of the characteristic polynomial of $\operatorname{ad} h$. The 0-eigenspace is $\mathfrak{h}$ itself. Thus the eigenspace decomposition takes the form

$$\mathfrak{g} = \mathfrak{h} \oplus \bigoplus_{\alpha \in \Phi} \mathfrak{g}_\alpha ,$$

where Φ is the set of roots and $\mathfrak{g}_\alpha$ is the α-eigenspace. One shows that each "root subspace" $\mathfrak{g}_\alpha$ is 1-dimensional. We have essentially seen all of this already in Section 7.9.3, in slightly different language.

It is a fact that the set Φ of roots is a crystallographic root system in $\mathfrak{h}^*$. More precisely, since we have defined root systems only in real vector spaces, we should say that Φ is a root system in the real vector space $V \subset \mathfrak{h}^*$ spanned by Φ. [Note: It is possible to use the Lie algebra structure to put a canonical inner product on V.] The coroots h_i discussed in Section 8.11.1 then live in V^*, which can be viewed as a real subspace of the complex vector space $\mathfrak{h} = \mathfrak{h}^{**}$.

Another basic fact is that one can reconstruct the Lie algebra $\mathfrak{g}$ from the root system Φ. There is more than one way to see this, but the one that is relevant to Kac–Moody theory is the following. Choose a system of simple roots $\alpha_i \in \mathfrak{h}^*$, $i = 1, \dots, n$, and let $h_i \in \mathfrak{h}$ be the coroot of α_i. This yields a Cartan matrix A as in Section 8.11.1, with $a_{ij} = \langle h_i, \alpha_j \rangle$, the eigenvalue of h_i acting on the α_j-root space. One can now recover the algebra $\mathfrak{g}$ from the matrix A by writing down a presentation by generators and relations due to

Chevalley, Harish-Chandra, and Serre. Namely, $\mathfrak{g}$ is generated by the vector space $\mathfrak{h}$ and $2n$ additional elements e_i, f_i [eigenvectors for the roots $\pm\alpha_i$], subject to the relations

$$[\mathfrak{h}, \mathfrak{h}] = 0 ,$$
$$[h, e_i] = \langle h, \alpha_i \rangle e_i \quad \text{for } h \in \mathfrak{h} ,$$
$$[h, f_i] = -\langle h, \alpha_i \rangle f_i \quad \text{for } h \in \mathfrak{h} ,$$
$$[e_i, f_i] = h_i ,$$
$$[e_i, f_j] = 0 \quad \text{if } i \neq j ,$$
$$(\operatorname{ad} e_i)^{-a_{ij}+1} e_j = 0 \quad \text{if } i \neq j ,$$
$$(\operatorname{ad} f_i)^{-a_{ij}+1} f_j = 0 \quad \text{if } i \neq j .$$

See Serre [213, p. VI-13] for details.

8.11.3 Kac–Moody Algebras

Suppose now that we are given a pair of dual finite-dimensional complex vector spaces $\mathfrak{h}, \mathfrak{h}^*$ with elements $h_i \in \mathfrak{h}$ and $\alpha_i \in \mathfrak{h}^*$. Define $A = (a_{ij})$ by $a_{ij} = \langle h_i, \alpha_j \rangle$, and assume that A is a *generalized Cartan matrix*; this means, by definition, that $a_{ij} \in \mathbb{Z}$, $a_{ii} = 2$, $a_{ij} \leq 0$ for $i \neq j$, and $a_{ij} = 0 \iff a_{ji} = 0$. Then the presentation above still makes sense, but it defines a Lie algebra that is infinite-dimensional in general.

The *Kac–Moody algebras* that one obtains by this construction have turned out to be quite important and to have unexpected connections with many areas of mathematics and physics. See [141, 164, 165] for more information and a more general formulation. [Note: Our definition is not precisely the same as the one that appears in some of the literature. In the symmetrizable case, for instance, our Kac–Moody algebra is the commutator subalgebra of the one introduced by Kac.]

Example 8.96. Let $\mathfrak{g} = \mathfrak{sl}_n(\mathbb{C})$ (matrices of trace 0). The standard Cartan subalgebra $\mathfrak{h}$ consists of the diagonal matrices in $\mathfrak{g}$, and the corresponding simple roots $\alpha_1, \ldots, \alpha_{n-1} \in \mathfrak{h}^*$ are given by $\alpha_i = x_i - x_{i+1}$, where x_i gives the ith diagonal entry of a matrix in $\mathfrak{h}$ (cf. Example 7.144). Let $\alpha_0 = -\tilde{\alpha}$, where $\tilde{\alpha} = \sum_{i=1}^{n-1} \alpha_i$. [This is the "highest root," although we do not need to know that here. It will arise naturally in Section 10.1.8 when we work out the structure of the fundamental chamber for a Euclidean reflection group.] Let $h_0, h_1, \ldots, h_{n-1}$ be the coroots of the α_i, and let $\tilde{a}_{ij} = \langle h_i, \alpha_j \rangle$ for $i, j = 0, \ldots, n-1$. Then $\tilde{A} = (\tilde{a}_{ij})$ is a generalized Cartan matrix, and the resulting Kac–Moody algebra turns out to be the "loop algebra"

$$\tilde{\mathfrak{g}} = \mathfrak{g} \otimes \mathbb{C}[t, t^{-1}] = \mathfrak{sl}_n\left(\mathbb{C}[t, t^{-1}]\right) .$$

If we think of $\mathbb{C}[t, t^{-1}]$ as the ring of trigonometric polynomials, we can view $\tilde{\mathfrak{g}}$ as the algebra of polynomial maps from the circle to $\mathfrak{g}$.

Notice that the α_i are dependent: $\sum_{i=0}^{n-1} \alpha_i = 0$. Similarly, $\sum_{i=0}^{n-1} h_i = 0$. There is a slight variation on this example. Replace $\mathfrak{h}$ by an n-dimensional vector space $\hat{\mathfrak{h}}$ with basis denoted by $h_0, h_1, \ldots, h_{n-1}$, and define α_i in the dual so that we get the same matrix $\tilde{a}_{ij} = \langle h_i, \alpha_j \rangle$. Thus we have a surjection $\hat{\mathfrak{h}} \twoheadrightarrow \mathfrak{h}$ with a 1-dimensional kernel, generated by $z := \sum_{i=0}^{n-1} h_i$. This time the Kac–Moody algebra is a central extension $\hat{\mathfrak{g}}$ of $\tilde{\mathfrak{g}}$ (with kernel $\mathbb{C}z$), which may be more familiar to those who have seen Kac–Moody algebras before.

8.11.4 The Weyl Group

Before proceeding further, let's note that there is a Coxeter group associated with a generalized Cartan matrix A in a natural way. Set $m_{ii} = 1$ and, for $i \neq j$, define m_{ij} in terms of A by

$$
\begin{array}{rccccc}
m_{ij} & = & 2 & 3 & 4 & 6 & \infty \\
\text{if} \quad a_{ij}a_{ji} & = & 0 & 1 & 2 & 3 & \geq 4 \,.
\end{array}
$$

Now define the *Weyl group* of A to be the Coxeter group W with generators $s_1, \ldots, s_n$ and relations $(s_i s_j)^{m_{ij}} = 1$. The definition of m_{ij} is motivated by Section 8.11.1, where the case $a_{ij}a_{ji} \geq 4$ does not occur. But this case is no harder to explain: Suppose A comes from vectors and covectors h_i, α_i as above, with the α_i pairwise independent. [This is forced by the conditions on A unless $a_{ij}a_{ji} = 4$.] Define linear reflections s_i on $\mathfrak{h}^*$ by the usual formula:

$$
s_i(x) = x - \langle h_i, x \rangle \alpha_i
$$

for $x \in \mathfrak{h}^*$. Then a straightforward computation (as in Remark 8.95(a)) shows that $s_i s_j$ has infinite order if $a_{ij}a_{ji} \geq 4$.

Note that the Weyl group is infinite in general. For the matrix $\tilde{A}$ in Example 8.96, W is the Weyl group that arose in Section 6.9 when we studied the BN-pair associated to SL_n over a field with a discrete valuation.

As in Remark 8.95(b), we call attention to the simply laced case, in which $a_{ij} \in \{0, -1\}$ for $i \neq j$ or, equivalently, $m_{ij} \in \{2, 3\}$ for $i \neq j$. The Coxeter diagram is then just a finite graph with no further decoration, and it determines the generalized Cartan matrix and hence the Kac–Moody algebra.

8.11.5 Kac–Moody Groups

If there are infinite-dimensional Lie algebras $\mathfrak{g}$, there should be corresponding "infinite-dimensional Lie groups" G. Tits [260] gave a general construction of such groups.

Start with a pair $\Lambda, \Lambda^{\vee}$ of free $\mathbb{Z}$-modules of finite rank that are dual to one another, together with families $\alpha_i \in \Lambda$ and $h_i \in \Lambda^{\vee}$, $i = 1, \ldots, n$. Let $a_{ij} = \langle h_i, \alpha_j \rangle$, and assume that the matrix $A = (a_{ij})$ is a generalized Cartan matrix. This is a refinement of what we had in Section 8.11.3; we can set $\mathfrak{h}^* = \mathbb{C} \otimes \Lambda$ and $\mathfrak{h} = \mathbb{C} \otimes \Lambda^\vee$ to get back to the previous situation. This refinement allows one to distinguish between different groups with the same Lie algebra, just as in the finite-dimensional theory. [Think of Λ as the character group X of a maximal torus; see Section 7.9.3.]

Tits associates to these data not just a group, but a *group functor* G on the category of commutative rings. The restriction of this functor to fields yields *Kac–Moody groups* $G(k)$ (where k is a field), generalizing the Chevalley groups discussed in Sections 6.4 and 7.9.2. For example, there are various group functors corresponding to the Kac–Moody algebra $\tilde{\mathfrak{g}}$ of Example 8.96, depending on the choice of the lattices $\Lambda, \Lambda^\vee$. One natural choice is to let $\Lambda^\vee$ be the lattice spanned by the coroots, in which case Λ is necessarily its dual (called the *weight lattice*). Then G is given by $G(k) = \mathrm{SL}_n\big(k[t, t^{-1}]\big)$. For instance, $G(\mathbb{C})$ is the "loop group" $\mathrm{SL}_n\big(\mathbb{C}[t, t^{-1}]\big)$ of polynomial maps from the circle to $\mathrm{SL}_n(\mathbb{C})$.

The details of Tits's construction in [260] are quite technical. (One can also find an expository account in [257].) All we will say here is that $G(k)$ is generated by a torus $T(k) = \mathrm{Hom}(\Lambda, k^*)$ and a family of "root groups" $U_\alpha(k)$, $\alpha \in \Phi$, where Φ is the set of roots of $\Sigma(W, S)$. Here W is the Weyl group associated with A as in Section 8.11.4. Each root group is isomorphic to the additive group k. Tits proves that the torus and root groups form an RGD system in $G(k)$. Such RGD systems were, historically, the motivating examples for the theory of RGD systems and twin buildings. In our canonical example, with $G(k) = \mathrm{SL}_n\big(k[t, t^{-1}]\big)$, we recover the RGD system of Example 8.47(b).

The groups given by Tits's construction are now usually called "split" Kac–Moody groups (of minimal type). There are also nonsplit versions of Kac–Moody groups (see Rémy [195]), which still have RGD systems and hence give rise to (Moufang) twin buildings.

Remark 8.97. Note that there is a great deal of freedom in choosing the generalized Cartan matrix A. In particular, one has a huge variety of twin BN-pairs, and hence thick twin buildings. The Weyl group can be any Coxeter group such that the entries m_{ij} of the Coxeter matrix are in $\{2, 3, 4, 6, \infty\}$ for $i \neq j$. The theory of Euclidean and hyperbolic reflection groups to be treated in Chapter 10 provides many concrete examples of such Coxeter groups, but the reader can also make up examples at random. See also Rémy [195, Section 13.3] for a detailed discussion of some specific hyperbolic examples.

8.11.6 The Simply Laced Case

The Kac–Moody groups $G(k)$ are easy to describe in the simply laced case mentioned at the end of Section 8.11.4 (for a suitable choice of $\Lambda, \Lambda^\vee$). For each index $i = 1, \ldots, n$, $G(k)$ has a subgroup $G_i(k)$ isomorphic to $\mathrm{SL}_2(k)$. For each pair of indices $i < j$, $G(k)$ has a subgroup $G_{ij}(k)$, which is isomorphic to $\mathrm{SL}_3(k)$ if there is an edge joining i and j in the Coxeter diagram and to $SL_2(k) \times SL_2(k)$ otherwise. The group $G_{ij}(k)$ contains $G_i(k)$ and $G_j(k)$ as subgroups. These are the subgroups

$$\begin{pmatrix} * & * & \\ * & * & \\ & & 1 \end{pmatrix} \quad \text{and} \quad \begin{pmatrix} 1 & & \\ & * & * \\ & * & * \end{pmatrix}$$

of $G_{ij}(k) = SL_3(k)$ if there is an edge joining i and j; otherwise, they are the two obvious copies of $SL_2(k)$ in $G_{ij}(k) = SL_2(k) \times SL_2(k)$.

The result, then, is that $G(k)$ is the direct limit (or amalgam) of the system consisting of the groups $G_i(k)$ and $G_{ij}(k)$ and the inclusions

$$
\begin{array}{c}
G_i(k) \\
\searrow \\
\qquad\qquad G_{ij}(k) \\
\nearrow \\
G_j(k)
\end{array}
\qquad (8.26)
$$

This means, concretely, that one can get a presentation for $G(k)$ by taking generators and relations for the groups $G_i(k)$ and $G_{ij}(k)$ and then adding further relations to express the inclusions in (8.26).

We explain briefly how this description of $G(k)$ relates to Tits's construction mentioned in Section 8.11.5. The latter, whether the diagram is simply laced or not, involves a Steinberg-type presentation of $G(k)$. If the twin building $\mathcal{C}$ associated to $G(k)$ is 2-spherical and satisfies condition (co) of Section 5.11 (which is always the case by Remark 5.212 if the diagram is simply laced), then the action of $G(k)$ on $\mathcal{C}$ can be used to reduce considerably the set of Steinberg relations needed in the presentation; see Abramenko and Mühlherr [12]. [We will explain this further in Section 14.1.2.] For instance, one always gets a finite presentation of $G(k)$ if the field k is finite. And in the simply laced case, we obtain (for arbitrary k) the amalgam presentation described above. See Caprace [69] for more details about presentations of 2-spherical Kac–Moody groups.

Example 8.98. The reader can get some feeling for the amalgam above by looking at the familiar example of the A_{n-1} diagram ($n \geq 3$), where the Kac–Moody functor G (restricted to fields) reduces to SL_n. The result, then, is that $SL_n(k)$ is the amalgam of the copies of $SL_2(k)$, $SL_3(k)$, and $SL_2(k) \times SL_2(k)$ embedded block diagonally in $SL_n(k)$.

Returning to the general (simply laced) case, suppose we take k to be a finite field $\mathbb{F}_q$. Then, as we noted above, we always obtain a finitely presented group $G(\mathbb{F}_q)$. A striking result of Caprace and Rémy [71,72] is that this finitely presented group is quite often simple, or at least simple modulo its center, if (W, S) is irreducible and not Euclidean. [Note that the group is automatically perfect, as a consequence of the amalgam presentation described above.] In the Euclidean case, on the other hand, one can get linear groups such as $SL_n\big(\mathbb{F}_q[t, t^{-1}]\big)$, which are residually finite and hence far from simple.

Remark 8.99. Although we have emphasized Kac–Moody theory as a source of RGD systems, there are also other examples. See, for instance, Ronan–Rémy [197] and further references cited there.

The Classification of Spherical Buildings

9.1 Introduction

One of Tits's greatest achievements is the classification of thick, irreducible, spherical buildings of rank at least 3, proved in [247]. Roughly speaking, the result is that such buildings correspond to simple algebraic groups (of relative rank at least 3) defined over an arbitrary field. We have seen several examples of this correspondence in Chapters 6 and 7. And we have also seen a proof (modulo the extension theorem) of the remarkable fact that all of the buildings in question come from groups with BN-pairs; see Proposition 7.11 and Theorem 7.59. It is equally remarkable that one can specify precisely the class of groups that can occur.

Our rough statement above captures the spirit of the classification theorem but is inaccurate in two respects:

(1) There are classical groups that are not algebraic groups but still give rise to buildings. For example, if D is a noncommutative division ring, then $\mathrm{GL}_n(D)$ is a classical group (with a closely associated simple group), but it is not an algebraic group unless D is a finite-dimensional algebra over a field. The associated building is nothing exotic; it is simply the flag complex of proper nontrivial subspaces of an n-dimensional vector space over D (see Section 4.3, including Exercise 4.32). For a second example, consider a quadratic form of finite Witt index on an infinite-dimensional vector space (see Remark 6.103); the orthogonal group is not an algebraic group, but it still has an associated building gotten from totally isotropic subspaces.

(2) There is a family of buildings of type F_4 associated to groups that again are something like algebraic groups but are not algebraic groups. There is a group of this type for every pair of fields k, K of characteristic 2 with $K^2 < k < K$. These groups, and the associated buildings, are said to be *mixed*. [Note: The construction actually produces a building with a

specific type function, having values in $\{1, 2, 3, 4\}$. Replacing (k, K) by $(K, k^{1/2})$ yields the same building, with the types reversed.]

So a more accurate statement of the classification system says that there are three classes of thick, irreducible, spherical buildings of rank ≥ 3:

(a) Classical buildings (associated to classical groups).
(b) Algebraic buildings (associated to algebraic groups).
(c) Mixed buildings (associated to mixed groups).

Note that "most" algebraic groups are classical, so there is a big overlap between classes (a) and (b). If we wanted to avoid this, we could confine ourselves to *exceptional* (i.e., nonclassical) algebraic groups in (b). Class (c) can be thought of as "truly exceptional"; it consists of a single family of buildings of type F_4, involving imperfect fields of characteristic 2.

Remarks 9.1. (a) The terminology here is potentially confusing. A building can be exceptional, in the sense that it corresponds to a nonclassical algebraic group, but its type (determined by its Coxeter diagram) need not be one of the exceptional types E_6, E_7, E_8, F_4. [We have omitted types H_3 and H_4 from this list because they cannot occur; see Corollary 7.61.] For example, we will see below that there is a family of (exceptional) algebraic buildings of type C_3, corresponding to a family of nonclassical algebraic groups. This confusion can occur only for groups defined over fields that are not algebraically closed.

(b) The remaining sections of this chapter will give a more detailed statement of the classification theorem, from which it will follow that all thick, irreducible, spherical buildings of rank ≥ 9 are classical.

The restriction to rank ≥ 3 in the classification theorem cannot be avoided. Indeed, the "free construction" given by Tits [251] suggests that there is no hope of classifying buildings of rank 2. And even if we consider only finite buildings of type A_2, for example, the problem is equivalent to classifying finite projective planes; this is a well-known problem that seems to be out of reach.

If, however, we impose the Moufang property, then the classification extends to rank 2. This result is due to Tits and Weiss and can be found in [262], which also includes a simplified proof of the original classification theorem. The simplification is possible because buildings of rank ≥ 3 can be treated much more systematically once the rank-2 Moufang buildings have been classified. We will say a few more words about it in Section 9.11.

The statement of the classification theorem for thick, irreducible, Moufang, spherical buildings of rank ≥ 2 is, in outline, exactly the same as in the case of rank ≥ 3 above: Every such building is classical, algebraic, or mixed. And as above, one can replace "algebraic" by "exceptional" in order to avoid overlap between the first two classes. There are five families of mixed buildings, including those of type F_4 already mentioned plus four families of rank-2 buildings. Three of these four involve fields k, K of characteristic 2 as in the

F$_4$ case, while the fourth involves a pair of fields of characteristic 3, with $K^3 \leq k < K$.

At the center of the proof of the classification theorem is Tits's extension theorem, which we stated without proof in Section 5.10. As we showed in Section 7.6, the extension theorem implies (after a little work) that all thick, irreducible, spherical buildings of rank ≥ 3 have the Moufang property; this explains why Tits did not need to add the latter as a hypothesis in his classification theorem.

Our purpose in this chapter is to survey, in slightly more detail than above, the thick, irreducible, spherical buildings of rank ≥ 3. We will omit some technicalities, but we hope that the reader will get the flavor of the classification. We will also give a few hints about rank-2 Moufang buildings, but we will be even more sketchy there. For more precise (but more technical) statements, see [200, 207, 247, 262, 281].

We now proceed case by case according to type, as determined by the Coxeter diagram; see Definition 4.37.

9.2 Type A$_n$

By a *building of type A$_n$* we will mean a (simplicial) building that comes equipped with a coloring having values in $\{1, \ldots, n\}$, in such a way that the resulting Coxeter diagram is the standard A$_n$-diagram with the vertices numbered $1, 2, \ldots, n$ from left to right. Any such building is isomorphic to the flag complex of proper subspaces of an n-dimensional projective space X. Vertices of type 1 correspond to the points of X, vertices of type 2 correspond to the lines of X, and so on. The building is thick if and only if every line in X has at least 3 points. For $n \geq 3$, any such X comes from an $(n+1)$-dimensional vector space over a division ring; thus the building comes from a vector space as in Section 4.3 (and Exercise 4.32). The building can also be described in terms of a BN-pair in a general linear group, as in Section 6.5. In particular, thick buildings of type A$_n$ for $n \geq 3$ are all classical, and there is one such (up to type-preserving isomorphism) for every isomorphism class of division rings.

If we want to classify our buildings up to arbitrary simplicial isomorphism (not necessarily type-preserving), then the result is that thick buildings of type A$_n$ ($n \geq 3$) correspond to equivalence classes of division rings, where two division rings are equivalent if they are isomorphic or if one is isomorphic to the opposite of the other (see Exercise 4.32 and its solution).

Remarks 9.2. (a) The equivalence between buildings of type A$_n$ and projective spaces is fairly straightforward. So the classification of such buildings is equivalent to the classification of projective spaces. In particular, the fact that a building of type A$_n$ (for $n \geq 3$) uniquely determines a division ring follows from the fundamental theorem of projective geometry.

(b) In view of (a), we can think of Tits's classification of spherical buildings as a generalization of the fundamental theorem of projective geometry.

Finally, if we allow $n = 2$ but impose the Moufang condition, then the classification is similar except that a new family of buildings arises, coming from Cayley division algebras. Many readers will have heard of the Cayley numbers, also called the octonions, which form an 8-dimensional "nonassociative division algebra" over the field $\mathbb{R}$ of real numbers. Analogues, called *Cayley division algebras*, exist over fields k other than $\mathbb{R}$, and it is possible to construct a Moufang projective plane, and hence a Moufang building of type A_2, for each Cayley division algebra. In terms of the "classical/algebraic/mixed" trichotomy in the introduction to this chapter, these buildings are algebraic but not classical, i.e., they are exceptional.

Remark 9.3. The classification simplifies if we consider only finite buildings, since there are no finite Cayley division algebras or finite noncommutative division rings. The result, then, is simply that the finite, thick Moufang buildings of type A_n ($n \geq 2$) are the buildings associated with the standard projective spaces over finite fields. Equivalently, they are the buildings associated with the linear groups $GL_{n+1}(k)$ (or $SL_{n+1}(k)$ or $PSL_{n+1}(k)$), where k is a finite field $\mathbb{F}_q$. In particular, they are all classical.

9.3 Type C_n

Just as buildings of type A_n correspond to projective spaces, buildings of type C_n correspond to polar spaces. These were first introduced by Veldkamp [268] in order to give an axiomatic description of the geometry of isotropic subspaces relative to a polarity in a projective space. The most familiar examples are given by totally isotropic subspaces in a vector space with a suitable form: symmetric bilinear, alternating bilinear, Hermitian, or quadratic. Tits [247] simplified the axioms and also introduced a new class of forms, called "pseudoquadratic," which are needed in characteristic 2. [The division ring of scalars for the underlying vector space is allowed to be noncommutative in this case.] A fundamental theorem of Buekenhout and Shult [63] provides an even simpler set of axioms.

Although we will not give the axioms for polar geometry here, the reader can find many examples in Chapter 6 in our discussions of symplectic, orthogonal, and unitary groups. [But recall from Section 6.7 that the resulting building of type C_n is not necessarily thick; there is one family of quadratic forms for which one needs to form an "oriflamme complex" in order to get a thick building, which is then of type D_n rather than C_n.] In all cases the rank n of the building is the Witt index of the corresponding form; recall that this is the common dimension of the maximal totally isotropic subspaces.

The precise conditions on the forms that arise in the theory of polar spaces are technical and will be omitted, but a convenient summary is given in Scharlau [207, Section 4.2]. See also Hahn–O'Meara [127] and Bruhat–Tits [59, Section 10], and see Aschbacher [24, Chapter 7] for a simplified treatment, sufficient to cover the case that the division ring is a finite field $k = \mathbb{F}_q$.

The theory of forms on vector spaces gives a complete classification of thick buildings of type C_n if $n \geq 4$. For $n = 3$, there is another family of thick buildings of type C_3 that we will not attempt to describe, one for each Cayley division algebra. They correspond to polar spaces that are not embeddable in projective spaces, where "embeddable" is understood in the sense of Tits [247, Chapter 8]. These buildings are algebraic (and exceptional).

Finally, if $n = 2$ and we impose the Moufang condition, then there are six families of thick buildings of type C_2, summarized in [262, Chapter 16, Figure 3]. They all admit algebraic descriptions, which are more complicated than in the higher-rank cases. Two of the six families consist of mixed buildings, and the other four are algebraic or classical.

Remark 9.4. As in the A_n case, the classification simplifies drastically if the buildings are finite. First, all finite, thick Moufang buildings of type C_n for $n \geq 2$ come from the geometry of totally isotropic subspaces associated to forms of Witt index n on vector spaces over finite fields k. In particular, the buildings are all classical. Second, the forms that yield thick C_n-buildings, as well as the corresponding groups, can easily be listed:

- There is an alternating form on k^{2n} as in Section 6.6, corresponding to the symplectic group $\mathrm{Sp}_{2n}(k)$.
- If char $k \neq 2$, there is a quadratic form on k^{2n+1}, described in Section 6.7.1, corresponding to the special orthogonal group $\mathrm{SO}_{2n+1}(k)$.
- There is a quadratic form on k^{2n+2} of Witt index n, corresponding to a special orthogonal group $\mathrm{SO}_{2n+2}^-(k)$. [This exists regardless of the characteristic of k; we described it in Section 6.7.2 only when char $k \neq 2$.]
- Given a quadratic extension $k < K$ of finite fields, there are Hermitian forms of Witt index n on K^{2n} and K^{2n+1} as in Section 6.8, corresponding to the special unitary groups $\mathrm{SU}_{2n}(K)$ and $\mathrm{SU}_{2n+1}(K)$.

9.4 Type D_n

The thick buildings of type D_n ($n \geq 4$) are all classical. They are precisely the ones studied in Section 6.7.1 (with appropriate modifications for the characteristic-2 case), corresponding to the group $\mathrm{SO}_{2n}(k)$ for an arbitrary field k. Recall that this group is defined via the standard quadratic form; if k is finite, it is usually denoted by $\mathrm{SO}_{2n}^+(k)$ as we explained in Section 6.7.2.

9.5 Type E_n

The thick buildings of type E_n ($n = 6, 7, 8$) are precisely the buildings associated to the Chevalley groups of those types over an arbitrary field k. (Recall that Chevalley groups are the groups constructed by Chevalley [81] that we mentioned in Section 6.4.) In particular, these buildings are all algebraic (and exceptional).

9.6 Digression: Twisted Chevalley Groups

In the finite case, all of the groups that arise in connection with Moufang buildings are either Chevalley groups or *twisted Chevalley groups*. We digress to explain what these are, starting with a familiar example. Consider the classical special unitary group $SU_n(\mathbb{C})$, which can be described as the subgroup of $SL_n(\mathbb{C})$ fixed by the automorphism $g \mapsto \bar{g}^{-t}$, where the bar denotes complex conjugation and the superscript $-t$ denotes transpose inverse. The existence of the transpose-inverse automorphism can be traced to the fact that the Coxeter diagram of type A_{n-1} has a symmetry of order 2. Thus we can view $SU_n(\mathbb{C})$ as arising from an automorphism of $SL_n(\mathbb{C})$ obtained by combining a field automorphism and a diagram automorphism.

Similar constructions exist for some of the other types of Chevalley groups whose Coxeter diagrams have symmetry, and, like the Chevalley groups themselves, the twisted groups have BN-pairs and hence buildings. Here is the list of diagrams with symmetry and the corresponding twisted groups:

- Type B_2 (or C_2): The diagram has an automorphism of order 2. Let K be a field of characteristic 2 with an automorphism σ such that σ^2 is the Frobenius automorphism $x \mapsto x^2$. [If K is finite, this means that $K = \mathbb{F}_q$ with q an odd power of 2.] Then one can combine σ with the diagram automorphism to construct a twisted Chevalley group, said to be of type $^2B_2$. The left superscript 2 indicates the order of the diagram automorphism. The associated building in this case has rank 1, so it will play no role in what follows.

- Type D_n: The diagram again has an automorphism of order 2, but the resulting twisted groups are orthogonal groups, so we get nothing new.

- Type D_4: The diagram has an automorphism of order 3. If K is any field with an automorphism of order 3, then we get a twisted Chevalley group, said to be of type $^3D_4$. It gives rise to an (exceptional) algebraic building of type G_2.

- Type E_6: The diagram has an automorphism of order 2. If K is any field with an automorphism of order 2, then we get a twisted Chevalley group, said to be of type $^2E_6$. It gives rise to an (exceptional) algebraic building of type F_4.

- Type F_4: The diagram has an automorphism of order 2. Let K be a field of characteristic 2 with an automorphism σ such that σ^2 is the Frobenius automorphism. Then we get a twisted Chevalley group, said to be of type $^2F_4$. It gives rise to a building of type $I_2(8)$, which is "mixed" in the terminology of [262].

- Type G_2: The diagram has an automorphism of order 2. Let K be a field of characteristic 3 with an automorphism σ such that σ^2 is the Frobenius automorphism $x \mapsto x^3$. [If K is finite, this means that $K = \mathbb{F}_q$ with q an odd power of 3.] Then we get a twisted Chevalley group, said to be of type $^2G_2$. The associated building has rank 1, so it will play no role in what follows.

The notion of a twisted Chevalley group is a special case of a more general construction that is needed to account for all absolutely simple algebraic groups and hence most Moufang spherical buildings. In the general (possibly infinite) case, one can have "twists" that act trivially on the Coxeter diagram; see Tits [245].

Chevalley groups and twisted Chevalley groups are often called groups of *Lie type*. We will have more to say about them in Section 9.10. See [75; 110, Chapters 2–4] for detailed treatments, with emphasis on the finite case, and see [107, Section 2.1] for an outline.

We now return to our survey of spherical buildings.

9.7 Type F_4

One description of the buildings of type F_4 is geometric: They are flag complexes of "metasymplectic spaces," which are incidence geometries with points, lines, planes, and "symplecta." We will say no more about these.

There is also an algebraic description of the buildings, which classifies them into five families. Let k be a field, and let K be a k-algebra that is either a division algebra (possibly commutative) or a Cayley division algebra. Assume that one of the following five conditions is satisfied:

(a) $K = k$.
(b) K is a separable quadratic extension of k.
(c) K is a quaternion division algebra.
(d) K is a Cayley division algebra.
(e) char $k = 2$, and K is an extension of k such that $K^2 < k < K$.

There is then a thick building of type F_4 associated to the pair (k, K). The buildings in cases (a)–(d) are algebraic (exceptional). Case (a) corresponds to the Chevalley group of type F_4 over k, and case (b) corresponds to the twisted Chevalley group of type $^2E_6$. The buildings that arise in case (e) are the mixed buildings mentioned in the introduction to this chapter. Cases (a) and (b) are the only ones that can yield finite buildings.

In summary, thick buildings of type F_4 are classified by pairs (k, K) as above, with one proviso: In case (e) the pairs (k, K) and $(K, k^{1/2})$ give isomorphic buildings, as we already noted in the introduction to this chapter.

9.8 Type G_2

Thick Moufang buildings of type G_2 are classified by algebraic objects called "hexagonal systems." See Tits–Weiss [262, Chapter 15] for the definition. There are six families of these, summarized in [262, Chapter 15, Figure 2]. These include the buildings associated to the Chevalley group of type G_2 (over an arbitrary field) and the twisted Chevalley group of type $^3D_4$. These are the only two cases that can arise if the building is finite. Of the remaining four families, three are (exceptional) algebraic and the fourth is mixed; it is the mixed family mentioned in Section 9.1 that involves imperfect fields of characteristic 3.

9.9 Type $I_2(8)$

Thick Moufang buildings of type $I_2(8)$ are classified by "octagonal sets" [256; 262, Chapter 10]. Here an *octagonal set* is a field K of characteristic 2 together with an endomorphism $\sigma\colon K \to K$ such that σ^2 is the Frobenius map $x \mapsto x^2$. The corresponding group is the twisted Chevalley group of type $^2F_4$ if σ is an automorphism (i.e., if K is a perfect field). The building is of mixed type in the terminology of [262].

The finite fields K that can arise here are the fields $\mathbb{F}_q$ with q an odd power of 2. The group $^2F_4(\mathbb{F}_q)$ has order

$$q^{12}(q^6 + 1)(q^4 - 1)(q^3 + 1)(q - 1) ,$$

and it is simple unless $q = 2$. In this case its commutator subgroup is of index 2 and is simple; it is the *Tits group* mentioned in Section 6.4.

9.10 Finite Simple Groups and Finite Buildings

In about 1980, Daniel Gorenstein announced that the classification of finite simple groups was complete. This appeared to mark the end of an unprecedented long-term cooperative effort involving many mathematicians and over 10,000 pages of journal articles. Many parts of the proof had not yet been published at the time of the announcement, and some parts had not even been written up completely. It is not surprising, therefore, that gaps were discovered over the next few years.

Most of the gaps were relatively easy to fix, but one of them, involving the "quasithin" case, turned out to be quite serious. By the mid-1990s the classification had reverted to being a conjecture rather than a theorem. But the quasithin case has now been dealt with by Aschbacher and Smith [25, 26] in about 1200 pages, and at this writing most experts seem to believe the classification theorem as originally stated in 1980.

Shortly after the first announcement, Gorenstein, Lyons, and Solomon began a "revision" project, hoping to write a proof that is clear, careful, and, as much as possible, self-contained. It is currently expected that this proof will occupy 11 or 12 volumes, of which 6 have been completed as of this writing [108–113]. There is also a second revision project underway, by Meierfrankenfeld, Stellmacher, Stroth, and others. This is not as far along, but there is an overview in [162].

The statement of the classification theorem is remarkably clean. It says that every nonabelian finite simple group is isomorphic to one of the following:

- An alternating group.
- A finite group of Lie type.
- One of 26 sporadic groups.

Recall from Section 9.6 that the finite groups of Lie type are the Chevalley groups and twisted Chevalley groups defined over finite fields. With a few exceptions that are easy to list, finite groups of Lie type are either simple or else have a closely related subquotient that is simple. For example, $\mathrm{SL}_n(\mathbb{F}_q)$ is a group of Lie type, as is its quotient $\mathrm{PSL}_n(\mathbb{F}_q)$ $(n \geq 2)$. As we know from Section 6.2.7, the latter is simple unless $n = 2$ and $q = 2$ or 3. Another example was mentioned at the end of Section 9.9.

All finite groups of Lie type have BN-pairs and hence associated buildings. For a few of the families the building is of rank 1 and hence provides a permutation representation but no geometry. In the "generic" case of rank ≥ 2, however, we get a Moufang building Δ, and the corresponding simple group of Lie type is almost always the subgroup G_1 of Aut Δ generated by the root groups, which we discussed in Section 7.8.7. [There are three exceptions, all in rank 2, in which the simple group is not G_1 but rather a subgroup of index 2. These are listed in [262, Proposition 37.3]; see also the paragraph following the proof of that proposition.] Thus an informal statement of the classification theorem is that the generic finite simple group is essentially the automorphism group of a finite, thick, irreducible Moufang building of rank ≥ 2. The list of such buildings and the corresponding groups has been given in the previous sections.

Remarks 9.5. (a) This correspondence between buildings and groups (for the "generic" case of rank ≥ 2) is almost 1–1. The only exception is the "coincidence" that the projective symplectic group $\mathrm{PSp}_4(\mathbb{F}_3)$ is isomorphic to the projective special unitary group $\mathrm{PSU}_4(\mathbb{F}_4)$, but the corresponding buildings of type C_2 are different. (This coincidence is called a *sporadic isomorphism* by finite-group theorists.)

(b) It is easy to describe the building Δ associated to a finite simple group G of Lie type, once one knows the characteristic p of the finite field over which G is defined. In fact, it turns out that the minimal parabolics (i.e., the conjugates of the B of the BN-pair) are the normalizers of the Sylow p-subgroups of G. So Δ can be identified with the set of subgroups that contain the normalizer of a Sylow p-subgroup, ordered by reverse inclusion.

(c) The classification of finite, rank-2, thick Moufang buildings, for which we have cited Tits–Weiss [262], can also be deduced from the classification by Fong and Seitz [103] of finite, rank-2, split BN-pairs. (We omit the definition of "split.")

*9.11 Remarks on the Simplified Proof

We mentioned in Section 9.1 that it was possible to simplify the proof of the classification theorem once the rank-2 Moufang buildings (also called Moufang polygons) were classified. In this section we briefly elaborate on that statement.

Tits's extension theorem (5.209) says that a building Δ (irreducible, spherical, of rank ≥ 3) is uniquely determined by $E_2(C)$, the union of the rank-2 residues containing a fixed chamber C. Here $E_2(C)$, which is sometimes called a *foundation* for the building, should be viewed as a set with some structure (e.g., an adjacency relation). Now we also know, again as a consequence of the extension theorem, that Δ is Moufang, so the rank-2 residues that are glued together to form the foundation are Moufang polygons; see Theorem 7.59 and Proposition 7.32. Since Moufang polygons have been classified [262], one can now systematically work out which Moufang polygons fit together to form a foundation and in how many different ways. See [262, Chapter 40] for details and [281, Chapter 12] for an overview. Tits had to work much harder in his original proof of the classification theorem, since he could not list a priori the possible rank-2 residues that might occur in a foundation.

One final remark: Given the extension theorem and the classification of Moufang polygons, one can break the classification problem into two steps. First, one has to figure out all ways to glue Moufang polygons together to form a structure that looks as if it could be the foundation for a spherical building. Next, one has to prove that every candidate for a foundation is in fact realized by a spherical building (which is then unique). For this second step, the *existence* of enough spherical buildings, Tits originally relied on the theory of algebraic groups. Later, Ronan and Tits [204] introduced *blueprints*. The results of that paper, which apply to general buildings, lead in particular to a uniform construction of all spherical Moufang buildings that is independent of the theory of algebraic groups. A write-up of this construction (just in the spherical case) can be found in [262, Sections 40.53–40.56].

*9.12 The Classification of Twin Buildings

Since there is an extension theorem for (most) 2-spherical twin buildings (Theorem 5.213), one might hope that the methods used in the spherical case would lead to a classification theorem for thick 2-spherical twin buildings. A program for doing this was proposed by Tits [261]. There were some highly nontrivial technical details that he did not anticipate, but the program has in fact been completed due to the efforts of Tits, Ronan, and Mühlherr. The complete proof has not been published yet, but we refer to Mühlherr [176] for a survey and for references to the literature. We confine ourselves here to a few very brief comments.

First, the proof of the classification theorem is similar in broad outline to the proof for the spherical case sketched in the previous section. Namely, there is first a uniqueness assertion, which says that a 2-spherical twin building is uniquely determined by a foundation, which is again a union of Moufang polygons. This was completely settled, at least under the hypothesis (co) of Section 5.11, in the papers [177, 203, 261].

Next, there is an existence question, in which one must determine which foundations are realizable by twin buildings. This step involves new difficulties that did not arise in the spherical case, in which (at least in Tits's original proof) algebraic groups essentially provided all the buildings. In the twin case, by contrast, Kac–Moody theory is not (or at least not yet) rich enough to directly yield the desired twin buildings. Moreover, the theory of blueprints mentioned above has not been extended to twin buildings. Indeed, it is an interesting open question whether this can be done.

What is true, however, is that (very roughly speaking) all thick, irreducible, 2-spherical twin buildings of rank at least 3 can be obtained by starting with twin buildings that come from Kac–Moody groups and then doing certain geometric constructions. One construction, for example, involves letting a group act on a known twin building and then taking the fixed-point set. The details of the existence proof for a special class of twin buildings can be found in [175]. The general case is treated in Mühlherr's Habilitationsschrift [174].

We close by mentioning one interesting and surprising aspect of Mühlherr's work. Recall that we have a standing convention that the generating set S of a Coxeter group W is always assumed to be finite. In particular, all of our buildings are of finite rank. But in order to carry out the existence part of the classification theorem, Mühlherr had to introduce certain twin buildings of *infinite* rank with group actions. He then obtained the desired twin buildings of finite rank by taking fixed points.

Euclidean and Hyperbolic Reflection Groups

Our study of the building associated to SL_n over a discretely valued field (Section 6.9) led to examples of buildings in which the apartments are Euclidean spaces. And readers of Chapter 8 saw that the theory of Kac–Moody groups leads to examples of buildings in which the apartments are hyperbolic spaces (Section 8.11). These examples motivate a systematic study of such buildings. But first we need to understand the apartments themselves, which means we need to start with the corresponding reflection groups. We will study these in the present chapter and then return to buildings in Chapters 11 and 12.

We begin with the Euclidean case, where our first task is to redo some of Chapter 1 in a more general setting.

10.1 Euclidean Reflection Groups

Let V be a real vector space of finite dimension $n \geq 1$. All of the geometric notions introduced in Chapter 1 treated the origin of V as a distinguished point. Our hyperplanes, for instance, were required to go through the origin. Our reflection groups therefore fixed the origin, and our cells were all cones with the origin as cone point. By the end of that chapter it had become clear that we were not doing Euclidean geometry at all, but rather spherical geometry. In Euclidean geometry, there is nothing special about the origin. So let's introduce the appropriate language for talking about reflections whose fixed hyperplane does not necessarily pass through the origin.

10.1.1 Affine Concepts

An *affine subspace* of V is a subset of the form $x + V_0$ with $x \in V$ and V_0 a linear subspace of V. In other words, it is a coset of a linear subspace. The *dimension* of $x + V_0$ is defined to be the dimension of V_0. If the dimension is $n - 1$ (i.e., if V_0 is a linear hyperplane), then $x + V_0$ is called an *affine hyperplane*. Equivalently, an affine hyperplane is a subset defined by a linear equation of the form $f = c$, where $f \colon V \to \mathbb{R}$ is a nonzero linear map and c is a constant.

For any nonempty subset $X \subseteq V$, there is a smallest affine subspace containing X, called the *affine span* of X. An *affine frame* for V is a subset X such that V is the affine span of X but not of any proper subset of X. Such a frame necessarily has exactly $n + 1$ elements; for we may assume that one of the elements is the origin, in which case the remaining elements form a basis for V. When $n = 2$, for instance, an affine frame is simply a set of 3 noncollinear points.

An *affine map* from V to a vector space V' is a map a of the form $a(x) = g(x) + v'$, where $g \colon V \to V'$ is linear and v' is a vector in V'. In other words, a is the composite $\tau_{v'} g$, where $\tau_{v'}$ is the translation $x' \mapsto x' + v'$. We will mainly be interested in the case that $V' = V$ and g is an automorphism, in which case we will call a an *affine automorphism* of V. Such an a is uniquely expressible as $\tau_v g$ with $v \in V$ and $g \in \mathrm{GL}(V)$, where $\mathrm{GL}(V)$ is the group of linear automorphisms of V. One deduces easily that the group $\mathrm{Aff}(V)$ of affine automorphisms of V is the semidirect product

$$\mathrm{Aff}(V) = V \rtimes \mathrm{GL}(V),$$

where V is identified with the (normal) subgroup consisting of translations. The $\mathrm{GL}(V)$-component g of an element $a \in \mathrm{Aff}(V)$ will be called the *linear part* of a.

Assume, now, that V is *Euclidean*, by which we mean that it comes equipped with a positive definite inner product $\langle -, - \rangle$. We then have a distance function on V given by $d(x, y) := \|x - y\|$, where $\|v\| := \sqrt{\langle v, v \rangle}$, and we will be interested in affine maps that are isometries. With the notation above, the affine map a is an isometry if and only if its linear part g is in the orthogonal group $\mathrm{O}(V) < \mathrm{GL}(V)$, consisting of automorphisms that preserve the inner product. Thus the group of affine isometries of V is $V \rtimes \mathrm{O}(V)$. It is worth noting that the word "affine" is redundant here; see Exercise 10.3 below.

Let H be an affine hyperplane and let H_0 be the linear hyperplane parallel to H, i.e., satisfying $H = x_1 + H_0$ for some $x_1 \in V$. Let s_0 be the orthogonal reflection s_{H_0}, and let $s_1 := \tau_{x_1} s_0 \tau_{-x_1}$; in other words, s_1 is the conjugate of s_0 by some translation taking H_0 to H. Explicitly, we have

$$s_1(x) = x_1 + s_0(x - x_1) = s_0(x) + (1 - s_0)(x_1) \tag{10.1}$$

for any $x \in V$. In particular, s_1 is an affine isometry whose linear part is s_0. It is easy to check that s_1 depends only on H, and not on the choice of the representative x_1; this follows, for instance, from (10.1) together with the observation that x_1 is unique mod $H_0 = \ker(1 - s_0)$. We can therefore write $s_1 = s_H$ and call s_1 the *reflection* with respect to H.

Exercises

10.1. Let $x_0, x_1, \ldots, x_n$ be an affine frame. Show that a point $x \in V$ is completely determined by the $n + 1$ numbers $d(x, x_i)$, i.e., if $d(x, x_i) = d(x', x_i)$ for all i, then $x = x'$.

10.2. Let $x_0, \ldots, x_n$ be an affine frame and let $y_0, \ldots, y_n$ be points such that $d(y_i, y_j) = d(x_i, x_j)$ for all i, j. Show that there is an affine isometry a such that $a(x_i) = y_i$ for all i. In particular, $y_0, \ldots, y_n$ is an affine frame.

10.3. Deduce from Exercises 10.1 and 10.2 that every isometry $V \to V$ is affine.

10.4. Show that the reflection s_H can be characterized as the unique nontrivial isometry of V that fixes H pointwise.

10.5. Let X and X' be isometric subsets of V. Show that every isometry $X \to X'$ extends to an isometry of V.

10.1.2 Formulas for Affine Reflections

In this brief subsection we elaborate on formula (10.1), to pave the way for a connection between Euclidean reflection groups and root systems that we will give in Section 10.1.8. We will need to use some notation defined in Appendix B. The reader may prefer to skip this subsection (and the appendix) and refer back to them as necessary.

Let H be a linear hyperplane and let α be a nonzero vector in $H^\perp$. Recall from the appendix just cited that the reflection s_0 with respect to H is given by

$$s_0(x) = x - \langle \alpha^\vee, x \rangle \alpha, \tag{10.2}$$

where $\alpha^\vee$ is the scalar multiple of α satisfying

$$\langle \alpha^\vee, \alpha \rangle = 2. \tag{10.3}$$

Note that we could have used $\alpha^\vee$ instead of α as our nonzero vector in $H^\perp$, so we can interchange α and $\alpha^\vee$ in (10.2) and write

$$s_0(x) = x - \langle \alpha, x \rangle \alpha^\vee. \tag{10.4}$$

Hence the rank-1 linear map $1 - s_0$ that occurs in formula (10.1) is given by

$$(1 - s_0)(x) = \langle \alpha, x \rangle \alpha^\vee. \tag{10.5}$$

Let H_1 be the affine hyperplane given by $\langle \alpha, x \rangle = 1$. Substituting (10.5) into (10.1), we obtain the following formula for the reflection s_1 with respect to H_1:

$$s_1(x) = s_0(x) + \alpha^\vee. \tag{10.6}$$

Note, in particular, that the composite $s_0 s_1$ is the translation $x \mapsto x - \alpha^\vee$.
Finally, we remark that equation (10.3) can also be written as

$$\alpha^\vee/2 \in H_1 \, ,$$

so the relation between α and $\alpha^\vee$ can be illustrated as in Figure 10.1.

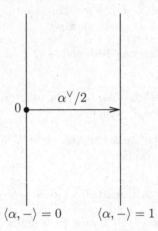

$$\langle \alpha, - \rangle = 0 \qquad \langle \alpha, - \rangle = 1$$

Fig. 10.1. The relation between $\alpha^\vee$ and α.

Exercise 10.6. For any constant c, let H_c be the hyperplane given by $\langle \alpha, x \rangle = c$. Show that the reflection s_c with respect to H_c is given by

$$s_c(x) = s_0(x) + c\alpha^\vee \, .$$

Hence

$$s_c(x) = x - (\langle \alpha, x \rangle - c)\alpha^\vee \, .$$

10.1.3 Affine Reflection Groups

We continue to assume that V is a Euclidean vector space of dimension $n \geq 1$. Let W be a group of affine isometries of V.

Definition 10.7. We say that W is an *affine reflection group* if there is a set $\mathcal{H}$ of affine hyperplanes in V satisfying:

(a) W is generated by the reflections s_H for $H \in \mathcal{H}$.
(b) $\mathcal{H}$ is W-invariant.
(c) $\mathcal{H}$ is locally finite, in the sense that every point of V has a neighborhood that meets only finitely many $H \in \mathcal{H}$.

Note that if we were only to assume that W is generated by reflections, then we could find a set $\mathcal{H}$ satisfying (a) and (b), but there is no reason to expect (c) to hold. Condition (c) plays the role in the affine setting of the requirement in the linear case (Chapter 1) that W be finite.

Much of what we did in Chapter 1 goes through with little or no change for affine reflection groups, although some of the arguments get slightly longer, since we now have only local finiteness of $\mathcal{H}$ instead of finiteness. We will sketch the theory, including proofs only for those results that require new ideas or are special to the affine case. Assume for the rest of this subsection that W is an affine reflection group and $\mathcal{H}$ is as in Definition 10.7.

The hyperplanes $H \in \mathcal{H}$ yield a partition of V into convex *cells*, these being nonempty sets A defined by linear equalities or strict inequalities, one for each $H \in \mathcal{H}$. More precisely, if H is defined by a linear equality $f = c$, then the definition of A will involve either the same equality or else an inequalitiy $f > c$ or $f < c$. A cell A has a *support*, defined as in Section 1.4.1; the support is an affine subspace of V, and A is open relative to its support. The *dimension* of A is the dimension of its support. The cells of maximal dimension n are called *chambers*; they are the connected components of the complement in V of $\bigcup_{H \in \mathcal{H}} H$. The cells of dimension $n - 1$ are called *panels*. Cells have faces, with properties similar to those proven in Chapter 1.

The supports of the panels of a chamber C are called the *walls* of C, and C is defined by its walls (i.e., by the inequalities corresponding to the walls). This is the first assertion we have made for which it is not entirely a routine matter to generalize the proof given in Chapter 1. To prove it, go back to the proof of Proposition 1.32. In the locally finite case, the same argument allows us to remove finitely many nonwalls from $\mathcal{H}$ and still have a defining set of hyperplanes. To finish the proof, let C' be the subset of V defined by the inequalities corresponding to the walls of C. It might, a priori, be bigger than C, so we must show that $C' \subseteq C$. Given $x \in C'$, consider the closed line segment joining x to a point of C. By local finiteness of $\mathcal{H}$ and compactness of the line segment, the latter can meet only finitely many hyperplanes. It remains to show that x is on the C-side of these. None of them are walls, since x is on the C-side of every wall. So x is on the C-side of all hyperplanes except perhaps finitely many nonwalls. Since we can remove these and still have a defining set of hyperplanes for C, we have $x \in C$, as required. [Alternatively, one can argue as in Exercise 1.34.]

Everything we have said so far applies to any locally finite collection of affine hyperplanes. Now let's bring W into the picture. Choose a chamber C and let S be the set of reflections with respect to the walls of C. Note that S, a priori, might be infinite; we will return to this question below. Let H_s for $s \in S$ be the hyperplane fixed by s, and let e_s be the unit vector perpendicular to H_s and pointing to the side of H_s containing C. Thus $e_s^{\perp}$ is the linear hyperplane parallel to H_s, and one of the defining inequalities for C has the form $\langle e_s, - \rangle > c$. Let $m(s, t)$ for $s, t \in S$ be the (possibly infinite) order of st. The following basic facts proved in the finite case remain valid:

(a) W is simply transitive on the chambers.

(b) W is generated by S.

(c) $\mathcal{H}$ necessarily consists of all affine hyperplanes H with $s_H \in W$.

(d) (W, S) is a Coxeter system.

(e) $\langle e_s, e_t \rangle = -\cos(\pi/m(s,t))$ for all $s, t \in S$. Moreover, $m(s,t) = \infty$ if and only if $s \neq t$ and H_s is parallel to H_t.

(f) $\overline{C}$ is a strict fundamental domain for the action of W on V, and the stabilizer of a point $x \in \overline{C}$ is the standard subgroup of W generated by $S_x := \{s \in S \mid sx = x\}$.

Everything except (d) is proved as in Chapter 1. For (d), one can argue as in Section 2.2, or one can verify the exchange condition as a byproduct of the proof of (a). Incidentally, the possibility $m(s,t) = \infty$ mentioned in (e) hardly ever occurs. In fact, we will see in Section 10.1.5 that the infinite dihedral group provides the only irreducible example whose Coxeter matrix involves ∞. [Alternatively, instead of appealing to Section 10.1.5, one can apply Corollary 3.27; for we will see below that the Coxeter complex $\Sigma(W, S)$ triangulates V if W is infinite and irreducible.]

10.1.4 Finiteness Results

Let's now settle the question of the finiteness of S, along with some related questions:

Theorem 10.8.

(1) C has only finitely many walls, and hence S is finite.

(2) The hyperplanes $H \in \mathcal{H}$ fall into finitely many classes under the relation of parallelism; in other words, there are only finitely many linear hyperplanes H_0 such that $\mathcal{H}$ contains a translate of H_0.

(3) Let $\overline{W} \leq \mathrm{GL}(V)$ be the set of linear parts of the elements $w \in W$, i.e., the image of W under the projection $\mathrm{Aff}(V) \twoheadrightarrow \mathrm{GL}(V)$. Then $\overline{W}$ is a finite reflection group.

Proof. (1) The inner product formula in (e) above shows that the angle $\angle(e_s, e_t)$ between e_s and e_t satisfies $\angle(e_s, e_t) \geq \pi/2$ for $s \neq t$. But if S were infinite, then the e_s would have a cluster point on the unit sphere and hence there would be s, t with $\angle(e_s, e_t)$ very small.

(2) Let $\Phi := \{\pm e_H \mid H \in \mathcal{H}\}$, where e_H is the unit vector perpendicular to H and pointing to the side containing C. We must show that Φ is finite. As in the proof of (1), it suffices to show that $\angle(e, e')$ is bounded away from 0 for $e \neq e'$ in Φ. We will show that in fact, there are only finitely many possibilities for this angle.

Let H and H' be elements of $\mathcal{H}$ perpendicular to e and e', respectively. If H and H' are parallel, then $\angle(e, e') = \pi$ [since $e \neq e'$]. Otherwise, choose

$x \in H \cap H'$ and choose $w \in W$ with $wx \in \overline{C}$; this is possible by (f) above. Then wH and wH' are elements of $\mathcal{H}$ that meet $\overline{C}$, and they are perpendicular to the vectors $\bar{w}e, \bar{w}e'$, where $\bar{w}$ is the linear part of w. Since $\measuredangle(e, e') = \measuredangle(\bar{w}e, \bar{w}e')$, the proof will be complete if we show that only finitely many elements of $\mathcal{H}$ meet $\overline{C}$. Now C has only finitely many walls by (1); so it is defined by finitely many inequalities and hence has only finitely many faces. And each face meets only finitely many elements of $\mathcal{H}$ by local finiteness and the definition of "cell." The union $\overline{C}$ of the faces therefore meets only finitely many elements of $\mathcal{H}$.

(3) The set Φ defined in the proof of (2) is $\overline{W}$-invariant. Since it was proven to be finite, $\overline{W}$ is a finite reflection group by Lemma 1.4.   □

10.1.5 The Structure of C

Call W *essential* if the associated finite reflection group $\overline{W}$ is essential. It is easy to reduce the general case to the essential case, as in Chapter 1. Similarly, we can decompose V according to the irreducible components of the Coxeter diagram of (W, S) and thereby reduce to the irreducible case. In this case we will prove the following theorem:

Theorem 10.9. *Assume that W is essential and irreducible. Then C is either a simplex or a simplicial cone. More precisely, one of the following holds:*

(1) *W is finite and has a fixed point. In this case C has exactly n walls, where $n = \dim V$, and C is a simplicial cone.*

(2) *W is infinite, C has exactly $n + 1$ walls, and any n of the $n + 1$ vectors e_s are linearly independent. The essentially unique linear relation $\sum_{s \in S} \lambda_s e_s = 0$ among the e_s has all of its coefficients λ_s nonzero and of the same sign. The chamber C is a simplex.*

Remarks 10.10. (a) If W has a fixed point x, then we can always assume that $x = 0$; for we can replace W by its conjugate $\tau_{-x} W \tau_x$. But if W fixes 0, then W is linear. Thus case (1), for practical purposes, is precisely the "spherical" case treated in Chapter 1. Case (2), then, describes the "genuinely Euclidean" irreducible reflection groups.

(b) If we drop the assumption that W is essential and irreducible, then C is a product of a vector space [corresponding to the inessential part of V], a simplicial cone [corresponding to the product of the finite irreducible factors of W], and simplices [one for each infinite irreducible factor of W].

(c) It follows from the theorem that the numbers $m(s, t)$ are always finite when W is irreducible, unless $n = 1$ and $W = D_\infty$; for an n-simplex with $n \geq 2$ cannot have parallel walls. In fact, a stronger statement is true:

Corollary 10.11. *With W as in case (2) of the theorem, the stabilizers of the points of $\overline{C}$ are precisely the proper standard subgroups of W, and all of these are finite.*

Proof. Given $x \in \overline{C}$, let S_x be as in statement (f) in Section 10.1.3. If A is the face of C containing x, then S_x can also be described as the set of $s \in S$ such that the H_s-component of the sign vector of A is 0. Since C is a simplex, the possible sets S_x that can occur are precisely the proper subsets of S, whence the first assertion of the corollary. To prove that the stabilizer W_x is finite, consider the homomorphism $W \to \overline{W}$ that sends each $w \in W$ to its linear part. The kernel of this homomorphism is the set of translations in W; hence the restriction of the homomorphism to W_x is trivial. [A nontrivial translation has no fixed points.] Thus W_x injects into the finite reflection group $\overline{W}$ and hence is finite. $\square$

Proof of the theorem. Let $H_1, \ldots, H_r$ be the walls of C, and let $e_1, \ldots, e_r$ be the corresponding unit vectors (pointing to the side containing C). Then $e_1, \ldots, e_r$ span V; for $[e_1, \ldots, e_r]^{\perp}$ is the fixed-point set of $\overline{W}$, and hence it is the trivial subspace. We therefore have $r \geq n$.

Suppose $r = n$, so that the vectors $e_1, \ldots, e_r$ form a basis. Then C is defined by n inequalities $\langle e_i, - \rangle > c_i$ and is therefore a simplicial cone; its cone point is the unique $x \in V$ such that $\langle e_i, x \rangle = c_i$ for all i. Replacing W by a conjugate, we may assume that $x = 0$. Then the generating reflections of W are linear, so $W = \overline{W}$ and we are in case (1).

Now suppose $r > n$, so that $e_1, \ldots, e_r$ are linearly dependent. Choose a linear relation

$$\sum_{i \in I} \lambda_i e_i = 0$$

with $\emptyset \neq I \subseteq \{1, \ldots, r\}$ and $\lambda_i \neq 0$ for all $i \in I$. Since $\langle e_i, e_j \rangle \leq 0$ for $i \neq j$, we can replace I by a subset, if necessary, to get a relation with $\lambda_i > 0$ for all $i \in I$; this follows from the first paragraph of the proof of Proposition 1.37. We can now deduce from the irreducibility assumption that I is the entire set of indices $\{1, \ldots, r\}$. For suppose it is not, and let J be the complementary set. Then for any $j \in J$ we have $\sum_{i \in I} \lambda_i \langle e_i, e_j \rangle = 0$, which implies that $\langle e_i, e_j \rangle = 0$ for all $i \in I$. But then the parts of the Coxeter diagram corresponding to I and J are disjoint, contradicting irreducibility.

Our relation now has the form $\sum_{i=1}^r \lambda_i e_i = 0$, with $\lambda_i > 0$ for all i. But we arrived at this relation by starting with an arbitrary relation among a subset of the e_i and then possibly passing to a further subset. It follows that every proper subset of the e_i is linearly independent; hence $r = n + 1$.

Since $e_1, \ldots, e_n$ form a basis for V, the intersection $\bigcap_{i=1}^n H_i$ consists of a single point, which we may take to be the origin. The chamber C is therefore defined by inequalities $\langle e_i, - \rangle > 0$ for $i = 1, \ldots, n$ and $\langle e_{n+1}, - \rangle > c$ for some constant c. Since e_{n+1} is a negative linear combination of $e_1, \ldots, e_n$, the last inequality can be rewritten in the form $\sum_{i=1}^n \mu_i \langle e_i, - \rangle < c'$ with $\mu_i > 0$. The constant c' is necessarily positive [and hence our original c was negative], since otherwise C would be empty. So we may multiply the inequality by a scalar in order to arrange that $c' = 1$. Thus C is defined by inequalities $f_i > 0$ for

$i = 1, \ldots, n$ and $\sum f_i < 1$, where $f_i := \mu_i \langle e_i, - \rangle$. Since the f_i form a basis for the dual space V^*, these inequalities define an open n-simplex.

Everything in (2) has now been proved except for the assertion that W is infinite. But this follows from the fact that W has compact fundamental domain $\overline{C}$; for if W were finite, then $V = \bigcup_{w \in W} w\overline{C}$ would be compact.   □

Definition 10.12. By a *Euclidean reflection group* we will mean an essential, irreducible, infinite, affine reflection group, as in case (2) of Theorem 10.9.

Notice that in contrast to the situation for finite reflection groups, the poset of cells has no smallest element and so cannot be a simplicial complex. In particular, it cannot be isomorphic to $\Sigma(W, S)$. We therefore adjoin to the poset of cells a smallest element, which one can think of as corresponding to the empty set, and we denote this new poset by Σ or $\Sigma(W, V)$. Since the chambers are simplices, it is easy to see (as in Section 1.5.8) that Σ is an abstract simplicial complex and that there is a canonical homeomorphism $|\Sigma| \cong V$.

It is immediate from property (f) in Section 10.1.3, together with Corollary 10.11, that $\Sigma(W, V) \cong \Sigma(W, S)$. We summarize some of these observations for future reference:

Proposition 10.13. *Let W be a Euclidean reflection group.*

(1) *The simplicial complex $\Sigma = \Sigma(W, V)$ is isomorphic to the Coxeter complex $\Sigma(W, S)$.*

(2) *Σ triangulates V.*

(3) *The poset of cells in V is isomorphic to the subposet $\Sigma_f(W, S)$ of $\Sigma(W, S)$ consisting of the finite standard cosets (ordered by reverse inclusion).*   □

We reiterate that the finite standard cosets are the same as the proper standard cosets, i.e., only the coset W is excluded, so statement (3) may seem superfluous. But we have stated it explicitly for comparison with the hyperbolic case that we will treat in Section 10.3.4.

Examples 10.14. We will treat some examples of Euclidean reflection groups rigorously in Section 10.1.7, and we will show how to construct the general Euclidean reflection group in Section 10.1.8. In the meantime, the reader might find it helpful to visualize the 2-dimensional examples. There are in fact precisely three of these, corresponding to the three possible shapes of a Euclidean triangle with angles of the form π/m, where m is an integer ≥ 2.

(a) There is a Euclidean reflection group whose chambers are equilateral triangles, which we have discussed in Example 3.7. Its Coxeter diagram is

and it is said to be of type $\tilde{A}_2$.

(b) There is a Euclidean reflection group whose chambers are isosceles right triangles. Its Coxeter diagram is

$$\circ \!\!\!\xrightarrow{4}\!\!\! \circ \!\!\!\xrightarrow{4}\!\!\! \circ$$

and it is said to be of type $\tilde{B}_2$ or $\tilde{C}_2$. The corresponding Coxeter complex can be obtained by first tiling the plane by squares and then drawing all of the diagonals. Alternatively, first tile the plane by squares and then barycentrically subdivide all the squares. [One has to rotate by 45 degrees in order to see that these two constructions give the same result.]

(c) There is a Euclidean reflection group whose chambers are 30-60-90 triangles. Its Coxeter diagram is

$$\circ \!\!\!\xrightarrow{}\!\!\! \circ \!\!\!\xrightarrow{6}\!\!\! \circ$$

and it is said to be of type $\tilde{G}_2$. The corresponding Coxeter complex can be obtained by first tiling the plane by equilateral triangles and then barycentrically subdividing all the triangles.

Our requirement that (W, V) be irreducible in order to be called a Euclidean reflection group was made for convenience, so that we would not have to leave the world of simplicial complexes. Another possible definition would replace irreducibility by the requirement that each irreducible component be infinite; this is equivalent to requiring that the action have a compact fundamental domain (which is then a product of simplices). We will have a few occasions to talk about such reflection groups, so we give them a name.

Definition 10.15. By a *semi-Euclidean* reflection group we will mean an essential affine reflection group such that each of its irreducible components is infinite.

Note that the rank of the Coxeter system (W, S) is then $n + k$, where k is the number of irreducible factors of W and n is the dimension of V. (We may also call n the *dimension* of (W, S).) Note also that the poset of cells associated to a semi-Euclidean reflection group is still isomorphic to the poset of finite standard cosets (ordered by reverse inclusion), but this is no longer a simplicial complex in the reducible case. It is the poset of cells of a cell complex in which the cells are *products* of simplices. Such complexes are sometimes said to be *polysimplicial*.

Exercise 10.16. Let (W, V) be a Euclidean reflection group. In the proof of Theorem 10.9 we arranged, for simplicity, for the origin to be one of the vertices of the simplex C. What would the $n + 1$ inequalities defining C look like if we had not done this?

10.1.6 The Structure of W, Part I

Let W be a Euclidean reflection group as above. Recall that we have a surjection $W \twoheadrightarrow \overline{W}$, where $\overline{W}$ is the finite reflection group consisting of the linear parts of the elements of W. The kernel of this surjection is the group T of translations in W. The first result about the structure of W is that the short exact sequence

$$1 \to T \to W \to \overline{W} \to 1$$

always splits:

Proposition 10.17. *There exist points $x \in V$ such that the stabilizer W_x maps isomorphically onto $\overline{W}$, so that $W = T \rtimes W_x$.*

Proof. Let $\overline{\mathcal{H}}$ be the set of linear hyperplanes H such that H is parallel to some element of $\mathcal{H}$. Then $\overline{W}$ is generated by $\{s_H \mid H \in \overline{\mathcal{H}}\}$. Since $\overline{W}$ is essential, it follows from Chapter 1 that $\overline{W}$ is actually generated by n such reflections s_H, whose hyperplanes H form the walls of a simplicial cone. Choose $H_1, \dots, H_n \in \mathcal{H}$ parallel to these walls, and let $s_i := s_{H_i}$. Then $\bigcap_{i=1}^{n} H_i$ consists of a single point x, which is fixed by each s_i, and the linear parts of the s_i generate $\overline{W}$. This shows that W_x surjects onto $\overline{W}$. But W_x also injects into $\overline{W}$, as we saw in the proof of Corollary 10.11. $\qquad\square$

Definition 10.18. A point x as in the proposition is called a *special point*.

It is easy to characterize the special points:

Proposition 10.19. *A point $x \in V$ is special if and only if every hyperplane in $\mathcal{H}$ is parallel to a hyperplane in $\mathcal{H}$ passing through x.*

Proof. If the condition on hyperplanes is satisfied, then the map $W_x \to \overline{W}$ is surjective and hence an isomorphism, as in the proof of Proposition 10.17. So x is special. For the converse, it is convenient to assume (without loss of generality) that $x = 0$. Suppose, then, that 0 is a special point. Since the elements of $W_x = W_0$ are linear, the isomorphism $W_0 \xrightarrow{\sim} \overline{W}$ is the identity map. Thus W contains the linear part of each of its elements. Now for any $H \in \mathcal{H}$, formula (10.1) shows that the linear part of s_H is the reflection s_0 with respect to the hyperplane H_0 through the origin that is parallel to H. So s_0 belongs to W and hence H_0 is in $\mathcal{H}$, as required. $\qquad\square$

Remarks 10.20. (a) The criterion of the proposition makes it simple to locate the special points if one has a picture of the Coxeter complex. In Example 10.14(a), for instance, all vertices are special. In Example 10.14(b), two types of vertices are special and one is not. In Example 10.14(c), only one type of vertex is special.

(b) The criterion of the proposition has a geometric consequence that will be useful later. Take the special vertex x to be the origin, as in the proof, and let C be any chamber with the origin as a vertex. Then C, being a simplex, has $n+1$ walls, n of which pass through the origin and are the walls of a simplicial cone $\mathfrak{C}$. The remaining wall cuts off C as a "corner" of $\mathfrak{C}$. [This is precisely what we showed in the proof of Theorem 10.9.] We will call $\mathfrak{C}$ the *conical extension* of C from the vertex x. Because the origin is a special point, the cone $\mathfrak{C}$ is in fact a $\overline{W}$-chamber. Indeed, $\mathfrak{C}$ is defined by some of the $\overline{W}$-walls, and C lies on one side of every $\overline{W}$-wall. [Why?] So a scaling argument shows that $\mathfrak{C}$ lies on one side of every $\overline{W}$-wall and hence is a $\overline{W}$-chamber, as claimed. Note further that every $\overline{W}$-chamber $\mathfrak{C}$ arises in this way from a W-chamber C having the origin as a vertex. For if we move along a straight line from the origin into $\mathfrak{C}$, then we enter a chamber C having the origin as a vertex, and $\mathfrak{C}$ is the unique $\overline{W}$-chamber containing C. It must therefore be the conical extension of C.

From now on we assume, as in the proof of Proposition 10.19, that 0 is a special point, so that $W_0 = \overline{W}$ and $W = T \rtimes \overline{W}$. Identifying T with the additive group

$$L := \{v \in V \mid \tau_v \in W\} \leq V,$$

we then have

$$W = L \rtimes \overline{W} \leq V \rtimes \mathrm{GL}(V) = \mathrm{Aff}(V).$$

Our next step in the analysis of the structure of W is to show that L is a *lattice* in V, by which we mean a subgroup of the form $\mathbb{Z}e_1 \oplus \cdots \oplus \mathbb{Z}e_n$ for some $\mathbb{R}$-basis $e_1, \ldots, e_n$ of V.

Lemma 10.21. *L is a discrete subgroup of the additive group of V, i.e., there is a neighborhood U of 0 in V such that $U \cap L = \{0\}$. The quotient group V/L, with the quotient topology, is compact.*

Proof. Pick a chamber C and a point $y \in C$. Since the transforms wC for $w \in W$ are all disjoint, the same is true of the translates $C + l$ for $l \in L$. So if we set $U := \{v \in V \mid y + v \in C\}$, then U is a neighborhood of 0 in V such that $U \cap L = \{0\}$. Recall now that the closed simplex $\overline{C}$ is a strict fundamental domain for the action of W. It follows that every point of V is equivalent mod L to a point of the compact set $\bigcup_{w \in \overline{W}} w\overline{C}$; hence V/L is compact.   □

Lemma 10.22. *If L is a discrete subgroup of the additive group of a finite-dimensional vector space V, then $L = \mathbb{Z}e_1 \oplus \cdots \oplus \mathbb{Z}e_r$ for some linearly independent vectors $e_1, \ldots, e_r$. If, in addition, V/L is compact, then L is a lattice.*

Proof. The second assertion follows immediately from the first. The following proof of the first assertion is taken from Pontryagin [189, Chapter 3, Section 19, Example 33], where a more general result is proved. We argue by induction on $\dim V$. If $L = 0$ there is nothing to prove, so assume $L \neq 0$. Give V

an arbitrary inner product and choose (by discreteness) a nonzero vector $e \in L$ of minimal length δ. Then any $l \in L$ that is not in $\mathbb{Z}e$ has distance at least $\delta/2$ from the line $\mathbb{R}e$. For if $y \in \mathbb{R}e$, then we can find $l' \in \mathbb{Z}e$ with $d(y, l') \le \delta/2$; since $d(l, l') = \|l - l'\| \ge \delta$, the triangle inequality implies that $d(l, y) \ge \delta/2$, as claimed. Consider now the subgroup $L/(L \cap \mathbb{R}e) = L/\mathbb{Z}e \le V/\mathbb{R}e$. If we give $V/\mathbb{R}e$ the metric induced by the canonical isomorphism $e^{\perp} \cong V/\mathbb{R}e$, then what we have just proven is that every nonzero element of $L/\mathbb{Z}e$ has distance at least $\delta/2$ from the origin of $V/\mathbb{R}e$. So $L/\mathbb{Z}e$ is a discrete subgroup of $V/\mathbb{R}e$. The lemma now follows easily from the induction hypothesis.   □

Returning now to our group $L := \{v \in V \mid \tau_v \in W\}$, the lemmas yield the following result on the structure of L and W:

Proposition 10.23. *L is a lattice in V. In particular, the Euclidean reflection group W is isomorphic to a semidirect product $\mathbb{Z}^n \rtimes \overline{W}$.*   □

Remarks 10.24. (a) Recall that V can be identified with the geometric realization of a Coxeter complex $\Sigma = \Sigma(W, V)$. So the vertices have *types*, and these are precisely the W-orbits. If x is a special vertex and t is its type, we can therefore identify the set of vertices of type t with W/W_x. Proposition 10.17 now implies that the lattice L acts simply transitively on the set of vertices of type t. So if we want to "see" L, we need only color the vertices in the essentially unique way, locate a special vertex x, and then look at all vertices of the same color. (More precisely, the arrows from x to the vertices of the same color represent translations, and these are the elements of L.)

(b) There is a fairly obvious way to topologize the group $\mathrm{Aff}(V)$, and the proof of Lemma 10.21 shows that an affine reflection group W is a discrete subgroup of $\mathrm{Aff}(V)$. Conversely, any discrete subgroup W of $\mathrm{Aff}(V)$ generated by reflections is an affine reflection group. For one can use the discreteness assumption to prove that the set of hyperplanes H with $s_H \in W$ is locally finite. Details are left to the interested reader.

Proposition 10.23 shows that there is a nontrivial condition satisfied by the finite reflection group $\overline{W}$, namely, it leaves a lattice invariant. One says that $\overline{W}$ is *crystallographic*. One can go much further, with only a little additional work, and show that there is an associated (crystallographic) root system Φ such that $\overline{W}$ is the Weyl group of Φ and W is completely determined by Φ. More precisely, W is the so-called *affine Weyl group* of Φ. Since the root systems are classified, this gives a classification of Euclidean reflection groups.

We will return to this in Section 10.1.8, after working out an example in Section 10.1.7.

10.1.7 Example

Let $\overline{W}$ be the symmetric group on n letters ($n \ge 2$). Recall that $\overline{W}$ is an essential, irreducible, finite reflection group acting on the $(n-1)$-dimensional space

$$V := \{(x_1, \ldots, x_n) \in \mathbb{R}^n \mid \sum x_i = 0\} .$$

The associated linear hyperplanes in V are defined by the equations

$$x_i - x_j = 0 \quad (i \neq j) .$$

We wish to construct an affine analogue of this example by introducing a lattice of translations.

Let $L := \mathbb{Z}^n \cap V$. Then L is a $\overline{W}$-invariant lattice in V. It is generated, as an abelian group, by the vectors $e_i - e_j$ $(i \neq j)$, where $\{e_1, \ldots, e_n\}$ is the standard basis for $\mathbb{R}^n$. We now set

$$W := L \rtimes \overline{W} \leq V \rtimes \mathrm{GL}(V) = \mathrm{Aff}(V) .$$

Note that W, as an abstract group, is the Weyl group that arose in Section 6.9. To see that W is a Euclidean reflection group, let $\mathcal{H}$ be the set of affine hyperplanes in V of the form $x_i - x_j = k$ with $i \neq j$ and $k \in \mathbb{Z}$. It is easy to check that $\mathcal{H}$ is locally finite and W-invariant. It is also easy to compute the reflection with respect to the hyperplane $x_i - x_j = k$; one finds that it is given by

$$x \mapsto s_{ij}(x) + k(e_i - e_j) ,$$

where s_{ij} is the transposition that interchanges the ith and jth coordinates. (This is a special case of the first formula in Exercise 10.6.) Thus W contains the reflection s_H for all $H \in \mathcal{H}$.

The subgroup generated by these reflections contains $\overline{W}$, and hence it contains the translations $x \mapsto x + k(e_i - e_j)$. It follows that this subgroup is the whole group W, which is therefore an affine reflection group. We will compute its Coxeter diagram below and see that W is irreducible; alternatively, the irreducibility of W follows from that of $\overline{W}$. Since W is obviously infinite, it is a Euclidean reflection group.

As fundamental chamber C we take the subset of V defined by

$$x_1 > \cdots > x_n > x_1 - 1 . \tag{10.7}$$

This is an intersection of half-spaces associated to n of the elements of $\mathcal{H}$, and it lies on one side of every $H \in \mathcal{H}$. So it is indeed a chamber. The reflections with respect to the walls $x_i = x_{i+1}$ are the basic transpositions $s_i := s_{i,i+1}$ $(i = 1, \ldots, n-1)$. And the reflection s_n with respect to the wall $x_n = x_1 - 1$ is the map $x \mapsto s_{n,1}(x) + (e_1 - e_n)$. The canonical unit vectors $f_1, \ldots, f_n$ associated to C are given by

$$f_i = \begin{cases} \dfrac{e_i - e_{i+1}}{\sqrt{2}} & \text{if } i \leq n-1, \\ \dfrac{e_n - e_1}{\sqrt{2}} & \text{if } i = n. \end{cases}$$

Notice that they satisfy the linear relation $\sum f_i = 0$, which has positive coefficients.

One can now find the Coxeter diagram of W, either by computing the orders of the products $s_i s_j$ or by computing the inner products $\langle f_i, f_j \rangle$. The diagram is o——∞——o if $n = 2$ and

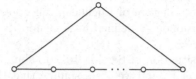

(with n vertices) if $n \geq 3$. In case $n = 3$, the diagram shows that the fundamental triangle C has all of its angles equal to $\pi/3$. Thus the Coxeter complex Σ in this case is the Euclidean plane tiled by equilateral triangles, which we discussed in Example 10.14(a).

We finish this example by indicating briefly how to get an explicit isomorphism between $\Sigma(W, V)$ and the fundamental apartment for $\mathrm{SL}_n(K)$ (Section 6.9). We will need such an isomorphism in Section 11.8.6. For this purpose it is convenient to replace V by the canonically isomorphic vector space $\mathbb{R}^n / V^\perp = \mathbb{R}^n / \mathbb{R} \cdot (1, 1, \ldots, 1)$. One can check that the set of vertices of $\Sigma(W, V)$ is the subset $\mathbb{Z}^n / \mathbb{Z} \cdot (1, \ldots, 1)$ of this quotient. A further calculation yields the following description of $\Sigma(W, V)$:

Given $u = (u_1, \ldots, u_n)$ and $v = (v_1, \ldots, v_n)$ in $\mathbb{Z}^n$, write $u \preceq v$ if

$$u_i \leq v_i \leq u_i + 1$$

for all i. Call two elements of $\mathbb{Z}^n / \mathbb{Z} \cdot (1, \ldots, 1)$ *incident* if they admit representatives u, v with $u \preceq v$. This incidence relation is symmetric, and one can check that the resulting flag complex is $\Sigma(W, V)$. Recall now that we gave a description of the building for $\mathrm{SL}_n(K)$ in terms of classes of A-lattices in K^n. The fundamental apartment Σ of this building has as vertices the classes $[[\pi^{a_1} e_1, \ldots, \pi^{a_n} e_n]]$, where $e_1, \ldots, e_n$ is the standard basis for K^n. In view of the definition of the incidence relation on the set of lattice classes, it is now evident that there is an isomorphism $\Sigma \xrightarrow{\sim} \Sigma(W, V)$ that sends the vertex $[[\pi^{a_1} e_1, \ldots, \pi^{a_n} e_n]]$ to the class of $(a_1, \ldots, a_n)$ mod $(1, \ldots, 1)$.

The group W we have been discussing is called the *affine Weyl group* of the root system of type A_{n-1}. Recall that this root system consists of the vectors $e_i - e_j$ that played such a prominent role above. We show next that every Euclidean reflection group admits a similar description in terms of a root system.

Exercises

10.25. What are the vertices of the fundamental chamber (10.7)? What are the corresponding elements of $\mathbb{Z}^n / \mathbb{Z} \cdot (1, \ldots, 1)$?

10.26. When $n = 3$, we remarked above that the Coxeter complex Σ is the plane tiled by equilateral triangles. Use this to get a picture of the lattice L.

*10.1.8 The Structure of W, Part II; Affine Weyl Groups

This subsection is optional and will not be referred to later in the book except for a very brief mention at the end of Chapter 11. But it is included in order to give a more complete picture of Euclidean reflection groups. We will use some notation and terminology defined in Appendix B, and we will use the results of Section 10.1.2.

We return to the situation of Section 10.1.6. Thus W is a Euclidean reflection group acting on V, and we are assuming that the origin is a special point so that $W = L \rtimes \overline{W}$ for some $\overline{W}$-invariant lattice L. Recall that the subset of $\mathcal{H}$ consisting of linear hyperplanes is precisely the set $\overline{\mathcal{H}}$ of hyperplanes associated with the finite reflection group $\overline{W}$. Moreover, $\overline{\mathcal{H}}$ contains a representative of every parallelism class of hyperplanes in $\mathcal{H}$.

What we will see in this subsection is that there is a root system Φ canonically associated with (W, V) such that:

(1) $\overline{W}$ is the Weyl group W_Φ.
(2) The lattice L is the coroot lattice of Φ.

Thus W is completely determined by Φ. Note that (1) and (2) are consistent with the example in Section 10.1.7, where Φ is the root system of type A_{n-1}. (We did not need to call attention to the occurrence of the coroot lattice in that example, since the root system of type A_{n-1} is self-dual.)

We begin by examining $\mathcal{H}$ in more detail. Consider an arbitrary $H \in \overline{\mathcal{H}}$, and let $\mathcal{H}' \subseteq \mathcal{H}$ be the set of hyperplanes in $\mathcal{H}$ that are parallel to H. Note first that $\mathcal{H}'$ consists of more than just H. For it contains $l + H$ for any $l \in L$, and these cannot all equal H, since then the rank-n lattice L would be contained in H. We claim that in fact, the hyperplanes in $\mathcal{H}'$ form a doubly infinite equally spaced family, i.e., there is a nonzero vector α such that $\mathcal{H}'$ consists of the hyperplanes $H_{\alpha,k}$ for $k \in \mathbb{Z}$, where $H_{\alpha,k}$ is defined by the equation $\langle \alpha, x \rangle = k$. See Figure 10.2. One can prove this by an easy direct argument (Bourbaki [44, Section VI.2.5, proof of Proposition 8]). But one can also deduce it quickly from what we have already done:

Let $W' \le W$ be the subgroup generated by the reflections with respect to the hyperplanes in $\mathcal{H}'$. Then W' is infinite [why?], and $\mathcal{H}'$ is W'-invariant. Since all the reflections generating W' have the same linear part, W' is essentially 1-dimensional. (More precisely, the essential part consists of W' acting on the line $H^\perp$.) Now the results we have proven earlier in the chapter, when specialized to the 1-dimensional case, show that $(W', H^\perp)$ must look like D_∞ acting on $\mathbb{R}$ as in Section 2.2.2, up to rescaling. The existence of the vector α follows at once.

The pair of vectors $\pm\alpha$ is uniquely determined by H, and the collection Φ of all such vectors $\pm\alpha$ for $H \in \overline{\mathcal{H}}$ is a $\overline{W}$-invariant set of vectors whose reflections generate $\overline{W}$. Thus Φ is a (possibly generalized) root system with $W_\Phi = \overline{W}$.

Lemma 10.27. Φ is a root system and L is the coroot lattice.

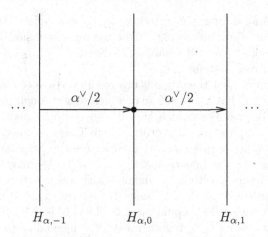

Fig. 10.2. The hyperplanes parallel to H.

Proof. Let L' be the additive group generated by the $\alpha^\vee$ for $\alpha \in \Phi$. Formula (10.6) shows that $\alpha^\vee$ is the translation component of the reflection with respect to $H_{\alpha,1}$. So $L' \le L$. On the other hand, L' contains the translation component $k\alpha^\vee$ of the reflection with respect to $H_{\alpha,k}$ (see Exercise 10.6). So $W' := L' \rtimes \overline{W}$ contains all the generating reflections of $W = L \rtimes \overline{W}$; hence $W' = W$ and $L' = L$.

We now make use of the fact that $\mathcal{H}$ is L-invariant. The content of this, in view of the previous paragraph, is that for each $\alpha, \beta \in \Phi$ and $k \in \mathbb{Z}$, the translate $\alpha^\vee + H_{\beta,k}$ is again in $\mathcal{H}$. Now this translate is defined by the equation $\langle \beta, x - \alpha^\vee \rangle = k$, or $\langle \beta, x \rangle = k + \langle \beta, \alpha^\vee \rangle$. So $\langle \beta, \alpha^\vee \rangle$ must be an integer, and we have proved the crystallographic condition (Cry) stated in Section B; thus Φ is a root system. $\square$

Remark 10.28. Note how naturally condition (Cry) and the coroot lattice arose. Even if we had never heard of root systems before, an analysis of Euclidean reflection groups would have forced us to consider them.

The lemma motivates the following definition:

Definition 10.29. The *affine Weyl group* of a root system Φ is the group

$$L \rtimes W_\Phi \le V \rtimes \mathrm{GL}(V) = \mathrm{Aff}(V) \,,$$

where L is the coroot lattice of Φ.

It is not hard to show that the affine Weyl group is in fact a Euclidean reflection group if Φ is irreducible; the proof is similar to the proof in Section 10.1.7 for the A_{n-1} case. In view of the lemma, we obtain every Euclidean reflection group in this way:

Theorem 10.30. *Every Euclidean reflection group with the origin as a special point is the affine Weyl group of an irreducible root system.* $\square$

For more information about root systems and affine Weyl groups, see Bourbaki [44] or Humphreys [133]. We close our discussion by returning to the description of a typical W-chamber C (Section 10.1.5), and restating it in the language of root systems.

We may assume that the origin is one of the vertices of C. As we noted in Remark 10.20(b), C has $n+1$ walls, n of which pass through the origin and are the walls of a $\overline{W}$-chamber (which we will call the *fundamental $\overline{W}$-chamber*) containing the simplex C as a "corner." Thus C is a truncated $\overline{W}$-chamber, obtained by cutting across the latter with some hyperplane $H_{\alpha,k}$. [In the A_{n-1}-case, this was the hyperplane $x_1 - x_n = 1$.] We may take α to be a positive root (where positivity is defined with respect to the fundamental $\overline{W}$-chamber); in this case $k > 0$, since $H_{\alpha,k}$ meets C. We must then have $k = 1$, since otherwise $H_{\alpha,1}$ would meet C. Thus C is defined by inequalities $\langle \alpha_i, x \rangle > 0$ for $i = 1, \ldots, n$, together with the inequality $\langle \alpha, x \rangle < 1$, where $\alpha_1, \ldots, \alpha_n$ are the simple roots and α is some particular positive root that we wish to determine.

To this end, note that for any root $\beta \in \Phi$ we have

$$\langle \beta, x \rangle \leq \langle \alpha, x \rangle \tag{10.8}$$

for all $x \in C$; for if this failed, then we would be able to scale x to arrange $\langle \beta, x \rangle = 1 > \langle \alpha, x \rangle$, so that $H_{\beta,1}$ would meet C. The inequality (10.8) extends to the entire fundamental $\overline{W}$-chamber by scaling, so α is the largest element of Φ with respect to the partial order defined by

$$\beta \leq \alpha \iff \langle \beta, - \rangle \leq \langle \alpha, - \rangle \text{ on the fundamental } \overline{W}\text{-chamber}$$

or, equivalently,

$$\beta \leq \alpha \iff \alpha - \beta = \sum_{i=1}^{n} \lambda_i \alpha_i \text{ with all } \lambda_i \geq 0.$$

(The last equivalence is proved as in the solution to Exercise 1.120.) This largest root α is often denoted by $\tilde{\alpha}$ and called the *highest root*.

See Figure 10.3 for an illustration of the description of C that we have arrived at. The following result restates the description in the language of root systems and affine Weyl groups.

Proposition 10.31. *Let W be the affine Weyl group of an irreducible root system Φ, and let $\alpha_1, \ldots, \alpha_n$ be the system of simple roots associated with a choice of fundamental $\overline{W}$-chamber. Then there is a W-chamber given by the inequalities $\langle \alpha_i, x \rangle > 0$ $(i = 1, \ldots, n)$ and $\langle \tilde{\alpha}, x \rangle < 1$, where $\tilde{\alpha}$ is the highest root.* □

Example 10.32. Let Φ be the root system of type A_{n-1}, with fundamental $\overline{W}$-chamber

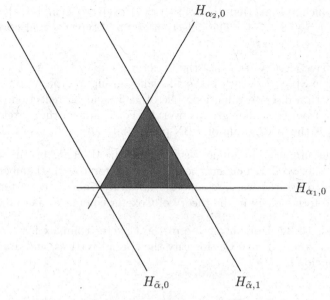

Fig. 10.3. A W-chamber as a truncated $\overline{W}$-chamber.

$$x_1 > \cdots > x_n . \tag{10.9}$$

The simple roots are $\alpha_i := e_i - e_{i+1}$ $(i = 1, \ldots, n-1)$, and the highest root is

$$\tilde{\alpha} = \alpha_1 + \cdots + \alpha_{n-1} = e_1 - e_n .$$

The resulting W-chamber as in Proposition 10.31 is gotten by adjoining to (10.9) the inequality $x_1 - x_n < 1$. This is the same chamber we gave in Section 10.1.7.

Remarks 10.33. (a) Note that our analysis of W-chambers has led to a proof that a root system has a highest root. This is a standard fact; direct proofs can be found in Bourbaki [44, Section VI.1.8] and Humphreys [131, Lemma 10.4A]. One can also explain the highest root in terms of the representation theory of the Lie algebra associated to the root system: It is the highest weight of the adjoint representation.

(b) In view of Theorem 10.30 and the classification of root systems, we know exactly what the Euclidean reflection groups are. Up to isometry and rescaling, they are the affine Weyl groups of the root systems of type A_n $(n \geq 1)$, B_n $(n \geq 3)$, C_n $(n \geq 2)$, D_n $(n \geq 4)$, E_n $(n = 6, 7, 8)$, F_4, and G_2. The corresponding Coxeter systems (W, S) are said to be of type $\tilde{A}_n$, $\tilde{B}_n$, and so on. For example, the infinite dihedral group is the Coxeter group of type $\tilde{A}_1$, and the group of isometries of the plane generated by the reflections with respect to the sides of an equilateral triangle is the Coxeter group of type $\tilde{A}_2$ (see Example 10.14(a)). The subscript denotes the dimension of the Euclidean space, which is 1 less than the rank of the Coxeter system, i.e., the size of the generating set S. (Note: The reflection group of type $\tilde{C}_2$ is also said to be of type $\tilde{B}_2$; see Exercise 1.17.)

(c) We have already seen that the Coxeter group of type $\tilde{A}_{n-1}$ arises from a BN-pair in $\mathrm{SL}_n(K)$ (which is the Chevalley group of type A_{n-1}), where K is a field with a discrete valuation. The other Euclidean reflection groups arise similarly from Chevalley groups over K. For example, the Coxeter group of type $\tilde{C}_n$ is the Weyl group of a BN-pair in $\mathrm{Sp}_{2n}(K)$.

(d) In the literature on affine Weyl groups, the W-chambers are often called *alcoves*, the word "chamber" being reserved for the $\overline{W}$-chambers. We will continue to use the word "chamber," however, which is consistent with the standard terminology in the theory of Coxeter complexes and buildings.

Exercise 10.34. Draw a picture of the Coxeter complex for the Euclidean reflection group of type $\tilde{C}_2$. Identify the special vertices, and draw a picture of the lattice L.

10.2 Euclidean Coxeter Groups and Complexes

Definition 10.35. A Coxeter system (W, S) will be called *Euclidean* if it is isomorphic to the Coxeter system associated with a Euclidean reflection group and a choice of fundamental chamber. Similarly, a Coxeter complex Σ will be called *Euclidean* if it is isomorphic to $\Sigma(W, S)$ for some Euclidean Coxeter system (W, S) or, equivalently, if it is isomorphic to $\Sigma(W, V)$ for some Euclidean reflection group (W, V).

The main point of this section is that a Euclidean Coxeter complex Σ has a well-defined Euclidean structure in a sense that will be made precise. Note that there is an obvious way to define a *semi-Euclidean Coxeter system*, based on the notion of semi-Euclidean reflection group (Definition 10.15). We will *not*, however, define semi-Euclidean Coxeter complexes. The problem is that the "natural" space on which a semi-Euclidean Coxeter group acts is a polysimplicial complex that in the reducible case is not simplicial. It could therefore be confusing to speak of semi-Euclidean Coxeter complexes.

10.2.1 A Euclidean Metric on $|\Sigma|$

Let Σ be a Euclidean Coxeter complex. Choose a Euclidean reflection group (W, V) and an isomorphism $\Sigma \cong \Sigma(W, V)$. Since there is a canonical bijection between $|\Sigma(W, V)|$ and V (Section 10.1.3), we obtain a bijection $|\Sigma| \cong V$. We wish to use this to transport to $|\Sigma|$ the notions of Euclidean geometry. The lemma below will enable us to show that these notions are independent of the choice of (W, V) and the choice of isomorphism $\Sigma \cong \Sigma(W, V)$.

The intuitive content of the lemma is that one can reconstruct the Euclidean space V (up to a dilation of its metric) from the abstract simplicial complex $\Sigma(W, V)$. Here is the precise statement:

Lemma 10.36. *Let (W, V) and (W', V') be Euclidean reflection groups. Let $\phi\colon \Sigma(W, V) \to \Sigma(W', V')$ be a simplicial isomorphism. Then the composite bijection*

$$V \cong |\Sigma(W, V)| \xrightarrow{\;|\phi|\;} |\Sigma(W', V')| \cong V'$$

is a similarity map, i.e., an affine isomorphism whose linear part g satisfies $\langle gv, gv' \rangle = \lambda \langle v, v' \rangle$ for some positive constant λ and all $v, v' \in V$.

Proof. Choose a chamber C in V, with panels A_i $(i = 1, \ldots, n + 1)$. Let H_i be the support of A_i, let e_i be the canonical unit normal to H_i (pointing to the side containing C), and let s_i be the reflection with respect to H_i. As in the proof of Theorem 10.9, we may conjugate W by a translation in order to arrange that H_i is defined by $\langle e_i, - \rangle = 0$ if $i \leq n$ and by $\langle e_{n+1}, - \rangle = c$ if $i = n + 1$, where $c < 0$. Conjugating W by a dilation, we can further arrange that $c = -1$. Let $C' := \phi(C)$ and $A'_i := \phi(A_i)$. Let H'_i, e'_i, and s'_i be the associated wall, unit vector, and reflection, respectively. We may assume that H'_i is defined by $\langle e'_i, - \rangle = 0$ if $i \leq n$ and by $\langle e'_{n+1}, - \rangle = -1$ if $i = n + 1$.

Recall now that the Coxeter matrix $M := (m_{ij})_{1 \leq i,j \leq n+1}$ of W is a combinatorial invariant of the Coxeter complex (Section 3.2); namely, m_{ij} is the diameter of the link of $A_i \cap A_j$. In view of the isomorphism ϕ, it follows that M is also the Coxeter matrix of W'. Consequently, we have

$$\langle e_i, e_j \rangle = \langle e'_i, e'_j \rangle$$

for all i, j. We can now construct a linear isometry $\psi\colon V \to V'$ such that $\psi(e_i) = e'_i$ for all i. For if we take ψ to be the linear map such that $\psi(e_i) = e'_i$ for $i \leq n$, then ψ preserves inner products and satisfies $\langle \psi(e_{n+1}), - \rangle = \langle e'_{n+1}, - \rangle$, whence $\psi(e_{n+1}) = e'_{n+1}$.

Note that $\psi(H_i) = H'_i$, so that $\psi s_i \psi^{-1} = s'_i$ and $\psi W \psi^{-1} = W'$. Thus ψ induces an isomorphism of pairs $(W, V) \to (W', V')$, and hence a simplicial isomorphism $\Sigma(W, V) \to \Sigma(W', V')$. This isomorphism takes C to C' and A_i to A'_i, so it coincides with our original isomorphism ϕ by the standard uniqueness argument. The lemma now follows from the commutative diagram

$$
\begin{array}{ccc}
|\Sigma(W, V)| & \longrightarrow & |\Sigma(W', V')| \\
\downarrow & & \downarrow \\
V & \longrightarrow & V'
\end{array}
$$

where the vertical arrows denote the canonical bijections and the horizontal arrows are induced by ψ. □

We now have a "Euclidean structure" on $|\Sigma|$ for any Euclidean Coxeter complex Σ, by which we mean that we can apply to $|\Sigma|$ any notion of Euclidean geometry that is invariant under similarity maps. In particular, $|\Sigma|$ has a well-defined equivalence class of metrics, where two metrics are equivalent if one is a positive scalar multiple of the other.

It will be convenient in what follows to have a canonical representative of this equivalence class. We can achieve this in many ways; for example, we could normalize the metric so that the chambers have diameter 1 or volume 1, or we could use the normalization that was suggested in the solution to Exercise 10.16. (The reader who works through the next subsection will see that this same normalization arises in connection with the canonical linear representation of (W, S).) The particular method of normalizing the metric choice makes no difference in what follows, so let's just agree that we have chosen one that we will stick with throughout the book. Thus $|\Sigma|$ is now a metric space, and any abstract isomorphism $\phi\colon \Sigma \to \Sigma'$ of Euclidean Coxeter complexes induces an isometry $|\Sigma| \to |\Sigma'|$.

We close this section by introducing one last bit of terminology.

Definition 10.37. By a *Euclidean space* we mean a metric space E that is isometric to $\mathbb{R}^n$ for some n.

What we have done above, then, is to give $|\Sigma|$ a canonical Euclidean space structure.

Note that we are making a somewhat pedantic distinction between the notions of *Euclidean space* and *Euclidean vector space*; recall that we have defined the latter to mean "vector space with an inner product." As a practical matter, the only difference between the two concepts is that a Euclidean vector space comes equipped with a preferred origin. More precisely, suppose E is a Euclidean space and x_0 is an arbitrary point in E. Then we can give E the structure of Euclidean vector space with x_0 as origin by choosing an isometry $\psi\colon \mathbb{R}^n \to E$ with $\psi(0) = x_0$ and using ψ to transport the vector space structure and inner product from $\mathbb{R}^n$ to E. It follows from Exercise 10.3 that this structure is independent of the choice of ψ.

Exercises

10.38. Convince yourself that the following assertion is meaningful and true: If Σ is a Euclidean Coxeter complex, then the Euclidean space $E := |\Sigma|$ contains a canonical locally finite collection $\mathcal{H}$ of hyperplanes, from which one can recover the decomposition of E into simplices.

10.39. State and prove an analogue of Exercise 1.68 for subcomplexes of Euclidean Coxeter complexes.

*10.2.2 Connection with the Tits Cone

The results of this somewhat technical subsection will not be used in any serious way in what follows, but they provide some motivation for the hyperbolic theory to be discussed in the next section. Assume throughout this subsection that (W, S) is a Euclidean Coxeter system of rank $n + 1$.

We have two geometric objects on which W acts with a good "chamber geometry." The first is $|\Sigma(W, S)|$; it is an n-dimensional Euclidean space, as we have just explained, decomposed into chambers that are simplices. The second is the Tits cone in V^*, where V is the space on which W acts by the canonical linear representation (see Sections 2.5 and 2.6); it is $(n + 1)$-dimensional and is decomposed into chambers that are simplicial cones. We will show here that the first model can be obtained from the second by cutting across the Tits cone with a suitable hyperplane. We have already seen this in Section 2.2.2 for the Coxeter group of type $\tilde{A}_1$.

Recall that the construction of the canonical linear representation of W (Section 2.5) started with a bilinear form B on the vector space $V := \mathbb{R}^S$, given by

$$B(e_s, e_t) := -\cos(\pi/m(s, t))$$

for $s, t \in S$. Recall further that B is positive definite if (and only if) W is finite. Here is the analogous result in the present Euclidean setup:

Proposition 10.40. *The bilinear form B is positive semidefinite with a 1-dimensional radical, spanned by a vector $v = \sum_{s \in S} \lambda_s e_s$ such that $\lambda_s > 0$ for all s.*

Here the radical of B is $V^\perp := \{v \in V \mid B(v, u) = 0 \text{ for all } u \in V\}$.

Proof. We may assume that W is given as a Euclidean reflection group acting on a Euclidean vector space V', and that S is the set of reflections with respect to the walls of a fundamental chamber C'. Let $(e'_s)_{s \in S}$ be the canonical unit normals to the walls of C'. Then we have a linear surjection $\phi \colon V \to V'$ given by $e_s \mapsto e'_s$ for $s \in S$. It satisfies $\langle \phi(x), \phi(y) \rangle = B(x, y)$ for all $x, y \in V$. Since the inner product in V' is positive definite, it follows immediately that B is positive semidefinite and that its radical is the kernel of ϕ. This kernel is 1-dimensional, since $\dim V' = n = \dim V - 1$. A nonzero vector $\sum \lambda_s e_s$ in $\ker \phi$ corresponds to a relation $\sum_{s \in S} \lambda_s e'_s = 0$ in V'. The proposition now follows from the fact that the coefficients of such a relation all have the same sign, as we saw in the proof of Theorem 10.9. $\qquad \square$

For definiteness, we choose a canonical v as in the proposition by requiring $\sum_{s \in S} \lambda_s = 1$. The proposition implies that the quotient $V/\mathbb{R}v$ inherits from V a positive definite bilinear form B', making $V/\mathbb{R}v$ an n-dimensional Euclidean vector space [which can be identified with the space called V' in the proof of the proposition]. Note next that in the action of W on V, the fixed-point set is the intersection of the hyperplanes $B(e_s, -) = 0$, which is precisely the radical of B. In particular, W fixes the vector v, and hence W leaves invariant the hyperplane in V^* that annihilates v, i.e., the n-dimensional space

$$E_0 := \{\xi \in V^* \mid \langle \xi, v \rangle = 0\} \ .$$

Here, as in Section 2.5, we use angle brackets to denote the canonical pairing between a vector space and its dual.

Note that E_0 can be identified with the dual of $V/\mathbb{R}v$ and hence with $V/\mathbb{R}v$ itself, since the bilinear form B' induces an isomorphism between $V/\mathbb{R}v$ and its dual. In particular, E_0 inherits via this isomorphism the structure of a Euclidean vector space, given by a positive definite bilinear form B'' computed as follows: Given $\xi, \eta \in E_0$, write $\xi = B(x, -)$ and $\eta = B(y, -)$ with $x, y \in V$; then $B''(\xi, \eta) = B(x, y)$. Let's work this out with $\eta = \eta_s := B(e_s, -)$ for $s \in S$. Then we have

$$B''(\xi, \eta_s) = B(x, e_s) = \langle \xi, e_s \rangle \ . \tag{10.10}$$

The vectors η_s arise when we compute the action of s on V^*. Namely, equation (2.14) says precisely that

$$s(\xi) = \xi - 2\langle \xi, e_s \rangle \eta_s \tag{10.11}$$

for $\xi \in V^*$. In particular, if we take $\xi \in E_0$ and use (10.10), this becomes

$$s(\xi) = \xi - 2B''(\xi, \eta_s)\eta_s \ .$$

In other words:

Lemma 10.41. *The generator s acts on E_0 as the orthogonal reflection with respect to $\eta_s^{\perp}$, where*

$$\eta_s^{\perp} = \{\xi \in E_0 \mid B''(\xi, \eta_s) = 0\} = \{\xi \in E_0 \mid \langle \xi, e_s \rangle = 0\} \ . \qquad \square$$

We now introduce the affine hyperplane E in V^* that we will use to cut across the chambers in the Tits cone. It is parallel to E_0 and is defined by

$$E := \{\xi \in V^* \mid \langle \xi, v \rangle = 1\} \ .$$

Note that E is invariant under the action of W and that it has a natural Euclidean metric, since it is a translate of E_0. Indeed, we need only choose an "origin" $\xi_0 \in E$ and use the bijection $\xi \mapsto \xi_0 + \xi$ from E_0 to E to transport the metric from E_0 to E. The resulting metric is independent of the choice of ξ_0.

We claim that any $s \in S$ acts on E as the reflection with respect to the hyperplane $H_s := \{\xi \in E \mid \langle \xi, e_s \rangle = 0\}$. Note first that H_s is indeed an affine hyperplane in E. To see this, we must check that the linear hyperplane in V^* given by $\langle -, e_s \rangle = 0$ is not parallel to E, i.e., is not E_0. In other words, we must check that e_s is not parallel to v, and this is immediate from Proposition 10.40

and the fact that $|S| \geq 2$ (since W is infinite). Returning now to the claim, choose $\xi_0 \in H_s$. Then a typical element of E has the form $\xi_0 + \xi$ for $\xi \in E_0$. Now equation (10.11) shows that $s(\xi_0) = \xi_0$, so we have $s(\xi_0 + \xi) = \xi_0 + s(\xi)$. In view of Lemma 10.41 and the definition of the Euclidean structure on E, it follows that s acts on E as the reflection with respect to the hyperplane $\xi_0 + \eta_s^{\perp}$ in E. This hyperplane is precisely H_s by the second description of $\eta_s^{\perp}$ in Lemma 10.41, whence the claim.

We have done almost all the work needed to realize W as a Euclidean reflection group acting on E. There is more than one way to carry out the remaining details, but the simplest method, given what we have done already, is to make use of the fact that W is already known to be realizable as a Euclidean reflection group, acting on a space that we called V' in the proof of Proposition 10.40. With the notation of that proof, we have a fundamental chamber C' in V' whose walls are given by $\langle e_s', - \rangle = c_s$ for some constants c_s ($s \in S$). Since the inequalities $\langle e_s, - \rangle > 0$ define a nonempty set, we must have $\sum_s \lambda_s c_s < 0$ (see the solution to Exercise 10.16), and we normalize the metric on V' to make $\lambda_s c_s = -1$.

Note that everything about the realization of W as a Euclidean reflection group acting on V' is completely determined by the Gram matrix calculation

$$\langle e_s', e_t' \rangle = -\cos \frac{\pi}{m(s,t)} \tag{10.12}$$

together with the equalities $\langle e_s', - \rangle = c_s$ defining the walls of C'. In fact, (10.12) allows us to reconstruct V' as a Euclidean vector space, and the description of the walls lets us reconstruct the action of the generators of W on V'.

We now show that the action of W on E has a similar description. Note first that the vectors η_s in E_0 have the same Gram matrix as the e_s'; this follows from (10.10) or directly from the original definition of the metric on E_0. Next, we need to choose a suitable origin $\xi_0 \in E$ so that we can view E as a Euclidean vector space (via the bijection $E_0 \to E$ given by $\xi \mapsto \xi_0 + \xi$). To this end we take $\xi_0 := -\sum_s c_s e_s^*$, where $(e_s^*)_{s \in S}$ is the basis of V^* dual to $(e_s)_{s \in S}$. Now our hyperplane H_s in E, when translated back to E_0, is the affine hyperplane $-\xi_0 + H_s$, on which $B''(\eta_s, -)$ takes the constant value $\langle -\xi_0, e_s \rangle = c_s$. Thus the generators of W acting on E are defined by the same formulas as the generators of W acting on V'. It follows that W is indeed a Euclidean reflection group acting on E, with $\mathcal{H} := \{wH_s \mid w \in W, s \in S\}$ as the set of walls.

To summarize what we have done so far, we started with an abstract Euclidean Coxeter system (W, S), and we realized W as a Euclidean reflection group in a completely canonical way. Note that the construction actually yields an action of W on a Euclidean metric space E that does not come with a preferred origin, as is appropriate for a canonical construction. It does come with a fundamental chamber C, however, as it should, since we are given the set S of fundamental reflections, and a choice of origin is equivalent to a choice of constants c_s determining the walls of C as above, with $\sum_s \lambda_s c_s = -1$.

The following theorem recaps the main points:

Theorem 10.42. *Any Euclidean Coxeter system (W, S) has a canonical realization as a Euclidean reflection group, acting on an affine hyperplane E in V^*, where $V = \mathbb{R}^S$. The cells into which E is decomposed by the reflecting hyperplanes are precisely the intersections with E of the cells other than $\{0\}$ in the Tits cone.*

Proof. The only thing that we have not yet proved is the description of the cells in terms of those of the Tits cone. Recall that the fundamental chamber C in the Tits cone is a simplicial cone with walls given by $\langle -, e_s \rangle = 0$ ($s \in S$). Thus our hyperplanes H_s in E are precisely the intersections with E of the walls of C. Using the W-action, we conclude that the set $\mathcal{H}$ of walls in E consists of the intersections with E of the walls defining the decomposition of the Tits cone. The proof will be complete once we observe that every ray $\mathbb{R}_+ \xi$ in the Tits cone ($\xi \neq 0$) intersects E. We may assume $\xi \in \overline{C}$, so that $\langle \xi, e_s \rangle \geq 0$ for all $s \in S$, and strict inequality must hold for at least one s. Then $\langle \xi, v \rangle = \sum_s \lambda_s \langle \xi, e_s \rangle > 0$, so we may scale ξ to get a vector in $\mathbb{R}_+ \xi \cap E$. □

Remark 10.43. Our argument based on V' could have been replaced by a more detailed analysis of the action of W on E. We would then have had to prove directly that $\mathcal{H}$ is locally finite, a fact that can be deduced from the discreteness of W in the group of isometries of E (Theorem 2.59). The advantage of this approach is that it leads to the proposition below, which we already stated without proof in Section 2.5.5. We have now given the main ideas behind its proof, and the missing details can be found in Bourbaki [44, Chapter V, Section 4.9].

Proposition 10.44. *Let (W, S) be an irreducible Coxeter system with Coxeter matrix M, and let B be the canonical bilinear form on $\mathbb{R}^S$, given by*

$$B(e_s, e_t) := -\cos \frac{\pi}{m(s, t)} . \tag{10.13}$$

Then (W, S) is Euclidean if and only if B is positive semidefinite and degenerate. □

Remark 10.45. Combining the proposition with the analogous characterization of finite reflection groups (Corollary 2.68), we see that a (possibly reducible) Coxeter system (W, S) can be realized as an affine reflection group if and only if it is of positive type. Each irreducible component of the Coxeter diagram then corresponds to either a finite reflection group or a Euclidean reflection group. Note that if the Euclidean space on which W acts has dimension n, then (W, S) has rank $n + k$, where k is the number of infinite irreducible factors of W. The number n will be called the *dimension* of (W, S). One can also speak of the *rank* of the bilinear form B (which is simply the rank of the matrix in (10.13)), and this is precisely the dimension n (it is *not* the rank of the Coxeter system (W, S)).

*10.3 Hyperbolic Reflection Groups

It would take us too far afield to give a complete treatment of hyperbolic reflection groups analogous to our treatment of Euclidean reflection groups. But the reader who wants to read the current literature on buildings needs at least a passing acquaintance with them. We will therefore give a sketch of the theory in this section. The standard reference for readers who want to know more is Vinberg [274], which is summarized by Margulis [158, Appendix C, Section 1]. See also Ratcliffe [193, Chapter 7], Maskit [159, Chapter IV], Davis [89, Chapter 6], and de la Harpe [95]. And there is a useful introduction to hyperbolic space in Bridson–Haefliger [48, Section I.1.2 and Chapter I.6].

10.3.1 Hyperbolic Space; Hyperplanes and Reflections

We will use the hyperboloid model of hyperbolic space (cf. Exercise 2.9), defined as follows. Fix an integer $n \geq 2$, and consider Lorentz space $\mathbb{R}^{n,1}$. By this we mean the vector space $\mathbb{R}^{n+1}$ endowed with the standard symmetric bilinear form of signature $(n, 1)$:

$$(x, y) \mapsto \langle x, y \rangle := x_1 y_1 + \cdots + x_n y_n - x_{n+1} y_{n+1} , \tag{10.14}$$

where $x = (x_1, \ldots, x_{n+1})$ and $y = (y_1, \ldots, y_{n+1})$. The associated quadratic form Q is given by

$$Q(x) := \langle x, x \rangle = x_1^2 + \cdots + x_n^2 - x_{n+1}^2 .$$

We define n-dimensional *hyperbolic space*, denoted by $\mathbb{H}^n$, to be the upper sheet of the hyperboloid $Q(x) = -1$, i.e.,

$$\mathbb{H}^n := \left\{ x \in \mathbb{R}^{n,1} \mid x_1^2 + \cdots + x_n^2 - x_{n+1}^2 = -1 \text{ and } x_{n+1} > 0 \right\} .$$

Alternatively, we can identify $\mathbb{H}^n$ with the set of rays in the cone U_+ defined by $Q(x) < 0$ and $x_{n+1} > 0$. This identification is useful when we want to talk about points at infinity, which correspond to rays in the boundary of U_+.

Although we will not make use of this fact, $\mathbb{H}^n$ has a canonical Riemannian metric making it a complete simply connected manifold of constant sectional curvature -1. The resulting distance function d is characterized by the equation

$$\cosh d(x, y) = -\langle x, y \rangle . \tag{10.15}$$

By a *hyperplane* in $\mathbb{H}^n$ we mean a nonempty subset H gotten by intersecting $\mathbb{H}^n$ with a linear hyperplane in $\mathbb{R}^{n,1}$. Equivalently, H is a subset of $\mathbb{H}^n$ defined by an equation of the form $\langle \alpha, x \rangle = 0$, where α is a vector such that $Q(\alpha) > 0$ (see Exercise 10.46 below). We can normalize α to make it a unit vector (i.e., to make $Q(\alpha) = 1$), in which case we will usually denote it by e and call it a *unit normal* to H. It is uniquely determined by H up to sign. A choice of one of the two unit normals will be called an *orientation* of H, and we will then call H an *oriented hyperplane*. Giving an orientation of H is the same as singling out one of the two half-spaces determined by H as the *positive* half-space.

Let H_1 and H_2 be distinct hyperplanes, oriented by unit normals e_1, e_2. If $H_1 \cap H_2 \neq \emptyset$, then there is a well-defined *angle* θ between H_1 and H_2, with $0 < \theta < \pi$. One can of course define this by pure geometry (it is the dihedral angle formed by the two positive half-spaces), but one can also take the equation

$$\langle e_1, e_2 \rangle = -\cos\theta \tag{10.16}$$

as the definition. Exercise 10.47 below implies that this equation determines a well-defined angle θ. If $H_1 \cap H_2 = \emptyset$, on the other hand, then the exercise just cited implies that $|\langle e_1, e_2 \rangle| \geq 1$. In this case there is still a geometric interpretation of $\langle e_1, e_2 \rangle$. To describe it, we may assume that the positive half-spaces of H_1 and H_2 are not nested (i.e., neither is contained in the other; for if they are nested, we can simply reverse the orientation of one of them. Then $\langle e_1, e_2 \rangle \leq -1$, with equality if and only if H_1 and H_2 are *parallel*, i.e., they meet at infinity; see Exercise 10.48. Otherwise, they are said to be *ultraparallel*, and there is a unique line segment going from one to the other and orthogonal to both. The length of this segment is the distance $d(H_1, H_2)$ between them, and one has

$$\langle e_1, e_2 \rangle = -\cosh d(H_1, H_2) . \tag{10.17}$$

The *reflection* $s = s_H$ of $\mathbb{H}^n$ with respect to a hyperbolic hyperplane H is given by

$$s(x) = x - 2\langle e, x \rangle e ,$$

where e is a unit normal to H. It is an isometry of order 2, and it is the unique nontrivial isometry of $\mathbb{H}^n$ that fixes H pointwise. As in Euclidean space, one can talk about the two half-spaces determined by a hyperplane H, and these are interchanged by the reflection s_H.

Finally, we remark that any two points of $\mathbb{H}^n$ can be connected by a unique geodesic, so one has notions of convexity, convex hull, and so on. The geodesic can be defined via differential geometry or directly [48].

Exercises

10.46. Given a nonzero vector $\alpha \in \mathbb{R}^{n,1}$, show that $\alpha^\perp$ intersects $\mathbb{H}^n$ if and only if $Q(\alpha) > 0$. Here $\alpha^\perp$ is defined with respect to the bilinear form $\langle -, - \rangle$.

10.47. Let $H_1 = \alpha_1^\perp \cap \mathbb{H}^n$ and $H_2 = \alpha_2^\perp \cap \mathbb{H}^n$ be distinct hyperbolic hyperplanes. Show that $H_1 \cap H_2 \neq \emptyset$ if and only if the bilinear form $\langle -, - \rangle$ is positive definite on the 2-dimensional space spanned by α_1 and α_2.

10.48. Let H_1 and H_2 be disjoint hyperbolic hyperplanes, oriented by unit vectors e_1 and e_2. Show that $|\langle e_1, e_2 \rangle| \geq 1$, with equality if and only if $e_1^\perp \cap e_2^\perp$ contains a nonzero vector x with $Q(x) \leq 0$. [Geometrically, this says that H_1 and H_2 meet at infinity.]

10.3.2 Reflection Groups in $\mathbb{H}^n$

Our starting point is the same as in the spherical and Euclidean cases. Let W be a group of isometries of $\mathbb{H}^n$ generated by reflections s_H, where H ranges over a locally finite W-invariant collection $\mathcal{H}$ of (hyperbolic) hyperplanes. As before, $\mathcal{H}$ induces a partition of $\mathbb{H}^n$ into convex "cells," with those of top dimension called chambers. In the Euclidean case, we were able to quickly reduce to a situation in which we could prove finiteness results and, ultimately, prove that the chambers were simplices. This does not work in hyperbolic space. The first problem is that the finiteness results might not hold. For example, a closed chamber (which is still a strict fundamental domain for the W-action) might have infinitely many walls. We therefore have to build the desired finiteness into the definition. We introduce some terminology first.

By a *polyhedron* in hyperbolic space we mean a subset P that has nonempty interior and is an intersection of finitely many closed half-spaces. As in Euclidean space, there is a unique minimal family of such half-spaces, and the bounding hyperplanes are called *walls*. Each wall comes with a canonical orientation, with the positive side being the side that contains P. Note that the positive half-spaces corresponding to two walls cannot be nested, so equation (10.17) holds whenever two walls are ultraparallel.

We denote by $\hat{P}$ the intersection of the closed half-spaces in $\mathbb{R}^{n,1}$ corresponding to the defining half-spaces of P. It is a closed polyhedral cone of the sort that we studied in Section 1.4.

We call a polyhedron P a *polytope* if it is compact. It can be shown, as in the theory of Euclidean polytopes, that a polytope P is the convex hull of finitely many vertices. The vertices of P correspond to the faces of $\hat{P}$ that are rays. Moreover, $\hat{P} - \{0\}$ is contained in the open cone U_+ that we defined in Section 10.3.1. Thus $\hat{P}$ is the cone over P, and we recover P from $\hat{P}$ by cutting the latter with the upper sheet of the hyperboloid $Q(x) = -1$. The faces of P (including the empty face) are in 1–1 correspondence with the faces of $\hat{P}$.

We call a polyhedron P a *generalized polytope* if it has finite volume. A generalized polytope P is again the convex hull of its vertices, but some of these might be "at infinity." This is less mysterious than it sounds. What happens is that $\hat{P}$ is not contained in U_+ in the noncompact case, but it is contained in the closure of U_+. Vertices at infinity correspond to faces of $\hat{P}$

that are rays in the boundary of U_+. But there are no higher-dimensional faces of $\hat{P}$ in the boundary of U_+. Thus the faces of P are in 1–1 correspondence with those of $\hat{P}$ other than the rays in ∂U_+.

We have already seen an example of a generalized polytope in our discussion of a fundamental domain for $\mathrm{PGL}_2(\mathbb{Z})$ in Section 2.2.3. This fundamental domain is a triangle with one vertex at infinity, and we noted the corresponding ray in ∂U_+ in our treatment from the point of view of linear algebra.

Remark 10.49. The terminology in the literature concerning polytopes in hyperbolic space is not consistent. Our terminology agrees with that of Ratcliffe [193, Section 6.5], but other authors, such as Vinberg [274], use the terms differently.

Returning now to our group W generated by reflections s_H, we introduce the central concept of this section.

Definition 10.50. we say that W is a *hyperbolic reflection group* if some (or every) closed chamber is a generalized polytope.

Thus the finiteness results that were *proved* in the Euclidean case have been taken as part of the definition in the hyperbolic case.

There is a second major difference between the Euclidean and hyperbolic theories. Namely, the chambers need not be simplices, even if we extend the notion of "simplex" to allow vertices at infinity. For example, there exists a regular right-angled pentagon in $\mathbb{H}^2$, and this can occur as the fundamental chamber for a hyperbolic reflection group. It is still true, however, that if we choose a fundamental chamber P and let S be the corresponding set of fundamental reflections, then (W, S) is a Coxeter system, which always turns out to be irreducible. In the pentagon example that we just mentioned, we get a Coxeter system (W, S) of rank 5.

A third difference between the hyperbolic theory and the Euclidean theory is that the connection between the Coxeter matrix M and the *Gram matrix* A is more complicated. Here $A = \big(a(s,t)\big)_{s,t \in S}$ is defined by

$$a(s,t) := \langle e_s, e_t \rangle \,,$$

where e_s is the canonical unit normal to the wall H_s of P fixed by s (corresponding to the orientation of the wall described above, so that $\langle e_s, - \rangle > 0$ on P). The equation

$$a(s,t) = -\cos \frac{\pi}{m(s,t)} \tag{10.18}$$

still holds if $m(s,t) < \infty$, which is the case if and only if $H_s \cap H_t \neq \emptyset$. If $H_s \cap H_t = \emptyset$, on the other hand, then $m(s,t) = \infty$, but as we noted in our discussion of oriented hyperplanes in Section 10.3.1, we do not necessarily have $a(s,t) = -1$. This happens if and only H_s and H_t are parallel; otherwise, they are ultraparallel, and

$$a(s,t) = -\cosh d(H_s, H_t) < -1 . \tag{10.19}$$

When $s \neq t$ and $m(s,t) < \infty$, equation (10.18) says that the angle between the oriented hyperplanes H_s and H_t is $\pi/m(s,t)$ (cf. (10.16)). In fact, more is true: The closed panels P_s and P_t of P supported by H_s and H_t form a dihedral angle of $\pi/m(s,t)$. The extra content in this statement is that $P_s \cap P_t$ has codimension 2, so that it makes sense to speak of the dihedral angle that they form. [An arbitrary polyhedron, or even polytope, can have panels whose intersection has codimension < 2 even if their walls intersect.]

In view of (10.19), the Gram matrix contains more information in general than the Coxeter matrix. To capture the extra information, we modify the Coxeter diagram as follows: Whenever $a(s,t) < -1$, we replace the corresponding edge in the Coxeter diagram (labeled ∞) by a dashed edge labeled by the number $-a(s,t) = \cosh d(H_s, H_t) > 1$. Thus a hyperbolic reflection group has a modified Coxeter diagram, from which we can reconstruct $(W, \mathbb{H}^n)$ up to isomorphism, and which in some cases contains more information than the Coxeter diagram of the underlying Coxeter system (W, S). We will give an example in the next subsection. See [158, Appendix C] for further examples.

Motivated by the special nature of the fundamental chamber P, we define a *Coxeter polyhedron* to be a polyhedron whose dihedral angles all have the form π/m for some integer $m \geq 2$. The importance of this notion comes from the following fact, known as *Poincaré's theorem*: A generalized polytope P in $\mathbb{H}^n$ can occur as a closed chamber for a hyperbolic reflection group if and only if it is a Coxeter polyhedron; see [95, Chapter 3; 159, Section IV.F; 193, Section 7.1].

10.3.3 Example

Consider a quadrilateral P in the hyperbolic plane as in Figure 10.4, with right angles at the base and vertical sides of equal length. The study of these quadrilaterals goes back to Omar Khayyám in the eleventh century [130, pp. 170–171], but they are usually called *Saccheri quadrilaterals*. Let b be the base length, θ the angle at the upper vertices, and h the "height," i.e., the length of the line segment from the midpoint of the upper edge to the midpoint of the base. The quadrilateral P is determined up to isometry by any two of the

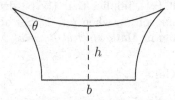

Fig. 10.4. A Saccheri (Khayyám) quadrilateral.

three parameters b, h, θ, and there are formulas that relate these parameters to one another (and to others) in [118, p. 415]. One can deduce, in particular, the equation

$$\cosh^2 h = \frac{\cosh b + \cos 2\theta}{\cosh b - 1},$$

$$(10.20)$$

giving h as a function of b and θ.

If we now take $\theta = \pi/m$ for some integer $m \geq 3$, then the hypotheses of Poincaré's theorem are satisfied, so there is a hyperbolic reflection group with P as fundamental chamber. The underlying Coxeter system (W, S) of rank 4 depends only on m. Figure 10.5 shows the Coxeter diagram, and Figure 10.6 shows the modified Coxeter diagram, where $c = \cosh h$ and $d = \cosh b$. Note that d can be any number > 1 and c is then determined by (10.20), which becomes

$$c^2 = \frac{d + \cos(2\pi/m)}{d - 1}.$$

Thus for each fixed $m \geq 3$ we have a 1-parameter family of inequivalent hyperbolic reflection groups with the same underlying Coxeter group.

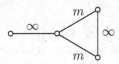

Fig. 10.5. The Coxeter diagram associated to a Saccheri quadrilateral.

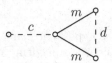

Fig. 10.6. The modified Coxeter diagram.

Remark 10.51. This phenomenon does not occur for hyperbolic reflection groups in dimensions $n > 2$. Indeed, Mostow's rigidity theorem [166, 167], as strengthened by Prasad [191], implies that (W, S) determines the reflection group $(W, \mathbb{H}^n)$ up to isomorphism when $n > 2$. [See also the reviews of [167] and [191] by Raghunathan in *Mathematical Reviews*, MR52 #5874–5.]

10.3.4 The Poset of Cells

Let W be a hyperbolic reflection group of dimension n (i.e., acting on $\mathbb{H}^n$), let P be a fundamental chamber, and let S be the set of fundamental reflections, one for each wall of P. As in the Euclidean case (Corollary 10.11), the stabilizers of the nonempty faces of P (not including the vertices at infinity) are precisely the finite standard subgroups of W; this follows from Vinberg [274, Theorem 3.1]. In contrast to the Euclidean case, however, not all the proper standard subgroups are finite unless the chambers happen to be simplices with no vertices at infinity. But the stabilizer calculation still yields the following analogue of Proposition 10.13(3):

Proposition 10.52. *If W is a hyperbolic reflection group, then the poset of cells in $\mathbb{H}^n$ is isomorphic to the subposet $\Sigma_f(W, S)$ of $\Sigma(W, S)$ consisting of the finite standard cosets (ordered by reverse inclusion).* $\square$

In order to incorporate the vertices at infinity, and therefore have a genuine cell complex, we cite another theorem, which characterizes the stabilizers of the vertices of P at infinity [274, Theorem 3.2]. The characterization requires the modified Coxeter diagram:

Theorem 10.53. *Let W be a hyperbolic reflection group acting on $\mathbb{H}^n$ with fundamental chamber P and modified Coxeter diagram D. Then the stabilizers of the vertices of P at infinity are the standard subgroups W_J $(J \subseteq S)$ such that the induced subdiagram D_J with vertex set J is the Coxeter diagram of a semi-Euclidean reflection group of dimension $n - 1$.*

(See Definition 10.15 for the notion of semi-Euclidean reflection group.)

The Coxeter diagram of a semi-Euclidean reflection group is an ordinary (not modified) diagram, so part of the requirement on D_J in the theorem is that it not contain any dashed edges. The example in Section 10.3.3 is instructive. Note that W does contain two standard subgroups that are 1-dimensional Euclidean reflection groups, corresponding to the two edges labeled ∞ in the Coxeter diagram. But those edges are dashed in the modified Coxeter diagram, so they do not correspond to vertices at infinity. And of course, we know from the construction of W that there are four ordinary vertices but none at infinity. The four vertex stabilizers are the finite standard subgroups of rank 2, corresponding to the two edges labeled m and the two nonexistent edges (for which $m(s, t) = 2$).

It might seem strange that a hyperbolic reflection group can have subgroups that act as reflection groups on Euclidean space, but this has a geometric explanation: If x is a vertex at infinity, then its stabilizer acts on a "horosphere" at x, which is an $(n - 1)$-dimensional Euclidean space (plus a point at infinity). [To explain this geometrically, we switch to Poincaré's unit ball model of $\mathbb{H}^n$. The point x at infinity is then a point on the bounding sphere of the ball, and the *horosphere* at x is the Euclidean sphere inside the ball that is tangent at x to the bounding sphere. If remove x, we get a submanifold of $\mathbb{H}^n$, which, with the induced Riemannian metric, is a Euclidean space.]

In view of Theorem 10.53, we have the following:

Proposition 10.54. *Let W be as in Theorem 10.53. Then the poset of cells, including the vertices at infinity, is isomorphic to the subposet of $\Sigma(W, S)$ consisting of the finite standard cosets together with the cosets wW_J with J as in the theorem.*   □

We repeat, for emphasis, that there is no easy way, in general, to find the subgroups W_J unless one knows the modified Coxeter diagram. More precisely, one needs to know which edges are dashed.

10.3.5 The Simplicial Case

Let (W, S) be the Coxeter system associated with a hyperbolic reflection group acting on $\mathbb{H}^n$ with fundamental chamber P. The simplest case to understand is the case in which P is a simplex, possibly with some vertices at infinity. This happens if and only if P has $n + 1$ panels, i.e., if and only if (W, S) has rank $n+1$. This case is simpler than the general case for several reasons. First, the modified Coxeter diagram coincides with the ordinary Coxeter diagram. Indeed, any two walls must intersect (possibly at infinity) since $n \geq 2$, so there are no ultraparallel pairs of walls. Second, there is an easy criterion for deciding whether a Coxeter system can arise as the Coxeter system of a hyperbolic reflection group with simplicial chambers:

Theorem 10.55. *An irreducible Coxeter system (W, S) arises from a hyperbolic reflection group with simplicial chambers if and only if it has the following properties:*

(1) *The canonical bilinear form B on $\mathbb{R}^S$ is nondegenerate but not positive definite.*

(2) *Each proper standard subgroup $W_J < W$ $(J \subset S)$ is of positive type.*

(See Section 2.5.5 for the terminology.)

Note that it suffices to consider J of the form $S \smallsetminus \{s\}$ in (2), in which case W_J is a vertex stabilizer. Thus the discussion in the previous subsection explains geometrically why it is of positive type. For a proof of the theorem, see Humphreys [133, Section 6.8] or Bourbaki [44, Chapter V, Section 4, Exercises 12 and 13]. Note also that one can check (2) by glancing at the Coxeter diagram of (W, S), since the diagrams of the finite and Euclidean Coxeter groups are known.

Theorem 10.55 has been used to classify hyperbolic reflection groups with simplicial chambers. In dimension $n = 2$ (so that the rank of (W, S) is 3), the result is that almost all irreducible (W, S) are hyperbolic reflection groups of this type; the only exceptions are the three that are finite (types A_3, C_3, and H_3) and the three that are Euclidean (types $\tilde{A}_2$, $\tilde{C}_2$, and $\tilde{G}_2$). For $n \geq 3$, however, it turns out that there are only finitely many possibilities (72 of them). The list can be found in [133, Section 6.9]. There are 32 in dimension three, 14 in dimension four, 12 in dimension five, 3 in dimension six, 4 in dimension seven, 4 in dimension eight, and 3 in dimension nine. Only 14 of these groups have compact fundamental chambers, 9 in dimension three and 5 in dimension four.

As in the Euclidean case (see Section 10.2.2), there is a close connection with the canonical linear representation and the Tits cone. In fact, the canonical bilinear form B turns out to have signature $(n, 1)$, so that the space $V := \mathbb{R}^S$ (which is canonically isomorphic to its dual, since B is nondegenerate) can be identified with the space $\mathbb{R}^{n,1}$ that we used in our definition of $\mathbb{H}^n$. The hyperbolic space on which W acts is then one sheet of the hyperboloid $B(x, x) = -1$, and the cells into which it is decomposed (including the empty cell and the vertices at infinity) correspond to the cells of the Tits cone.

10.3.6 The General Case

Much less is known in the general case. It is still possible to give criteria for recognizing the (possibly modified) Coxeter diagram of a hyperbolic reflection group, but the criteria are more complicated, and such diagrams have not been classified. In stating the criteria, we will use the term "modified diagram" for a finite graph D with labeled edges (possibly dashed), of the sort that arose in Section 10.3.2. Thus an ordinary edge joining vertices s, t may be labeled by $m \in \{4, 5, \ldots, \infty\}$, and a dashed edge is labeled by a real number $c > 1$. Any such diagram has an associated Coxeter matrix $M = \big(m(s, t)\big)_{s,t \in S}$, and the diagram then encodes a symmetric matrix $A = (a(s, t))$ with $a(s, t) = -\cos\big(\pi/m(s, t)\big)$ if $m(s, t) < \infty$ and $a(s, t) \leq -1$ if $m(s, t) = \infty$.

Theorem 10.56. *Let D be a connected modified diagram with vertex set S, and let A be the corresponding symmetric matrix. Then D arises from a hyperbolic reflection group acting on $\mathbb{H}^n$ if and only if the following conditions hold:*

(1) *Let B be the bilinear form on $V = \mathbb{R}^S$ with matrix A. Then the induced form on the quotient of V by the radical of B has signature $(n, 1)$.*

(2) *There is at least one subset J of S such that the induced subdiagram D_J of D is the Coxeter diagram of a spherical or semi-Euclidean Coxeter group W_J of dimension $n - 1$.*

(3) *For any J_1 satisfying the condition in (2) and any $K \subset S$ such that W_K is a spherical reflection group of dimension $n - 2$, there is a set $J_2 \supset K$ distinct from J_1 and also satisfying the condition in (2).*

This criterion, which is due to Vinberg, follows from [274, Theorem 2.1 and Proposition 4.2]. The geometric meaning of (2) is that the fundamental chamber P must have at least one vertex, possibly at infinity, and the geometric meaning of (3) is that any edge emanating from a vertex must lead to another vertex. Note that as in the simplicial case, one can check conditions

(2) and (3) by glancing at the diagram D. But in contrast to the simplicial case, one really needs D, and not just the Coxeter diagram.

Hyperbolic reflection groups, even those with compact fundamental chamber, have not been classified, in spite of Vinberg's criterion 10.56. One major problem is that for a given n, there is no a priori upper bound on the size of S. It is known, however, as in the simplicial case, that the dimension n cannot be arbitrarily large. For example, Vinberg [273] showed that there are no such groups with compact fundamental domain in dimensions $n > 29$. [This bound is probably not sharp. The largest known compact example was constructed by Bugaenko [64], and it has dimension 7.] If we drop the compactness assumption, then there is an even bigger gap between the examples and the known upper bound, which is 995 (see [143]). [Again the bound is probably not sharp. The known examples of highest dimension [271, 272, 275] have $n = 19$.]

*10.4 Hyperbolic Coxeter Groups and Complexes

There are at least three different notions of "hyperbolic Coxeter group" in the literature. For example, Bourbaki [44] and Humphreys [133] use this term for hyperbolic reflection groups whose chambers are simplices, as in Section 10.3.5. Another possible definition is that a hyperbolic Coxeter group is one that is hyperbolic in the sense of Gromov [119]. Such groups were characterized by Moussong [170]; see Section 12.3.9 below for a statement of his result. We will adopt a definition that is different from both of these:

Definition 10.57. We call a Coxeter system (W, S) *hyperbolic* if it arises from a hyperbolic reflection group as defined in Section 10.3.2.

This class of Coxeter groups properly includes the hyperbolic Coxeter groups in the sense of Bourbaki and Humphreys. But a hyperbolic Coxeter group as just defined is *not* necessarily hyperbolic in the sense of Gromov if the chambers are not compact. In fact, we have seen that such groups can contain Euclidean reflection groups of dimension ≥ 2, and hence they can contain free abelian subgroups of rank ≥ 2. But hyperbolic groups in the sense of Gromov cannot have such subgroups.

It is less clear what we should mean by a "hyperbolic Coxeter complex." One could, for instance, use this term for a Coxeter complex Σ that is isomorphic to $\Sigma(W, S)$ for some hyperbolic Coxeter system (W, S). With this definition, a hyperbolic Coxeter complex would always be simplicial, but it would generally have higher dimension than that of the hyperbolic space on which W acts as a reflection group.

On the other hand, it is usually more useful to work with the "metric polyhedral complex" consisting of hyperbolic space decomposed into cells by the reflecting hyperplanes. This is a more complicated object than a simplicial complex or even a cell complex, because it can have vertices at infinity. See Bridson–Haefliger [48] for a comprehensive treatment of ordinary metric polyhedral complexes (with no vertices at infinity), and see Gaboriau–Paulin [104] for the more general theory. We indicate briefly how to construct such an object for a given hyperbolic Coxeter group W.

We already know from the examples in Section 10.3.3 that the desired complex (taking the metric into account) is not in general uniquely determined by (W, S), at least in the important case $n = 2$. So we must first *choose* the modified Coxeter diagram that we want to use, or, equivalently, the generalized polytope P that is to be the fundamental chamber, with panels P_s indexed by S. Next, note that the space X we are trying to construct should consist of translates wx ($w \in W$, $x \in P$), with $wx = w'x'$ if and only if $x' = x$ and $w' \in w\langle S_x \rangle$, where $S_x := \{s \in S \mid x \in P_s\}$. So we simply *define* X to be the quotient of $W \times P$ by the equivalence relation $(w, x) \sim (w', x') \iff x' = x'$ and $w' \in w\langle S_x \rangle$. Equivalently, this relation is generated by declaring $(w, x) \sim (ws, x)$ if $x \in P_s$. Thus we are taking a copy of P for each $w \in W$, and we are identifying the faces P_s in copies of P that correspond to s-adjacent elements of W. This construction is due to Vinberg [270].

The resulting set X has an obvious action of W, with a copy of P as strict fundamental domain. We have a hyperbolic metric on P, and we can metrize X by minimizing the lengths of piecewise geodesic paths between two given points. In this way we have reconstructed the hyperbolic space on which W acts. We will return to this in Section 12.1, where we will be able to explain the construction of the metric in detail in a more general setting.

11

Euclidean Buildings

Having introduced Euclidean Coxeter complexes in Section 10.2, we can now develop a systematic theory of buildings whose apartments are Euclidean spaces. Recall that we gave an example of such a building in Section 6.9.3.

Definition 11.1. By a *Euclidean building* we mean a building Δ whose apartments are Euclidean Coxeter complexes.

One also says that Δ is an *affine building*, or a building of *affine type*. But we use the word "Euclidean," as Tits did in [248], in order to emphasize the fact that there is a well-defined Euclidean metric in every apartment (Section 10.2.1). In this chapter, which is based on Bruhat–Tits [59], we investigate the geometry that these Euclidean metrics impose on the building as a whole.

An important fact that will emerge is that the geometric realization X of a Euclidean building Δ has a natural metric that satisfies a curvature condition called the CAT(0) *property*. The theory of CAT(0) spaces has become quite important in recent years, and many readers will already be familiar with it. For the sake of those who are not, we begin this chapter by giving the relevant definitions, so that we do not have to interrupt the flow of ideas when the CAT(0) property arises in connection with buildings.

11.1 CAT(0) Spaces

The standard reference for this section is Bridson–Haefliger [48]. See also Ballmann [28] for a more concise treatment.

Definition 11.2. Let X be a metric space with metric denoted by $d(-,-)$ or $d_X(-,-)$. A *geodesic*, or *geodesic segment*, in X is a subset isometric to a closed interval of real numbers. We say that X is a *geodesic metric space* if any two points of X can be connected by a geodesic. We denote by $[x,y]$ any geodesic joining x and y, even though we are not assuming that this geodesic is unique. We will always parametrize $[x,y]$ by $t \mapsto p_t$ $(0 \le t \le 1)$, where $d(x, p_t) = t d(x, y)$ for all t.

Given three points x, y, z in a metric space X, the triangle inequality implies that there is a *comparison triangle* in the Euclidean plane $\mathbb{R}^2$, whose vertices $\bar{x}, \bar{y}, \bar{z}$ have the same pairwise distances as x, y, z. It is unique up to an isometry of $\mathbb{R}^2$. Given a geodesic $[x, y]$ and a point $p = p_t \in [x, y]$, there is a corresponding point $\bar{p} = \bar{p}_t$ $[= (1-t)\bar{x} + t\bar{y}]$ on the line segment $[\bar{x}, \bar{y}]$ in $\mathbb{R}^2$; see Figure 11.1, which illustrates this for $t = 1/2$. The picture also indicates

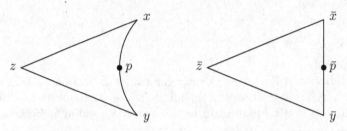

Fig. 11.1. A geodesic triangle in X and its comparison triangle in $\mathbb{R}^2$.

choices of geodesics from z to x and y as an aid to the intuition, so that we have a "geodesic triangle" in X; but the comparison triangle does not depend on these choices.

Definition 11.3. A metric space X is called a CAT(0) *space* if for any x, y in X there is a geodesic $[x, y]$ with the following property: For all $p \in [x, y]$ and all $z \in X$, we have

$$d_X(z, p) \le d_{\mathbb{R}^2}(\bar{z}, \bar{p}),\tag{11.1}$$

with $\bar{z}$ and $\bar{p}$ as above.

This should be thought of intuitively as being related to nonpositive curvature, as we have tried to suggest in Figure 11.1. The intuition is justified by a theorem in differential geometry that says that a Riemannian manifold has sectional curvature ≤ 0 if and only if it is locally a CAT(0) space; a proof can be found in Bridson–Haefliger [48, Appendix to Chapter II.1]. See also Exercise 11.22 below. The "CAT" terminology is explained in [48, p. 159; 119, Section 2.4]; we will explain the "0" below.

We have used one of several possible equivalent definitions of CAT(0). For example, there is a variant of the definition that does not explicitly mention the comparison triangle. Such a definition exists because there is a formula for the distance from a vertex of a Euclidean triangle to any point on the opposite side: Given $x, y, z \in \mathbb{R}^2$ and $t \in [0, 1]$, let $p_t = (1-t)x + ty$; then the formula is

$$d^2(z, p_t) = (1-t)d^2(z, x) + t d^2(z, y) - t(1-t)d^2(x, y).\tag{11.2}$$

This is best known when $t = 1/2$, in which case it can be found in Pappus [185, Book VII, Proposition 122].* It is essentially the parallelogram law in that case, so we will call (11.2) the *generalized parallelogram law*. To prove it, we may assume that $z = 0$. Then

$$d^2(z, p_t) = \|p_t\|^2 = (1-t)^2\|x\|^2 + t^2\|y\|^2 + 2t(1-t)\langle x, y \rangle \; ;$$

one obtains (11.2) from this by using the formula

$$d^2(x, y) = \|x\|^2 + \|y\|^2 - 2\langle x, y \rangle$$

to eliminate $\langle x, y \rangle$.

The following reformulation of the CAT(0) property is now immediate:

Proposition 11.4. *A metric space X is a* CAT(0) *space if and only if for any $x, y \in X$ there is a geodesic $[x, y]$ with the following property: For any point $p = p_t \in [x, y]$ and any $z \in X$,*

$$d^2(z, p) \le (1-t)d^2(z, x) + td^2(z, y) - t(1-t)d^2(x, y) \; . \tag{11.3}$$

$\square$

Note that both versions of the definition allow for the possibility that there is more than one geodesic joining two given points. But we can quickly deduce that in fact this does not happen:

Proposition 11.5. *For any two points x, y in a* CAT(0) *space X, there is a unique geodesic $[x, y]$ joining them. It is characterized by*

$$[x, y] = \{z \in X \mid d(x, y) = d(x, z) + d(z, y)\} \; . \tag{11.4}$$

Proof. Any geodesic joining x and y is contained in the set on the right side of (11.4). It therefore suffices to show that the right side is contained in the left, with $[x, y]$ as in the definition of "CAT(0) space." Suppose z satisfies $d(x, y) = d(x, z) + d(z, y)$, and let $p = p_t \in [x, y]$, where $t := d(x, z)/d(x, y)$, so that $d(x, p) = d(x, z)$ and $d(p, y) = d(z, y)$. Then the comparison triangle in $\mathbb{R}^2$ with vertices $\bar{x}, \bar{y}, \bar{z}$ degenerates to the line segment $[\bar{x}, \bar{y}]$, and $\bar{z} = \bar{p}$. Hence $d(z, p) = 0$ by (11.1), so $z = p \in [x, y]$. (Alternatively, check that the right side of (11.3) is equal to 0 under our assumptions.) $\square$

Remarks 11.6. (a) Our definition of "CAT(0)" is similar to one of several equivalent conditions given in Bridson–Haefliger [48, I.1.7(2)], but it is superficially different from the latter. Where we stated "there is a geodesic $[x, y]$" with a certain property, they require the property to hold for *all* geodesics. But the two definitions are in fact equivalent, since our definition implies uniqueness of geodesics.

---

* Pappus's version is $d^2(z, x) + d^2(z, y) = 2\big(d^2(z, m) + d^2(m, y)\big)$, where $m := p_{1/2}$ is the midpoint of $[x, y]$.

(b) The "0" in "CAT(0)" refers to the fact that the comparison space $\mathbb{R}^2$ has curvature 0. More generally, there is a notion of CAT(κ) space for any real number κ, where $\mathbb{R}^2$ is replaced by the complete simply connected 2-manifold of constant curvature κ. This is the sphere of radius $1/\sqrt{\kappa}$ if $\kappa > 0$, and it is the hyperbolic plane with metric scaled by $1/\sqrt{-\kappa}$ if $\kappa < 0$. In case $\kappa > 0$, some care is needed in formulating the definition because comparison triangles in the sphere will exist only if the points x, y, z are sufficiently close together. When $\kappa = 1$, for example, one assumes in the definition that $d(x, y) < \pi$, and one considers only points z such that $d(x, y) + d(y, z) + d(z, x) < 2\pi$. In particular, X is not required to be a geodesic metric space; we require only that geodesics exist between points at distance $< \pi$ from one another. [And these geodesics then turn out to be unique.]

We will use Proposition 11.4 to prove that geodesic segments in a CAT(0) space vary continuously with their endpoints; see [48, Proposition II.1.4] for a different proof. In the precise statement of the result, we denote by $p_t(x, y)$ the point p_t as above on the unique geodesic $[x, y]$.

Proposition 11.7. *Let X be a CAT(0) space. Then the map $(x, y, t) \mapsto p_t(x, y)$ is continuous as a function of x, y, t. In particular, X is contractible.*

Proof. Fix x, y, t and apply the inequality (11.3) to $z = p_{t'}(x', y')$ for (x', y', t') close to (x, y, t). Since $d(z, x)$ is close to $d(z, x') = t'd(x', y')$, it is clear that $d(z, x) \to td(x, y)$ as $(x', y', t') \to (x, y, t)$; hence the first term on the right side of (11.3) approaches $(1 - t)t^2d^2(x, y)$. Similarly, the second term approaches $t(1 - t)^2d^2(x, y)$. The right side of (11.3) therefore approaches

$$(1 - t)t^2d^2(x, y) + t(1 - t)^2d^2(x, y) - t(1 - t)d^2(x, y) = 0 ,$$

whence $d(z, p_t(x, y)) \to 0$.   □

Next, we return to Euclidean geometry and note a simple qualitative consequence of the generalized parallelogram law (11.2): If we move z so that it gets closer to x and y, then it also gets closer to p. More precisely:

Proposition 11.8. *Consider two triangles in the Euclidean plane, with vertices x, y, z and x, y, z'. Let p be an arbitrary point on the common side $[x, y]$. If $d(z', x) \leq d(z, x)$ and $d(z', y) \leq d(z, y)$, then $d(z', p) \leq d(z, p)$.*   □

See Figure 11.2 for an illustration, and see Exercise 11.11 for an alternative proof that is slightly longer but provides better intuition as to why the result is true.

Finally, we record the special case $t = 1/2$ of the inequality (11.3), which suffices for many purposes. Letting m be the midpoint $p_{1/2}$ of $[x, y]$, we can write this special case as

$$d^2(z, m) \leq \frac{1}{2}\left(d^2(z, x) + d^2(z, y)\right) - \frac{1}{4}d^2(x, y) . \tag{NC}$$

["(NC)" is intended to suggest nonpositive curvature.] Thus every CAT(0) space has the following property, first introduced by Bruhat–Tits [59]:

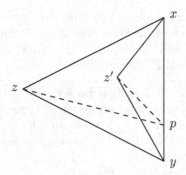

Fig. 11.2. Two triangles in the Euclidean plane.

(NC) *For any two points* $x, y \in X$ *there is a point* $m \in X$ *such that inequality* (NC) *holds for all* $z \in X$.

Conversely, property (NC) "almost" implies CAT(0), and it does imply CAT(0) if the metric space is complete (see Exercise 11.14 below).

Exercises

11.9. Let $[x, y]$ and $[y, z]$ be geodesics in an arbitrary metric space X. If $d(x, z) = d(x, y) + d(y, z)$, show that there is a geodesic $[x, z]$ obtained by concatenating $[x, y]$ and $[y, z]$. More precisely, let $\alpha \colon [a, b] \to [x, y]$ and $\beta \colon [b, c] \to [y, z]$ be isometries with $\alpha(a) = x$, $\alpha(b) = \beta(b) = y$, and $\beta(c) = z$, where $a \leq b \leq c$ in $\mathbb{R}$. Let $\gamma \colon [a, c] \to X$ be given by $\gamma|_{[a,b]} = \alpha$ and $\gamma|_{[b,c]} = \beta$. Then γ is an isometry from $[a, c]$ onto $[x, y] \cup [y, z]$.

11.10. Give an alternative proof of the generalized parallelogram law by deducing it from the law of cosines.

11.11. Consider a triangle with side lengths a, b, c in the Euclidean plane. Let θ be the angle opposite the side of length c. If a and b are fixed while θ varies from 0 to π, then c strictly increases from $|a - b|$ to $a + b$. (One can give a rigorous proof of this intuitively obvious fact via the law of cosines.) Use this fact to give an alternative proof of Proposition 11.8.

Assume throughout the remaining exercises that X is a metric space with property (NC).

11.12. Show that the point m necessarily satisfies

$$d(x, m) = d(y, m) = \frac{1}{2} d(x, y) .$$

Moreover, m is the only point satisfying these equations, so we may call m the *midpoint* of the pair $\{x, y\}$.

11.13. Given three points $a, b, c \in X$, let m_1 be the midpoint of $\{a, b\}$ and let m_2 be the midpoint of $\{a, c\}$. Show that $d(b, c) \geq 2d(m_1, m_2)$. Intuitively, this says that two geodesics emanating from a given point separate at least as quickly as they would in Euclidean space.

11.14. Fix $x, y \in X$ and let $t \in [0, 1]$ be a dyadic rational number, i.e., a rational number whose denominator is a power of 2. Show that there is a point $p_t \in X$ such that the inequality (11.3) holds for all $z \in X$. Show further that any such p_t can be characterized as the unique point satisfying the equations

$$d(x, p_t) = td(x, y) \,,$$
$$d(y, p_t) = (1 - t)d(x, y) \,.$$

Extend all this to arbitrary $t \in [0, 1]$ if X is complete.

11.15. A subset Y of X will be called *midpoint convex* if it contains the midpoint of any pair of its points. Suppose that Y is midpoint convex, let $x \in X$ be arbitrary, and let $d = d(x, Y) := \inf_{y \in Y} d(x, y)$. If there is a point $y \in Y$ such that $d(x, y) = d$, show that $d(y, y') \leq d(x, y')$ for all $y' \in Y$.

11.2 Euclidean Buildings as Metric Spaces

Assume throughout this section that Δ is a Euclidean building, equipped with an arbitrary system of apartments $\mathcal{A}$. Nothing we do will depend on the choice of $\mathcal{A}$.

Let X be the geometric realization $|\Delta|$ of the simplicial complex Δ (Section A.1.1). For the moment, we view X as a set, with no topology. [We will soon endow X with a metric, and hence a topology.] The set X is the union of open simplices $|A|$, one for each nonempty simplex $A \in \Delta$. To avoid cumbersome notation, we will omit the vertical bars and simply denote by A this open simplex and by $\overline{A}$ the corresponding closed simplex. Thus $\overline{A}$ now denotes the geometric realization of the subcomplex $\Delta_{\leq A}$ that we have sometimes called $\overline{A}$ in earlier chapters.

It will be convenient to apply to X terminology that we have previously used for the abstract complex Δ. In particular, we will refer to X itself as a building and to the subsets $E = |\Sigma|$ as apartments ($\Sigma \in \mathcal{A}$). For any such apartment E and any chamber C of E, the geometric realization of $\rho_{\Sigma, C} : \Delta \to \Sigma$ is a retraction $X \to E$, denoted by $\rho_{E, C}$.

In view of Section 10.2.1, each apartment E of X is a Euclidean space, with a metric d_E. Moreover, the isomorphisms between apartments given by the building axiom (B2) can be taken to be isometries. We now wish to piece the metrics d_E together to make the entire building X a metric space.

Given two points $x, y \in X$, axiom (B1) implies that there is an apartment E containing both x and y. Choose such an E and set

$$d(x, y) := d_E(x, y) \ .$$

If E' is another apartment containing x and y, then (B2) gives us an isometry $E \to E'$ fixing x and y, so $d(x, y)$ is independent of the choice of apartment. We therefore have a well-defined distance function

$$d \colon X \times X \to \mathbb{R} \ .$$

This should not be confused with the combinatorial distance function that we have used before (defined via galleries), although, as we will see, the two kinds of distance functions have some similar properties. To avoid confusion, we will write $\mathbf{d}(-, -)$ from now on for the combinatorial distance function.

Theorem 11.16.

(1) *The distance function* $d \colon X \times X \to \mathbb{R}$ *is a metric.*

(2) *The metric space* X *is complete and is a* CAT(0) *space.*

(3) *The retraction* $\rho = \rho_{E,C} \colon X \to E$ *is distance-decreasing for any apartment* E *and chamber* C *of* E, *i.e.,*

$$d(\rho(x), \rho(y)) \leq d(x, y)$$

for all $x, y \in X$. *Equality holds if* $x \in \overline{C}$.

(4) *For any* $x, y \in X$, *choose an apartment* E *containing* x *and* y *and let* $[x, y]$ *be the line segment joining them in the Euclidean space* E. *Then* $[x, y]$ *is independent of the choice of* E *and can be characterized by*

$$[x, y] = \{z \in X \mid d(x, y) = d(x, z) + d(z, y)\} \ .$$

Proof. We begin by proving (3), which makes sense even before we know that d is a metric. The second assertion of (3) follows immediately from the fact that ρ maps every apartment containing C isometrically onto E. This fact also implies that for any chamber C' of X, ρ maps $\overline{C'}$ isometrically onto its image. Suppose now that x and y are arbitrary points of X. Choose an apartment E' containing them, and let $[x, y]$ be the line segment joining them in E'. It is easy to see that we can subdivide this segment in such a way that each subinterval is contained in a closed chamber. [Use the fact that the decomposition of E' into simplices is induced by a locally finite collection $\mathcal{H}$ of hyperplanes.] Let the subdivision points be $x = x_0, x_1, \dots, x_m = y$. We then have

$$d(\rho(x), \rho(y)) \leq \sum_{i=1}^{m} d(\rho(x_{i-1}), \rho(x_i)) = \sum_{i=1}^{m} d(x_{i-1}, x_i) = d(x, y) \ ,$$

where the inequality follows from the triangle inequality in the Euclidean space E, and the first equality follows from the fact that ρ is an isometry on each closed chamber. This proves the first assertion of (3).

It is now easy to prove (1), the content of which is that d satisfies the triangle inequality: Given $x, y, z \in X$, choose an apartment E containing x

and y, and let $\rho = \rho_{E,C}$ for some chamber C of E. Using (3) and the triangle inequality in E, we find that

$$d(x,y) \leq d(x,\rho(z)) + d(\rho(z),y) \leq d(x,z) + d(z,y) ,$$

as required.

The same circle of ideas leads easily to the CAT(0) property: Given $x,y \in X$, choose an apartment E containing them and let $[x,y]$ be the line segment joining them in E; it is a geodesic in X. Given $z \in X$ and $p = p_t \in [x,y]$ $(0 \leq t \leq 1)$, choose a chamber C of E with $p \in \overline{C}$, and let ρ be the retraction $\rho_{E,C}$. By (3) we have $d(\rho(z),x) \leq d(z,x)$, $d(\rho(z),y) \leq d(z,y)$, and $d(\rho(z),p) = d(z,p)$. There are now two ways to finish. Method 1: According to the criterion of Proposition 11.4, we must verify

$$d^2(z,p) \leq (1-t)d^2(z,x) + td^2(z,y) - t(1-t)d^2(x,y) .$$

This follows from the generalized parallelogram law in E, which gives

$$d^2(\rho(z),p) = (1-t)d^2(\rho(z),x) + td^2(\rho(z),y) - t(1-t)d^2(x,y) .$$

Method 2: We use our original definition of CAT(0), taking the comparison triangle $\bar{x}, \bar{y}, \bar{z}$ to be in E, with $\bar{x} = x$ and $\bar{y} = y$. Then $\bar{p} = p$, and we must prove $d(z,p) \leq d(\bar{z},p)$. This follows from Proposition 11.8, applied to the Euclidean triangles $x,y,\bar{z}$ and $x,y,\rho(z)$. This completes the proof that X is a CAT(0) space, and assertion (4) follows immediately via Proposition 11.5.

Finally, we must show that X is complete. Fix a chamber C and let $\tau \colon X \to \overline{C}$ be the geometric realization of the unique retraction $\Delta \to \Delta_{\leq C}$ (Proposition A.13). For any chamber C', we claim that τ maps $\overline{C'}$ isometrically onto $\overline{C}$. To see this, choose an apartment $E = |\Sigma|$ containing C and C' and let w be the unique type-preserving automorphism of Σ such that $wC' = C$. Then $\tau|_{\overline{C'}}$ is induced by the restriction of w to C' and its faces, and the assertion now follows from the fact that w is an isometry of E.

It now follows easily that τ is distance-decreasing; the proof is the same as the proof of the analogous fact about retractions onto apartments. So if we are given a Cauchy sequence $(x_m)_{m \geq 1}$ in X, then the image sequence $(\tau(x_m))$ is a Cauchy sequence in $\overline{C}$. The latter being a closed subset of a Euclidean space, it follows that there is a point $y \in \overline{C}$ such that $\tau(x_m) \to y$ as $m \to \infty$. Choose for each m a chamber C_m with $x_m \in \overline{C}_m$, and let y_m be the unique point in $\overline{C}_m$ such that $\tau(y_m) = y$. (We will then say that y_m is of type y.) Since $\tau|_{\overline{C}_m}$ is an isometry, we have

$$d(x_m, y_m) = d(\tau(x_m), y) \to 0 \quad \text{as } m \to \infty ;$$

hence (y_m) is also a Cauchy sequence. On the other hand, we claim that the set of points of a given type y is discrete. [Draw a picture of the tree case to see why this is intuitively plausible.] Hence the Cauchy sequence (y_m) is eventually constant, and the fact that $d(x_m, y_m)$ tends to 0 now says that $x_m \to y'$, where $y' := y_m$ for large m.

It remains to prove the discreteness claim. Recall that the *star* of a point $x \in X$, denoted by $\mathrm{st}\, x$ or $\mathrm{st}_X\, x$, is the union of the closed simplices $\overline{A}$ containing x. If our metric on X is at all reasonable, we expect the star of x to be a neighborhood of x. In fact, a more precise statement is true:

Lemma 11.17. *Given $y \in \overline{C}$ there is a $\delta > 0$ with the following property: For any $x \in X$ of type y, $\mathrm{st}\, x$ contains the closed ball of radius δ centered at x.*

Now $\mathrm{st}\, x$ contains no point distinct from x and having the same type as x. So the lemma implies that $d(x, x') > \delta$ for any two distinct points x, x' of type y. This proves the claim and completes the proof of the theorem, modulo the lemma. □

Proof of the lemma. Choose an apartment E containing C, and let $\mathcal{H}$ be the locally finite collection of walls that defines the simplicial decomposition of E. Let δ be the minimum distance from y to a wall $H \in \mathcal{H}$ not containing y. Then for any $y' \in E$ with $d(y, y') \le \delta$, the open segment (y, y') does not cross any wall. We therefore have $(y, y') \subseteq A$ for some open cell A; hence $y, y' \in \overline{A}$ and $y' \in \mathrm{st}_E\, y$.

Now suppose x and x' are points of X with x of type y and $d(x, x') \le \delta$. We can find an apartment E' containing x and x' and an isomorphism $\phi \colon E' \to E$ such that $\phi(x) = y$. Then $d(y, \phi(x')) \le \delta$, so $\phi(x') \in \mathrm{st}_E\, y$ by the previous paragraph and hence $x' \in \mathrm{st}_{E'}\, x \subseteq \mathrm{st}_X\, x$. □

Exercises

11.18. The *open star* of x is the union of the open simplices A such that $x \in \overline{A}$. State and prove an analogue of the lemma for open stars.

11.19. Let X' be a subcomplex of X, i.e., $X' = |\Delta'|$ for some subcomplex Δ' of Δ. [Equivalently, X' is a subset of X that is a union of closed simplices.] Deduce from Exercise 11.18 that X' is a closed subset of X.

11.20. Deduce from Exercise 11.18 or 11.19 that any chamber C is an open subset of X.

11.21. Let Δ' be a subcomplex of Δ, and let $X' = |\Delta'|$. Show that Δ' is a convex subcomplex as defined in Section 4.11 if and only if X' is convex in the following sense: For any points $x, y \in X'$, the geodesic $[x, y]$ is contained in X'.

The next exercise assumes some knowledge of differential geometry.

11.22. Use the ideas in the proof of Theorem 11.16 to show that a complete simply connected Riemannian manifold M with sectional curvature ≤ 0 is a CAT(0) space.

11.3 The Bruhat–Tits Fixed-Point Theorem

If G is a compact group of isometries of a complete simply connected Riemannian manifold M of nonpositive curvature, then a famous theorem of É. Cartan says that G fixes a point of M. Cartan's fixed-point theorem is a fundamental tool in the theory of Lie groups. In this section we will prove a generalization of Cartan's theorem to complete CAT(0) spaces. This generalization, due to Bruhat and Tits [59], then applies to Euclidean buildings as well as to complete simply connected manifolds of nonpositive curvature. Like Cartan's theorem, the Bruhat–Tits theorem has applications to group theory; we will explore these in the next section.

Theorem 11.23. *Let G be a group of isometries of a complete CAT(0) space X. If G stabilizes a nonempty bounded subset of X, then G fixes a point of X.*

Remarks 11.24. (a) In the situation of Cartan's theorem, where G is compact, any orbit Gx is a bounded set stabilized by G. So Cartan's theorem is indeed a special case of the Bruhat–Tits theorem.

(b) The geometric realization of a tree is a complete CAT(0) space. The theorem in this case can be found in Serre [217, Section I.4.3, Proposition 19], which also contains applications to group theory.

(c) Theorem 11.23 is of interest even when X is a Euclidean space or, more generally, a Hilbert space (real or complex, possibly infinite-dimensional). The result in this case has a cohomological interpretation, which is often used in representation theory; see Exercise 11.30 below.

The Bruhat–Tits proof of Theorem 11.23 consists in associating to every nonempty bounded subset $A \subseteq X$ a point $c = c(A) \in X$ which, intuitively, is some sort of "center" of A. The construction of c depends only on the metric on X, so it is compatible with isometries. In particular, if A is invariant under a group G of isometries of X, then c is fixed by G.

We will give a variant of this proof due to Serre [private communication]. The basic idea remains the same, but Serre's definition of $c(A)$ is different from that of Bruhat and Tits. Namely, $c(A)$ is defined to be the center of the sphere circumscribed about A. Here are the details.

Let X be an arbitrary metric space and A a nonempty bounded subset. For any $x \in X$, let $r(x, A)$ be the smallest real number r such that A is contained in the closed ball of radius r centered at x; equivalently,

$$r(x, A) = \sup_{a \in A} d(x, a) .$$

Definition 11.25. The *circumradius* of A, denoted by $r(A)$, is defined by

$$r(A) := \inf_{x \in X} r(x, A).$$

If $r(A) = r(x, A)$ for some $x \in X$, then any such x will be called a *circumcenter* of A.

See Figure 11.3 for an example. For a simpler example, note that the midpoint m that occurs in property (NC) is a circumcenter (and even the unique circumcenter) of the two-point set $\{x, y\}$.

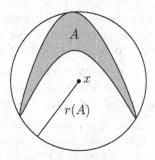

Fig. 11.3. x is a circumcenter of A.

If X is a sphere, which is a manifold of positive curvature, then circumcenters always exist but are not necessarily unique (see Exercise 11.29 below). In Euclidean space, however, it is known that circumcenters exist and are unique. More generally, we have the following observation of Serre:

Theorem 11.26. *If X is a complete CAT(0) space, then every nonempty bounded subset A admits one and only one circumcenter.*

As we explained above, Theorem 11.23 follows immediately from Theorem 11.26; for the circumcenter of A will clearly be fixed by any group of isometries of X that stabilizes A. It remains to prove Theorem 11.26.

Proof of Theorem 11.26. For any two points $x, y \in X$, we can apply the inequality (NC) with $z \in A$ to get

$$r^2(m, A) \leq \frac{1}{2}\left(r^2(x, A) + r^2(y, A)\right) - \frac{1}{4}d^2(x, y),$$

where m is the midpoint of $\{x, y\}$. Hence

$$d^2(x, y) \leq 2\left(r^2(x, A) + r^2(y, A) - 2r^2(m, A)\right).$$

Since $r(m, A) \geq r(A)$, this implies

$$d^2(x, y) \leq 2\left(r^2(x, A) + r^2(y, A) - 2r^2(A)\right). \tag{11.5}$$

Uniqueness of the circumcenter is now immediate; for if x and y are both circumcenters, then the right side of (11.5) is 0, and hence $x = y$. To prove existence, take a sequence of points $x_n \in X$ such that $r(x_n, A) \to r(A)$, and apply (11.5) with $x = x_n$ and $y = x_m$. Then the right side can be made arbitrarily small by taking n and m sufficiently large, so (x_n) is a Cauchy sequence. Hence (x_n) has a limit $x \in X$, and it is easy to check that $r(x, A) = r(A)$. $\qquad\square$

We close this section by proving one more result about circumcenters. This will not be needed in what follows, but it provides a nice illustration of the inequality (NC). In Euclidean geometry, it is known that the circumcenter of a bounded set A is contained in the closure of the convex hull of A. We will show that this too generalizes to complete CAT(0) spaces.

Theorem 11.27. *Let X be a complete* CAT(0) *space, let A be a nonempty bounded subset, and let Y be the smallest closed convex subset of X that contains A. Then the circumcenter of A is contained in Y.*

Here a subset of X is *convex* if it contains the (unique) geodesic joining any two of its points. At first glance it might seem that the theorem is a formal consequence of Theorem 11.26. For Y, in its own right, is a complete CAT(0) space, so Theorem 11.26 implies that A has a circumcenter in Y. What is not obvious, however, is that the circumcenter of A in Y is the same as its circumcenter in X. In other words, it is conceivable that the circumradius $r_Y(A)$ of A in Y is bigger than the circumradius $r_X(A)$ of A in X. Theorem 11.27 will follow as soon as we prove that this cannot happen. This result does not require geodesics or completeness, so we will formulate it in terms of property (NC) and the notion of "midpoint convexity" defined in Exercise 11.15.

Lemma 11.28. *Let X be a metric space with property* (NC), *and let Y be a midpoint-convex subset. Then $r_Y(A) = r_X(A)$ for any nonempty bounded subset $A \subseteq Y$.*

Proof. Given any $x \in X$ and any $r > r(x, A)$, we must find $y \in Y$ such that $r(y, A) \le r$. This is easy if there is a $y \in Y$ with $d(x, y) = d(x, Y)$; in this case we have $r(y, A) \le r(x, A) < r$ by Exercise 11.15. In the general case, let $d = d(x, Y)$, and choose a sequence of points $y_n \in Y$ such that $d_n := d(x, y_n) \to d$. We will show that $r(y_n, A) \le r$ for some n. Suppose this is false. Then for each n we can find a point $a_n \in A$ such that $d(y_n, a_n) > r$. As in the solution to Exercise 11.15, consider the points p_t between y_n and a_n, where t ranges over the dyadic rationals in $[0, 1]$. These points are in Y, and we have

$$d^2 \le d^2(x, p_t) \le (1 - t)d_n^2 + tr^2(x, A) - t(1 - t)r^2 .$$

Fixing t and letting $n \to \infty$, we conclude that

$$d^2 \le (1 - t)d^2 + tr^2(x, A) - t(1 - t)r^2 = d^2 + \alpha t + r^2 t^2 ,$$

where $\alpha := -d^2 + r^2(x, A) - r^2$. But this is absurd; for α is negative, so $\alpha t + r^2 t^2 < 0$ for small $t > 0$. This contradiction shows that $r(y_n, A) \le r$ for some n, as required. $\qquad\square$

Exercises

11.29. (a) If X is a compact metric space, show that every nonempty subset admits a circumcenter.

(b) If X is a sphere (of any dimension ≥ 0), show that X has subsets with more than one circumcenter. In fact, there is even a subset such that *every* point of X is a circumcenter. (More generally, this happens whenever X is a metric space of finite diameter that admits a transitive group of isometries.)

11.30. Let V be a real or complex Hilbert space on which a group G acts by linear isometries. A 1-*cocycle* on G with values in V is a function $c \colon G \to V$ such that $c(gh) = c(g) + gc(h)$ for all $g, h \in G$. It is called a *coboundary* if there is a vector $v \in V$ such that $c(g) = gv - v$ for all $g \in G$. Deduce from Theorem 11.23 that a cocycle is a coboundary if and only if it is bounded.

11.4 Application: Bounded Subgroups

There is a classical application of Cartan's fixed-point theorem to the study of compact subgroups of a Lie group G: Under suitable hypotheses on G, one constructs a complete simply connected Riemannian manifold X of nonpositive curvature on which G acts as a group of isometries; the fixed-point theorem then implies that any compact subgroup of G must be contained in the stabilizer G_x of some point $x \in X$. If $G = \mathrm{SL}_n(\mathbb{R})$, for instance, then G acts transitively on the associated X and has the special orthogonal group $\mathrm{SO}_n(\mathbb{R})$ as one of the stabilizers. The conclusion, then, is that every compact subgroup of $\mathrm{SL}_n(\mathbb{R})$ is conjugate to a subgroup of $\mathrm{SO}_n(\mathbb{R})$. [*Note:* The group SO_n here is defined with respect to the standard Euclidean inner product on $\mathbb{R}^n$, not the "standard quadratic form" used in Section 6.7.1.] In this section we will use the Bruhat–Tits fixed-point theorem to prove similar results for groups acting on Euclidean buildings. These results then apply to certain "p-adic groups" such as $\mathrm{SL}_n(\mathbb{Q}_p)$.

Definition 11.31. Let G be a group with a BN-pair, and let $\Delta = \Delta(G, B)$ be the associated building. We say that the BN-pair is *Euclidean* if Δ is a Euclidean building.

Assume throughout this section that G is a group with a Euclidean BN-pair. It is then immediate from the definition of the metric on $X := |\Delta|$ that G acts as a group of isometries of X. In many cases G has a natural topology, so that the notion of compact subgroup makes sense. In general, however, it is more convenient to deal with "bounded" subgroups. We begin by figuring out what that should mean.

Lemma 11.32. *Let G be a group with a Euclidean BN-pair. The following conditions on a subset $F \subseteq G$ are equivalent:*

(i) *F is contained in a finite union of double cosets BwB.*

(ii) *For some $x \in X$, the set $Fx := \{gx \mid g \in F\}$ is a bounded subset of the metric space X.*

(iii) *For every bounded set $Y \subseteq X$, the set $FY := \bigcup_{y \in Y} Fy$ is a bounded subset of X.*

Proof. (i) $\implies$ (ii): It suffices to consider the case that F is a double coset BwB. Let C be the fundamental chamber of X; it is fixed pointwise by B. Let $\tilde{w}$ be a representative of w in N. Then for any $g = b\tilde{w}b' \in F$ and any $x \in \overline{C}$, we have

$$d(x, gx) = d(bx, gx) = d(x, \tilde{w}b'x) = d(x, wx) .$$

Hence Fx is contained in the sphere of radius $r := d(x, wx)$ centered at x.

(ii) $\implies$ (iii): This is left as an exercise; it is valid for any set of isometries of any metric space.

(iii) $\implies$ (i): By (iii) applied with Y equal to the fundamental chamber C, the set FC is bounded. So there is a bound on the metric distance between C and gC ($g \in F$), and it follows easily that there are only finitely many possibilities for the Weyl distance $\delta(C, gC)$. [All Weyl distances can be computed by means of galleries in a fixed bounded set in the model Coxeter complex $\Sigma(W, V)$. Such a set meets only finitely many chambers because of the local finiteness of the set of hyperplanes defining the simplicial decomposition.] Hence F is contained in a finite union of double cosets by the interpretation of the Bruhat decomposition given in Section 6.1.4.

Alternatively, we could argue as follows: Let E be an apartment containing C, and let $\rho = \rho_{E,C} \colon X \to E$. Since ρ is distance-decreasing, $\rho(FC)$ is a bounded subset of E. As above, this set meets only finitely many chambers. The interpretation of the Bruhat decomposition given in Exercise 6.18 now implies that F is contained in a finite union of double cosets. $\square$

Definition 11.33. We will call a set $F \subseteq G$ *bounded* if it satisfies the equivalent conditions of the lemma.

Exercises 11.40 and 11.41 below should convince you that this definition is reasonable.

We are now ready to apply the fixed-point theorem. Note that the stabilizers of the points of X are the same as the stabilizers of the nonempty simplices; hence they are the proper parabolic subgroups. In particular, the maximal elements among these stabilizers are the maximal (proper) parabolic subgroups, which are the stabilizers of the vertices. We will omit the word "proper" in what follows, since the notion of "maximal parabolic subgroup" would be of no interest otherwise.

Theorem 11.34. *Let G be a group with a Euclidean BN-pair. The following conditions on a subgroup $H \leq G$ are equivalent:*

(i) H *is bounded.*

(ii) H *fixes a point of X.*

(iii) H *fixes a vertex of X.*

(iv) H *is contained in a maximal parabolic subgroup.*

Proof. It is immediate that (iv) $\iff$ (iii) $\iff$ (ii) $\implies$ (i). The content of the theorem, then, is that (i) $\implies$ (ii), and this follows from the fixed-point theorem. $\square$

Corollary 11.35. *Every bounded subgroup is contained in a maximal bounded subgroup, and the maximal bounded subgroups are the maximal parabolic subgroups. G contains precisely $n+1$ conjugacy classes of maximal bounded subgroups, where $n := \dim X$; they are represented by the standard parabolic subgroups $BW_J B$ with $J = S \smallsetminus \{s\}$ for some $s \in S$.* $\square$

Remark 11.36. Suppose we are in the situation that G is a topological group and "bounded" is the same as "relatively compact" (e.g., $G = \mathrm{SL}_n(\mathbb{Q}_p)$). Then a maximal bounded subgroup is necessarily compact, since otherwise its closure would be a bigger bounded subgroup. Consequently, the corollary remains valid with "bounded" replaced by "compact."

We close this section by proving that the building Δ can be entirely reconstructed from the group G, viewed simply as a bornological group in the sense of Exercise 11.41 below. In particular, in the situation of Remark 11.36, Δ can be reconstructed from G as a topological group. The precise statement will be given in Theorem 11.38 below.

Recall that Δ can be identified with the poset of parabolic subgroups of G, ordered by reverse inclusion. Recall also that all buildings are flag complexes (Exercise 4.50). Hence Δ is the flag complex of the incidence geometry consisting of the maximal parabolic subgroups, with two maximal parabolics P, Q incident if and only if $P \cap Q$ contains a parabolic subgroup. [This says precisely that the corresponding vertices of Δ are joinable.] Since any subgroup of G containing a parabolic subgroup is itself parabolic, we can state this more simply: P and Q are incident if and only if $P \cap Q$ is parabolic.

Lemma 11.37. *If P and Q are distinct maximal parabolics, then $P \cap Q$ is parabolic if and only if $P \cap Q$ is a maximal (proper) subgroup of P.*

Proof. Let x (resp. y) be the vertex fixed by P (resp. Q). If $P \cap Q$ is parabolic, then x and y are joinable and $P \cap Q$ is the stabilizer of the edge A that they determine. Any subgroup P' with $P > P' > P \cap Q$ would be parabolic and would therefore correspond to a simplex A' with $\{x\} < A' < A$. So no such P' can exist, i.e., $P \cap Q$ is maximal in P.

Conversely, suppose that $P \cap Q$ is a maximal subgroup of P, and consider the geodesic $[x, y]$. It is fixed pointwise by $P \cap Q$, since the latter is a group of isometries fixing x and y. For any $x' \in (x, y)$ sufficiently close to x, the segment $(x, x']$ is contained in some simplex A of positive dimension having x

as a vertex; hence the stabilizer P' of x' (which is the same as the stabilizer of A) is properly contained in P. We therefore have $P > P' \geq P \cap Q$, which implies that $P \cap Q$ is equal to the parabolic subgroup P'.   □

We have now obtained the following description of Δ in terms of G:

Theorem 11.38. *The building Δ associated to a group G with a Euclidean BN-pair is isomorphic to the flag complex of the incidence geometry consisting of the maximal bounded subgroups of G, where two distinct such subgroups P, Q are incident if and only if $P \cap Q$ is a maximal subgroup of P.*   □

There is an analogous theorem about Lie groups. Under suitable hypotheses on a Lie group G, the associated manifold X of nonpositive curvature can be identified with the set of maximal compact subgroups of G, and the Riemannian-manifold structure on X depends only on G as a topological group.

Remark 11.39. Notice that we have made no essential use of the N in the BN-pair. In fact, we could simply have assumed that we were given a Tits subgroup B of G such that $\Delta(G, B)$ is Euclidean.

Exercises

11.40. Suppose $G = \mathrm{SL}_n(K)$ as in Section 6.9. Show that a set F is bounded if and only if there is an upper bound on the absolute values of the matrix entries of the elements of F. If K is complete and the residue field k is finite (e.g., $K = \mathbb{Q}_p$), show further that F is bounded if and only if it is relatively compact. Here G is topologized as a subset of the vector space of $n \times n$ matrices, and a set is called *relatively compact* if its closure is compact.

11.41. Show that our notion of "bounded set" satisfies the following conditions, which are the axioms for a *bornology* on a set:

(1) Every singleton is bounded.
(2) If $F' \subseteq F$ and F is bounded, then F' is bounded.
(3) A finite union of bounded sets is bounded.

Show further that the following two axioms for a *bornological group* are satisfied:

(4) If F_1 and F_2 are bounded, then so is their product $F_1 F_2$.
(5) If F is bounded, then so is F^{-1}.

11.42. Describe the n conjugacy classes of maximal bounded subgroups of $\mathrm{SL}_n(K)$.

11.5 Bounded Subsets of Apartments

We return to the study of a general Euclidean building $X = |\Delta|$, equipped with an arbitrary system of apartments. The theorem of this section is the analogue for Euclidean buildings of the fact that a spherical building admits a unique system of apartments, consisting of the convex hulls of pairs of opposite chambers (Theorem 4.70).

Recall that a subset $X' \subseteq X$ is called *convex* if it contains the geodesic $[x, y]$ connecting any two of its points x, y. If X' is a chamber subcomplex of X (by which we mean that it is the geometric realization of a chamber subcomplex of Δ), then it is convex if and only if it contains every minimal gallery joining any two of its chambers; see Exercise 11.21.

Given chambers C, C' of X, let $B(C, C')$ be the smallest convex subcomplex containing C and C'. In other words, $B(C, C')$ is the geometric realization of the combinatorial convex hull $\Gamma(C, C')$ discussed in Section 4.11.1 and illustrated in Figure 3.11. In particular, $B(C, C')$ is the union of the closed chambers that can occur in a minimal gallery from C to C'.

Theorem 11.43. *Let $\mathcal{B}$ be the collection of bounded subsets $Y \subseteq X$ such that Y is contained in an apartment. Then $\mathcal{B}$ is independent of the system of apartments $\mathcal{A}$. In fact, $\mathcal{B}$ consists of all subsets $Y \subseteq X$ such that Y is contained in $B(C, C')$ for some pair C, C' of chambers.*

The proof of the theorem requires a result about Euclidean Coxeter complexes analogous to Lemma 4.69. We need some terminology before we can state it.

Definition 11.44. Let $E = |\Sigma|$ be the geometric realization of a Euclidean Coxeter complex, and let $\mathcal{H}$ be the associated set of hyperplanes in E. Fix $x \in E$ and let $\overline{\mathcal{H}}$ be the set of hyperplanes through x and parallel to some element of $\mathcal{H}$. Then $\overline{\mathcal{H}}$ is finite (Theorem 10.8) and defines a decomposition of E into conical cells $\mathfrak{A}$ with x as the cone point. These cells will simply be referred to as *conical cells based at x*.

For example, the open sector bounded by the heavy lines in Figure 11.4 is a conical cell. The open rays bounding the sector are also conical cells, as is the singleton $\{x\}$.

Here is another description of the conical cells that we will often use: Choose an identification of Σ with the complex $\Sigma(W, V)$ associated to a Euclidean reflection group. This yields an identification of E with V. Let $\overline{W}$ be the finite reflection group consisting of the linear parts of the elements of W. The conical cells based at x, then, are simply the translates $\mathfrak{A} = x + \mathfrak{D}$, where $\mathfrak{D}$ is a cell associated to $\overline{W}$. We will call $\mathfrak{D}$ the *direction* of $\mathfrak{A}$.

Definition 11.45. If the $\overline{W}$-cell $\mathfrak{D}$ is a chamber (hence a simplicial cone), then the conical cell $x + \mathfrak{D}$ will be called a *sector* ("quartier" in [59]).

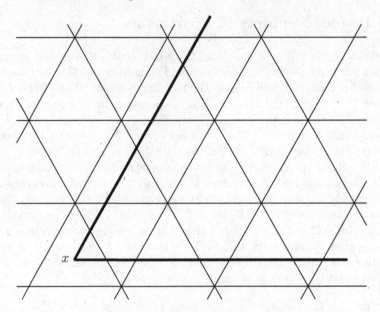

Fig. 11.4. Conical cells based at x.

If the n walls of $\mathfrak{D}$ are defined by linear equations $f_i = 0$, where $f_i > 0$ on $\mathfrak{D}$, then a sector $\mathfrak{C}$ with direction $\mathfrak{D}$ is given by linear inequalities of the form $f_i > c_i$ $(i = 1, \ldots, n)$. It is clear from this that the intersection of two sectors with direction $\mathfrak{D}$ is again a sector with direction $\mathfrak{D}$. See Figure 11.5.

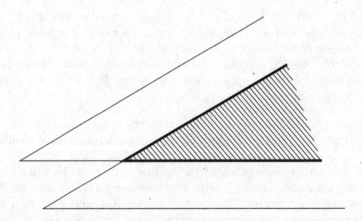

Fig. 11.5. The intersection of two sectors with the same direction.

If $\mathfrak{C}$ and $\mathfrak{C}'$ are sectors with $\mathfrak{C}' \subseteq \mathfrak{C}$, then we will say that $\mathfrak{C}'$ is a *subsector* of $\mathfrak{C}$. Note that $\mathfrak{C}$ and $\mathfrak{C}'$ then necessarily have the same direction. For suppose $\mathfrak{C} = x + \mathfrak{D}$ and $\mathfrak{C}' = x' + \mathfrak{D}'$. Letting $\mathfrak{D}$ be defined by inequalities $f_i > 0$ as above, we conclude that the f_i are bounded below on $\mathfrak{D}'$; hence no f_i can be negative on the cone $\mathfrak{D}'$. Thus $f_i > 0$ on $\mathfrak{D}'$ for all i, which implies that $\mathfrak{D}' \subseteq \mathfrak{D}$ and hence $\mathfrak{D}' = \mathfrak{D}$.

Consider now two sectors $\mathfrak{C}_1 := x + \mathfrak{D}$ and $\mathfrak{C}_2 := y - \mathfrak{D}$ having opposite directions $\pm\mathfrak{D}$. Let $\overline{\mathfrak{C}}_1$ and $\overline{\mathfrak{C}}_2$ be the closures $x + \overline{\mathfrak{D}}$ and $y - \overline{\mathfrak{D}}$. Assume that $x \in \overline{\mathfrak{C}}_2$ and $y \in \overline{\mathfrak{C}}_1$, so that the two closed sectors $\overline{\mathfrak{C}}_1, \overline{\mathfrak{C}}_2$ overlap, as in Figure 11.6.

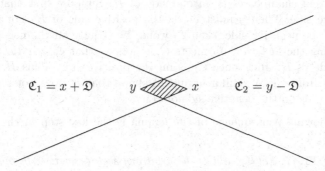

$\mathfrak{C}_1 = x + \mathfrak{D}$ y x $\mathfrak{C}_2 = y - \mathfrak{D}$

Fig. 11.6. Two sectors with opposite directions.

We will show that if C_1 and C_2 are chambers that are "sufficiently far out" in $\mathfrak{C}_1$ and $\mathfrak{C}_2$, respectively, then $B(C_1, C_2)$ contains the overlap $\overline{\mathfrak{C}}_1 \cap \overline{\mathfrak{C}}_2$. Let $\mathfrak{C}_1'$ be the subsector of $\mathfrak{C}_1$ based at y and let $\mathfrak{C}_2'$ be the subsector of $\mathfrak{C}_2$ based at x, as indicated by the dotted lines in Figure 11.7; in other words, $\mathfrak{C}_1' = y + \mathfrak{D}$ and $\mathfrak{C}_2' = x - \mathfrak{D}$.

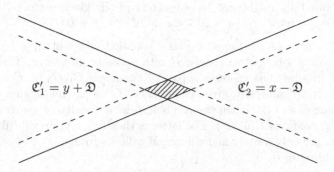

$\mathfrak{C}_1' = y + \mathfrak{D}$ $\mathfrak{C}_2' = x - \mathfrak{D}$

Fig. 11.7. Subsectors.

Lemma 11.46. *With the notation above, suppose C_1 and C_2 are chambers in E such that $\overline{C}_1$ meets $\mathfrak{C}_1'$ and $\overline{C}_2$ meets $\mathfrak{C}_2'$. If C is any chamber of E such that $\overline{C}$ meets $\overline{\mathfrak{C}}_1 \cap \overline{\mathfrak{C}}_2$, then $C \subseteq B(C_1, C_2)$.*

(*Note*: Since sectors are open sets, the hypothesis that $\overline{C}_i$ meets $\mathfrak{C}'_i$ implies that C_i meets $\mathfrak{C}'_i$.)

Proof. We must show that no wall (i.e., element of $\mathcal{H}$) separates C from both C_1 and C_2. Let H be a wall, defined by a linear equation $f = c$. We may choose f such that $f > 0$ on $\mathfrak{D}$, in which case we will say that the closed half-space $f \geq c$ (resp. $f \leq c$) is the positive (resp. negative) side of H. (In Figures 11.6 and 11.7, think of $\mathfrak{C}_1$ and $\mathfrak{C}'_1$ as opening in the positive direction.)

The closed chamber $\overline{C}$ is on one side of H. Suppose first that it is on the positive side. Then y must be on the positive side of H. For if y were strictly on the negative side, then $\overline{\mathfrak{C}}_2$ would be strictly on the negative side, contradicting the fact that $\overline{\mathfrak{C}}_2$ meets $\overline{C}$. It follows that $\mathfrak{C}'_1$ is strictly on the positive side of H, and hence C_1 is on the positive side. Thus H does not separate C from C_1. A similar argument shows that H does not separate C from C_2 if C is on the negative side of H. □

The following consequence of the lemma is the key step in the proof of Theorem 11.43:

Corollary 11.47. *Let $\mathfrak{C}_1$ and $\mathfrak{C}_2$ be arbitrary sectors in E with opposite directions. Given any bounded subset Y of E, there are subsectors $\mathfrak{C}'_1 \subseteq \mathfrak{C}_1$ and $\mathfrak{C}'_2 \subseteq \mathfrak{C}_2$ with the following property: If C_1 and C_2 are chambers in E such that $\overline{C}_1$ meets $\mathfrak{C}'_1$ and $\overline{C}_2$ meets $\mathfrak{C}'_2$, then $B(C_1, C_2)$ contains Y.*

Proof. Let $\mathfrak{D}$ be the direction of $\mathfrak{C}_1$, so that $-\mathfrak{D}$ is the direction of $\mathfrak{C}_2$. Observe first that we can find sectors $x + \mathfrak{D}$ and $y - \mathfrak{D}$ as in Lemma 11.46, with $Y \subseteq \mathfrak{C}_1 \cap \mathfrak{C}_2$. In fact, with the notation above, we need only choose constants c_i, c'_i ($i = 1, \ldots, n$) such that $c_i < f_i < c'_i$ on Y for all i. The lemma therefore implies that the subsectors $y + \mathfrak{D}$ and $x - \mathfrak{D}$ have the property stated in the corollary, but they might not be subsectors of the given sectors $\mathfrak{C}_1, \mathfrak{C}_2$. To achieve this, we set $\mathfrak{C}'_1 := (y + \mathfrak{D}) \cap \mathfrak{C}_1$ and $\mathfrak{C}'_2 := (x - \mathfrak{D}) \cap \mathfrak{C}_2$. □

Proof of Theorem 11.43. Suppose Y is a bounded subset of an apartment E. Take an arbitrary pair of sectors in E with opposite directions. Then Corollary 11.47 implies that there is a pair of chambers C_1, C_2 in E such that $Y \subseteq B(C_1, C_2)$. Conversely, given chambers C, C' of X, choose an apartment E containing C and C'. Then the combinatorial convexity of apartments implies that E contains $B(C, C')$. The latter is therefore a bounded subset of E [since there are only finitely many minimal galleries from C to C' in E]; hence so is any subset of it. □

We close this section with a variant of Corollary 11.47 that will be useful later.

Lemma 11.48. *Let E be the geometric realization of a Euclidean Coxeter complex, and let C and D be chambers of E. Then there are sectors $\mathfrak{C}_1, \mathfrak{C}_2$ in E with the following property: For any subsectors $\mathfrak{C}'_1 \subseteq \mathfrak{C}_1$ and $\mathfrak{C}'_2 \subseteq \mathfrak{C}_2$, there is a gallery that starts at a chamber meeting $\mathfrak{C}'_1$, ends at a chamber meeting $\mathfrak{C}'_2$, and passes through both C and D (in that order).*

Proof. Choose a point $x \in D$ and a direction $\mathfrak{D}$ such that $x + d$ belongs to C for some $d \in \mathfrak{D}$. Let $\mathfrak{C}_1$ be a sector with direction $\mathfrak{D}$, and let $\mathfrak{C}_2$ be a sector with direction $-\mathfrak{D}$. Then any subsector $\mathfrak{C}_1' \subseteq \mathfrak{C}_1$ contains $x + td$ for sufficiently large $t > 0$, and any subsector $\mathfrak{C}_2' \subseteq \mathfrak{C}_2$ contains $x - td$ for all sufficiently large $t > 0$. We can therefore find points $y_i \in \mathfrak{C}_i'$ $(i = 1, 2)$ such that the line segment $[y_1, y_2]$ passes through both C and D. Moving y_1 and y_2 slightly if necessary, we may assume that they are contained in open chambers of E and that the line segment never crosses two walls of E simultaneously. The successive chambers that it passes through therefore form the desired gallery. $\square$

Exercises

In these exercises E continues to denote the geometric realization of a Euclidean Coxeter complex.

11.49. Take $x = y$ in Lemma 11.46, so that $\overline{\mathfrak{C}}_1$ and $\overline{\mathfrak{C}}_2$ meet only at the basepoint x. Deduce that $B(C_1, C_2)$ contains a neighborhood of x in E if C_1 meets $\mathfrak{C}_1$ and C_2 meets $\mathfrak{C}_2$.

11.50. Let H be a wall and $\mathfrak{C}$ a sector in E. Show that one of the roots of E determined by H contains a subsector of $\mathfrak{C}$.

11.51. Fill in the missing details in the proof of Lemma 11.48.

11.6 A Metric Characterization of the Apartments

Assume now that $\mathcal{A}$ is the complete system of apartments in our Euclidean building $X = |\Delta|$. In view of Section 11.5, of course, anything we say concerning bounded subsets of apartments will then apply to an arbitrary apartment system. Recall that we stated two characterizations of the apartments as simplicial complexes in Remark 4.56, and we characterized them as W-metric spaces in Section 5.5.2. We now characterize them as ordinary metric spaces. The ideas in this section are very similar to those in Section 5.5.

Theorem 11.52. *Let $n = \dim X$. Then a subset $E \subseteq X$ is an apartment if and only if E is isometric to $\mathbb{R}^n$.*

Another way to say this is that (a) a subset isometric to $\mathbb{R}^n$ is necessarily a subcomplex, and (b) the collection of all such subcomplexes is a system of apartments. These assertions can be viewed as generalizations to arbitrary X of elementary facts about trees. We will deduce Theorem 11.52 from the more precise Theorem 11.53 below.

Theorem 11.53. *Let Y be a subset of X. Assume either that Y is convex or that Y has nonempty interior. If Y is isometric to a subset of $\mathbb{R}^n$, then Y is contained in an apartment.*

To deduce Theorem 11.52 from Theorem 11.53, suppose E is isometric to $\mathbb{R}^n$. Then E is easily seen to be convex in X; this follows from the characterization of geodesics $[x, y]$ given in Theorem 11.16. So Theorem 11.53 implies that E is contained in an apartment E'. But E' cannot be isometric to a proper subset of itself, so E must be the entire apartment E'.

The rest of this section will be devoted to the proof of Theorem 11.53. Choose a type function on Δ with values in a set S, let W be the Weyl group, and let $E := |\Sigma(W, S)|$ be the "model apartment." We will make heavy use of certain canonical maps $\rho_D \colon X \to E$. These were discussed in Exercises 4.51 and 4.94, but we repeat the relevant facts.

For any chamber D of X there is a canonical type-preserving chamber map $\rho = \rho_D \colon X \to E$ such that $\rho(D)$ is the fundamental chamber of E and ρ maps any apartment containing D isomorphically onto E. It can be described in terms of retractions onto apartments: If E_1 is an apartment containing D and $\iota \colon E \to E_1$ is the unique type-preserving isomorphism that takes the fundamental chamber of E to D, then the composite $\iota\rho$ is the retraction $\rho_{E,D} \colon X \to E_1$. This observation, together with Theorem 11.16(3), implies that ρ_D is distance-decreasing and preserves distances from points of $\overline{D}$, i.e., $d(\rho_D(x), \rho_D(y)) \leq d(x, y)$ for $x, y \in X$, with equality if $x \in \overline{D}$.

On the level of chambers, we can describe ρ by $\rho(D') = \delta(D, D') \in W = \mathcal{C}(E)$ for any chamber D' of X. Since the fundamental chamber of E is the identity element of W, we can also write this as

$$\rho(D') = wC \,, \tag{11.6}$$

where C is the fundamental chamber of E and $w = \delta(D, D')$.

For any chambers D of X and C of E (not necessarily the fundamental chamber) let $\tau_{D,C} \colon \overline{D} \to \overline{C}$ be the unique type-preserving simplicial isomorphism. One can easily check that it is an isometry. [We may assume that C is the fundamental chamber of E, in which case $\tau_{D,C}$ is the restriction to $\overline{D}$ of ρ_D; the latter maps any apartment containing D isometrically onto E.] In case C is the fundamental chamber of E, we will write τ_D instead of $\tau_{D,C}$.

Lemma 11.54. *Let Y be a subset of X that contains a nonempty open subset U of a chamber D. If Y is isometric to a subset of $\mathbb{R}^n$, then there is a unique isometry α from Y into E such that $\alpha|_U = \tau_D|_U$. Moreover, $\alpha = \rho|_Y$, where $\rho = \rho_D \colon X \to E$.*

Proof. Suppose first that there exists an isometry α from Y into E such that $\alpha|_U = \tau|_U$, where $\tau := \tau_D$. Then $\rho\alpha^{-1} \colon \alpha(Y) \to E$ fixes the open set $\tau(U) = \alpha(U)$ pointwise and preserves distances from points of $\tau(U)$. Hence $\rho\alpha^{-1} = \mathrm{id}_{\alpha(Y)}$ by Exercise 10.1. This proves the last assertion of the lemma, as well as the uniqueness of α. To prove existence, start with an arbitrary isometry β from Y into E. Then $\beta(U)$ and $\tau(U)$ are isometric subsets of E, and the isometry $\tau\beta^{-1} \colon \beta(U) \to \tau(U)$ extends to an isometry $\gamma \colon E \to E$ by Exercise 10.5 (or [48, Chapter I.2, Proposition 2.20]). So we may take $\alpha = \gamma\beta$. $\qquad\square$

The next lemma is the crucial step in the proof.

Lemma 11.55. *Let Y be a subset of X that contains a closed chamber $\overline{D}$, and suppose there is an isometry α from Y into E such that α maps $\overline{D}$ onto a closed chamber $\overline{C}$ of E by the map $\tau_{D,C}$. Let C' be a chamber of E adjacent to C. Then there is a chamber D' adjacent to D such that α extends to an isometry from $Y \cup \overline{D'}$ into E taking $\overline{D'}$ to $\overline{C'}$ by the map $\tau_{D',C'}$.*

Proof. We may assume that C is the fundamental chamber of E, in which case $\alpha = \rho|_Y$ by Lemma 11.54, where $\rho = \rho_D \colon X \to E$. Then $C' = sC$ for some fundamental reflection $s \in S$. We will take D' to be a suitably chosen chamber D' of X such that $\delta(D', D) = s$. Consider, for the moment, an arbitrary such D', and let $\rho' \colon X \to E$ be the composite $s\rho_{D'}$. [In the terminology of Exercise 4.51, ρ' is the canonical map taking D' to C'.] We will show that D' can be chosen such that $\rho'|_Y = \rho|_Y$ ($= \alpha$). This will yield the lemma. For then $\rho'|_{(Y \cup \overline{D'})}$ will extend α, will be $\tau_{C',D'}$ on $\overline{D'}$, and will be an isometry because ρ' preserves distances from points of $\overline{D'}$.

We claim that ρ and ρ' are related as follows: Let H be the wall of E separating C from C', i.e., the wall fixed by the reflection s. Let U be the open half-space bounded by H and containing C'. Then for any $x \in X$ we have $\rho(x) = \rho'(x)$ or $s\rho'(x)$. We always have $\rho'(x) = \rho(x)$ except possibly if $\rho(x) \in U$. In this case there is a choice of D' (depending on x) such that $\rho'(x) = \rho(x)$.

Given the claim, we can choose D' such that $\rho'(y) = \rho(y)$ for some $y \in Y \cap \rho^{-1}(U)$ (unless the latter is empty, in which case we are already done). We will show that we then have $\rho'(y) = \rho(y)$ ($= \alpha(y)$) for all $y \in Y \cap \rho^{-1}(U)$. Let $Z = \rho(Y) \cap U$ and let $f = \rho'\alpha^{-1} \colon Z \to E$. Then f is distance-decreasing, $f(z) = z$ or sz for all $z \in Z$, and $f(z) = z$ for at least one $z \in Z$. We want to conclude that $f(z) = z$ for all $z \in Z$. To this end we use the fact that $d(z, sz') > d(z, z')$ for any $z, z' \in U$. [This has a proof similar to that of the analogous combinatorial fact, given in Exercise 3.57. Namely, consider the line segment $[z, sz']$, and fold it back onto $\overline{U}$ to obtain a path from z to z' that has the same length but is not straight.] Since $f(z) = z$ for some $z \in Z$, this fact implies that $f(z') = z'$ for all $z' \in Z$, as required; for otherwise we would have $f(z') = sz'$ for some z', contradicting the fact that f is distance-decreasing.

It remains to prove the claim. Let D'' be an arbitrary chamber of X, let $w = \delta(D, D'')$, and let $w' = \delta(D', D'')$:

Using equation (11.6) twice, we see that $\rho(D'') = wC$ and $\rho'(D'') = sw'C$. Recall now that $w' = sw$ or w; hence $\rho'(D'') = \rho(D'')$ or $s\rho(D'')$. More precisely, if $l(sw) = l(w) + 1$, then $w' = sw$ (hence $\rho'(D'') = \rho(D'')$); if $l(sw) = l(w) - 1$, on the other hand, then there is a unique choice of D' (depending on D'') for which $w' = sw$. The claim follows at once, as soon as one recalls that $l(sw) = l(w) - 1$ if and only if wC ($= \rho(D'')$) is in U. $\square$

Lemma 11.56. *If Y is a subset of X that contains a closed chamber and is isometric to a subset of $\mathbb{R}^n$, then Y is contained in an apartment.*

(This is a special case of the theorem we are trying to prove; for a closed chamber has nonempty interior by Exercise 11.20.)

Proof. Let $\overline{D}$ be a closed chamber contained in Y. By Lemma 11.54 we can find an isometry α from Y into E that maps $\overline{D}$ onto the closed fundamental chamber of E by the map τ_D. By repeated applications of Lemma 11.55 we can successively adjoin closed chambers to $\alpha(Y)$ and extend α^{-1} to an isometry β from E into X that is simplicial on each closed chamber of E. But then β is a chamber map, and its image $\beta(E)$ is an apartment containing Y by Proposition 4.59. $\square$

Proof of Theorem 11.53. In view of Lemma 11.56, it suffices to show that we can enlarge the given Y to a set that contains a closed chamber and is still isometric to a subset of $\mathbb{R}^n$. Suppose first that Y has nonempty interior. Then Y contains a nonempty open subset of a chamber D, and Lemma 11.54 implies that $\rho = \rho_D$ maps Y isometrically into E. But then ρ also maps $Y \cup \overline{D}$ isometrically into E, since ρ preserves distances from points of $\overline{D}$. So we are done in this case.

Suppose now that Y is convex. Choose a simplex A that is maximal among the simplices meeting Y, and let D be a chamber having A as a face. We claim that $\rho = \rho_D$ maps $Y \cup \overline{D}$ isometrically into E. As above, it suffices to show that $\rho|_Y$ is an isometry. For this purpose it will be convenient to replace the abstract Coxeter complex E by an apartment containing D and to replace ρ by the retraction onto that apartment centered at D.

We must show that $d(\rho(y), \rho(z)) = d(y, z)$ for all $y, z \in Y$. We may assume $y, z \notin \overline{D}$. Choose $x \in Y \cap A$, and let $T \subseteq Y$ be the convex hull of $\{x, y, z\}$. Note that any isometry α from Y into $\mathbb{R}^n$ must take T to the convex hull of $\{\alpha(x), \alpha(y), \alpha(z)\}$; this follows from the characterization of geodesics in Theorem 11.16. So T is, in an obvious sense, a *Euclidean triangle*. Since $y \neq x$ and $z \neq x$, it follows that there is a well-defined angle θ at the vertex x, with $0 \leq \theta \leq \pi$.

If we take any $y' \in (x, y]$ and $z' \in (x, z]$, then the triangle T' determined by $\{x, y', z'\}$ has the same angle at x. In particular, we will take y' and z' close enough to x that they are in A and hence in $\overline{D}$. [This is possible because of the maximality of A; for if $[x, y]$, say, does not stay in A for a little while, then it enters a simplex having A as a proper face.] Now ρ maps $[x, y]$ (resp. $[x, z]$) isometrically onto $[x, \rho(y)]$ (resp. $[x, \rho(z)]$) and fixes T'. The angle θ is therefore equal to the angle between $[x, \rho(y)]$ and $[x, \rho(z)]$. By elementary geometry ("two sides and the included angle") we conclude that T is congruent to the triangle in E with vertices $x, \rho(y), \rho(z)$. Hence $d(\rho(y), \rho(z)) = d(y, z)$, as required.   □

Exercises

11.57. Extract the following fact from the proof above: Given $x, y, z \in X$ with $x \neq y$ and $x \neq z$, there is a well-defined angle θ between $[x, y]$ and $[x, z]$. [One can actually define angles in much greater generality; see Bridson–Haefliger [48].]

11.58. With x, y, z, θ as in Exercise 11.57, prove the following *cosine inequality*:

$$d^2(y, z) \geq d^2(x, y) + d^2(x, z) - 2d(x, y)d(x, z)\cos\theta \,,$$

with equality if and only if $\{x, y, z\}$ is contained in an apartment. This reinforces the nonpositive curvature intuition again: Two geodesics emanating from x tend to separate at least as fast as they would in Euclidean space.

11.59. Let $x, y, z \in X$ be three distinct points. Suppose that the union $[y, x] \cup [x, z]$ is locally a geodesic at x, in the sense that $[y', x] \cup [x, z'] = [y', z']$ for some $y' \in [y, x]$ and $z' \in (x, z]$. Show that $[y, x] \cup [x, z] = [y, z]$.

11.60. If Δ is thick, show that the simplicial decomposition of $X = |\Delta|$ is completely determined by the metric.

11.7 Construction of Apartments

We continue to denote by X a Euclidean building, equipped with its complete system of apartments. As an illustration of our techniques for constructing apartments (Theorems 11.53 and 5.73), we will prove two results asserting the existence of apartments containing given subsets of X. The results are stated as the two parts of Theorem 11.63 below.

Definition 11.61. By a *conical cell* in X we will mean a subset $\mathfrak{A}$ that is contained in some apartment E and is a conical cell in E in the sense of Definition 11.44. Similarly, a *sector* in X is a subset $\mathfrak{C}$ that is contained in some apartment E and is a sector in E in the sense of Definition 11.45. Equivalently, a sector is a conical cell of maximal dimension.

If $\mathfrak{A}$ is a conical cell (or sector) as just defined, then $\mathfrak{A}$ is a conical cell (or sector) in *any* apartment E' that contains it:

Proposition 11.62. *If $\mathfrak{A}$ is a conical cell in some apartment E, then $\mathfrak{A}$ is a conical cell in every apartment E' that contains it.*

Proof. This follows easily from the fact that there is an isomorphism $E \to E'$ fixing $E \cap E'$ pointwise (Proposition 4.101). In the present Euclidean context, we can give an alternative proof of that proposition as follows. Take an isomorphism $\phi\colon E \to E'$ that fixes pointwise a maximal simplex A of the subcomplex $E \cap E'$. We will show that ϕ fixes $E \cap E'$ pointwise. Choose any $x \in A$. Given $y \neq x$ in $E \cap E'$, consider the geodesic $[x, y]$. It is contained in $E \cap E'$, and a nontrivial initial segment of it is contained in A by maximality; hence $[x, y]$ and its image $[x, \phi(y)]$ have a common initial segment. But these are geodesics of the same length in the Euclidean space E, so they must coincide. In particular, $\phi(y) = y$. □

Let's return now to sectors, which are the only conical cells that will concern us in this section.

Theorem 11.63.

(1) *Given a sector $\mathfrak{C}$ and a chamber C in X, there is an apartment containing C and a subsector of $\mathfrak{C}$.*
(2) *Given two sectors $\mathfrak{C}_1$ and $\mathfrak{C}_2$, there is an apartment containing subsectors of $\mathfrak{C}_1$ and $\mathfrak{C}_2$.*

(Think about trees to see why the theorem is plausible; a sector, in this case, is simply a ray tending toward an "end" of the tree.)

The crux of the proof of (1) is the following result. Recall that we are now using $\mathbf{d}(-, -)$ to denote combinatorial distances between chambers.

Lemma 11.64. *Let $\mathfrak{C}$ be a sector in an apartment E, and let C be a chamber in X. Then we can find a subsector $\mathfrak{C}' \subseteq \mathfrak{C}$ and a chamber C_0 in E such that the retraction $\rho = \rho_{E, C_0}\colon X \to E$ has the following property: For any chamber C' of E that meets $\mathfrak{C}'$,*

$$\mathbf{d}(C', \rho(C)) = \mathbf{d}(C', C) . \tag{11.7}$$

Proof. Note first that there is a bounded subset Z of E such that for any choice of C_0, we will have $\rho(C) \subseteq Z$. In fact, let z be any point of E and let Y be any ball in X centered at z and containing C; then we can take $Z = Y \cap E$. [This works because our retractions are distance-decreasing.] Let $\mathfrak{C}''$ be a sector in E containing Z and having direction opposite to that of $\mathfrak{C}$. We can choose $\mathfrak{C}''$ such that its basepoint x is in $\mathfrak{C}$. Now let $\mathfrak{C}'$ be the subsector of $\mathfrak{C}$ based at x and let C_0 be any chamber of E whose closure contains x.

Consider any chamber C' of E that meets $\mathfrak{C}'$. Applying Lemma 11.46 to the two opposite sectors $\mathfrak{C}', \mathfrak{C}''$, and recalling that $\rho(C) \subseteq Z \subseteq \mathfrak{C}''$, we get

$$\begin{aligned} \mathbf{d}(C', \rho(C)) &= \mathbf{d}(C', C_0) + \mathbf{d}(C_0, \rho(C)) \\ &= \mathbf{d}(C', C_0) + \mathbf{d}(C_0, C) . \end{aligned} \tag{11.8}$$

If we now take a minimal gallery from C' to C_0 and compose it with a minimal gallery from C_0 to C, we get a gallery Γ from C' to C whose image under ρ is minimal; hence Γ is minimal and (11.7) holds. □

In the proof that follows we will have occasion to speak of isometries in the sense of ordinary metric spaces as well as isometries in the W-metric sense (Definition 5.59). To avoid confusion, we will refer to the latter as *Weyl isometries*.

Proof of Theorem 11.63(1). Let $\mathfrak{C}'$ and ρ be as in the lemma. There are two ways, both of which are instructive, to deduce that there is an apartment containing $\mathfrak{C}'$ and C.

Method 1: Let C' be the set of chambers of E that meet $\mathfrak{C}'$. The lemma, together with the fact that ρ is type-preserving, implies that $C' \cup \{C\}$ is Weyl isometric to a subset of $\mathcal{C}(E)$. So there is an apartment containing it by Theorem 5.73. Since the closures of the chambers in C' cover $\mathfrak{C}'$, this proves (1).

Method 2: In view of Theorem 11.53, it suffices to show that $\rho\colon X \to E$ is an isometry on $\mathfrak{C}' \cup C$. And this will follow if we show that ρ is an isometry on $\overline{C'} \cup C$ for any chamber C' that meets $\mathfrak{C}'$. Now (11.7) implies that $\rho(C) = \rho_{E,C'}(C)$, so the result follows from the fact that $\rho_{E,C'}$ is an isometry on $\overline{C'} \cup C$. $\qquad \square$

Exercise 11.65. Give an alternative proof of Theorem 11.63(1) based on Exercises 5.83 and 11.50.

As a consequence of (1) we can define a new kind of *retraction* onto an apartment, which will be useful in the proof of Theorem 11.63(2). Given an apartment E and a sector $\mathfrak{C}$ in E, (1) implies that X is the union of the apartments E' that contain a subsector of $\mathfrak{C}$. We now define $\rho = \rho_{E,\mathfrak{C}}\colon X \to E$ to be the map whose restriction to any such E' is the isomorphism $\phi_{E'}\colon E' \to E$ that fixes $E \cap E'$ pointwise. It is easy to check, as in the construction of the "ordinary" retractions $\rho_{E,C}$ in Chapter 4, that ρ is well defined.

Note, for future reference, the following property of $\rho = \rho_{E,\mathfrak{C}}$: For any chamber C of X there is a subsector $\mathfrak{C}'$ of $\mathfrak{C}$ such that

$$\rho(C) = \rho_{E,C'}(C) \tag{11.9}$$

for any chamber C' of E that meets $\mathfrak{C}'$. Indeed, if we take $\mathfrak{C}'$ to be a subsector such that $\mathfrak{C}'$ and C are contained in an apartment E', then both sides of the equality to be proved are equal to $\phi_{E'}(C)$. We can also formulate (11.9) in terms of Weyl distances:

$$\delta(C', \rho(C)) = \delta(C', C) \tag{11.10}$$

for any chamber C' of E that meets $\mathfrak{C}'$, i.e., ρ is a Weyl isometry on $C' \cup \{C\}$, where C' is the set of chambers in E that meet $\mathfrak{C}'$. Before proceeding further, the reader might find it useful to look at Exercises 11.66–11.68 below to get some intuition about this new kind of retraction.

We will use retractions of the form $\rho = \rho_{E,\mathfrak{C}}$ in the proof of part (2) of the theorem. More precisely, we will choose an apartment E_1 containing $\mathfrak{C}_1$, and we will find subsectors $\mathfrak{C}'_i$ of $\mathfrak{C}_i$ ($i = 1, 2$) such that $\rho = \rho_{E_1,\mathfrak{C}_1}$ is an isometry on $\mathfrak{C}'_1 \cup \mathfrak{C}'_2$ (or a Weyl isometry on the set of chambers meeting this union). The key step in the proof is to figure out how to choose $\mathfrak{C}'_2$. Consider, for example, the case that X is a tree. Figure 11.8 shows a typical configuration. Note that as we move out along the ray $\mathfrak{C}_2$, the image under ρ heads toward

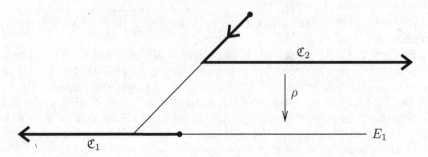

Fig. 11.8. The retraction onto E_1 based at $\mathfrak{C}_1$.

the same end of E_1 as $\mathfrak{C}_1$ at first, but then it reverses direction and heads toward the opposite end. So if we want to find a subsector $\mathfrak{C}'_2$ on which ρ is an isometry, we need only start $\mathfrak{C}'_2$ at any chamber C in $\mathfrak{C}_2$ such that $\rho(C)$ is already heading in the opposite direction from $\mathfrak{C}_1$.

It turns out that a similar idea works in higher dimensions also. Here are the details:

Proof of Theorem 11.63(2). Choose apartments E_1 and E_2 containing $\mathfrak{C}_1$ and $\mathfrak{C}_2$, respectively, and let $\rho = \rho_{E_1,\mathfrak{C}_1}$ as above. Identify E_1 with a vector space V as in Section 11.5, so that $\mathfrak{C}_1$ has a direction $\mathfrak{D}$; the latter is a chamber in the Coxeter complex associated to a finite reflection group $\overline{W}$. We will use combinatorial distances $\mathbf{d}(-, -)$ in this Coxeter complex in order to compare directions. The intuitive idea to keep in mind is that the bigger $\mathbf{d}(\mathfrak{D}, \mathfrak{D}')$ is, the more nearly opposite $\mathfrak{D}$ and $\mathfrak{D}'$ are.

We associate to any chamber C of E_2 a $\overline{W}$-chamber $\mathfrak{D}'$ in E_1, which we think of as the *direction of* $\rho(\mathfrak{C}_2)$ *at* $\rho(C)$, as follows: $\mathfrak{D}'$ is the direction of the sector $\phi(\mathfrak{C}_2)$ in E_1, where $\phi \colon E_2 \to E_1$ is the type-preserving isomorphism taking C to $\rho(C)$. Alternatively, we can characterize $\mathfrak{D}'$ as follows: Choose a directed line segment $\overrightarrow{xy}$ in C that is *parallel* to $\mathfrak{C}_2$, in the sense that it is a translate of a segment going from the cone point of $\mathfrak{C}_2$ to some point of $\mathfrak{C}_2$. Then $\mathfrak{D}'$ is the unique $\overline{W}$-chamber such that $\rho(\overrightarrow{xy})$ is parallel to $\mathfrak{D}'$.

Let's focus now on those chambers of E_2 that meet $\mathfrak{C}_2$. Choose among these a chamber C_0 such that the resulting direction $\mathfrak{D}'$ makes $\mathbf{d}(\mathfrak{D}, \mathfrak{D}')$ as big as possible. Such a C_0 exists because $\mathbf{d}(\mathfrak{D}, \mathfrak{D}')$ is bounded by the diameter of the spherical Coxeter complex associated to $\overline{W}$. Let x_0 be any point in $C_0 \cap \mathfrak{C}_2$, and let $\mathfrak{C}'_2$ be the subsector of $\mathfrak{C}_2$ based at x_0. As we noted above while defining ρ, we can find a subsector $\mathfrak{C}'_1 \subseteq \mathfrak{C}_1$ such that $\rho(C_0) = \rho_{E_1,C}(C_0)$ for any chamber C that meets $\mathfrak{C}'_1$. Passing to a further subsector if necessary, we can also arrange that $\mathfrak{C}_1' \subseteq \rho(x_0) + \mathfrak{D}$. We will show that ρ is an isometry on $\mathfrak{C}_1' \cup \mathfrak{C}_2'$. In view of Theorem 11.53, this will complete the proof. [Note: The interested reader can easily make minor changes in what follows so that Theorem 5.73 can be applied instead of Theorem 11.53.]

Let C' be any chamber of E_2 that meets $\mathfrak{C}_2'$. Choose $x \in C' \cap \mathfrak{C}_2'$, and consider the directed line $\overrightarrow{x_0 x}$. It crosses exactly those walls $H_1, \dots, H_l$ of E_2 that separate C_0 from C'. By moving x slightly, if necessary, we can make sure that $\overrightarrow{x_0 x}$ does not simultaneously cross two walls. For if $\overrightarrow{x_0 x}$ meets $H_i \cap H_j$ for some $i \neq j$, then x is in the affine span of x_0 and $H_i \cap H_j$; this affine span is a hyperplane, so we need only choose x so as to miss finitely many hyperplanes. With such a choice of x, then, $\overrightarrow{x_0 x}$ passes through chambers $C_0, \dots, C_l = C'$ that form a minimal gallery from C_0 to C'.

We claim that ρ maps $C_0, \dots, C_l$ to a gallery (not just a pregallery) in E_1 and that in addition, $\rho(C_i) = \rho_{E_1, C}(C_i)$ for each $i = 1, \dots, l$ and any chamber C that meets $\mathfrak{C}_1'$. Once the claim is proved, we will be done. For the first assertion of the claim implies [by the standard uniqueness argument] that ρ coincides on $\mathfrak{C}_2'$ with the type-preserving isomorphism $E_2 \to E_1$ taking C_0 to $\rho(C_0)$. Hence ρ is an isometry on $\mathfrak{C}_2'$. And the second assertion of the claim implies that ρ preserves distances between points of $\mathfrak{C}_1'$ and points of $\mathfrak{C}_2'$. It remains to prove the claim.

Arguing by induction on l, we may assume that $l > 0$ and that the claim is known for the subgallery $C_0, \dots, C_{l-1}$. Hence ρ is an isometry on $\mathfrak{C}_1' \cup C_0 \cup \cdots \cup C_{l-1}$, and this union is therefore contained in an apartment E'. Note, then, that ρ maps E' isomorphically onto E_1. Moreover, $\rho|_{E'} = \rho_{E_1, C}|_{E'}$ for any chamber C of $E_1 \cap E'$ and hence, in particular, for any chamber C that meets $\mathfrak{C}_1'$. Let A be the common panel of C_{l-1} and C_l, and let C_l' be the chamber of E' adjacent to C_{l-1} along A. Let H be the support of A in E', i.e., the wall of E' separating C_{l-1} from C_l'. As in the proof of Lemma 11.46, it will be convenient to refer to the two closed half-spaces of E' determined by H as the positive and negative sides, the positive side being defined by a linear inequality $f \geq c$ with f bounded below on $\mathfrak{C}_1'$. We can similarly define the positive and negative sides of the wall $\rho(H)$ in E_1.

We now consider three cases:

(a) $C_l' = C_l$. In other words, E' contains C_l. Since $\rho|_{E'}$ is an isomorphism and coincides with $\rho_{E_1, C}|_{E'}$ for any C meeting $\mathfrak{C}_1'$, the claim is trivial in this case.

(b) $C_l' \neq C_l$, and C_0 is on the positive side of H. Figure 11.9 illustrates this case when X is a tree. Since x_0 is on the positive side of H, $\rho(x_0)$ is on the positive side of $\rho(H)$; hence all of $\rho(x_0) + \mathfrak{D}$ is on the positive side of $\rho(H)$. In view of our choice of $\mathfrak{C}_1'$, it follows that $\mathfrak{C}_1'$ is on the positive side of $\rho(H)$ in E_1, whence $\mathfrak{C}_1'$ is on the positive side of H in E'.

Suppose now that C is any chamber that meets $\mathfrak{C}_1'$. Then C is on the positive side of H, so there is a minimal gallery in E' of the form $C, \dots, C_{l-1}, C_l'$. Replacing C_l' by C_l, we obtain a gallery of the same type, so it is still minimal (Proposition 4.41). Hence $\rho_{E_1, C}$ maps this gallery to a minimal gallery in E_1;

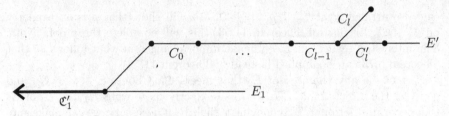

Fig. 11.9. Case (b): C_0 is on the positive side of H.

in particular, $\rho_{E_1,C}$ maps our original gallery $C_0, \ldots, C_l$ to a gallery. Since $\rho_{E_1,C}(C_i) = \rho(C_i)$ for $i < l$, it follows that $\rho_{E_1,C}(C_l)$ is independent of C. This common value of $\rho_{E_1,C}(C_l)$ for C meeting $\mathfrak{C}_1'$ must equal $\rho(C_l)$, and the claim is now proved in case (b).

(c) $C_l' \neq C_l$, and C_0 is on the negative side of H. Figure 11.10 illustrates this case when X is a tree; it suggests that ρ changes direction as one goes from C_{l-1} to C_l. We will show that our choice of C_0 prohibits this case from

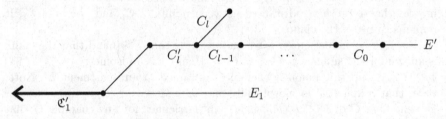

Fig. 11.10. Case (c): C_0 is on the negative side of H.

occurring. Let $\mathfrak{C}_1'' \subseteq \mathfrak{C}_1'$ be the subsector based at some point of $\mathfrak{C}_1'$ on the positive side of H. Then $\mathfrak{C}_1''$ lies entirely on the positive side of H. Considering galleries as in case (b), we conclude that $\rho_{E,C}(C_l) = \rho_{E,C}(C_{l-1})$ for any chamber C that meets $\mathfrak{C}_1''$. Consequently, $\rho(C_l) = \rho(C_{l-1})$.

Recall now that we have a directed line segment $\overrightarrow{x_0 x}$ that is parallel to $\mathfrak{C}_2$ and that passes through $C_0, \ldots, C_l$. The initial portion of this segment is mapped by ρ to a directed segment in E parallel to a $\overline{W}$-chamber that we called $\mathfrak{D}'$. Let x_1 be the point where $\overrightarrow{x_0 x}$ crosses A, and let y_0, y_1, and y be, respectively, $\rho(x_0)$, $\rho(x_1)$, and $\rho(x)$. Let $z = sy$, where s is the reflection of E_1 with respect to $\rho(H)$. See Figure 11.11 for a picture of E_1 in the 2-dimensional case. One can now check that ρ maps $\overrightarrow{x_0 x}$ to the path obtained from $\overrightarrow{y_0 z}$ by folding it onto the negative side of $\rho(H)$. In particular, $\overrightarrow{y_1 z}$ has the same direction as $\overrightarrow{y_0 y_1} = \rho(\overrightarrow{x_0 x_1})$; hence it is parallel to $\mathfrak{D}'$. Thus $y_1 + \mathfrak{D}'$ is on the positive side of $\rho(H)$, i.e., on the same side as $y_1 + \mathfrak{D}$. Moreover, since $\rho(\overrightarrow{x_1 x}) = s(\overrightarrow{y_1 z})$, the direction of $\rho(\mathfrak{C}_2)$ at C_l is $\bar{s}\mathfrak{D}'$, where $\bar{s}$ is the linear part of s. But $\mathbf{d}(\mathfrak{D}, \bar{s}\mathfrak{D}') > \mathbf{d}(\mathfrak{D}, \mathfrak{D}')$ by Exercise 3.57, contradicting the

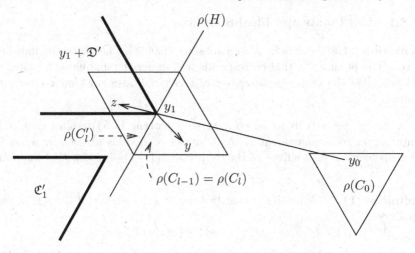

Fig. 11.11. The apartment E_1 in case (c).

maximality of $\mathbf{d}(\mathfrak{D}, \mathfrak{D}')$. This contradiction completes the proof of the claim and hence of the theorem.   □

Exercises

11.66. Describe $\rho_{E,\mathfrak{c}}$ if X is a tree.

11.67. Prove that for any bounded subset Y of X there is a subsector $\mathfrak{C}'$ of $\mathfrak{C}$ such that $\rho_{E,\mathfrak{c}}|_Y = \rho_{E,C}|_Y$ for any chamber C of E that meets $\mathfrak{C}'$.

11.68. Prove or disprove the following purported generalization of Theorem 11.63(1): *Given a sector $\mathfrak{C}$ and a bounded subset Y of an apartment, there is a subsector $\mathfrak{C}' \subseteq \mathfrak{C}$ such that $\mathfrak{C}'$ and Y are contained in some apartment.*

11.8 The Spherical Building at Infinity

At the end of Section 6.9 we suggested the possibility of constructing a spherical building by attaching a "sphere at infinity" to every apartment of a Euclidean building. In this section we carry out the details of that construction. Some of what we do here generalizes to complete CAT(0) spaces; see Bridson–Haefliger [48, Chapters II.8 and II.9].

We continue to denote by X an arbitrary Euclidean building, and "apartment" continues to refer to the complete apartment system unless the contrary is explicitly stated.

11.8.1 Ideal Points and Ideal Simplices

Definition 11.69. A *ray* in X is a subset $\mathfrak{r}$ that is isometric to the half-line $[0, \infty)$. The point $x \in \mathfrak{r}$ that corresponds to 0 under the unique such isometry will be called the *origin* or *basepoint* of $\mathfrak{r}$. We will also say that $\mathfrak{r}$ *emanates* from x.

A ray is easily seen to be convex, so Theorem 11.53 implies that it is contained in some apartment E. As a subset of E, it is necessarily a ray in the usual sense, i.e., a subset of the form $\mathfrak{r} := \{(1 - t)x + ty \mid t \geq 0\}$ for some $y \neq x$.

Definition 11.70. We will say that two rays $\mathfrak{r}, \mathfrak{s}$ are *parallel* if the sets of real numbers

$$\{d(y, \mathfrak{s}) \mid y \in \mathfrak{r}\} \quad \text{and} \quad \{d(z, \mathfrak{r}) \mid z \in \mathfrak{s}\}$$

are bounded. In other words, we require that there be a number M such that for each $y \in \mathfrak{r}$ there is a $z \in \mathfrak{s}$ with $d(y, z) < M$, and similarly with the roles of $\mathfrak{r}$ and $\mathfrak{s}$ reversed.

If $\mathfrak{r}$ and $\mathfrak{s}$ are subsets of some apartment E, one can easily check that they are parallel if and only if there is a translation of E taking one to the other. It is also easy to check that the relation of parallelism is an equivalence relation.

Definition 11.71. An equivalence class of rays will be called an *ideal point* of X.

If $\mathfrak{r}$ is a ray emanating from x and representing an ideal point e, one thinks of e as sitting "at infinity," or at the "end" of $\mathfrak{r}$. To reinforce this intuition, we will write $\mathfrak{r} =: [x, e)$; this notation is justified by the following lemma, which shows that an ideal point admits a unique representative ray emanating from a given point x.

Lemma 11.72. *Given a point x and a ray $\mathfrak{s}$, there is a unique ray $\mathfrak{r}$ that is based at x and parallel to $\mathfrak{s}$.*

Proof. To prove existence, let E be an apartment containing $\mathfrak{s}$. Let $\mathfrak{C}$ be a sector in E, based at the origin of $\mathfrak{s}$, such that the closure of $\mathfrak{C}$ contains $\mathfrak{s}$. By Theorem 11.63 we can find a subsector $\mathfrak{C}'$ of $\mathfrak{C}$ such that $\mathfrak{C}'$ and x are contained in some apartment E'. Since $\mathfrak{C}'$ is a translate of $\mathfrak{C}$ in E, its closure contains a ray $\mathfrak{s}'$ parallel to $\mathfrak{s}$; we can now translate $\mathfrak{s}'$ in E' to obtain the desired $\mathfrak{r}$.

We must now show that there cannot exist distinct parallel rays with the same origin. Suppose, to the contrary, that $\mathfrak{r}_1$ and $\mathfrak{r}_2$ are distinct parallel rays based at x. Then $\mathfrak{r}_1 \cap \mathfrak{r}_2$, being a closed, convex subset of a line, must be an interval $[x, x']$. Replacing $\mathfrak{r}_1$ and $\mathfrak{r}_2$ by the subrays based at x', we are reduced to the case $\mathfrak{r}_1 \cap \mathfrak{r}_2 = \{x\}$. By Exercises 11.57 and 11.58, there is then a well-defined angle $\theta > 0$ between $\mathfrak{r}_1$ and $\mathfrak{r}_2$ at x, and the cosine inequality holds:

For any $s, t \geq 0$ let p_s (resp. q_t) be the point of $\mathfrak{r}_1$ (resp. $\mathfrak{r}_2$) at distance s (resp. t) from x; then

$$d^2(p_s, q_t) \geq s^2 + t^2 - 2st \cos\theta .$$

Fix s and consider the right side of this inequality as t varies. If $\theta \geq \pi/2$, then the minimum value of the right side is s^2, which is achieved when $t = 0$. If $\theta < \pi/2$, then the minimum value is $s^2 \sin^2\theta$, which is achieved when $t = s \cos\theta$. In either case, $\lim_{s\to\infty} d(p_s, \mathfrak{r}_2) = \infty$, contradicting the assumption that $\mathfrak{r}_1$ and $\mathfrak{r}_2$ are parallel. $\square$

Let X_∞ be the set of ideal points. We wish to decompose X_∞ into "ideal simplices."

Definition 11.73. Let $\mathfrak{A}$ be a conical cell in X in the sense of Definition 11.61. The *face of* $\mathfrak{A}$ *at infinity*, denoted by $\mathfrak{A}_\infty$, is defined to be the set of ideal points e such that $\mathfrak{A}$ contains the open ray $(x, e) := [x, e) \smallsetminus \{x\}$, where x is the cone point of $\mathfrak{A}$.

Note that we can recover $\mathfrak{A}$ from its cone point x and its face at infinity $F := \mathfrak{A}_\infty$; namely, $\mathfrak{A}$ is the "open join" $x * F$, where the latter is defined as follows:

$$x * F := \begin{cases} \{x\} & \text{if } F = \emptyset, \\ \bigcup_{e \in F} (x, e) & \text{otherwise.} \end{cases}$$

Definition 11.74. An *ideal simplex* of X is a subset F of X_∞ such that $F = \mathfrak{A}_\infty$ for some conical cell $\mathfrak{A}$.

Lemma 11.75. *If F is an ideal simplex and x is an arbitrary point of X, then there is a conical cell $\mathfrak{A}$ based at x such that $F = \mathfrak{A}_\infty$. Consequently, there is a 1–1 correspondence between the set of ideal simplices of X and the set of conical cells based at any given point $x \in X$.*

Proof. The proof is similar to that of Lemma 11.72: Write $F = \mathfrak{B}_\infty$ for some conical cell $\mathfrak{B}$, let E be an apartment containing $\mathfrak{B}$, and choose a sector $\mathfrak{C}$ in E (based at the cone point of $\mathfrak{B}$) such that $\mathfrak{B}$ is a face of $\mathfrak{C}$. Replacing $\mathfrak{C}$ by a subsector and $\mathfrak{B}$ by a translate, we may assume that $\mathfrak{B}$ and x are contained in an apartment E'. The desired $\mathfrak{A}$ is then a translate of $\mathfrak{B}$ in E'. This proves the first assertion, and the second follows at once. $\square$

Lemma 11.76. *The ideal simplices partition X_∞.*

Proof. Any open ray (x, e) is contained in an apartment E. It is therefore contained in some conical cell $\mathfrak{A}$ in E based at x, whence $e \in \mathfrak{A}_\infty$. This shows that X_∞ is the union of the ideal simplices. Suppose now that we have two distinct ideal simplices $F = \mathfrak{A}_\infty$ and $F' = \mathfrak{A}'_\infty$. Then $\mathfrak{A}$ (resp. $\mathfrak{A}'$) is a face of a sector $\mathfrak{C}$ (resp. $\mathfrak{C}'$) in some apartment E (resp. E'). By Theorem 11.63, there is an apartment E'' containing subsectors of $\mathfrak{C}$ and $\mathfrak{C}'$. We may therefore replace $\mathfrak{A}$ and $\mathfrak{A}'$ by translates (in E and E'), in order to reduce to the case that F and F' are represented by conical cells in E''. But now it is evident that F and F' are disjoint; in fact, we can represent F and F' by conical cells $x * F$ and $x * F'$ with $x \in E''$; so our assertion follows from the fact that the conical cells in E'' based at a given point x partition E''. □

In the theory of trees, one usually defines an *end* of a tree X to be an equivalence class of rays, where two rays are equivalent if they have a common subray. Readers familiar with this theory have probably already wondered how it relates to the theory being developed here. The answer is that if our Euclidean building X is a tree, then the ideal points of X are the same as its ends. In other words, two rays represent the same ideal point if and only if they have a common subray. Here is a generalization of this fact to Euclidean buildings of arbitrary dimension:

Lemma 11.77. *Two sectors of X have the same face at infinity if and only if they have a common subsector.*

Proof. It is obvious that a sector has the same face at infinity as any subsector, whence the "if" part. Conversely, suppose $\mathfrak{C}_1$ and $\mathfrak{C}_2$ are sectors with the same face at infinity. Let E be an apartment containing subsectors $\mathfrak{C}'_1$ and $\mathfrak{C}'_2$. Then these subsectors have the same face at infinity, so they have the same direction $\mathfrak{D}$ (defined with respect to some vector space structure on E). The intersection $\mathfrak{C}'_1 \cap \mathfrak{C}'_2$ is then a sector with direction $\mathfrak{D}$, so it is a common subsector of $\mathfrak{C}_1$ and $\mathfrak{C}_2$. □

11.8.2 Construction of the Building at Infinity

Let Δ_∞ be the set of ideal simplices of X. The first step is to define a face relation on this set. Recall that there is a face relation on the set of conical cells in an apartment E based at a given point x. We extend this to conical cells in X based at x by saying that $\mathfrak{A}'$ is a *face* of $\mathfrak{A}$ if $\mathfrak{A}'$ is contained in the closure of $\mathfrak{A}$ and is a face of $\mathfrak{A}$ in some apartment containing $\mathfrak{A}$. In this case $\mathfrak{A}'$ is a face of $\mathfrak{A}$ in every apartment containing $\mathfrak{A}$.

We can now use the 1–1 correspondence in Lemma 11.75 to define a *face relation* on Δ_∞, so that it becomes a poset.

Definition 11.78. Given ideal simplices F', F, we say that F' is a *face* of F if $x * F'$ is a face of $x * F$ for some $x \in X$.

A glance at the proof of Lemma 11.75 shows that $x * F'$ is then a face of $x * F$ for every $x \in X$. For any apartment $E = |\Sigma|$ of X, let Σ_∞ be the set of ideal simplices F such that $F = \mathfrak{A}_\infty$ for some conical cell $\mathfrak{A}$ in E. Note that Σ_∞ is a subset of Δ_∞ closed under passage to faces. (So we can call it a *subcomplex*, as soon as we have proven that Δ_∞ is a simplicial complex.) Note further that Σ_∞, with the face relation that it inherits from Δ_∞, is a finite Coxeter complex. In fact, if we identify Σ with $\Sigma(W, V)$ as we have done before, then Σ_∞ is isomorphic to $\Sigma(\overline{W}, V)$. We will call Σ_∞ an *apartment* of Δ_∞.

Theorem 11.79. *The poset Δ_∞ is a spherical building. Its apartments are in 1–1 correspondence with those of X.*

Proof. The proof of Lemma 11.76 showed that any two elements F, F' of Δ_∞ are contained in an apartment Σ_∞. Since the latter is a simplicial complex and is closed under passage to faces, it follows that (a) any two elements of Δ_∞ have a greatest lower bound and (b) for any $F \in \Delta_\infty$, the poset $(\Delta_\infty)_{\leq F}$ is a Boolean lattice. Thus Δ_∞ is a simplicial complex (Definition A.1), and each Σ_∞ is a subcomplex. Moreover, in the course of proving this we have already verified the building axioms (B0) and (B1). To complete the proof that Δ_∞ is a building, we will prove the variant (B2″) of (B2).

Suppose $E = |\Sigma|$ and $E' = |\Sigma'|$ are two apartments of X such that Σ_∞ and Σ'_∞ have a common chamber. This means that there are sectors $\mathfrak{C} \subset E$ and $\mathfrak{C}' \subset E'$ such that $\mathfrak{C}_\infty = \mathfrak{C}'_\infty$. By Lemma 11.77, $\mathfrak{C}$ and $\mathfrak{C}'$ have a common subsector; in particular, $E \cap E' \neq \emptyset$. Let $\phi \colon E' \to E$ be the isomorphism that fixes this intersection pointwise. Then ϕ induces an isomorphism $\phi_\infty \colon \Sigma'_\infty \to \Sigma_\infty$, and we will show that ϕ_∞ fixes every simplex $F \in \Sigma_\infty \cap \Sigma'_\infty$. Choose any $x \in E \cap E'$. Then $x * F$ is the unique conical cell in E based at x with F as its face at infinity, and similarly for E'. So $x * F \subseteq E \cap E'$. Hence ϕ fixes $x * F$ pointwise, and ϕ_∞ therefore fixes F. This completes the proof that Δ_∞ is a spherical building.

Continuing with the same notation, suppose that $\Sigma_\infty = \Sigma'_\infty$. Then the previous paragraph shows that $x * F \subseteq E \cap E'$ for any $F \in \Sigma_\infty = \Sigma'_\infty$, and hence $E = E'$. The function $E \mapsto \Sigma_\infty$ is therefore a bijection from the set of apartments of X to the set of apartments of Δ_∞. $\square$

Definition 11.80. We will call Δ_∞ the *building at infinity* associated to Δ.

Note that the geometric realization of the building at infinity is a union of spheres, one for each apartment E of X.

Remark 11.81. Under mild hypotheses, there is a reasonable way to topologize X_∞. If X is a tree, for example, one gets the space of ends, which is typically a Cantor set. We will say a little more about this in Section 13.2.1, and we will mention an analogous construction in differential geometry in Section 14.3.1.

More generally, there is often a reasonable topology on $X \amalg X_\infty$, so that the latter is a compactification of X. The tree case, as in Figure 6.3, is again easy to understand. In fact, the tree in that figure was drawn with its edges getting smaller and smaller, so as to suggest the possibility of compactifying it. See [48,

Chapter II.8] for a general discussion of this compactification in the setting of CAT(0) spaces. And see [126, 151, 284] for much more information about compactifications of Euclidean buildings. Rémy [196] has given applications of these compactifications to the theory of Kac–Moody groups.

Exercises

11.82. Show that there is a bijection $|\Delta_\infty| \cong X_\infty$. [In view of Remark 11.81, one should not expect this to be a homeomorphism when both sides are given their natural topologies.]

11.83. Let $E = |\Sigma|$ be an apartment, and let $x \in E$ be a special vertex. Show that there is a simplicial isomorphism $\mathrm{lk}_E(x) \xrightarrow{\sim} \Sigma_\infty$. [One can visualize this as "radial projection" from x.]

11.84. Assume that Δ is thick.

(a) Any automorphism of Δ induces an automorphism of Δ_∞, so we have a canonical homomorphism $\mathrm{Aut}\,\Delta \to \mathrm{Aut}\,\Delta_\infty$. Show that this homomorphism is injective.
(b) Deduce that every nontrivial automorphism ϕ of Δ has unbounded displacement, i.e., $d(x, \phi(x))$ is unbounded for $x \in X$.
(c) Give examples to show that thickness is necessary in (a) and (b).

11.85. Assume that Δ is thick, and let G be a subgroup of $\mathrm{Aut}\,\Delta$ such that there is a bounded set containing representatives for the G-orbits in $X = |\Delta|$. Deduce from Exercise 11.84 that G has trivial center. [In typical applications, Δ is locally finite and $G \backslash X$ is compact.]

Remark 11.86. Exercises 11.84(b) and 11.85 have combinatorial analogues, valid for any thick irreducible building whose Weyl group is infinite. The results are phrased in terms of the metric space $\mathcal{C}$ of chambers. See [11].

11.8.3 Type-Preserving Maps

Since buildings are colorable, it makes sense to ask whether a map between subcomplexes of X (or of Δ_∞) is type-preserving. The following result will be used when we look at type-preserving automorphism groups below.

Proposition 11.87. *Let* $\phi\colon E \to E'$ *be a type-preserving isomorphism between apartments* $E = |\Sigma|$ *and* $E' = |\Sigma'|$ *of* X. *Then the induced isomorphism* $\phi_\infty\colon \Sigma_\infty \to \Sigma'_\infty$ *is type-preserving. In particular, a type-preserving automorphism of* Δ *induces a type-preserving automorphism of* Δ_∞.

Proof. Assume first that E and E' have a common sector $\mathfrak{C}$ and that ϕ is the isomorphism that fixes $E \cap E'$ pointwise. Then ϕ_∞ fixes $\mathfrak{C}_\infty$ and all its faces, so it is type-preserving. For arbitrary E, E', choose an apartment E''

such that $E \cap E''$ and $E' \cap E''$ each contain a sector. Let $\psi \colon E \to E'$ be the composite of the isomorphisms $E \to E'' \to E'$ of the type just considered. Then ψ_∞ is type-preserving. Now the automorphism $w := \psi^{-1}\phi$ of E is in the Coxeter group $W := \mathrm{Aut}_0\, \Sigma$ of type-preserving automorphisms of Σ, and w_∞ is simply the image of w in the finite reflection group $\overline{W} := \mathrm{Aut}_0\, \Sigma_\infty$. So w_∞ is type-preserving and hence so is $\phi_\infty = \psi_\infty w_\infty$. $\hspace{1em}\square$

Exercise 11.88. Let g be an automorphism of Δ. Is it true that g is type-preserving if and only if the induced automorphism of Δ_∞ is type-preserving?

11.8.4 Incomplete Apartment Systems

If $\mathcal{A}$ is now an incomplete apartment system, it is reasonable to try to define a building at infinity using only the apartments in $\mathcal{A}$. This is of interest because incomplete apartment systems arise quite often (e.g., from BN-pairs). Let $\mathcal{A}_\infty$ be the set of apartments Σ_∞ in Δ_∞ with $\Sigma \in \mathcal{A}$. Let $\Delta_\infty(\mathcal{A})$ be the union of these apartments Σ_∞. It is a subcomplex of Δ_∞, and we would like to know whether it is a building, with $\mathcal{A}_\infty$ as system of apartments. Since Δ_∞ is already known to be a building, the only issue is whether axiom (B1) holds. It is easy to characterize the apartment systems $\mathcal{A}$ for which this is true. Call a sector in X an $\mathcal{A}$-*sector* if it is contained in an apartment in $\mathcal{A}$. The following result is immediate:

Proposition 11.89. $\Delta_\infty(\mathcal{A})$ *is a building with* $\mathcal{A}_\infty$ *as system of apartments if and only if* $\mathcal{A}$ *has the following property: For any two* $\mathcal{A}$-*sectors, there is an apartment in* $\mathcal{A}$ *containing a subsector of each of them.* $\hspace{1em}\square$

By thinking about trees, one can easily see that there are many apartment systems besides the complete one that satisfy the condition in the proposition, but that not all apartment systems do.

Definition 11.90. We will say that $\mathcal{A}$ is *good* if it satisfies the condition of the proposition.

What we have done so far, then, is to make the (trivial) observation that certain subbuildings of Δ_∞ arise from good apartment systems. Our goal for the remainder of this subsection is to show, more generally, that *every* subbuilding of Δ_∞ arises from a good apartment system in some subbuilding of X. To this end, we start with an arbitrary nonempty subset $\mathcal{A}$ of the complete apartment system and set

$$ X' := \bigcup_{E \in \mathcal{A}} E . $$

Consider the following conditions that $\mathcal{A}$ might or might not satisfy:

(0) Given any two chambers of X', there is an apartment in $\mathcal{A}$ containing both of them.

(1) Given a chamber C in X' and an $\mathcal{A}$-sector $\mathfrak{C}$, there is an apartment in $\mathcal{A}$ containing C and a subsector of $\mathfrak{C}$.
(2) Given any two $\mathcal{A}$-sectors, there is an apartment in $\mathcal{A}$ containing a subsector of each of them.

Note that (1) and (2) are the properties that were shown to hold for the complete apartment system in Theorem 11.63, while (0) holds trivially in that setting. Indeed, (0) says precisely that X' is a subbuilding of X with $\mathcal{A}$ as a system of apartments.

Proposition 11.91. *If* (2) *holds, then so do* (0) *and* (1). *In particular, X' is then a subbuilding of X with $\mathcal{A}$ as a good system of apartments.*

Proof. To prove (0), consider any two chambers C_1, C_2 in X', and choose an apartment E in the complete apartment system that contains them. According to Lemma 11.48, there are sectors $\mathfrak{C}_1, \mathfrak{C}_2$ in E with the following property: For any subsectors $\mathfrak{C}_1' \subseteq \mathfrak{C}_1$ and $\mathfrak{C}_2' \subseteq \mathfrak{C}_2$, there is a gallery that starts at a chamber meeting $\mathfrak{C}_1'$, ends at a chamber meeting $\mathfrak{C}_2'$, and passes through both C_1 and C_2 (in that order). Choose, for $i = 1, 2$, an apartment $E_i \in \mathcal{A}$ containing C_i, and let $\phi_i \colon E \to E_i$ be the isomorphism fixing $E \cap E_i$ pointwise. Then $\phi_1(\mathfrak{C}_1)$ and $\phi_2(\mathfrak{C}_2)$ are $\mathcal{A}$-sectors, so there is an apartment $E' \in \mathcal{A}$ containing a subsector of each of them. We may assume that E' in fact contains $\phi_i(\mathfrak{C}_i)$ for each i. We will show that E' contains C_1 and C_2.

Choose a minimal gallery Γ in E that starts at a chamber D_1 meeting $\mathfrak{C}_1$, ends at a chamber D_2 meeting $\mathfrak{C}_2$, and passes through both C_1 and C_2, in that order. Then we can form a new gallery Γ' by replacing the first part of Γ (from D_1 to C_1) by its image under ϕ_1 and the last part of Γ (from C_2 to D_2) by its image under ϕ_2. This has the same type as Γ, so it is still minimal. Since Γ' starts and ends in E', it is contained in E'; hence C_1 and C_2 are contained in E', and (0) is proved.

To prove (1), let C be a chamber in X' and let $\mathfrak{C}$ be an $\mathcal{A}$-sector. Since the complete apartment system satisfies (1), there is an apartment E that contains C and a subsector $\mathfrak{C}_1$ of $\mathfrak{C}$. Let $\mathfrak{C}_2$ be a sector in E with direction opposite to that of $\mathfrak{C}_1$. In view of Corollary 11.47, we may arrange (after replacing $\mathfrak{C}_1$ and $\mathfrak{C}_2$ by subsectors if necessary) that the following condition holds: If C_1 and C_2 are chambers in E such that C_i meets $\mathfrak{C}_i$ for $i = 1, 2$, then the convex hull of C_1 and C_2 contains C.

Now choose an apartment E' in $\mathcal{A}$ containing C, and let $\phi \colon E \to E'$ be the isomorphism fixing $E \cap E'$ pointwise. Then $\phi(\mathfrak{C}_2)$ is an $\mathcal{A}$-sector, so our hypothesis implies that there is an apartment E'' in $\mathcal{A}$ containing subsectors of $\mathfrak{C}_1$ and $\phi(\mathfrak{C}_2)$ (which may be assumed to be $\mathfrak{C}_1$ and $\phi(\mathfrak{C}_2)$ themselves). We claim that E'' contains C, so that we are done.

To prove the claim, choose chambers C_i meeting $\mathfrak{C}_i$ for $i = 1, 2$, and choose a minimal gallery Γ from C_1 to C_2 in E passing through C. Replace the second part of Γ (from C to C_2) by its image under ϕ to get a gallery Γ' from C_1 to $\phi(C_2)$ passing through C. Then Γ' has the same type as Γ, so it is still minimal. (Alternatively, Γ' is minimal because its image under the retraction $\rho_{E,C}$ is minimal.) Hence Γ' is contained in E'', and therefore C is contained in E''.   □

The implication (2) $\implies$ (0) is essentially due to Kleiner–Leeb [147, Lemma 3.3], though our proof is superficially different. (See Exercise 11.94 for their proof.) This implication can be restated as follows:

Corollary 11.92. *There is a 1–1 correspondence between subbuildings of Δ_∞ and pairs $(X', \mathcal{A})$, with X' a subbuilding of X and $\mathcal{A}$ a good apartment system for X'. It is given by $\Delta' \mapsto (X', \mathcal{A})$, where Δ' is a subbuilding of Δ_∞, $\mathcal{A}$ is the set of apartments E of X such that E_∞ is an apartment of Δ', and $X' = \bigcup_{E \in \mathcal{A}} E$.*   □

Finally, we can use the implication (2) $\implies$ (1) to prove a useful thickness result. Our proof will also use Remark 10.20(b), which the reader may need to review before proceeding.

Proposition 11.93. *Let $\mathcal{A}$ be a good system of apartments for a Euclidean building X. If X is thick, then the building at infinity $\Delta_\infty(\mathcal{A})$ is thick.*

Proof. Any two adjacent chambers of $\Delta_\infty(\mathcal{A})$ can be represented by adjacent sectors $\mathfrak{C}_1, \mathfrak{C}_2$ (with a common cone point) in an apartment E in $\mathcal{A}$. We may take their cone point x to be a special vertex of E, in which case each $\mathfrak{C}_i$ is the conical extension in E of a chamber C_i having x as a vertex (see Remark 10.20(b)). It is easy to check (by a scaling argument) that C_1 and C_2 are adjacent. We wish to find a third sector $\mathfrak{C}_3$ based at x that is adjacent to $\mathfrak{C}_1$ and $\mathfrak{C}_2$ along their common (conical) face and does not have a subsector in common with either of them. By thickness of X, there is a chamber C_3 of X adjacent to C_1 and C_2 along their common panel. Since $\mathcal{A}$ has property (1), there is an apartment E' in $\mathcal{A}$ containing C_3 and a subsector of $\mathfrak{C}_1$. Then E' necessarily contains the entire sector $\mathfrak{C}_1$, since the latter is contained in the convex hull of x and any subsector. We can now take the desired $\mathfrak{C}_3$ to be the conical extension of C_3 in E'.   □

Exercises

11.94. Given an alternative proof of Proposition 11.91 that uses geodesics instead of minimal galleries.

11.95. Give an example to show that the converse of Proposition 11.93 is false.

11.8.5 Group-Theoretic Consequences

In this subsection we assume that our Euclidean building $X = |\Delta|$ comes equipped with a type-preserving action of a group G. Assume further that

G has a subgroup N that stabilizes an apartment $E = |\Sigma|$ (in the complete apartment system) and is chamber transitive on E. We can then introduce the set of apartments

$$\mathcal{A} := \{gE \mid g \in G\}\,,$$

and the action of G on X is strongly transitive on $\mathcal{A}$. [This makes sense even if $\mathcal{A}$ is not a *system* of apartments; see the discussion preceding Lemma 6.4.] Choose a chamber C and a sector $\mathfrak{C}$ in E, let B be the stabilizer of C, and let $\mathfrak{B}$ be the stabilizer of $\mathfrak{C}_\infty$. We now ask whether $\mathcal{A}$ satisfies condition (2) of the previous subsection.

Proposition 11.96. $\mathcal{A}$ *satisfies* (2) *if and only if* $G = \mathfrak{B}N\mathfrak{B}$.

Proof. The group G acts transitively on $\mathcal{A}_\infty$, and N stabilizes Σ_∞ and acts on the latter via the quotient map $N \twoheadrightarrow W \twoheadrightarrow \overline{W}$. So N is chamber transitive on Σ_∞. The result now follows from Lemma 6.4. $\square$

If $\mathcal{A}$ satisfies (2), then we know from Proposition 11.91 that it also satisfies (0) and (1). These also have group-theoretic interpretations. The interpretation of (0) is immediate via another application of Lemma 6.4:

Corollary 11.97. *If* $\mathcal{A}$ *satisfies* (2), *then* $G = BNB$. *In this case,* $X' := \bigcup_{g \in G} gE$ *is a Euclidean building with* $\mathcal{A}$ *as a good apartment system. The action of* G *on* X' *is strongly transitive with respect to* $\mathcal{A}$, *and the action of* G *on* $\Delta_\infty(\mathcal{A})$ *is strongly transitive with respect to its unique apartment system* $\mathcal{A}_\infty$. $\square$

Corollary 11.98. *If* $\mathcal{A}$ *satisfies* (2) *and* Δ *is thick, then* (B, N) *and* $(\mathfrak{B}, N)$ *are BN-pairs in* G

Proof. $\Delta_\infty(\mathcal{A})$ is thick by Proposition 11.93. Now apply Theorem 6.56(2) twice. $\square$

We turn now to the implication (2) $\implies$ (1) of Proposition 11.91. The group-theoretic interpretation of this is called the *Iwasawa decomposition*. Choose a special vertex $x \in E$ and let K be its stabilizer. In familiar examples, K is a maximal compact subgroup of G (see Remark 11.36). For example, if $G = \mathrm{SL}_n(\mathbb{Q}_p)$ as in Section 6.9, then one choice of x yields $K = \mathrm{SL}_n(\mathbb{Z}_p)$.

Proposition 11.99. *If* $\mathcal{A}$ *satisfies* (2), *then* $G = \mathfrak{B}K$.

Proof. Since G acts strongly transitively on $\Delta_\infty(\mathcal{A})$, every apartment in $\mathcal{A}$ containing a subsector of $\mathfrak{C}$ has the form bE for some $b \in \mathfrak{B}$; hence, since we know that $\mathcal{A}$ satisfies (1),

$$X' = \bigcup_{b \in \mathfrak{B}} bE\,. \tag{11.11}$$

On the other hand, the $\mathfrak{B}$-orbit of the special vertex x contains all vertices of E of type t, where t is the type of x; see Remark 10.24(a). [In order to apply that remark, identify the Weyl group $W = N/T$ with a Euclidean reflection group acting on E, and observe that the elements of N that act as translations on E are contained in $\mathfrak{B}$.] Combining this fact with (11.11), we conclude that $\mathfrak{B}$ acts transitively on the set of all vertices of X' of type t. Equivalently, $\mathfrak{B}$ acts transitively on G/K, which says precisely that $G = \mathfrak{B}K$.   □

We now specialize to the most important situation in which the considerations above are applicable. Assume for the remainder of this subsection that Δ is *given* to us as the building $\Delta(G, B)$ associated to a group G with a Euclidean BN-pair. Let $\mathcal{A}$ be the corresponding apartment system (so that the G-action is strongly transitive with respect to $\mathcal{A}$). The results above in this case can be summarized as follows:

Proposition 11.100. *The apartment system $\mathcal{A}$ is good if and only if $G = \mathfrak{B}N\mathfrak{B}$. If this holds, then $(\mathfrak{B}, N)$ is a BN-pair in G whose associated building is $\Delta_\infty(\mathcal{A})$. Moreover, one then has the Iwasawa decomposition $G = \mathfrak{B}K$.*   □

In order to apply this, one needs to be able to compute the group $\mathfrak{B}$ and decide whether $G = \mathfrak{B}N\mathfrak{B}$. We close this subsection with a computation of $\mathfrak{B}$. We will illustrate it in the next subsection in the canonical example, where $G = \mathrm{SL}_n(K)$.

Suppose we are given $g \in \mathfrak{B}$. Then $g\mathfrak{C}$ and $\mathfrak{C}$ have a common subsector. In particular, gE and E have a common chamber. Since G is strongly transitive and type-preserving on X, it follows that there is an element $g' \in G$ that maps gE to E by the unique isomorphism that fixes $E \cap gE$ pointwise. Then g' is in the subgroup $\mathfrak{B}' \le \mathfrak{B}$ consisting of those elements of G that fix some subsector of $\mathfrak{C}$ pointwise. Now $g'g$ stabilizes E and $\mathfrak{C}_\infty$, so its action on E is given by an element $w \in W' := \ker\{W \twoheadrightarrow \overline{W}\}$. Let T^* be the set of elements of G that fix E pointwise, and let $\tilde{w}$ be a representative of w in N. Then we have $g'g\tilde{w}^{-1} \in T^* \le \mathfrak{B}'$, whence $g \in \mathfrak{B}'\tilde{w}$. As in Chapter 6, we will write $\mathfrak{B}'w$ instead of $\mathfrak{B}'\tilde{w}$ (this being independent of the choice of $\tilde{w}$). So what we have proven so far may be written as

$$\mathfrak{B} = \mathfrak{B}'W'.\tag{11.12}$$

Now let's look more carefully at $\mathfrak{B}'$. Let $\mathfrak{B}_0 \le \mathfrak{B}'$ be the set of elements of G that fix $\mathfrak{C}$ pointwise. We will show that

$$\mathfrak{B}' = \bigcup_{w \in W'} w\mathfrak{B}_0w^{-1}.\tag{11.13}$$

Note first that $w\mathfrak{B}_0w^{-1}$ is the set of elements of G that fix $w\mathfrak{C}$ pointwise. So $w\mathfrak{B}_0w^{-1}$ fixes the subsector $\mathfrak{C} \cap w\mathfrak{C}$ of $\mathfrak{C}$, and the right side of (11.13) is therefore contained in the left side. For the opposite inclusion, it suffices to show that any subsector $\mathfrak{C}'$ of $\mathfrak{C}$ contains a translate $w\mathfrak{C}$ for some $w \in W'$. Now it is easy to check that $\mathfrak{C}'$ contains a set of representatives for the orbits of the (translation) action of W' on E; this follows from the fact that there is a bounded set of representatives (a "fundamental parallelepiped"). In particular, we can find a $w \in W'$ that maps the cone point of $\mathfrak{C}$ into $\mathfrak{C}'$, whence $w\mathfrak{C} \subseteq \mathfrak{C}'$. This completes the proof of (11.13). We have the following immediate consequence of (11.12) and (11.13):

Proposition 11.101. $\mathfrak{B}$ *is generated by* $\mathfrak{B}_0$ *and any set of representatives for* W' *in* N. $\square$

Exercises

11.102. Show that there is a short exact sequence $1 \to \mathfrak{B}' \to \mathfrak{B} \to W' \to 1$.

11.103. Let W'' be the submonoid of W' consisting of those $w \in W'$ such that $w\mathfrak{C} \subseteq \mathfrak{C}$. [Equivalently, if we identify E with a vector space V and W' with a lattice L in V, then $W'' = L \cap \overline{\mathfrak{D}}$, where $\mathfrak{D}$ is the direction of $\mathfrak{C}$.] Show that (11.13) remains valid if W' is replaced by W''.

Remark 11.104. Readers familiar with ascending HNN extensions (see [228, Section 1.2]) should find the situation in these exercises familiar. This suggests that $\mathfrak{B}$ is, in some sense that we will not make precise, a "generalized ascending HNN extension with base group $\mathfrak{B}_0$."

11.8.6 Example

Let K be a field with a discrete valuation, and consider the Euclidean BN-pair in $G := \mathrm{SL}_n(K)$ constructed in Section 6.9. Its Weyl group W is the Euclidean reflection group studied in Section 10.1.7. And as we saw in Section 10.1.8, W is the Coxeter group of type $\tilde{\mathrm{A}}_{n-1}$ and is the affine Weyl group of the root system of type A_{n-1}. We already know that there is a second BN-pair in G, obtained by forgetting that K has a valuation and applying Section 6.5; its Weyl group is the symmetric group on n letters, which is the finite reflection group $\overline{W}$ associated to W. Recall that $\overline{W}$ is the (ordinary) Weyl group of the root system of type A_{n-1}. In view of the relationship between the Weyl groups of our two BN-pairs, the following result is not surprising:

Proposition 11.105. *Let X be the Euclidean building $|\Delta(G, B)|$ associated to $G = \mathrm{SL}_n(K)$.*

(1) *There is a sector $\mathfrak{C}$ in the fundamental apartment $E = |\Sigma|$ such that the stabilizer $\mathfrak{B}$ of $\mathfrak{C}_\infty$ is the upper-triangular subgroup of G.*

(2) *The apartment system $\mathcal{A}$ associated to (G, B, N) is good. The subcomplex $\Delta_\infty(\mathcal{A})$ of Δ_∞ is therefore isomorphic to the spherical building associated to G in Section 6.5.*

(3) *$\mathcal{A}$ is the complete apartment system if and only if K is complete with respect to the given valuation.*

Sketch of proof. Identify the fundamental apartment Σ with the complex $\Sigma(W, V)$ studied in Section 10.1.6. [Recall that we gave an explicit way of making this identification in Section 10.1.7.] Then $E = |\Sigma|$ is identified with V. As "fundamental sector" $\mathfrak{C}$ we take the subset of V defined by $x_1 < \cdots < x_n$. Its closure is a subcomplex of E whose vertices, from the point of view of A-lattices, are the classes $[[\pi^{a_1} e_1, \ldots, \pi^{a_n} e_n]]$ with $a_1 \leq \cdots \leq a_n$. The group $\mathfrak{B}_0$ that fixes $\mathfrak{C}$ pointwise is therefore given by

$$\mathfrak{B}_0 = \bigcap_{d \in D} d \cdot \mathrm{SL}_n(A) \cdot d^{-1},$$

where D is the set of matrices in $\mathrm{GL}_n(K)$ of the form $\mathrm{diag}(\pi^{a_1}, \ldots, \pi^{a_n})$ with $a_1 \leq \cdots \leq a_n$. An easy computation shows that this intersection is the upper-triangular subgroup of $\mathrm{SL}_n(A)$.

Now apply formula (11.13) to get $\mathfrak{B}' = \bigcup_{t \in T} t \mathfrak{B}_0 t^{-1}$, where T is the diagonal subgroup of G. Another easy computation then shows that this union consists of all upper-triangular matrices in G whose diagonal entries are units in A. In view of (11.12), we conclude that $\mathfrak{B}$ is indeed the full upper-triangular subgroup of G.

Statement (1) is now proved, and (2) follows immediately from (1) and Proposition 11.100. Turning now to (3), suppose first that K is not complete. Let $\hat{K}$ be the completion of K, and let $\hat{G}$, $\hat{B}$, and $\hat{N}$ be the analogues of G, B, and N over $\hat{K}$. Then G is dense in $\hat{G}$, and $\hat{B}$ is an open subgroup of $\hat{G}$; it follows that $\Delta(G, B) = \Delta(\hat{G}, \hat{B})$ (Section 6.10.1). On the other hand, it is easy to see that $\hat{G}/\hat{N}$ is strictly bigger than G/N, so we definitely get more apartments using $(\hat{G}, \hat{B}, \hat{N})$ than we get from (G, B, N). The apartment system associated to (G, B, N) is therefore not complete.

Finally, suppose K is complete, and let E' be an arbitrary apartment in the complete system. We will show that $E' \in \mathcal{A}$ by constructing a $g \in G$ such that $E' = gE$, where E is the fundamental apartment. We may assume that E' contains the fundamental chamber C, in which case we will find the desired g in B.

Let $\phi \colon E \to E'$ be the isomorphism that fixes $E \cap E'$. In view of Theorem 11.43, every bounded subset of E' is contained in an apartment in $\mathcal{A}$. So if we exhaust E by an increasing sequence of bounded sets F_i containing C, then we can find $b_i \in B$ such that b_i maps F_i into E' by the map $\phi|_{F_i}$. We claim that the F_i and b_i can be chosen such that $b_{i+1} \equiv b_i \mod \pi^i$. Accepting this for the moment, we can easily complete the proof. For the completeness of K implies that b_i converges to some $b \in B$ as $i \to \infty$, whence $E' = bE$ and we are done. It remains to prove the claim.

By looking at the stabilizers of the vertices of E, one sees first that the F_i can be chosen such that any element of G that fixes F_i pointwise is in $\mathrm{SL}_n(A)$ and is congruent to a diagonal matrix mod π^i. In particular, for any choice of the b_i we will have $b_{i+1}^{-1} b_i$ congruent to a diagonal matrix mod π^i. Now assume inductively that $b_1, \ldots, b_i$ have been chosen and that they satisfy the required congruences. Let b be any element of B such that $b|_{F_{i+1}} = \phi|_{F_{i+1}}$. Then there is a diagonal matrix $t \in \mathrm{SL}_n(A)$ such that $b^{-1}b_i \equiv t \mod \pi^i$. Since t fixes E pointwise, we can complete the inductive step by setting $b_{i+1} := bt$.   □

11.9 Classification

Recall from Chapter 9 that Tits classified thick, irreducible, spherical buildings of rank ≥ 3, and that they are (roughly) in 1–1 correspondence with simple algebraic groups or classical groups or mixed groups over an arbitrary field. [More generally, the classification extends to Moufang buildings of rank 2.] Tits later [258] classified thick Euclidean buildings of rank ≥ 4. [More generally, the classification extends to thick Euclidean buildings of rank 3 such that the spherical building at infinity is Moufang.] The result this time is that they are (roughly) in 1–1 correspondence with absolutely simple algebraic groups or classical groups or mixed groups defined over a field that is complete with respect to a discrete valuation. See Tits [252, Section 4] or Weiss [283, Chapter 28] for the list of such groups in case the residue field is finite (which holds if and only if the corresponding building is locally finite).

The existence of a building associated to a group as above had already been proved by Bruhat and Tits [59, 60]; see Section C.11 below. To achieve the classification, then, Tits essentially had to find an inverse to this construction and produce, from a Euclidean building, an algebraic group over a field with discrete valuation.

For overviews of the classification theorem, see Ronan [200, Section 10.5] or Weiss [282]. For a complete proof, see Weiss [283]. We will not try to say much about the proof except to indicate very briefly why one might expect a Euclidean building X to yield an algebraic group over a field with discrete valuation.

The starting point for the proof is the consideration of the spherical building at infinity. This has rank ≥ 3 [or else it has rank 2 and is Moufang], so one can assume that it is known and, typically, corresponds to an algebraic group over a field K. This field is visible in the root groups U_α, many of which are isomorphic to the additive group of K (see Section 7.9). The desired valuation on K will come from a function $v \colon U_\alpha^* \to \mathbb{Z}$, defined as follows.

Choose a fundamental apartment $E = |\Sigma|$ in the given Euclidean building $X = |\Delta|$. We may identify E with a vector space on which the Weyl group W of X acts as a Euclidean reflection group, with the origin as a special point. As we explained in Section 10.1.8, this leads to an isomorphism of W with the affine Weyl group of a crystallographic root system Φ. We can identify Φ with the set of roots of the corresponding apartment Σ_∞ of Δ_∞. Now any root α of Σ_∞ comes from a root of E by passage to faces at infinity, but this root of E is uniquely determined only up to parallelism. Indeed, there is an infinite family of parallel walls $H_{\alpha,m}$ in E ($m \in \mathbb{Z}$), and any one of these can bound a root that gives α at infinity. This family of walls is the source of the valuation v on the root group $U_\alpha \leq \mathrm{Aut}\,\Delta_\infty$.

In slightly more detail, one first shows that each U_α extends to a group of automorphisms of X. The fixed-point set in E of an element $g \in U_\alpha^*$ then turns out to be one of the infinitely many possible roots of E corresponding to α as above. The bounding wall of this root is therefore $H_{\alpha,m}$ for some $m \in \mathbb{Z}$, and one sets $v(g) = m$. See the cited references for further details.

Remark 11.106. Tits [258] and Bruhat–Tits [59, 60] actually generalize the concept of Euclidean building in order to allow fields with nondiscrete valuations. The resulting geometric objects might be called "nondiscrete Euclidean buildings" or "$\mathbb{R}$-buildings." Many readers are probably familiar with $\mathbb{R}$-trees, which are the 1-dimensional $\mathbb{R}$-buildings.

*11.10 Moufang Euclidean Buildings

In this section the reader is assumed to be familiar with the general theory of (not necessarily spherical) Moufang buildings as treated in Chapter 8. Here we are talking a priori about ordinary buildings as opposed to twin buildings, and we use the term "Moufang" in the sense of Remark 8.57(b). But as we noted in Remark 8.57(c), a Moufang building in this sense gives rise to an RGD system, so it is indeed part of a Moufang twin building (see Theorem 8.81).

The classification of Euclidean buildings that we sketched in the previous section depends heavily on the Moufang property of their buildings at infinity. So it is natural to ask whether this Moufang property is related to the Moufang property for the Euclidean buildings themselves. One could in principle answer this question by comparing the classification of Euclidean buildings with that of 2-spherical twin buildings (Section 9.12), but this is not easy to carry out in practice. Instead, we will be content to derive an important necessary condition for a Euclidean building to be Moufang or to be part of a twin building. This will in particular explain why, as claimed in Remark 6.134, the construction of a twin BN-pair in $\mathrm{SL}_n\left(k[t, t^{-1}]\right)$ does not generalize to $\mathrm{SL}_n(\mathbb{Q}_p)$ (if $n \geq 3$); see Example 11.109 below.

We start by deriving (independently of the Moufang property) a necessary condition for a Euclidean building X to be part of a twin building. The result, Proposition 11.107 below, is a special case of a more general result of Mühlherr and Van Maldeghem [178, Theorem 17]. Our proof is different and requires only methods that we have already developed in this book.

Modifying slightly the notation we have used earlier, we write X_∞ for the spherical building at infinity (with respect to the complete apartment system of X). Our proof will use σ-isometries, as introduced in Section 5.5, so we have to relate types of vertices of X_∞ to the types of vertices of X. We do this as follows: Let $I_0 = \{0, 1, \ldots, n\}$ be the set of types of X (where $n + 1$ is the rank of X), and fix a special vertex v of X. We may assume that v has type 0. A vertex z of $Z := X_\infty$ is represented by the ray $[v, z)$, which is a closed conical cell based at v. The initial segment of this ray is an edge of

X having v as a vertex. If x is the other vertex, we define the type of z to be the type of x. (In particular, $I := \{1, \ldots, n\}$ is the set of types of Z.) To understand our assignment of types intuitively, note that if E is an apartment of X containing v, then the "radial projection" isomorphism $\mathrm{lk}_E(v) \xrightarrow{\sim} E_\infty$ (Exercise 11.83) is type-preserving.

The necessary condition for X to be part of a twin building that we want to prove is the following:

Proposition 11.107. *Let X be a Euclidean building and $Z := X_\infty$ its spherical building at infinity. If there exists a twin building (X_+, X_-) with $X_+ = X$, then for any special vertex v of X, $\mathrm{lk}_X(v)$ is isomorphic to a subbuilding of Z.*

We will combine the simplicial approach with the W-metric approach in the proof that follows. If A is a simplex of X, we denote by $\mathcal{C}_A$ the corresponding residue.

Proof. We assume that (X_+, X_-) is a twin building with $X_+ = X$. Let a special vertex $v = v_+$ of X be given. We choose any vertex $v_- \in X_-$ that is opposite v_+, which means that the residues $\mathcal{R}_+ := \mathcal{C}_{v_+}$ and $\mathcal{R}_- := \mathcal{C}_{v_-}$ are opposite. By Proposition 5.152, the projection maps between $\mathcal{R}_+$ and $\mathcal{R}_-$ are σ_I-isometries, where I is the type of $\mathcal{R}_+$ (and the cotype of v_+). We want to construct a (canonical) σ_I-isometry of $\mathcal{R}_-$ into Z, which then immediately yields an isometry of $\mathcal{R}_+$ into Z. [Note: To simplify notation, we are not distinguishing between Z and its set of chambers when the context makes it clear what we mean.]

For any $C_- \in \mathcal{R}_-$, we set $C_+ := \mathrm{proj}_{\mathcal{R}_+} C_-$ and consider the convex hull $\Gamma(C_+, C_-)$, which has two components $\Gamma_+(C_+, C_-)$ and $\Gamma_-(C_+, C_-)$. If (Σ_+, Σ_-) is any twin apartment containing C_+ and C_-, then, by Proposition 5.193, $\Gamma(C_+, C_-)$ is the intersection of all twin roots (α_+, α_-) of (Σ_+, Σ_-) that contain C_+ and C_-. Now the first halves of these twin roots are precisely the roots α_+ of Σ_+ such that α_+ contains C_+ and has v_+ in its boundary. (If v_+ were not in the boundary of α_+, then v_- would be in the interior of the root of Σ_- opposite α_+; and if v_+ is in the boundary of α_+, then v_- is in the boundary of α_- and $C_- = \mathrm{proj}_{\mathcal{R}_-} C_+$ is in α_-, since twin roots are convex.) So $\Gamma_+(C_+, C_-)$ is the intersection of all roots α_+ of Σ_+ such that α_+ contains C_+ and $\partial\alpha_+$ passes through v_+. Since v_+ is a special vertex, this is a sector of Σ_+. (Here we identify roots with half-spaces.) We set $\mathfrak{C}(C_-) := \Gamma_+(C_+, C_-)$, and we denote by $\widetilde{C}_-$ the corresponding chamber $\mathfrak{C}(C_-)_\infty$ of Z. Define a map $f: \mathcal{R}_- \to Z$ by $f(C_-) := \widetilde{C}_-$ for $C_- \in \mathcal{R}_-$. We claim that f is a σ_I-isometry.

By Lemma 5.61 it suffices to show that f takes i-adjacent chambers to i'-adjacent chambers for $i \in I$, where $i' := \sigma_I(i)$. Let C_- and D_- be i-adjacent chambers of $\mathcal{R}_-$. Recall first that C_+ and D_+ are i'-adjacent, where $C_+ := \mathrm{proj}_{\mathcal{R}_+} C_-$ and $D_+ := \mathrm{proj}_{\mathcal{R}_+} D_-$. Denote by P_+ the panel (codimension-1 simplex) they share, and note that P_+ has cotype i'. For studying the passage to chambers of Z, it is crucial that there exists a twin apartment (Σ_+, Σ_-) that contains all four chambers C_+, D_+, C_-, D_-. Just choose (Σ_+, Σ_-) such that it contains (say) C_+ and D_-. Then Σ_- meets $\mathcal{R}_-$ and hence contains (in view of the convexity of twin apartments) $C_- = \mathrm{proj}_{\mathcal{R}_-} C_+$; and Σ_+ meets $\mathcal{R}_+$ and hence contains $D_+ = \mathrm{proj}_{\mathcal{R}_+} D_-$. So the sectors $\mathfrak{C}(C_-)$ and $\mathfrak{C}(D_-)$ constructed above are both contained in Σ_+. (It turns out to be very useful that the convex hull of two chambers can be constructed in any twin apartment that contains the two chambers and is independent of the choice of the twin apartment.) And $\mathfrak{C}(C_-)$ and $\mathfrak{C}(D_-)$ both contain P_+ in their boundary. Their closures therefore intersect in a codimension-1 face. Since the panel P_+ is of cotype i', it follows from our definition of types in Z that the chambers $\widetilde{C}_-$ and $\widetilde{D}_-$ represented by $\mathfrak{C}(C_-)$ and $\mathfrak{C}(D_-)$ have a common panel of cotype i'. They are easily seen to be distinct, so they are i'-adjacent in Z. This proves the claim.

We may now apply Lemma 5.62 to conclude that $f(\mathcal{R}_-)$ is a subbuilding of Z. Hence (in simplicial language) $\mathrm{lk}_{X_-}(v_-)$ and thus also $\mathrm{lk}_X(v)$ can be identified with a subbuilding of Z. $\qquad\square$

Corollary 11.108. *Let X be a thick Euclidean building of rank at least 3. Assume that X_∞ is Moufang and that there exists a twin building (X_+, X_-) with $X = X_+$. Then $\mathrm{lk}_X(v)$ is Moufang for any special vertex v of X, and the root groups of $\mathrm{lk}_X(v)$ are subgroups of the root groups of X_∞.*

Recall that X_∞ is always thick if X is thick (Proposition 11.93) and that it is automatically Moufang by Theorem 7.59 if the rank of X is at least 4.

Proof. By Proposition 11.107, $L_v := \mathrm{lk}_X(v)$ can be embedded in X_∞ as a subbuilding for any special vertex v of X. So the corollary follows immediately from Proposition 7.37. $\qquad\square$

Corollary 11.108 easily implies that thick p-adic buildings of rank at least 3 cannot be twinned. We discuss this here only for a special case:

Example 11.109. Consider our standard example, where X is the Bruhat–Tits building associated to $\mathrm{SL}_n(K)$, K being a field with a discrete valuation. If k is the residue field, then the link of a vertex is the spherical building associated to $\mathrm{SL}_n(k)$, whereas the building at infinity is the spherical building associated to $\mathrm{SL}_n(\hat{K})$ (where $\hat{K}$ is the completion of K). The root groups of the former are isomorphic to the additive group of k, and the root groups of the latter are isomorphic to the additive group of $\hat{K}$. We can therefore conclude from Corollary 11.108 (if $n \geq 3$) that X cannot be part of a twin building unless K and k have the same characteristic. In particular, the Bruhat–Tits building associated to $\mathrm{SL}_n(\mathbb{Q}_p)$ cannot be part of a twin building for $n \geq 3$.

We now return to the situation described at the beginning of this section and assume that X is Moufang in the sense of Remark 8.57(b). Fix an apartment E of X.

Proposition 11.110. *Suppose that the Euclidean building X is Moufang of rank at least 3 with root group system (U_α), where α ranges over the roots of E. Suppose further that the spherical building X_∞ is Moufang. Then for any root α of E, U_α embeds in U_a, where $a := \alpha_\infty$ is the corresponding root of E_∞.*

Remark 11.111. As we already noted in connection with Corollary 11.108, we do not need to require that X_∞ be Moufang if the rank of X is at least 4. In the present situation, we also do not need to require this for rank 3, since Van Maldeghem and Van Steen [265] proved that the Moufang property of X implies the Moufang property of X_∞ if X has rank 3. In this case, they also prove that U_α is a subgroup of U_a. Their argument is different and does not refer to twin buildings.

Proof of Proposition 11.110. As we mentioned earlier, the assumption that X is Moufang implies that X is part of a Moufang twin building (X, X_-). Given α, we first choose a panel $P \in \partial\alpha$ that contains a special vertex v. It is easy to see that this is always possible. [In most cases X has at least two types of special vertices. Where this is not true (i.e., for $\tilde{E}_8$, $\tilde{F}_4$, and $\tilde{G}_2$), one can instead use the fact that there is a type t in the Coxeter diagram, adjacent to the unique special type t', such that $m(t, t') = 3$; the details are left as an exercise.] Now $L_v := \mathrm{lk}_X(v)$ is a Moufang spherical building, and U_α can be identified with a root group of L_v (see Section 7.3.2 and Proposition 8.21). By Proposition 11.107, L_v is a subbuilding of $Z := X_\infty$. This identification can be chosen such that $E \cap L_v$ is mapped onto E_∞ and $\alpha_v := \alpha \cap L_v$ is mapped onto a: By construction of the Moufang twin building (X, X_-), E is part of a twin apartment (E, E_-) of that twin building. Now choose the vertex v_- in the proof of Proposition 11.107 to be in E_-. Then the map f defined there identifies $\mathrm{lk}_{E_-}(v_-)$ with E_∞. Composing it with the projection isomorphism between L_v and $\mathrm{lk}_{X_-}(v_-)$ yields the desired additional properties of the embedding $L_v \hookrightarrow Z$. So Proposition 7.37 implies that U_α, which can be identified with the root group U_{α_v} of L_v, is a subgroup of U_a.   □

Thus if X is Moufang, its root groups need to be subgroups of the root groups of X_∞. In particular, X cannot be Moufang if it is locally finite and all root groups of X_∞ are torsion-free. This again shows that p-adic buildings cannot be Moufang.

Buildings as Metric Spaces

In the introduction to this book we mentioned three approaches to buildings: the simplicial approach (Chapter 4), the combinatorial approach (Chapter 5), and the metric approach. The present chapter is devoted to the last of these. We do not give a new definition of "building" here, just a different way of thinking about buildings and working with them. The intuitive idea is that one can specify, more or less arbitrarily, a metric model for a closed chamber, and one can then glue copies of this model chamber together to get a realization of the building as a metric space.

The ideas that we present here have not yet reached maturity. But our hope is that by giving a unified treatment of several constructions that have appeared in the literature, we will stimulate further research.

We have already seen two examples to motivate the present chapter. First, we hinted in Section 4.7 at the possibility of putting a spherical metric on each apartment of a spherical building; as we noted there, one can prove the Solomon–Tits theorem in this way. Second, we saw in Chapter 11 the usefulness of thinking of Euclidean buildings (or rather their geometric realizations) as metric spaces. In these two cases, one can in fact characterize the buildings by the metric properties of their geometric realizations; see Charney and Lytchak [80].

In this chapter we will set up a general framework for constructing metric realizations of buildings. The most important example at the time of this writing is a construction due to Davis [88] that provides a CAT(0) realization for *every* building. (See [89, Chapters 7, 12, and 18] for a recent treatment of the subject by Davis.) This has become a standard tool in current research, and it should be viewed as the main result of the chapter.

We begin by giving in Section 12.1 the general method for constructing metric realizations alluded to above. Then in Section 12.2 we illustrate the method by giving several special cases. We are able to include at this point Davis's realization, provided we accept one technical ingredient as a black box.

In Section 12.3 we fill in the missing details of the construction. We state without proof a fundamental theorem of Moussong [170], which says that the Davis realization of an apartment is a CAT(0) space. It is then a simple matter to go from apartments to buildings, which we do in Section 12.4.

We will need to use some results from the theory of regular cell complexes in this chapter. Everything we need is summarized in Appendix A. But we have tried to write this chapter in such a way that readers can omit the appendix (except for some terminology) and still follow the main ideas.

Conventions: We will have several occasions to deal with flag complexes of posets, for which we have generally used the notation $\Delta(-)$ (see Definition A.4). But since the letter Δ will usually denote a building in what follows, we will write $K(P)$ instead of $\Delta(P)$ in this chapter for the flag complex of a poset P. Recall, finally, our convention from Section 2.5 that *the generating set S of a Coxeter group is always understood to be finite.*

12.1 Metric Realizations of Buildings

Throughout this section Δ will denote a building of type (W, S), and $\mathcal{C} = \mathcal{C}(\Delta)$ will be its set of chambers with Weyl distance function $\delta \colon \mathcal{C} \times \mathcal{C} \to W$. The construction we are about to give is most naturally described in terms of $(\mathcal{C}, \delta)$, though we will see that simplices of Δ other than chambers play a role also. Readers who prefer to stay entirely in the framework of W-metric spaces can identify these simplices with residues.

12.1.1 The Z-Realization as a Set

Let Z be a set with a family of nonempty subsets Z_s ($s \in S$). Intuitively, Z is a model for a closed chamber, and Z_s is its s-panel. Starting in the next subsection we will assume that Z has a metric and that each Z_s is a closed subset. We will then construct a metric space $X = Z(\Delta)$, which we call the *Z-realization* of Δ, by gluing together copies $Z(C)$ of Z, one for each chamber $C \in \mathcal{C}$. For the moment, however, there is no metric.

If C and D are s-adjacent chambers of Δ, then we want our gluing to identify the copy of Z_s in $Z(C)$ with the copy of Z_s in $Z(D)$. This leads to the following construction. Start with the product $\mathcal{C} \times Z$, which is simply a disjoint union of copies of Z, one for each $C \in \mathcal{C}$, and let "$\sim$" be the smallest equivalence relation on $\mathcal{C} \times Z$ such that $(C, z) \sim (D, z)$ if $\delta(C, D) = s$ and $z \in Z_s$. Thus $(C, z) \sim (C', z')$ if and only if $z' = z$ and there is a gallery $C = C_0, \ldots, C_l = C'$ such that $\delta(C_{i-1}, C_i) = s_i \in S$ and $z \in Z_{s_i}$ for $i = 1, \ldots, l$. If we set

$$S_z := \{s \in S \mid z \in Z_s\}$$

and $W_z := \langle S_z \rangle$, then we can describe the relation more concisely as follows: $(C, z) \sim (C', z')$ if and only if $z' = z$ and $\delta(C, C') \in W_z$.

Definition 12.1. The Z-*realization* of Δ, denoted by $Z(\Delta)$ or $Z(\mathcal{C})$ or $Z(\mathcal{C}, \delta)$, is the quotient of $\mathcal{C} \times Z$ by the equivalence relation defined above. We denote by $[C, z]$ the equivalence class of (C, z), and we set $Z(C) := \{[C, z] \mid z \in Z\}$. There is a well-defined function $\tau \colon X \to Z$, given by $[C, z] \mapsto z$, and we call $\tau(x)$ for $x \in X$ the *type* of x.

Note that the *type function* τ maps each $Z(C)$ bijectively onto Z.

Examples 12.2. (a) Suppose Z is a geometric closed simplex with vertex set S. For each $s \in S$, let Z_s be the face of Z not containing the vertex s. Then $Z(\Delta)$ is simply the usual geometric realization $|\Delta|$ (except that we are temporarily viewing the latter as a set with no topology). For any $C \in \mathcal{C}$, the set $Z(C)$ is the closed simplex that has often been denoted by $\overline{C}$ in previous chapters.

(b) Suppose Δ consists of a single apartment Σ. Choose a fundamental chamber and identify Σ with the standard Coxeter complex $\Sigma(W, S)$. Thus $\mathcal{C} = W$ and $\delta(w, w') = w^{-1}w'$ for $w, w' \in W$. The resulting Z-realization $X = Z(\Sigma)$, also denoted by $Z(W)$ or $Z(W, S)$, is then the quotient of $W \times Z$ by the equivalence relation generated by declaring $(w, z) \sim (ws, z)$ if $z \in Z_s$. Explicitly,

$$(w, z) \sim (w', z') \iff z' = z \text{ and } w' \in w \langle S_z \rangle.$$

When we describe $Z(W, S)$ in this way, we see that it is cooked up so that W acts on it with a copy of Z as strict fundamental domain and with stabilizers $W_z = \langle S_z \rangle$ for z in the fundamental domain (which is the set $Z(1)$). This is Vinberg's construction that we described in Section 10.4 in the hyperbolic setting.

(c) Let W be the Coxeter group of type $\tilde{A}_1 \times \tilde{A}_1$, i.e., W is a product of two infinite dihedral groups. It is generated by four fundamental reflections s, t, u, v, where s and t commute with u and v. Let Z be a solid square, let Z_s and Z_t be one pair of opposite sides, and let Z_u and Z_v be the other pair. See Figure 12.1. Then the Z-realization $Z(W, S)$ is the plane, tiled by

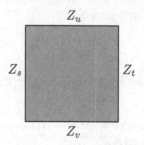

Fig. 12.1. $W = D_\infty \times D_\infty$, and Z is a square.

squares. [Note that $Z(W, S)$ in this case is a much more natural geometric model for (W, S) than the (simplicial) Coxeter complex $\Sigma(W, S)$, which is 3-dimensional.] This example generalizes in an obvious way to semi-Euclidean reflection groups (Section 10.2), with Z being a product of simplices.

(d) Let W be an essential finite reflection group with fundamental chamber C, and let S be the set of fundamental reflections. If we take Z to be the closed simplicial cone $\overline{C}$ and Z_s to be the closed panel $Z \cap sZ$ of C fixed by s, then the Z-realization $Z(W, S)$ is the vector space on which W acts. More generally, let (W, S) be arbitrary, and consider the dual of the canonical linear representation of W as in Section 2.6. Let Z be the closure of the fundamental chamber. Then, with the obvious definition of the subspaces Z_s, the Z-realization $Z(W, S)$ is the Tits cone.

(e) Let (W, S) be an arbitrary Coxeter system, and let $\mathcal{S}$ be the poset of spherical subsets of S, ordered by inclusion. Here a subset of S is called *spherical* if the Coxeter group it generates is finite. Whenever it is convenient, we will identify $\mathcal{S}$ with the poset of finite standard subgroups of W. Let Z be the geometric realization of the flag complex of $\mathcal{S}$, i.e., $Z = |K(\mathcal{S})|$. For $s \in S$, let $\mathcal{S}_s$ be the set of spherical subsets containing s, and let $Z_s \subseteq Z$ be the geometric realization of its flag complex, i.e., $Z_s = |K(\mathcal{S}_s)|$. Note that a typical point $z \in Z$ lies in an open simplex corresponding to a chain $W_0 < \cdots < W_p$ of finite standard subgroups, and $z \in Z_s$ if and only if $s \in W_0$. So the subset S_z defined above is $W_0 \cap S$; hence $W_z = W_0$. In particular, the various groups W_z ($z \in Z$) are precisely the finite standard subgroups of W.

The Z-realization $Z(\Delta)$ in this case will be discussed in Section 12.4. It is the underlying set of the Davis realization mentioned in the introduction to this chapter.

Remark 12.3. In view of constructions that have occurred elsewhere in this book, the reader may be surprised that we ordered $\mathcal{S}$ by inclusion rather than reverse inclusion. In fact, the choice is completely irrelevant in the present context, since reversing the order on a poset does not change the flag complex. But we will see in Section 12.3 that the inclusion order has a geometric interpretation. (Briefly, there is a naturally occurring regular cell complex whose poset of cells is isomorphic to $\mathcal{S}$.)

Exercises

12.4. Under what conditions are the sets $Z(C)$ ($C \in \mathcal{C}$) distinct from one another?

12.5. In Example 12.2(b), under what conditions is the stabilizer of the fundamental domain (as a set) trivial?

12.6. Continuing with Example 12.2(b), under what conditions does W have a fixed point in X?

12.1.2 A Metric on X

Assume from now on that Z has a metric, denoted by $d(-,-)$ or $d_Z(-,-)$, and that each Z_s is a closed subspace. We will put a metric on X in such a way that τ maps each $Z(C)$ isometrically onto Z. We do this via *chains*, which one can think of intuitively as corresponding to piecewise geodesic paths.

Definition 12.7. A *chain* in X is a finite sequence γ of points $x_0, x_1, \ldots, x_m$ ($m \geq 0$) such that for each $i = 1, \ldots, m$ there is a chamber $C_i \in \mathcal{C}$ such that x_{i-1} and x_i are both in $Z(C_i)$. We say that γ is an *m-chain*, and we call the subchain x_{i-1}, x_i the *i*th *segment* of γ for $i = 1, \ldots, m$. The *length* of γ is

$$l(\gamma) := \sum_{i=1}^{m} d_Z\big(\tau(x_{i-1}), \tau(x_i)\big) \, .$$

If we think of a chain γ as representing (intuitively) a piecewise geodesic path, then the length of γ is the sum of the lengths of the pieces. (Warning: Be careful not to confuse the length $l(\gamma)$ with m, which is the number of segments.)

Definition 12.8. We define a *distance function* on X by

$$d(x, y) := \inf_{\gamma} l(\gamma) \, , \tag{12.1}$$

where γ ranges over all chains from x to y.

Note that there is at least one such chain from x to y (hence $d(x, y) < \infty$) because of our assumption that each Z_s is nonempty.

Before proving that d is a metric, we make some remarks about chains. Given a chain $\gamma \colon x_0, \ldots, x_m$, choose chambers $C_1, \ldots, C_m \in \mathcal{C}$, as in the definition of "chain," such that $Z(C_i)$ contains the ith segment of γ. Letting $z_i := \tau(x_i)$, we then have

$$\begin{aligned}
x_0 &= [C_1, z_0] \, , \\
x_1 &= [C_1, z_1] = [C_2, z_1] \, , \\
x_2 &= [C_2, z_2] = [C_3, z_2] \, , \\
&\ \ \vdots \\
x_{m-1} &= [C_{m-1}, z_{m-1}] = [C_m, z_{m-1}] \, , \\
x_m &= [C_m, z_m] \, .
\end{aligned} \tag{12.2}$$

See Figure 12.2 for an example with $m = 3$. Note that the chambers C_1, C_2, C_3 happen to form a gallery in this example, but we do not insist that the chambers $C_1, \ldots, C_m$ in (12.2) necessarily form a gallery in general. [On the other hand, we will see in Section 12.1.6 below that this can always be arranged.]

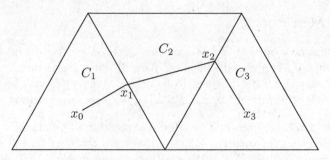

Fig. 12.2. A chain with three segments.

Note further that for the purpose of computing distances in X, we may assume that no three consecutive points x_{i-1}, x_i, x_{i+1} lie in a single $Z(C)$; for if this happened, then we could delete x_i from γ to get a chain γ' with $l(\gamma') \le l(\gamma)$.

We will prove shortly that the distance function d is a metric. The only property of a metric that is not obvious is that $d(x, y) > 0$ if $x \ne y$. The proof of this will make use of an analogue of Lemma 11.17. For any $x \in X$, we define the *star* of x, denoted by $\operatorname{st} x$, to be the union of the sets $Z(C)$ containing x. And for any $z \in Z$, we define $\epsilon(z) > 0$ by

$$\epsilon(z) := \min_{z \notin Z_s} d_Z(z, Z_s)\,, \tag{12.3}$$

where this minimum is taken to be $+\infty$ if $z \in Z_s$ for all $s \in S$. The minimum exists because S is finite, and it is positive because each Z_s is closed. Heuristically, $\epsilon(z)$ is the minimum distance from z to a wall not containing it; this is exactly the quantity δ that occurred in the proof of Lemma 11.17.

Lemma 12.9. *Given $x \in X$ of type $\tau(x) = z$, the star of x contains the open ball $\{y \in X \mid d(x, y) < \epsilon(z)\}$ of radius $\epsilon(z)$ centered at x.*

Proof. We must show that $y \notin \operatorname{st} x \implies d(x, y) \ge \epsilon(z)$. Suppose $y \notin \operatorname{st} x$, and consider a chain from x to y represented as in (12.2). Assume, as we may, that x_0, x_1, x_2 do not lie in a single $Z(C)$. [Note that there is indeed an x_2, i.e., that $m \ge 2$, by our assumption that $y \notin \operatorname{st} x$.] We claim that $d_Z(z_0, z_1) \ge \epsilon(z)$, which implies $d(x, y) \ge \epsilon(z)$. To prove the claim, observe first that $\delta(C_1, C_2) \in W_{z_1}$, since $[C_1, z_1] = [C_2, z_1]$. Now we cannot have $W_{z_1} \le W_z$, since that would imply $x_0 = [C_1, z] = [C_2, z]$ and hence $x_0, x_1, x_2 \in Z(C_2)$. So there is at least one $s \in S$ such that $z_1 \in Z_s$ but $z \notin Z_s$, and then

$$d_Z(z_0, z_1) = d_Z(z, z_1) \ge d(z, Z_s) \ge \epsilon(z)\,,$$

as claimed. $\qquad\square$

We can now prove, among other things, that d is a metric.

Proposition 12.10.

(1) *The distance function* $d\colon X \times X \to \mathbb{R}$ *in Definition 12.8 is a metric.*
(2) *The type function* $\tau\colon X \to Z$ *maps* $Z(C)$ *isometrically onto* Z *for every chamber* $C \in \mathcal{C}$.
(3) *If* Z *is complete, then so is* X.

Proof. We begin by observing that the triangle inequality in Z implies that

$$d(x, y) \geq d_Z\big(\tau(x), \tau(y)\big) \tag{12.4}$$

for any $x, y \in X$. If x and y both lie in $Z(C)$, then the opposite inequality also holds, since we then have a 1-chain from x to y of length $d_Z\big(\tau(x), \tau(y)\big)$. This proves (2). Turning now to (1), we must show that $d(x, y) > 0$ if $x \neq y$. If $\tau(x) \neq \tau(y)$, this follows from (12.4). And if $\tau(x) = \tau(y)$, it follows from Lemma 12.9. [Note that $y \notin \operatorname{st} X$, since τ is bijective on each $Z(C)$.]

Finally, we prove (3) by imitating the proof of the corresponding fact for Euclidean buildings (Theorem 11.16). Assume that Z is complete, and consider a Cauchy sequence $(x_n)_{n\geq 1}$ in X. Then the image sequence $(\tau(x_n))$ is a Cauchy sequence in Z by (12.4), so it converges to a point $z \in Z$. Choose for each n a chamber C_n with $x_n \in Z(C_n)$, and let y_n be the unique point in $Z(C_n)$ such that $\tau(y_n) = z$. In view of (2),

$$d(x_n, y_n) = d(\tau(x_n), \tau(y_n)) = d(\tau(x_n), z) \to 0 \quad \text{as } n \to \infty\,;$$

hence (y_n) is also a Cauchy sequence. On the other hand, we know that distinct points of type z are at positive distance $\geq \epsilon(z)$ from one another. Hence the Cauchy sequence (y_n) is eventually constant, and the fact that $d(x_n, y_n)$ tends to 0 now says that $x_n \to y$, where $y := y_n$ for large n. $\qquad\square$

Corollary 12.11. *Suppose* Δ *consists of a single apartment* $\Sigma = \Sigma(W, S)$ *as in Example 12.2(b). Then* W *acts on* $X = Z(W, S)$ *by isometries, and the identification of the fundamental domain with* Z *is compatible with the original metric on* Z.

(The meaning of the second statement is that the map $z \mapsto [1, z]$ is an isometry from Z onto the fundamental domain $Z(1)$, where the latter is metrized as a subspace of X.)

Proof. The first assertion is immediate from the definitions, and the second follows from part (2) of the proposition. $\qquad\square$

Exercise 12.12. There is a different way to define the metric on a Z-realization that is more in the spirit of Section 11.2. Namely, one first treats the case of an apartment and then defines $d(x, y)$ to be the distance between x and y in any apartment containing them. Carry out the details of this approach, and show that it leads to the same metric as the one defined above using chains.

12.1.3 From Apartments to Chambers and Back Again

In this subsection we focus on the situation of Corollary 12.11, but from a slightly different point of view. In many examples one *starts* with an action of W on an "apartment" X, which is already given as a metric space. One then finds suitable $(Z, (Z_s)_{s \in S})$ inside X and forms the Z-realization $Z(W, S)$. We wish to give sufficient conditions under which $Z(W, S)$ can be identified with the original X.

Suppose we are given a Coxeter system (W, S) and an action of W on a metric space X by isometries. Assume:

(1) The action has a strict fundamental domain Z (Definition 1.103).
(2) For any $z \in Z$ the stabilizer W_z is the standard subgroup $\langle S_z \rangle$, where
    $S_z := \{s \in S \mid sz = z\}$.

We then metrize Z as a subspace of X, and we set $Z_s := Z \cap sZ = Z \cap X^s$. For the sake of intuition, we will call the sets wZ ($w \in W$) *closed chambers*. Consider the Z-realization $Z(W, S)$. We have a W-equivariant bijection $\phi \colon Z(W, S) \to X$ given by $[w, z] \mapsto wz$ for $w \in W$ and $z \in Z$ (Example 12.2(b)). It maps $Z(w)$ isometrically onto the closed chamber wZ for each $w \in W$, and we seek conditions under which ϕ is a global isometry. Here is another way to formulate the problem:

Consider "chains" in X, i.e., finite sequences $x_0, x_1, \ldots, x_m$ such that any two consecutive points x_{i-1}, x_i lie in a closed chamber. The *length* of such a chain is defined to be $\sum_{i=1}^{n} d(x_{i-1}, x_i)$, where d is the given metric on X. For any $x, y \in X$, we set $d'(x, y)$ equal to the infimum of the lengths of the chains from x to y. Then ϕ is an isometry if and only if $d' = d$. In view of Lemma 12.9, there is an obvious necessary condition:

(3) For each $x \in X$, the *star* of x, defined to be the union of the closed chambers containing x, is a neighborhood of x.

It is also reasonable to assume that the following condition holds:

(4) X is a geodesic metric space, and Z is a convex subset of X in the sense that it contains all geodesics joining any two of its points.

We now show that these conditions are sufficient.

Proposition 12.13. *Under the hypotheses* (1)–(4), *the map* $\phi \colon Z(W, S) \to X$ *is an isometry.*

Proof. We claim that every geodesic segment $[x, y]$ in X can be subdivided so that each piece is contained in a closed chamber. Accepting this for the moment, we obtain a chain from x to y of length $d(x, y)$, whence $d'(x, y) \leq d(x, y)$. The triangle inequality in X implies the opposite inequality, so $d' = d$ and ϕ is an isometry. It remains to prove the claim.

For each point $u \in [x, y]$, assumption (3) implies that there is a neighborhood N_u of u relative to $[x, y]$ contained in $\operatorname{st} u$. We can take N_u to be an interval with endpoints in st u, and then the convexity of closed chambers implies that N_u is contained in a union of two chambers. Since finitely many of the N_u cover $[x, y]$, the claim follows.   □

Examples 12.14. (a) Let W be an essential finite reflection group acting on a Euclidean vector space V as in Chapter 1, and let S be the set of simple reflections with respect to a chosen fundamental chamber C. Let X be the unit sphere in V, let Z be the spherical simplex $\overline{C} \cap X$, and let Z_s be the panel of Z fixed by s. Then the proposition applies and shows that the Z-realization $Z(W, S)$ is canonically isometric to the sphere X.

(b) Let W be a Euclidean reflection group (Definition 10.12) acting on a Euclidean space V, and let S be the set of simple reflections with respect to a chosen fundamental chamber C. Let Z be the Euclidean simplex $\overline{C}$, and let Z_s be the panel of Z fixed by s. Then the proposition applies with $X = V$ and shows that $Z(W, S)$ is canonically isometric to the Euclidean space V. This generalizes to semi-Euclidean reflection groups (see Example 12.2(c)).

(c) Let W be a hyperbolic reflection group (Definition 10.50) acting on a hyperbolic space $\mathbb{H} = \mathbb{H}^n$, and let S be the set of simple reflections with respect to a chosen fundamental chamber P. Let $Z = P$ and, for $s \in S$, let Z_s be the panel of Z fixed by s. Then the proposition applies with $X = \mathbb{H}$ and shows that $Z(W, S)$ is canonically isometric to the hyperbolic space $\mathbb{H}$.

Remark 12.15. An interesting feature of the last example is that $Z(W, S)$, which is the most natural metric realization of $\Sigma = \Sigma(W, S)$, might have much lower dimension than the simplicial complex Σ. For example, if the fundamental polytope P is a pentagon in the hyperbolic plane, then $Z(W, S)$ is 2-dimensional but Σ is 4-dimensional. This phenomenon does not occur in the spherical and Euclidean cases (although it does occur in the semi-Euclidean case, as we already mentioned in connection with Example 12.2(c)).

12.1.4 The Effect of a Chamber Map

We return now to the general theory, where Z is a metric space with closed nonempty subspaces Z_s ($s \in S$). Let Δ and Δ' be buildings of type (W, S) and let $X = Z(\Delta)$ and $X' = Z(\Delta')$ be their Z-realizations. Let $\phi \colon \Delta \to \Delta'$ be a type-preserving chamber map. Then ϕ induces a map $Z(\phi) \colon X \to X'$ given by $[C, z] \mapsto [\phi(C), z]$ for all $C \in \mathcal{C}$ and $z \in Z$. We will sometimes simply write ϕ instead of $Z(\phi)$ when the meaning is clear from the context. Important examples for us are inclusions of apartments and retractions onto apartments.

Note that $Z(\phi)$ takes a chain in X to a chain in X' of the same length. This immediately implies the following:

Proposition 12.16. *Given a type-preserving chamber map* $\phi \colon \Delta \to \Delta'$, *the induced map* $\phi \colon X \to X'$ *is distance-decreasing, i.e.,*

$$d(\phi(x), \phi(y)) \leq d(x, y)$$

for all $x, y \in X$.   □

Corollary 12.17. *Let Δ be a building and Σ an apartment in Δ, viewed as a building in its own right. Then the inclusion $\Sigma \hookrightarrow \Delta$ induces an isometric embedding of $Z(\Sigma)$ into $Z(\Delta)$.*

Proof. If $\iota\colon \Sigma \hookrightarrow \Delta$ is the inclusion and $\rho\colon \Delta \to \Sigma$ is a retraction, then the induced maps on Z-realizations are distance-decreasing. Since $\rho\iota = \mathrm{id}_\Sigma$, it follows that no distances can be strictly decreased by ι.   □

A less formal way of stating the corollary is that the subspace metric on an apartment in X coincides with its intrinsic metric. (Here, for the sake of intuition, we are referring to $Z(\Sigma)$ as an *apartment* in X.) The significance of this for us is that if we want to compute the distance in X between two points x, y, then we can choose an apartment containing them and confine our attention to chains from x to y in the apartment.

Finally, we return to the case that ϕ is a retraction and prove the following sharper version of Proposition 12.16, as in the Euclidean case (Theorem 11.16).

Proposition 12.18. *Let $\rho = \rho_{\Sigma,C}$, where Σ is an apartment and C is a chamber of Σ. Then*

$$d(\rho(x), \rho(y)) \leq d(x, y)$$

for all $x, y \in X$, with equality if $x \in Z(C)$.

Proof. We need only prove the assertion about equality. Choose a chamber D with $y \in Z(D)$, and let Σ' be an apartment containing C and D. Recall that the restriction of ρ to Σ' is given by a type-preserving isomorphism $\phi\colon \Sigma' \to \Sigma$. Then ϕ induces an isometry $Z(\Sigma') \to Z(\Sigma)$, and the result follows at once.   □

Our next goal is to find sufficient conditions under which X is a geodesic metric space. This requires some preliminary results, which we give in the following two subsections.

12.1.5 The Carrier of a Point of X

In the ordinary geometric realization of Δ, each point is in a unique open simplex $|A|$. For a general Z-realization $X = Z(\Delta)$, we will similarly associate a simplex $A \in \Delta$ to each point $x \in X$. The identification of simplices with residues makes this particularly easy to do. Let $z = \tau(x)$ and note that by the definition of X, the set of chambers C such that $x \in Z(C)$ is a residue of type S_z. This set therefore equals $\mathcal{C}_{\geq A}$ for a unique simplex $A \in \Delta$ (Corollary 4.11).

Definition 12.19. Given $x \in X = Z(\Delta)$, let $\mathcal{R} = \{C \in \mathcal{C} \mid x \in Z(C)\}$. The *carrier* of x is the unique simplex $A \in \Delta$ such that $\mathcal{R} = \mathcal{C}_{\geq A}$. Equivalently, if C is any chamber with $x \in Z(C)$, the carrier A of x is the face of C of cotype S_z, where $z = \tau(x)$.

Note, by way of illustration, that if C and D are adjacent chambers and $x \in Z(C) \cap Z(D)$, then the carrier A of x is a face of the common panel $P := C \cap D$. This is almost a tautology; for we have $C, D \in \mathcal{C}_{\geq A}$ by definition, so $A \leq C \cap D$.

The rest of this subsection is devoted to some simple observations, all of which are motivated by the ordinary geometric realization.

Proposition 12.20. *Two points of X with the same carrier and the same type are equal.*

Proof. Let z be the common type of the two points, and write them as $[C, z]$ and $[D, z]$. By hypothesis, the chambers C and D have the same face of cotype S_z and hence are in the same S_z-residue. But then $[C, z] = [D, z]$ by the definition of X. □

Consider now a type-preserving chamber map $\phi \colon \Delta \to \Delta'$ between buildings of type (W, S). As in Section 12.1.4, there is an induced map $\phi \colon X \to X'$ on Z-realizations.

Proposition 12.21. *If $x \in X$ has carrier $A \in \Delta$, then $\phi(x)$ has carrier $\phi(A)$.*

Proof. Write $x = [C, z]$, so that A is the face of C of cotype S_z. Then $\phi(x) = [\phi(C), z]$, so the carrier of $\phi(x)$ is the face of $\phi(C)$ of cotype S_z; but this face is $\phi(A)$. □

Corollary 12.22. *Let $\phi_1, \phi_2 \colon \Delta \to \Delta'$ be type-preserving chamber maps between buildings of type (W, S). Let $x \in X$ have carrier $A \in \Delta$. Then $\phi_1(x) = \phi_2(x)$ if and only if $\phi_1(A) = \phi_2(A)$. In particular, a type-preserving endomorphism of Δ fixes x if and only if it fixes A.*

Proof. The second assertion follows from the first, applied with one map equal to the identity. The first assertion is an immediate consequence of Propositions 12.20 and 12.21. □

Exercise 12.23. Under what circumstances is the carrier of a point the empty simplex of Δ? Give a naturally occurring example in which this happens.

12.1.6 Chains and Galleries

Let's return to the representation (12.2) of a chain in X. Such representations are easier to work with if the chambers C_i form a gallery, as in Figure 12.2. We will show that we can always achieve this by repeating some points x_i, without

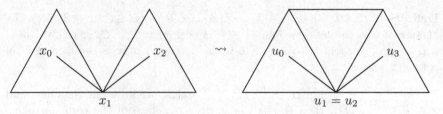

Fig. 12.3. Repeating a point to create a gallery.

changing the length $l(\gamma)$ of the chain. This simply introduces some segments of length 0. See Figure 12.3, which illustrates a 2-chain being replaced by a 3-chain in which the second point is repeated, so that the middle segment of the resulting 3-chain has length 0.

To see that this is always possible, consider two consecutive chambers $C = C_i$ and $D = C_{i+1}$ in the representation (12.2) ($1 \le i < m$), and let $x = x_i$. The equation corresponding to this point in (12.2) has the form

$$x = [C, z] = [D, z] , \tag{12.5}$$

where $z = z_i$. Then $\delta(C, D) \in W_z$, so we can join C and D by a gallery $C = E_0, \ldots, E_r = D$ ($r \ge 1$) in which each adjacency type is given by an element of S_z. We then add $r - 1$ copies of the point x to our chain, so that the equation (12.5) in the representation of the chain is replaced by r equations:

$$x = [E_0, z] = [E_1, z] ,$$
$$x = [E_1, z] = [E_2, z] ,$$
$$\vdots$$
$$x = [E_{r-1}, z] = [E_r, z] .$$

This proves our assertion.

Definition 12.24. We will say that the representation (12.2) of an m-chain is in *standard form* if $C_1, \ldots, C_m$ is a gallery.

Writing $\delta(C_i, C_{i+1}) = s_i$, we then have $s_i \in W_{z_i}$ for $i = 1, \ldots, m - 1$, and hence $z_i \in Z_{s_i}$, as in Figure 12.2. For emphasis, we add this information to our representation, setting $Z_i = Z_{s_i}$:

$$
\begin{aligned}
x_0 &= [C_1, z_0] , \\
x_1 &= [C_1, z_1] = [C_2, z_1], & z_1 &\in Z_1 , \\
x_2 &= [C_2, z_2] = [C_3, z_2], & z_2 &\in Z_2 , \\
&\;\;\vdots \\
x_{m-1} &= [C_{m-1}, z_{m-1}] = [C_m, z_{m-1}], & z_{m-1} &\in Z_{m-1} , \\
x_m &= [C_m, z_m] .
\end{aligned}
\tag{12.6}
$$

Next, we show that the gallery $\Gamma\colon C_1,\dots,C_m$ can be taken to be minimal. More precisely:

Proposition 12.25. *Let γ be a chain from x to y, represented in standard form as in (12.6). Let A be the carrier of x and let B be the carrier of y. Then there is a chain γ' from x to y with $l(\gamma') \le l(\gamma)$ such that γ' can be represented in standard form with the associated gallery being a minimal gallery from A to B.*

Proof. We may assume, as we remarked near the end of Section 12.1.4, that there is an apartment Σ such that the chain γ lies in $Y := Z(\Sigma)$ and the associated gallery Γ is in Σ. We proceed in two steps. First we show that we can find a γ' whose gallery is minimal; then we show that we can make it minimal from A to B.

Suppose Γ is not minimal. Then, as in the proof of Lemma 3.70, we can find a root α in Σ and indices i, j, with $2 \le i < j \le m$, such that C_{i-1} and C_j are in α but $C_k \in -\alpha$ for $i \le k < j$. Assume for simplicity that $i < j - 1$; if $i = j - 1$, the argument that follows needs trivial modifications.

Let $\phi\colon \Sigma \to \Sigma$ be the folding with image α, and recall from Section 12.1.4 that ϕ induces a distance-decreasing map, still denoted by ϕ, from Y to itself. If we apply ϕ to the portion $x_i, \dots, x_{j-2}$ of γ, we obtain an m-chain

$$\gamma_1\colon x_0, \dots, x_{i-1}, \phi(x_i), \dots, \phi(x_{j-2}), x_{j-1}, \dots, x_m$$

from x to y. Note that $\phi(x_{i-1}) = x_{i-1}$ and $\phi(x_{j-1}) = x_{j-1}$, as illustrated in Figure 12.4. (In that example $i = 2$, $j = m = 5$, and we have set $y_k = \phi(x_k)$ for $k = 2, 3$.) So γ_1 is indeed a chain. The consecutive pairs of points lie in

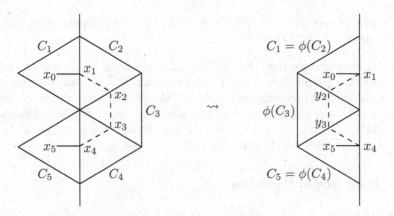

Fig. 12.4. Folding a chain.

the copies of Z corresponding to the chambers

$$C_1, \dots, C_{i-1}, \phi(C_i), \dots, \phi(C_{j-1}), C_j, \dots, C_m \; ,$$

which form a pregallery with exactly two repetitions. There are three consecutive points $x_{i-2}, x_{i-1}, \phi(x_i)$ in $Z(C_{i-1}) = Z(\phi(C_i))$ and three consecutive points $\phi(x_{j-2}), x_{j-1}, x_j$ in $Z(\phi(C_{j-1})) = Z(C_j)$. We may therefore delete the points x_{i-1} and x_{j-1} from γ_1 to obtain an $(m-2)$-chain γ' with $l(\gamma') \leq l(\gamma_1) = l(\gamma)$, and γ' admits a standard representation with gallery

$$C_1, \ldots, C_{i-1}, \phi(C_{i+1}), \ldots, \phi(C_{j-2}), C_j, \ldots, C_m .$$

If this new gallery is not minimal, we can repeat the process and reduce the number of segments again. After finitely many steps, we arrive at a chain of length $\leq l(\gamma)$ whose gallery is minimal.

We may now assume that our original m-chain γ has an associated gallery Γ that is minimal. If it is not minimal from A to B, then Proposition 3.78 implies that there is a wall $H \notin \mathcal{S}(A, B)$ such that Γ crosses H (exactly once). In other words, there is a root α with $A, B \in \alpha$ such that $C_1, \ldots, C_i \in \alpha$ and $C_{i+1}, \ldots, C_m \in -\alpha$, where $1 \leq i < m$. Let ϕ be the folding of Σ onto α. Applying ϕ to γ, we obtain an m-chain $\phi(\gamma)$ that admits a representation as in (12.2) with the associated chambers forming the pregallery $\phi(\Gamma)$. Note that ϕ fixes A and B, so $\phi(\gamma)$ is again a chain from x to y by Corollary 12.22. Since $\phi(C_{i+1}) = \phi(C_i)$, we can delete a point from $\phi(\gamma)$ to get an $(m-1)$-chain γ' from x to y with $l(\gamma') \leq l(\phi(\gamma)) = l(\gamma)$. Iterating this process, we arrive after finitely many steps at a chain of length $\leq l(\gamma)$ whose chain is minimal from A to B.   □

Corollary 12.26. *Given* $x, y \in X$, *there is a finite collection* $\mathcal{G}$ *of galleries such that*

$$d(x, y) = \inf_{\gamma} l(\gamma) ,$$

where γ *ranges over the chains from* x *to* y *that can be represented as in* (12.6) *with the associated gallery in* $\mathcal{G}$.

Proof. We may assume that $\Delta = \Sigma(W, S)$. Let A and B be the carriers of x and y as in the proposition. Then we can compute $d(x, y)$ using only chains whose galleries are minimal from A to B. Recall now that there are only finitely many such galleries modulo the action of $W_A \cap W_B$ (see Proposition 3.124). Since $W_A \cap W_B$ acts on X by isometries, we can take $\mathcal{G}$ to be a set of representatives for the orbits.   □

12.1.7 Existence of Geodesics

We begin by giving mild hypotheses under which X contains chains of minimal length joining any two points. This is a discrete analogue of the existence of geodesics. Recall that a metric space is *proper* if closed bounded sets are compact. In most examples of the present theory that arise in practice, Z is a proper metric space, and hence so is each Z_s.

Theorem 12.27. *Suppose that Z_s is a proper metric space for each $s \in S$. Then for any $x, y \in X$, there is a chain from x to y of length $d(x, y)$.*

Proof. In view of Corollary 12.26, it suffices to show that among the chains γ representable as in (12.6) with a given associated gallery Γ, there is one of minimal length. Now the length of γ is given by

$$l(\gamma) = \sum_{i=1}^{m} d_Z(z_{i-1}, z_i) \,, \tag{12.7}$$

where $z_0 = \tau(x)$, $z_m = \tau(y)$, and z_i ranges over Z_i for $i = 1, \ldots, m - 1$. The right side of (12.7) may be viewed as a continuous function

$$f \colon Z_1 \times \cdots \times Z_{m-1} \to \mathbb{R} \,,$$

and the set $\{f \leq M\}$ is compact for any real number M. This follows from the properness of Z_i together with the observation that the inequality $f(z_1, \ldots, z_{m-1}) \leq M$ forces z_i to lie in a bounded subset of Z_i for each i. [Check this by induction on i.] Hence f takes on a minimum value. □

Corollary 12.28. *If Z is a geodesic metric space and each Z_s is a proper metric space, then $X = Z(\Delta)$ is a geodesic metric space.*

Proof. Given $x, y \in X$, choose a chain $x = x_0, x_1, \ldots, x_m = y$ of length $d(x, y)$. Then there are geodesics $[x_{i-1}, x_i]$ for each $i = 1, \ldots, m$, and we can concatenate these to obtain a geodesic $[x, y]$ by Exercise 11.9. □

12.1.8 Curvature

Results like those in the previous subsection will be familiar to readers who know the theory of piecewise polyhedral complexes [47, 48], but the proofs above are considerably easier than those in the piecewise polyhedral setting. One might therefore hope that it will also be easy to prove curvature properties of X (such as the CAT(0) property) under appropriate hypotheses on Z and the subspaces Z_s. Unfortunately, we do not know of any general results in this direction. All we can say is that in view of the following proposition, it suffices to understand apartments.

Proposition 12.29. *Suppose that $Z(W, S)$ is a CAT(κ) space for some real number κ. Then the Z-realization of any building Δ of type (W, S) is a CAT(κ) space.*

(See Remark 11.6(b) for the definition of "CAT(κ) space.")

The proof is essentially the same as the proof that Euclidean buildings are CAT(0) spaces (Theorem 11.16), once we establish some basic properties of the CAT(κ) property for $\kappa \neq 0$. Note that we can scale the metric to reduce to the cases $\kappa = \pm 1$, so we will confine ourselves to those two cases. We begin with the spherical and hyperbolic analogues of the generalized parallelogram law (equation (11.2) in Section 11.1).

Lemma 12.30. *Let X be the 2-sphere S^2 or the hyperbolic plane $\mathbb{H}^2$, and let x, y, z be three points in X. If $X = S^2$, assume that x and y are not antipodal. Set $c := d(x, y)$, and for $0 \leq t \leq 1$, let $p = p_t$ be the point on the geodesic $[x, y]$ such that $d(x, p) = td(x, y)$.*

(1) *If $X = S^2$, then*

$$\cos d(z, p) = \frac{\sin(1 - t)c}{\sin c} \cos d(z, x) + \frac{\sin tc}{\sin c} \cos d(z, y) \, .$$

(2) *If $X = \mathbb{H}^2$, then*

$$\cosh d(z, p) = \frac{\sinh(1 - t)c}{\sinh c} \cosh d(z, x) + \frac{\sinh tc}{\sinh c} \cosh d(z, y) \, .$$

(In both equations we allow the uninteresting possibility that $c = 0$, in which case we interpret the fractions in terms of their limiting values $1 - t$ and t.)

Proof. Assume first that $X = S^2$, and recall (from [48, Chapter I.2], for instance) that the distance function is given by $\cos d(u, v) = \langle u, v \rangle$, where the right side denotes the inner product of unit vectors in $\mathbb{R}^3$. Our task, then, is to compute $\langle z, p \rangle$, for which the key step is to observe that

$$p = \frac{\sin(1 - t)c}{\sin c} x + \frac{\sin tc}{\sin c} y \, . \tag{12.8}$$

Taking the inner product with z, we immediately obtain the desired formula. To prove (12.8), note that there is a unit vector x' orthogonal to x such that

$$y = (\cos c)x + (\sin c)x' \, ,$$

and then

$$p_t = (\cos tc)x + (\sin tc)x'$$

for $0 \leq t \leq 1$. Eliminating x' from these two equations, one obtains (12.8) after a short computation.

If $X = \mathbb{H}^2$, the same proof is valid with no essential change if one uses the hyperboloid model of the hyperbolic plane; see equation (10.15) in Section 10.3.1.   □

This leads at once to the following $\mathrm{CAT}(\kappa)$ criterion, analogous to Proposition 11.4:

Proposition 12.31. *Let X be a metric space.*

(1) *X is a $\mathrm{CAT}(1)$ space if and only if for any $x, y \in X$ with $d(x, y) < \pi$, there is a geodesic $[x, y]$ with the following property: For any $p = p_t \in [x, y]$ and any $z \in X$ with $d(x, y) + d(y, z) + d(z, x) < 2\pi$, we have $d(z, p) \leq \pi$ and*

$$\cos d(z, p) \geq \frac{\sin(1 - t)c}{\sin c} \cos d(z, x) + \frac{\sin tc}{\sin c} \cos d(z, y) \, ,$$

where $c = d(x, y)$.

(2) X *is a* $\mathrm{CAT}(-1)$ *space if and only if for any* $x, y \in X$, *there is a geodesic* $[x, y]$ *with the following property: For any* $p = p_t \in [x, y]$ *and any* $z \in X$,

$$\cosh d(z, p) \leq \frac{\sinh(1 - t)c}{\sinh c} \cosh d(z, x) + \frac{\sinh tc}{\sinh c} \cosh d(z, y) ,$$

where $c = d(x, y)$. □

There are also analogues of Propositions 11.5, 11.7, and 11.8, which the interested reader can formulate and prove. We now return to Z-realizations.

Proof of Proposition 12.29. The proof is essentially the same as in the case of Euclidean buildings (Theorem 11.16), but we will repeat it for the convenience of the reader. Assume, for example, that $\kappa < 0$, in which case we may assume that $\kappa = -1$. The same method works for the other cases.

Given $x, y \in X := Z(\Delta)$, choose an "apartment" $E = Z(\Sigma)$ containing them. It is a $\mathrm{CAT}(-1)$ space by hypothesis. Let $[x, y]$ be the geodesic joining them in E, and note that it is also a geodesic in X by Corollary 12.17. Given $z \in X$ and $p = p_t \in [x, y]$, choose a chamber C of Σ with $p \in Z(C)$, and let ρ be the retraction $\rho_{\Sigma, C}$. By Proposition 12.18 we have $d(\rho(z), x) \leq d(z, x)$, $d(\rho(z), y) \leq d(z, y)$, and $d(\rho(z), p) = d(z, p)$. There are now two ways to finish. Method 1: According to the criterion of Proposition 12.31, we must verify

$$\cosh d(z, p) \leq \frac{\sinh(1 - t)c}{\sinh c} \cosh d(z, x) + \frac{\sinh tc}{\sinh c} \cosh d(z, y) ,$$

where $c = d(x, y)$. This follows from the $\mathrm{CAT}(-1)$ property for E, which implies

$$\cosh d(\rho(z), p) \leq \frac{\sinh(1 - t)c}{\sinh c} \cosh d(\rho(z), x) + \frac{\sinh tc}{\sinh c} \cosh d(\rho(z), y) .$$

Method 2: Use the original definition of $\mathrm{CAT}(-1)$. Choose a comparison triangle $\bar{x}, \bar{y}, \bar{z} \in \mathbb{H}^2$ for x, y, z, and choose $\bar{z}' \in \mathbb{H}^2$ such that $\bar{x}, \bar{y}, \bar{z}'$ is a comparison triangle for $x, y, \rho(z)$. We now have

$$\begin{aligned} d(z, p) = d(\rho(z), p) \quad &\text{as noted above} \\ \leq d(\bar{z}', \bar{p}) \quad &\text{by the } \mathrm{CAT}(-1) \text{ property for } E \\ \leq d(\bar{z}, \bar{p}) , \end{aligned}$$

where the last inequality follows from the hyperbolic analogue of Proposition 11.8. □

Curvature properties of apartments themselves are not yet well understood. In other words, we do not know of any general conditions on Z and the Z_s under which we can conclude that $Z(W, S)$ is a $\mathrm{CAT}(\kappa)$ space. In particular, we would like an answer to the following question:

Question 12.32. If Z is a CAT(0) space, under what conditions is $Z(W, S)$ a CAT(0) space?

In attempting to answer this question, one might try first to answer the following two questions, which are motivated by results in Bridson–Haefliger [48, Chapters II.4 and II.5]. For simplicity, assume that Z is a CAT(0) space as in Question 12.32. Further hypotheses on Z and the Z_s might be necessary in order for the questions to be reasonable.

Question 12.33. Is $Z(W, S)$ locally CAT(0) if it locally has unique geodesics?

Question 12.34. Under what conditions does $Z(W, S)$ locally have unique geodesics?

Remark 12.35. Easy examples show that some condition is required in Question 12.34. Consider, for example, the Euclidean reflection group

$$W := \left\langle\, s, t, u \; ; \; s^2 = t^2 = u^2 = (st)^4 = (tu)^4 = (su)^2 = 1 \,\right\rangle\,,$$

generated by the reflections of the plane in the sides of an isosceles right triangle. [This is the Coxeter group of type $\tilde{C}_2$; see Example 10.14(b).] The "natural" Z for this group is an isosceles right triangle, and $Z(W, S)$ is then isometric to the plane (Example 12.14(b)). A portion of $Z(W, S)$ is shown in Figure 12.5. Suppose now that we use the same combinatorial data but

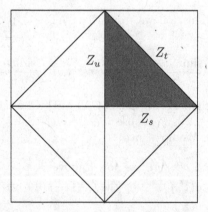

Fig. 12.5. The group of type $\tilde{C}_2$ and its fundamental chamber.

change the metric on Z so that it is an equilateral triangle. Then we get the same $Z(W, S)$ as a set, but now we have positive curvature at the point in the center of the figure. The new $Z(W, S)$, near that point, looks metrically like the boundary of a regular octahedron near one of its vertices, and geodesics are not locally unique.

Exercises

12.36. Give an alternative proof of Lemma 12.30 based on the law of cosines, as in Exercise 11.10.

12.37. Write down the $\mathrm{CAT}(\kappa)$ analogue of (NC) for $\kappa = \pm 1$.

12.38. Generalize Exercise 11.11 to the spherical and hyperbolic cases.

12.2 Special Cases

In this section we list some examples of metric realizations that have occurred in the literature and that fit into our framework. We also mention one example that does not fit.

Example 12.39. For every spherical building Δ, the ordinary geometric realization $|\Delta|$ admits a canonical CAT(1) metric, obtained as follows. Let Δ have type (W, S); then W is finite. Let Z be a simplex with vertex set S as in Example 12.2(a), so that the Z-realization X of Δ can be identified with $|\Delta|$ as a set. We metrize Z as a spherical simplex using the canonical representation of W as a finite reflection group (Section 1.5.5) and intersecting the fundamental chamber with the unit sphere. This gives us a metric on X such that every apartment is isometric to the standard unit sphere of dimension $|S| - 1$ (see Example 12.14(a)). In view of Proposition 12.29, X is a (complete) CAT(1) space.

The metric on $|\Delta|$ that we have just described is the one alluded to in Remark 4.74, which can be used to give a geometric proof of the Solomon–Tits theorem.

Example 12.40. For every Euclidean building Δ, the CAT(0) metric on the geometric realization $|\Delta|$ that we discussed in detail in Chapter 11 is a special case of the construction in the present chapter. Since we get nothing new from this, we will just give a brief indication of why this is true, leaving the details to the interested reader:

As in the previous example, take Z to be a simplex with vertex set S. It has a canonical Euclidean metric. To show that the resulting metric on the Z-realization is the same as the metric constructed in Chapter 11, it suffices to consider the case of an apartment, since any two points are contained in an apartment. Now apply Example 12.14(b).

Before treating our next example, we introduce the notion of *hyperbolic building*, with a warning that there is more than one possible definition.

Definition 12.41. A *hyperbolic building* is a building whose Weyl group (W, S) is a hyperbolic Coxeter system in the sense of Definition 10.57.

Example 12.42. Let Δ be a hyperbolic building of type (W, S). Choose a realization of (W, S) as a hyperbolic reflection group acting on $\mathbb{H}^n$ with fundamental domain P. Let $Z = P$ with its hyperbolic metric, and let Z_s for $s \in S$ be the face of P fixed by s. In view of Example 12.14(c), the Z-realization of any apartment is the hyperbolic space $\mathbb{H}^n$, with its canonical metric. The Z-realization X of Δ is then a (complete) CAT(-1) space by Proposition 12.29.

The study of hyperbolic buildings via their metric realizations is a very active area of research at the time of this writing, largely because many examples arise from the theory of Kac–Moody groups (see Remark 8.97). In fact, it is this realization itself that is often called a "hyperbolic building." See Gaboriau–Paulin [104] for an extensive study of hyperbolic buildings from this point of view, along with many examples. See also [45, 46, 266, 267].

Example 12.43. Let Δ be an arbitrary building, with Weyl group (W, S). Let Z be the geometric realization of the flag complex of the poset of spherical subsets of S as in Example 12.2(e). There is a piecewise Euclidean metric on Z obtained by identifying each simplex with a suitable Euclidean simplex. We will explain this in detail in Section 12.3, where we will quote a theorem of Moussong, according to which the Z-realization of an apartment is always a CAT(0) space. The Z-realization of Δ is then a (complete) CAT(0) space by Proposition 12.29; see Theorem 12.66 below. This is the *Davis realization* mentioned in the introduction to this chapter.

If Δ is spherical, then the Davis realization, as a set, is the cone over the ordinary geometric realization; each apartment is the cone over a sphere and is metrically a ball. If Δ is Euclidean, the Davis realization is the same as the usual one that we studied extensively in Chapter 11. If Δ is hyperbolic, however, the Davis realization is *not* in general the same as the one in Example 12.42, even as a set. For example, suppose Δ is a single apartment $\Sigma(W, S)$, where $W = \mathrm{PGL}_2(\mathbb{Z})$. Then the realization in Example 12.42 is the hyperbolic plane with its standard metric, whereas the Davis realization is the subset of the plane obtained by "cutting off the cusps"; see Figure 12.11, where the Davis realization is the shaded part. Moreover, the metric on the Davis realization is not the hyperbolic one, but rather a piecewise Euclidean metric (which is Lipschitz equivalent to the hyperbolic metric).

Example 12.44. Let Δ be a building of type (W, S), where W is hyperbolic in the sense of Gromov. Moussong has characterized such Coxeter groups W; see Section 12.3.9 below. Moussong has also shown that the Davis realization of an apartment admits a different metric in this case, obtained by using the same set Z but giving it a piecewise hyperbolic metric instead of a piecewise-Euclidean metric. The new metric makes the Z-realization X of Δ a (complete) CAT(-1) space; see Section 12.3.9 below.

Example 12.45. Our final example is actually a nonexample, i.e., it is a metric realization that can be defined for an arbitrary Coxeter group W but that does *not* fit into our framework. Namely, Niblo and Reeves [180] have constructed a finite-dimensional CAT(0) cubical complex on which W acts, where the cubes are metrically equivalent to the standard cube. (See Section A.2.1 for the notion of "cubical complex.") The Niblo–Reeves complex has had useful applications to the study of Coxeter groups; see [68, 70], for example. But the construction works only for Coxeter groups and does not lead to realizations of arbitrary buildings.

The rest of this chapter is devoted to filling in some of the missing details in the construction of the Davis realization.

12.3 The Dual Coxeter Complex

In Example 12.43 above we described the choice of $(Z, \{Z_s\}_{s \in S})$ that will give the Davis realization of a building as soon as we explain the metric on Z. It would be possible to do that fairly quickly as in [88], but we will instead give a more long-winded treatment that reveals some interesting combinatorial geometry, beginning with the case of a single apartment. Readers who just want to get the main ideas may wish to concentrate on the introductory remarks and examples (Sections 12.3.1 and 12.3.2) and ignore the rigorous construction (Section 12.3.3).

12.3.1 Introduction

Given a Coxeter system (W, S), recall that the Coxeter complex $\Sigma = \Sigma(W, S)$ is the poset of standard cosets in W, ordered by reverse inclusion. It is a simplicial complex on which W acts, and the stabilizers of the nonempty simplices are the proper parabolic subgroups of W. Now when one is using geometric methods to study a group, it is often convenient to have an action with finite stabilizers. This leads one to try to construct a modified Coxeter complex, in which the stabilizers are the *finite* parabolic subgroups.

A natural starting point is the subposet $\Sigma_f = \Sigma_f(W, S)$ of Σ consisting of the finite standard cosets. Note, however, that Σ_f is not a subcomplex of Σ in general, i.e., it is not closed under passage to faces. In fact, it has the dual property that if $A < B$ in Σ and $A \in \Sigma_f$, then $B \in \Sigma_f$. So perhaps we should reverse the ordering, i.e., we should consider the poset of finite standard cosets ordered by inclusion instead of reverse inclusion. This poset does indeed turn out to be the poset of cells of a cell complex (generally not simplicial), which we will call the *dual Coxeter complex of* (W, S) and denote by Σ_d or $\Sigma_d(W, S)$.

The upshot of all this will be the following. Given (W, S), there are two complexes on which W acts. The first is the ordinary Coxeter complex Σ, which is a simplicial complex whose nonempty simplices correspond to the proper standard cosets in W, ordered by reverse inclusion. The stabilizers of the nonempty simplices are the proper parabolic subgroups of W. The second is the dual Coxeter complex Σ_d, which is a regular cell complex whose (nonempty) cells correspond to the *finite* standard cosets in W, ordered by inclusion. The stabilizers are the *finite* parabolic subgroups of W.

Most of this section will be devoted to a combinatorial discussion of Σ_d, independent of Section 12.1. At the end, however, we will make the connection with metric realizations. We proceed now to the details, starting with some motivating examples. The reader may find it helpful to glance first at Section A.2 for terminology regarding cell complexes.

12.3.2 Examples

Example 12.46. Let W be the rank-1 Coxeter group $\langle s \rangle$, thought of as a reflection group acting on $\mathbb{R}$. Its ordinary Coxeter complex Σ is combinatorially the 0-sphere, with two 0-simplices and the empty simplex. But it is more convenient for us to identify Σ with the poset of conical cells (two open half-lines, and their common face $\{0\}$). If we draw the chamber graph on top of a picture of $\mathbb{R}$, we see a regular cell complex with one 1-cell and two 0-cells, which are faces of the 1-cell (Figure 12.6). This will be the dual Coxeter complex Σ_d. Note that the poset of (nonempty) cells is isomorphic to Σ, with the order

$$0$$

Fig. 12.6. The dual Coxeter complex of type A_1.

reversed. Note also that the action of s flips the 1-cell, interchanging its two vertices.

Example 12.47. Let W be the dihedral group of order 6, i.e., the Coxeter group of type A_2, viewed as a reflection group acting on $\mathbb{R}^2$. Its ordinary Coxeter complex Σ is a circle decomposed into six arcs by the three reflecting hyperplanes. Equivalently, it is the hexagon pictured in Figure 12.7. It has 13 simplices, counting the empty simplex: six of rank 2, six of rank 1, and one of rank 0. The dual Coxeter complex Σ_d will turn out to be the solid hexagon shown in Figure 12.8, which appears naturally if one draws the chamber graph of Σ on top of the picture of $\mathbb{R}^2$ and then fills it in. It is a regular cell complex with six 0-cells, six 1-cells, and one 2-cell. Note that the stabilizer of a 1-cell is a rank-1 Coxeter group, acting on that edge as in Example 12.46. The stabilizer of a 0-cell, however, is trivial. In fact, the 0-cells correspond to the chambers of Σ, and W permutes these simply transitively.

This example generalizes in an obvious way to the dihedral group of order $2m$, in which case the dual Coxeter complex is a solid $2m$-gon, with a W-action that permutes the vertices simply transitively.

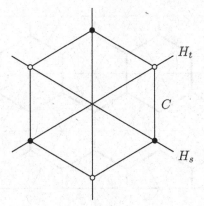

Fig. 12.7. The Coxeter complex of type A_2.

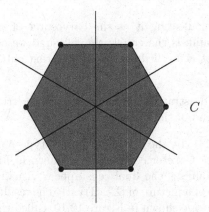

Fig. 12.8. The dual Coxeter complex of type A_2.

Example 12.48. Let W be the Coxeter group of type $\tilde{A}_2$. Its Coxeter complex Σ is the plane tiled by equilateral triangles. If we draw the chamber graph (or Cayley graph) on top of a picture of Σ as in Figure 12.9, we see the honeycomb tiling of the plane by hexagons. This yields a decomposition of the plane as a regular cell complex, with the 2-cells being solid hexagons. This will turn out to be the dual Coxeter complex Σ_d. Note that the stabilizer of each 2-cell is a dihedral group of order 6, acting on that cell as in Example 12.47.

We can describe the cells of Σ_d in the following way, which will be familiar to readers who have seen dual cell decompositions of triangulated manifolds as in [129, p. 232; 179, Section 64]: For each vertex $v \in \Sigma$, its link in Σ is a hexagon. Inside the barycentric subdivision of Σ we see a smaller copy of that link, barycentrically subdivided, and we cone it off from v to get a solid hexagon (said to be *dual* to v). For each edge $e \in \Sigma$, its link is a 0-sphere (two vertices), and we cone off a smaller copy of that link in the barycentric subdivision, using the barycenter of e as the cone point, to get an edge dual

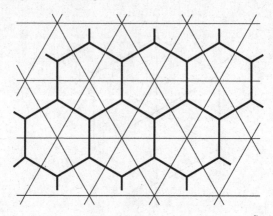

Fig. 12.9. The dual Coxeter complex of type $\tilde{A}_2$.

to e. It cuts across e, meeting it at the barycenter of e. Finally, for each 2-simplex $\sigma \in \Sigma$, its link is the empty simplex, and we cone it off from the barycenter of σ to get a 0-cell dual to σ, consisting only of the barycenter of σ.

Our final example illustrates the same idea, with the triangulated manifold being the sphere instead of the plane.

Example 12.49. Let W be the Coxeter group of type A_3 (symmetric group on four letters), whose Coxeter complex Σ is shown in Figures 0.1 and 1.3. As in the previous example, the dual cells appear visually if one draws the chamber graph on top of a picture of $|\Sigma|$. [Try it!] The result is combinatorially equivalent to the polytope shown in Figure 12.10, called the *permutahedron*.[*] The interior of this polytope is included as a cell in the dual Coxeter complex; it can be thought of as dual to the empty simplex of Σ. [This phenomenon did not occur in Example 12.48, because the empty simplex there is not in Σ_f.]

Note that the 2-cells are solid squares or hexagons, which are the dual Coxeter complexes of the rank-2 parabolic subgroups of W (see Example 12.47).

12.3.3 Construction of the Dual Coxeter Complex

Let Σ be an arbitrary Coxeter complex. The link of every simplex is again a Coxeter complex; hence if it is finite, it triangulates a sphere. Intuitively, the construction that follows consists in coning off smaller copies of these spheres in the barycentric subdivision of Σ to get cells, as in Examples 12.48 and 12.49.

Let Σ_f be the subposet of Σ consisting of the *spherical simplices* in Σ, i.e., the simplices $A \in \Sigma$ whose link $\mathrm{lk}_\Sigma A$ is finite. If Σ is given to us

---

[*] The permutahedron in Figure 12.10 was drawn by Frank Sottile and is reproduced here with his permission.

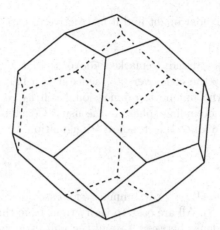

Fig. 12.10. The permutahedron.

as $\Sigma(W, S)$, then Σ_f is the set of simplices whose stabilizer W_A is finite or, equivalently, it is the set of finite standard cosets, ordered by reverse inclusion. Let $X = |K(\Sigma_f)|$, the geometric realization of the flag complex of Σ_f. Thus X is decomposed into simplices, one for each chain $A_0 < \cdots < A_p$ of spherical simplices in Σ. To make a connection with Section 12.3.2 above, we remark that if Σ is infinite, everything here can be viewed as taking place in the barycentric subdivision of Σ (see Example A.5). If Σ is finite, on the other hand, then the poset Σ_f ($= \Sigma$) contains the empty simplex, and so we are working in the cone over the barycentric subdivision; one can think of the cone point as the barycenter of the empty simplex.

For each $A \in \Sigma_f$, let $e_A := |K(\Sigma_{\geq A})| \subseteq X$. Thus e_A is homeomorphic to the cone over the barycentric subdivision of $\mathrm{lk}_\Sigma A$. This link is a finite Coxeter complex, hence a sphere, so e_A is a topological ball. Its relative interior is the union of the open simplices in X corresponding to the chains $A_0 < \cdots < A_p$ with $A_0 = A$, while its boundary corresponds to the chains with $A_0 > A$. It is now a routine matter to verify that the set of cells e_A ($A \in \Sigma_f$) is a regular cell complex with underlying space X.

Definition 12.50. We denote by Σ_d the regular cell complex consisting of the cells e_A (where A ranges over the spherical simplices of Σ), and we call Σ_d the *dual* of Σ. If $\Sigma = \Sigma(W, S)$ for some Coxeter system (W, S), then we set $\Sigma_d = \Sigma_d(W, S)$ and call it the *dual Coxeter complex* of (W, S).

Note that the correspondence $A \mapsto e_A$ is order-reversing, so there is an *order-preserving* bijection between the (nonempty) cells of $\Sigma_d(W, S)$ and the finite standard cosets in W, ordered by inclusion.

Remark 12.51. If we had wanted to rely on Appendix A, we could have simply applied Proposition A.25 to the poset of finite standard cosets (ordered by inclusion). The reader might find it instructive to carry out this approach as an exercise.

We repeat, as we already remarked above, that if Σ is finite then $X = |\Sigma_d|$ is topologically the cone over $|\Sigma|$; hence it is a ball. In fact, the whole space is equal to the unique top-dimensional cell $e_\emptyset$ corresponding to the empty simplex. Its bounding sphere is the usual Coxeter complex, with the cell decomposition dual to the standard triangulation.

12.3.4 Properties

Fix a Coxeter system (W, S). We summarize for ease of reference some properties of $\Sigma_d = \Sigma_d(W, S)$. All are easy to verify right from the definition and/or from the correspondence between Σ_d and the finite standard cosets. Some of these properties were already stated in the course of the construction of Σ_d.

(1) W operates on Σ_d, and the cell stabilizers are the finite parabolic subgroups of W. [Warning: In contrast to the situation for the ordinary Coxeter complex, the cell stabilizers do not fix the cells pointwise. For example, the stabilizer of every edge is a group of order 2 whose generator flips the edge.]
(2) The 1-skeleton of Σ_d is the Cayley graph of (W, S).
(3) There is a set of cells $L \subseteq \Sigma_d$ such that the cell stabilizers W_e for $e \in L$ are the finite standard parabolic subgroups of W. More precisely, the function $e \mapsto W_e$ is an order-preserving bijection from L to the set of finite standard parabolic subgroups of W, ordered by inclusion. Every cell in Σ_d is W-equivalent to a unique cell in L.
(4) The bijection in (3) extends to an order-preserving, W-equivariant bijection from the cells of Σ_d to the set of finite standard cosets wW_J in W.
(5) If W is finite, then $|\Sigma_d|$ is a ball of dimension equal to the rank of (W, S).
(6) For any standard subgroup W_J of W ($J \subseteq S$), the dual Coxeter complex $\Sigma_d^J := \Sigma(W_J, J)$ is a subcomplex of Σ_d. In case W_J is finite, the ball $|\Sigma_d^J|$ is the cell of Σ_d corresponding to W_J under the bijection in (3).

Note that L in (3) is simply the set of cells e_A such that A is a face of the fundamental chamber in Σ.

The reader may find it helpful to explicitly work through the definition of Σ_d in the case of Example 12.48 (starting with the barycentric subdivision of Σ) to see how it leads to the picture in Figure 12.8. Example 12.49 is also instructive; here one starts with the cone over the barycentric subdivision of Σ.

We close this subsection with one nontrivial property of the dual Coxeter complex.

Proposition 12.52. *The underlying space $X = |\Sigma_d|$ is always contractible.*

Sketch of proof. As in Section 4.12, our proof will be complete except for (routine) homotopy-theoretic details. The proposition is trivial if Σ is finite, so we may assume that it is infinite and hence contractible (Theorem 4.127). Let Σ' be the set of nonempty simplices. Then the flag complex $K(\Sigma')$ is the barycentric subdivision of Σ and hence is also contractible. So it suffices to show that the inclusion of $X = |K(\Sigma_f)|$ into $|K(\Sigma')|$ is a homotopy equivalence.

We can build the poset Σ' from Σ_f by successively adjoining elements $A \in \Sigma' \smallsetminus \Sigma_f$ in order of increasing codimension of A as a simplex of Σ. Since Σ_f already contains all simplices of codimensions 0 and 1 (chambers and panels), this means that we start with the simplices A of codimension 2, then those of codimension 3, and so on. For fixed codimension, the adjunctions can be done in any order. Each time we adjoin an element A of $\Sigma' \smallsetminus \Sigma_f$, all $B > A$ are already present, but no $B \le A$ is present. On the level of flag complexes, then, the effect of the adjunction is to cone off the subcomplex $K(\Sigma_{>A})$. But $K(\Sigma_{>A})$ is the barycentric subdivision of $\mathrm{lk}_\Sigma A$, which is an infinite Coxeter complex and hence is contractible. It follows that none of the adjunctions change the homotopy type. $\square$

Exercises

12.53. What is L in Example 12.48? Locate the cells of L in Figure 12.9 and describe their stabilizers.

12.54. If (W, S) has irreducible components $(W_1, S_1), \ldots, (W_k, S_k)$, describe $\Sigma_d(W, S)$ in terms of the cell complexes $\Sigma_d(W_i, S_i)$. Contrast this with the behavior of the ordinary (simplicial) Coxeter complex $\Sigma(W, S)$.

12.3.5 Remarks on the Spherical Case

Let (W, S) be a Coxeter system with W finite. We outline here an alternative construction of $\Sigma_d = \Sigma(W, S)$ based on the duality theory for polytopes as given, for instance, in Ziegler [286]. To apply this theory one needs to know that $\Sigma(W, S)$ can be realized as the boundary of a polytope $\widehat{\Sigma}$, whose facets cut across the chambers. We have not mentioned this before, but it is a general fact about the cell decomposition of the sphere induced by an essential hyperplane arrangement $\mathcal{H}$. A proof is given in Appendix A (see Section A.2.3). Other proofs can be found in [37, Example 4.1.7; 286, Corollary 7.18].

A simple illustration of this fact is given in Figure A.4. For a second example, consider the group of type A_3 discussed in Example 12.49, and imagine "flattening out" each spherical triangle in Figure 0.1 or 1.3, i.e., replacing it by the (Euclidean) convex hull of its vertices. The resulting polytope is $\widehat{\Sigma}$.

By the well-known duality theory of polytopes, which we briefly review in Section A.2.2, $\widehat{\Sigma}$ has a dual polytope P, called the *zonotope* associated with $\mathcal{H}$. For the reflection arrangement of type A_3, P is the permutahedron in Figure 12.10. See Ziegler [286, Example 0.10] for more information about the permutahedron and for further pictures.

Returning now to the case of a general reflection arrangement, duality theory gives an order-reversing correspondence between the faces of $\widehat{\Sigma}$ and those of P, including the entire polytope and the empty face in both cases, so P provides a model for Σ_d. [Recall that a polytope is naturally the underlying space of a regular cell complex; see Section A.2.2.] In particular, the W-action is simply transitive on the vertices of P. So we can describe P as the convex hull of a generic W-orbit in the ambient space of the reflection representation of W. For definiteness, we could take the W-orbit of the point of the fundamental chamber at distance 1 from each wall. [Such a point exists and is unique because C is a simplicial cone.] Thus we could have dispensed with duality theory and simply taken this description of P as the definition of the dual Coxeter complex in the finite case. This approach is taken in Charney–Davis [79, Lemma 2.1.3].

12.3.6 The Euclidean and Hyperbolic Cases

If W is a Euclidean reflection group acting on a Euclidean space V, then $|\Sigma_d|$ can be identified with V, decomposed into cells dual to the (nonempty) simplices of Σ. Here we can appeal to the theory of manifolds or simply to Proposition 10.13. Example 12.48 described a special case of this.

The situation is similar if W is a hyperbolic reflection group (see Proposition 10.52), except that $|\Sigma_d|$ is not the entire hyperbolic space if the fundamental polyhedron has vertices at infinity. The Coxeter group $W = \mathrm{PGL}_2(\mathbb{Z})$, whose Coxeter complex was shown in Figure 2.4, provides an instructive example. If one draws the chamber graph on top of that picture, one sees a pattern of hyperbolic hexagons and squares. The cell complex Σ_d is obtained by filling these in. One can view it as a truncated hyperbolic plane obtained by cutting off neighborhoods of the cusps. The result is a "thickened tree," decomposed into solid hexagons and squares, as in Figure 12.11.* (The underlying picture of Σ in that figure looks slightly different from the one in Figure 2.4, but it is equivalent under a hyperbolic isometry.)

12.3.7 A Fundamental Domain

We return to an arbitrary Coxeter system (W, S) and its dual Coxeter complex Σ_d. The action of W on $X = |\Sigma_d|$, unlike the action on the ordinary Coxeter complex, has the property that a cell stabilizer generally does *not* fix that cell pointwise. So we should not expect to find a strict fundamental domain that is a union of cells. On the other hand, it is easy to describe a strict fundamental domain if we go back to our original definition of X in

---

* Figure 12.11 was drawn by Kai-Uwe Bux; we are grateful to him for permission to reproduce it here.

Fig. 12.11. The Coxeter complex and its dual for $W = \mathrm{PGL}_2(\mathbb{Z})$.

Section 12.3.3 as the geometric realization of the simplicial complex $K(\Sigma_f)$. Indeed, it is immediate that the subcomplex $K(\mathcal{S})$ is a simplicial fundamental domain for the action of W on $K(\Sigma_f)$, where $\mathcal{S}$ is the poset of spherical subsets of S as in Example 12.2(e). So the space Z of that example sits inside X and is a strict fundamental domain for the action of W on X.

To understand this geometrically, recall that $K(\Sigma_f)$ is a subcomplex of the barycentric subdivision of Σ (or, if W is finite, the cone over this barycentric subdivision). Our fundamental domain Z, then, is obtained by taking the union of the simplices that lie in the closed fundamental chamber $\overline{C}$ (or the cone over it). In other words, Z is the union of the simplices with finite stabilizer in the barycentric subdivision of $\overline{C}$ (or the cone over that barycentric subdivision).

Consider, for example, the group

$$W = \mathrm{PGL}_2(\mathbb{Z}) = \left\langle\, s, t, u \,;\, s^2 = t^2 = u^2 = (st)^3 = (su)^2 = 1 \,\right\rangle$$

as in Figure 12.11. A schematic picture of the fundamental chamber is shown in Figure 12.12, with the fundamental domain Z shaded. Each vertex is labeled by the fundamental reflections that fix it, and for easier comparison with Figure 12.11, we have drawn two of the edges of the barycentric subdivision as dashed lines. Note that the unlabeled vertex in the figure is *not* part of the fundamental domain (or even of X), because it is fixed by t and u, which generate an infinite dihedral group.

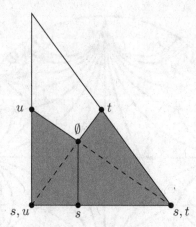

Fig. 12.12. A fundamental domain for $\mathrm{PGL}_2(\mathbb{Z})$ acting on $|\Sigma_d|$.

Returning to the general case, note that the part of Z fixed by a fundamental reflection s is precisely the subset called Z_s in Example 12.2(e). The following result is now immediate.

Proposition 12.55. *As a set, X is canonically in 1–1 correspondence with the Z-realization of Σ.* $\square$

Exercise 12.56. Draw a picture of the fundamental domain Z if $W = D_6$ as in Figure 12.8. Label the vertices in a way that indicates the identification of Z with $|K(\mathcal{S})|$.

12.3.8 A CAT(0) Metric on X

We continue to consider an arbitrary Coxeter system (W, S) and its dual Coxeter complex Σ_d. In view of Section 12.3.5, every cell of Σ_d can be identified with a polytope in Euclidean space and hence has a canonical metric. The complex Σ_d is therefore an example of a *piecewise Euclidean polyhedral complex*. There is a well-developed theory of such complexes; see, for example, [48, Chapter I.7], which is based on Bridson's thesis [47]. The theory implies, in particular, that one can put a metric on $X = |\Sigma_d|$ by minimizing lengths of piecewise linear paths. In this way X becomes a complete geodesic metric space.

Remark 12.57. In some of the literature one sees a slight variant of this, in which X is subdivided into metric simplices. Here the simplicial decomposition is given by the original definition of X as $|K(\Sigma_f)|$, and each simplex inherits a metric from a cell of Σ_d that contains it. This approach yields the same metric on X. Another variant is to give the fundamental domain Z its canonical metric as a finite piecewise Euclidean simplicial complex and then to extend this metric to X by the theory developed in Section 12.1 (which is applicable by Proposition 12.55). This approach again yields the same metric, and it has the technical advantage that we need to appeal to Bridson's theory only in the case of a finite complex. Note that the assumption of properness in some of the results of Section 12.1 is automatically satisfied here, since Z is a finite union of simplices and hence is compact.

If (W, S) is Euclidean, it follows from the remarks in Section 12.3.6 that we recover the ambient Euclidean space with its Euclidean metric (up to a constant factor). In the hyperbolic case, X is a subspace of the ambient hyperbolic space, and the piecewise Euclidean metric is Lipschitz equivalent to the hyperbolic metric (i.e., each is bounded by a constant multiple of the other).

Returning now to the general case, a remarkable result of Moussong asserts that X is always a CAT(0) space:

Theorem 12.58. *For any Coxeter system (W, S), the space $X = |\Sigma_d(W, S)|$ with its piecewise Euclidean metric is a CAT(0) space.*

The proof is long and we will not give it here. Moussong's original proof can be found in [170, Theorem 14.1]. See [89, Section 12.3; 90, Corollary 6.7.5; 149, Appendix B] for other proofs.

Moussong's theorem has the following purely algebraic corollary, which can also be proved using the Tits cone (see Proposition 2.87):

Corollary 12.59. *If (W, S) is an arbitrary Coxeter system, then every finite subgroup of W is conjugate to a subgroup of a finite standard subgroup.*

Proof. The point stabilizers W_x for the W-action on X are the conjugates of the finite standard subgroups. The corollary now follows from the Bruhat–Tits fixed-point theorem (Theorem 11.23).   □

12.3.9 The Gromov Hyperbolic Case: A CAT(−1) Metric

We mentioned in Section 10.4 that Moussong characterized the Coxeter groups that are hyperbolic in the sense of Gromov. Readers who do not know what this means can still appreciate the theorem of Moussong to be stated in this subsection. Before stating it, we need to go back to the construction of the piecewise Euclidean metric on $X = |\Sigma_d|$ above. We based this construction on the fact that the cells of Σ_d can be realized as convex subsets of Euclidean space. It turns out that they can also be realized as convex subsets of hyperbolic space [170, Section 13], so we have a second metric on X, which is piecewise hyperbolic.

We make one further observation in order to motivate Moussong's theorem. A well-known property of Gromov-hyperbolic groups is that they do not contain subgroups isomorphic to $\mathbb{Z} \times \mathbb{Z}$. In other words, if a group has a free abelian subgroup of rank ≥ 2, then it cannot be Gromov hyperbolic. Now there is an obvious way to construct free abelian subgroups of Coxeter groups: Take a family $J_1, \ldots, J_k$ of pairwise disjoint subsets of S such that $W_i := W_{J_i}$ is infinite for $i = 1, \ldots, k$. If W_i is a Euclidean reflection group, let T_i be its group of translations; otherwise, let T_i be an arbitrary infinite cyclic subgroup (see Exercise 12.62 below). Then $T := T_1 \times \cdots \times T_k$ is free abelian. We will call a subgroup T of this form a *standard* free abelian subgroup. Note that it will have rank ≥ 2 if and only if either (a) $k \geq 2$ or (b) $k = 1$ and W_1 has rank ≥ 3.

Moussong's theorem says, among other things, that the existence of a standard free abelian subgroup of rank ≥ 2 is the only obstruction to hyperbolicity for a Coxeter group. More precisely:

Theorem 12.60. *The following conditions on a Coxeter system (W, S) are equivalent:*

(i) *W is hyperbolic in the sense of Gromov.*
(ii) *W does not contain a free abelian subgroup of rank 2.*
(iii) *For all $J \subseteq S$, the standard parabolic subgroup W_J has at most one infinite irreducible factor, and this factor is not a Euclidean Coxeter group unless it has rank 2 (in which case it is D_∞).*
(iv) *The space $X = |\Sigma_d(W, S)|$, with its piecewise hyperbolic metric, is a $\mathrm{CAT}(-1)$ space.*

Moussong's original proof can be found in [170, proof of Theorem 17.1]. See also [89, Section 12.6; 90, Theorem 11.1].

Remarks 12.61. (a) Note that condition (iii) is simply a concise way of ruling out the existence of a standard free abelian subgroup of rank ≥ 2 (see (a) and (b) above).

(b) Krammer [149, Theorem 6.8.3] proved a stronger version of the equivalence (ii) $\iff$ (iii): For any Coxeter system (W, S), every free abelian subgroup of W has a finite-index subgroup that is conjugate to a subgroup of a standard free abelian subgroup.

Exercise 12.62. In the discussion of standard free abelian subgroups above, we needed the following fact: If W is a (finitely generated) infinite Coxeter group, then W has elements of infinite order. Prove this.

12.3.10 A Cubical Subdivision of Σ_d

In Figures 12.8, 12.9, and 12.11, the walls of Σ cut across the cells of Σ_d, inducing a subdivision of the latter into combinatorial cubes. This is true in general. For completeness, we give a precise statement of the result, although we will not make any use of it in what follows. We will give only a brief indication of the proof, which relies on the technical Section A.3; but it is quite easy, given the results of that section. We emphasize, before stating the result, that our notion of *cubical complex* (Definition A.24) is purely combinatorial; the cells are *not* required to be isometric to cubes. In fact, endowing each cell with its cubical metric will almost *never* yield a CAT(0) metric on Σ_d.

Proposition 12.63. *Let (W, S) be an arbitrary Coxeter system. There is a cubical complex $\Sigma_c(W, S)$ that subdivides $\Sigma_d(W, S)$. Its poset of cells can be identified with the set of closed intervals $[wW_J, wW_K]$ ($J \subseteq K \subseteq S$) in the poset of finite standard cosets, these intervals being ordered by inclusion. The action of W on $X = |\Sigma_d| = |\Sigma_c|$ permutes the cubes, and the stabilizer of each cube fixes the cube pointwise. The union of the cubes corresponding to the intervals $[W_J, W_K]$ is a strict fundamental domain for the action of W on X.*

Sketch of proof. It is is easy to verify that the poset of finite standard cosets has a cubical realization, as defined in Section A.3. The existence of Σ_c now follows easily from Proposition A.38. (See also Remark A.39.) The remaining assertions are equally easy and are left to the interested reader. Note, for instance, that the stabilizer of the cube corresponding to the interval $[W_J, W_K]$ is W_J. □

The fundamental domain described in the last sentence of the proposition is the same as the one given in Section 12.3.7, with some of the simplices lumped together to form cubes. In Figure 12.12, for example, one need only ignore the dashed lines to see the cubes. See also the figure in the solution to Exercise 12.56. The correspondence between cubes and intervals should be clear from the vertex labels in those figures.

Remark 12.64. We phrased the proposition in terms of the Coxeter system (W, S), but it is clear that every Coxeter complex Σ gives rise to a well-defined cubical complex Σ_c. It can be defined as the cubical realization of the poset Σ_d (or Σ_f) in the sense of Section A.3.

12.4 The Davis Realization of a Building

We are now ready for the main result of this chapter. All of the work has been done, and it is just a matter of putting the pieces together. We have already stated the result in Example 12.43, but we repeat it here.

Let (W, S) be an arbitrary Coxeter system, and let $(Z, \{Z_s\})$ be as in Example 12.2(e). Thus $Z = |K(\mathcal{S})|$ and $Z_s = |K(\mathcal{S}_s)|$, where $\mathcal{S}$ is the poset of spherical subsets of S, and $\mathcal{S}_s$ is the poset of spherical subsets containing s. As we explained in Section 12.3.8, Z is a compact geodesic metric space.

Definition 12.65. With Z as above, the Z-realization of a building Δ of type (W, S) is called the *Davis realization* of Δ.

The main result of this section is the following theorem of Davis [88, Theorem 11.1], which Davis says was also known to Moussong:

Theorem 12.66. *For any building Δ, its Davis realization $X = Z(\Delta)$ is a complete* CAT(0) *space.*

Proof. X is a geodesic metric space by Corollary 12.28, and it is complete by Proposition 12.10. The hard part is that X is a CAT(0) space. In view of Proposition 12.29, we need only check this when Δ consists of a single apartment, and for this we have Moussong's Theorem 12.58. □

As in the case of a single apartment (Section 12.3), there are a number of variants on the definition of the Davis realization X. For example, we could take X (as a set) to be the ordinary or cubical geometric realization of the subposet Δ_s of Δ consisting of the *spherical simplices*, i.e., those whose link is a spherical building. [If we are in the setting of Chapter 5, we would instead describe Δ_s as the set of spherical residues, where a residue is viewed as a building in its own right (Corollary 5.30).]

And as in Section 12.3.8, we can immediately get nontrivial consequences of the CAT(0) property by applying the Bruhat–Tits fixed-point theorem (Theorem 11.23). For example:

Corollary 12.67. *Let H be a group of type-preserving automorphisms of a building Δ. If H stabilizes a bounded set of chambers, then H fixes a spherical simplex. Equivalently, H stabilizes a spherical residue in $\mathcal{C}(\Delta)$.*

Note: The word "bounded" here refers to the gallery metric on $\mathcal{C}(\Delta)$.

Proof. Since Z is compact, a bound on the combinatorial distance between two chambers C, D yields a bound on the distance in X between points of $Z(C)$ and points of $Z(D)$. So H stabilizes a bounded subset of X and therefore has a fixed point x. In view of Corollary 12.22, it follows that H fixes the carrier A of x. This carrier is a simplex of cotype S_z, where $z = \tau(x)$. Since $W_z := \langle S_z \rangle$ is finite, A is a spherical simplex. □

If we apply this to $\Delta = \Delta(G, B)$, where G is a group with a BN-pair, the result is the following: If $H \leq G$ is a subgroup that stabilizes a bounded set of chambers of Δ, then H is contained in a parabolic subgroup of spherical type, i.e., a conjugate of a subgroup P_J such that W_J is finite. Here is a concrete special case of this.

Corollary 12.68. *Let G be a group with a BN-pair. Then every finite subgroup of G is a subgroup of a parabolic subgroup of spherical type.* □

Finally, we record the following consequence of the CAT(0) property:

Corollary 12.69. *The Davis realization of a building is contractible.* □

Remarks 12.70. (a) Modulo technicalities involving the difference between the metric topology and the weak topology on $X = |K(\Delta_s)|$, one can give a more elementary proof of Corollary 12.69 in the spirit of the proof of Proposition 12.52. Details are omitted.

(b) It is natural to ask whether there is a version of the Davis realization that is a regular cell complex reducing to Σ_d in the case of a single apartment. The answer is no. One has to at least pass to the cubical subdivision Σ_c before the gluing of apartments respects the cell structure. (One can see this already in the case of a tree.) The intuitive explanation for this is that apartments branch along walls, and the walls cut across the cells of Σ_d to create the subdivision Σ_c.

(c) If the underlying Coxeter system is hyperbolic in the sense of Gromov, then Moussong's Theorem 12.60 leads to a slight variant of Theorem 12.66, in which the same space X has a CAT(-1) metric, as we mentioned in Example 12.44.

We close this chapter by posing a question about metric realizations that is motivated by Theorem 11.52.

Question 12.71. Under what conditions on (W, S) and Z is the following assertion true: If X is the Z-realization of a building of type (W, S), then every subset of X isometric to an apartment is an apartment?

Examples due to Caprace [private communication] show that this does not hold in complete generality, even if one is interested only in the Davis realization.

13

Applications to the Cohomology of Groups

This chapter is a survey, without proofs, of a few of the applications of buildings to the cohomology theory of groups, especially arithmetic groups. A prerequisite for this chapter is familiarity with the basic facts about group cohomology, as given for instance in [50]. We will also use some algebraic topology (fundamental group, covering spaces, homology theory of manifolds, . . .).

A less serious prerequisite involves the theory of algebraic groups. In order to make accurate statements, we will need to use standard terminology about linear algebraic groups. But we hope that the reader can get the flavor of the results by thinking of familiar examples. For example, if you see "Let G be a linear algebraic group," you can think "Let $G = \mathrm{SL}_n$." Symbols like $G(\mathbb{Q})$ or $G(\mathbb{R})$ can then be interpreted as $\mathrm{SL}_n(\mathbb{Q})$ or $\mathrm{SL}_n(\mathbb{R})$. Any technical assumptions about G (semisimplicity, simple connectivity, . . .) can be ignored, since they are all satisfied by the example $G = \mathrm{SL}_n$.

The reader who is not content to think about SL_n can consult Appendix C, where we define most of the terms that are used.

13.1 Arithmetic Groups over the Rationals

13.1.1 Definition

An arithmetic group, roughly speaking, is a group of integral matrices defined by polynomial equations. For example, $\mathrm{SL}_n(\mathbb{Z})$ is an arithmetic group, defined by the single equation $\det(a_{ij}) = 1$. For the precise definition of "arithmetic group," start with a linear algebraic group G defined over $\mathbb{Q}$ (e.g., $G = \mathrm{SL}_n$). We can think of G as a subgroup of GL_n for some n, defined by polynomial equations with rational coefficients in the n^2 matrix entries. The rational matrices satisfying the given equations then form a group $G(\mathbb{Q})$ (e.g., $\mathrm{SL}_n(\mathbb{Q})$). And for any extension field $K \geq \mathbb{Q}$, the matrices in $\mathrm{GL}_n(K)$ satisfying the defining equations form a group $G(K)$ (e.g., $\mathrm{SL}_n(\mathbb{R})$, $\mathrm{SL}_n(\mathbb{C})$, $\mathrm{SL}_n(\mathbb{Q}_p)$).

We can also consider the invertible integral matrices satisfying the defining equations. These form a group $G(\mathbb{Z}) := G(\mathbb{Q}) \cap \mathrm{GL}_n(\mathbb{Z})$.

Definition 13.1. The group $G(\mathbb{Z})$ is said to be an *arithmetic group*. More generally, any subgroup $\Gamma \leq G(\mathbb{Q})$ that is commensurable with $G(\mathbb{Z})$ is said to be arithmetic, where two subgroups are *commensurable* if their intersection is of finite index in both of them.

For example, if $\Gamma \leq \mathrm{SL}_n(\mathbb{Z})$ is the subgroup consisting of matrices that are congruent to the identity matrix mod m for some integer $m \geq 2$, then Γ has finite index in $\mathrm{SL}_n(\mathbb{Z})$ and hence is arithmetic. For future reference, we remark that a group Γ of this form is torsion-free if $m \geq 3$ [50, Section II.4, Exercise 3].

Remark 13.2. Readers familiar with the theory of algebraic groups as presented in Appendix C, for instance, might legitimately object to the notation $G(\mathbb{Z})$. For G is assumed to be defined over $\mathbb{Q}$, and $\mathbb{Z}$ is not a $\mathbb{Q}$-algebra. Indeed, we were able to define the group $G(\mathbb{Z})$ above only because we assumed that we were given a specific embedding of G in a linear group GL_n, and a different embedding can lead to a different group $G(\mathbb{Z})$. It turns out, however, that the new $G(\mathbb{Z})$ is commensurable with the old one. So the notion of *arithmetic subgroup of $G(\mathbb{Q})$* is well defined in spite of our abuse of notation.

13.1.2 The Symmetric Space

The way to get homological information about an arithmetic group Γ is to view it as a subgroup of $L := G(\mathbb{R})$. The latter is a closed subgroup of $\mathrm{GL}_n(\mathbb{R})$ and hence is a locally compact topological group. More precisely, L is a Lie group with only finitely many connected components, and Γ is a discrete subgroup. (It suffices to verify this last assertion for $\Gamma = G(\mathbb{Z})$, in which case it is an immediate consequence of the fact that $\mathbb{Z}$ is discrete in $\mathbb{R}$.) The significance of having Γ embedded as a discrete subgroup of a Lie group is that it enables us to exhibit an Eilenberg–Mac Lane space of type $K(\Gamma, 1)$ for computing the cohomology of Γ, provided Γ is torsion-free.

The starting point for constructing a $K(\Gamma, 1)$ is the existence of a contractible manifold X associated to L, on which L acts by diffeomorphisms. If $G = \mathrm{SL}_2$, for example, then X is the hyperbolic plane, which we may take to be the upper half-plane with $\mathrm{SL}_2(\mathbb{R})$ acting by linear fractional transformations (see Section 2.2.3). In general, X can be constructed as the homogeneous space L/H, where H is a maximal compact subgroup of L. (Such an H exists and is unique up to conjugacy.)

Remark 13.3. We will soon specialize to the case that the algebraic group G is connected and semisimple. The space X is then a complete simply connected Riemannian manifold of nonpositive curvature, and L acts by isometries. The manifold X is called the *symmetric space* associated to L. The fact that all maximal compact subgroups are conjugate to H follows from Cartan's fixed-point theorem in this case (see Section 11.3). It is also known that H is equal to its own normalizer in L. Hence X can be identified with the set of maximal compact subgroups of L, with L acting by conjugation.

The compactness of the subgroup H implies that the action of L on X is *proper*. This means that for every compact subset $C \subseteq X$, the set $\{g \in L \mid gC \cap C \neq \emptyset\}$ is a compact subset of L. The action remains proper if we restrict it to any closed subgroup of L. In particular, the discrete subgroup $\Gamma \leq L$ acts properly on X. But then the compact subsets of Γ that occur in the definition of "proper" are finite. One easily deduces that the Γ-action satisfies the following condition, which is sometimes taken as the definition of properness for a discrete group action: For every $x \in X$, the stabilizer Γ_x of x in Γ is finite, and x has a Γ_x-invariant neighborhood U such that $gU \cap U = \emptyset$ for all $g \in \Gamma \smallsetminus \Gamma_x$.

Suppose now that the arithmetic group Γ is torsion-free. [This assumption is relatively harmless, since we can always achieve it by passing to a subgroup of finite index; we have already seen this for $\mathrm{SL}_n(\mathbb{Z})$.] The finite stabilizers Γ_x are then trivial, and properness reduces to a familiar condition from covering space theory. Thus if we form the quotient space $Y := \Gamma \backslash X$, then X is a regular covering space of Y, with Γ as group of deck transformations. Since X is contractible, it follows that Y is an Eilenberg–Mac Lane space of type $K(\Gamma, 1)$. Consequently, the homology and cohomology groups of Γ are the same as those of the manifold Y. A Riemannian manifold Y of this form is said to be a *locally symmetric space*.

Now it is no easy matter to actually calculate the cohomology of the manifold Y. But it is at least possible to get some qualitative results. For example, since Y is finite-dimensional, we immediately conclude that the *cohomological dimension* $\mathrm{cd}\,\Gamma$ is finite:

$$\mathrm{cd}\,\Gamma \leq d := \dim Y = \dim L - \dim H . \tag{13.1}$$

This means that $H^i(\Gamma) = 0$ for $i > d$ and any coefficient module.

If Y happens to be compact, we can say a lot more. We will spell this out now, for motivation, before returning to the more typical noncompact case.

13.1.3 The Cocompact Case

Assume that Y is compact, in which case Γ is said to be *cocompact*. Then Y is a closed manifold, so it has nonzero cohomology in the top dimension d (with $\mathbb{Z}/2\mathbb{Z}$ coefficients, for instance). Thus equality holds in (13.1).

Another consequence of compactness is that the groups $H^i(\Gamma, M)$ are finitely generated whenever the coefficient module M is finitely generated as an abelian group; for one knows that $H^i(Y, M)$ is finitely generated. And if we use the triangulability of Y, then we can deduce a stronger homological finiteness property of Γ. Namely, the $\mathbb{Z}\Gamma$-module $\mathbb{Z}$ admits a free resolution $(F_i)_{i \geq 0}$

such that F_i is finitely generated for all i and zero for all sufficiently large i; here $\mathbb{Z}\Gamma$ is the group ring of Γ over $\mathbb{Z}$. One expresses this by saying that Γ is of *type FL*.

Note next that there is a *Poincaré duality* isomorphism between the homology and cohomology of Γ. More precisely,

$$H^i(\Gamma, M) \cong H_{d-i}(\Gamma, \Omega \otimes M)$$

for any Γ-module M, where Ω is the orientation module of X, i.e., Ω is a free abelian group of rank 1 whose two generators correspond to the two orientations of X. The tensor product above is over $\mathbb{Z}$ and is given the diagonal Γ-action. We can get rid of Ω by replacing Γ by its subgroup (of index 1 or 2) consisting of the elements whose action on X is orientation-preserving.

Finally, the compactness of Y implies that Γ is a finitely presented group. This is not really a homological result, but it is closed related to homological finiteness properties such as the FL property.

All of these results are very nice, but it is relatively rare that they are applicable (i.e., that Γ is cocompact). If $G = \mathrm{SL}_n$, for instance, then Γ is not cocompact except in the trivial case $n = 1$. In the case of SL_2, the non-cocompactness follows from the discussion in Section 2.2.3. For if Γ were cocompact, then $W \backslash X$ would be compact, where $W = \mathrm{PGL}_2(\mathbb{Z})$; but we know that $W \backslash X$ is homeomorphic to a closed 2-simplex with one vertex removed.

13.1.4 The General Case

It is remarkable that all of the properties mentioned above generalize to the case that Γ is not cocompact. Most surprising, perhaps, is that there is a generalization of the duality theorem. It is in proving this that buildings come into the picture.

To avoid uninteresting technicalities, we will state the results under the assumption that the algebraic group G is connected and semisimple. We denote by l the $\mathbb{Q}$-rank of G, i.e., the rank of a maximal $\mathbb{Q}$-split torus (see Appendix C). The number l is significant for us for two reasons: (a) Γ is cocompact if and only if $l = 0$. Thus Section 13.1.3 was really a discussion of a very special case (where G is said to be anisotropic; see Section C.10). (b) The group $G(\mathbb{Q})$ has a BN-pair and an associated (spherical) building, and l is the rank of that building. In other words, l is the number of vertices of a chamber, or, equivalently, the number of generating reflections of the Weyl group W. If $G = \mathrm{SL}_n$, for example, then $l = n - 1$.

The first step in dealing with the general case is to prove that the manifold Y can be compactified by the adjunction of a boundary, i.e., Y is diffeomorphic to the interior of a compact manifold $\overline{Y}$ with boundary. This was first proved by Raghunathan [192]. The spaces Y and $\overline{Y}$ are homotopy equivalent by a standard result from topology, so $\overline{Y}$ is still a $K(\Gamma, 1)$ manifold. This implies as above that Γ is finitely presented and of type FL. Raghunathan's proof, however, yields no information about the boundary $\partial \overline{Y}$ that is adjoined to Y, so we get no further homological properties of Γ. In particular, we do not get a calculation of cd Γ or a duality theorem yet.

Borel and Serre [41] gave a more explicit construction of $\overline{Y}$. They worked directly with X (independent of any particular Γ) and adjoined a boundary to it. Their construction is canonical enough that the action of $G(\mathbb{Q})$ on X extends to the resulting manifold $\overline{X}$ with boundary, and the action of any arithmetic subgroup Γ is still proper, but now the quotient is compact. This quotient is then the desired $\overline{Y}$ when Γ is torsion-free.

If $G = \mathrm{SL}_2$, for example, $\overline{X}$ is obtained from X by adjoining a disjoint union of lines, one for each cusp in the pictures in Section 2.2.3. This may seem hard to visualize, but a picture can be found in [216, p. 216]. [If you have trouble visualizing *any* example of a 2-dimensional manifold whose boundary is a disjoint union of infinitely many lines, you can draw one yourself: First draw a picture of an infinite tree in the plane that branches at every vertex. Now trace over that tree using a marker with a very wide tip. The result is a picture of a surface whose boundary consists of infinitely many lines.]

The crucial feature of the Borel–Serre construction is that one is able to understand the algebraic topology of the boundary $\partial \overline{X}$: It is homotopy equivalent to the spherical building Δ associated to the BN-pair in $G(\mathbb{Q})$. The idea behind the proof of this is that $\partial \overline{X}$ is constructed as a disjoint union of contractible pieces e_P, indexed by the proper parabolic subgroups $P < G(\mathbb{Q})$. These pieces fit together in a manner that reflects the inclusion relations among the parabolic subgroups, and the desired homotopy equivalence then follows from a consideration of nerves of covers.

In view of the Solomon–Tits theorem (Sections 4.7 and 4.12), we now know that $\partial \overline{X}$ has the homotopy type of a bouquet of $(l-1)$-spheres. This leads to the calculation of cd Γ and to the duality theorem. A detailed explanation of the method can be found in [50, Sections VIII.7–10], so we will be brief. Let H_c^* denote cohomology with compact supports and let $\widetilde{H}_*$ denote reduced homology. We take $\mathbb{Z}$ coefficients in both cases. Combining Poincaré–Lefschetz duality and the contractibility of $\overline{X}$, one obtains

$$H_c^i(\overline{X}) \cong H_{d-i}(\overline{X}, \partial \overline{X}) \cong \widetilde{H}_{d-i-1}(\partial \overline{X}) \ .$$

Hence $H_c^i(\overline{X}) = 0$ unless $i = d-l$, and $H_c^{d-l}(\overline{X})$ is free abelian. Since $H_c^*(\overline{X}) \cong H^*(\Gamma, \mathbb{Z}\Gamma)$, one concludes, first, that

$$\mathrm{cd}\,\Gamma = d - l \ .$$

The point here is that the cohomological dimension of a group Γ of type FL can be computed as the top dimension in which $H^*(\Gamma, \mathbb{Z}\Gamma)$ is nontrivial.

In the present case, the top dimension is the *only* dimension in which $H^*(\Gamma, \mathbb{Z}\Gamma)$ is nontrivial. Using this, together with the fact that the nontrivial cohomology group is $\mathbb{Z}$-torsion-free, one deduces that Γ satisfies *Bieri–Eckmann duality*:

$$H^i(\Gamma, M) \cong H_{d-l-i}(\Gamma, D \otimes M)$$

for any Γ-module M and any i. Here D, the *dualizing module*, is a fixed Γ-module, independent of M. In the present situation, D is simply the Γ-module $H_c^{d-l}(\overline{X})$; it is isomorphic to $\widetilde{H}_{l-1}(\Delta) \otimes \Omega$, where Ω is the orientation module of X as in Section 13.1.3.

If $l = 0$, then $\widetilde{H}_{l-1}(\Delta) = \widetilde{H}_{-1}(\emptyset) = \mathbb{Z}$, so $D = \Omega$, and Bieri–Eckmann duality reduces to Poincaré duality. If $l > 0$, on the other hand, then D is a free abelian group of infinite rank.

To summarize, we have the following result of Borel and Serre:

Theorem 13.4. *Let G be a connected semisimple linear algebraic group defined over $\mathbb{Q}$. Let d be the dimension of the symmetric space associated to $G(\mathbb{R})$, and let l be the $\mathbb{Q}$-rank of G. Let Γ be a torsion-free arithmetic subgroup of $G(\mathbb{Q})$. Then Γ is finitely presented and of type FL and is a $(d-l)$-dimensional duality group. It is a Poincaré duality group if and only if $l = 0$, i.e., if and only if Γ is cocompact.*

Remarks 13.5. (a) We have said practically nothing about the actual construction of $\overline{X}$, which is extremely difficult. Grayson [116, 117] has given an alternative approach that avoids some of the technical problems faced by Borel and Serre. Instead of explicitly attaching a boundary to X, he finds his $\overline{X}$ inside of X. In other words, he constructs the sort of manifold one would get from the Borel–Serre $\overline{X}$ by removing an open collar neighborhood of the boundary. Grayson's method was based on a new approach to reduction theory introduced by Stuhler [229, 230] instead of the classical reduction theory used by Borel and Serre.

(b) The theorem generalizes to an arbitrary linear algebraic group defined over $\mathbb{Q}$, but one has to define the integers d and l slightly differently in the general case.

13.1.5 Virtual Notions

One says that a group *virtually* has a certain property if a subgroup of finite index has that property. It is sometimes convenient to use this language in order to avoid the assumption that Γ is torsion-free. For example, any arithmetic subgroup $\Gamma \leq G(\mathbb{Q})$ is *virtually torsion-free*, and it has a well-defined *virtual cohomological dimension* vcd Γ; this is the common cohomological dimension $d - l$ of its torsion-free subgroups of finite index. Similarly, Γ is *virtually of type FL*, or, more briefly, it is of *type VFL*, because it has a subgroup of finite index that is of type FL. Finally, the arithmetic group Γ is a *virtual duality group*. Note that we can dispense with "virtual" when talking about finite presentation: Γ is itself finitely presented since it has a finitely presented subgroup of finite index.

For our canonical example of SL_n, one has $d = (n(n+1)/2) - 1$ and $l = n - 1$, so

$$\mathrm{vcd}\,(\mathrm{SL}_n(\mathbb{Z})) = \frac{n(n-1)}{2}\,.$$

There is an easy way to remember this result—it says that $\mathrm{vcd}(\mathrm{SL}_n(\mathbb{Z}))$ is equal to the "obvious" lower bound on this vcd that one gets by looking at the strict upper-triangular subgroup of $\mathrm{SL}_n(\mathbb{Z})$.

In order to explain this, we need to recall some facts about solvable groups. Let Γ be an abstract solvable group. Choose a normal series

$$1 = \Gamma_0 \lhd \Gamma_1 \lhd \cdots \lhd \Gamma_r = \Gamma$$

with abelian quotients Γ_i/Γ_{i-1}, and set

$$h := \sum_{i=1}^{r} \dim_{\mathbb{Q}} \mathbb{Q} \otimes (\Gamma_i/\Gamma_{i-1})\,.$$

Definition 13.6. The number h, which is independent of the choice of normal series, is called the *Hirsch rank* of Γ.

The Hirsch rank is closely related to the homological and cohomological dimensions of Γ. In particular,

$$h \le \mathrm{cd}\,\Gamma \le h + 1$$

if Γ is torsion-free [33, Section 7.3].

Returning now to the strict upper-triangular group, it is a torsion-free nilpotent group of Hirsch rank $n(n-1)/2$, whence the "obvious" inequality $\mathrm{vcd}(\mathrm{SL}_n(\mathbb{Z})) \ge n(n-1)/2$.

Here is another example to illustrate this principle. Consider $\mathrm{SL}_n(\mathbb{Z}[1/p])$, where p is a prime number. (This is not arithmetic, but we are about to enlarge our framework so as to allow groups like this.) Its full upper-triangular subgroup is a virtually torsion-free solvable group of Hirsch rank $n(n-1)/2 + n - 1$, so we get

$$\mathrm{vcd}\big(\mathrm{SL}_n\,(\mathbb{Z}[1/p])\big) \ge \frac{n(n-1)}{2} + n - 1\,.$$

We will see in the next section how to use a Euclidean building to prove that equality holds.

13.2 *S*-Arithmetic Groups

Let S be a finite set of prime numbers, and let $\mathbb{Z}_S < \mathbb{Q}$ be the ring of rational numbers a/b $(a, b \in \mathbb{Z})$ such that the primes dividing b are in S. Thus the elements of $\mathbb{Z}_S$ are *integral except possibly at S*. Using $\mathbb{Z}_S$ instead of $\mathbb{Z}$, we get the notion of *S-arithmetic group*:

Definition 13.7. Let G be a linear algebraic group defined over $\mathbb{Q}$, and let S be a finite set of primes. Then any subgroup of $G(\mathbb{Q})$ commensurable with $G(\mathbb{Z}_S)$ is said to be an S-*arithmetic group*.

For example, $\mathrm{SL}_n(\mathbb{Z}[1/p])$ is an S-arithmetic subgroup of $\mathrm{SL}_n(\mathbb{Q})$, with $S = \{p\}$. In this section we will describe work of Borel and Serre [42] that extends their results about arithmetic groups to the S-arithmetic case.

13.2.1 A p-adic Analogue of the Symmetric Space

Fix a prime number p, and let L now denote the group $G(\mathbb{Q}_p)$. Groups of this type are often called p-*adic groups*, and are thought of as analogues of real or complex Lie groups. Assume that the algebraic group G is simply connected and absolutely almost simple. These assumptions guarantee that L admits a Euclidean BN-pair, analogous to the one we studied for $G = \mathrm{SL}_n$ (Section 6.9). We therefore have a (Euclidean) Bruhat–Tits building X on which L acts.

The building X is a locally finite simplicial complex and is contractible by Theorem 4.127 or by Theorem 11.16 and Proposition 11.7. Its dimension is the $\mathbb{Q}_p$-rank of G (or, more precisely, of the linear algebraic group over $\mathbb{Q}_p$ obtained from G by extension of scalars). Recall that this dimension is $n - 1$ for the case $G = \mathrm{SL}_n$. The stabilizers of the simplices of X are compact open subgroups of L, and it follows easily that the action of L on X is proper.

For applications to the cohomology of discrete groups, we will want to know $H_c^*(X)$. This was computed by Borel and Serre [42], using a method remarkably similar to the method used for the symmetric space associated to $G(\mathbb{R})$. The first step is to embed X as a dense open subspace of a compact contractible space $\overline{X} = X \cup \partial X$. The compact space ∂X that is adjoined to X is, as a set, the geometric realization of the (spherical) building at infinity. Remark 11.81 and Exercise 11.82 hinted at the possibility of doing this and also suggested that the topology on ∂X should *not* be expected to be the usual simplicial topology. We will say more about the significance of having a second topology on the geometric realization of a building in Section 14.3.1.

If X is 1-dimensional, for example, then it is a tree and $\overline{X}$ is its endpoint compactification. Thus ∂X is the space of ends of X in this case; it is a Cantor set, whose points are in 1–1 correspondence with the vertices of the (0-dimensional) spherical building at infinity. If this spherical building were given the weak topology, however, then it would be discrete.

The Borel–Serre compactification $\overline{X}$ enables one to compute $H_c^*(X)$. Using a suitable cohomology theory (e.g., Alexander–Spanier cohomology), one finds that

$$H_c^i(X) \cong H^i(\overline{X}, \partial X) \cong \widetilde{H}^{i-1}(\partial X) .$$

Borel and Serre go on to prove an analogue of the Solomon–Tits theorem for the reduced Alexander–Spanier cohomology of the compactly topologized spherical building ∂X: This cohomology is zero except in the top dimension (which is $\dim X - 1$), and it is free abelian in that dimension. The end result, then, is that $H_c^i(X)$ vanishes for $i \neq \dim X$ and is free abelian for $i = \dim X$.

13.2.2 Cohomology of *S*-Arithmetic Groups: Method 1

In Section 13.1 our emphasis was on using *proper* actions of discrete groups to get homological information. We will return to that point of view in Section 13.2.3 below. But first, for the sake of variety, we will show how to get the same kind of information from an action that is not proper. The method we will follow here is based on [214] and [57]. To keep the discussion as simple as possible, we begin with the familiar case $G = \mathrm{SL}_n$, and we assume that S is a singleton $\{p\}$. Thus an *S*-arithmetic group, for the moment, is simply a subgroup of $\mathrm{SL}_n(\mathbb{Q})$ commensurable with $\mathrm{SL}_n(\mathbb{Z}[1/p])$.

Let $\Gamma = \mathrm{SL}_n(\mathbb{Z}[1/p])$, viewed as a subgroup of $L := \mathrm{SL}_n(\mathbb{Q}_p)$. Note that Γ is not discrete in L; in fact, it is dense in L. But we can still consider the (nonproper) simplicial action of Γ on the Bruhat–Tits building X associated to L. Because of the density of Γ (and the fact that the stabilizers of the simplices are open in L), a fundamental domain for the L-action on X will still be a fundamental domain for the Γ-action. Hence Γ has a closed chamber $\overline{C}$ as strict fundamental domain. Moreover, the stabilizers of the faces of C are commensurable with $\mathrm{SL}_n(\mathbb{Z})$. For example, if v is the vertex corresponding to the standard lattice, then we know that the stabilizer of v in L is $\mathrm{SL}_n(\mathbb{Z}_p)$, where $\mathbb{Z}_p$ is the ring of p-adic integers; hence the stabilizer of v in Γ is $\mathrm{SL}_n(\mathbb{Z}[1/p]) \cap \mathrm{SL}_n(\mathbb{Z}_p) = \mathrm{SL}_n(\mathbb{Z})$. Thus the stabilizers are not finite, as they would be for a proper action, but they are groups that are known to have good finiteness properties.

Now let Γ be a torsion-free subgroup of $\mathrm{SL}_n(\mathbb{Q})$ commensurable with $\mathrm{SL}_n(\mathbb{Z}[1/p])$, e.g., a torsion-free subgroup of $\mathrm{SL}_n(\mathbb{Z}[1/p])$ of finite index. Then Γ acts on X with compact quotient and torsion-free arithmetic stabilizers. Since torsion-free arithmetic groups are finitely presented and of type FL, it follows that Γ is finitely presented [51] and of type FL [214, Proposition 11].

To calculate cd Γ and prove duality, we use the equivariant cohomology spectral sequence for (Γ, X) with coefficients in $\mathbb{Z}\Gamma$ [50, VII.7.10]. All the stabilizers are duality groups of the same dimension $m := \mathrm{vcd}(\mathrm{SL}_n(\mathbb{Z}))$, so the spectral sequence is concentrated on the line $q = m$. Moreover, one can calculate $E_1^{*,m}$ by the method of [57, Sections 2 and 3], and one finds that it is $C_c^*(X) \otimes D$; here C_c^* denotes simplicial cochains with compact support, and D is the dualizing module for the torsion-free arithmetic subgroups of $\mathrm{SL}_n(\mathbb{Z})$, i.e., $D = \widetilde{H}_{l-1}(\Delta) \otimes \Omega$ in the notation of Section 13.1. In view of the calculation of $H_c^*(X)$ stated in Section 13.2.1, the spectral sequence collapses at E_2 and gives the following result: $H^*(\Gamma, \mathbb{Z}\Gamma)$ is concentrated in dimension $\mathrm{vcd}(\mathrm{SL}_n(\mathbb{Z})) + \dim X$, and in that dimension it is the Γ-module $H_c^{\dim X}(X) \otimes D$. Thus Γ is a duality group, and we have calculated its dimension and its dualizing module. In particular,

$$\mathrm{vcd}\big(\mathrm{SL}_n(\mathbb{Z}[1/p])\big) = \frac{n(n-1)}{2} + n - 1.$$

The method works equally well if S consists of more than one prime. Simply pick some $p \in S$, let Γ act on the corresponding X, and note that the stabilizers are $(S \smallsetminus \{p\})$-arithmetic. So the analysis can be done by induction on the number of primes in S. The method also works equally well if SL_n is replaced by any G that is simply connected and absolutely almost simple. To state the result, let X_p be the Bruhat–Tits building associated to $G(\mathbb{Q}_p)$, let $d_p := \dim X_p$, and let $D_p := H_c^{d_p}(X_p)$. It is convenient to introduce a fictitious prime ∞ and to set X_∞ equal to the symmetric space associated to $G(\mathbb{R})$. Let $d_\infty := \dim X_\infty - l$, where l is the $\mathbb{Q}$-rank of G, and let $D_\infty := H_c^{d_\infty}(\overline{X}_\infty)$. Let $S' = S \cup \{\infty\}$, and set

$$d := \sum_{p \in S'} d_p \qquad \text{and} \qquad D := \bigotimes_{p \in S'} D_p \, .$$

Theorem 13.8. *Let G be a simply connected and absolutely almost simple linear algebraic group defined over $\mathbb{Q}$. Then, with the notation above, any torsion-free S-arithmetic subgroup of $G(\mathbb{Q})$ is finitely presented and of type FL and is a duality group of dimension d with dualizing module D.*

13.2.3 Cohomology of S-Arithmetic Groups: Method 2

We now sketch the method actually used by Borel and Serre [42] to prove Theorem 13.8. Instead of letting the torsion-free S-arithmetic group Γ act on the various X_p one at a time, they let it act on them simultaneously. More precisely, let $L_p := G(\mathbb{Q}_p)$ for $p \in S'$, where $\mathbb{Q}_\infty := \mathbb{R}$. Let $L := \prod_{p \in S'} L_p$. Then Γ can be embedded diagonally in the locally compact group L, and it is discrete in L. The point here is that $\mathbb{Z}_S$ is a discrete subring of $\prod_{p \in S'} \mathbb{Q}_p$, since a sequence of nonzero elements of $\mathbb{Z}_S$ that converges to 0 p-adically for all $p \in S$ will not converge to 0 in $\mathbb{R}$. Now Γ acts properly on the contractible space $X := \overline{X}_\infty \times \prod_{p \in S} X_p$. As in the arithmetic case, the quotient $Y := \Gamma \backslash X$ is a compact $K(\Gamma, 1)$-space.

A suitable triangulation theorem now implies that Γ is finitely presented and of type FL. Moreover, letting d and D be as above, we can apply the Künneth theorem to calculate that $H_c^*(X) = H^*(\Gamma, \mathbb{Z}\Gamma)$ is concentrated in dimension d and is isomorphic to D in that dimension. Thus Γ is a d-dimensional duality group with dualizing module D.

Remark 13.9. Borel and Serre proved a more general theorem than the one stated above. First, they worked over an arbitrary algebraic number field F, not just $\mathbb{Q}$. Their L involved the groups $G(\hat{F})$ for various completions $\hat{F}$ of F, which may include several copies of $\mathbb{R}$, several copies of $\mathbb{C}$, and several $\mathfrak{p}$-adic completions. Second, their hypothesis on G was weaker than the one stated above; it suffices to assume that G is a linear algebraic group (defined over F) such that the connected component of the identity is reductive.

13.2.4 The Nonreductive Case

The finiteness properties proven by Borel and Serre in the reductive case hold for some nonreductive groups, but not for all. Consider, for example, the following subgroups of the 2×2 upper-triangular group:

$$G := \begin{pmatrix} 1 & * \\ & * \end{pmatrix}, \qquad G_0 := \begin{pmatrix} 1 & * \\ & 1 \end{pmatrix}.$$

Then G is not reductive because it has the connected unipotent group G_0 as a normal subgroup. (See Sections C.7 and C.9 for the relevant definitions.) Nevertheless, it is not hard to show that $G(\mathbb{Z}[1/p])$ is finitely presented and of type VFL. On the other hand, $G_0(\mathbb{Z}[1/p])$ is isomorphic to the additive group $\mathbb{Z}[1/p]$, so it is not even finitely generated. Another interesting example is the 3×3 group

$$G_1 := \begin{pmatrix} 1 & * & * \\ & * & * \\ & & 1 \end{pmatrix};$$

one can show that $G_1(\mathbb{Z}[1/p])$ is finitely generated but not finitely presented.

The groups G_0 and G_1 are part of an infinite sequence of groups whose study was initiated by H. Abels [1]. The next one in the sequence is

$$G_2 := \begin{pmatrix} 1 & * & * & * \\ & * & * & * \\ & & * & * \\ & & & 1 \end{pmatrix}.$$

In general, G_n is the subgroup of GL_{n+2} consisting of upper-triangular matrices such that the diagonal entries in the upper left corner and lower right corner are 1. (The subscript n indicates the rank of the torus consisting of the diagonal matrices in G_n.)

We have already noted that $G_0(\mathbb{Z}[1/p])$ is not finitely generated, whereas $G_1(\mathbb{Z}[1/p])$ is finitely generated but not finitely presented. And Abels [1] proved that $G_2(\mathbb{Z}[1/p])$ is finitely presented. In order to describe the situation for arbitrary n, we need to introduce finiteness conditions F_n that generalize finite generation and finite presentation.

Definition 13.10. We say that a group Γ is of type F_1 if it is finitely generated. For any $n \geq 2$, we say that Γ is of type F_n if it is finitely presented and if the $\mathbb{Z}\Gamma$-module $\mathbb{Z}$ (with trivial Γ-action) admits a free resolution $(P_i)_{i \geq 0}$ with P_i finitely generated for $i \leq n$. If Γ is of type F_n for all n, then we say that Γ is of type F_∞; this is equivalent to saying that Γ is finitely presented and that there is a resolution as above with P_i finitely generated for all i [50, VIII.4.5]. Finally, we make the convention that every group Γ is of type F_0. Let $\phi(\Gamma)$ be the largest n $(0 \leq n \leq \infty)$ such that Γ is of type F_n. We call $\phi(\Gamma)$ the *finiteness length* of Γ.

It is easy to see that any group of type VFL is of type F_∞ [50, VIII.5.1]. Hence $\phi(\Gamma) = \infty$ if Γ is arithmetic, and the same is true if Γ is S-arithmetic and G is reductive.

Let's return now to our sequence of nonreductive groups G_n, and consider the S-arithmetic groups $\Gamma_n := G_n(\mathbb{Z}[1/p])$. The results stated above can be expressed by saying that $\phi(\Gamma_0) = 0$, $\phi(\Gamma_1) = 1$, and $\phi(\Gamma_2) \geq 2$. Abels and Brown [5] generalized these results by showing that $\phi(\Gamma_n) = n$ for all n. A slightly different proof was later given by Brown [52]. Both proofs involve an analysis of the action of Γ_n on the Bruhat–Tits building X associated to $\mathrm{SL}_{n+2}(\mathbb{Q}_p)$. (Recall that $\mathrm{GL}_{n+2}(\mathbb{Q}_p)$ acts on this building, so Γ_n also acts.) As in Section 13.2.2, the stabilizers are arithmetic groups. The problem, however, is that the quotient is not compact; so it takes some work to deduce finiteness properties (or the lack thereof) from the action.

At the moment, this application of buildings is not understood in any systematic way. In other words, one does not know how to find a suitable building to use for the study of the finiteness properties of an arbitrary S-arithmetic group.

Remark 13.11. Given the action of Γ_n on the $(n+1)$-dimensional building X, we can interpret the result that $\phi(\Gamma_n) = n$ as saying that Γ_n just barely fails to be of type F_∞. For one has the following general result, which is a consequence of [52, Theorems 2.2 and 3.2]: Suppose a group Γ acts on a d-dimensional contractible complex X. If the stabilizer of every simplex is of type F_∞, then Γ is of type F_∞ if and only if it is of type F_d.

It should be mentioned, finally, that the F_1 and F_2 conditions (i.e., finite generation and finite presentation) are understood for an arbitrary S-arithmetic group Γ. The results, which are due to Kneser, Borel–Tits, and Abels, are too complicated to state here. See the introduction to [2] for a survey.

13.3 Discrete Subgroups of p-adic Groups

In Section 13.2.1 we mentioned that Euclidean buildings play a role in the theory of p-adic groups analogous to the role played by symmetric spaces in the the theory of real Lie groups. We made use of such buildings in our discussion of the cohomology of S-arithmetic groups. One can also use the Euclidean building associated with a p-adic group L to study the cohomology of discrete subgroups $\Gamma < L$. Garland [105] gave a striking application of this sort. To avoid technicalities, we state a special case of his result, as improved by Casselman [77]:

Theorem 13.12. *Let G be a simple algebraic group defined over $\mathbb{Q}_p$ and let l be its $\mathbb{Q}_p$-rank. If Γ is a discrete cocompact subgroup of $G(\mathbb{Q}_p)$, then $H^i(\Gamma, \mathbb{R}) = 0$ for $0 < i < l$.*

Remark 13.13. Here is a simple way to get examples of such discrete subgroups Γ. Start with a simple algebraic group G_0 defined over $\mathbb{Q}$ and $\mathbb{R}$-anisotropic. This is equivalent to saying that $G_0(\mathbb{R})$ is compact. On the other hand, $G_0(\mathbb{Z}[1/p])$ is discrete in $G_0(\mathbb{R}) \times G_0(\mathbb{Q}_p)$ (Section 13.2.3), and it follows easily that $G_0(\mathbb{Z}[1/p])$ is discrete in $G_0(\mathbb{Q}_p)$. Thus we can take G in the theorem to be the group obtained from G_0 by extension of scalars to $\mathbb{Q}_p$, and we can take $\Gamma = G_0(\mathbb{Z}[1/p])$. An easy example of such a G_0 is the norm-1 group of a quaternion algebra over $\mathbb{Q}$ whose norm form is positive definite (see Section 6.11).

Garland's theorem can be viewed as a p-adic analogue of a vanishing theorem of Matsushima [161] for the cohomology of locally symmetric spaces. Matsushima's proof was based on estimates of the eigenvalues of certain operators related to the curvature tensor. The most interesting feature of Garland's proof of Theorem 13.12 is that it is similarly based on eigenvalue estimates, where the operators are now *p-adic curvature* operators. The underlying space here is the quotient $Y = \Gamma\backslash X$, where X is a Euclidean building; one can think of Y as a p-adic analogue of a locally symmetric space.

The formal similarity between Garland's method and Matsushima's is explained geometrically by the work of Pansu [183]. See also [30, 98, 138, 153, 184] for further results related to Garland's ideas.

13.4 Cohomological Dimension of Linear Groups

A special case of the results of Section 13.2 is that $\mathrm{vcd}(\mathrm{SL}_n(\mathbb{Z}_S)) < \infty$ for any n and S. The proof of this does not require the full force of the arguments we sketched as long as we do not care about the precise value of the virtual cohomological dimension. In particular, we need the proper action of $\mathrm{SL}_n(\mathbb{Z}_S)$ on the contractible finite-dimensional space $X = \prod_{p \in S'} X_p$, but we do not need the spaces $\overline{X}_p$. As a consequence, we have the following theorem, first pointed out by Serre [214, Théorème 5]:

Theorem 13.14. *Let F be an algebraic extension of $\mathbb{Q}$ and let Γ be an arbitrary finitely generated subgroup of $\mathrm{GL}_n(F)$. Then $\mathrm{vcd}\,\Gamma < \infty$.*

To prove this, we may assume that F is finite over $\mathbb{Q}$, and then we can easily reduce to the case $F = \mathbb{Q}$. [An n-dimensional vector space over F is a finite-dimensional vector space over $\mathbb{Q}$.] Then $\Gamma \leq \mathrm{GL}_n(A)$ for some finitely generated subring $A < \mathbb{Q}$; hence $\Gamma \leq \mathrm{GL}_n(\mathbb{Z}_S)$ for some S. Finally, we may replace Γ by $\Gamma \cap \mathrm{SL}_n(\mathbb{Z}_S)$, since $\det(\Gamma)$ is a finitely generated abelian group. The theorem now follows from the result about $\mathrm{SL}_n(\mathbb{Z}_S)$ stated above.

It is natural to ask what can be said if F is not assumed to be algebraic. For example, what if F is a rational function field $\mathbb{Q}(X)$? Easy examples show that finitely generated subgroups do not necessarily have finite vcd in this case. Suppose, for instance, that $\Gamma = G(\mathbb{Z}[X])$, where G is the 2×2 matrix group defined at the beginning of Section 13.2.4. Then Γ is finitely generated, but $\mathrm{vcd}\,\Gamma = \infty$ because the unipotent subgroup $G_0(\mathbb{Z}[X])$ is free abelian of infinite rank.

It turns out that this example is essentially the only kind of counterexample. In other words, the failure of vcd Γ to be finite can always be explained in terms of the unipotent subgroups of Γ. This is the content of the following theorem of Alperin and Shalen [21]:

Theorem 13.15. *Let Γ be a finitely generated subgroup of $\mathrm{GL}_n(F)$, where F is a field of characteristic 0. Then* vcd $\Gamma < \infty$ *if and only if there is an upper bound on the Hirsch ranks of the unipotent subgroups of Γ.*

Recall that any unipotent subgroup U of $\mathrm{GL}_n(F)$ is torsion-free and nilpotent by Kolchin's theorem (Section C.7). So the Hirsch rank of U is indeed defined and differs from cd U by at most 1 (see Section 13.1.5). Thus "Hirsch rank" could be replaced by "cohomological dimension" in the statement of the theorem. The "only if" part is now obvious. We will say a few words about the proof of the "if" part, in order to indicate how buildings enable one to reduce the question of finite vcd to the question of finding a bound on cd U, where U ranges over the unipotent subgroups of Γ.

As in the proof of Serre's theorem, the finite generation of Γ guarantees that $\Gamma \leq \mathrm{GL}_n(A)$ for some finitely generated subring $A \leq F$. We may assume that F is the field of fractions of A and that $\Gamma \leq \mathrm{SL}_n(A)$. The first step in the proof is pure commutative algebra. One shows that there is a finite collection $\{v_i\}$ of discrete valuations on F that can be used to test integrality of elements of A, in the following sense: Let A_i be the valuation ring of v_i and let B be the ring of algebraic integers in F; then $A \cap \bigcap_i A_i \leq B$.

Let X_i be the Bruhat–Tits building associated to $\mathrm{SL}_n(F)$ and the valuation v_i (Section 6.9). Then Γ acts on X_i for all i. We can either analyze one of these actions at a time, as in Section 13.2.2, or we can let Γ act on the product as in Section 13.2.3. Either way, we are reduced to finding a bound on vcd Γ', where Γ' ranges over the subgroups of Γ that stabilize a vertex in each X_i. Now the stabilizer of a vertex of X_i stabilizes an A_i-lattice in F^n; hence its characteristic polynomial has coefficients in A_i. Consequently, the characteristic polynomial of each element of Γ' has coefficients in the ring of integers B. One says that Γ' has *integral characteristic*.

Intuitively, Γ' resembles a subgroup of the arithmetic group $\mathrm{SL}_n(B)$, and so one might hope to be able to bound vcd Γ' by embedding Γ' as a discrete subgroup of a product $\mathrm{SL}_n(\mathbb{R})^{r_1} \times \mathrm{SL}_n(\mathbb{C})^{r_2}$. Now this is certainly not possible in general—unipotent groups again provide counterexamples. (Note that any unipotent group has integral characteristic.) But Alperin and Shalen, using techniques introduced by Bass for studying groups of integral characteristic, show that unipotent groups are the only obstruction. More precisely, there is a unipotent normal subgroup $U \trianglelefteq \Gamma'$ such that Γ'/U is a discrete subgroup of a Lie group as above. Since vcd $\Gamma' \leq$ cd $U +$ vcd Γ'/U, we are done by the hypothesis on the unipotent subgroups.

13.5 S-Arithmetic Groups over Function Fields

We close with a discussion of the finiteness properties of some matrix groups in characteristic $\neq 0$. Let K be the function field of an irreducible projective smooth curve C defined over a finite field $k := \mathbb{F}_q$. Readers not familiar with these concepts can think of C as something like a Riemann surface and K as the field of meromorphic functions on C. The canonical example is the rational function field $K = k(t)$, which corresponds to the case that C is the projective line (analogue of the Riemann sphere).

Let S be a finite nonempty set of (closed) points of C, and let $\mathcal{O}_S < K$ be the ring of functions that have no poles except possibly at points in S. We can also describe $\mathcal{O}_S$ in terms of discrete valuations. Each point p of the curve C gives rise to a discrete valuation v_p on K such that $v_p(f)$ is the order of vanishing of f at p. Thus $v_p(f) < 0$ if and only if f has a pole at p; hence the valuation ring $\mathcal{O}_p$ associated to v_p is the set of functions that do not have a pole at p. The definition of $\mathcal{O}_S$ can now be rewritten as

$$\mathcal{O}_S = \bigcap_{p \notin S} \mathcal{O}_p .$$

Suppose, for example, that $K = k(t)$. Then the curve has one point at infinity together with one point for every irreducible polynomial in $k[t]$. If S consists only of the point at infinity, then $\mathcal{O}_S$ is the polynomial ring $k[t]$.

Let G be a linear algebraic group defined over K. We can then define the notion of S-arithmetic subgroup of $G(K)$ exactly as in Definition 13.7, with $\mathbb{Q}$ replaced by K and $\mathbb{Z}_S$ replaced by $\mathcal{O}_S$. For example, $\mathrm{SL}_n(k[t])$ is an S-arithmetic subgroup of $\mathrm{SL}_n(k(t))$ when $S = \{\infty\}$ as above.

Assume now that G is simply connected and absolutely almost simple. For each $p \in S$ we have a locally compact group $L_p := G(K_p)$, where K_p is the completion of K with respect to v_p, and we have a Bruhat–Tits building X_p on which L_p acts properly. Set

$$L := \prod_{p \in S} L_p \quad \text{and} \quad X := \prod_{p \in S} X_p .$$

As in the number field case, the group $\Gamma := G(\mathcal{O}_S)$ is a discrete subgroup of L. It therefore acts properly on the contractible space X. And, as in the number field case again, the quotient $\Gamma \backslash X$ is compact if and only if $l = 0$, where l is the K-rank of G. One can deduce that Γ is of type VFL, and hence of type F_∞, when $l = 0$. (*Note:* Part of what has to be proved here is that Γ is virtually torsion-free, which is not automatic in characteristic $\neq 0$. The proof in the present case uses the compactness of $\Gamma \backslash X$ [214, Théorème 4].)

If $l > 0$, on the other hand, the situation becomes different from that in the number field case, at least as far as the F_n properties are concerned. Indeed, Γ need not even be finitely generated. The simplest example is $\Gamma = \mathrm{SL}_2(k[t])$. The space X is a tree in this case, and there is a half-line that is a strict fundamental domain for the action. By analyzing the stabilizers along this half-line [217, Section II.1.6] one can see that Γ is not finitely generated. [This is a theorem of Nagao, which had been proved earlier without the aid of the tree.] More generally, $SL_n(k[t])$ acts on its $(n-1)$-dimensional building with a sector as strict fundamental domain, and one suspects that this group is of type F_{n-2} but not of type F_{n-1}. We will return to this example below.

In the rest of this section we summarize what is known about the finiteness properties of $\Gamma = G(\mathcal{O}_S)$ for arbitrary G, K, and S. Assume throughout this discussion that $l > 0$, and set

$$d := \dim X = \sum_{p \in S} d_p , \tag{13.2}$$

where $d_p = \dim X_p$ (= the K_p-rank of G). Note that $d_p \geq l$, so we always have $d \geq 1$. The first result is that finite generation and finite presentation are completely understood in terms of d (see Behr [32]):

Theorem 13.16.

(1) Γ is finitely generated if and only if $d \geq 2$.
(2) Γ is finitely presented if and only if $d \geq 3$.

For example, $SL_2(\mathcal{O}_S)$ is finitely generated if and only if $|S| \geq 2$ and finitely presented if and only if $|S| \geq 3$. Similarly, $SL_3(\mathcal{O}_S)$ is finitely generated for any S and is finitely presented if and only if $|S| \geq 2$. And for $n \geq 4$, $SL_n(\mathcal{O}_S)$ is finitely presented for any S.

For SL_2, not only does one know when Γ is finitely generated or finitely presented, but one in fact knows the precise finiteness length (see Definition 13.10). This is given by the following theorem of Stuhler [231]:

Theorem 13.17. $SL_2(\mathcal{O}_S)$ has finiteness length $d - 1 = |S| - 1$.

Suppose, next, that $G = SL_n$ for arbitrary n and that we are in the simplest possible case: $\mathcal{O}_S = k[t]$. Then $\Gamma = SL_n(k[t])$, and X has only one factor, which is a Euclidean building of dimension $d = n - 1$. We know from the results stated above that the finiteness length of Γ is 0 if $n = 2$, is 1 if $n = 3$, and is at least 2 if $n \geq 4$. The following result, due independently to Abels and Abramenko (see [3, 4, 6]), almost settles the question for arbitrary n:

Theorem 13.18. For any n there is an integer N such that $SL_n(\mathbb{F}_q[t])$ has finiteness length $n - 2$ if $q \geq N$.

If $n \leq 5$, Abramenko has shown that one can take $N = 2$, i.e., there is no restriction on q. If $n \geq 6$, however, the best known value of N is $N = \max_{1 \leq i \leq n-2} \binom{n-2}{i}$, again due to Abramenko, but this is not believed to be sharp. Taking $n = 6$, for example, one does not know the finiteness length of $SL_6(\mathbb{F}_q[t])$ for $q = 2, 3, 4, 5$. Abramenko [9, Section III.2] went on to generalize Theorem 13.18 to arbitrary classical groups (including symplectic, orthogonal, and unitary groups). Here is one version of the result:

Theorem 13.19. *Let G be an absolutely almost simple $\mathbb{F}_q$-group that is not of exceptional type, and let l be its $\mathbb{F}_q$-rank. If $q \geq 2^{2l-1}$, then $G\left(\mathbb{F}_q[t]\right)$ and $G\left(\mathbb{F}_q[t, t^{-1}]\right)$ are both of type F_{l-1}, and $G\left(\mathbb{F}_q[t]\right)$ is not of type F_l (so it has finiteness length precisely $l - 1$).*

If we turn now to a general G, K, S (with $l > 0$), the theorems stated above suggest the following question (with d as in equation (13.2)):

Question 13.20. Is the finiteness length $\phi(\Gamma)$ always equal to $d - 1$?

Several people have conjectured that the answer is yes. At the time of this writing, however, all that is known (aside from the special cases mentioned earlier) is the following result of Bux and Wortman [67]:

Theorem 13.21. $\phi(\Gamma) < d$.

(In view of Remark 13.11, this is equivalent to saying that Γ is not of type F_∞.)

The simplest group for which one does not know the precise finiteness length is the group $\Gamma = \mathrm{SL}_3\left(\mathbb{F}_q[t, t^{-1}]\right)$. Here $|S| = 2$, and X is a product of two Euclidean buildings of dimension 2, so $d = 4$. We know that Γ is finitely presented by Theorem 13.16, and we know that Γ is not of type F_4 by Theorem 13.21. So the finiteness length $\phi(\Gamma)$ is either 2 or 3, and the conjecture mentioned above predicts that $\phi(\Gamma) = 3$.

Remark 13.22. In this section we have focused on a particular kind of finiteness property, for which the function field case seems very different from the number field case. But if one asks slightly different questions, then the two cases do not seem quite so different. In fact, Grayson's version of the Borel–Serre construction for number fields [116, 117] was inspired by homological finiteness results in the function field case proved by Serre for SL_2 and Quillen for SL_n. See [217, Sections II.2.8 and II.2.9] and [115]. Similar ideas can be found in the work of Stuhler [229, 230].

14

Other Applications

In the last chapter we surveyed some of the applications of buildings to the cohomology theory of groups. We have also mentioned applications of the Bruhat–Tits fixed-point theorem to the structure of groups acting on buildings. As we saw in Section 12.4, this has very wide applicability via the Davis realization. And of course, the theory of buildings has been deeply intertwined with the theory of algebraic groups since the beginning of the subject, as we have noted many times in this book. More recently, the connection between twin buildings and Kac–Moody groups has been discovered (see Section 8.11).

The purpose of the present chapter is to mention briefly some additional applications of buildings, with pointers to the literature for readers who want to know more. There has been an explosion of such applications, far beyond what could have been envisioned in the early days of the theory of buildings, and we make no claim of completeness. See also Tits's survey [248] for the applications known to him as of the early 1970s.

We could not have written this chapter without the help of experts in the various application areas. In particular, we wish to thank Michael Aschbacher, Werner Ballmann, Dan Barbasch, Francis Buekenhout, Stephen DeBacker, Ralf Gramlich, Lizhen Ji, Bill Kantor, Bruce Kleiner, Dimitri Leemans, Enrico Leuzinger, Antonio Pasini, Bertrand Rémy, Peter Schneider, Ernie Shult, Steve Smith, Ron Solomon, Ulrich Stuhler, Gudlauger Thorbergsson, and Richard Weiss.

14.1 Presentations of Groups

It has been known for a long time that one can often obtain a presentation for a group G by considering a fundamental domain for the action of G on a suitable space. This idea goes back to the work of Poincaré, Klein, and Fricke. Here is an easily stated result of this type; it is an improvement by Brown [51] of a result of Soulé [222]:

Proposition 14.1. *Suppose a group G acts on a simply connected simplicial complex Δ, and suppose $F \subseteq \Delta$ is a simplicial fundamental domain in the sense of Definition 3.74. Then G is the sum of the vertex stabilizers G_v amalgamated along the edge stabilizers G_e, where v ranges over the vertices of F and e ranges over the edges of F.*

This means, by definition, that G is the direct limit of the system consisting of the groups G_v and G_e, together with the inclusions

whenever e is an edge of F with vertices v, v'. [See, for example, Serre [217, Section I.1.1] for the definition of "direct limit."] An equivalent formulation is that one can get a presentation for G by combining presentations of the groups G_v and then introducing relations to identify the copy of G_e in G_v with the copy of G_e in $G_{v'}$, where e, v, and v' are as above.

A variant of this result (which is actually a special case of it) can be found in Tits [259]. See also [73; 195, Chapter 3]. It is stated in terms of group actions on posets and is formulated with applications to buildings in mind. Suppose a group G acts on a poset X, and suppose that F is a fundamental domain for the action. By this we mean that F is closed under passage to predecessors and that every element of X is G-equivalent to a unique element of F. Note that the existence of such a fundamental domain implies that $G_y \leq G_x$ whenever $x \leq y$. Here, as usual, G_x is the stabilizer of x.

Proposition 14.2. *If the poset X is simply connected, then G is the direct limit of the system consisting of the stabilizers G_x $(x \in F)$ and the inclusions $G_y \hookrightarrow G_x$ for $x < y$ in F.*

(Simple connectivity of a poset is defined combinatorially in the cited references. It is equivalent to simple connectivity of the flag complex of X in the usual sense of topology.)

As in Proposition 14.1, we may express the conclusion of Proposition 14.2 in the language of amalgamations: G is the amalgam (or amalgamated sum) of the subgroups G_x relative to the inclusions $G_y \hookrightarrow G_x$ for $x < y$.

We now proceed to some specific presentations that can be obtained when a group acts on a building. All of them exhibit the group as an amalgam of various subgroups, though in some cases the proofs in the literature use the algebraic theory of BN-pairs instead of directly applying one of the propositions above.

14.1.1 Chamber-Transitive Actions

Suppose a group G admits a chamber-transitive, type-preserving action on a building Δ of type (W, S). Choose a fundamental chamber C, and let G_J

$(J \subseteq S)$ be the stabilizer of the face of C of cotype J. For example, $G_\emptyset$ is the stabilizer of C.

In this subsection we will be concerned only with sets J such that $|J| \leq 2$. We set $G_s := G_{\{s\}}$ $(s \in S)$ and $G_{s,t} := G_{\{s,t\}}$ $(s, t \in S, \ s \neq t)$. As a first application of Proposition 14.2, Tits [259, Section 14] proves that G is the amalgam of the the groups $G_{s,t}$ with $m(s,t) < \infty$, the groups G_s for $s \in S$, and the group $G_\emptyset$; the amalgam is formed relative to the inclusions $G_s \hookrightarrow G_{s,t}$ and $G_\emptyset \hookrightarrow G_s$.

Note, for plausibility, that this result reduces to the Coxeter presentation of W if $\Delta = \Sigma(W,S)$ and $G = W$. For another familiar example, suppose W is the infinite dihedral group. Then Δ is a tree on which G acts with an edge as fundamental domain. The conclusion in this case is that G is the amalgamated free product $G_s *_{G_\emptyset} G_t$, where $S = \{s, t\}$. This could also be deduced from the theory of groups acting on trees [217, Section I.4].

14.1.2 Further Results for BN-Pairs

We specialize now to the case that G has a BN-pair. We can then apply the results of Section 14.1.1 to the action of G on $\Delta(G, B)$. The groups G_J in this case are the standard parabolics P_J, so we obtain a description of G as an amalgam of spherical standard parabolic subgroups of rank ≤ 2.

There are two further descriptions of G as an amalgam, as well as a third one in the spherical case. The first result says that G is the sum of the rank-1 parabolics P_s $(s \in S)$ and the group N, amalgamated along B (which is a subgroup of each P_s) and the intersections $N \cap P_s$. This is proved in [247, Proposition 13.3]; see also [217, Section II.1.7, Theorem 8].

For the second result, which can be deduced easily from the first, assume that the rank of (W, S) is at least 3. Then G is the sum of the maximal standard parabolics $P_{S \smallsetminus \{s\}}$, amalgamated along their intersections. See [217, Section II.1.7, Corollary 3].

For the final result, known as the *Curtis–Tits theorem*, assume that we are in the spherical case, i.e., that W is finite. For simplicity, assume further that our BN-pair comes from an RGD system as in Section 7.8.5. We then have a family of "rank-1" groups $H_s := T\langle U_s, U_{-s}\rangle$ $(s \in S)$ as in Section 7.8.3, and we set $H_{s,t} := \langle H_s, H_t\rangle$ for $s \neq t$ in S. The Curtis–Tits theorem [247, Theorem 13.32] then says that G is the sum of the groups $H_{s,t}$, amalgamated along the subgroups H_s. Equivalently, G is the direct limit of the system consisting of the groups H_s and $H_{s,t}$ and the inclusion maps

For general (nonspherical) RGD systems as defined in Chapter 8, Abramenko and Mühlherr [12] proved that the Curtis–Tits theorem remains valid provided

(W, S) is 2-spherical and the associated twin building satisfies condition (co) of Section 5.11. It is this generalization of the Curtis–Tits theorem that enabled us to give a simple description of certain Kac–Moody groups in Section 8.11.6.

14.1.3 The Group U_+

Still in the setting of a general RGD system, we proved in Theorem 8.85 that U_+ is an amalgam of its subgroups U_w ($w \in W$). Our proof was based on work that we had already done in Section 8.7, but we could instead have deduced the theorem from Proposition 14.2. This is done in [259, Section 15]; see also [73]. The starting point is that U_+ acts on Δ_- with Σ_- as fundamental domain (cf. Remark 8.35). Here (Δ_+, Δ_-) is the twin building associated to the RGD system, and (Σ_+, Σ_-) is the fundamental twin apartment.

14.1.4 S-Arithmetic Groups

We already saw in Chapter 13 that (Euclidean) buildings are very useful for deriving results about the finite presentability of S-arithmetic groups; see in particular Theorem 13.8 (where the part referring to finite presentability does not require the assumption that the S-arithmetic subgroup is torsion-free) and Theorem 13.16. These general results are qualitative in the sense that they do not provide explicit (finite) presentations or at least not any "nice" ones. This changes in special situations, where simplicial fundamental domains are available and one can therefore apply Proposition 14.1.

As a first example, consider the group $\Gamma := \mathrm{SL}_2(\mathbb{Z}[1/p])$. It acts on the building Δ associated to $G := \mathrm{SL}_2(\mathbb{Q}_p)$, which is a tree; see Section 6.9. Since Γ is dense in G, it acts chamber transitively (even Weyl transitively) on Δ, as we discussed in Section 6.10.2. If we choose the fundamental edge $e = \{v, w\}$ so that the vertex stabilizers in G are, respectively, $G_v = \mathrm{SL}_2(\mathbb{Z}_p)$ and $G_w = d\,\mathrm{SL}_2(\mathbb{Z}_p)d^{-1}$ with $d := \mathrm{diag}(1, p)$, then an immediate application of Proposition 14.1 yields that Γ is the free product of $\mathrm{SL}_2(\mathbb{Z})$ and $d\,\mathrm{SL}_2(\mathbb{Z})d^{-1}$, amalgamated along their intersection (see also [217, Section II.1.4]).

But Proposition 14.1 can also be applied when the fundamental domain F is infinite. One of the best-known examples of this is Serre's interpretation of Nagao's theorem (see [217, Section II.1.6]): Here we consider the subgroup $\Gamma := \mathrm{SL}_2(k[t])$ of $G := \mathrm{SL}_2(K)$, where k is a field and $K := k((1/t))$ is the field of formal Laurent series over k; this is the completion of $k(t)$ with respect to the discrete valuation $v := v_\infty$ of $k(t)$ introduced in Example 6.108(b). The group Γ is a discrete subgroup of G, and it is an S-arithmetic subgroup of G (with $S = \{v\}$) if k is finite.

Let Δ again be the tree associated to G. It is shown in [217] that the action of Γ on this tree admits a ray F as simplicial fundamental domain. Moreover, this ray can be chosen such that if the vertices of F are consecutively numbered $x_0, x_1, \dots$, then $\Gamma_{x_0} = \mathrm{SL}_2(k)$ and $\Gamma_{x_1} < \Gamma_{x_2} < \cdots$ is a strictly increasing sequence whose union is equal to $\mathrm{SB}_2(k[t])$, the group of all upper-triangular matrices in $\mathrm{SL}_2(k[t])$. It now follows from Proposition 14.1 that Γ is the free product of $\mathrm{SL}_2(k)$ and $\mathrm{SB}_2(k[t])$, amalgamated along their intersection $\mathrm{SB}_2(k)$. Since $\mathrm{SB}_2(k[t])$ is an infinite, strictly increasing union of subgroups, it is not finitely generated. Combining this with the presentation of Γ as an amalgam, one deduces that Γ is *not* finitely generated (not even for finite k). This is strikingly different from the situation for arithmetic groups and was quite a surprise when it was first discovered by Nagao.

The example discussed in the previous paragraph was generalized by Soulé [223] to the following situation: Let $\mathcal{G}$ be a simply connected, almost-simple Chevalley group (e.g., $\mathcal{G} = \mathrm{SL}_n$), set $\Gamma := \mathcal{G}(k[t])$ and $G := \mathcal{G}(K)$, and consider the action of Γ on the Euclidean Bruhat–Tits building Δ associated to G. Soulé proves that this action again admits a simplicial fundamental domain F, which can be chosen to be a sector in the fundamental apartment of Δ. Applying Proposition 14.1, one again gets a presentation of Γ as an amalgam of vertex stabilizers. Since F is infinite, this amalgam first involves infinitely many subgroups of Γ. As in the previous paragraph, however, these vertex stabilizers can be organized into finitely many families of increasing sequences so that the original amalgam can be transformed into one involving only finitely many subgroups of Γ. For details, we refer to [223] or to [259, Section 15.7], where the fundamental domain F is derived differently.

The presentations we discussed in the last two paragraphs involve infinitely many generators and relations. But the group $\Gamma := \mathrm{SL}_n\left(\mathbb{F}_q[t]\right)$ (to name one example) is finitely presented if $n \geq 4$, as was shown by Rehmann and Soulé [194]. So it is natural to look for different actions of Γ that yield nice *finite* presentations. This is indeed possible, and it again involves the Euclidean building Δ of $G := \mathrm{SL}_n(K)$, where $K := \mathbb{F}_q((1/t))$. Consider the fundamental chamber C of Δ (suitably chosen with respect to the Γ-action), and remove the vertex x_0 of C with $\Gamma_{x_0} = \mathrm{SL}_n(\mathbb{F}_q)$ to obtain a panel $P := C \smallsetminus \{x_0\}$ in Δ. It can be shown (see Abramenko [6]) that the Γ-invariant complex $\Gamma \cdot P$ is simply connected if $(q, n) \neq (2, 4)$. This yields (by Proposition 14.1 again) a presentation of Γ as an amalgam of the finite stabilizers Γ_x, where x ranges over the $n - 1$ vertices of P. One can derive similar amalgam presentations (using some results from [9]) for groups of the form $\mathcal{G}\left(\mathbb{F}_q[t]\right)$, where $\mathcal{G}$ is a classical $\mathbb{F}_q$-group of rank ≥ 3 and q is "sufficiently big."

14.2 Finite Groups

Recall that finite buildings are very closely related to finite simple groups, as we indicated in Section 9.10. It is therefore not surprising that the theory of buildings plays a role in the proof of the classification theorem for finite simple groups that we stated in that section (both the existing "proof" and the various revised proofs being developed). The first connection involves "recognition theorems" for simple groups, which are needed at several places in the proof. Typically one has an abstract simple group G with a collection of subgroups resembling subgroups of a known simple group $\bar{G}$, and one wants to conclude that $G \cong \bar{G}$. A possible approach if $\bar{G}$ is a group of Lie type is to construct a simplicial complex with G-action from the collection of subgroups and prove that it is isomorphic to the building associated with $\bar{G}$.

Suzuki [232] was one of the first to give such a proof. More recent proofs along these lines, for rank-2 groups of Lie type, can be found in [25, Section F.4; 109, Section 30]. And Aschbacher and Smith [25, Section I.5] mention another case in which they could have used this method but chose to give a more elementary proof instead. It is not clear what role the method will have in future revised proofs.

In order to even state a recognition theorem, one has to thoroughly understand the structure of the target group $\bar{G}$. Here again, buildings can be useful. For example, the proof of the *Curtis–Tits theorem* in [110, Section 2.9] essentially uses the simple connectivity of spherical buildings of rank ≥ 3 to express certain groups of Lie type as amalgams of parabolic subgroups. (See also Section 14.1.2 above.)

There is a big generalization of the Curtis–Tits theorem, which also generalizes some results of Phan. The resulting "Phan theory," which makes serious use of buildings, is still under active development and is being used in the ongoing revision of the proof of the classification theorem. See [113, p. 333] for an example of this, and see Gramlich [114] for a comprehensive survey of Phan theory.

Next, we mention that the theory of buildings, especially the "local approach" in [255], inspired Buekenhout to invent "diagram geometries." These are generalizations of buildings that can be used for simple groups other than those of Lie type. For example, they have been used to prove recognition theorems for some sporadic groups; see [23, 135, 136] for examples. There is now a vast literature on diagram geometries, with several successes, but there is still no general picture that includes all of the sporadic groups. In addition to the references already cited, the reader can consult [61, 62, 187] and the forthcoming book by Smith [219, Chapter 2].

Root subgroups are used as tools in the classification and elsewhere, and these subgroups, as we know from Chapter 7, have interpretations in terms of buildings. There are generalizations to some of the sporadic groups, which are related to the generalizations of buildings mentioned above. See Timmesfeld [238].

Finally, moving away from the classification, which involves applications of spherical buildings, we mention that the theory of Euclidean buildings has also had applications to finite group theory due to Kantor, Liebler, and Tits [142]. See also Ronan [200, Section 10.6].

14.3 Differential Geometry

L. Ji [140] has written a survey of the applications of buildings to geometry and topology, with a comprehensive bibliography. We confine ourselves here to mentioning a few of the high points.

14.3.1 Mostow Rigidity

Mostow [167] gave the first application of buildings to differential geometry in his proof of strong rigidity of locally symmetric spaces. We will not give a precise description of the locally symmetric spaces that occur in Mostow's theorem, but we have already seen some typical examples in Section 13.1.2. For our present purposes, the reader needs to know only that we are talking about certain Riemannian manifolds of the form $Y = \Gamma \backslash X$, where X is a complete simply connected Riemannian manifold of nonpositive curvature associated with a noncompact semisimple real Lie group L, and Γ is a torsion-free discrete subgroup of L.

We assume that Y has finite volume (which is automatic in the arithmetic setting of Section 13.1.2). Since Y is an Eilenberg–Mac Lane space of type $K(\Gamma, 1)$, it is uniquely determined up to homotopy type by its fundamental group Γ, viewed as an abstract group. The remarkable fact proved by Mostow is that under suitable rank and irreducibility assumptions, Y is actually determined *as a Riemannian manifold* by Γ, up to scaling of the metric.

Mostow's proof makes essential use of Tits's rigidity theorem for spherical buildings (Theorem 5.205). Such buildings arise in differential geometry as boundaries at infinity of symmetric spaces. Recall that as we suggested at the beginning of Section 11.8, every complete CAT(0) space has a boundary at infinity that can be constructed from equivalence classes of geodesic rays. In the present setting, the CAT(0) space is the symmetric space X associated with a Lie group $L = G(\mathbb{R})$, where G is a semisimple algebraic group defined over $\mathbb{R}$, and the boundary X_∞ turns out to be in 1–1 correspondence (as a set) with the geometric realization of the spherical building Δ associated with G; see [29, 167]. This explains why there is a connection with buildings. Further details can be found in [140] and the references cited there.

Remarks 14.3. (a) A geodesic ray in the symmetric space X starting at a given $x \in X$ is determined by its initial direction vector; so X_∞ is topologically a sphere, decomposed into (infinitely many) simplices. In view of the bijection between X_∞ and $|\Delta|$ mentioned above, we get a new topology on $|\Delta|$, quite different from the simplicial topology, under which $|\Delta|$ is homeomorphic to a sphere. We can view this new topology as imposing some extra structure on the building Δ. We saw hints of this in a different context in Remark 11.81, Exercise 11.82, and Section 13.2.1. This situation has been axiomatized by Burns and Spatzier [65], and further applications of such "topological buildings" can be found in [140]. See also [199].

(b) The "geodesic compactification" $X \cup X_\infty$ is one of many possible compactifications or partial compactifications of a symmetric space in which the space at infinity can be described in terms of a spherical building. The general theory is surveyed in Ji [140], and details are given in [40, 125].

14.3.2 Further Rigidity Theorems

Mostow's rigidity theorem inspired a tremendous amount of further work, including many extensions of the theorem. See, for instance, Ballmann et al. [28, 29], Burns–Spatzier [66], Gromov–Schoen [122], Kleiner–Leeb [146], Margulis [158], Pansu [184], Prasad [190], and further references cited in [140]. We will not attempt to summarize this work, but one interesting feature is that Euclidean buildings are used in addition to spherical buildings. For example, Gromov and Schoen [122] develop a theory of harmonic maps into Euclidean buildings. And Kleiner and Leeb [146] even have occasion to use $\mathbb{R}$-buildings (cf. Remark 11.106); these arise in connection with the asymptotic geometry of symmetric spaces.

14.3.3 Isoparametric Submanifolds

The basic reference for this subsection is Thorbergsson's survey [237], which includes a detailed history of the subject.

Given a surface Σ in $\mathbb{R}^3$ and a point $x \in \Sigma$, consider the curves in Σ cut by planes normal to Σ at x. The curvature at x, as we run through all these curves, achieves a maximum and minimum value. These extreme values are called the *principal curvatures* of Σ at x. They can also be described as the eigenvalues of the "shape operator" (also called the Weingarten operator), on the tangent plane to Σ at x. The product of the principal curvatures is the Gaussian curvature. The surface is called *isoparametric* if the principal curvatures are constant. For example, spheres, planes, and circular cylinders are isoparametric. The study of isoparametric surfaces goes back at least as far as 1918, when they arose in a problem in geometric optics.

More generally, one can consider *isoparametric hypersurfaces* in Euclidean space $\mathbb{R}^{n+1}$, the sphere S^{n+1}, or hyperbolic space $\mathbb{H}^{n+1}$. There are now n principal curvatures at each point (counted with multiplicity), and one again requires that they be constant. Such hypersurfaces were classified early on in the Euclidean and hyperbolic cases, but É. Cartan showed in the 1930s that the theory is much more difficult in the spherical case. He constructed several interesting families of isoparametric hypersurfaces in spheres, and such hypersurfaces are still not classified today. As we will see, they are related to rank-2 buildings, which are also unclassified. (The buildings that arise are not Moufang in general, so the classification results stated in Chapter 9 do not apply.)

In the 1980s the concept was generalized to submanifolds of codimension bigger than 1, and buildings entered the subject in a 1991 paper of Thorbergsson [235] classifying compact isoparametric submanifolds of Euclidean space of codimension at least 3. The result is that under mild hypotheses, any such submanifold is a principal orbit of the isotropy representation of a symmetric space, also called a *generalized flag manifold*. We will not state the hypotheses or define the technical terms used in the conclusion; the main point for our purposes is the connection with Lie groups, and this is where buildings come in. Namely, Thorbergsson constructed a simplicial complex from his isoparametric submanifold, and he showed that it is a (topological) spherical building when the codimension in Euclidean space is at least 3. He then appealed to the results of [65].

In a follow-up paper [236], he considered some codimension-2 examples. In codimension 2, the isoparametric submanifolds in question are the same as isoparametric hypersurfaces in a sphere. The examples considered by Thorbergsson had four distinct principal curvatures, and he again showed that his complex was a building, of type C_2, i.e., the flag complex of a polar plane. Immervoll [134] later generalized this result to arbitrary isoparametric hypersurfaces with four principal curvatures. It is conjectured that the complex is always a building, and the result is known except possibly if there are exactly six distinct principal curvatures, each of multiplicity 2. In this case there is precisely one known isoparametric hypersurface, which is conjectured to be the only one, and the complex for this example is indeed known to be a building (of type G_2).

Further useful references for this material beyond those cited above are Kramer [148, Chapter 8] and Ji [140].

14.3.4 Singular Spaces and p-adic Groups

Differential geometry concerns smooth spaces, and the fundamental concept is curvature. We have seen in this book that some curvature notions can be extended to singular spaces, and to buildings in particular, via $\mathrm{CAT}(\kappa)$ theory. See Section 12.2 for a list of examples. A different sort of extension of curvature to buildings is Garland's p-adic curvature (Section 13.3).

For geometers interested in extending concepts from smooth manifolds to singular spaces, buildings often play the role of test examples. This is analogous to the special role played by symmetric spaces in the theory of simply connected smooth manifolds of nonpositive curvature.

In the same spirit, Euclidean buildings have been used to prove structure theorems (Iwasawa and Cartan decompositions, etc.) for p-adic groups, in much the same way that symmetric spaces are used for real Lie groups. See, for instance, Ronan [201, 202].

Note, finally, that singular spaces arise naturally in many areas of mathematics, including differential geometry itself, where they occur as Gromov–Hausdorff limits of smooth manifolds; see Gromov [120, 121].

14.4 Representation Theory and Harmonic Analysis

L. Solomon [220] was the first to notice a connection between buildings and representation theory, in the context of finite groups. Let G be a finite group with a BN-pair, and let Δ be the corresponding building. Then the action of G on the unique nonvanishing reduced homology group of Δ yields a representation of G. Solomon observed that this is the *Steinberg representation*, first constructed by Steinberg [226] for the groups $\mathrm{GL}_n(\mathbb{F}_q)$. It is interesting that Steinberg did not actually construct the representation explicitly, but only its character. He then appealed to abstract results from representation theory to prove that the function he constructed was in fact the character of a representation. The homology of the building provided the first concrete realization of Steinberg's character.

Further applications of spherical buildings to the representation theory of finite groups of Lie type were later given by Lusztig [155] and others. See Smith [218] for a survey.

Next, Euclidean buildings have played a major role in the representation theory of p-adic groups, often through harmonic analysis. For an early introduction to this theory, see the lecture notes of Macdonald [157]. There is, unfortunately, only a limited amount of expository material on the more recent developments. See, for instance DeBacker [93, 94] and Schneider [209]. Here are a few of the research papers; the list is not meant to be complete, but only to provide some entry points into the vast literature:

- Adler and DeBacker [20]
- Barbasch and Moy [31]
- Cunningham and Hales [87]
- DeBacker [91, 92]
- Hakim and Murnaghan [128]
- Kim and Murnaghan [144]
- Moy and Prasad [171, 172]
- Rogawski [198]
- Schneider and Stuhler [210, 211]
- Teitelbaum [233]
- Vigneras [269]
- Yu [285], based on Adler [19]

In a somewhat different direction, there has been a great deal of recent work on harmonic analysis on Euclidean buildings with a view toward applications to random walks. See, for example, [76, 186] and the references cited in those papers.

A

Cell Complexes

A.1 Simplicial Complexes

A.1.1 Definitions

Recall that a *simplicial complex* with vertex set $\mathcal{V}$ is a nonempty collection Δ of finite subsets of $\mathcal{V}$ (called *simplices*) such that every singleton $\{v\}$ is a simplex and every subset of a simplex A is a simplex (called a *face* of A). The cardinality r of A is called the *rank* of A, and $r-1$ is called the *dimension* of A. There are different conventions in the literature as to whether the empty set is considered a simplex. Our convention, as forced by the definition above, is that the empty set *is* a simplex; it has rank 0 and dimension -1. A *subcomplex* of Δ is a subset Δ' that contains, for each of its elements A, all the faces of A; thus Δ' is a simplicial complex in its own right, with vertex set equal to some subset of $\mathcal{V}$.

Note that Δ is a *poset*, ordered by the face relation. As a poset, it has the following formal properties:

(a) Any two elements $A, B \in \Delta$ have a greatest lower bound $A \cap B$.
(b) For any $A \in \Delta$, the poset $\Delta_{\leq A}$ of faces of A is isomorphic to the poset of subsets of $\{1, \ldots, r\}$ for some integer $r \geq 0$.

Note. We will often express (b) more concisely by saying that $\Delta_{\leq A}$ is a *Boolean lattice* of rank r.

Conditions (a) and (b) actually characterize simplicial complexes, in the sense that any nonempty poset Δ satisfying (a) and (b) can be identified with the poset of simplices of a simplicial complex. Namely, take the vertex set $\mathcal{V}$ to be the set of rank-1 elements of Δ [where the rank of A is defined to be the rank of the Boolean lattice $\Delta_{\leq A}$]; then we can associate to each $A \in \Delta$ the set $A' := \{v \in \mathcal{V} \mid v \leq A\}$. It is easy to check that $A \mapsto A'$ defines a poset isomorphism of Δ onto a simplicial complex with vertex set $\mathcal{V}$.

We will therefore extend the standard terminology as follows:

Definition A.1. Any nonempty poset Δ satisfying (a) and (b) will be called a *simplicial complex*. The elements of Δ will be called *simplices*, and those of rank-1 will be called *vertices*.

Note that in contrast to (a), two simplices $A, B \in \Delta$ do *not* necessarily have a least upper bound (or any upper bound). We say that A and B are *joinable* if they have an upper bound; in this case they have a least upper bound $A \cup B$, which is simply the set-theoretic union when Δ is viewed as a set of subsets of its vertex set. More generally, an arbitrary set of simplices in a simplicial complex is said to be *joinable* if it has an upper bound, in which case it has a least upper bound.

We visualize a simplex A of rank r as a geometric $(r-1)$-simplex, the convex hull of its r vertices. One makes this precise by forming the *geometric realization* $|\Delta|$ of Δ, which is a topological space partitioned into (open) simplices $|A|$, one for each nonempty $A \in \Delta$. To construct this topological space, start with an abstract real vector space with $\mathcal{V}$ as a basis. Let $|A|$ be the interior of the simplex in this vector space spanned by the vertices of A, i.e., $|A|$ consists of the linear combinations $\sum_{v \in A} \lambda_v v$ with $\lambda_v > 0$ for all v and $\sum_{v \in A} \lambda_v = 1$. We then set

$$|\Delta| := \bigcup_{A \in \Delta} |A| .$$

If Δ is finite, then all of this is going on in $\mathbb{R}^N$, where N is the number of vertices of Δ, and we simply topologize $|\Delta|$ as a subspace of $\mathbb{R}^N$. The question of how to topologize $|\Delta|$ in the general case is more subtle, but the most commonly used topology is the weak topology with respect to the closed simplices: One first topologizes each closed simplex as a subspace of Euclidean space, and one then declares a subset of $|\Delta|$ to be closed if and only if its intersection with each closed simplex is closed. For more details see any standard topology text, such as [129, p. 103; 179, Sections 2 and 3; 224, Section 3.1].

Exercises

A.2. (a) Why is Δ required to be nonempty in Definition A.1?
(b) What should we mean by the "empty simplicial complex"?

A.3. Show that condition (a) in Definition A.1 can be replaced by the following two conditions:

(a_1) Δ has a smallest element.
(a_2) If two elements $A, B \in \Delta$ have an upper bound, then they have a least upper bound $A \cup B$.

A.1.2 Flag Complexes

Let P be a set with a binary relation called "incidence," which is reflexive and symmetric. For example, P might consist of the points and lines of a projective plane, with the usual notion of incidence; or P might consist of the points, lines, and planes of a 3-dimensional projective space; or P might be a poset, with x and y incident if they are comparable (i.e., if $x \leq y$ or $y \leq x$).

A *flag* in P is a set of pairwise incident elements of P. For example, if P is a poset with incidence defined as above, a flag is simply a chain, i.e., a linearly ordered subset.

Definition A.4. The *flag complex* associated to P is the simplicial complex $\Delta(P)$ with P as vertex set and the finite flags as simplices. A flag complex of rank 2 (dimension 1) is also called an *incidence graph*.

In the case of a poset, the flag complex is also called the *order complex* of the poset. One example of this construction appears naturally in the foundations of the theory of simplicial complexes:

Example A.5. If P is the poset of nonempty simplices of a simplicial complex Σ, then $\Delta(P)$ is the *barycentric subdivision* of Σ [179, Lemma 15.3; 224, 3.3]. See Figure A.1 for an example, where Σ consists of a 2-simplex and its faces, and the vertices of the barycentric subdivision are labeled with the corresponding element of P. Note how the six 2-simplices in the barycentric

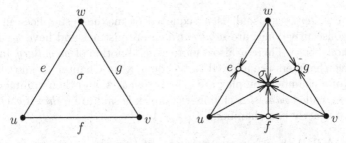

Fig. A.1. A 2-simplex and its barycentric subdivision.

subdivision correspond to the maximal chains of (nonempty) simplices of Σ. From the point of view of geometric realizations, an element of the poset P is a simplex, and the corresponding vertex of $\Delta(P)$ is its barycenter. Thus there are three types of barycenters, corresponding to the three types of simplices of Σ (vertex, edge, 2-simplex). These types are indicated by the three colors (black, white, gray) in the picture of $\Delta(P)$.

Remark A.6. The requirement that P be simplicial in this example is not necessary; P could just as easily be the poset of cells of a more general kind of complex, called a *regular cell complex* (see Section A.2), where the cells are not necessarily simplices. The interested reader can draw some low-dimensional examples to illustrate this.

Not all simplicial complexes are flag complexes. For example, the boundary of a triangle is not a flag complex. [If it were, then the three vertices would be pairwise incident and hence would be the vertices of a simplex of the complex.] Here is a useful characterization of flag complexes:

Proposition A.7. *The following conditions on a simplicial complex Δ are equivalent:*

(i) *Δ is a flag complex.*
(ii) *Every finite set of pairwise joinable simplices is joinable.*
(iii) *Every set of three pairwise joinable simplices is joinable.*
(iv) *Every finite set of pairwise joinable vertices is joinable.*

Proof. It is immediate that (i) $\implies$ (ii) $\implies$ (iv) $\implies$ (i) and that (ii) $\implies$ (iii). The proof that (iii) $\implies$ (ii) is a straightforward induction and is left as an exercise.   □

A.1.3 Chamber Complexes and Type Functions

Definition A.8. Let Δ be a finite-dimensional simplicial complex. We say that Δ is a *chamber complex* if all maximal simplices have the same dimension and any two can be connected by a gallery.

Here a *gallery*, as usual, is a sequence of maximal simplices in which any two consecutive ones are *adjacent*, i.e., are distinct and have a common codimension-1 face. There is also a more general notion of *pregallery*, in which consecutive chambers are allowed to be equal, as in Chapter 1. The maximal simplices of a chamber complex are called *chambers*, and their codimension-1 faces will be called *panels*. A chamber complex is said to be *thin* if each panel is a face of exactly two chambers.

Example A.9. Although many examples of thin chamber complexes occur throughout this book, we mention here the classical examples: If $|\Delta|$ is a connected manifold without boundary, then Δ is a thin chamber complex. Proofs can be found in many topology textbooks. For the convenience of the reader, we outline a proof in Exercise A.17 below, based on local homology calculations. [Note: In the topological literature, thin chamber complexes are often called *pseudomanifolds*.]

Note that the set $\mathcal{C} = \mathcal{C}(\Delta)$ of chambers has a well-defined *distance function* $d(-,-)$, defined to be the minimal length of a gallery joining two chambers. In other words, $d(-,-)$ is the standard metric on the vertices of the chamber graph, the latter being defined in the obvious way. If there is any danger of confusion, we will also refer to the distance just defined as the *gallery distance.* The diameter of Δ, denoted by diam Δ, is the diameter of the metric space of chambers.

We can also consider galleries between arbitrary simplices A and B, i.e., galleries $C_0, \dots, C_l$ with $A \leq C_0$ and $B \leq C_l$, and we define $d(A, B)$ to be the minimal length of such a gallery. Equivalently, $d(A, B)$ is the distance between the sets $\mathcal{C}_{\geq A}$ and $\mathcal{C}_{\geq B}$ in the metric space $\mathcal{C}$. We will again call $d(A, B)$ the *distance*, or *gallery distance*, between A and B. The reader is warned, however, that this distance function on simplices is not a metric; one can, for instance, have $d(A, B) = 0$ with $A \neq B$. In fact, $d(A, B) = 0$ if and only if there is a chamber with both A and B as faces or, equivalently, if and only if A and B are joinable.

Let Δ be a chamber complex of rank n (dimension $n-1$), and let I be a set with n elements (which can be thought of as colors). A *type function* on Δ with values in I is a function τ that assigns to each vertex v an element $\tau(v) \in I$, in such a way that the vertices of every chamber are mapped bijectively onto I. Less formally, each vertex is colored, vertices that are joined by an edge have different colors, and the total number of colors is equal to the rank of Δ (so that a chamber has exactly one vertex of each color). Given a vertex v, we will call $\tau(v)$ the *type* of v.

Definition A.10. We say that Δ is *colorable*, or *vertex colorable*, if it admits a type function.

Colorable complexes are often called "balanced" complexes in the combinatorics literature.

A type function, if it exists, is essentially unique: Any two type functions (with values in sets I and I') differ by a bijection $I \cong I'$. Equivalently, the partition of vertices into types is unique. To see this, just note that if the type function is known on a chamber, then it is uniquely determined on any adjacent chamber. See Figure A.2 for an illustration of this in the rank-3 case. In principle, then, one could try to construct a type function by assigning types to the vertices of an arbitrarily chosen "fundamental chamber" and then moving out along galleries. If one knows a priori that Δ is colorable, then this method is guaranteed to succeed.

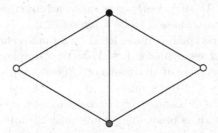

Fig. A.2. The type function on adjacent chambers.

If $n = 2$, then Δ is a graph; it is colorable if and only if it is *bipartite* in the usual sense of graph theory. (The vertices are partitioned into two subsets, and every edge has one vertex in each subset.) This is the case if and only if Δ contains no circuits of odd length. For another familiar example, suppose the chamber complex Δ is a barycentric subdivision as in Example A.5. Then every vertex of Δ corresponds to a simplex of the original complex, and we may declare its type to be the dimension of that simplex. (See Figures 1.3 and A.1 for illustrations of this.) Similarly, if Δ is the flag complex of a projective space or affine space, then the vertices naturally fall into types (point, line, plane,...).

It is useful to think of type functions as chamber maps. Recall first that a *simplicial map* from a simplicial complex Δ to a simplicial complex Δ' is a function ϕ from the vertices of Δ to those of Δ' that takes simplices to simplices. If the image $\phi(A)$ of a simplex A always has the same dimension as A, then ϕ is called *nondegenerate*. A nondegenerate simplicial map is the same as a poset map $\phi: \Delta \to \Delta'$ such that ϕ maps $\Delta_{\le A}$ isomorphically to $\Delta'_{\le \phi(A)}$ for every $A \in \Delta$.

Definition A.11. If Δ and Δ' are chamber complexes of the same dimension, then a simplicial map $\phi: \Delta \to \Delta'$ is called a *chamber map* if it takes chambers to chambers or, equivalently, if it is nondegenerate.

Note that a chamber map takes adjacent chambers to chambers that are either equal or adjacent, and hence it takes galleries to pregalleries. This is why we need the concept of "pregallery." Note further that the image of a chamber map $\Delta \to \Delta'$ is always a chamber subcomplex of Δ', in the sense of the following definition.

Definition A.12. Given a chamber complex Δ, a *chamber subcomplex* of Δ is a simplicial subcomplex Σ that is a chamber complex in its own right and has the same dimension as Δ. Equivalently, the maximal simplices in Σ are chambers in Δ, and any two can be connected by a gallery in Σ.

Note that if Σ is a chamber subcomplex of Δ, then the inclusion $\Sigma \hookrightarrow \Delta$ is a chamber map. Another important kind of chamber map that arises fairly often in this book is a chamber map ϕ from Δ to a subcomplex Σ such that ϕ is the identity on Σ; such a map ϕ is called a *retraction* of Δ onto Σ. If a retraction exists, then Σ is said to be a *retract* of Δ.

Returning now to type functions, let Δ be a rank-n chamber complex and let I be an n-element set as above. Let Δ_I be the "simplex with vertex set I," i.e., the complex consisting of all subsets of I. Then a type function on Δ with values in I is simply a chamber map $\tau: \Delta \to \Delta_I$. Given $A \in \Delta$ we will call $\tau(A)$ the *type* of A; it is a subset of I, consisting of the types of the vertices of A. As we will see, it is often more convenient to work with the *cotype* of a simplex A, defined to be the complement $I \smallsetminus \tau(A)$. Its cardinality is the codimension of A. For example, the cotype of a chamber is the empty set, and the cotype of a panel is a singleton. A panel of cotype $\{i\}$ will be called an *i-panel*.

For any chamber C of Δ, the type function τ maps the subcomplex $\Delta_{\leq C}$ generated by C isomorphically onto Δ_I; hence we may compose τ with the inverse isomorphism to get a retraction $\phi \colon \Delta \to \Delta_{\leq C}$. In concrete terms, $\phi(A)$ is simply the unique face of C having the same type as A. Conversely, a retraction onto $\Delta_{\leq C}$ can be viewed as a type function on Δ, with the set I of types being the set of vertices of C. Thus we have another characterization of colorability:

Proposition A.13. Δ *is colorable if and only if* $\Delta_{\leq C}$ *is a retract of* Δ. $\qquad\square$

Next, we record the easy observation that a chamber map ϕ between colorable chamber complexes induces a *type-change map* that describes the effect of ϕ on the colors. More precisely:

Proposition A.14. *Let* Δ *and* Δ' *be colorable chamber complexes with type functions* τ *and* τ', *respectively, having values in sets* I *and* I'. *If* $\phi \colon \Delta \to \Delta'$ *is a chamber map, then there is a bijection* $\phi_* \colon I \to I'$ *such that for any simplex* $A \in \Delta$,

$$\tau'(\phi(A)) = \phi_*(\tau(A)) \,.$$

Proof. View τ as a chamber map $\Delta \to \Delta_I$ as above, and similarly for τ'. Then the composite $\tau' \circ \phi$ gives a type function on Δ with values in I', so it must differ from τ by a bijection $I \to I'$. This bijection is the desired ϕ_*. $\qquad\square$

Finally, we note a useful property of retracts. Recall that a subcomplex Δ' of a simplicial complex Δ is said to be *full* if it contains every simplex of Δ whose vertices are all in Δ'. The reader can prove the following result as an exercise:

Lemma A.15. *If* Δ' *is a retract of* Δ, *then it is a full subcomplex.* $\qquad\square$

Exercises

A.16. (a) Give an example of a thin chamber complex that has three mutually adjacent chambers.

(b) Show that a colorable thin chamber complex cannot contain three mutually adjacent chambers.

A.17. The purpose of this exercise is to outline a proof that every connected triangulated manifold (without boundary) is a thin chamber complex. To facilitate certain inductive arguments, we work more generally with *homology manifolds*. Recall that if X is a topological space, then the *local homology* of X at a point $x \in X$ is defined to be $H_*(X, X \smallsetminus \{x\})$. By excision, this does not change if we replace X by a neigborhood of x. A topological space X is called a *homology n-manifold* if its local homology at every point is the same as that of $\mathbb{R}^n$, i.e., $H_i(X, X \smallsetminus \{x\}) \cong \mathbb{Z}$ if $i = n$ and 0 otherwise. Let X be the geometric realization $|\Delta|$ of a finite-dimensional simplicial complex Δ.

(a) Let A be a nonempty simplex, and let x be a point in the corresponding geometric open simplex $|A| \subseteq X$. Show that $H_i(X, X \smallsetminus \{x\}) \cong \tilde{H}_{i-k-1}(\mathrm{lk}\, A)$ for all i, where $k := \dim A$. Here the tilde denotes reduced homology, and $\mathrm{lk}\, A$ is the link of A in Δ (Definition A.19 below). [Reduced homology is obtained by forming the chain complex in the usual way, but including the empty simplex in dimension -1.]

(b) Show that X is a homology n-manifold if and only if for every $k \geq 0$ and every k-simplex A, $\mathrm{lk}\, A$ has the same homology as the sphere S^{n-k-1}.

Assume from now on that X is a homology n-manifold.

(c) Show that every maximal simplex of Δ has dimension n.

(d) Show that every $(n-1)$-simplex of Δ is a face of exactly two n-simplices.

(e) With A as in (a), show that $|\mathrm{lk}\, A|$ is a homology $(n - k - 1)$-manifold.

(f) If X is connected, show that Δ is gallery connected and hence a thin chamber complex.

A.1.4 Chamber Systems

We wish to show that if Δ is a sufficiently nice chamber complex, then Δ is completely determined by the system consisting of its chambers together with a suitable refinement of the adjacency relation. Thus we can forget about the vertices and, indeed, all the nonmaximal simplices, when it is convenient to do so.

To refine the adjacency relation, we assume that Δ is colorable. Choose a type function with values in a set I. Then any panel of Δ has cotype $\{i\}$ for some $i \in I$, i.e., it is an i-panel. Given $i \in I$, two adjacent chambers of Δ will be called *i-adjacent* if their common panel is an i-panel. The *chamber system* associated to Δ is the set $\mathcal{C} = \mathcal{C}(\Delta)$ of chambers together with the relations of i-adjacency, one for each i. One can view this chamber system as the chamber graph of Δ, together with a coloring of the edges; the edge between two adjacent chambers gets color i if the chambers are i-adjacent.

It will be convenient to introduce further relations on chambers, called J-equivalence, one for each subset $J \subseteq I$. We say that two chambers are *J-equivalent* if they can be connected by a gallery $C_0, \ldots, C_l$ $(l \geq 0)$ such that any two consecutive chambers C_{k-1}, C_k are j-adjacent for some $j \in J$.

Definition A.18. The equivalence classes of chambers under J-equivalence are called *J-residues*, or residues of *type J*. If J is a singleton $\{i\}$, then we simply say "i-equivalent" and "i-residue."

Thus two chambers C, D are i-equivalent if and only if they have the same i-panel or, in other words, if and only if either $C = D$ or C is i-adjacent to D.

In order to state conditions under which we can recover a chamber complex Δ from its chamber system, we need to recall some more terminology.

Definition A.19. The *link* of a simplex A, denoted by $\operatorname{lk} A$ or $\operatorname{lk}_\Delta A$, is the subcomplex of Δ consisting of the simplices B that are disjoint from A [i.e., $A \cap B$ is the empty simplex] and joinable to A.

Note that as a poset, $\operatorname{lk} A$ is isomorphic to the poset $\Delta_{\geq A}$ via $B \mapsto B \cup A$ ($B \in \operatorname{lk} A$). In particular, the maximal simplices of $\operatorname{lk} A$ are in 1–1 correspondence with the chambers of Δ having A as a face. But $\operatorname{lk} A$ need not be a chamber complex. For it might not be possible to connect two chambers in $\Delta_{\geq A}$ by a gallery in $\Delta_{\geq A}$.

Proposition A.20. *Let Δ be a colorable chamber complex. Assume that the link of every simplex is a chamber complex, and assume further that every panel is a face of at least two chambers. Then Δ is determined up to isomorphism by its chamber system. More precisely:*

(1) *For every simplex A, the set $\mathcal{C}_{\geq A}$ of chambers having A as a face is a J-residue, where J is the cotype of A.*

(2) *Every residue has the form $\mathcal{C}_{\geq A}$ for some simplex A.*

(3) *For any simplex A, we can recover A from $\mathcal{C}_{\geq A}$ via*

$$A = \bigcap_{C \geq A} C \,.$$

(4) *Δ, as a poset, is isomorphic to the set of residues in $\mathcal{C}$, ordered by reverse inclusion.*

Proof. Note first that (4) implies that Δ is determined by its chamber system, since the residues are defined entirely in terms of the chamber system. We proceed now to (1)–(4).

(1) Our first assumption implies that the chambers in $\mathcal{C}_{\geq A}$ are all J-equivalent to one another, so there is a J-residue $\mathcal{R}$ with $\mathcal{C}_{\geq A} \subseteq \mathcal{R}$. Since J-equivalent chambers have the same face of cotype J, we also have the opposite inclusion; hence $\mathcal{C}_{\geq A} = \mathcal{R}$.

(2) Let $\mathcal{R}$ be an arbitrary J-residue. Then the chambers in $\mathcal{R}$ all have the same face A of cotype J, so $\mathcal{R} \subseteq \mathcal{C}_{\geq A}$. In view of (1), equality must hold.

(3) We can think of simplices as sets of vertices, so that the greatest lower bound in the statement is simply the set-theoretic intersection. If v is a vertex of Δ such that $v \notin A$, we must find a chamber $C \geq A$ with $v \notin C$. Start with an arbitrary chamber $C \geq A$. If $v \notin C$, we are done. Otherwise, let P be the panel of C not containing v. Then there is a chamber $D \neq C$ with $P < D$. We then have $D > A$ but $v \notin D$.

(4) It is immediate from (1)–(3) that there is a bijection from Δ to the set of residues, given by

$$A \mapsto \mathcal{C}_{\geq A} \,,$$

with inverse

$$\mathcal{R} \mapsto \bigcap_{C \in \mathcal{R}} C \,.$$

These bijections are order-reversing if the residues are ordered by inclusion.

$\square$

Exercise A.21. Let Δ be a colorable chamber complex, and assume only that the link of every vertex is again a chamber complex. Show that Δ is still determined up to isomorphism by its chamber system.

*A.2 Regular Cell Complexes

In this section we outline the theory of regular cell complexes, with emphasis on the aspects that are needed in Chapter 12. For further information and a more detailed account, see [34; 37, Section 4.7; 84; 154].

A.2.1 Definitions and First Properties

By a *ball* we mean a topological space homeomorphic to a closed ball in Euclidean space of some dimension $d \geq 0$. A ball e has a (relative) *interior* $\operatorname{int} e$, homeomorphic to an open d-ball, and a *boundary* ∂e, homeomorphic to the sphere S^{d-1} (which is empty if $d = 0$).

Definition A.22. A *regular cell complex* is a collection K of balls in a Hausdorff space $|K|$ such that:

(1) The interiors $\operatorname{int} e$ for $e \in K$ partition $|K|$.
(2) For each $e \in K$, the boundary ∂e is a union of finitely many elements of K (necessarily of dimension less than that of e).
(3) $|K|$ has the weak topology with respect to the collection K, i.e., a subset of $|K|$ is closed if its intersection with each $e \in K$ is closed. [This is redundant if K is finite.]

The elements of K are called *closed cells*, and their relative interiors are called *open cells*.

A regular cell complex K has a natural *face relation*, corresponding to the inclusion relation on closed cells. Thus K is a poset. For each cell $e \in K$, the set $K_{\leq e}$ of faces of e is again a regular cell complex, whose underlying space is e itself. Similarly, the set $K_{<e}$ of proper faces of e is a regular cell complex whose underlying space is the sphere ∂e.

Two regular cell complexes K, K' are said to be *isomorphic* if there is a homeomorphism $f \colon |K| \to |K'|$ that maps every cell of K onto a cell of K'. (It then follows that f induces a poset isomorphism $K \to K'$.)

The canonical example of a regular cell complex is the set K of nonempty simplices of a simplicial complex, with $|K|$ being the geometric realization of that complex as defined in Section A.1.1. Here is a simple example of a regular cell complex that is not simplicial:

Example A.23. Let X be the n-cube $[0,1]^n$, and let K consist of the (closed) faces of X in the usual sense, i.e., the subsets gotten by freezing zero or more coordinates at 0 or 1. Then K is a regular cell complex with $|K| = X$. See Exercise A.27 below for a description of K as a poset.

Example A.23 is the foundation for an important class of regular cell complexes:

Definition A.24. A *cubical complex* is a regular cell complex with the following two properties:

(1) For each cell e, the subcomplex $K_{\leq e}$ is isomorphic to a standard cube, viewed as a regular cell complex as in Example A.23.
(2) The intersection of two cells, if nonempty, is again a cell.

One can also characterize cubical complexes abstractly, in the spirit of Definition A.1; see Exercises A.30 and A.31.

Cubical complexes have become quite important in recent years, largely because of a theorem of Gromov [48, Theorem II.5.20; 119, 4.2.C]: Suppose K is a finite-dimensional cubical complex in which each cell comes equipped with a metric. Assume that the isomorphisms in (1) above can be taken to be isometries, and assume that the metric assigned to a face of a cell e agrees with the one it inherits as a subspace of e. Then there is a metric on $X = |K|$ obtained by minimizing lengths of piecewise linear paths, and Gromov's theorem says that X is a CAT(0) space if and only if the link of every vertex is a flag complex. [See Exercise A.28 below for the notion of "link" in the present context.]

Returning now to a general regular cell complex K, recall that K is a poset and hence has an associated flag complex $\Delta = \Delta(K)$. As we suggested in Remark A.6, Δ is the *barycentric subdivision* of K. In particular, $|\Delta|$ is homeomorphic to $|K|$. To get a specific homeomorphism, we have to choose for each cell e a homeomorphism between e and the cone over ∂e; the cone point then corresponds to a chosen "barycenter" in $\mathrm{int}\, e$. See, for instance, [37, proof of Proposition 4.7.8]. Note that if we then identify $|\Delta|$ with $|K|$, we recover the closed cell e as the union of the open simplices in $|\Delta|$ corresponding to the flags $e_0 < e_1 < \cdots < e_d$, with $e_d \leq e$.

This gives us a way of reconstructing the topological space $|K|$ from the abstract poset K: Simply *define* $|K|$ to be $|\Delta(K)|$ and, for each $e \in K$, define the corresponding closed cell to be $|\Delta(K_{\leq e})|$. This circle of ideas leads to the following result [34, Proposition 3.1; 37, Proposition 4.7.23]:

Proposition A.25.

(1) *A poset P is isomorphic to the poset of cells of a regular cell complex if and only if $|\Delta(P_{<x})|$ is homeomorphic to a sphere for each $x \in P$.*

(2) *If P satisfies the condition in (1), then there is a unique (up to isomorphism) regular cell complex whose poset of cells is isomorphic to P. Its underlying space is $|\Delta(P)|$, and its cells are the subspaces $e_x := |\Delta(P_{\leq x})|$ for $x \in P$.*

(3) *Two regular cell complexes K, K' are isomorphic as regular cell complexes if and only if they are isomorphic as posets.* □

Exercises

A.26. Let P be the poset of cells of a hyperplane arrangement (Section 1.4), with the smallest cell removed. Show that P is a regular cell complex whose underlying space is a sphere. [The intention is that this should be done directly; it will also follow from the results of Section A.2.3 below.]

A.27. Let K be the standard cube, as in Example A.23. Note that the set of vertices (0-cells) of K is $P = \{0,1\}^n$, which is itself a poset in an obvious way. [Declare $0 < 1$, and order P componentwise. Equivalently, identify P with the Boolean lattice of subsets of $\{1, \ldots, n\}$.] Describe the poset structure on K in terms of the poset P. Figure A.3 might provide a clue, where we have written a, b, c instead of $1, 2, 3$, i.e., we are identifying P with the poset of subsets of $\{a, b, c\}$.

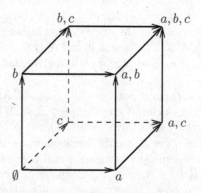

Fig. A.3. The faces of a cube.

A.28. If e is a cell of a cubical complex K, show that the poset $K_{\geq e}$ is a simplicial complex. It is called the *link* of e.

Remark A.29. This terminology is motivated by the simplicial case; see the comment following Definition A.19. There is also a geometric explanation [48, Chapter I.7]. Briefly, one should think of the link of e as the space of directions orthogonal to e. It is decomposed into spherical pieces, one for each cell $f > e$. To understand intuitively why the link is simplicial in the cubical case, it suffices to consider the link of a vertex, in which case one need only note that a cube, in the neighborhood of a vertex, looks like a simplicial cone.

A.30. Show that a poset P is isomorphic to the poset of cells of a cubical complex if and only if it satisfies the following two conditions:

(a) If two elements $x, y \in P$ have a lower bound, then they have a greatest lower bound.

(b) For any $x \in P$, the poset $P_{\leq x}$ is isomorphic to the poset of faces of a cube.

A.31. Show that condition (a) in Exercise A.30 can be replaced by:

(a') If two elements $x, y \in P$ have an upper bound, then they have a least upper bound.

[This exercise is the cubical analogue of Exercise A.3.]

A.32. Let K be a regular cell complex, let e be an n-cell for some $n \geq 2$, and let $e' < e$ be a face of dimension $n - 2$. Show that there are exactly two $(n-1)$-cells f such that $e' < f < e$.

A.33. A regular cell complex is said to have the *intersection property* if the intersection of two cells is either empty or is again a cell. For example, simplicial and cubical complexes have the intersection property, as does the poset of cells of a hyperplane arrangement.

(a) If K has the intersection property, show that every cell is the least upper bound of its vertices.

(b) If K has the intersection property and $f < e$ in K, show that f is an intersection of codimension-1 faces of e.

A.2.2 Regular Cell Complexes from Polytopes

A subset X of a Euclidean space V is called a *polytope* if X is compact, has nonempty interior in V, and is the intersection of finitely many closed halfspaces. We will assume for simplicity (and without loss of generality) that the origin is an interior point of X. It follows that X can be defined by finitely many linear inequalities of the form $g_j \leq 1$, where $(g_j)_{j \in J}$ is a family of linear functions indexed by a finite set J. The assumption that X has nonempty interior is made for convenience; if it failed, we could replace V by the span of X.

By a slight variant of what we did in Section 1.4, the polytope X has a natural decomposition as a regular cell complex, where the open cells are the nonempty sets defined by equalities $g_j = 1$ or strict inequalities $g_j < 1$, one for each $j \in J$. The unique top-dimensional cell is the interior of X, which is defined by $g_j < 1$ for all j. For example, the cube $[0, 1]^n$ is a polytope, and the cell structure that we have just defined agrees with the one in Example A.23.

The set of proper faces of the polytope X forms a subcomplex, which is a regular cell complex in its own right, whose underlying space is topologically a sphere. The top-dimensional faces of this subcomplex, i.e., the codimension-1 faces of X, are called the *facets* of X.

The *face lattice* of X is the poset of cells of X, together with the empty set. As the terminology suggests, it is indeed a lattice, i.e., it has least upper bounds and greatest lower bounds. [Since it is finite and has a largest element, we need only check the existence of greatest lower bounds; these are given by set-theoretic intersections of closed cells.] For example, Exercise A.27 showed that the face lattice of a cube is the poset of closed intervals in a Boolean lattice, together with the empty set, ordered by inclusion.

Every polytope X has a *dual polytope* X^*, sometimes called the *polar* of X [286, Section 2.3]. With the hypotheses and notation above, the dual of X is the subset of the dual vector space V^* consisting of all linear functions $g \in V^*$ such that $g \leq 1$ on X. The face lattice of the dual is isomorphic to the opposite of the face lattice of X. The empty face of X^* corresponds to the top-dimensional cell of X, the vertices of X^* correspond to the facets of X, and so on. These vertices are in fact the linear functions g_j above if the defining inequalities $g_j \leq 1$ form a minimal set of inequalities defining X. (By extension of the terminology we used in Section 1.4, the hyperplanes $g_j = 1$ are then the *walls* of X.)

A.2.3 Regular Cell Complexes from Arrangements

Let $\mathcal{H} = \{H_i\}_{i \in I}$ be an essential hyperplane arrangement in V, and let Σ be the poset of conical cells as in Section 1.4.1. The nontrivial elements of Σ, i.e., those different from $\{0\}$, intersect the unit sphere in subsets that are the cells of a regular cell decomposition of the sphere. In Chapter 1 we treated this in detail for reflection arrangements (i.e., arrangements coming from finite reflection groups), in which case the cell complex is simplicial. It is not hard to check directly that the assertion remains true in general, but we will instead deduce it from a stronger assertion. Namely, we will construct a polytope X whose proper faces correspond to the nontrivial conical cells; the boundary of X then gives, by radial projection, the desired cell decomposition of the sphere.

Roughly speaking, we get X by choosing suitable affine hyperplanes that cut across the chambers. More precisely, choose for each $i \in I$ a linear function $f_i \colon V \to \mathbb{R}$ such that H_i is given by $f_i = 0$. Since $\mathcal{H}$ is essential, the f_i span the dual space V^*. For any sequence $\tau = (\tau_i)_{i \in I}$ with $\tau_i = \pm 1$, set $g_\tau := \sum_{i \in I} \tau_i f_i$. We then define $X \subseteq V$ by the $2^{|I|}$ inequalities $g_\tau \leq 1$, one for each τ. Note that the defining inequalities imply that $|f_i| \leq 1$ on X, so X is compact [because the f_i span V^*] and hence a polytope.

It will follow from what we do below that the inequalities $g_\tau \leq 1$ are redundant in general and that X is actually defined by the inequalities $g_\sigma \leq 1$ in which σ is the sign sequence of a chamber. Since $g_\sigma > 0$ on C if $\sigma = \sigma(C)$ for a chamber C, this implies that we can visualize the facets of X (given by $g_\sigma = 1$) as cutting across the chambers and matching up correctly along faces. For example, let $\mathcal{H}$ consist of three lines in the plane as in Figure 1.4. Then the line $f_1 + f_2 + f_3 = 1$ cuts across the chamber $+++$, the line $-f_1 + f_2 + f_3 = 1$

cuts across the chamber $-++$, and they agree on the common panel, where $f_1 = 0$; see Figure A.4. The polytope X in this case is a solid hexagon.

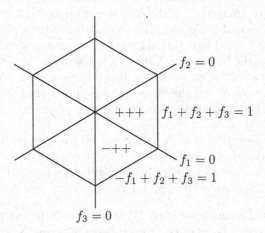

Fig. A.4. The facets of X cut across the chambers.

Returning now to the general theory, we wish to show that the proper faces of X are in 1–1 correspondence with the cells $A \neq \{0\}$ in Σ. The discussion that follows is adapted from [56, Appendix].

Given $A \neq \{0\}$ in Σ, let $\sigma = \sigma(A)$ and let $h_\sigma := \sum_{i \in I} \sigma_i f_i$. [Note that $h_\sigma = g_\sigma$ if A is a chamber. But if A is not a chamber, then h_σ is *not* one of the linear functions g_τ occurring in the definition of X.] We have $h_\sigma > 0$ on A, and A is the cone over

$$A_1 := A \cap \{h_\sigma = 1\} \ .$$

Note that A_1 is defined by the following linear equalities and inequalities, where $I_0 := \{i \in I \mid \sigma_i = 0\}$:

$$f_i = 0 \quad \text{for } i \in I_0, \tag{A.1}$$
$$\sigma_i f_i > 0 \quad \text{for } i \notin I_0, \tag{A.2}$$
$$h_\sigma = 1. \tag{A.3}$$

We claim that A_1 is a (relatively open) face of X. In fact, we will show that A_1 is the face defined by

$$g_\tau = 1 \quad \text{if } \tau \text{ is consistent with } \sigma, \tag{A.4}$$
$$g_\tau < 1 \quad \text{otherwise.} \tag{A.5}$$

Here τ is *consistent* with σ if $\tau_i = \sigma_i$ for all i such that $\sigma_i \neq 0$. [Recall that τ_i is required to be ± 1 for all i.]

For any τ that is consistent with σ, we can write $g_\tau = h_\sigma + \sum_{i \in I_0} \tau_i f_i$. Moreover, we can change the sign of any τ_i with $i \in I_0$ and still have a sequence consistent with σ. It follows that (A.4) is equivalent to (A.1) and (A.3). And in the presence of (A.1) and (A.3), (A.5) is equivalent to (A.2). Thus we have transformed (A.4) and (A.5) to (A.1)–(A.3), whence the claim.

We now have $V \smallsetminus \{0\}$ partitioned into the cones over some of the (relatively open) faces A_1 of the boundary ∂X of X. It follows that the cells A_1 are in fact all of the faces of X, and we have established the desired 1–1 correspondence between the faces of X and the elements of Σ (other than $\{0\}$). It is easy to check that this correspondence is a poset isomorphism, i.e., it preserves the face relation.

Finally, our assertion that X is defined by the inequalities $g_\sigma \leq 1$ corresponding to the chambers follows from the general fact that a polytope with nonempty interior can always be defined by one inequality for each facet [286, Theorem 2.15]. Alternatively, one can check the assertion directly.

Remark A.34. The dual polytope X^* is the convex hull of the linear functions g_τ. It is a special kind of polytope called a *zonotope*. In some treatments of the subject one starts with this zonotope and then *defines* X to be its dual. See [37, Example 4.1.7; 286, Corollary 7.18]. Zonotopes associated to reflection arrangements play an important role in the theory of geometric realizations of Coxeter complexes; see Section 12.3.5.

A.3 Cubical Realizations of Posets

Every poset P gives rise to a flag complex $\Delta(P)$. It is a simplicial complex with P as vertex set. Our purpose here is to call attention to a class of posets for which one can lump certain simplices of $\Delta(P)$ together to create a *cubical* complex. The motivating example is the case that P is the Boolean lattice $\{0,1\}^n$; we saw in Exercise A.27 how to construct the standard cube (viewed as a regular cell complex) from P. But we did not explain there the connection between this cube and $\Delta(P)$, so let's go back to the beginning and do that. Our discussion will make use of the *Hasse diagram* of a poset P. This is a graph with one vertex for each element of P and one edge for each "cover relation" $x \lessdot y$, where the latter means that $x < y$ and there is no z with $x < z < y$. By convention, y is drawn higher than x in the diagram.

Examples A.35. (a) Let P be the poset $\{0,1\}$ with $0 < 1$, i.e., the Boolean lattice of rank 1. The flag complex $\Delta(P)$ has four simplices: the empty chain, the two singletons, and the chain $0 < 1$. Its geometric realization can be identified with the unit interval $[0,1]$, which is also the 1-cube.

(b) Let P be the direct product $\{0,1\} \times \{0,1\}$ with the componentwise ordering, i.e., P is the Boolean lattice of rank 2. Its Hasse diagram is shown on the left and its flag complex on the right in Figure A.5. Note that $\Delta(P)$

triangulates the unit square $[0,1] \times [0,1]$, or the 2-cube. If we delete the dotted edge so that the two 2-simplices are lumped together, we see the standard cell structure on the 2-cube.

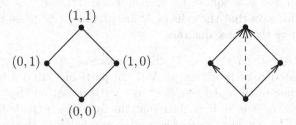

Fig. A.5. The Hasse diagram and flag complex of $\{0,1\} \times \{0,1\}$.

Example A.35(b) illustrates the following classical fact:

Proposition A.36. *If P and Q are posets, then the canonical map*

$$|\Delta(P \times Q)| \to |\Delta(P)| \times |\Delta(Q)|$$

is a homeomorphism. Here $P \times Q$ is ordered componentwise.

Note. If the posets are infinite, one has to be careful about how the product on the right is topologized.

We omit the proof, which can be found in [100, II.8.9; 154, pp. 96–97; 163, Theorem 2].

Example A.37. Let $P = \{0,1\}^n$, i.e., the Boolean lattice of rank n. Then Proposition A.36 implies that $\Delta(P)$ triangulates the n-cube. The reader is encouraged to draw the Hasse diagram when $n = 3$; the cube will be visible in the picture (which should be compared with Figure A.3).

We can now give conditions on a poset P under which we can construct a cubical complex with vertex set P. The answer is very simple and is in the spirit of Proposition A.25. The result is probably well known, but we are not aware of a reference for it.

Proposition A.38. *Let P be a poset in which every closed interval $[x,y]$ ($x,y \in P$, $x \le y$) is a Boolean lattice. Then one can construct a cubical complex $K = \Delta_c(P)$ with the following properties:*

(1) *The underlying space of K is $|\Delta(P)|$.*
(2) *The cells of K are the subspaces $e_{x,y} := \big|\Delta([x,y])\big|$, where $x \le y$ in P. In particular, the vertices of K are the same as those of $\Delta(P)$, i.e., they are the elements of P.*

(3) *As a poset, K is isomorphic to the set of closed intervals in P, ordered by inclusion.*

Note, then, that K has a vertex for each $x \in P$, an edge from x to y for each cover relation $x \lessdot y$, a square for each interval $[x, y]$ of rank 2, and so on. Intuitively, this says that the cells of K are precisely the cubes that one sees when one draws the Hasse diagram.

Sketch of proof. Given $x \le y$ in P, the subspace $e_{x,y} \subseteq |\Delta(P)|$ defined in (2) is a triangulated cube by Example A.37. In particular, it is a ball. Its interior is the union of the open simplices in $|\Delta(P)|$ corresponding to the chains $x = x_0 < \cdots < x_d = y$, so it is clear that the interiors of the balls $e_{x,y}$ partition $|\Delta(P)|$. The rest of the proof is routine and is left to the reader.   □

Remark A.39. It may happen that P is itself the poset of (nonempty) cells of a regular cell complex L. Then $\Delta(P)$ is the barycentric subdivision of L, while $\Delta_c(P)$ is a coarser subdivision into cubes. It has a vertex b_e for each cell e of L, which we can visualize as the barycenter of e, and it has one cube for each face relation $e \le f$ between cells of L. The vertices of this cube are the barycenters of the faces g with $e \le g \le f$.

Examples A.40. (a) Let P be the poset of nonempty simplices of a simplicial complex Σ. Then P satisfies the hypotheses of Proposition A.38 and Remark A.39. We therefore obtain a cubical subdivision of $|\Sigma|$, with one cube for every pair consisting of a simplex of Σ and a nonempty face of it. See Figure A.6 for the case that Σ consists of a 2-simplex and its faces. Note that one can see, from the orientations on the edges, the smallest and largest vertex of each cube and hence the corresponding interval in P.

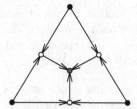 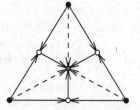

Fig. A.6. The cubical and barycentric subdivisions of a 2-simplex.

(b) Now let P be the poset of *all* simplices of Σ, including the empty simplex; more briefly, $P = \Sigma$. The hypotheses of the proposition are again satisfied, and we obtain a cubical subdivision of the cone over Σ. It has one cube for every pair consisting of a simplex of Σ and a face of it, one or both of which might be empty. (The cube corresponding to the case that both are empty is the cone point.) Figure A.7 illustrates this when Σ is the boundary of a square. Combinatorially, K is a solid octagon, decomposed into four quadrilaterals (corresponding to the intervals from the empty simplex to the four original edges).

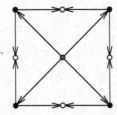

Fig. A.7. The cubical subdivision of the cone over a square.

B

Root Systems

The reader can consult Bourbaki [44] or Humphreys [131, 133] for a detailed treatment of root systems. Here we will just present the definition and one or two consequences.

B.1 Notation

Let V be a Euclidean vector space with inner product $\langle -, - \rangle$. Recall from Section 1.1 that the orthogonal reflection s with respect to a hyperplane H in V is given by

$$s(x) = x - 2 \frac{\langle \alpha, x \rangle}{\langle \alpha, \alpha \rangle} \alpha \, , \tag{B.1}$$

where α is any nonzero vector in $H^{\perp}$. If we set

$$\alpha^{\vee} := \frac{2\alpha}{\langle \alpha, \alpha \rangle} \, , \tag{B.2}$$

this takes the simpler form

$$s(x) = x - \langle \alpha^{\vee}, x \rangle \alpha \, . \tag{B.3}$$

The relation between $\alpha^{\vee}$ and α is probably best remembered by the property that $\alpha^{\vee}$ is the scalar multiple of α satisfying

$$\langle \alpha^{\vee}, \alpha \rangle = 2 \, . \tag{B.4}$$

This shows that there is a symmetric relationship between α and $\alpha^{\vee}$, so that $(\alpha^{\vee})^{\vee} = \alpha$.

B.2 Definition and First Properties

Let Φ be a generalized root system in V (Definition 1.5). Thus Φ is a finite set of nonzero vectors, and Φ is W_Φ-invariant, where W_Φ is the (finite) group generated by the reflections s_α $(\alpha \in \Phi)$.

Definition B.1. We say that Φ is a *root system*, or a *crystallographic root system*, if it satisfies the following condition:

(Cry) $\langle \alpha^\vee, \beta \rangle$ *is an integer for all* $\alpha, \beta \in \Phi$.

One sometimes assumes further that Φ spans V, which is equivalent to saying that the reflection group W_Φ is essential, but this is less important; it can always be achieved by replacing V by the span of Φ. Condition (Cry) arises naturally in the theory of Lie algebras, and it also arises naturally in the theory of infinite Euclidean reflection groups (see Section 10.1.8).

Assume now that we have chosen a fundamental chamber for the Weyl group $W = W_\Phi$. Recall from Section 1.5.10 that this gives rise to a notion of *positive root* and to a system of *simple roots* $\alpha_1, \dots, \alpha_n$, corresponding to the fundamental reflections $s_1, \dots, s_n$ that generate W. By Exercise 1.120, every positive root is a nonnegative linear combination of the α_i. In the presence of (Cry), we can say more:

Proposition B.2. *Let Φ be a root system that spans V.*

(1) *Every positive root is a nonnegative linear combination of the simple roots, with integer coefficients.*

(2) *The additive subgroup of V generated by Φ is a lattice in V, i.e., it is the free abelian group generated by a vector-space basis.*

Proof. The content of (1), in view of what we already know, is that every root is an integral linear combination of the simple roots. To see this, recall from Exercise 1.121 that every root α is W-equivalent to a simple root α_j. Now as we repeatedly apply the generating reflections s_i to get from α_j to α, condition (Cry) and formula (B.3) show that we repeatedly subtract integral multiples of simple roots. Hence α is an integral linear combination of simple roots. Statement (2) follows immediately from (1), since the α_i form a basis for V. $\square$

Exercise B.3. The *Cartan matrix* of Φ relative to the system of simple roots $\alpha_1, \dots, \alpha_n$ is the $n \times n$ matrix $A = (a_{ij})$ given by $a_{ij} = \langle \alpha_i^\vee, \alpha_j \rangle$. Show that it is related to the Coxeter matrix $M = (m_{ij})$ by

$$a_{ij} = -2 \frac{\|\alpha_j\|}{\|\alpha_i\|} \cos \frac{\pi}{m_{ij}} .$$

Consequently,

$$a_{ij} a_{ji} = 4 \cos^2 \frac{\pi}{m_{ij}} .$$

B.3 The Dual Root System

If Φ is a generalized root system, then

$$\Phi^\vee := \{\alpha^\vee \mid \alpha \in \Phi\}$$

is again a generalized root system, said to be *dual* to Φ. It is crystallographic (and hence a root system) if and only if Φ is. In particular, a root system Φ naturally gives rise to two lattices, the *root lattice*, generated by Φ, and the *coroot lattice*, generated by $\Phi^\vee$. Note that Φ and $\Phi^\vee$ have the same Weyl group W, and that both lattices are W-invariant.

B.4 Examples

The examples that follow are based on those in Section 1.2, and they use the same notation.

Example B.4. Let Φ be the root system of type A_{n-1}, consisting of the vectors $e_i - e_j$ $(i \neq j)$. Since each root α satisfies $\langle \alpha, \alpha \rangle = 2$, we have $\alpha^\vee = \alpha$ and $\Phi^\vee = \Phi$. The root lattice and coroot lattice (in the subspace V of $\mathbb{R}^n$ spanned by the roots) are equal to

$$\mathbb{Z}^n \cap V = \{(x_1, \ldots, x_n) \in \mathbb{Z}^n \mid \textstyle\sum x_i = 0\} \ .$$

It has a basis $\alpha_1, \ldots, \alpha_{n-1}$, where $\alpha_i = e_i - e_{i+1}$. This is the system of simple roots corresponding to the fundamental chamber $x_1 > \cdots > x_n$.

Example B.5. Let Φ be the root system of type B_n, consisting of the vectors $\pm e_i \pm e_j$ $(i \neq j)$ together with the vectors $\pm e_i$. The roots $\alpha = \pm e_i \pm e_j$ satisfy $\langle \alpha, \alpha \rangle = 2$, so $\alpha^\vee = \alpha$. The roots $\alpha = \pm e_i$ satisfy $\langle \alpha, \alpha \rangle = 1$, so $\alpha^\vee = 2\alpha$. Thus $\Phi^\vee$ is the root system of type C_n. The root lattice is the standard lattice $\mathbb{Z}^n$, while the coroot lattice is the sublattice

$$\{(x_1, \ldots, x_n) \in \mathbb{Z}^n \mid \textstyle\sum x_i \equiv 0 \bmod 2\} \ , \tag{B.5}$$

which is of index 2 in $\mathbb{Z}^n$.

The root lattice, of course, has its standard basis $e_1, \ldots, e_n$. But the present point of view leads to a different basis, consisting of the vectors

$$e_1 - e_2, \ e_2 - e_3, \ \ldots, \ e_{n-1} - e_n, \ e_n \ .$$

These are the simple roots corresponding to the fundamental chamber

$$x_1 > \cdots > x_n > 0$$

that we wrote down in Example 1.82. The same fundamental chamber leads to the basis

$$e_1 - e_2, \ e_2 - e_3, \ \ldots, \ e_{n-1} - e_n, \ 2e_n$$

for the root lattice of type C_n, and hence for the coroot lattice of type B_n, given in (B.5).

Example B.6. If Φ is the root system of type D_n, consisting of the vectors $\pm e_i \pm e_j$ $(i \neq j)$, then $\Phi^\vee = \Phi$. The root lattice and the coroot lattice are equal to the lattice in (B.5) again. From the point of view of D_n, however, the natural basis to write down is

$$e_1 - e_2, \, e_2 - e_3, \, \ldots, \, e_{n-1} - e_n, \, e_{n-1} + e_n \, ;$$

this arises from the fundamental chamber given in Example 1.83.

*B.5 Nonreduced Root Systems

Recall from Section 1.1 our convention that (generalized) root systems Φ are always assumed to be *reduced*, which means that for any $\alpha \in \Phi$, the only scalar multiples of α that are in Φ are $\pm \alpha$. But nonreduced root systems arise naturally in connection with algebraic groups defined over fields that are not algebraically closed (see Section 7.9.3), so we will mention them briefly. Note first that the only way a crystallographic root system can fail to be reduced is if it contains a pair $\{\alpha, 2\alpha\}$. More precisely, if α and $\lambda\alpha$ are both roots for some $\lambda \in \mathbb{R}$, then necessarily $\lambda \in \{\pm 1, \pm 2, \pm\frac{1}{2}\}$. Indeed, we have $\langle \lambda\alpha, \alpha^\vee \rangle = 2\lambda$ and $\langle \alpha, (\lambda\alpha)^\vee \rangle = 2/\lambda$, so our assertion follows from the fact that 2λ and $2/\lambda$ are both integers.

We close by mentioning the unique family of irreducible, nonreduced, crystallographic root systems:

Example B.7. For each $n \geq 1$, the root system of type BC_n is defined to be the union of the systems of type B_n and C_n in $\mathbb{R}^n$, i.e., it consists of the vectors $\pm e_n$, $\pm 2e_n$, and $\pm e_i \pm e_j$ for $1 \leq i, j \leq n$, $i \neq j$.

C

Algebraic Groups

A typical example of what we want to talk about in this appendix is the group $SL_n(k)$, where k is a field. As a set, this is defined by the equation $\det(a_{ij}) = 1$, which is a polynomial equation of degree n in the n^2 variables a_{ij}. And the group structure is given by the matrix-multiplication map

$$SL_n(k) \times SL_n(k) \to SL_n(k) \ ,$$

which is also describable by polynomials. In fact, each of the n^2 components of this map is a quadratic function of $2n^2$ variables. As this example suggests, we will be looking at groups G of the following form: G is a set defined by polynomial equations, with a group structure defined by a polynomial map.

We will see later that there are many properties of such groups that are revealed only when we pass from k to a bigger field k'. Examples are given in Sections C.6–C.9 below. Thus we will want to take the defining equations for G and look at their solutions over various extensions k' of k. More generally, there are reasons for looking at solutions over k-algebras R that are not necessarily fields. [By a k-*algebra* here we mean a commutative ring with identity that comes equipped with a homomorphism $k \to R$. Since k is a field, the given homomorphism is necessarily 1–1, and we may think of R as a ring that contains k as a subring.] These considerations lead to the notion of *group scheme*. There are many references for this material, but we will mostly follow Waterhouse [278].

C.1 Group Schemes

Suppose we are given a collection of polynomial equations in m variables with coefficients in a field k. For any k-algebra R, let $G(R) \subseteq R^m$ be the solution set of the given equations. Assume further that we are given m polynomials in $2m$ variables such that the map $R^m \times R^m \to R^m$ that they define sends $G(R) \times G(R)$ into $G(R)$ for every R and makes $G(R)$ a group. What we have, then, is not just a single group, but rather a group-valued functor on the category of k-algebras. [Take a moment to verify this assertion; the essential point is that the formulas defining the group structure are compatible with k-algebra maps.]

Definition C.1. A functor G defined by polynomials as above is called an *algebraic affine group scheme* over k, and the group $G(R)$ is called the group of R-*points* of G.

The canonical example is $G = \mathrm{SL}_n$, viewed now as the functor $R \mapsto \mathrm{SL}_n(R)$.

Remark C.2. Readers familiar with topological groups might be surprised that we did not require the inversion map $g \mapsto g^{-1}$ to be a polynomial map. The reason for not requiring it is that it turns out to be a formal consequence of our definition [278, Chapter 1]. In our SL_n example, for instance, Cramer's rule provides a polynomial formula for the inverse.

Examples C.3. (a) Fix a matrix $g \in \mathrm{SL}_n(k)$. For any k-algebra R, let $G(R)$ be the centralizer of g in $\mathrm{SL}_n(R)$. Then G is defined by polynomial equations (which depend on the given g), and the group law is given by matrix multiplication. Thus G is an algebraic affine group scheme over k.

(b) The *multiplicative group* is the functor $R \mapsto R^*$, the latter being the group of invertible elements of R. To describe it by an equation, note that we have $x \in R^*$ if and only if there is a $y \in R$ with $xy = 1$. Since y is uniquely determined by x, we can identify R^* with the "hyperbola" $xy = 1$ in the plane R^2. The group structure is given by

$$(x, y) \cdot (x', y') = (xx', yy') \,.$$

[Note, incidentally, that we have the polynomial formula $(x, y)^{-1} = (y, x)$ for inversion.] Another way to describe this group by equations is to identify it with the diagonal subgroup of SL_2, i.e., with the matrix group defined by the equations $a_{12} = a_{21} = 0$, $\det(a_{ij}) = 1$.

(c) The general linear group GL_n can be treated similarly: We identify it with the set of solutions of the equation $\det(a_{ij}) \cdot y = 1$ in $n^2 + 1$ variables a_{ij}, y. [Exercise: Write down a polynomial formula for inversion.] Alternatively, we can identify GL_n with the matrix group

$$\begin{pmatrix} * & \cdots & * & \\ \vdots & \ddots & \vdots & \\ * & \cdots & * & \\ & & & * \end{pmatrix} < \mathrm{SL}_{n+1} \,,$$

which is defined by adding $2n$ equations to the determinant equation defining SL_{n+1}. When $n = 1$, this example reduces to Example (b).

(d) The *additive group* is defined by $G(R) = R$, with addition as the group law. It is the set of solutions of the empty set of equations in one variable, and the group structure is clearly given by a polynomial map. Alternatively, G can be identified with the matrix group $\left(\begin{smallmatrix} 1 & * \\ 0 & 1 \end{smallmatrix}\right)$.

(e) The *circle group* is the curve $x^2 + y^2 = 1$ in the plane, with group structure given by imitating the familiar rule for multiplying complex numbers of norm 1:

$$(x, y) \cdot (x', y') = (xx' - yy', xy' + x'y) .$$

Once again, our group can be identified with a matrix group; namely, it is isomorphic to the *rotation group*, consisting of the matrices $\left(\begin{smallmatrix} a & b \\ c & d \end{smallmatrix}\right) \in \mathrm{SL}_2$ satisfying the equations $a = d$ and $b = -c$.

(f) Similarly, the 3-sphere $x_1^2 + x_2^2 + x_3^2 + x_4^2 = 1$ has a polynomial group law given by imitating the rule for multiplying quaternions $x = x_1 + x_2 i + x_3 j + x_4 k$ of norm 1. Alternatively, we get the same group law by formally writing $z = x_1 + i x_2$ and $w = x_3 - i x_4$ and identifying the quaternion x with the 2×2 matrix

$$\begin{pmatrix} z & w \\ -\bar{w} & \bar{z} \end{pmatrix} ;$$

see formula (6.27) in Section 6.11. Here the bars denote "complex conjugation" in the ring $R[i] = R \oplus Ri$ obtained by formally adjoining $i = \sqrt{-1}$ to R. Thus the 3-sphere group is isomorphic to the special unitary group SU_2, whose group of R-points is the subgroup of $\mathrm{SL}_2(R[i])$ consisting of the matrices $\left(\begin{smallmatrix} a & b \\ c & d \end{smallmatrix}\right)$ satisfying $a = \bar{d}$ and $b = -\bar{c}$.

(g) For any integer $n \geq 2$, there is a group scheme μ_n, called the *group of nth roots of unity*, defined by $\mu_n(R) := \{x \in R \mid x^n = 1\}$ with group structure given by multiplication. This is a group of 1×1 matrices.

Remark C.4. It is no accident that we were able to represent every example as a matrix group. In fact, one of the first theorems of the subject is that every affine algebraic group scheme is isomorphic to a closed subgroup of some GL_n, i.e., a subgroup defined by polynomial equations in the n^2 matrix entries [278, Section 3.4].

Exercise C.5. Let G be the 3-sphere group as in Example C.3(f). Note that we identified $G(R)$ with a group of 2×2 matrices, but the matrices did not have entries in R. Thus our matrix representation does not, on the face of it, have the form described in Remark C.4. Exhibit G as a closed subgroup of some GL_n.

C.2 The Affine Algebra of G

Let G be an algebraic affine group scheme defined by a set of polynomial equations in m variables. Write the equations in the form $f(x_1, \ldots, x_m) = 0$, and let I be the ideal in the polynomial ring $k[X_1, \ldots, X_m]$ generated by the given f's. The *affine algebra* of G is the quotient $A := k[X_1, \ldots, X_m]/I$. For example, the affine algebra of the circle is $k[X, Y]/(X^2 + Y^2 - 1)$.

The affine algebra A of G *represents* G in the following sense: For any k-algebra R, the set $G(R)$ is in 1–1 correspondence with the set $\text{Hom}(A, R)$ of algebra homomorphisms $A \to R$. More concisely, $G = \text{Hom}(A, -)$. Of course this describes G only as a set-valued functor. To describe the group structure on G we need to impose some extra structure on A, consisting of a *comultiplication* $c: A \to A \otimes A$ satisfying certain axioms. The algebra A with this extra structure is called a *Hopf algebra*. See [278, Chapter 1] for details.

C.3 Extension of Scalars

Suppose we have a field extension $k' \geq k$. Then any polynomial with coefficients in k also has coefficients in k'. So if G is a group scheme defined as above by polynomials with coefficients in k, then G yields a group scheme G' over k' defined by the same formulas. In other words, we simply "restrict" G from the category of k-algebras to the subcategory of k'-algebras. [Any k'-algebra can be viewed as a k-algebra.] The group scheme G' is said to be obtained from G by *extension of scalars* from k to k'. If G is represented by a Hopf algebra A over k, then G' is represented by the Hopf algebra $A' := k' \otimes_k A$ over k'.

Here is an example to show what can happen when one extends scalars. Let G be the circle group over $\mathbb{R}$. After extending scalars to $\mathbb{C}$, the resulting G' is still the circle group, viewed now as a group scheme over $\mathbb{C}$. But the defining equation for the circle can be written as $(x + iy)(x - iy) = 1$ over $\mathbb{C}$, and it follows easily that there is an isomorphism $G' \to GL_1$ given by $(x, y) \mapsto x + iy$. Thus G becomes isomorphic to the multiplicative group after extension of scalars, but the two group schemes are easily seen to be nonisomorphic over $\mathbb{R}$.

C.4 Group Schemes from Groups

Let's go back to the naïve point of view, as in the first paragraph of this appendix. Thus we assume that we are given a group $G_0 \subseteq k^m$ that is the solution set of a collection of polynomial equations and that has a group law $G_0 \times G_0 \to G_0$ defined by a polynomial map. Assume further that the inversion map $G_0 \to G_0$ is a polynomial map. There is then a canonical way to "extend" G_0 to a group scheme G, with G_0 as the group of k-points $G(k)$. Namely, consider *all* polynomial equations that are satisfied by G_0, and define $G(R)$ to be the set of solutions of the same equations in R^m. It is not hard to show that the polynomial formula defining the group structure on G_0 works for arbitrary R and makes G a group scheme [278, Section 4.4].

This passage from groups to group schemes has a simple interpretation in terms of Hopf algebras: Given G_0, let A be the ring of functions $G_0 \rightarrow k$ given by polynomials. Equivalently, $A = k[X_1, \ldots, X_m]/I$, where I is the ideal consisting of all polynomials that vanish on G_0. Then the group structure on G_0 yields a Hopf algebra structure on A, and the group scheme G is simply the functor $\mathrm{Hom}(A, -)$ represented by A.

A group scheme G over k that arises from a group G_0 in this way will be said to be *determined by its k-points*. For example, one can show that GL_n and SL_n are determined by their k-points as long as k is infinite [278, Section 4.5]. On the other hand, the group scheme μ_3 over $\mathbb{Q}$ is not determined by its group of $\mathbb{Q}$-points, which is the trivial group.

It is easy to characterize the group schemes G that are determined by their k-points: If A is the affine algebra of G, then G is determined by its k-points if and only if no nonzero element of A goes to zero under all k-algebra homomorphisms $A \rightarrow k$. In case k is algebraically closed, Hilbert's Nullstellensatz allows us to restate the criterion as follows [278, Section 4.5]: G is determined by its k-points if and only if A is *reduced*, i.e., if and only if A has no nonzero nilpotent elements.

C.5 Linear Algebraic Groups

We are ready, finally, for the main definition. Let k be a field and let $\bar{k}$ be its algebraic closure. Let G be an algebraic affine group scheme over k, and let $\bar{G}$ be the group scheme over $\bar{k}$ obtained from G by extension of scalars.

Definition C.6. We say that G is a *linear algebraic group defined over k* if the group scheme $\bar{G}$ is determined by its group of $\bar{k}$-points.

For example, GL_n and SL_n are linear algebraic groups over k for any k, and μ_3 is a linear algebraic group over k unless k has characteristic 3. In characteristic 3, on the other hand, μ_3 over $\bar{k}$ is not determined by its group of $\bar{k}$-points, which is the trivial group; so μ_3 is not a linear algebraic group defined over k in this case.

Remarks C.7. (a) Our primary interest here is in actual groups rather than group functors. This is why we insist that we should get an actual group (i.e., the group scheme associated to an actual group) after extension of scalars. But it would be too restrictive to demand that G itself be the group scheme associated to a group, since that would exclude such examples as μ_3 over $\mathbb{Q}$ or SL_n over a finite field.

(b) The word "linear" in Definition C.6 serves as a reminder of the fact, mentioned at the end of Section C.1, that G is isomorphic to a closed subgroup of a general linear group.

(c) In view of Hilbert's Nullstellensatz, we can restate the definition of "linear algebraic group defined over k" in terms of the affine algebra A of G (see the last paragraph of Section C.4): The group scheme G is a linear algebraic group defined over k if and only if $\bar{k} \otimes_k A$ is reduced. This is equivalent to a condition called *smoothness* [278, Chapter 11], and there are techniques for checking it. In characteristic 0 it is known that *all* algebraic affine group schemes are smooth, so there is nothing to check. In characteristic p, however, we have already seen that smoothness can fail (e.g., μ_3 in characteristic 3).

C.6 Tori

Let G be the *n-dimensional torus* over $\mathbb{R}$, i.e., the product of n copies of the circle group. The example in Section C.3 shows that G becomes isomorphic to the direct product $(\mathrm{GL}_1)^n$ of n copies of the multiplicative group after extension of scalars to $\mathbb{C}$. This motivates the following terminology.

Definition C.8. A linear algebraic group G is a *torus* of rank n if G becomes isomorphic to $(\mathrm{GL}_1)^n$ after extension of scalars to $\bar{k}$. The torus is said to be *split* (or k-split) if it is already isomorphic to $(\mathrm{GL}_1)^n$ over k.

We saw in Chapter 6 the canonical examples in which split tori arise "in nature"; namely, the diagonal groups called T in Sections 6.5–6.7 are split tori. And we saw examples of nonsplit tori in Section 6.11.3. Other natural examples of nonsplit tori can be found in orthogonal and unitary groups (with respect to quadratic and Hermitian forms other than the standard ones).

C.7 Unipotent Groups

An element $g \in \mathrm{GL}_n(k)$ is called *unipotent* if $g - 1$ is nilpotent. This is equivalent to saying that g is conjugate to an element of $U_n(k)$, where U_n here is the strict upper-triangular group (*not* the unitary group), i.e., the group of upper-triangular matrices with 1's on the diagonal. A group of $n \times n$ matrices is called *unipotent* if each of its elements is unipotent. This is equivalent, by a theorem of Kolchin [278, Section 8.1], to saying that the group is conjugate to a subgroup of $U_n(k)$. Finally, if G is a linear algebraic group over k, choose an embedding of G as a closed subgroup of some GL_n, and call G *unipotent* if $G(\bar{k})$ is a unipotent subgroup of $\mathrm{GL}_n(\bar{k})$. This is equivalent to saying that there is an element of $\mathrm{GL}_n(k)$ that conjugates G into U_n [278, Section 8.3]. Moreover, this notion is independent of the choice of embedding of G in a general linear group.

C.8 Connected Groups

There is a topology on k^m, called the *Zariski topology*, in which the closed sets are the subsets defined by polynomial equations. The subset $G(k) \subseteq k^m$ inherits a Zariski topology, and we can therefore apply topological concepts, such as connectivity, to $G(k)$. More useful for us is the Zariski topology on $G(\bar{k}) \subseteq \bar{k}^m$. In particular, we will say that the linear algebraic group G is *connected* if $G(\bar{k})$ is connected in the Zariski topology. For example, GL_n and SL_n are connected (for any k), but μ_3 over $\mathbb{Q}$ is not. Note, however, that $GL_n(k)$ is disconnected if k is a finite field, whereas the disconnected group μ_3 has the property that $\mu_3(\mathbb{Q})$ is connected. Thus it is important to look at $G(\bar{k})$ rather than $G(k)$ in order to get the "right" answer.

C.9 Reductive, Semisimple, and Simple Groups

Let G be a connected linear algebraic group over k.

Definition C.9. G is called *reductive* if $G(\bar{k})$ contains no nontrivial connected normal unipotent subgroup.

For example, GL_n and SL_n are reductive.

Definition C.10. G is called *semisimple* if $G(\bar{k})$ contains no nontrivial connected normal solvable subgroup.

For example, SL_n is semisimple but GL_n is not (because of its center). Note that any semisimple group is reductive, since unipotent matrix groups are solvable (and even nilpotent) by Kolchin's theorem.

If G is semisimple, then $G(\bar{k})$ is "almost" a finite direct product of simple groups. More precisely, $G(\bar{k})$ has the following properties: (a) it has only finitely many minimal nontrivial closed connected normal subgroups N_i; (b) the N_i commute and generate $G(\bar{k})$; (c) the canonical surjection $\prod_i N_i \twoheadrightarrow G(\bar{k})$ has finite kernel; and (d) each N_i is *almost simple*, which means that its center Z_i is finite and that the quotient N_i/Z_i is simple as an abstract group. Proofs can be found in [132, Sections 27.5 and 29.5]. If there is only one N_i, i.e., if $G(\bar{k})$ is almost simple, then G is said to be *absolutely almost simple*. For example, SL_n is absolutely almost simple.

C.10 BN-Pairs and Spherical Buildings

If G is reductive, then the group $G(k)$ has a BN-pair whose associated building, sometimes called the *Tits building* of G, is spherical. We will give a brief description of this in the semisimple case. (See also Section 7.9.3 where the BN-pair is described from the point of view of RGD systems.) For more details, see [247] and the references cited there. See also [132, Section 28.3] for the case that k is algebraically closed, and see Warner [277, Section 1.2] for an analytic approach in case $k = \mathbb{R}$.

A *Borel subgroup* of $G(\bar{k})$ is a maximal connected solvable subgroup of $G(\bar{k})$. Borel subgroups exist and are unique up to conjugacy. If k is algebraically closed, any Borel subgroup can serve as the B of the BN-pair in $G(k) = G(\bar{k})$. Let T be a maximal torus in G. These also exist and are unique up to conjugacy. Since $T(k)$ is connected and solvable, we can choose the Borel subgroup B to contain $T(k)$. Still assuming that $k = \bar{k}$, we can then take the N of the BN-pair to be the normalizer of $T(k)$ in $G(k)$. The resulting spherical building has rank l (dimension $l - 1$), where l is the rank of T. Its chambers are in 1–1 correspondence with the Borel subgroups of G, and its apartments are in 1–1 correspondence with the maximal tori in G. This construction is described in Chapter 6 for several groups G, and further information is given in Sections 7.9.2 and 7.9.3.

It is immediate from the definitions above that the parabolic subgroups with respect to this BN-pair are the subgroups of $G(k)$ that contain a Borel subgroup. There is another characterization of them, whose statement involves concepts that we have not defined (and will not define): They are the subgroups of the form $P(k)$, where P is a closed subgroup of G such that G/P is a projective variety. For example, let $G = \mathrm{SL}_n$ and let P be the subgroup defined by $a_{i1} = 0$ for $i > 1$ (i.e., P is the stabilizer of the line $[e_1]$); then G/P is $(n - 1)$-dimensional projective space.

When $k \neq \bar{k}$, the situation is more complicated, the problem being that $G(\bar{k})$ might not have a Borel subgroup that is defined over k (i.e., which is the group of $\bar{k}$-points of a linear algebraic subgroup of G defined over k). Orthogonal groups provide examples of this phenomenon (see Section 6.7.2).

To get a BN-pair, in general, one has to forget about Borel subgroups and instead take B to be the group $P(k)$ for some minimal parabolic subgroup P of G, where now "parabolic" is *defined* by the property that G/P is a projective variety. B is again unique up to conjugacy. We can choose B to contain $T(k)$, where T is now a maximal k-split torus in G, and we take N to be the normalizer of $T(k)$ in $G(k)$. The rank of the resulting spherical building is again equal to the rank l of T. This rank l is also called the *k-rank* of G. It can be strictly smaller than the rank of the building associated to $G(\bar{k})$. In fact, it can be zero, in which case the building is empty (i.e., it consists only of the empty simplex). The group G is said to be *anisotropic*, or *k-anisotropic*, if $l = 0$. The terminology is motivated by the case of the orthogonal group (Section 6.7.2). More generally, if k' is an extension of k, we say that G is k'-anisotropic if the group over k' obtained by extension of scalars is anisotropic.

C.11 BN-Pairs and Euclidean Buildings

Here we will be even briefer. Suppose G is absolutely almost simple and isotropic and is defined over a field K with a discrete valuation. Assume that K is complete with respect to the valuation and, to avoid technicalities, that the residue field is perfect. Assume further that G is simply connected. (This is another term that we have not defined; an example is SL_n.) Then there is a Euclidean BN-pair in $G(K)$, analogous to the one in Section 6.9 for $\mathrm{SL}_n(K)$. The associated Euclidean building, often called the *Bruhat–Tits building* of G, has dimension l, where l is again the K-rank of G; thus the dimension of this building exceeds by 1 the dimension of the spherical building associated to $G(K)$. When K is locally compact (e.g., $K = \mathbb{Q}_p$), the parabolic subgroups of $G(K)$ with respect to this BN-pair are open and compact. Theorem 11.38 therefore provides a description of the Bruhat–Tits building in terms of the maximal compact subgroups of $G(K)$. Here, of course, we use the locally compact topology that comes from the valuation, not the Zariski topology. For more information see Bruhat–Tits [59, 60], Tits [252], or Weiss [283].

Remark C.11. The assumption that the residue field is perfect is made only to handle a few special cases, and it may not be necessary even then. More precisely, it is not necessary if the conjecture in Weiss [283, 21.16] is true.

C.12 Group Schemes versus Groups

We have insisted on thinking of algebraic groups as functors because we find this point of view useful. But in so doing, we may have given the misleading impression that a "typical" linear algebraic group G defined over k is not determined by its group of k-points. We will therefore close by stating a theorem that says that G is determined by its k-points much more often than one might expect. Let G be a linear algebraic group defined over k. Then G is determined by its k-points whenever the following three conditions are satisfied: (a) G is connected; (b) k is infinite; and (c) either k is perfect or G is reductive. (Note that (b) and (c) hold automatically in characteristic 0.) For a proof of this theorem see Borel [39, Corollary 18.3], where the result is stated in the following equivalent form: If (a), (b), and (c) hold, then $G(k)$ is Zariski-dense in $G(\bar{k})$.

Hints/Solutions/Answers to Selected Exercises

Chapter 1

1.29. By considering sign sequences, check that this union is an intersection of open half-spaces.

1.45. Consider the set $\mathcal{H}_A$ of hyperplanes in $\mathcal{H}$ containing A, and let Σ_A be the set of $\mathcal{H}_A$-cells. There is an obvious 1–1 map $\Sigma_{\geq A} \to \Sigma_A$ that sends an $\mathcal{H}$-cell $B \geq A$ to the unique $\mathcal{H}_A$-cell containing it.

1.61. See Proposition 1.40.

1.62. Think in terms of sign sequences. Given chambers $C \geq A$ and $D \geq B$, $\sigma(C)$ and $\sigma(D)$ must disagree wherever A and B have opposite (nonzero) signs, i.e., in $d(A,B)$ places. To minimize the length, we try to make them agree elsewhere. For example, if $\sigma_i(A) = 0$ and $\sigma_i(B) = +$, then necessarily $\sigma_i(D) = +$, so we want $\sigma_i(C) = +$. This line of reasoning shows that we minimize $d(C,D)$ precisely by taking $C \geq AB$ and $D = BC$.

1.65. The implication (ii) $\implies$ (i) is immediate from Proposition 1.56. To prove (i) $\implies$ (ii), one has to show that if C is a chamber not in $\mathcal{D}$, then there is a hyperplane H separating C from $\mathcal{D}$. Choose $D \in \mathcal{D}$ at minimal distance from C, and let $D, D', \ldots, C$ be a minimal gallery from D to C. Then $D' \notin \mathcal{D}$, and the hyperplane H separating D from D' also separates D from C. To complete the proof, we will show that all chambers in $\mathcal{D}$ are on the D-side of H. Given $E \in \mathcal{D}$, we have $d(E,D) = d(E,D') \pm 1$. The sign cannot be $+$, because then there would be a minimal gallery from E to D passing through D', which would contradict (i). So the sign is $-$, which means that E and D are on the same side of H.

1.68. It is immediate from the definitions that (ii) $\implies$ (i). For the converse, assume first that Σ' contains a chamber, in which case we will prove (i) $\implies$ (v) $\implies$ (iv) $\implies$ (ii).

(i) $\implies$ (v): Consider a minimal gallery $A \leq C_0, C_1, \ldots, C_l = C$. If A_i is the common panel between C_{i-1} and C_i, then we have $C_0 = AC$, $C_1 = A_1C$, and so on. So if (i) holds and $A, C \in \Sigma'$, one sees inductively that all $C_i \in \Sigma'$.

(v) $\implies$ (iv): Trivial.

(iv) $\implies$ (ii): Consider the intersection of the closed half-spaces (bounded by hyperplanes in H) that contain all the cells in Σ'. Let Σ'' be the set of cells in this intersection. If (iv) holds, we know by Exercise 1.65 that Σ' and Σ'' have the same chambers. To show that they are equal, observe that Σ'' is a subsemigroup of Σ containing a chamber, which implies that every maximal simplex of Σ'' is a chamber.

Suppose now that Σ' does not necessarily contain a chamber. To prove (i) $\implies$ (ii), note that (i) implies that all maximal cells of Σ' have the same support L. [If A and B are maximal, then $A = AB$ and $B = BA$; now use the fact that AB and BA have the same support.] We can now replace the ambient vector space V by L and thereby reduce to the case already treated, where Σ' contains a chamber.

Finally, we show that (i) and (ii) are equivalent to (iii): If (iii) holds, then Σ' is a subsemigroup because the line segment from a point of A to a point of B passes through AB. So (iii) implies (i) and (ii). Conversely (ii) $\implies$ (iii) because an intersection of half-spaces is convex.

1.87. $W \cong A_5 \times \{\pm 1\}$.

1.101. (a) V' is invariant under the rank-1 operator $s - 1$, whose image is $\mathbb{R}e_s$ and whose kernel is H_s.

(b) If $m(s,t) > 2$, then the operator $t - 1$ is nonzero on e_s. So $e_s \in V' \implies e_t \in V'$. Now use connectivity of the Coxeter diagram.

(c) Either V' is contained in $\bigcap_{s \in S} H_s = 0$ or else V' contains some (and hence all) e_s.

(d) Suppose $u \colon V \to V$ commutes with all $w \in W$. It suffices to show that u has a real eigenvalue λ, since the λ-eigenspace is then a nonzero W-invariant subspace, hence is the whole space V. The existence of such an eigenvalue follows from the fact that u commutes with any reflection $s \in W$, and s has a 1-dimensional eigenspace (which is therefore u-invariant).

1.112. First, we need a description of the barycentric subdivision of the boundary of an $(n-1)$-simplex. The boundary of an $(n-1)$-simplex is the abstract simplicial complex consisting of the proper subsets of $\{1, 2, \ldots, n\}$. Its barycentric subdivision consists of chains of such subsets (see Section A.1.2). Here are two possible ways to identify this with Σ. Method 1: The symmetric group W acts on the barycentric subdivision in an obvious way; calculate a fundamental domain and stabilizers. Method 2: The cells of Σ correspond to ordered partitions of $\{1, 2, \ldots, n\}$ (Section 1.4.7). These are in 1–1 correspondence with chains of nonempty proper subsets.

1.120. We may assume that W is essential, so that the simple roots form a basis for the ambient vector space. The result is then equivalent to the (almost obvious) fact that in $\mathbb{R}^n$ a nonzero linear functional $\sum c_i x_i$ is positive on the positive orthant (given by $x_i > 0$ for all i) if and only if $c_i \geq 0$ for all i.

1.121. Recall from the proof of Theorem 1.69 that every hyperplane in $\mathcal{H}$ is W-equivalent to a wall of the fundamental chamber.

1.125. $l(w) = l(w^{-1}) =$ the number of walls separating C from $w^{-1}C$.

Chapter 2

2.9. The upper sheet of the hyperboloid $Q = -1$ is one of the standard models of the hyperbolic plane.

2.30. (b) This follows from Exercise 2.28 if $j > i$, so assume $j < i$. We may also assume $i \geq 2$, and we argue by induction on i. Suppose first that $j \neq i - 1$. Then w_{i-1} is s_i-reduced by (a) and s_j-reduced by induction. So it is $\{s_i, s_j\}$-reduced, and we have

$$l(w_i s_j) = l(w_{i-1} s_i s_j) = l(w_{i-1}) + 2 = i + 1$$

by Proposition 2.20. Now suppose $j = i-1$, and note that w_{i-2} is s_{i-1}-reduced and s_i-reduced. Hence it is $\{s_{i-1}, s_i\}$-reduced, and we have

$$l(w_i s_{i-1}) = l(w_{i-2} s_{i-1} s_i s_{i-1}) = l(w_{i-2}) + 3 = i + 1$$

by Proposition 2.20 again. [Readers who want to look ahead at Theorem 2.33 might find it instructive to think about this exercise from the point of view of that theorem.]

2.31. (a) The w_i are (right) J-reduced by Exercise 2.30, with a renumbering of the generators s_i. To see that they are the only J-reduced elements, note that W acts transitively on $\{1, 2, \ldots, n\}$ and W_J is the stabilizer of 1. Since w_i takes 1 to $i + 1$ [using a left action of W on $\{1, \ldots, n\}$], the w_i form a complete set of coset representatives for W/W_J. Alternatively, one can argue directly with S-words.

(b) As in (a), the left J-reduced elements are $s_1 s_2 \cdots s_i$, and 1 and s_1 are the only two of these that are also right J-reduced. [The fact that there are only two (W_J, W_J)-double cosets can also be deduced from the theory of permutation groups; it says that W acts doubly transitively on $\{1, 2, \ldots, n\}$.]

(c) Here W has infinitely many generators $s_1, s_2, \ldots$, with $m(s_i, s_{i+1}) = 3$ and $m(s_i, s_j) = 2$ if $j > i + 1$. The Coxeter diagram is an infinite path with unlabeled edges. There is also a permutation group interpretation: W is the group of permutations w of the natural numbers such that w fixes all but finitely many elements. Setting $J := S \setminus \{s_1\}$, we find as in (a) and (b) that the right J-reduced elements are the $s_i \cdots s_1$ ($i = 0, 1, 2, \ldots$), the left J-reduced elements are the $s_1 \cdots s_i$ ($i = 0, 1, 2, \ldots$), and the (J, J)-reduced are elements are 1 and s_1.

2.42. See [246].

2.47. As in the finite case, we may assume that $J = S \setminus \{s\}$. If there are infinitely many $t \in S$ with $m(s,t) > 2$, then the elements ts provide infinitely many J-reduced elements and we are done. Otherwise, the Coxeter diagram of (W_J, J) falls into finitely many connected components $J_1, J_2, \ldots$, at least one of which (say J_1) must be infinite. Choose $t \in J_1$ with $m(s,t) > 2$, and observe, as above, that it suffices to find infinitely many $(J_1 \setminus \{t\})$-reduced elements of W_{J_1} or, equivalently, to show that $W_{J_1 \setminus \{t\}}$ has infinite index in W_{J_1}. Thus our problem for W and s has been reduced to the same problem for W_{J_1} and t. Continuing in this way, we have two possibilities: Either the process terminates after finitely many steps and we are done, or else there is an infinite sequence $(s_1, s_2, \ldots) = (s, t, \ldots)$ of distinct elements of S such that $m(s_i, s_{i+1}) > 2$ for all $i \geq 1$. We can now set $w_i := s_i \cdots s_2 s_1$ and check that the w_i form an infinite set of J-reduced elements (see Exercise 2.30).

2.51. This is not obvious if one defines "Coxeter system" via (C), as is often done. But it is easy using (A) or (D).

2.89. We may assume that A is a face of the fundamental chamber C. Let C' be the intersection of the open half-spaces $U_+(s)$, where s ranges over the fundamental reflections that fix A. Then, as in the proof of Lemma 2.58, C' is essentially the fundamental chamber for the dual of the canonical linear representation of W_A. If W_A is finite, then it is a finite reflection group acting on V^*, with $\mathcal{H}_A$ as its set of walls. [Just note that $\mathcal{H}_A$ is a W_A-invariant set of hyperplanes whose reflections generate W_A.] So every $\mathcal{H}_A$-cell is W_A-equivalent to a face of C'. Since every face of C' contains a face of C, it follows that every $\mathcal{H}_A$-cell contains a cell of X.

2.90. If we can show that X_f is open, then it will automatically be the full interior of X by Lemma 2.86. [If y is in the interior of X, choose $x \in X_f$ and note that y is in $[x, y')$ for a suitable $y' \in X$.] Consider now a point $x \in X_f$, and let A be the cell containing x. Let $U = \bigcup_{B \geq A} B$. Everything in (a) and (b) will follow easily if we show that U is open in $\overline{V}^*$. By Lemma 2.85, $\mathcal{H}_A$ is finite. And, as we saw in the proof of that lemma, there are only finitely many chambers $\geq A$. So we may enlarge $\mathcal{H}_A$ to a finite set $\mathcal{H}'$ containing the walls of all chambers $\geq A$. We claim that the cells $\geq A$ are the same as the $\mathcal{H}'$-cells $\geq A$. This claim, together with Exercise 1.29, will complete the proof that U is open.

To prove the claim, we appeal to Exercises 1.45 and 2.89. Combining the bijections of those exercises, we get a bijection from the cells in X with A as a face to the $\mathcal{H}'$-cells with A as a face; it sends a cell $B \geq A$ to the unique $\mathcal{H}'$-cell B' containing B. But the definition of $\mathcal{H}'$ implies that any such B is already an $\mathcal{H}'$-cell, so $B' = B$ and the claim is proved.

Chapter 3

3.11. (a) We may assume that C is the fundamental chamber. Then $D = wC$ for some $w \in W$, and the hypothesis says that $l(ws) \le l(w)$ for all $s \in S$. Now apply Corollary 2.19.

(b) This follows trivially from (a). For a direct proof, observe that Σ has finite diameter if and only if the length function on W is bounded, and this is the case if and only if W is finite.

3.12. (a) The equivalence of the two formulations is obtained using the W-action. To prove the first version, suppose $d(C, y)$ is bounded as y ranges over all vertices of type s. Then there are finitely many chambers containing all these vertices.

(b) We may assume that x is a vertex of the fundamental chamber, i.e., x is a standard subgroup W_I, where $I = S \smallsetminus \{s'\}$ for some $s' \in S$. A vertex y of type s is a coset wW_J with $J := S \smallsetminus \{s\}$, and a gallery from x to y is a sequence $u, us_1, us_1s_2, \dots, us_1 \cdots s_l$ of elements of W with $u \in W_I$, $s_i \in S$ for all i, and $us_1 \cdots s_l \in wW_J$. Hence $d(x, y)$ is the minimum value of $l(u^{-1}v)$ with $u \in W_I$ and $v \in wW_J$, i.e., $d(x, y)$ is the length of the shortest representative of the double coset $W_I w W_J$ (see Proposition 2.23). There are infinitely many such double cosets by Proposition 2.45, so the length is unbounded.

3.13. We may assume that A is a face of the fundamental chamber. Then A *is* a standard subgroup, and the exercise reduces to the fact that A acts transitively on itself by left translation.

3.57. Argue as in the proof of Lemma 3.45: Take a minimal gallery from C to sC and fold it onto α.

3.58. (a) The canonical isomorphism $\Sigma \to \Sigma'$, on the level of chambers, associates to a permutation π the chain of subsets

$$\{\pi(1)\} \subset \{\pi(1), \pi(2)\} \subset \cdots \subset \{\pi(1), \dots, \pi(n-1)\} \ .$$

So a chamber $X_1 \subset \cdots \subset X_{n-1}$ of Σ' is in α'_{ij} if and only if the first set X_k that contains i does not contain j. The vertices of chambers of this form are exactly the sets described in the statement of (a).

(b) The root opposite to α'_{ij} is α'_{ji}. Hence $\partial \alpha'_{ij} = \alpha'_{ij} \cap \alpha'_{ji}$. Now apply (a).

(c) This follows from (a) and (b).

3.59. (a) Let $\Sigma = \Sigma(W, S)$, with fundamental chamber C, and let H be the wall corresponding to the reflection t. Then H separates C from wC; for otherwise we would have $d(C, twC) > d(C, wC)$ by Exercise 3.57, contradicting the hypothesis $l(tw) < l(w)$. Consider now the gallery $C = C_0, \dots, C_d = wC$ corresponding to the given decomposition of w. It must cross H, so there is an index i such that $C_{i-1} = tC_i$. There are now two ways to finish. Method 1: Since the wall separating C_{i-1} from C is the wall fixed by t, we have $t = us_iu^{-1}$, where $u = s_1 \cdots s_{i-1}$; now calculate tw. Method 2: We have a pregallery

$$C_0, \ldots, C_{i-1}, tC_i, \ldots, tC_d$$

from C to twC with exactly one repetition. The gallery obtained by deleting the repetition has type $(s_1, \ldots, \hat{s}_i, \ldots, s_d)$.

(b) (i) $\implies$ (ii) trivially. For (ii) $\implies$ (iii), take $t = us_iu^{-1}$ with $u = s_1 \cdots s_{i-1}$. Finally, (iii) $\implies$ (i) by the strong exchange condition.

(c) Acyclicity follows from the fact that the length $l(w)$ strictly increases along any directed path. This implies that if $w' \leq w$ and $w \leq w'$, then $w' = w$. Transitivity of "$\leq$" is trivial.

(d) Assume first that there is a directed edge $w' \to w$, and choose a reduced decomposition $w = s_1 \cdots s_d$. Then $w' = s_1 \cdots \hat{s}_i \cdots s_d$ for some $1 \leq i \leq d$, and $sw' = ss_1 \cdots \hat{s}_i \cdots s_d$. If $l(sw) > l(w)$, this yields an edge $sw' \to sw = \max\{sw, w\}$. Otherwise, we may assume that the decomposition of w has been chosen such that $s_1 = s$. If $i = 1$, then $sw' = w$ and there is nothing to prove, so assume $i > 1$. Then we have a path

$$sw' = s_2 \cdots \hat{s}_i \cdots s_d \to s_2 \cdots s_d \to w \; ;$$

hence $sw' < w = \max\{sw, w\}$.

In the general case, there is a path $w' = w_0 \to w_1 \to \cdots \to w_k = w$. By the special case already treated, we have

$$sw' = sw_0 \leq \max\{sw_0, w_0\} \leq \cdots \leq \max\{sw_k, w_k\} = \max\{sw, w\} \; ,$$

and the proof is complete.

(e) If $w' < w$ and we start with an arbitrary decomposition of w, then, by repeated applications of the strong exchange condition, we obtain w' by deleting some letters. Hence (i) $\implies$ (ii). It is trivial that (ii) $\implies$ (iii). Finally, to show that (iii) $\implies$ (i), suppose we have a reduced decomposition $w = s_1 \cdots s_d$, and let w' be obtained by deleting some letters. We will show by induction on $d = l(w)$ that $w' < w$. Set $u = s_2 \cdots s_d$, and note that $u < w$. If s_1 is one of the letters deleted in passing from w to w', then w' is obtained from u by deleting zero or more letters; hence $w' \leq u < w$ by induction. Otherwise, $w' = s_1u'$, where u' is obtained from u by deleting some letters. By induction, $u' < u$, and (d) now implies $w' \leq \max\{s_1u, u\} = w$. Since $l(w') < l(w)$, we have strict inequality.

3.62. The support of A is the intersection of the walls containing A; hence it is the fixed-point set of the reflections fixing A. Now recall that these reflections generate the stabilizer of A. [We may assume that A is a face of the fundamental chamber, i.e., A is a standard subgroup W_J, $J \subseteq S$. The stabilizer is W_J.]

3.81. If $A \in H$, choose an arbitrary chamber $> A$ and consider its image under the reflection s_H.

3.83. (a) If there were only finitely many walls, then Σ would have finite diameter and hence would be finite by Exercise 3.11(b).

(b) Proposition 3.78 implies that $d(x, C)$, for any chamber C, is bounded by the number of walls not containing x. Now apply Exercise 3.12(a).

3.97. (a) Consider a chamber $D' \notin \mathcal{D}$ such that D' is adjacent to some chamber in $\mathcal{D}$. It suffices to prove that there is a unique root α that contains $\mathcal{D}$ but not D'. Equivalently, we must show that there is a unique chamber $D \in \mathcal{D}$ adjacent to D'. Suppose, to the contrary, that D_1 and D_2 are two distinct such chambers. Then we have a gallery D_1, D', D_2, which is easily seen to be minimal (see, for instance, Exercise A.16). This contradicts the assumption that $\mathcal{D}$ is convex.

(b) We know from the solution to (a) that $D'_2 \in \alpha_1$; hence

$$d(D'_1, D'_2) = d(D_1, D'_2) + 1 .$$

And we have

$$d(D_1, D'_2) = d(D_1, D_2) + 1 ,$$

since $D_1 \in \mathcal{D} \subseteq \mathcal{C}(\alpha_2)$. Combining these two equations, we get $d(D'_1, D'_2) = d(D_1, D_2) + 2$. [Alternatively, observe that a gallery $D'_1, D_1, \ldots, D_2, D'_2$ as in the statement of the exercise does not cross any wall twice.]

3.117. Method 1: Apply Proposition 3.78. Method 2: The inequality is obvious, since any gallery from A_2 to B is also a gallery from A_1 to B. Now suppose $A_2 \geq A_1 B$, and consider a minimal gallery $C_0, \ldots, C_l$ from A_2 to B. Then this is also minimal from C_0 to B, so $d(A_2, B) = d(C_0, B)$. On the other hand, $C_0 \geq A_1 B$, so we know from Theorem 3.108 that C_0 can start a minimal gallery from A_1 to B. Hence $d(A_1, B) = d(C_0, B) = d(A_2, B)$.

3.120. (a) If $D \neq C$, then the first step of a minimal gallery from C to D crosses a wall $H \in \mathcal{H}_C$, so $\sigma_H(D) \neq \sigma_H(C)$.

(b) Consider the chamber AC. We have $\sigma_H(AC) = \sigma_H(C)$ for all $H \in \mathcal{H}_C$, so $AC = C$ by (a).

(c) Suppose $\sigma_H(A) \leq \sigma_H(B)$ for all $H \in \mathcal{H}_C$. Then every panel of C containing B also contains A. Now use the fact that B is the intersection of the panels of C containing it.

(d) Suppose $\sigma_H(B) = \sigma_H(A)$ for all $H \in \mathcal{H}_C$. Then $B \leq C$ by (b), and then $B = A$ by (c).

3.121. (a) Let P_s and P_t be the panels of C corresponding to the walls H_s and H_t, and consider the codimension-2 face $A := P_s \cap P_t$ of C. The first step is to replace D by AD in order to reduce to the case $D \geq A$. Since $\sigma_H(AD) = \sigma_H(D)$ for $H = H_s$ or H_t, this replacement is legitimate as long as we show that H_s and H_t are still walls of AD. Consider $H := H_s$, for instance. Set $P := P_s$ and let Q be the panel of D contained in H. Then a consideration of sign sequences shows that AP is a panel of AD and is contained in H; hence H is indeed a wall of AD.

We may now assume that $D \geq A$. The next step is to replace Σ by $\mathrm{lk}_\Sigma A$ in order to reduce to the case that Σ has rank 2. This is easily justified via

Proposition 3.79. Finally, the rank-2 case is easy. If $m < \infty$, then we have already given the proof as part of the "geometric proof" of Corollary 1.91. And if $m = \infty$, then Σ is a triangulated line, and there is no $D \neq C$ having the same two walls as C.

(b) If $m = 2$ and Σ has rank 2, then there are four chambers, all having the same two walls.

(c) This is immediate from (a), since irreducibility implies that there is a path in the Coxeter diagram connecting any two of the reflections s_i.

3.146. The assumption $d(C, D) = d - 1$ implies that C and D are separated by all walls but one, so there is a unique root containing them.

3.150. We may assume that the given simplices are chambers. Their sign sequences all agree except with respect to finitely many walls. We therefore have a collection of roots α, whose bounding walls include almost all $H \in \mathcal{H}$, such that the convex hull is contained in their intersection. This determines the sign sequence of a chamber in the convex hull except for finitely many coordinates.

3.156. (a) This is immediate from Exercise 3.83(b) and the finiteness of $d(x, y)$.

(b) This follows easily from (a). [First figure out what the walls look like in a join of Coxeter complexes.]

3.168. (a) For any simplex $B \in \Sigma$, we have $B \in H \iff AB \in H$. [For the implication $\impliedby$, use the fact that $B \in \operatorname{supp} AB$.] As in the proof of Lemma 3.162, we therefore have

$$B \in H \iff AB \in H \cap \Sigma_{\geq A} \iff (AB \smallsetminus A) \in H'.$$

(b) As above, we have

$$B \in \alpha \implies (AB \smallsetminus A) \in \alpha' \tag{$*$}$$

for any simplex $B \in \Sigma$. Similarly,

$$B \in -\alpha \implies (AB \smallsetminus A) \in (-\alpha)' = -\alpha'. \tag{$**$}$$

To prove the converse of $(*)$, suppose $AB \smallsetminus A$ is in α'. If $AB \smallsetminus A$ is also in $-\alpha'$, then it is in H'; hence B is in $H \subseteq \alpha$ by (a). Otherwise, $(**)$ implies $B \notin -\alpha$, so B is in α.

Chapter 4

4.13. Given chambers $C, D \geq A$, any minimal gallery from C to D is contained in an apartment. Now apply Proposition 3.93.

4.24. (a) V has $2^4 - 1 = 15$ nonzero vectors. Given such a nonzero vector v, the totally isotropic 2-dimensional subspaces containing v are the 2-dimensional subspaces of the 3-dimensional space $v^\perp$ containing v. These are in 1–1 correspondence with the 1-dimensional subspaces of the 2-dimensional space $v^\perp/\mathbb{F}_2 v$, and there are $2^2 - 1 = 3$ of these. So far we have proven that Q has 15 points, each contained in 3 lines. So there are 45 point–line pairs. Since it is immediate that every line has 3 points, it follows that there must be 15 lines. Alternatively, it is not hard to directly count the 2-dimensional totally isotropic subspaces of V. [A pair of orthogonal nonzero vectors determines one.]

(b) The vertices of the quadrilateral, in cyclic order, are e_1, e_2, f_1, f_2; note that any two cyclically consecutive ones are orthogonal in V, hence collinear in Q.

(c) This is straightforward and is also a special case of a result to be proved in Section 6.6.

(d) The description of the apartments in the statement of (a) implies that any pair of opposite vertices has the form (i) or (ii). It remains to show that any two vertices as in (i) or (ii) are opposite. Consider case (i), for instance. We are given two nonzero vectors e_1', f_1' with $\langle e_1', f_1' \rangle = 1$, and we wish to extend them to a symplectic basis e_1', e_2', f_1', f_2'. Let U be the 2-dimensional space spanned by e_1', f_1', and note that since our bilinear form is nondegenerate on U, we have an orthogonal decomposition $V = U \oplus U^\perp$. It follows that the form is nondegenerate on the 2-dimensional space $U^\perp$, so we can take e_2', f_2' to be any two nonorthogonal vectors in $U^\perp$.

(e) Q cannot contain 6 pairwise noncollinear points, since that would yield $6 \cdot 3 = 18$ distinct lines. Similarly, Q cannot contain 6 pairwise nonintersecting lines. We will construct 5 pairwise noncollinear points, i.e., 5 pairwise nonorthogonal vectors in V. [One can also construct 5 pairwise nonintersecting lines by a similar method.]

Start with two nonorthogonal vectors, say e_1, f_1. There are 4 vectors orthogonal to neither of these, namely, the vectors that have both a nonzero e_1-component and a nonzero f_1-component. One of these vectors $(e_1 + f_1)$ is orthogonal to the others. But if we delete this one, the remaining 3 are pairwise nonorthogonal. This gives us the following 5 pairwise nonorthogonal vectors:

$$e_1, \quad f_1, \quad e_1 + f_1 + e_2, \quad e_1 + f_1 + f_2, \quad e_1 + f_1 + e_2 + f_2 .$$

4.32. First, we need to distinguish between left vector spaces (i.e., left k-modules) and right vector spaces. Since a right vector space over k is the same thing as a left vector space over the opposite division ring k^{op}, there is no loss of generality in just considering left vector spaces. The only other change is in Exercise 4.31. Since the dual of a left vector space is naturally a right vector space, and hence a left vector space over k^{op}, the result is that the building associated with an n-dimensional (left) vector space over k is isomorphic in a type-reversing way to the building associated with an n-dimensional (left) vector space over k^{op}. Moreover, one cannot in general say that the building associated to a vector space admits a type-reversing automorphism, unless $k \cong k^{\mathrm{op}}$.

4.50. Given n pairwise joinable simplices, we may assume inductively that the first $n-1$ are joinable. Hence there is an apartment Σ containing all n simplices. They are still pairwise joinable in Σ because Σ is a retract of Δ, so they are joinable because Σ is a flag complex (Exercise 3.116). Alternatively, skip the induction and recall that it suffices to consider the case $n = 3$.

4.52. (a) Given a chamber $D \neq C$, we must find an apartment Σ containing C but not D. Choose a minimal gallery $C, C', \ldots, D$ from C to D. By the convexity of apartments, we are reduced to finding an apartment Σ containing C but not C'. Let C'' be a chamber different from C and C' that is adjacent to C along the panel $C \cap C'$. Then we can take Σ to be any apartment containing C and C''.

(b) Δ might contain only one apartment.

4.61. (a) Suppose α is a root in an apartment Σ, and let Σ' be another apartment containing α. Let $\phi \colon \Sigma \to \Sigma'$ be the isomorphism fixing the intersection pointwise (cf. (B2″)). Then ϕ takes roots of Σ to roots of Σ', since roots are defined intrinsically in terms of the simplicial structure. Hence $\alpha = \phi(\alpha)$ is a root of Σ'.

(b) This follows from the solution to (a).

4.77. Let Σ be an apartment containing C and D. Then the hypothesis implies that $d(C, D') < d(C, D)$ for every chamber $D' \in \Sigma$ adjacent to D. Now apply Exercise 3.11. See also Exercise 1.59 for some geometric insight.

4.89. Observe first that Δ' has the same Weyl group (W, S) as Δ by Proposition 4.63. Now use the fact that the Weyl distance between two chambers can be computed by using any apartment containing them.

4.90. Choose a chamber $D \in \Sigma$ at maximal distance from C. We must show that $d(C, D) = \operatorname{diam} \Delta$ or, equivalently, that $\delta(C, D) = w_0$, where w_0 is the longest element of W (Sections 1.5.2 and 2.3.2). Suppose not. Then $w := \delta(C, D)$ must satisfy $l(ws) > l(w)$ for some $s \in S$. If D' is the chamber of Σ that is s-adjacent to D, it follows that $\delta(C, D') = ws$ and hence $d(C, D') > d(C, D)$, contradicting the choice of D.

4.100. We need to be a little careful because of nonassociativity of the product. But we can deduce the assertion from the fact that it holds in any apartment (Proposition 3.112(4)). Note that in the spherical case, we need only the more elementary version of the theory given in Section 1.4.6.

4.108. (a) The assertions about Λ and Λ' follow from the results of Section 3.6.6 (see Proposition 3.136 and Corollary 3.141). Moreover, we have $K \subseteq \Lambda$ and $K \subseteq \Lambda'$ by Lemma 3.140; hence $K \subseteq \Lambda \cap \Lambda'$. The opposite inclusion is trivial.

(b) Walls (and hence supports) in a Coxeter complex are defined intrinsically in terms of the simplicial structure. So an isomorphism between Coxeter complexes takes supports to supports.

(c) View ϕ as a chamber map from Λ to Λ'. The standard uniqueness argument (in which we move out along Λ-galleries starting at M) then shows that ϕ is uniquely determined by the fact that it fixes M pointwise. Now recall that K is a chamber subcomplex of both Λ and Λ', and consider K-galleries starting at M.

4.109. Choose apartments Σ and Σ' containing α and α', respectively, and let $-\alpha$ and $-\alpha'$ be the opposite roots in the chosen apartments. Let $\phi\colon \Sigma \to \Sigma'$ be an isomorphism fixing $\Sigma \cap \Sigma'$ pointwise. In particular, ϕ fixes $H := \partial\alpha = \partial\alpha'$ pointwise. Then $\phi(\alpha)$ is a root of Σ' bounded by H, so $\phi(\alpha) = \pm\alpha'$. We may assume that $\phi(\alpha) = -\alpha'$. (Otherwise compose ϕ with the reflection of Σ' that interchanges $\pm\alpha'$.) There is then an isomorphism $\psi\colon \Sigma \to \alpha \cup \alpha'$ such that $\psi|_\alpha = \mathrm{id}$ and $\psi|_{-\alpha} = \phi|_{-\alpha}$. [Draw a picture.] Proposition 4.59 now implies that $\alpha \cup \alpha'$ is an apartment.

4.112. It is not hard to prove this using the ideas of the present section. But a more general result will be proved in the next chapter (Corollary 5.118), and the reader may prefer to look ahead and translate that proof into simplicial language.

4.114. Let L be a thick line and p a thick point. If p and L are incident, then the result follows from Exercise 4.113. Otherwise, consideration of an apartment containing p and L shows that there is a vertex p' opposite p such that p' and L are incident. To complete the proof, note that p' is thick by Exercise 4.100.

4.125. In view of Theorem 4.66, it suffices to show that Δ' is weak. Let P be a panel of Δ', and let Σ be an apartment of Δ contained in Δ'. By Exercise 4.90, Σ contains a panel P' opposite P. We can now get two chambers of Δ' containing P by projecting from $\mathcal{C}_{P'}$ to $\mathcal{C}_P$ and applying Exercise 4.100 (and the closure of Δ' under projections).

Chapter 5

5.10. Assuming (WD2a) and (WD2b) instead of (WD2), we have to show that (WD2) holds when $l(sw) < l(w)$. Note first that by (WD2a), s-equivalence is an equivalence relation. [Symmetry follows from the second proof of Lemma 5.3(1).] Now suppose that $\delta(C',C) = s \in S$ and $\delta(C,D) = w$, where $l(sw) < l(w)$. Apply (WD3) to get a chamber C'' such that $\delta(C'',C) = s$ and $\delta(C'',D) = sw$. Then $C'' \sim_s C'$. If $C'' = C'$, we are done. Otherwise, $\delta(C',C'') = s$, and then $\delta(C',D) = s(sw) = w$ by (WD2b).

5.20. (a) We have to show that every minimal gallery has reduced type. Suppose there is a minimal gallery from C to D of type $\mathbf{s} = (s_1,\ldots,s_n)$, and set $w = \delta(C, D)$. By Lemma 5.16 there is a decomposition of the form $w = s_{i_1} \ldots s_{i_m}$ with $1 \leq s_{i_1} < \cdots < s_{i_m} < n$. In view of Corollary 5.17(1), we then have

$$n = d(C, D) = l(w) \leq m \leq n .$$

Hence $l(w) = m = n$, and $\mathbf{s}$ is a reduced decomposition of w.

(b) This is immediate from (a) and Lemma 5.16.

5.40. (a) The projection $C_0 := \operatorname{proj}_{\mathcal{P}} D$ has the property that $d(C, D) = d(C_0, D) + 1$ for all $C \neq C_0$ in $\mathcal{P}$ by the gate property (Proposition 5.34(3)). So if two chambers in $\mathcal{P}$ have different distances from D, one of them must be C_0.

5.63. Given $\phi \in \operatorname{Iso}(W)$ and $w \in W$, we have

$$\phi(1)^{-1}\phi(w) = \delta(\phi(1), \phi(w)) = \delta(1, w) = w .$$

Hence $\phi(w) = \phi(1)w$. Thus the set of isometries can be identified with W acting on itself by left translation.

5.83. (a) By definition, there is an isometry $\phi \colon \beta \to C$ for some root $\beta \subset W$. Then $C = \phi(v)$ for some $v \in \beta$ such that v is adjacent to a chamber $v' \notin \beta$. It follows, as in Remark 5.82, that $\beta = v\alpha_s$ for some $s \in S$. So we may assume that $\beta = \alpha_s$ and $\phi(1) = C$. Since $l(sw) = l(w) + 1$ for all $w \in \alpha_s$, we can extend ϕ to an isometry $\alpha_s \cup \{s\} \to C$ by setting $\phi(s) = D$ as in the proof of Theorem 5.73. That theorem now yields an extension of ϕ to an isometry $\phi \colon W \to C$, whose image is the desired apartment $\mathcal{A}$.

(b) Continuing with the notation of the solution to (a), consider $V := \phi^{-1}(\mathcal{A} \cap \mathcal{A}')$. It is a convex subset of W containing α_s but not s. This implies that $V = \alpha_s$ (see Exercise 3.96); hence $\mathcal{A} \cap \mathcal{A}' = \phi(V) = \alpha$.

(c) Every root in a thick building is an intersection of two apartments by (a) and (b). We can now appeal to the fact that every convex subset of an apartment $\mathcal{A}$ is an intersection of roots (Proposition 3.94).

5.84. Identify $\mathcal{A}$ with the set of chambers of a Coxeter complex Σ, so that we can apply the results of Chapter 3. Specifically, we will use the concepts and results of Exercise 3.97 and its solution (where $\mathcal{M}$ was called $\mathcal{D}$). Thus $\mathcal{M}$ has *walls*, which are subcomplexes of Σ, each wall H determines a *root* α_H (called $\mathcal{C}(\alpha_H)$ in the exercise just cited), and we have

$$\mathcal{M} = \bigcap_H \alpha_H ,$$

where H ranges over the walls of $\mathcal{M}$. For each such wall H, choose a chamber $D^- \in \mathcal{A} \smallsetminus \mathcal{M}$ such that D^- is adjacent to a chamber $D^+ \in \mathcal{M}$, and let $\mathcal{M}^-$ be the set obtained from $\mathcal{M}$ by adjoining all these chambers D^-, one for each wall H of $\mathcal{M}$. By thickness, we can choose for each wall a chamber $D' \in \mathcal{C}$ that is adjacent to both D^+ and D^-. Let $\mathcal{M}'$ be the set obtained from $\mathcal{M}$ by adjoining all these chambers D'.

Using Exercise 3.97(b), one checks that $\mathcal{M}'$ is isometric to $\mathcal{M}^+$. In more detail, that exercise gave a minimal gallery that we would write as $D_1^-, D_1^+, \ldots, D_2^+, D_2^-$ in the present notation. Replacing the extremities by D_1' and D_2' yields a gallery of the same (reduced) type, which is therefore minimal and can be used to show that $\delta(D_1', D_2') = \delta(D_1^-, D_2^-)$. It now follows from Theorem 5.73 that $\mathcal{M}'$ is contained in an apartment $\mathcal{A}'$. For each wall H as above, the convex set $\mathcal{A} \cap \mathcal{A}'$ contains D^+ but not D^-, so it is contained in α_H. Hence $\mathcal{A} \cap \mathcal{A}' = \mathcal{M}$.

5.138. The first equation follows from the definition of δ^* and a standard property of w_0. Now suppose C' op D in $\mathcal{C}$. Then $d_0 - d(C, D) = d(C', D) - d(C, D) \geq d(C, C')$, with equality if and only if C is in the convex hull of C' and D.

5.163. $\mathcal{C}_+$ and $\mathcal{C}_-$ are opposite residues of (spherical) type S, so the projection maps $\mathrm{proj}_{\mathcal{C}_+}$ and $\mathrm{proj}_{\mathcal{C}_-}$ induce mutually inverse σ_0-isometries between $\mathcal{C}_+$ and $\mathcal{C}_-$ by Proposition 5.152.

5.168. Surprisingly, we do not have to assume that S is spherical. Let J be the type of $\mathcal{R}$, and let K be the type of S. Set $w_1 := \min\big(\delta^*(\mathcal{R}, S)\big)$. Then the characterization of $\mathrm{proj}_{\mathcal{R}} S$ is

$$\mathrm{proj}_{\mathcal{R}} S = \{C \in \mathcal{R} \mid w_0(J)w_1 \in \delta^*(C, S)\} . \qquad (*)$$

Moreover, if $C \in \mathcal{R}$ and $D \in S$ satisfy $\delta^*(C, D) = w_0(J)w_1$, then $C = \mathrm{proj}_{\mathcal{R}} D$. To prove this, suppose first that $\delta^*(C, D) = w_0(J)w_1$ for some $D \in S$. Then $\delta^*(C, D) = \max\big(\delta^*(\mathcal{R}, D)\big)$, so $C = \mathrm{proj}_{\mathcal{R}} D$. This proves the "moreover" assertion, which implies that the right side of $(*)$ is contained in the left. For the opposite inclusion, assume $C = \mathrm{proj}_{\mathcal{R}} D$ for some $D \in S$, and set $w := \delta^*(C, D)$. Then Lemma 5.149 and equation (5.11) imply that $l(w_0(J)w) = l(w) - l(w_0(J))$. After some further (slightly tricky) arguments, one concludes that w is in $w_0(J)w_1 W_K$; hence $w_0(J)w_1 \in wW_K = \delta^*(C, S)$.

For the second part of the exercise, one shows as in the proof of Lemma 5.36(2) that $\mathrm{proj}_{\mathcal{R}} S$ is a residue of type $w_0(J)(J \cap w_1 K w_1^{-1})w_0(J)^{-1}$. Start by choosing $C_1 \in \mathrm{proj}_{\mathcal{R}} S$ and $D_1 \in S$ such that $\delta^*(C_1, D_1) = w_0(J)w_1$. In particular, $C_1 = \mathrm{proj}_{\mathcal{R}} D_1$. For $C \in \mathcal{C}_\epsilon$ with $u := \delta_\epsilon(C, C_1)$, one now checks that

$$C \in \mathrm{proj}_{\mathcal{R}} S \iff u \in W_J \cap w_0(J)w_1 W_K w_1^{-1} w_0(J)^{-1} .$$

The proof uses $(*)$ together with the fact that $\delta^*(C, D_1) = \delta_\epsilon(C, C_1)\delta^*(C_1, D_1)$ for $C \in \mathcal{R}$, so that $\delta^*(C, S) = uw_0(J)w_1 W_K$.

5.169. Let J be the type of $\mathcal{R}$. Choose $C_0 \in \mathcal{R} \cap \mathcal{M}_\epsilon$ and set $w := \delta^*(C_0, D)$. If w is not the longest element w_1^* of $\delta^*(\mathcal{R}, D)$, then there exists $s \in J$ such that $l(sw) > l(w)$. Letting $\mathcal{P}$ be the s-panel containing C_0, we conclude that $C_1 := \mathrm{proj}_{\mathcal{P}} D$ is in $\mathcal{R} \cap \mathcal{M}_\epsilon$ and satisfies $d^*(C_1, D) > d^*(C_0, D)$. Continuing in this way, we obtain a gallery $C_0, C_1, \ldots$ in $\mathcal{R} \cap \mathcal{M}_\epsilon$ along which the numerical codistance to D is strictly increasing. Since J is spherical, the process must terminate after finitely many steps with $C_n = \mathrm{proj}_{\mathcal{R}} D \in \mathcal{R} \cap \mathcal{M}_\epsilon$.

5.189. Let J be the common type of $\mathcal{R}$ and $\mathcal{S}$, and set $w := \delta^*(C, D)$. Since C, C', and D are in Σ, we have $\delta_-(C', D) = w$, so $\delta^*(\mathcal{R}, D) = W_J w = \delta_-(\mathcal{S}, D)$. Set $w_1 := \min(W_J w)$ and $D_1 := \mathrm{proj}_{\mathcal{S}} D$. Thus $w_1 = \delta_-(D_1, D)$.

   (a) For any $X \in \mathcal{R}$, we now deduce

$$\delta^*(X, D) = w_1 = \delta_-(D_1, D) \implies \delta^*(X, D_1) = 1\,,$$

and, conversely,

$$\delta^*(X, D_1) = 1 \implies d^*(X, D) \le d(D_1, D) = l(w_1) \implies \delta^*(X, D) = w_1\,,$$

since the latter is minimal in $\delta^*(\mathcal{R}, D)$.

   (b) Set $C_0 := \mathrm{proj}_{\mathcal{R}} D$. Then $\delta^*(C_0, D) = \max(W_J w) = w_0(J) w_1 \implies d^*(C_0, D_1) \ge d^*(C_0, D) - d(D, D_1) = l(w_0(J))$. Since $\delta^*(\mathcal{R}, \mathcal{S}) = W_J$, this implies that $\delta^*(C_0, D_1) = w_0(J)$ and hence $C_0 = \mathrm{proj}_{\mathcal{R}} D_1$. The connection with Exercise 5.167 is now the following: For any $X \in \mathcal{R}$, we have $\delta^*(X, D_1) = 1 \iff \delta_+(X, \mathrm{proj}_{\mathcal{R}} D_1) = w_0(J) \iff X$ and C_0 are opposite in $\mathcal{R}$.

5.199. Use Corollary 5.194 and the convexity of twin apartments.

Chapter 6

6.37. Argue by induction on d. [Note that the case $d = 1$ is precisely (Bru3).] The induction is completely routine for the second inclusion. For the first, we may assume that the decomposition is reduced. Then

$$
\begin{aligned}
C(ww') &\subseteq C(s_1) C(s_2 \cdots s_d w') && \text{by (Bru3)} \\
&\subseteq C(s_1) C(s_2 \cdots s_d) C(w') && \text{by induction} \\
&= C(w) C(w') && \text{by part (3) of Proposition 6.36.}
\end{aligned}
$$

6.49. The first assertion follows from the fact that B is self-normalizing. The second assertion follows from the first and the fact that $Z(G)$ is normal in G, so that $Z(G) \le \bigcap_{g \in G} gBg^{-1}$. We can restate this assertion as follows: If a group G acts Weyl transitively on a thick building Δ, then $Z(G)$ acts trivially. Applying this statement to the image of G in $\mathrm{Aut}\, \Delta$, we obtain the third assertion of the exercise.

6.54. (a) Check the usual two conditions (see Definition A.1). Note that the standard parabolic subgroups $P_J = BW_J B$ are in 1–1 correspondence with the standard parabolic subgroups W_J of W. Note further that for $g \in G$ and $J, K \subseteq S$, the smallest (with respect to inclusion) standard parabolic subgroup of G containing g, P_J, P_K corresponds to the smallest standard parabolic subgroup of W containing w, J, K, where $g \in C(w)$. The existence of this was shown in the proof of Theorem 3.5.

(b) Define $\iota\colon \Sigma(W, S) \to \Delta$ by $\iota(wW_J) = wP_J$, where the notation is as in the solution to (a). Check that ι is well defined and is a simplicial map. Note that the image of ι is Σ. Define $\rho\colon \Delta \to \Sigma(W, S)$ by $\rho(gP_J) = wW_J$ if $g \in C(w)$. To see that ρ is well defined, suppose $g' = gh$ is another representative of the coset gP_J ($h \in P_J$). Then $h \in C(w_J)$ for some $w_J \in W_J$, and $g' \in C(w)C(w_J)$. Now apply Exercise 6.37 to show that the representative g' yields the same coset wW_J. Show that ρ is simplicial and that $\rho\iota = \mathrm{id}_{\Sigma(W,S)}$, so ι maps $\Sigma(W, S)$ isomorphically onto Σ.

Note, for use in the solution to (c), that the proof has given us a retraction $\rho' := \iota\rho\colon \Delta \to \Sigma$, given by $\rho'(gP) = wP$ if $g \in C(w)$.

(c) To verify (B1), we may assume that one of the two given simplices is a standard parabolic subgroup P. The other one is then gQ for some $g \in G$ and some standard parabolic Q. Writing $g = bnb'$ with $n \in N$ and $b, b' \in B$, we have $gQ = bnQ \in b\Sigma$, so $b\Sigma$ is an apartment containing P and gQ.

It follows from what we have done so far that Δ is a chamber complex. It has a G-invariant type function given by $gP_J \mapsto S \setminus J$. Strong transitivity is also immediate; for G is transitive on $\mathcal{A}$, and the subgroup N stabilizes Σ and is transitive on $\mathcal{C}(\Sigma)$.

Finally, we verify (B2''). Given two apartments with a common chamber C, we must construct an isomorphism between them fixing their intersection. By strong transitivity, we may assume that one of the two apartments is Σ and that C is the standard parabolic coset B. Let Σ' be the other apartment. Since the stabilizer of C is precisely the subgroup B, we can apply strong transitivity again to find an isomorphism $\phi\colon \Sigma' \to \Sigma$ given by the action of some element $b \in B$. We show that ϕ fixes $\Sigma' \cap \Sigma$ by showing that $\phi = \rho'|_{\Sigma'}$, where ρ' is the retraction mentioned above.

Every simplex of $\Sigma' = b^{-1}\Sigma$ has the form $b^{-1}wP$ for some $w \in W$ and some standard parabolic P; the definition of ρ' now gives

$$\rho'(b^{-1}wP) = wP = b \cdot b^{-1}wP = \phi(b^{-1}wP),$$

as required.

6.89. (a) This is the algebraic analogue of Example 5.136(a). It can be done algebraically, or it can be deduced from that example.

(b) This is almost immediate from the solution to Exercise 5.163. Namely, if $C_{\pm}$ are the fundamental chambers of $\mathcal{C}(G, B_+, B_-)$, then $\mathrm{proj}_{C_-} C_+ = w_0 C_-$. Since B_+ stabilizes C_+, it follows that B_+ stabilizes $w_0 C_-$; hence $B_+ \leq w_0 B_- w_0$. Similarly, $B_- \leq w_0 B_+ w_0$. [This solution is a nice example of the use of buildings to obtain short, transparent proofs of purely algebraic results.]

6.90. (a) In view of the Birkhoff decomposition, it suffices to show that for all $w \in W$, $B_+ \not\subseteq wB_- w^{-1}$. Equivalently, we will show that B_+ does not fix the chamber $D_- := wC_-$ of Σ_-. Here, as usual, Σ is the fundamental twin apartment of $\mathcal{C}(G, B_+, B_-)$. Our method will be to use strong transitivity to prove the existence of an element $b \in B_+$ such that $bD_- \neq D_-$.

Let $D_+ := wC_+$; it is the chamber of Σ_+ opposite D_-. Since Σ_+ has infinitely many walls (see Exercise 3.83(a)) and only finitely many of them separate C_+ from D_+, there is a root α_+ of Σ_+ containing both C_+ and D_+. The corresponding twin root $\alpha = (\alpha_+, \alpha_-)$ then contains C_+ but not D_-. By thickness and Lemma 5.198, there is a twin apartment $\Sigma' \ne \Sigma$ containing α, and strong transitivity implies that $\Sigma' = b\Sigma$ for some $b \in B_+$. Since $\Sigma \cap \Sigma' = \alpha$ by Exercise 5.202, we conclude that $bD_- \notin \Sigma$ and hence $bD_- \ne D_-$.

(b) Now the result to be proved is that B_+ does not fix any vertex v_- of Σ_-. [Here we work in the simplicial buildings associated to $\mathcal{C}\pm$, and we still denote by Σ the corresponding simplicial apartment.] Let v_+ be the vertex of Σ opposite v_-. As in the solution to (a) above, but using Exercise 3.83(b) instead of (a), we can find a root α_+ that contains both C_+ and v_+, with $v_+ \notin \partial\alpha_+$. The corresponding twin root contains C_+ but not v_-, and the argument in (a) gives an element $b \in B_+$ that does not fix v_-.

6.114. (b) The first assertion will be proved in Section 6.10.2.

(c) For the first assertion, see Serre [217, Section II.1.1].

6.115. (a) Every vertex in the link has a unique representative L with $\pi A^2 < L < A^2$. So the vertices of the link correspond to the 1-dimensional subspaces of the 2-dimensional k-vector space $A^2/\pi A^2$.

(b) Before specializing to $K = k(t)$, note that in general, a lattice L as in the solution to (a) has the form $L = \pi A^2 + Ax$ with $x \in A^2 \setminus \pi A^2$. For $L = [e_1, \pi e_2]$, we can take $x = e_1$. For all other L, we can take $x = \lambda e_1 + e_2$ with $\lambda \in A$, and λ is uniquely determined mod π. We denote by L_λ the resulting lattice $\pi A^2 + Ax$. Thus, for example, $L_0 = [[\pi e_1, e_2]]$, and L_λ for any λ is the image of L_0 under the action of the elementary matrix $E_{12}(\lambda) := \left(\begin{smallmatrix} 1 & \lambda \\ 0 & 1 \end{smallmatrix}\right)$. What is special about the case $K = k(t)$ is that the residue field k can also be viewed as a subring of A; the desired subgroup U is then $\{E_{12}(\lambda) \mid \lambda \in k\}$.

6.123. This result is given in Lam [150, Corollary VI.2.5(1)]. Here is a sketch of a direct proof: By Hilbert's criterion, it suffices to show that the equation $\alpha x^2 + \beta y^2 = 1$ has a solution over $\mathbb{Q}_p$. By standard results about lifting solutions of equations [215, Section II.2.2, Corollary 2], it suffices to solve the equation mod p. We now appeal to the fact that a nondegenerate binary quadratic form over a finite field of characteristic $\ne 2$ represents every nonzero element; see, for instance, [215, Section IV.1.7, Proposition 4], where this result is proved by an easy counting argument.

6.124. We try to replace the basis vector $e_2 \in D$ by a suitable linear combination $e_2' := \lambda e_2 + \mu e_4$ with $\lambda, \mu \in \mathbb{Q}$. Note that any such e_2' anticommutes with e_3 and that $(e_2')^2 = \lambda^2 \alpha - \mu^2 \alpha\beta =: \alpha' \in \mathbb{Q}$. We will show that λ, μ can be chosen such that $\alpha' \in U^2$, where U is the group $\mathbb{Z}_p^*$ of p-adic units. Setting $e_4' = e_2' e_3$, we will then have a "quaternion basis" $1, e_2', e_3, e_4'$, showing that $D \cong (\alpha', \beta)_\mathbb{Q}$.

The expression defining α' above is the binary quadratic form $\langle \alpha, -\alpha\beta \rangle$ in the variables λ, μ. As in the solution to Exercise 6.123, this form, viewed as a form over $\mathbb{Q}_p$, represents all elements of U. In particular, we can find $\lambda, \mu \in \mathbb{Q}_p$ such that $\lambda^2\alpha - \mu^2\alpha\beta = 1$, for instance. Since U^2 is an open subset of $\mathbb{Q}_p$, we can replace λ, μ by approximations in $\mathbb{Q}$ and still have $\lambda^2\alpha - \mu^2\alpha\beta \in U^2$.

Chapter 7

7.12. (a) Consider the projection $C_1 := AC_0 \in \Sigma$. Then $gC_1 = (gA)C_0 \in \Sigma$ and hence, since g preserves Weyl distance from C_0, $gC_1 = C_1$. Being type-preserving, g fixes the face A of C_1.

(b) The lemma says that every simplex of Δ is U-equivalent to a simplex of Σ. Since U fixes C_0, (a) implies that no two distinct simplices of Σ are U-equivalent.

7.77. Recall that a root vector α corresponds to the half-space $\langle \alpha, - \rangle \geq 0$. We can now appeal to the following elementary fact from linear algebra: Let $f, g, h: V \to \mathbb{R}$ be linear functions on a real vector space V. Then h is a nonnegative linear combination of f and g if and only if $h \geq 0$ on $\{v \in V \mid f(v) \geq 0 \text{ and } g(v) \geq 0\}$.

7.78. Method 1: Recall from Example 1.10 that the root system of type A_{n-1} consists of the vectors $\alpha_{ij} := e_i - e_j$ ($i \neq j$), where $e_1, \ldots, e_n$ is the standard basis for $\mathbb{R}^n$. The only strictly positive linear combination of α_{ij} and α_{jk} that is a root is their sum, which is α_{ik}, so the result follows from Exercise 7.77. Method 2: By Example 3.52, we can identify α_{ij} (or, more precisely, $\mathcal{C}(\alpha_{ij})$) with the set of permutations π of $\{1, \ldots, n\}$ such that $\pi^{-1}(i) < \pi^{-1}(j)$. If $\pi^{-1}(i) < \pi^{-1}(j)$ and $\pi^{-1}(j) < \pi^{-1}(k)$, then $\pi^{-1}(i) < \pi^{-1}(k)$, and no other such conclusion can be drawn. So $\alpha_{ij} \cap \alpha_{jk} \subseteq \alpha_{ik}$, and this is the only such inclusion.

7.97. Choose $s \in S$ such that $w \notin \alpha_s$. We then have $w \in -\alpha_s$; hence $1 \in w^{-1}(-\alpha_s) =: \alpha$, so $\alpha \in \Phi_+$ and $w\alpha = -\alpha_s$.

7.110. The expression is to be interpreted as $\tilde{w}V\tilde{w}^{-1}$ for some representative $\tilde{w}$ of w in N. It is legitimate if V is normalized by T. Note that we must use the *same* representative $\tilde{w}$ for both occurrences of w, unless $T \leq V$. In particular, we cannot write sVs instead of sVs^{-1} for $s \in S$, since $\tilde{s}V\tilde{s} = \tilde{s}V\tilde{s}^{-1}$ only if $\tilde{s}^2 \in V$.

7.112. In view of Proposition 7.106, it suffices to show that there is a minimal gallery in Σ whose associated sequence of roots contains both α and β. [Equivalently, there exists $w \in W$ such that $w\alpha$ and $w\beta$ are both in $\Phi(w')$ for some $w' \in W$.] Since Σ contains no nested roots, the intersections $\alpha \cap \beta$ and $(-\alpha) \cap (-\beta)$ each contain at least one chamber. Choose a chamber $C \in \alpha \cap \beta$ and a chamber $D \in (-\alpha) \cap (-\beta)$. Then any minimal gallery from C to D must cross both $\partial\alpha$ and $\partial\beta$, so α and β both occur in the associated sequence of roots.

7.129. Note that $\langle U_\alpha, U_{-\alpha}\rangle$ admits a BN-pair, and apply Exercise 6.50.

7.139. (a) This can of course be done via elementary row/column operations, but we have already done it: Apply equation (7.25) with $\mu = 1$.

(b) Equation (7.20) implies $[h(\lambda), E_{12}(\mu)] = E_{12}(\lambda\mu\lambda - \mu)$. This reduces to the formula to be proved if $\lambda\mu = \mu\lambda$.

(c) It suffices to show that every $x \in D$ can be written in the form $x = (\lambda^2 - 1)\mu$ for some commuting elements $\lambda, \mu \in D$ with $\lambda \neq 0$. If $x \neq 0, \pm 1$, we can take $\lambda = x$ and $\mu = (x^2 - 1)^{-1}x$. Otherwise, we can choose any $\lambda \neq 0, \pm 1$ and set $\mu = (\lambda^2 - 1)^{-1}x$. Note that such a λ exists precisely when $D \neq \mathbb{F}_2$ or $\mathbb{F}_3$.

Chapter 8

8.36. (a) We may assume that $d^*(C_+, D') \leq d^*(C_+, C)$ and, by Lemma 8.7, that $D = wC_- \in \Sigma_-$. Let α be the twin root of Σ containing $D = wC_-$ but not wsC_-, and let $\mathcal{P}$ be the s-panel containing D, D', and wsC_-. If $\delta^*(C_+, D') \neq \delta^*(C_+, D)$, then $\delta^*(C_+, D') = ws$, and $l(ws) < l(w)$. Then α does not contain C_-, since $d(C_-, wsC_-) < d(C_-, wC_-)$, so α is a positive root. Transitivity of U_α on $\mathcal{C}(\mathcal{P}, \alpha)$ now gives us an element $u \in U_\alpha \leq U_+$ such that $uD' = wsC_-$ and $uD = D = wC_-$. All assertions of (a) now follow in this case, with $x = u^{-1}$.

Suppose now that $\delta^*(C_+, D') = \delta^*(C_+, D)$ $[= w]$. Then we must have $l(ws) > l(w)$, so α is a negative root. Transitivity of $U_{-\alpha}$ on $\mathcal{C}(\mathcal{P}, -\alpha)$ now yields an element $y \in U_{-\alpha}$ such that $D' = yD$, and the assertions of (a) are satisfied with $x = 1$.

(b) Combine (a) and Remark 8.35.

8.46. By Corollary 3.166, there is a nested pair of roots $\alpha \subsetneq \beta$. Using the action of W, we can arrange that the fundamental chamber is in $\beta \smallsetminus \alpha$, so that β is positive and α is negative. Then $\{-\alpha, \beta\}$ is a positive nonprenilpotent pair.

8.60. The geometric interpretation is that $\Delta = \bigcup_{u \in U_-} u\Sigma_0$, where $\Delta = \Delta(G, B_+)$ and Σ_0 is the fundamental apartment. This is to be expected in view of Lemma 8.7. To prove the equation, note that it suffices to show that $B_-WB_+ = G$, since $WB_+ = WTB_+ = TWB_+$. The proof now uses ideas from the theory of twin BN-pairs, but one has to take care to use only those facts about (B_+, B_-, N) that we know at this point.

8.63. U_+ fixes the fundamental chamber, but U_{-s} does not. Since there is no opposite chamber to work with, there is no obvious way to interchange $+$ and $-$ in this argument.

8.94. The procedure described in the exercise would yield a presentation with many copies of each generating subgroup U_α, and it is not obvious that they are all equal in the direct limit. Now they are in fact all equal in the direct limit, but we know this only because of the isomorphism $\varinjlim U_w \xrightarrow{\sim} U_+$. To put it another way, there is no obvious way to construct the inverse of the map $q: \varinjlim U_w \longrightarrow V$ that occurred in our proof without making use of the isomorphism $\varinjlim U_w \xrightarrow{\sim} U_+$.

Chapter 10

10.1. We may assume $x_0 = 0$. Then if we know the numbers $d(x, x_i)$, we can compute the inner products $\langle x, x_i \rangle$ via the identity $d^2(x, y) = \|x\|^2 + \|y\|^2 - 2\langle x, y \rangle$, which is essentially the law of cosines.

10.5. We will need this result only when V is the affine span of X. In this case one can find an affine frame $x_0, \ldots, x_n$ in X and argue as in Exercise 10.3. The proof in the general case is similar but takes slightly more work. See [48, Chapter I.2, Proposition 2.20] for a detailed proof of a more general result.

10.16. The chamber C is defined by $\langle e_i, - \rangle > c_i$ $(i = 1, \ldots, n+1)$. We have no control over the c_i, but we can be sure that $\sum_{i=1}^{n+1} \lambda_i c_i < 0$. If we want to try to make our description of C (and hence of (W, V)) as canonical as possible, we can first normalize the λ_i by requiring $\sum_{i=1}^{n+1} \lambda_i = 1$. We can then scale the inner product to achieve $\sum_{i=1}^{n+1} \lambda_i c_i = -1$. This determines the vector $(c_1, \ldots, c_{n+1})$ up to translation by a vector in the n-dimensional subspace of $\mathbb{R}^{n+1}$ given by $\sum_{i=1}^{n+1} \lambda_i x_i = 0$.

10.34. See Example 10.14(b). The special vertices are those where four lines intersect rather than two. To get a picture of L, see Remark 10.24(a).

10.46. If $Q(\alpha) \neq 0$, then we have an orthogonal decomposition $\mathbb{R}^{n,1} = \mathbb{R}\alpha \oplus \alpha^\perp$. Suppose now that $Q(\alpha) > 0$. Then the bilinear form $\langle -, - \rangle$ must have signature $(n-1, 1)$ on $\alpha^\perp$; hence Q takes negative values on $\alpha^\perp$. It follows that $\alpha^\perp$ meets $\mathbb{H}^n$. Conversely, suppose $\alpha^\perp$ meets $\mathbb{H}^n$, and choose β in the intersection. Then $\mathbb{R}^{n,1} = \mathbb{R}\beta \oplus \beta^\perp$, and since $Q(\beta) = -1$, the bilinear form $\langle -, - \rangle$ must be positive definite on $\beta^\perp$. Since $\alpha \in \beta^\perp$, we have $Q(\alpha) > 0$.

10.47. Let $V = \mathbb{R}\alpha_1 + \mathbb{R}\alpha_2$. Then $H_1 \cap H_2 = V^\perp \cap \mathbb{H}^n$. Suppose this is nonempty, and choose β in it. Then, as in the previous solution, $\langle -, - \rangle$ is positive definite on $\beta^\perp$, which contains V, so $\langle -, - \rangle$ is positive definite on V. Conversely, suppose $\langle -, - \rangle$ is positive definite on V. Then, in particular, $\langle -, - \rangle$ is nondegenerate on V, so we have an orthogonal decomposition $\mathbb{R}^{n,1} = V \oplus V^\perp$; hence $\langle -, - \rangle$ has signature $(n-2, 1)$ on $V^\perp$ and Q takes negative values on $V^\perp$. It follows that $V^\perp$ meets $\mathbb{H}^n$, so $H_1 \cap H_2 \neq \emptyset$.

10.48. Consider the restriction of the bilinear form $\langle -, - \rangle$ to $V := \mathbb{R}e_1 + \mathbb{R}e_2$. It has matrix $\left(\begin{smallmatrix} 1 & c \\ c & 1 \end{smallmatrix}\right)$, where $c := \langle e_1, e_2 \rangle$. In view of the previous exercise, the hypothesis implies that this bilinear form is not positive definite, i.e., $|c| \geq 1$. If strict inequality holds, then the form is nondegenerate but indefinite, i.e., it has signature $(1, -1)$. We then have an orthogonal decomposition $\mathbb{R}^{n,1} = V \oplus V^{\perp}$, which implies that Q is positive definite on $V^{\perp} = e_1^{\perp} \cap e_2^{\perp}$. This says precisely that there is no nonzero vector x in $e_1^{\perp} \cap e_2^{\perp}$ with $Q(x) \leq 0$. Conversely, suppose there is no such vector x. Then Q is positive definite on $V^{\perp}$. In particular, it is nondegenerate on $V^{\perp}$, so we have an orthogonal decomposition $\mathbb{R}^{n,1} = V^{\perp} \oplus (V^{\perp})^{\perp} = V^{\perp} \oplus V$, and Q must have signature $(1, -1)$ on V. Thus $|c| > 1$.

Chapter 11

11.11. Assume first that equality holds in one of the two given inequalities, say $d(z', x) = d(z, x)$. Let θ (resp. θ') be the angle at x in the triangle x, y, z (resp. x, y, z'). By the fact quoted in the statement of the exercise, we have $\theta' \leq \theta$. Now use that fact again to compare the triangles x, p, z and x, p, z', and conclude that $d(z', p) \leq d(z, p)$.

In the general case, where both of the given inequalities are strict, introduce an intermediate triangle x, y, z'', where z'' is chosen such that $d(z'', x) = d(z, x)$ and $d(z'', y) = d(z', y)$. [Such a z'' exists because $d(x, y) \leq d(x, z') + d(z', y) \leq d(x, z) + d(z', y)$.] Now apply the first case twice.

11.15. First draw a picture to see why the assertion is plausible. For the proof, consider the points p_t "between" y and y' as in Exercise 11.14, where t ranges over the dyadic rationals in $[0, 1]$. Then

$$d^2(x, p_t) \leq (1 - t)d^2 + td^2(x, y') - t(1 - t)d^2(y, y') .$$

The right side of this inequality would be a decreasing function of t for small t if we had $d(y, y') > d(x, y')$.

11.22. Let V be the tangent space to M at p_t and let $\rho: M \to V$ be the inverse of the exponential map. Then ρ is distance-decreasing and preserves distances from p_t.

11.30. Use the given cocycle to define an action of G on V by affine isometries.

11.50. Let $\mathfrak{D}$ be the direction of $\mathfrak{C}$, and let the wall H be defined by an equation $f = c$. We may assume that $f > 0$ on $\mathfrak{D}$ and hence that $f(x) > c$ for some $x \in \mathfrak{C}$. The subsector $x + \mathfrak{D}$ of $\mathfrak{C}$ is then contained in the half-space $f > c$.

11.58. To prove the inequality, take a suitable apartment E containing $\{x, y', z'\}$, and consider $\rho = \rho_{E,C}$, where $x \in \overline{C}$. [See Bridson–Haefliger [48] for a proof that is valid in an arbitrary CAT(0) space.] If equality holds, then ρ is an isometry on $\{y, z\} \cup \overline{C}$.

11.59. It suffices to show that $x \in [y, z]$, i.e., that $d(y, z) = d(y, x) + d(x, z)$. This follows from the cosine inequality in Exercise 11.58. (Note that our hypothesis says precisely that $\theta = \pi$.)

11.84. (a) Let ϕ be an automorphism of Δ such that the induced map ϕ_∞ on Δ_∞ is the identity. In view of Exercise 4.52, it suffices to show that ϕ stabilizes every apartment $E = |\Sigma|$. But this is immediate from Theorem 11.79 and the fact that ϕ stabilizes every apartment of Δ_∞.

(b) An automorphism of bounded displacement maps every ray to a parallel ray and hence induces the identity map on X_∞.

(c) Euclidean Coxeter complexes admit nontrivial translations.

11.94. To prove (0) we must show that any two chambers C_1, C_2 in X' are contained in an apartment in $\mathcal{A}$. Choose points $x_1 \in C_1$ and $x_2 \in C_2$, and consider the directed line segment from x_1 to x_2. Working in an apartment E_2 in $\mathcal{A}$ containing C_2, we can continue the line segment in the same direction so as to obtain a ray starting at x_1 that is eventually in E_2. (See Exercise 11.59.) Similarly, we can continue the ray in the opposite direction in an apartment E_1 in $\mathcal{A}$. The result is a line (i.e., a subset isometric to $\mathbb{R}$) such that each end is eventually in an apartment in $\mathcal{A}$. Perturbing x_1 and x_2 slightly, if necessary, we can arrange that the two ends of this line are contained in open sectors. Our hypothesis now implies that the entire line is contained in an apartment in $\mathcal{A}$, which then also contains C_1 and C_2.

The proof of (1) is similar: Given a chamber C of X' and an $\mathcal{A}$-sector $\mathfrak{C}$, choose $x \in C$ and construct a ray starting at x that is eventually in $\mathfrak{C}$. [This is possible because there is an apartment in the complete system containing C and a subsector $\mathfrak{C}'$ of $\mathfrak{C}$. If $\mathfrak{C}'$ has direction $\mathfrak{D}$, we can use the ray given by $x + td$ $(t \geq 0)$ for any $d \in \mathfrak{D}$.] Now piece this together with a ray in an apartment in $\mathcal{A}$ containing C, where the direction is chosen so as to match up with the first ray. After a small perturbation if necessary, we get a line in X whose two ends are contained in $\mathcal{A}$-sectors, and we can finish as in the proof of (0).

Chapter 12

12.12. This exercise is not trivial at the moment. You can try it to see what the issues are and then return to it after reading Sections 12.1.4 and 12.1.5. At that point it should be easy.

12.36. The spherical and hyperbolic laws of cosines can be found in [48, Chapter I.2]. In the spherical case, for example, the law of cosines says the following: Given a geodesic triangle with side lengths a, b, c and with vertex angle θ opposite the side of length c, we have

$$\cos c = \cos a \cos b + \sin a \sin b \cos \theta .$$

12.37. For $\kappa = -1$, it is

$$\cosh d(z, m) \le \frac{\cosh d(z, x) + \cosh d(z, y)}{2 \cosh(d(x, y)/2)} .$$

For $\kappa = 1$, assume $d(x, y) + d(y, z) + d(z, x) < 2\pi$, and conclude that $d(z, m) \le \pi$ and

$$\cos d(z, m) \ge \frac{\cos d(z, x) + \cos d(z, y)}{2 \cos(d(x, y)/2)} .$$

12.56. See the figure below.

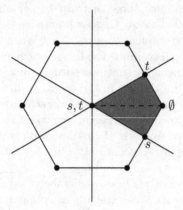

12.62. A classical theorem of Schur [212] implies that finitely generated infinite linear groups cannot be torsion groups. (See Wehrfritz [279, Corollary 4.9] for a modern treatment.) Alternatively, there is a longer but more elementary proof based on Tits's solution to the word problem, as in Exercise 2.42. A third alternative is to apply Proposition 2.74.

Appendix A

A.2. (a) Δ must contain the empty simplex as an element.

(b) The "empty simplicial complex" Δ has one element, which is the empty simplex. The vertex set of Δ is empty, as is the geometric realization $|\Delta|$.

A.16. (a) The boundary of a triangle has this property. More generally, the boundary of an n-simplex for any $n \ge 2$ has $n+1$ mutually adjacent chambers.

(b) Let C, D, E be distinct chambers in a colorable thin chamber complex. Suppose C is adjacent to D and D is adjacent to E. We will show that C is not adjacent to E. Choose a type function with values in a set I. Then $C \cap D$ is an i-panel for some $i \in I$, and $D \cap E$ is a j-panel for some $j \in I$. By thinness, $i \ne j$. One can now check that $C \cap E$ has cotype $\{i, j\}$ and hence is not a panel, so C is not adjacent to E.

A.17. (a) A detailed proof can be found in Munkres [179, Lemma 63.1]. The essential point is that the local homology can be computed using the geometric realization of the star of A, which is a neighborhood of each point of $|A|$. Here the *star* of a simplex A consists of all simplices joinable to A; it is the simplicial join $\Delta_{\leq A} * \mathrm{lk}_\Delta A$.

(b) This is immediate from (a).

(c) Apply (b). Note that $\mathrm{lk}\, A$ is the empty simplicial complex (whose only simplex is the empty simplex) if A is maximal, so we must have $n-k-1 = -1$. [More directly: $|A|$ is an open subset of X homeomorphic to $\mathbb{R}^k$, so the local homology at each of its points is the same as that of $\mathbb{R}^k$.]

(d) Apply (b). Note that $\mathrm{lk}\, A$ is 0-dimensional, with one vertex for every n-simplex having A as a face.

(e) Apply (b), noting that the link in $\mathrm{lk}\, A$ of a simplex B is equal to the link in Δ of $A \cup B$.

(f) Arguing by induction on n, we may assume that the link of every nonempty simplex is gallery connected. Hence any two chambers with nonempty intersection can be connected by a gallery. To complete the proof, use the connectivity of X to show that any two chambers can be joined by a sequence $C_0, \ldots, C_m$ such that $C_{i-1} \cap C_i$ is nonempty for each $i = 1, \ldots, m$. [Partition the chambers into suitable equivalence classes, and note that X would be disconnected if there were more than one class.]

A.27. Every face of the cube has a smallest vertex x and a largest vertex y with respect to the ordering on P, and the set of vertices of that face is the closed interval $[x, y]$ in P. [In Figure A.3, we have oriented each edge toward its larger vertex in order to make this visually obvious.] Thus the cells of K correspond to intervals in P, ordered by inclusion.

A.30. If these conditions hold, then P can be identified with the poset of cells of a regular cell complex by Proposition A.25(1). The subcomplex $P_{\leq x}$ is isomorphic to a cube by part (2) of the same proposition. And condition (a) is exactly what is needed to guarantee that a nonempty intersection of closed cells is a closed cell.

A.31. Note that (b) implies that $P_{\leq x}$ is finite for each $x \in P$. In view of this, the implication (a$'$) $\implies$ a is almost immediate. [x and y have finitely many common lower bounds, and their supremum, which exists by (a$'$), is the largest of them.] The opposite implication is similar but slightly trickier, because there might be infinitely many upper bounds. To get around this, consider a finite collection of upper bounds whose infimum is of minimal dimension.

A.32. The flag complex $\Delta(K_{<e})$ triangulates an $(n-1)$-sphere; hence it is an $(n-1)$-dimensional thin chamber complex by Example A.9. The result follows easily.

Appendix B

B.3. Recall that the angle between α_i and $-\alpha_j$ is π/m_{ij} (see Theorem 1.88 or its proof). Hence

$$a_{ij} = -\|\alpha_i^\vee\|\,\|\alpha_j\|\cos\frac{\pi}{m_{ij}}\,.$$

Now note that $\|\alpha^\vee\| = 2/\|\alpha\|$ for any root α.

Appendix C

C.5. G is isomorphic to a closed subgroup of GL_4 over k. To see this, identify G with SU_2 and note that $\mathrm{SU}_2(R[i])$ acts on $R[i]^2 = R^4$. Alternatively, the group of unit quaternions acts on the (4-dimensional) quaternion algebra by multiplication.

References

The numbers following the uparrow (↑) at the end of
each entry indicate the pages on which the item is cited.

[1] H. Abels, *An example of a finitely presented solvable group*, Homological group theory (C. T. C. Wall, ed.), London Mathematical Society Lecture Note Series, vol. 36, Cambridge University Press, Cambridge, 1979, pp. 205–211. MR564423 (82b:20047) ↑643

[2] _____, *Finite presentability of S-arithmetic groups. Compact presentability of solvable groups*, Lecture Notes in Mathematics, vol. 1261, Springer-Verlag, Berlin, 1987. MR903449 (89b:22017) ↑644

[3] _____, *Finiteness properties of certain arithmetic groups in the function field case*, Israel J. Math. **76** (1991), no. 1-2, 113–128. MR1177335 (94a:20077) ↑648

[4] H. Abels and P. Abramenko, *On the homotopy type of subcomplexes of Tits buildings*, Adv. Math. **101** (1993), no. 1, 78–86. MR1239453 (95a:55013) ↑648

[5] H. Abels and K. S. Brown, *Finiteness properties of solvable S-arithmetic groups: an example*, J. Pure Appl. Algebra **44** (1987), no. 1-3, 77–83. MR885096 (88g:20104) ↑644

[6] P. Abramenko, *Endlichkeitseigenschaften der Gruppen* $SL_n(\mathbf{F}_q[t])$, Ph.D. Thesis, Frankfurt am Main, 1987. ↑648, 655

[7] _____, *Reduktive Gruppen über lokalen Körpern und Bruhat-Tits-Gebäude*, Bayreuth. Math. Schr. **47** (1994), 1–69. MR1285203 (95k:20074) ↑443

[8] _____, *Walls in Coxeter complexes*, Geom. Dedicata **49** (1994), no. 1, 71–84. MR1261574 (95a:20043) ↑163

[9] _____, *Twin buildings and applications to S-arithmetic groups*, Lecture Notes in Mathematics, vol. 1641, Springer-Verlag, Berlin, 1996. MR1624276 (99k:20060) ↑290, 434, 648, 655

[10] P. Abramenko and K. S. Brown, *Transitivity properties for group actions on buildings*, J. Group Theory **10** (2007), no. 3, 267–277. MR2320966 ↑365, 371

[11] _____, *Automorphisms of buildings have unbounded displacement*, preprint, 2007, available at arXiv:0710.1426[math.GR]. ↑584

[12] P. Abramenko and B. Mühlherr, *Présentations de certaines BN-paires jumelées comme sommes amalgamées*, C. R. Acad. Sci. Paris Sér. I Math. **325** (1997), no. 7, 701–706. MR1483702 (98h:20043) ↑497, 653

[13] P. Abramenko and G. Nebe, *Lattice chain models for affine buildings of classical type*, Math. Ann. **322** (2002), no. 3, 537–562. MR1895706 (2003a:20048) ↑363

[14] P. Abramenko and H. Van Maldeghem, *Connectedness of opposite-flag geome-tries in Moufang polygons*, European J. Combin. **20** (1999), no. 6, 461–468. MR1703593 (2000e:51007) ↑290

[15] ———, *On opposition in spherical buildings and twin buildings*, Ann. Comb. **4** (2000), no. 2, 125–137. MR1770684 (2001m:51016) ↑196

[16] ———, *1-twinnings of buildings*, Math. Z. **238** (2001), no. 1, 187–203. MR1860741 (2002g:51013) ↑266, 272, 372, 373

[17] ———, *Combinatorial characterizations of convexity and apartments in build-ings*, Australas. J. Combin. **34** (2006), 89–104. MR2195312 (2006j:51008) ↑237

[18] M. Abramson and C. D. Bennett, *Embeddings of twin trees*, Geom. Dedicata **75** (1999), no. 2, 209–215. MR1686759 (2000c:51010) ↑291

[19] J. D. Adler, *Refined anisotropic K-types and supercuspidal representations*, Pacific J. Math. **185** (1998), no. 1, 1–32. MR1653184 (2000f:22019) ↑660

[20] J. D. Adler and S. DeBacker, *Some applications of Bruhat-Tits theory to har-monic analysis on the Lie algebra of a reductive p-adic group*, Michigan Math. J. **50** (2002), no. 2, 263–286. MR1914065 (2003g:22016) ↑660

[21] R. C. Alperin and P. B. Shalen, *Linear groups of finite cohomological dimen-sion*, Invent. Math. **66** (1982), no. 1, 89–98. MR652648 (84a:20052) ↑646

[22] E. Artin, *Geometric algebra*, Interscience Publishers, Inc., New York-London, 1957. MR0082463 (18,553e) ↑343, 344, 348

[23] M. Aschbacher, *Sporadic groups*, Cambridge Tracts in Mathematics, vol. 104, Cambridge University Press, Cambridge, 1994. MR1269103 (96e:20020) ↑656

[24] ———, *Finite group theory*, 2nd ed., Cambridge Studies in Advanced Math-ematics, vol. 10, Cambridge University Press, Cambridge, 2000. MR1777008 (2001c:20001) ↑343, 344, 348, 350, 503

[25] M. Aschbacher and S. D. Smith, *The classification of quasithin groups. I*, Mathematical Surveys and Monographs, vol. 111, American Mathematical Society, Providence, RI, 2004. Structure of strongly quasithin K-groups. MR2097623 (2005m:20038a) ↑507, 656

[26] ———, *The classification of quasithin groups. II*, Mathematical Surveys and Monographs, vol. 112, American Mathematical Society, Providence, RI, 2004. Main theorems: the classification of simple QTKE-groups. MR2097624 (2005m:20038b) ↑507

[27] M. F. Atiyah and I. G. Macdonald, *Introduction to commutative algebra*, Addison-Wesley Publishing Co., Reading, Mass.-London-Don Mills, Ont, 1969. MR0242802 (39 #4129) ↑354

[28] W. Ballmann, *Lectures on spaces of nonpositive curvature*, DMV Seminar, vol. 25, Birkhäuser Verlag, Basel, 1995. With an appendix by Misha Brin. MR1377265 (97a:53053) ↑549, 658

[29] W. Ballmann, M. Gromov, and V. Schroeder, *Manifolds of nonpositive curva-ture*, Progress in Mathematics, vol. 61, Birkhäuser Boston Inc., Boston, MA, 1985. MR823981 (87h:53050) ↑657, 658

[30] W. Ballmann and J. Świątkowski, *On L^2-cohomology and property (T) for au-tomorphism groups of polyhedral cell complexes*, Geom. Funct. Anal. **7** (1997), no. 4, 615–645. MR1465598 (98m:20043) ↑645

[31] D. Barbasch and A. Moy, *A new proof of the Howe conjecture*, J. Amer. Math. Soc. **13** (2000), no. 3, 639–650 (electronic). MR1758757 (2001f:22053) ↑660

[32]   H. Behr, *Arithmetic groups over function fields. I. A complete characterization of finitely generated and finitely presented arithmetic subgroups of reductive algebraic groups*, J. Reine Angew. Math. **495** (1998), 79–118. MR1603845 (99g:20088) ↑648

[33]   R. Bieri, *Homological dimension of discrete groups*, 2nd ed., Queen Mary College Mathematical Notes, Queen Mary College Department of Pure Mathematics, London, 1981. MR715779 (84h:20047) ↑639

[34]   A. Björner, *Posets, regular CW complexes and Bruhat order*, European J. Combin. **5** (1984), no. 1, 7–16. MR746039 (86e:06002) ↑670, 671

[35]   _____, *Some combinatorial and algebraic properties of Coxeter complexes and Tits buildings*, Adv. in Math. **52** (1984), no. 3, 173–212. MR744856 (85m:52003) ↑214

[36]   _____, *Topological methods*, Handbook of combinatorics. Vol. 1, 2 (R. L. Graham, M. Grötschel, and L. Lovász, eds.), Elsevier Science B.V., Amsterdam, 1995, pp. 1819–1872. MR1373690 (96m:52012) ↑197, 214

[37]   A. Björner, M. Las Vergnas, B. Sturmfels, N. White, and G. M. Ziegler, *Oriented matroids*, 2nd ed., Encyclopedia of Mathematics and its Applications, vol. 46, Cambridge University Press, Cambridge, 1999. MR1744046 (2000j:52016) ↑25, 623, 670, 671, 676

[38]   A. Borel, *Groupes linéaires algébriques*, Ann. of Math. (2) **64** (1956), 20–82. MR0093006 (19,1195h) ↑335

[39]   _____, *Linear algebraic groups*, 2nd ed., Graduate Texts in Mathematics, vol. 126, Springer-Verlag, New York, 1991. MR1102012 (92d:20001) ↑443, 693

[40]   A. Borel and L. Ji, *Compactifications of symmetric and locally symmetric spaces*, Mathematics: Theory & Applications, Birkhäuser Boston Inc., Boston, MA, 2006. MR2189882 (2007d:22030) ↑658

[41]   A. Borel and J.-P. Serre, *Corners and arithmetic groups*, with an appendix by A. Douady and L. Hérault, Comment. Math. Helv. **48** (1973), 436–491. MR0387495 (52 #8337) ↑637

[42]   _____, *Cohomologie d'immeubles et de groupes S-arithmétiques*, Topology **15** (1976), no. 3, 211–232. MR0447474 (56 #5786) ↑640, 642

[43]   N. Bourbaki, *Éléments de mathématique. 23. Première partie: Les structures fondamentales de l'analyse. Livre II: Algèbre. Chapitre 8: Modules et anneaux semi-simples*, Actualités Sci. Ind. no. 1261, Hermann, Paris, 1958. MR0098114 (20 #4576) ↑367

[44]   _____, *Éléments de mathématique. Fasc. XXXIV. Groupes et algèbres de Lie. Chapitre IV: Groupes de Coxeter et systèmes de Tits. Chapitre V: Groupes engendrés par des réflexions. Chapitre VI: Systèmes de racines*, Actualités Scientifiques et Industrielles, No. 1337, Hermann, Paris, 1968. MR0240238 (39 #1590) ↑15, 48, 100, 103, 106, 259, 315, 316, 337, 347, 526, 528, 529, 536, 544, 546, 681

[45]   M. Bourdon, *Immeubles hyperboliques, dimension conforme et rigidité de Mostow*, Geom. Funct. Anal. **7** (1997), no. 2, 245–268. MR1445387 (98c:20056) ↑616

[46]   _____, *Sur les immeubles fuchsiens et leur type de quasi-isométrie*, Ergodic Theory Dynam. Systems **20** (2000), no. 2, 343–364. MR1756974 (2001g:20056) ↑616

[47]   M. R. Bridson, *Geodesics and curvature in metric simplicial complexes*, Group theory from a geometrical viewpoint (É. Ghys, A. Haefliger, and A. Verjovsky, eds.), World Scientific Publishing Co. Inc., River Edge, NJ, 1991, pp. 373–463. MR1170372 (94c:57040) ↑611, 626

[48]   M. R. Bridson and A. Haefliger, *Metric spaces of non-positive curvature*, Grundlehren der Mathematischen Wissenschaften, vol. 319, Springer-Verlag, Berlin, 1999. MR1744486 (2000k:53038) ↑537, 538, 547, 549, 550, 551, 552, 570, 573, 579, 583, 611, 612, 614, 626, 671, 672, 713, 714, 715

[49]   B. Brink and R. B. Howlett, *A finiteness property and an automatic structure for Coxeter groups*, Math. Ann. **296** (1993), no. 1, 179–190. MR1213378 (94d:20045) ↑101, 106

[50]   K. S. Brown, *Cohomology of groups*, Graduate Texts in Mathematics, vol. 87, Springer-Verlag, New York, 1982. MR672956 (83k:20002) ↑633, 634, 637, 641, 643

[51]   ———, *Presentations for groups acting on simply-connected complexes*, J. Pure Appl. Algebra **32** (1984), no. 1, 1–10. MR739633 (85k:20100) ↑641, 651

[52]   ———, *Finiteness properties of groups*, J. Pure Appl. Algebra **44** (1987), 45–75. MR885095 (88m:20110) ↑644

[53]   ———, *Buildings*, Springer-Verlag, New York, 1989. MR969123 (90e:20001) ↑vii, 6, 8, 32

[54]   ———, *WHAT IS... a building*, Notices Amer. Math. Soc. **49** (2002), no. 10, 1244–1245. ↑3

[55]   ———, *Semigroup and ring theoretical methods in probability*, Representations of finite dimensional algebras and related topics in Lie theory and geometry (V. Dlab and C. M. Ringel, eds.), Fields Institute Communications, vol. 40, American Mathematical Society, Providence, RI, 2004, pp. 3–26. MR2057147 (2005b:60118) ↑26, 29

[56]   K. S. Brown and P. Diaconis, *Random walks and hyperplane arrangements*, Ann. Probab. **26** (1998), no. 4, 1813–1854. MR1675083 (2000k:60138) ↑675

[57]   K. S. Brown and R. Geoghegan, *Cohomology with free coefficients of the fundamental group of a graph of groups*, Comment. Math. Helv. **60** (1985), no. 1, 31–45. MR787660 (87b:20066) ↑641

[58]   F. Bruhat, *Représentations induites des groupes de Lie semi-simples complexes*, C. R. Acad. Sci. Paris **238** (1954), 437–439. MR0059283 (15,504h) ↑335

[59]   F. Bruhat and J. Tits, *Groupes réductifs sur un corps local*, Inst. Hautes Études Sci. Publ. Math. No. 41 (1972), 5–251. MR0327923 (48 #6265) ↑7, 351, 363, 413, 434, 447, 503, 549, 552, 558, 565, 592, 593, 693

[60]   ———, *Groupes réductifs sur un corps local. II. Schémas en groupes. Existence d'une donnée radicielle valuée*, Inst. Hautes Études Sci. Publ. Math. No. 60 (1984), 197–376. MR756316 (86c:20042) ↑592, 593, 693

[61]   F. Buekenhout, P. Cara, M. Dehon, and D. Leemans, *Residually weakly primitive geometries of small sporadic and almost simple groups: a synthesis*, Topics in diagram geometry, 2003, pp. 1–27. MR2066521 (2005e:51015) ↑656

[62]   F. Buekenhout and A. Pasini, *Finite diagram geometries extending buildings*, Handbook of incidence geometry (F. Buekenhout, ed.), North-Holland, Amsterdam, 1995, pp. 1143–1254. MR1360737 (96i:51012) ↑656

[63]   F. Buekenhout and E. Shult, *On the foundations of polar geometry*, Geometriae Dedicata **3** (1974), 155–170. MR0350599 (50 #3091) ↑502

[64] V. O. Bugaenko, *Groups of automorphisms of unimodular hyperbolic quadratic forms over the ring* $\mathbf{Z}[(\sqrt{5}+1)/2]$, Vestnik Moskov. Univ. Ser. I Mat. Mekh. **5** (1984), 6–12. MR764026 (86d:11030) ↑546

[65] K. Burns and R. Spatzier, *On topological Tits buildings and their classification*, Inst. Hautes Études Sci. Publ. Math. No. 65 (1987), 5–34. MR908214 (88g:53049) ↑657, 659

[66] _____, *Manifolds of nonpositive curvature and their buildings*, Inst. Hautes Études Sci. Publ. Math. No. 65 (1987), 35–59. MR908215 (88g:53050) ↑658

[67] K.-U. Bux and K. Wortman, *Finiteness properties of arithmetic groups over function fields*, Invent. Math. **167** (2007), no. 2, 355–378. MR2270455 (2007k:11082) ↑649

[68] P.-E. Caprace, *Conjugacy of 2-spherical subgroups of Coxeter groups and parallel walls*, Algebr. Geom. Topol. **6** (2006), 1987–2029 (electronic). MR2263057 (2007j:20059) ↑617

[69] _____, *On 2-spherical Kac-Moody groups and their central extensions*, Forum Math. **19** (2007), no. 5, 763–781. MR2350773 ↑497

[70] P.-E. Caprace and B. Mühlherr, *Reflection triangles in Coxeter groups and biautomaticity*, J. Group Theory **8** (2005), no. 4, 467–489. MR2152693 (2006b:20053) ↑617

[71] P.-E. Caprace and B. Rémy, *Simplicity and superrigidity of twin building lattices*, preprint, 2006, available at arXiv:math/0607664[math.GR]. ↑324, 497

[72] _____, *Simplicité abstraite des groupes de Kac-Moody non affines*, C. R. Math. Acad. Sci. Paris **342** (2006), no. 8, 539–544. MR2217912 (2006k:20102) ↑497

[73] _____, *Groups with a root group datum*, preprint, 2007. ↑488, 492, 652, 654

[74] L. Carbone, M. Ershov, and G. Ritter, *Abstract simplicity of complete Kac-Moody groups over finite fields*, preprint, 2006, available at arXiv:math/0612772[math.GR]. ↑324

[75] R. W. Carter, *Simple groups of Lie type*, John Wiley & Sons, London-New York-Sydney, 1972. Pure and Applied Mathematics, Vol. 28. MR0407163 (53 #10946) ↑505

[76] D. I. Cartwright and W. Woess, *Isotropic random walks in a building of type* $\tilde{A}_d$, Math. Z. **247** (2004), no. 1, 101–135. MR2054522 (2005d:60011) ↑660

[77] W. Casselman, *On a p-adic vanishing theorem of Garland*, Bull. Amer. Math. Soc. **80** (1974), 1001–1004. MR0354933 (50 #7410) ↑644

[78] A. Cayley, *On the theory of groups*, Amer. J. Math. **11** (1889), no. 2, 139–157. ↑39

[79] R. Charney and M. W. Davis, *Finite* $K(\pi, 1)s$ *for Artin groups*, Prospects in topology (F. Quinn, ed.), Annals of Mathematics Studies, vol. 138, Princeton University Press, Princeton, NJ, 1995, pp. 110–124. MR1368655 (97a:57001) ↑624

[80] R. Charney and A. Lytchak, *Metric characterizations of spherical and Euclidean buildings*, Geom. Topol. **5** (2001), 521–550 (electronic). MR1833752 (2002h:51008) ↑597

[81] C. Chevalley, *Sur certains groupes simples*, Tôhoku Math. J. (2) **7** (1955), 14–66. MR0073602 (17,457c) ↑335, 504

[82] _____, *Sur les décompositions cellulaires des espaces G/B*, Algebraic groups and their generalizations: classical methods (W. J. Haboush and B. J. Parshall, eds.), Proceedings of Symposia in Pure Mathematics, vol. 56, American Mathematical Society, Providence, RI, 1994, pp. 1–23. With a foreword by Armand Borel. MR1278698 (95e:14041) ↑137

[83] _____, *Classification des groupes algébriques semi-simples*, Springer-Verlag, Berlin, 2005. Collected works. Vol. 3; edited and with a preface by P. Cartier. With the collaboration of P. Cartier, A. Grothendieck, and M. Lazard. MR2124841 (2006b:20068) ↑335

[84] G. E. Cooke and R. L. Finney, *Homology of cell complexes*, Based on lectures by Norman E. Steenrod, Princeton University Press, Princeton, N.J., 1967. MR0219059 (36 #2142) ↑670

[85] H. S. M. Coxeter, *The complete enumeration of finite groups of the form $R_i^2 = (R_iR_j)^{k_{ij}} = 1$*, J. London Math. Soc. **10** (1935), 21–25. ↑91, 98

[86] _____, *Regular polytopes*, 3rd ed., Dover Publications Inc., New York, 1973. MR0370327 (51 #6554) ↑17

[87] C. Cunningham and T. C. Hales, *Good orbital integrals*, Represent. Theory **8** (2004), 414–457 (electronic). MR2084489 (2006d:22021) ↑660

[88] M. W. Davis, *Buildings are* CAT(0), Geometry and cohomology in group theory (P. H. Kropholler, G. A. Niblo, and R. Stöhr, eds.), London Mathematical Society Lecture Note Series, vol. 252, Cambridge University Press, Cambridge, 1998, pp. 108–123. MR1709955 (2000i:20068) ↑1, 7, 597, 617, 630

[89] _____, *The geometry and topology of Coxeter groups*, London Mathematical Society Monographs Series, vol. 32, Princeton University Press, Princeton, NJ, 2007. ↑537, 597, 627, 628

[90] M. W. Davis and G. Moussong, *Notes on nonpositively curved polyhedra*, Low dimensional topology (K. Böröczky Jr. and W. Neumann, eds.), Bolyai Society Mathematical Studies, vol. 8, János Bolyai Mathematical Society, Budapest, 1999, pp. 11–94. MR1747268 (2001b:57040) ↑627, 628

[91] S. DeBacker, *Parametrizing nilpotent orbits via Bruhat-Tits theory*, Ann. of Math. (2) **156** (2002), no. 1, 295–332. MR1935848 (2003i:20086) ↑660

[92] _____, *Some applications of Bruhat-Tits theory to harmonic analysis on a reductive p-adic group*, Michigan Math. J. **50** (2002), no. 2, 241–261. MR1914064 (2003g:22018) ↑660

[93] _____, *Lectures on harmonic analysis for reductive p-adic groups*, Representations of real and p-adic groups (E.-C. Tan and C.-B. Zhu, eds.), Lecture Notes Series, Institute for Mathematical Sciences, National University of Singapore, vol. 2, Singapore University Press, Singapore, 2004, pp. 47–94. MR2090869 (2005g:22009) ↑660

[94] _____, *Homogeneity for reductive p-adic groups: an introduction*, Harmonic analysis, the trace formula, and Shimura varieties (J. Arthur, D. Ellwood, and R. Kottwitz, eds.), Clay Mathematics Proceedings, vol. 4, American Mathematical Society, Providence, RI, 2005, pp. 523–550. MR2192015 (2006m:22024) ↑660

[95] P. de la Harpe, *An invitation to Coxeter groups*, Group theory from a geometrical viewpoint (É. Ghys, A. Haefliger, and A. Verjovsky, eds.), World Scientific Publishing Co. Inc., River Edge, NJ, 1991, pp. 193–253. MR1170367 (93g:20080) ↑537, 541

[96] V. V. Deodhar, *On the root system of a Coxeter group*, Comm. Algebra **10** (1982), no. 6, 611–630. MR647210 (83j:20052a) ↑88, 97, 100

[97] A. W. M. Dress and R. Scharlau, *Gated sets in metric spaces*, Aequationes Math. **34** (1987), no. 1, 112–120. MR915878 (89c:54057) ↑27, 34, 151

[98] J. Dymara and T. Januszkiewicz, *Cohomology of buildings and their automorphism groups*, Invent. Math. **150** (2002), no. 3, 579–627. MR1946553 (2003j:20052) ↑645

[99] M. J. Dyer, *Hecke algebras and reflections in Coxeter groups*, Ph.D. Thesis, University of Sydney, 1987. ↑101

[100] S. Eilenberg and N. Steenrod, *Foundations of algebraic topology*, Princeton University Press, Princeton, New Jersey, 1952. MR0050886 (14,398b) ↑677

[101] W. Feit and G. Higman, *The nonexistence of certain generalized polygons*, J. Algebra **1** (1964), 114–131. MR0170955 (30 #1189) ↑337

[102] D. G. Fon-Der-Flaass, *A combinatorial construction for twin trees*, European J. Combin. **17** (1996), no. 2-3, 177–189. Discrete metric spaces (Bielefeld, 1994). MR1379370 (97c:05047) ↑291

[103] P. Fong and G. M. Seitz, *Groups with a (B, N)-pair of rank 2. I, II*, Invent. Math. **21** (1973), 1–57; ibid. **24** (1974), 191–239. MR0354858 (50 #7335) ↑508

[104] D. Gaboriau and F. Paulin, *Sur les immeubles hyperboliques*, Geom. Dedicata **88** (2001), no. 1-3, 153–197. MR1877215 (2002k:51014) ↑547, 616

[105] H. Garland, *p-adic curvature and the cohomology of discrete subgroups of p-adic groups*, Ann. of Math. (2) **97** (1973), 375–423. MR0320180 (47 #8719) ↑644

[106] P. Garrett, *Buildings and classical groups*, Chapman & Hall, London, 1997. MR1449872 (98k:20081) ↑363

[107] D. Gorenstein, *Finite simple groups*, University Series in Mathematics, Plenum Publishing Corp., New York, 1982. An introduction to their classification. MR698782 (84j:20002) ↑505

[108] D. Gorenstein, R. Lyons, and R. Solomon, *The classification of the finite simple groups*, Mathematical Surveys and Monographs, vol. 40.1, American Mathematical Society, Providence, RI, 1994. MR1303592 (95m:20014) ↑507

[109] _____, *The classification of the finite simple groups. Number 2. Part I. Chapter G*, Mathematical Surveys and Monographs, vol. 40.2, American Mathematical Society, Providence, RI, 1996. General group theory. MR1358135 (96h:20032) ↑507, 656

[110] _____, *The classification of the finite simple groups. Number 3. Part I. Chapter A*, Mathematical Surveys and Monographs, vol. 40.3, American Mathematical Society, Providence, RI, 1998. Almost simple K-groups. MR1490581 (98j:20011) ↑505, 507, 656

[111] _____, *The classification of the finite simple groups. Number 4. Part II. Chapters 1–4*, Mathematical Surveys and Monographs, vol. 40.4, Providence, RI, 1999. Uniqueness theorems. MR1675976 (2000c:20028) ↑507

[112] _____, *The classification of the finite simple groups. Number 5. Part III. Chapters 1–6*, Mathematical Surveys and Monographs, vol. 40.5, American Mathematical Society, Providence, RI, 2002. The generic case, stages 1–3a. MR1923000 (2003h:20028) ↑507

[113] _____, *The classification of the finite simple groups. Number 6. Part IV*, Mathematical Surveys and Monographs, vol. 40.6, American Mathematical Society, Providence, RI, 2005. The special odd case. MR2104668 (2005m:20039) ↑507, 656

[114] R. Gramlich, *Developments in finite Phan theory*, preprint, 2007. ↑656

[115] D. R. Grayson, *Finite generation of K-groups of a curve over a finite field (after Daniel Quillen)*, Algebraic K-theory. Part I (R. K. Dennis, ed.), Lecture Notes in Mathematics, vol. 966, Springer-Verlag, Berlin, 1982, pp. 69–90. MR689367 (84f:14018) ↑649

[116] _____, *Reduction theory using semistability*, Comment. Math. Helv. **59** (1984), no. 4, 600–634. MR780079 (86h:22018) ↑638, 649

[117] _____, *Reduction theory using semistability. II*, Comment. Math. Helv. **61** (1986), no. 4, 661–676. MR870711 (88e:22015) ↑638, 649

[118] M. J. Greenberg, *Euclidean and non-Euclidean geometries*, 3rd ed., W. H. Freeman and Company, New York, 1993. Development and history. MR1261866 (94k:51001) ↑542

[119] M. Gromov, *Hyperbolic groups*, Essays in group theory (S. M. Gersten, ed.), Mathematical Sciences Research Institute Publications, vol. 8, Springer-Verlag, New York, 1987, pp. 75–263. MR919829 (89e:20070) ↑546, 550, 671

[120] _____, *Asymptotic invariants of infinite groups*, Geometric group theory. Vol. 2 (G. A. Niblo and M. A. Roller, eds.), London Mathematical Society Lecture Note Series, vol. 182, Cambridge University Press, Cambridge, 1993, pp. 1–295. MR1253544 (95m:20041) ↑659

[121] _____, *Metric structures for Riemannian and non-Riemannian spaces*, Progress in Mathematics, vol. 152, Birkhäuser Boston Inc., Boston, MA, 1999. Based on the 1981 French original. With appendices by M. Katz, P. Pansu, and S. Semmes. Translated from the French by Sean Michael Bates. MR1699320 (2000d:53065) ↑659

[122] M. Gromov and R. Schoen, *Harmonic maps into singular spaces and p-adic superrigidity for lattices in groups of rank one*, Inst. Hautes Études Sci. Publ. Math. No. 76 (1992), 165–246. MR1215595 (94e:58032) ↑658

[123] L. C. Grove, *Classical groups and geometric algebra*, Graduate Studies in Mathematics, vol. 39, American Mathematical Society, Providence, RI, 2002. MR1859189 (2002m:20071) ↑344, 348, 350

[124] L. C. Grove and C. T. Benson, *Finite reflection groups*, 2nd ed., Graduate Texts in Mathematics, vol. 99, Springer-Verlag, New York, 1985. MR777684 (85m:20001) ↑15, 48

[125] Y. Guivarc'h, L. Ji, and J. C. Taylor, *Compactifications of symmetric spaces*, Progress in Mathematics, vol. 156, Birkhäuser Boston Inc., Boston, MA, 1998. MR1633171 (2000c:31006) ↑658

[126] Y. Guivarc'h and B. Rémy, *Group-theoretic compactification of Bruhat-Tits buildings*, Ann. Sci. École Norm. Sup. (4) **39** (2006), no. 6, 871–920. MR2316977 ↑584

[127] A. J. Hahn and O. T. O'Meara, *The classical groups and K-theory*, Grundlehren der Mathematischen Wissenschaften, vol. 291, Springer-Verlag, Berlin, 1989. With a foreword by J. Dieudonné. MR1007302 (90i:20002) ↑503

[128] J. Hakim and F. Murnaghan, *Distinguished tame supercuspidal representations*, preprint, 2007, available at arXiv:0709.3506[math.RT]. ↑660

[129] A. Hatcher, *Algebraic topology*, Cambridge University Press, Cambridge, 2002. MR1867354 (2002k:55001) ↑197, 619, 662

[130] D. W. Henderson and D. Taimina, *Experiencing Geometry*, 3rd ed., Prentice Hall, 2005. ↑541

[131] J. E. Humphreys, *Introduction to Lie algebras and representation theory*, Graduate Texts in Mathematics, vol. 9, Springer-Verlag, New York, 1972. MR0323842 (48 #2197) ↑529, 681

[132] ———, *Linear algebraic groups*, Graduate Texts in Mathematics, vol. 21, Springer-Verlag, New York, 1975. MR0396773 (53 #633) ↑691, 692

[133] ———, *Reflection groups and Coxeter groups*, Cambridge Studies in Advanced Mathematics, vol. 29, Cambridge University Press, Cambridge, 1990. MR1066460 (92h:20002) ↑15, 48, 97, 103, 138, 528, 544, 546, 681

[134] S. Immervoll, *Isoparametric hypersurfaces and smooth generalized quadrangles*, J. Reine Angew. Math. **554** (2003), 1–17. MR1952166 (2003m:51018) ↑659

[135] A. A. Ivanov, *Geometry of sporadic groups. I*, Encyclopedia of Mathematics and its Applications, vol. 76, Cambridge University Press, Cambridge, 1999. Petersen and tilde geometries. MR1705272 (2000i:20028) ↑656

[136] A. A. Ivanov and S. V. Shpectorov, *Geometry of sporadic groups. II*, Encyclopedia of Mathematics and its Applications, vol. 91, Cambridge University Press, Cambridge, 2002. Representations and amalgams. MR1894594 (2003i:20027) ↑656

[137] N. Iwahori and H. Matsumoto, *On some Bruhat decomposition and the structure of the Hecke rings of p-adic Chevalley groups*, Inst. Hautes Études Sci. Publ. Math. No. 25 (1965), 5–48. MR0185016 (32 #2486) ↑351

[138] H. Izeki and S. Nayatani, *Combinatorial harmonic maps and discrete-group actions on Hadamard spaces*, Geom. Dedicata **114** (2005), 147–188. MR2174098 (2006k:58024) ↑645

[139] N. Jacobson, *Basic algebra. II*, 2nd ed., W. H. Freeman and Company, New York, 1989. MR1009787 (90m:00007) ↑184

[140] L. Ji, *Buildings and their applications in geometry and topology*, Asian J. Math. **10** (2006), no. 1, 11–80. MR2213684 (2007j:22033) ↑657, 658, 659

[141] V. G. Kac, *Infinite-dimensional Lie algebras*, 3rd ed., Cambridge University Press, Cambridge, 1990. MR1104219 (92k:17038) ↑494

[142] W. M. Kantor, R. A. Liebler, and J. Tits, *On discrete chamber-transitive automorphism groups of affine buildings*, Bull. Amer. Math. Soc. (N.S.) **16** (1987), no. 1, 129–133. MR866031 (87m:20124) ↑656

[143] A. G. Khovanskiĭ, *Hyperplane sections of polyhedra, toric varieties and discrete groups in Lobachevskiĭ space*, Funktsional. Anal. i Prilozhen. **20** (1986), no. 1, 50–61, 96. MR831049 (87k:22015) ↑546

[144] J.-L. Kim and F. Murnaghan, *Character expansions and unrefined minimal K-types*, Amer. J. Math. **125** (2003), no. 6, 1199–1234. MR2018660 (2004k:22024) ↑660

[145] F. Klein, *Vorlesungen über die Theorie der elliptischen Modulfunctionen, ausgearbeitet und vervollständigt von Dr. Robert Fricke*, Vol. I, B. G. Teubner, Leipzig, 1890. ↑44, 73

[146] B. Kleiner and B. Leeb, *Rigidity of quasi-isometries for symmetric spaces and Euclidean buildings*, Inst. Hautes Études Sci. Publ. Math. No. 86 (1997), 115–197 (1998). MR1608566 (98m:53068) ↑658

[147] ———, *Rigidity of invariant convex sets in symmetric spaces*, Invent. Math. **163** (2006), no. 3, 657–676. MR2207236 (2006k:53064) ↑587

[148] L. Kramer, *Homogeneous spaces, Tits buildings, and isoparametric hypersurfaces*, Mem. Amer. Math. Soc. **158** (2002), no. 752, xvi+114. MR1904708 (2003m:53078) ↑659

[149] D. Krammer, *The conjugacy problem for Coxeter groups*, Ph.D. Thesis, Universiteit Utrecht, 1994. ↑107, 627, 628

[150] T. Y. Lam, *Introduction to quadratic forms over fields*, Graduate Studies in Mathematics, vol. 67, American Mathematical Society, Providence, RI, 2005. MR2104929 (2005h:11075) ↑348, 365, 366, 367, 368, 370, 710

[151] E. Landvogt, *A compactification of the Bruhat-Tits building*, Lecture Notes in Mathematics, vol. 1619, Springer-Verlag, Berlin, 1996. MR1441308 (98h: 20081) ↑584

[152] J. Lehner, *Discontinuous groups and automorphic functions*, Mathematical Surveys, No. VIII, American Mathematical Society, Providence, R.I., 1964. MR0164033 (29 #1332) ↑72, 75

[153] E. Leuzinger, *Isoperimetric inequalities and random walks on quotients of graphs and buildings*, Math. Z. **248** (2004), no. 1, 101–112. MR2092723 (2005j:11030) ↑645

[154] A. T. Lundell and S. Weingram, *The Topology of CW Complexes*, Van Nostrand Reinhold, New York, 1969. ↑670, 677

[155] G. Lusztig, *The discrete series of GL_n over a finite field*, Princeton University Press, Princeton, N.J., 1974. Annals of Mathematics Studies, No. 81. MR0382419 (52 #3303) ↑660

[156] R. C. Lyndon, *Groups and geometry*, London Mathematical Society Lecture Note Series, vol. 101, Cambridge University Press, Cambridge, 1985. Revised from "Groupes et géométrie" by A. Boidin, A. Fromageot, and Lyndon. MR794131 (87i:20068) ↑17

[157] I. G. Macdonald, *Spherical functions on a group of p-adic type*, Ramanujan Institute, Centre for Advanced Study in Mathematics, University of Madras, Madras, 1971. Publications of the Ramanujan Institute, No. 2. MR0435301 (55 #8261) ↑660

[158] G. A. Margulis, *Discrete subgroups of semisimple Lie groups*, Ergebnisse der Mathematik und ihrer Grenzgebiete, vol. 17, Springer-Verlag, Berlin, 1991. MR1090825 (92h:22021) ↑537, 541, 658

[159] B. Maskit, *Kleinian groups*, Grundlehren der Mathematischen Wissenschaften, vol. 287, Springer-Verlag, Berlin, 1988. MR959135 (90a:30132) ↑537, 541

[160] H. Matsumoto, *Générateurs et relations des groupes de Weyl généralisés*, C. R. Acad. Sci. Paris **258** (1964), 3419–3422. MR0183818 (32 #1294) ↑336

[161] Y. Matsushima, *On Betti numbers of compact, locally sysmmetric Riemannian manifolds*, Osaka Math. J. **14** (1962), 1–20. MR0141138 (25 #4549) ↑645

[162] U. Meierfrankenfeld, B. Stellmacher, and G. Stroth, *Finite groups of local characteristic p: an overview*, Groups, combinatorics & geometry (A. A. Ivanov, M. W. Liebeck, and J. Saxl, eds.), World Scientific Publishing Co. Inc., River Edge, NJ, 2003, pp. 155–192. MR1994966 (2004i:20020) ↑507

[163] J. Milnor, *The geometric realization of a semi-simplicial complex*, Ann. of Math. (2) **65** (1957), 357–362. MR0084138 (18,815d) ↑677

[164] R. V. Moody and A. Pianzola, *Lie algebras with triangular decompositions*, Canadian Mathematical Society Series of Monographs and Advanced Texts, John Wiley & Sons Inc., New York, 1995. MR1323858 (96d:17025) ↑494

[165] _____, *Kac-Moody Lie algebras: a survey of their algebraic development over the first 30 years*, Canadian Mathematical Society. 1945–1995, Vol. 3, Canadian Math. Soc., Ottawa, ON, 1996, pp. 211–245. MR1661617 (2000k:17033) ↑494

[166] G. D. Mostow, *Quasi-conformal mappings in n-space and the rigidity of hyperbolic space forms*, Inst. Hautes Études Sci. Publ. Math. No. 34 (1968), 53–104. MR0236383 (38 #4679) ↑542

[167] _____, *Strong rigidity of locally symmetric spaces*, Princeton University Press, Princeton, N.J., 1973. Annals of Mathematics Studies, No. 78. MR0385004 (52 #5874) ↑542, 657

[168] R. Moufang, *Die Schnittpunktsätze des projektiven speziellen Fünfecksnetzes in ihrer Abhängigkeit voneinander*, Math. Ann. **106** (1932), no. 1, 755–795. MR1512782 ↑393

[169] _____, *Alternativkörper und der Satz vom vollständigen Vierseit*, Abh. Math. Sem. Univ. Hamburg **9** (1933), 207–222. ↑393

[170] G. Moussong, *Hyperbolic Coxeter groups*, Ph.D. Thesis, Ohio State University, 1988. ↑546, 598, 627, 628

[171] A. Moy and G. Prasad, *Unrefined minimal K-types for p-adic groups*, Invent. Math. **116** (1994), no. 1-3, 393–408. MR1253198 (95f:22023) ↑660

[172] _____, *Jacquet functors and unrefined minimal K-types*, Comment. Math. Helv. **71** (1996), no. 1, 98–121. MR1371680 (97c:22021) ↑660

[173] B. Mühlherr, *A rank 2 characterization of twinnings*, European J. Combin. **19** (1998), no. 5, 603–612. MR1637764 (99g:20051) ↑272

[174] _____, *On the existence of 2-spherical twin buildings*, Habilitationsschrift, Dortmund, 1999. ↑509

[175] _____, *Locally split and locally finite twin buildings of 2-spherical type*, J. Reine Angew. Math. **511** (1999), 119–143. MR1695793 (2000f:51023) ↑509

[176] _____, *Twin buildings*, Tits buildings and the model theory of groups (K. Tent, ed.), London Mathematical Society Lecture Note Series, vol. 291, Cambridge University Press, Cambridge, 2002. Papers from the workshop held in Würzburg, September 2000, pp. 103–117. MR2018383 (2004m:51032) ↑509

[177] B. Mühlherr and M. A. Ronan, *Local to global structure in twin buildings*, Invent. Math. **122** (1995), no. 1, 71–81. MR1354954 (96h:20062) ↑290, 509

[178] B. Mühlherr and H. Van Maldeghem, *Codistances in buildings*, preprint, 2007. ↑593

[179] J. R. Munkres, *Elements of algebraic topology*, Addison-Wesley Publishing Company, Menlo Park, CA, 1984. MR755006 (85m:55001) ↑619, 662, 663, 717

[180] G. A. Niblo and L. D. Reeves, *Coxeter groups act on* CAT(0) *cube complexes*, J. Group Theory **6** (2003), no. 3, 399–413. MR1983376 (2004e:20072) ↑101, 617

[181] O. T. O'Meara, *Introduction to quadratic forms*, Classics in Mathematics, Springer-Verlag, Berlin, 2000. Reprint of the 1973 edition. MR1754311 (2000m:11032) ↑348, 365, 367, 370

[182] P. Orlik and H. Terao, *Arrangements of hyperplanes*, Grundlehren der Mathematischen Wissenschaften, vol. 300, Springer-Verlag, Berlin, 1992. MR1217488 (94e:52014) ↑28

[183] P. Pansu, *Formules de Matsushima, de Garland et propriété (T) pour des groupes agissant sur des espaces symétriques ou des immeubles*, Bull. Soc. Math. France **126** (1998), no. 1, 107–139. MR1651383 (2000d:53067) ↑645

[184] _____, *Super-rigidité géométrique et applications harmoniques*, Non-positively curved geometry, discrete groups and rigidity (L. Bessières, A. Parreau, and B. Rémy, eds.), Séminaires et Congrès, vol. 18, Société Mathématique de France, Paris. Proceedings of the 2004 Grenoble summer school, to appear. ↑645, 658

[185] Pappus of Alexandria, *Mathematicæ collectiones*, HH. de Duccijs, Bologna, 1660. ↑551

730   References

[186] J. Parkinson, *Spherical harmonic analysis on affine buildings*, Math. Z. **253** (2006), no. 3, 571–606. MR2221087 (2007j:20039) ↑660

[187] A. Pasini, *Diagram geometries*, Oxford Science Publications, The Clarendon Press Oxford University Press, New York, 1994. MR1318911 (96f:51018) ↑656

[188] V. Platonov and A. Rapinchuk, *Algebraic groups and number theory*, Pure and Applied Mathematics, vol. 139, Academic Press Inc., Boston, MA, 1994. Translated from the 1991 Russian original by Rachel Rowen. MR1278263 (95b:11039) ↑368

[189] L. S. Pontryagin, *Topological groups*, Gordon and Breach Science Publishers, Inc., New York, 1966. Translated from the second Russian edition by Arlen Brown. MR0201557 (34 #1439) ↑522

[190] G. Prasad, *Lattices in semisimple groups over local fields*, Studies in algebra and number theory (G.-C. Rota, ed.), Advances in Mathematics, Supplementary Studies, vol. 6, Academic Press Inc. [Harcourt Brace Jovanovich Publishers], New York, 1979, pp. 285–356. MR535769 (81g:22014) ↑658

[191] _____, *Strong rigidity of* $\mathbf{Q}$-*rank* 1 *lattices*, Invent. Math. **21** (1973), 255–286. MR0385005 (52 #5875) ↑542

[192] M. S. Raghunathan, *A note on quotients of real algebraic groups by arithmetic subgroups*, Invent. Math. **4** (1967/1968), 318–335. MR0230332 (37 #5894) ↑636

[193] J. G. Ratcliffe, *Foundations of hyperbolic manifolds*, 2nd ed., Graduate Texts in Mathematics, vol. 149, Springer, New York, 2006. MR2249478 (2007d:57029) ↑537, 540, 541

[194] U. Rehmann and C. Soulé, *Finitely presented groups of matrices*, Algebraic K-theory (M. R. Stein, ed.), Springer-Verlag, Berlin, 1976. Lecture Notes in Mathematics, Vol. 551, pp. 164–169. Lecture Notes in Math., Vol. 551. MR0486175 (58 #5955) ↑655

[195] B. Rémy, *Groupes de Kac-Moody déployés et presque déployés*, Astérisque No. 277 (2002), viii+348. MR1909671 (2003d:20036) ↑476, 496, 652

[196] _____, *Topological simplicity, commensurator super-rigidity and non-linearities of Kac-Moody groups*, Geom. Funct. Anal. **14** (2004), no. 4, 810–852. With an appendix by P. Bonvin. MR2084981 (2005g:22024) ↑584

[197] B. Rémy and M. Ronan, *Topological groups of Kac-Moody type, right-angled twinnings and their lattices*, Comment. Math. Helv. **81** (2006), no. 1, 191–219. MR2208804 (2007b:20063) ↑498

[198] J. D. Rogawski, *An application of the building to orbital integrals*, Compositio Math. **42** (1980/81), no. 3, 417–423. MR607380 (83g:22011) ↑660

[199] J. Rohlfs and T. A. Springer, *Applications of buildings*, Handbook of incidence geometry (F. Buekenhout, ed.), North-Holland, Amsterdam, 1995, pp. 1085–1114. MR1360735 (97b:20041) ↑657

[200] M. A. Ronan, *Lectures on buildings*, Perspectives in Mathematics, vol. 7, Academic Press Inc., Boston, MA, 1989. MR1005533 (90j:20001) ↑224, 403, 408, 409, 474, 501, 592, 656

[201] _____, *Buildings: main ideas and applications. I. Main ideas*, Bull. London Math. Soc. **24** (1992), no. 1, 1–51. MR1139056 (93b:51006a) ↑659

[202] _____, *Buildings: main ideas and applications. II. Arithmetic groups, buildings and symmetric spaces*, Bull. London Math. Soc. **24** (1992), no. 2, 97–126. MR1148671 (93b:51006b) ↑659

[203] _____, *Local isometries of twin buildings*, Math. Z. **234** (2000), no. 3, 435–455. MR1774092 (2001g:20030) ↑290, 459, 509

[204] M. A. Ronan and J. Tits, *Building buildings*, Math. Ann. **278** (1987), no. 1-4, 291–306. MR909229 (89e:51005) ↑508

[205] _____, *Twin trees. I*, Invent. Math. **116** (1994), no. 1-3, 463–479. MR1253201 (94k:20058) ↑455

[206] _____, *Twin trees. II. Local structure and a universal construction*, Israel J. Math. **109** (1999), 349–377. MR1679605 (2000f:05030) ↑291

[207] R. Scharlau, *Buildings*, Handbook of incidence geometry (F. Buekenhout, ed.), North-Holland, Amsterdam, 1995, pp. 477–645. MR1360726 (97h:51018) ↑180, 501, 503

[208] W. Scharlau, *Quadratic and Hermitian forms*, Grundlehren der Mathematischen Wissenschaften, vol. 270, Springer-Verlag, Berlin, 1985. MR770063 (86k:11022) ↑348, 365, 367

[209] P. Schneider, *Gebäude in der Darstellungstheorie über lokalen Zahlkörpern*, Jahresber. Deutsch. Math.-Verein. **98** (1996), no. 3, 135–145. MR1421022 (98d:22016) ↑660

[210] P. Schneider and U. Stuhler, *The cohomology of p-adic symmetric spaces*, Invent. Math. **105** (1991), no. 1, 47–122. MR1109620 (92k:11057) ↑660

[211] _____, *Representation theory and sheaves on the Bruhat-Tits building*, Inst. Hautes Études Sci. Publ. Math. No. 85 (1997), 97–191. MR1471867 (98m:22023) ↑660

[212] I. Schur, *Über Gruppen periodischer Substitutionen*, Sitzungsber. Preuss. Akad. Wiss. (1911), 619–627. ↑716

[213] J.-P. Serre, *Algèbres de Lie semi-simples complexes*, W. A. Benjamin, Inc., New York-Amsterdam, 1966. MR0215886 (35 #6721) ↑494

[214] _____, *Cohomologie des groupes discrets*, Prospects in mathematics, Annals of Mathematics Studies, vol. 70, Princeton University Press, Princeton, N.J., 1971, pp. 77–169. MR0385006 (52 #5876) ↑641, 645, 647

[215] _____, *A course in arithmetic*, Graduate Texts in Mathematics, vol. 7, Springer-Verlag, New York, 1973. Translated from the French. MR0344216 (49 #8956) ↑75, 348, 365, 370, 710

[216] _____, *Arithmetic groups*, Homological group theory (C. T. C. Wall, ed.), London Mathematical Society Lecture Note Series, vol. 36, Cambridge University Press, Cambridge, 1979, pp. 105–136. MR564421 (82b:22021) ↑637

[217] _____, *Trees*, Springer Monographs in Mathematics, Springer-Verlag, Berlin, 2003. Translated from the French original by John Stillwell, corrected 2nd printing of the 1980 English translation. MR1954121 (2003m:20032) ↑72, 196, 357, 358, 363, 370, 487, 558, 648, 649, 652, 653, 654, 710

[218] S. D. Smith, *Homology representations from group geometries*, Proceedings of the conference on groups and geometry. Part B (D. W. Crowe, J. M. Osborn, and L. Solomon, eds.), Hadronic Press Inc., Palm Harbor, FL, 1985, pp. 514–540. MR852422 (87m:20021) ↑660

[219] _____, *Subgroup complexes*, in preparation. ↑656

[220] L. Solomon, *The Steinberg character of a finite group with BN-pair*, Theory of finite groups: A symposium (R. Brauer and C.-H. Sah, eds.), W. A. Benjamin, Inc., New York-Amsterdam, 1969, pp. 213–221. MR0246951 (40 #220) ↑197, 660

[221] _____, *A Mackey formula in the group ring of a Coxeter group*, J. Algebra **41** (1976), no. 2, 255–264. MR0444756 (56 #3104) ↑26

[222] C. Soulé, *Groupes opérant sur un complexe simplicial avec domaine fondamental*, C. R. Acad. Sci. Paris Sér. A-B **276** (1973), A607–A609. MR0317307 (47 #5854) ↑651

[223] _____, *Chevalley groups over polynomial rings*, Homological group theory (C. T. C. Wall, ed.), London Mathematical Society Lecture Note Series, vol. 36, Cambridge University Press, Cambridge, 1979, pp. 359–367. MR564437 (81g:20080) ↑655

[224] E. H. Spanier, *Algebraic topology*, Springer-Verlag, New York, 1995. Corrected reprint of the 1966 original. MR1325242 (96a:55001) ↑197, 662, 663

[225] B. Srinivasan, *Ruth Moufang, 1905–1977*, Math. Intelligencer **6** (1984), no. 2, 51–55. MR738907 (85m:01084) ↑393

[226] R. Steinberg, *A geometric approach to the representations of the full linear group over a Galois field*, Trans. Amer. Math. Soc. **71** (1951), 274–282. MR0043784 (13,317d) ↑660

[227] _____, *Lectures on Chevalley groups*, Yale University, New Haven, Conn., 1968. Notes prepared by John Faulkner and Robert Wilson. MR0466335 (57 #6215) ↑440, 441, 442

[228] R. Strebel, *Finitely presented soluble groups*, Group theory (K. W. Gruenberg and J. E. Roseblade, eds.), Academic Press Inc. [Harcourt Brace Jovanovich Publishers], London, 1984, pp. 257–314. MR780572 (86g:20050) ↑590

[229] U. Stuhler, *Eine Bemerkung zur Reduktionstheorie quadratischer Formen*, Arch. Math. (Basel) **27** (1976), no. 6, 604–610. MR0424707 (54 #12666) ↑638, 649

[230] _____, *Zur Reduktionstheorie der positiven quadratischen Formen. II*, Arch. Math. (Basel) **28** (1977), no. 6, 611–619. MR0447126 (56 #5441) ↑638, 649

[231] _____, *Homological properties of certain arithmetic groups in the function field case*, Invent. Math. **57** (1980), no. 3, 263–281. MR568936 (81g:22016) ↑648

[232] M. Suzuki, *Characterizations of linear groups*, Bull. Amer. Math. Soc. **75** (1969), 1043–1091. MR0260889 (41 #5509) ↑656

[233] J. Teitelbaum, *The geometry of p-adic symmetric spaces*, Notices Amer. Math. Soc. **42** (1995), no. 10, 1120–1126. MR1350009 (96g:11064) ↑660

[234] K. Tent, *A short proof that root groups are nilpotent*, J. Algebra **277** (2004), no. 2, 765–768. MR2067630 (2005a:20043) ↑424

[235] G. Thorbergsson, *Isoparametric foliations and their buildings*, Ann. of Math. (2) **133** (1991), no. 2, 429–446. MR1097244 (92d:53053) ↑659

[236] _____, *Clifford algebras and polar planes*, Duke Math. J. **67** (1992), no. 3, 627–632. MR1181317 (93i:51033) ↑659

[237] _____, *A survey on isoparametric hypersurfaces and their generalizations*, Handbook of differential geometry. Vol. I (F. J. E. Dillen and L. C. A. Verstraelen, eds.), North-Holland, Amsterdam, 2000, pp. 963–995. MR1736861 (2001a:53097) ↑658

[238] F. G. Timmesfeld, *Abstract root subgroups and simple groups of Lie type*, Monographs in Mathematics, vol. 95, Birkhäuser, Basel, 2001. MR1852057 (2002f:20070) ↑656

[239] J. Tits, *Sur la trialité et certains groupes qui s'en déduisent*, Inst. Hautes Études Sci. Publ. Math. No. 2 (1959), 13–60. MR1557095 ↑337

[240] _____, *Groupes et géométries de Coxeter* (1961), unpublished. ↑65, 91, 94, 103

[241] _____, *Théorème de Bruhat et sous-groupes paraboliques*, C. R. Acad. Sci. Paris **254** (1962), 2910–2912. MR0138706 (25 #2149) ↑336

[242] _____, *Géométries polyédriques et groupes simples*, Atti della II riunione del groupement de mathématiciens d'expression latine (Florence, 1961), Edizioni Cremonese, Rome, 1963, pp. 66–88. ↑336

[243] _____, *Structures et groupes de Weyl*, Séminaire Bourbaki (1964/1965), Exp. No. 288, W. A. Benjamin, Inc., New York-Amsterdam, 1966. MR1608796 ↑1, 336

[244] _____, *Algebraic and abstract simple groups*, Ann. of Math. (2) **80** (1964), 313–329. MR0164968 (29 #2259) ↑322, 323, 324

[245] _____, *Classification of algebraic semisimple groups*, Proceedings of Symposia in Pure Mathematics. Vol. IX: Algebraic groups and discontinuous subgroups (A. Borel and G. D. Mostow, eds.), Proceedings of the Symposium in Pure Mathematics of the American Mathematical Society held at the University of Colorado, Boulder, Colorado (1965), American Mathematical Society, Providence, R.I., 1966, pp. 33–62. MR0224710 (37 #309) ↑505

[246] _____, *Le problème des mots dans les groupes de Coxeter*, Symposia mathematica. Vol. I. Convegni del Dicembre del 1967 e dell' Aprile del 1968, Istituto Nazionale di Alta Matematica Roma, Academic Press, London, 1969, pp. 175–185. MR0254129 (40 #7339) ↑86, 698

[247] _____, *Buildings of spherical type and finite BN-pairs*, Lecture Notes in Mathematics, vol. 386, Springer-Verlag, Berlin, 1974. MR0470099 (57 #9866) ↑1, 6, 25, 26, 128, 157, 180, 210, 286, 288, 289, 379, 380, 387, 402, 499, 501, 502, 503, 653, 692

[248] _____, *On buildings and their applications*, Proceedings of the International Congress of Mathematicians. Volume 1, Canadian Mathematical Congress, Montreal, Que., 1975, pp. 209–220. MR0439945 (55 #12826) ↑173, 549, 651

[249] _____, *Two properties of Coxeter complexes*, J. Algebra **41** (1976), no. 2, 265–268. Appendix to "A Mackey formula in the group ring of a Coxeter group" (J. Algebra **41** (1976), no. 2, 255–264) by Louis Solomon. MR0444757 (56 #3105) ↑26

[250] _____, *Non-existence de certains polygones généralisés. I*, Invent. Math. **36** (1976), 275–284. MR0435248 (55 #8208) ↑403

[251] _____, *Endliche Spiegelungsgruppen, die als Weylgruppen auftreten*, Invent. Math. **43** (1977), no. 3, 283–295. MR0460485 (57 #478) ↑393, 397, 402, 500

[252] _____, *Reductive groups over local fields*, Automorphic forms, representations and *L*-functions. Part 1 (A. Borel and W. Casselman, eds.), Proceedings of Symposia in Pure Mathematics, XXXIII, American Mathematical Society, Providence, R.I., 1979, pp. 29–69. MR546588 (80h:20064) ↑592, 693

[253] _____, *Non-existence de certains polygones généralisés. II*, Invent. Math. **51** (1979), no. 3, 267–269. MR530633 (83c:20055) ↑403

[254] _____, *Buildings and Buekenhout geometries*, Finite simple groups. II (M. J. Collins, ed.), Academic Press Inc. [Harcourt Brace Jovanovich Publishers], London, 1980, pp. 309–320. ↑336

[255] _____, *A local approach to buildings*, The geometric vein (C. Davis, B. Grünbaum, and F. A. Sherk, eds.), Springer-Verlag, New York, 1981, pp. 519–547. MR661801 (83k:51014) ↑1, 224, 337, 656

[256] _____, *Moufang octagons and the Ree groups of type $^2F_4$*, Amer. J. Math. **105** (1983), no. 2, 539–594. MR701569 (84m:20048) ↑459, 506

[257] _____, *Groups and group functors attached to Kac–Moody data*, Arbeitstagung Bonn 1984 (F. Hirzebruch, J. Schwermer, and S. Suter, eds.), Lecture Notes in Mathematics, vol. 1111, Springer-Verlag, Berlin, 1985, pp. 193–223. MR797422 (87a:22032) ↑496

[258] _____, *Immeubles de type affine*, Buildings and the geometry of diagrams (L. A. Rosati, ed.), Lecture Notes in Mathematics, vol. 1181, Springer-Verlag, Berlin, 1986, pp. 159–190. MR843391 (87h:20077) ↑592, 593

[259] _____, *Ensembles ordonnés, immeubles et sommes amalgamées*, Bull. Soc. Math. Belg. Sér. A **38** (1986), 367–387. MR885545 (88j:20041) ↑652, 653, 654, 655

[260] _____, *Uniqueness and presentation of Kac-Moody groups over fields*, J. Algebra **105** (1987), no. 2, 542–573. MR873684 (89b:17020) ↑266, 419, 462, 476, 489, 495, 496

[261] _____, *Twin buildings and groups of Kac-Moody type*, Groups, combinatorics & geometry (M. Liebeck and J. Saxl, eds.), London Mathematical Society Lecture Note Series, vol. 165, Cambridge University Press, Cambridge, 1992, pp. 249–286. MR1200265 (94d:20030) ↑1, 217, 266, 290, 337, 364, 371, 411, 413, 429, 455, 458, 459, 466, 476, 509

[262] J. Tits and R. M. Weiss, *Moufang polygons*, Springer Monographs in Mathematics, Springer-Verlag, Berlin, 2002. MR1938841 (2003m:51008) ↑6, 180, 375, 390, 393, 403, 424, 500, 501, 503, 505, 506, 507, 508

[263] B. L. van der Waerden, *Gruppen von linearen transformationen*, Springer-Verlag, Berlin, 1935. ↑347

[264] H. Van Maldeghem, *Generalized polygons*, Monographs in Mathematics, vol. 93, Birkhäuser Verlag, Basel, 1998. MR1725957 (2000k:51004) ↑180, 389

[265] H. Van Maldeghem and K. Van Steen, *Moufang affine buildings have Moufang spherical buildings at infinity*, Glasgow Math. J. **39** (1997), no. 3, 237–241. MR1484566 (99b:51009) ↑596

[266] A. Vdovina, *Combinatorial structure of some hyperbolic buildings*, Math. Z. **241** (2002), no. 3, 471–478. MR1938699 (2003m:20040) ↑616

[267] _____, *Polyhedra with specified links*, Séminaire de Théorie Spectrale et Géométrie. Vol. 21. Année 2002–2003, Université de Grenoble I Institut Fourier, Saint-Martin-d'Hères, pp. 37–42. MR2052822 (2005h:52019) ↑616

[268] F. D. Veldkamp, *Polar geometry. I, II, III, IV, V*, Nederl. Akad. Wetensch. Proc. Ser. A 62; 63 = Indag. Math. 21 (1959), 512-551 **22** (1959), 207–212. MR0125472 (23 #A2773) ↑502

[269] M.-F. Vignéras, *Dualité pour les représentations des groupes réductifs p-adiques et finis*, Bol. Acad. Nac. Cienc. (Córdoba) **65** (2000), 225–231. MR1840456 (2002m:22013) ↑660

[270] È. B. Vinberg, *Discrete linear groups that are generated by reflections*, Izv. Akad. Nauk SSSR Ser. Mat. **35** (1971), 1072–1112. MR0302779 (46 #1922) ↑103, 107, 547

[271] _____, *Some arithmetical discrete groups in Lobačevskiĭ spaces*, Discrete subgroups of Lie groups and applications to moduli, Published for the Tata Institute of Fundamental Research, Bombay, 1975, pp. 323–348. MR0422505 (54 #10492) ↑546

[272] _____, *The two most algebraic K3 surfaces*, Math. Ann. **265** (1983), no. 1, 1–21. MR719348 (85k:14020) ↑546

[273] _____, *Absence of crystallographic groups of reflections in Lobachevskiĭ spaces of large dimension*, Trudy Moskov. Mat. Obshch. **47** (1984), 68–102, 246. MR774946 (86i:22020) ↑546

[274] _____, *Hyperbolic groups of reflections*, Uspekhi Mat. Nauk **40** (1985), no. 1(241), 29–66, 255. MR783604 (86m:53059) ↑537, 540, 543, 545

[275] È. B. Vinberg and I. M. Kaplinskaja, *The groups $O_{18,1}(Z)$ and $O_{19,1}(Z)$*, Dokl. Akad. Nauk SSSR **238** (1978), no. 6, 1273–1275. MR0476640 (57 #16199) ↑546

[276] K. Vogtmann, *Spherical posets and homology stability for $O_{n,n}$*, Topology **20** (1981), no. 2, 119–132. MR605652 (82d:18016) ↑348

[277] G. Warner, *Harmonic analysis on semi-simple Lie groups. I*, Springer-Verlag, New York, 1972. Die Grundlehren der mathematischen Wissenschaften, Band 188. MR0498999 (58 #16979) ↑692

[278] W. C. Waterhouse, *Introduction to affine group schemes*, Graduate Texts in Mathematics, vol. 66, Springer-Verlag, New York, 1979. MR547117 (82e:14003) ↑685, 686, 687, 688, 689, 690

[279] B. A. F. Wehrfritz, *Infinite linear groups. An account of the group-theoretic properties of infinite groups of matrices*, Springer-Verlag, New York, 1973. Ergebnisse der Matematik und ihrer Grenzgebiete, Band 76. MR0335656 (49 #436) ↑716

[280] R. M. Weiss, *The nonexistence of certain Moufang polygons*, Invent. Math. **51** (1979), no. 3, 261–266. MR530632 (82k:05062) ↑403

[281] _____, *The structure of spherical buildings*, Princeton University Press, Princeton, NJ, 2003. MR2034361 (2005b:51027) ↑vii, 224, 289, 408, 501, 508

[282] _____, *The classification of affine buildings*, Proceedings of the Hermann Weyl conference, Bielefeld, September 2006, to appear. ↑592

[283] _____, *The structure of affine buildings*, Princeton University Press, Princeton, N.J., 2008. Annals of Mathematics Studies. ↑vii, 592, 693

[284] A. Werner, *Compactification of the Bruhat-Tits building of PGL by seminorms*, Math. Z. **248** (2004), no. 3, 511–526. MR2097372 (2005k:20075) ↑584

[285] J.-K. Yu, *Construction of tame supercuspidal representations*, J. Amer. Math. Soc. **14** (2001), no. 3, 579–622 (electronic). MR1824988 (2002f:22033) ↑660

[286] G. M. Ziegler, *Lectures on polytopes*, Graduate Texts in Mathematics, vol. 152, Springer-Verlag, New York, 1995. MR1311028 (96a:52011) ↑623, 624, 674, 676

Notation Index

Subject Index

Graduate Texts in Mathematics

(*continued from page* ii)